Soil Components

Volume 1

Organic Components

Soil Components

Volume 1

Organic Components

Edited by

John E. Gieseking

Springer-Verlag   New York   ·   Heidelberg   ·   Berlin

1975

John E. Gieseking
Professor of Soil Chemistry
University of Illinois
Urbana, Illinois

Library of Congress Cataloging in Publication Data

Gieseking, John Eldon, 1905–
  Soil components.

  CONTENTS: v. 1. Organic components.—

  1. Soils—Composition.   I. Title.
S592.5.G53   631.4′1.   73–14742

ISBN 0–387–06861–9 Springer-Verlag New York · Heidelberg · Berlin
ISBN 3–540–06861–9 Springer-Verlag Berlin · Heidelberg · New York

Preface

Organic substances returned to the soil by plants, animals and microorganisms go through biochemical cycles and subcycles that provide essential media for the growth of plants in the soil. These cycles involve numerous, complicated and interdependent chemical reactions.

Many books have been written to describe the genesis, the nature and the reactions of soil organic matter and have contributed much to organizing parts of the knowledge about soil organic matter. Each book is an important contribution but none has duplicated any of the others to any great extent; each has developed essential but loose fitting segments of the knowledge about soil organic matter. Each of these books complements the other.

The authors of chapters in this volume have done considerable, scholarly research in their areas and have reported a very comprehensive coverage of their topics. Their contributions will be important supplements to already published works on soil organic matter.

The editor wishes to express his appreciation for the vast library resources of the University of Illinois that were used during the production of this volume. He also wishes to express his appreciation for the help of his wife, Flossie Y. Gieseking, who assisted or promoted every operation until it was completed.

Urbana, Illinois
August 30, 1974

JOHN E. GIESEKING

Contributors

Anderson, G., The Macaulay Institute for Soil Research, Craigiebuckler, Aberdeen, Scotland.

Babel, U., Biopedologie, Universität Hohenheim, Stuttgart, Germany.

Beutelspacher, H., Institut für Biochemie des Bodens der Forschungsanstalt für Landwirtschaft, Braunschweig, Germany.

Braids, O. C., U.S. Geological Survey, Water Resources Division, Mineola, New York.

Flaig, W., Professor and Director, Institut für Biochemie des Bodens der Forschungsanstalt für Landwirtschaft, Braunschweig, Germany.

Greenland, D. J., Professor and Head, Department of Soil Science, The University, Reading, England.

Kononova, M. M., Professor, Dokuchaev Soil Institute, Moscow, U.S.S.R.

Miller, R. H., Professor, Agronomy Department, The Ohio State University, Columbus, Ohio.

Oades, J. M., Department of Agricultural Biochemistry and Soil Science, Waite Agricultural Research Institute, University of Adelaide, Adelaide, Australia.

Parsons, J. W., Department of Soil Science, University of Aberdeen, Aberdeen, Scotland.

Rietz, E., Institut für Biochemie des Bodens der Forschungsanstalt für Landwirtschaft, Braunschweig, Germany.

Tinsley, J., Professor and Head, Department of Soil Science, University of Aberdeen, Aberdeen, Scotland.

Contents

Chapter 1

Chemical Composition and Physical Properties of Humic Substances

W. Flaig, H. Beutelspacher, and E. Rietz

Contents

A. Humification

The processes of humification occur mainly under aerobic conditions. Animals that live in the soil may initially reduce the size of the fresh organic residues. Further transformations are promoted by the activity of the enzymes of bacteria and fungi living in the soil. Cellulose, proteins, and fats are readily available carbon sources for the microorganisms, whereas such compounds as lignin and other phenolic plant constituents are decomposed more slowly. These, as well as some of the new substances that are formed through oxidation of phenolic units, are toxic to different degrees. They can therefore serve as carbon sources only for special species of microorganisms. Besides the organic residues derived from higher animals and plants, many substances synthesized by microorganisms can serve as carbon sources.

Certain species of microorganisms also synthesize phenolic or quinonoid metabolic products that decompose slowly, by reason of their more or less toxic properties or their sorption on inorganic soil constituents. The resistance to decomposition in soil of antibiotics formed by various species of microorganisms varies greatly (PINCK, HOLTON, and ALLISON [1961], PINCK, SOULIDES, and ALLISON [1961], PINCK [1962]).

The transformation of nitrogen-containing substances plays an important role during humification. The microorganisms use, as a nitrogen source, the proteins and their decomposition products derived from dead animals and plants as well as from the microorganisms themselves. The ratio of carbon to nitrogen varies greatly in different plant residues. The average carbon/nitrogen ratio is about 100, but during humification it decreases due to the consumption of carbon as a source of energy by the microorganisms. At a carbon/nitrogen ratio of about 10 the humification processes are more or less completed. At this stage about 95 percent of the soil nitrogen is bound in organic form and only slowly becomes available for the microorganisms and the plants. The rate of availability depends upon the climatic conditions that influence the activity of the microorganisms.

The nitrogen compounds of the humus are the natural reservoir for the nitrogen nutrition of plants, because this element is not supplied in significant quantities by the weathering of the rocks or the minerals of the soil. Nor is the amount of nitrogen that reaches the soil in the rainwater or that is fixed by nitrogen-fixing organisms sufficient for the annual plant growth. The nitrogen-containing organic compounds of the soil humus are, therefore, very important for the fertility of the soil under natural conditions.

The transformations of carbohydrates, proteins, and other constituents of plants and animals during humification are described in other chapters. Only the processes of lignin degradation are described in detail in this chapter (Section B II). The possible participation of carbohydrates in the formation of humic acids is mentioned in Section B X.

The humification process involves not only the decomposition of high and low molecular weight plant, animal, and microbial cell constitutents and transformation products, but also the synthesis of low and high molecular weight compounds, which are not formed within living cells, but must be considered as typical organic soil constituents formed by humification processes. These compounds are dark-colored and are called humic substances. They can be separated into different fractions from the other substances of the humus by their characteristic solubilities or dispersibilities.

The methods for the isolation and separation of the humic substances were developed by SPRENGEL [1826, 1828] and BERZELIUS [1833]. Later, ODEN [1912, 1919, 1922] made further suggestions for the fractionation of humus. These proposals and others by HOPPE-SEYLERS [1889] for soil humus nomenclature (Table 1) are very old, but they are the best so far proposed.

Table 1. Fractionation of humic substances.

Fractions	Alkali	Acid	Alcohol
Fulvic acids	Soluble	Soluble	—
Hymatomelanic acids	Soluble	Insoluble	Soluble
Humic acids	Soluble	Insoluble	Insoluble
Humins	Insoluble	Insoluble	Insoluble

Briefly, the procedure for soil humus fractionation is as follows: Undecomposed organic residues are removed by hand and by sieving. The soil is usually acidified with a mineral acid, and washed until there are no more calcium ions in the wash water. It is then extracted with an alkaline solution. Upon acidification of the dark-colored extract a dark-colored precipitate is formed. The supernatant liquid remains yellow or orange, and contains low and high molecular weight organic compounds, which are called fulvic acids.

The dark-colored precipitate is extracted with alcohol. The residue of the extraction is called humic acids. The hymatomelanic acids are obtained by evaporation of the alcohol. The organic fractions that cannot be extracted by alkaline solutions are called humins. Not all authors subdivide the acid precipitate into humic and hymatomelanic acids by extraction with alcohol, but denote all portions that are precipitable with acids, as humic acids (DAVIES and COULSON [1959]). The specific methods used for the extraction of soil humic substances have varied greatly; these investigations are described in a separate section (B IV).

Examples of the distribution of the soil organic carbon in these various fractions are presented in Table 2, a summary of the papers of SCHREINER and SHOREY (S.a.S.) [1910], BERTHELOT and ANDRÈ (B) [1892] and SHMUK (S) [1930].

Table 2. Distribution of Soil Organic Carbon after Extraction by Schreiner and Shorey (1910) (S.a.S.), Berthelot and Andrè (1892) (B), and Shmuk (1930) (S) Procedures.

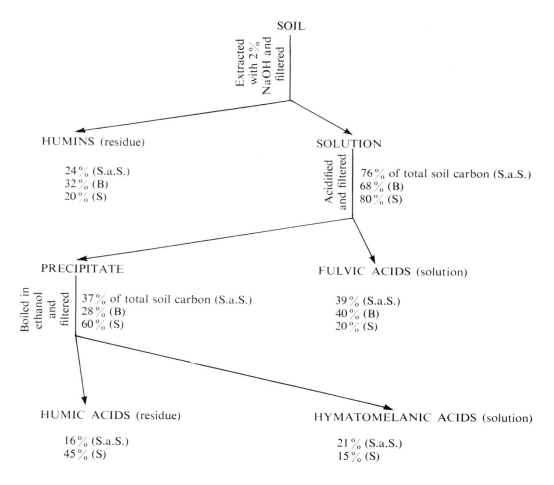

Results of the determination of the composition of the different fractions from a single soil sample depend to a considerable degree on the method used. The compositions of fractions of different origins obtained by isolation with a standardized procedure are also variable. The names of isolated fractions represent series of similar substances. As an example, the investigations of KONONOVA [1961] follow. The amounts of fulvic and humic acids of the soils of the USSR were investigated from north to south with a standardized procedure and their elemental

Table 3. Elemental composition of humic and fulvic acids of the main soils of the USSR (as percentage of absolute weight of ash-free material, according to KONONOVA [1961]).

Soils		Carbon %	Hydrogen %	Oxygen %	Nitrogen %	Carbon/ Hydrogen %
Northern podzol under forest;	I*	58.11	5.37	32.00	4.52	10.82
humus alluvial horizon 16–24 cm;	II†	52.37	3.53	42.89	1.21	14.84
Arkhangel region						
Sod-podzolic soil, arable;	I	57.63	5.23	35.33	4.81	11.02
0–20 cm; Moscow region	II	42.63	5.05	44.60	4.12	9.15
Dark gray forest soil under	I	61.20	3.60	31.32	3.88	17.00
oak; 12–19 cm; Shipov Forest;	II	47.46	3.64	45.87	3.03	13.04
Voronezh region						
Ordinary chernozem, arable;	I	62.13	2.91	31.38	3.58	21.35
0–20 cm; Kamennaya Steppe;	II	44.84	3.45	49.36	2.35	13.00
Voronezh region						
Chestnut soil; virgin land;	I	61.74	3.72	30.62	3.92	16.60
0–20 cm; Valuisk Exp. Sta.,	II	43.19	3.61	51.43	1.77	11.96
Stalingrad region						
Light serozem, arable;	I	61.94	3.93	29.46	4.67	15.76
0–20 cm; Pakhta-Aral,	II	45.80	4.30	46.00	3.90	10.65
Kazakhsk SSR						
Krasnozem arable; 0–20 cm;	I	59.65	4.37	31.54	4.44	13.65
Anaseuli, Georgian SSR	II	49.82	3.35	44.33	2.50	14.87

* I   represents humic acids.
† II   represents fulvic acids.

composition was determined (Table 3). The percentage of the elements depends upon the conditions of formation of the humic substances.

Nine soils of the USSR from the north to south are arranged in Figure 1. From north to south the average yearly temperature increases from $+2$ to $+11°C$, and the yearly rainfall decreases from about 600 to 100 mm. The curves indicate that a maximum and a minimum exist in the case of chernozems.

The different environmental factors and the inorganic initial materials lead in each case by the different ecological conditions to various biological processes, which finally influence the quantity and the properties of humic substances. These differences have an influence on soil productivity by interaction with the inorganic soil constituents.

It is evident from these investigations that the amount as well as the physical properties and the chemical constitution of the fulvic or humic acid fractions depends on the climatic conditions under which they were formed, and the plant species and inorganic constituents of the soils.

B. Chemical Composition

I. Degradation of Plant Constituents During Humification

Studies on the decomposition of plant constituents in a straw mulch in the field (SAUER-LANDT and GRAFF [1959]) show that cellulose decomposes much faster than lignin. The percentages of cellulose and lignin lost from the beginning of September to the end of November in the humid climate of central Europe are reported in Table 4.

The decomposition of lignin and the total organic matter was influenced in opposite directions under two moisture regimes.

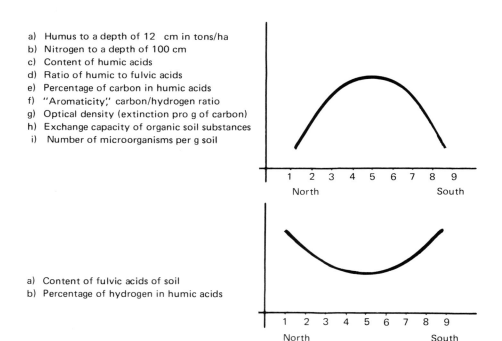

a) Humus to a depth of 12 cm in tons/ha
b) Nitrogen to a depth of 100 cm
c) Content of humic acids
d) Ratio of humic to fulvic acids
e) Percentage of carbon in humic acids
f) "Aromaticity," carbon/hydrogen ratio
g) Optical density (extinction pro g of carbon)
h) Exchange capacity of organic soil substances
i) Number of microorganisms per g soil

a) Content of fulvic acids of soil
b) Percentage of hydrogen in humic acids

1 = podzolic soil; 2 = light-gray soil; 3 = dark-gray, podzolic forest soil; 4 = degraded soil; 5 = deep soil;
6 = ordinary chernozem; 7 = dark soil; 8 = light chestnut soil; 9 = serozem.

Figure 1. Measurements of properties of soil organic matter of different soil types of the USSR (according to KONONOVA [1966]).

The humification of plant material of different composition has been investigated primarily in the laboratory (KOLENBRANDER [1955], SPRINGER [1944, 1945, 1955], SPRINGER and LEHNER [1952a, b], SPRINGER and SEISCHAB [1961]). WAKSMAN and TENNEY [1927a, b] called attention to the fact that young plants with a low methoyxl lignin content decompose easier than older ones with a higher methoxyl content.

Studies with carbon-14-labeled plant material have shown that glucose decomposes faster than hemicelluloses, and hemicelluloses decompose faster than cellulose. The degrada-

Table 4. Decrease of Plant Constituents during a three-month period (SAUERLANDT and GRAFF [1959]).

Constituent	Loss.	
Ether-soluble fraction	52.1%	
Water-soluble fraction	48.6%	
Cellulose	43.6%	
Lignin	4.7%	
	No additional water	Sprinkled
Organic material	36.5%	42.3%
Lignin	21.7%	7.3%

tion of lignin proceeds very slowly. The investigations showed, further, that the radioactivity of added carbohydrate is found mainly in the hydrolyzable parts of the humic acids, namely the amino acids, which are formed from the glucose through the metabolism of the microorganisms. Lignin or its degradation products, such as p-hydroxybenzaldehyde, vanillin, and syringaldehyde, are chiefly concentrated in the nonhydrolyzable fraction of the humic acids. (SIMONART and MAYAUDON [1958a, b], SIMONART, MAYAUDON, and BATISTIC [1959], MAYAUDON and SIMONART [1958, 1959a, b], SØRENSEN [1963, 1966], FÜHR [1962], FÜHR and SAUERBECK [1966]).

The most intensive studies on the processes of humification have been made with straw, a plant material that is relatively rich in cellulose and lignin and poor in protein (BARTLETT and NORMAN [1938], BARTLETT, SMITH, and BROWN [1937], BROADBENT [1954], FLAIG, SCHOBINGER, and DEUEL [1959], KAILA [1952], MAEDER [1960], MOHTADI [1962], PHILLIPS [1934], SMITH, STEVENSON, and BROWN [1930], SCHOBINGER [1958], SPRINGER and LEHNER [1952a, b], WAKSMAN, TENNEY, and DIEHM [1929]).

The degradation of holocellulose (determined according to ADAMS and CASTAGNE [1948] with sodium chlorite and acetic acid) occurs much faster than that of lignin (determined according to RITTER, SEBORG, and MITCHELL [1932] with 72% sulfuric acid).

Since the content of available nitrogen is a limiting factor for the activity of the microorganisms and therefore also for the rate of humification, the addition of nitrogen to low-nitrogen residues accelerates decomposition. The quotient of the amount of the cellulose remaining after a given time without addition of nitrogen divided by the amount of the cellulose remaining with addition of nitrogen increases more rapidly than that of lignin (Figure 2). This means that the degradation of the cellulose is accelerated much more than that of lignin by added nitrogen (FLAIG [1962]).

The method of proximate analysis of plant constituents according to WAKSMAN and STEVENS [1930] has been used to determine the rate of decomposition of single components such as cellulose, hemicellulose, lignin, proteins, and ether- and water-soluble substances.

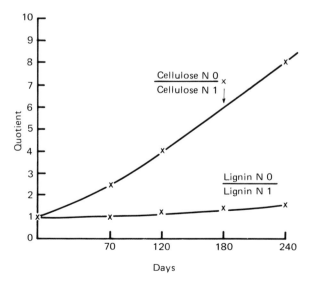

Figure 2. Acceleration of the degradation of cellulose and lignin of rye straw by addition of nitrogen. N 0 = no nitrogen added; N 1 = 1% as NH_4NO_3 per dry weight of straw (according to MAEDER [1960]).

Monomers of lignin

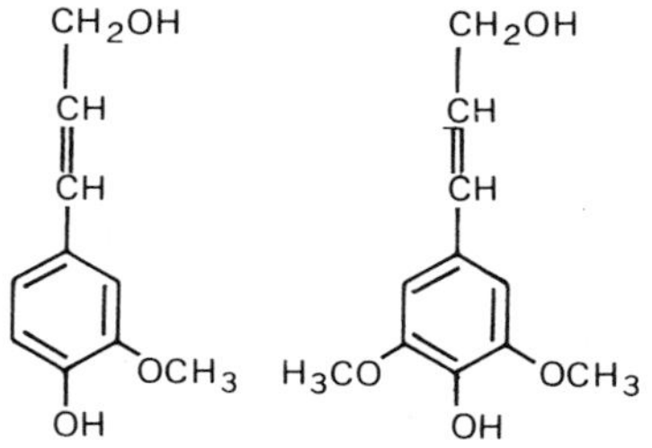

	Coniferyl alcohol	Sinapyl alcohol	*p*-Coumaryl alcohol
Conifers	+	0	0
Deciduous trees	+	+	0
Graminees	+	+	+

Polymeric structural scheme of coniferyl lignin

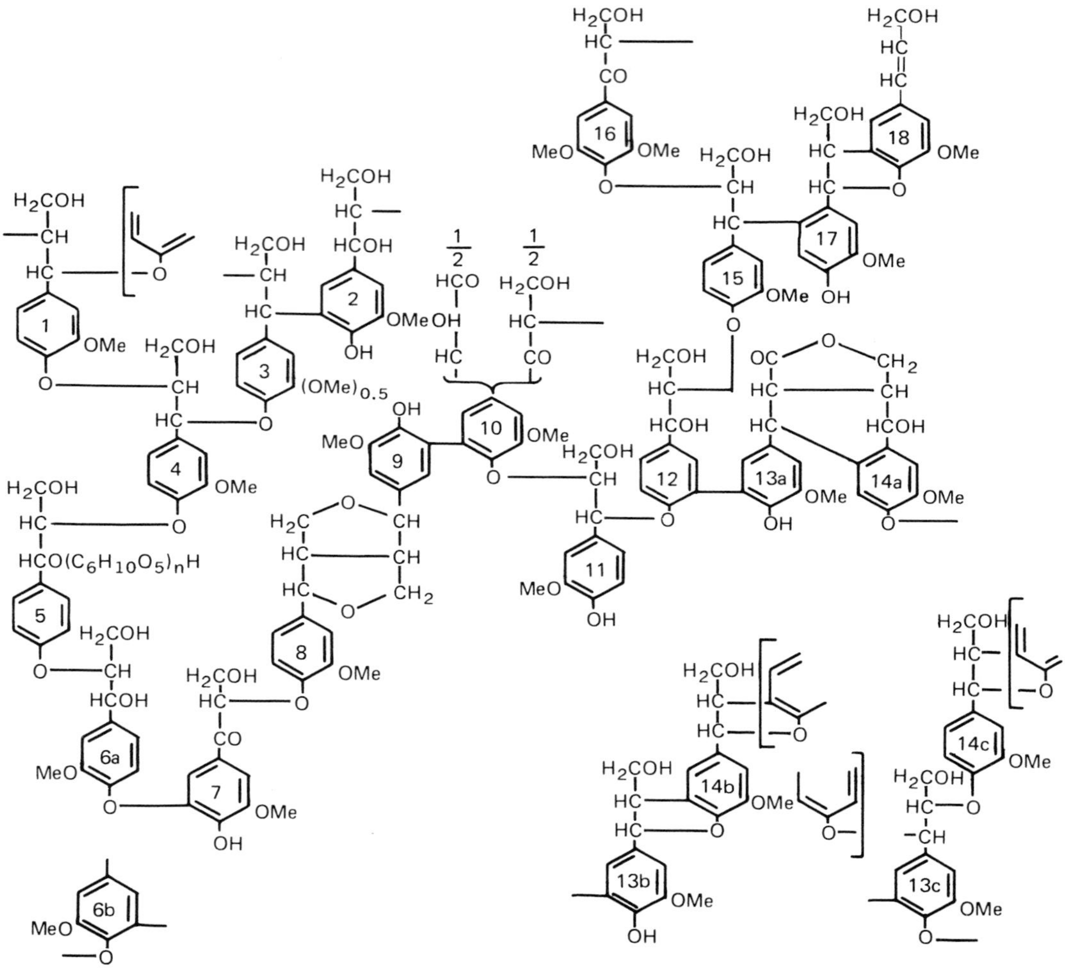

Figure 3. Structural scheme of coniferous lignin according to FREUDENBERG [1962, 1964].

This method worked out for undecomposed plant materials is not very reliable in the case of decomposed residues.

While estimating the amount of lignin, it is possible to simultaneously determine humic substances. In the ether-soluble fraction there are not only fats and waxes, but also lignin degradation products. In light of this, different investigators have altered the method. The determination of humic acids by other methods such as bromacetolysis or sulfacetolysis should also be mentioned (SPRINGER and LEHNER [1941/42], LEHNER and SPRINGER [1950], SPRINGER, KLEE, and WAGNER [1962]).

II. Lignin as an Initial Material for the Formation of Humic Substances

The lignins of different species of plants are polymers or copolymers of different phenyl-propenyl alcohols, which are substituted in the benzene ring with one hydroxyl group. The ring may have one, two, or no methoxyl groups. The lignin of conifers consists mainly of coniferyl alcohol; that of deciduous trees consists of a mixture of coniferyl and sinapyl alcohols; in the case of graminees, p-coumaryl alcohol copolymerizes as a third component.

The structural scheme of spruce lignin (see Figure 3) developed by Freudenberg and co-workers (FREUDENBERG [1962, 1964], FREUDENBERG and HARKIN [1964]) illustrates the trans-formations of lignin during the formation of humic substances. Coniferyl alcohol, as the building block, is connected through different C-C-linkages of the side chains and the rings, and through ether linkages to form a three-dimensional polymer. The different linkages and their frequencies are described by several authors (ADLER [1961], FREUDENBERG [1962, 1964], FREUDENBERG and HARKIN [1964]).

The relatively slow decomposition of the lignin and other phenolic plant constituents leads to the idea that these are important initial materials for the formation of humic sub-stances. Other findings support this idea: for example, the chemical degradation of humic acids to phenolic compounds, which may be derived from lignin and will be discussed later. For this reason the changes in the properties of lignins during humification have been intensively studied by various investigators.

1. Isolation of Lignin Fractions and Their Elemental Composition

Attempts have been made to determine the sources of error during the isolation of lignin with sulfuric acid. One error, coprecipitation of pentoses (NORMAN and JENKINS [1934a]), is reduced by pretreatment with dilute acids. The coprecipitation of proteins (NORMAN and JENKINS [1934b]), which causes errors in the nitrogen content of isolated fractions, has also been investigated. The conditions for the isolation of different lignins from different plant species and the necessary precautions for good yields with the sulfuric acid procedure are described by PLOETZ [1940]. Different methods for the isolation of lignin from rotted straw have been checked in extensive investigations (FLAIG, SCHOBINGER, and DEUEL [1959]). The most useful values were obtained with the 72% sulfuric acid method by careful treatment of the rotted straw (MAEDER [1960]). "Sulfuric acid" lignins can be obtained with a sulfur content of 0.6 to 0.8% and an ash content of 8 to 11%.

The "lignin fractions," which can be isolated from straw during its humification, differ in their elemental composition from the original lignin. Therefore, it is incorrect to call them lignin. The humified residue contains less nitrogen and methoxyl groups than the original lignin, as shown by fractions isolated with dioxane according to the method of BJÖRKMAN [1954, 1956], (BROADBENT [1954], BARTLETT and NORMAN [1938], BARTLETT [1939], FLAIG et al. [1959], FLAIG [1960a], WAKSMAN and SMITH [1934], STÖCKLI [1952], NEHRING and SCHIEMANN [1952a, b]).

Table 5. Elementary analysis of the lignin fractions of rye straw from experiments with and without added nitrogen (calculated for ash-free substances) (MAEDER [1960]).

Days	Nitrogen	Carbon, %	Hydrogen, %	Oxygen, % (Diff)	Nitrogen, %	Sulfur, %	OCH_3, %	Ash, %
0		62.73	5.48	30.55	0.53	0.55	17.08	6 15
70	N 0*	62.73	5.48	31.20	0.54	0.49	15.53	7.65
	N 1†	61.42	5.25	30.20	1.44	0.69	12.76	9.00
120	N 0	62.13	5.42	31.41	0.56	0.48	14.99	8.62
	N 1	60.93	5.41	31.20	1.68	0.78	11.33	10.02
180	N 0	62.20	5.41	31.30	0.56	0.53	14.37	9.31
	N 1	60.94	5.38	31.15	1.74	0.79	10.95	11.24
240	N 0	62.14	5.27	31.03	0.56	0.97	13.46	9.36
	N 1	59.61	5.13	32.77	1.88	0.61	13.46	11.69

* N 0 = No nitrogen added.
† N 1 = 1% as nitrogen as NH_4NO_3 per dry weight of straw.

Tables 5 and 6 show that the conditions during humification and the composition of the initial material (rye and wheat straw) influence the transformations of the lignin. But in all cases a correlation exists between the decrease of the methoxyl content and the increase of the nitrogen content (FLAIG [1960a]), which is taken up later.

2. Differences in the Degradation of Monomers of Lignins During Humification

The rate of decomposition of the monomers of various lignins has been investigated. KRATZL and CLAUS [1962] found by ethanolysis of rotted rye straw that the ratio of guaiacyl to syringyl component is shifted from 1:0.62 to 1:0.37 after 240 days (1% N per dry weight of straw was added as NH_4NO_3). This means that the guaiacyl component is enriched in the rotted straw; this may be caused by a favored degradation of the syringyl component. Syringyl components seem to be more readily utilized by the microorganisms than the guaiacyl components.

In water extracts of rotted straw the syringl compounds could never be identified. Only guaiacyl components such as ferulic acid, vanillic acid, or vanillin could be observed. p-Hydroxybenzoic acid and its aldehyde were also found (MAEDER [1960]). Lignin degradation

Table 6. Elementary analysis of isolated "sulfuric acid" and "Björkman" lignin fractions of wheat straw. (Addition of 0.5% nitrogen as $NaNO_3$ per dry weight of straw, SCHOBINGER [1958]).

	Days of humification	Carbon, %	Hydrogen, %	Oxygen, % (Diff)	Nitrogen, %	Sulfur, %	OCH_3, %
H_2SO_4 lignin	0	58.59	5.60	31.37	0.54	3.9	15.33
	70	58.78	5.93	29.48	2.01	3.8	11.14
	180	58.67	5.91	29.15	2.37	3.9	9.06
	260	58.00	5.83	29.93	3.14	3.9	7.84
	340	54.97	5.96	32.09	3.08	3.9	8.57
	410	54.29	5.86	32.88	2.97	4.0	7.32
Björkman lignin	0	60.68	5.79	33.11	0.42	—	16.76
	70	59.75	7.35	30.93	1.97	—	10.84
	180	58.81	6.56	31.27	3.36	—	8.19
	340	57.37	6.95	31.14	4.54	—	7.50

Substances identified in rotted straw

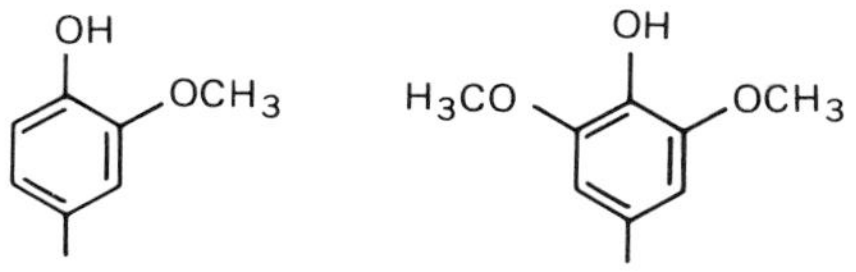

Fresh straw	1	:	0.6
Rotted straw			
(240 days + N)	1	:	0.4
Fresh lignin	1	:	0.6
Lignin fraction			
(240 days + N)	1	:	1.2

Substances identified in water extract of rotted straw:

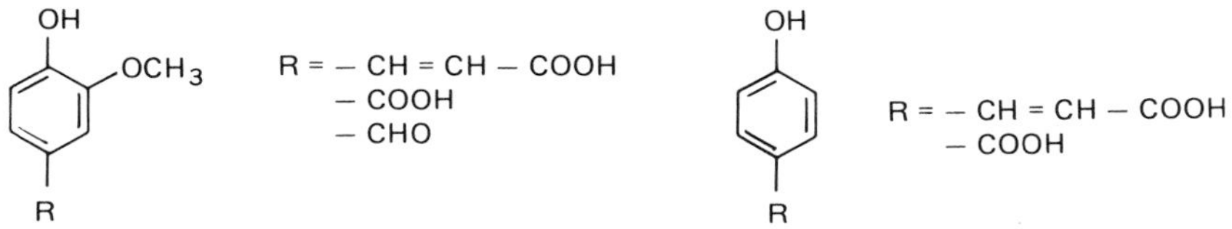

Figure 4. Alterations of the content of the guaiacyl or syringyl component of rye straw during humification (according to KRATZL and CLAUS [1962]).

products were also isolated during investigations of other humified materials (JACQUIN [1963]).

The ratio of guaiacyl to syringyl components changes from 1:0.62 to 1:1.2 in the lignin fraction of rye straw after humification. The guaiacyl component seems to be split off more than the syringyl component or, if the guaiacyl component remains in the lignin molecule, it is demethylated to a larger extent.

HIGUSHI *et al.* [1955, 1956] (Figure 5) investigated the degradation of milled and pre-extracted beechwood by different white and brown rot fungi. Brown rot fungi did not change the lignin remarkably after incubation time of five months at 25°C. But the methoxyl content of the lignin was lowered by the action of white rot fungi. The content of the guaiacyl component of the lignin was also decreased in the milled wood after inoculation.

Some other alterations of the lignin molecule must occur during the humification, especially in the presence of white rot fungi, because the amount of insoluble aldehydes after nitrobenzene oxidation decreases from 18% in the lignin of original wood to 2 to 6% in the lignin of the degraded wood.

The differential degradation of various lignins seems to be connected with the linkages of the monomers in the lignin molecule. It is suggested that these degradations take place mainly between the carbon atom-2 of the side chains and the oxygen on the ring in the adjoining chain. This type of linkage may be prevalent in lignins containing sinapyl alcohol units, because both o-positions of the hydroxyl group in the ring are substituted with methoxyl groups and therefore the linkages between the rings and the side chains or between two rings might be rare (compare with Figure 3).

The elementary analysis of lignin and humic acids (see Table 7) demonstrates that the carbon content of the lignin of different plants is higher than that of humic acids. The decrease of the carbon content and the increase of the oxygen content shows that oxidative processes play an important role during the transformation of lignin into humic acids. Furthermore, the

	Content of lignin*	Total aldehydes in % of lignin[†]	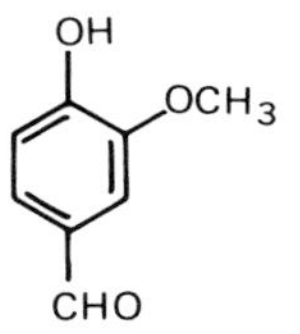		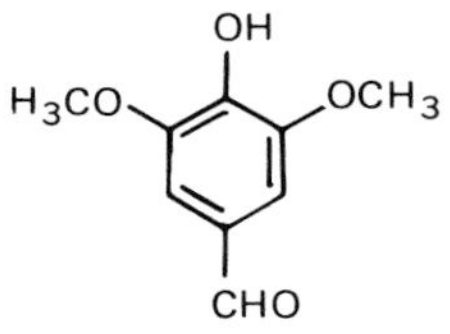
Fresh wood	23.7%	18.4	1 (4.7)	:	2.4 (13.7)
Wood degraded by white rot fungus (Fomes fomentarius)	9.5%	2.6	1 (0.4)	:	4.3 (2.2)
Wood degraded by brown rot fungus (Poria vaporaria)	25.8%	19.1	1 (3.4)	:	3.8 (15.7)

* Calculated for fresh wood.
[†] Content of lignin determined according to Klason.
( ) Calculated for content of Klason − lignin.

Figure 5. Nitrobenzene oxidation of lignin from beechwood degraded by different fungi (according to HIGUSHI et al. [1955]).

Table 7. Elemental composition of lignin, humified lignin, humic and fulvic acids.

		Carbon, %	Hydrogen, %	Oxygen, %	OCH_3, %	Nitrogen, %
Lignin						
(1) Beech (*Fagus silvaticus*)* $C_9H_{6,5}O_2(H_2O)_{0,64}(OCH_3)_{1,41}$		61.87	6.03	32.10	21.65	—
(2) Spruce (*Picea exelsa*)* $C_9H_{7,12}O_2(H_2O)_{0,40}(OCH_3)_{0,92}$		65.08	5.88	29.04	12.87	—
(3) Peat moss (*Sphagnum*)* $C_9H_{7,96}O_2(H_2O)_{0,90}(OCH_3)_{0,25}$		61.67	5.89	32.44	4.36	—
(4) Rye straw (*Secale cereale*)* $C_9H_{6,51}O_2(H_2O)_{1,05}(OCH_3)_{1,13}N_{0,08}$		63.10	5.92	30.67	17.20	0.54
Humified lignin						
5) Rye straw (decomposed 180 days)[†] $C_9H_{5,45}O_2(H_2O)_{1,42}(OCH_3)_{0,68}N_{0,24}$		61.15	5.42	32.42	11.05	1.75
Humic and fulvic acids						
(6) Northern podsol under forest[‡]	h. a.	58.11	5.37	32.00	—	4.52
	f. a.	52.37	3.33	42.89		1.21
(7) Podzolic soil§	h. a.	57.94	5.79	31.41	1.54	4.86
(8) Ordinary chernozem, arable[‡]	h. a.	62.13	2.91	31.38	—	3.58
	f. a.	44.84	3.45	49.36		2.35
(9) Chernozem§	h. a.	57.32	4.25	34.39	1.17	4.04

* FREUDENBERG and HARKIN [1964].   h. a. = humic acids.
[†] MAEDER [1960].   f. a. = fulvic acids.
[‡] KONONOVA [1966].
§ DRAGUNOV [1948].

content of nitrogen increases, whereas the content of methoxyl groups decreases remarkably during humification.

STADNIKOV [1963, 1965] concludes by oxygen uptake of peat constituents in alkaline solution before and after "demethoxylation" (methylether cleavage) in the presence of resorcinol and phloroglucinol that the syringyl component is split off at the beginning of peat formation and that the guaiacyl component is mainly present in the peat constituents.

KRATZL and RISNYOVSKY [1964] and KRATZL, RISNYOVSKY, CLAUS, and WITTMANN [1966] conclude by their investigations of sulfite liquors of cellulose production from coniferous wood that demethylation occurs through semiquinone formation during oxidation in alkaline medium. Thereby, the methylether group reacts like an ester and is saponified by alkali. Reactive groups are formed by demethylation, which are important for further reactions of lignin with other reactive compounds, such as amino acids, and finally for the degradation that occurs in nature by enzymatic processes.

3. *Function of Oxygen*

BARTLETT and NORMAN [1938] and BROADBENT [1954] (Table 8) have found that the number of hydroxyl groups tends to decrease upon degradation. The hydroxyl groups were estimated by acetylation with acetic anhydride in pyridine for 45 minutes and titration of the excess anhydride as acetic acid with alcoholic alkali (Br) after hydrolysis, according to the method of OGG, PORTER, and WILLITS [1945].

The content of carboxylic groups (determined by the increase of the methoxyl content according to Fischer-esterification with methanol and cold hydrochloric acid (Br)), the cation exchange capacity (determined by treatment with barium acetate and exchange with calcium acetate (Ba and N) or with hydrochloric acid (Br)), the exchangeable hydrogen (by the iodite-iodate method (Br)), and the neutralization equivalents (by suspension in dilute alkali during 24 hours and potentiometric back titration (Br)) progressively increase with humification (compare Table 8).

Also, the titration of lignin fractions, isolated from wheat straw according to the method of Björkman, in ethylendiamine with ethanolamine, according to BROCKMANN and MEYER

Table 8. Transformation of the lignin of oat straw at different stages of humification with (N 1) and without (N 0) addition of ammonium phosphate (N 1 = 225 ml ammonium phosphate with 100 ppm per 75 g fresh straw = 3 % N calculated per fresh straw; room temperature), according to BROADBENT [1954]. Determination of lignin according to NORMAN and JENKINS [1934a, b].

Days	Loss of weight of original substance		Some chemical properties											
			Carbon, %		Hydrogen, %		OCH_3, %		OH, %		COOH, meq/100g		CEC, meq/100g	
	N 0	N 1	N 0	N 1	N 0	N 1	N 0	N 1	N 0	N 1	N 0	N 1	N 0	N 1
0	0	0	61.5	63.0	6.67	6.25	11.40	11.48	6.82	7.84	91	81	29	44
14	27.4	22.7	61.9	63.0	6.30	5.88	10.32	11.42	7.53	7.24	98	85	42	47
40	38.8	37.7	61.6	63.9	6.21	6.14	9.52	11.23	7.77	6.63	105	101	62	74
88	44.6	47.6	61.2	63.1	6.35	6.04	10.39	10.37	5.05	5.32	112	107	82	81
135	58.0	56.4	59.7	61.7	6.30	9.24	9.24	8.66	4.13	6.77	122	117	101	110
180	60.1	58.0	60.2	60.7	6.34	5.65	8.33	8.96	4.01	6.70	117	127	100	109
244	67.2(?)	59.3	60.7	61.1	5.81	5.74	8.56	9.00	5.83	6.31	135	115	126	102
355	66.0	61.3	59.8	62.0	6.23	5.85	7.85	7.71	6.37	5.77	128	142	116	118
452	65.5	63.9	59.4	60.5	5.93	5.67	7.85	7.99	6.05	5.20	161	141	124	131

Table 9. Transformations of lignin fractions of wheat straw during different periods of humification, according to FLAIG, SCHOBINGER, and DEUEL [1959].

Days	Equivalent weight	Acid groups equiv./100 g	Readily decarboxylated groups in mol CO_2/100 g	CO_2 formation in % of the acidic groups
Fresh straw	560 ± 13	178	10.6	6
70	677 ± 11	173	14.5	8
180	429 ± 7	233	30.4	13
340	412 ± 9	243	—	—

[1953], shows a decrease in equivalent weight with increasing time of humification (see Table 9).

The titration curves indicate an increase of acidic groups, which is due primarily to an increase in carboxylic groups. A portion of these new acidic groups can be split off with 12% hydrochloric acid to form CO_2. Carbon dioxide may also originate from the side chain of the oxidized lignin building blocks or from acids, which are formed by the cleavage of the ring. The lignin of the fresh straw is altered by treatment with strong acids, whereby ring condensations occur (BRAUNS and BRAUNS [1960]). The products formed are decomposed by oxidation to benzene polycarboxylic acids.

The formation of carbon dioxide can be explained by these reactions as well as by the decomposition of pentosans that might be included. According to the investigations of DEUEL and DUBACH [1958a, b], and DEUEL, DUBACH, and BACH [1958], ketocarboxylic acids, unsaturated heterocyclic acids, and aromatic hydrocarboxylic acids are decarboxylated by treatment with 12% HCl.

The appreciable increase in exchange capacity is, according to BARTLETT and NORMAN [1933], not associated with proteins, which might be coprecipitated during the isolation of the lignin fractions. The addition of egg albumin to lignin isolated from fresh wheat or oat straw in an amount equivalent to the nitrogen content of rotted straw lignin had little effect on the exchange capacity. Furthermore, the authors believe that the increase in the exchange capacity of the lignin fractions isolated from the humified straw is not caused by the condensation of nitrogen compounds with the transformed lignin.

The lignin fractions of oat straw incubated for 180 days with ammonium phosphate were more readily oxidized in 20% sodium hydroxide solution than similar fractions isolated from composted straw without supplemental nitrogen. The increased oxygen uptake can be observed for only 180 days, after which it decreases in both cases, up to the 452-day period of humification. The activity of the microorganisms is accelerated by the addition of nitrogen salts, whereby the transformations of the lignin occur faster. The greatest change occurs during the first 180 days; afterward, there is little change. The alterations in many other properties of the lignin fractions were always greater during the first 180 days in all three experiments (BROADBENT [1954], SCHOBINGER [1958], MAEDER [1960]). The conditions of humification with the addition of nitrogen were generally comparable in all these studies. This observation is important with respect to the later discussion of the spectra of the lignin fractions.

Further investigations on functional oxygen in the lignin fractions isolated from wheat straw according to BJÖRKMAN [1954] showed an increase in carbonyl content from 0.3% in the fresh straw to 0.47% and 0.59% after 70 and 410 days, respectively. Carbonyl content was determined with 2,4-dinitrophenylhydrazine (TRAYNARD and EYMERY [1956]). This increase may be due to an increase of the carbonyl groups in the side chains and perhaps also to an oxidation of phenolic hydroxyl groups to quinones during humification. STEELINK and TOLLIN

Table 10. Methylation of lignin fractions of wheat straw isolated by the Björkman method (SCHOBINGER [1958]).

	OCH_3, %	Nitrogen content, %
Lignin from fresh straw	16.76	0.42
Lignin from fresh straw, methylated	21.76	
Lignin 70 days rotted	10.84	1.97
Lignin 70 days rotted, methylated	13.97	

[1962] have demonstrated that free radicals increase in the lignin fractions during humification and that the free radicals were probably related to the presence of quinonoid groups.

The decrease of the methoxyl content of the lignin fractions during humification (Table 10) could theoretically lead to an increase of phenolic hydroxyl groups by cleavage of the methylether. Because the ratio of the methoxyl content before and after methylation is in both cases about 3:4, no increase in the number of free phenolic hydroxyl groups in the lignin fractions occurs. The increase of nitrogen from 0.42 to 1.97% shows that the phenols have reacted with nitrogen compounds. Model investigations with 1-hydroxy-2-methoxybenzene compounds demonstrated that methoxyl-containing phenols do not condense with primary amino compounds such as amino acids (HAIDER, FREDERICK, and FLAIG [1965]).

4. Function of Nitrogen

It is noted in Table 11 that for fresh straw the sum of nitrogen determined as α-amino nitrogen and that which remains in the residue after hydrolysis is more than 90%, while the sum is lower for the decomposing straw (with the exception of Björkman lignin, 70 days). This result may be explained by the investigations of GALE [1946], who found that certain amino acids are decarboxylated to amines by the action of microorganisms, and therefore cannot be determined by the titrimetric method with ninhydrin (VAN SLYKE, MCFADYEN, and

Table 11. Distribution of the nitrogen of the lignin fractions of wheat straw after different periods of humification (calculated for ash-free substances) (FLAIG, SCHOBINGER, and DEUEL [1959]).

Days of humification	Residue of hydrolysis	Total nitrogen in %	α-Amino nitrogen in the hydrolysate	In residue of hydrolysis As NH$_3$	As "residual" nitrogen	Loss of weight by hydrolysis
Lignin fractions (isolated with sulfuric acid)						
0	44.0	0.55	58.1	40.0	0	17
70	44.6	2.09	39.2	27.8	18.7	20
180	55.7	2.46	21.1	25.2	18.7	24
260	42.4	3.26	20.8	17.2	25.1	24
340	52.1	3.20	20.3	25.6	26.5	32
Lignin fractions (isolated according to Björkman)						
0	43.0	0.42	47.6	43.0	0	24
70	57.5	1.97	41.1	—	—	46
180	40.7	3.36	33.9	—	—	43

In % of total nitrogen (= 100) before hydrolysis

Table 12. Elemental composition of the residues of lignin fractions of wheat straw after hydrolysis.

Days of humification	Carbon, % A*	B†	Hydrogen, % A	B	Oxygen, % A	B (Diff)	Nitrogen, % A	B
0	63.39	69.35	5.41	5.88	30.98	24.59	0.22	0.18
70	60.15	68.45	5.68	7.45	33.20	22.97	0.97	1.13
180	60.40	67.36	5.66	6.37	32.86	24.90	1.08	1.37
260	58.86	—	5.61	—	34.15	—	1.38	—
340	56.66	—	5.62	—	36.05	—	1.67	—

* Isolated with sulfuric acid.
† According to Björkman; % calculated for ash-free substances, without the addition of nitrogen salts.

HAMILTON [1941]). Thirteen amino acids could be identified in some cases with paper chromatography in the hydrolysates of fractions in which α-amino nitrogen was found.

An oxidative deamination of amino acids can occur in an oxidizing medium by o-diphenol or quinones that are formed from the lignin degradation products by cleavage of the methylethers. The increase in nitrogen content of the residue measured by hydrolysis and, the increase of NH_3 nitrogen and residual nitrogen measured by reductive fusion with potassium hydroxide and sodium acetate (BREYHAN [1956]) indicate that the nitrogen is bound in such a way that it can be split off only under the drastic conditions of the alkali fusion (NH_3-N) or is no longer hydrolyzable ("residual"-N). Also, indole compounds could be detected (FLAIG and BREYHAN [1956]).

A considerable loss of weight is observed upon recovery of the residues after hydrolysis; it cannot be explained on the basis of hydrolyzable amino acids. Therefore, other reactions, as yet unknown, must occur during the hydrolysis with 6 N hydrochloric acid.

The elementary analysis of the residue after hydrolysis shows increased carbon values (Table 12) in comparison with those of the nonhydrolyzed lignin fractions (Table 6). The observed increase of carbon and the decrease of oxygen following hydrolysis of the lignin fractions isolated by the Björkman method cannot be entirely explained by the splitting off of the amino acids. Presumably, alterations can occur by treatment with 6 N hydrochloric acid, which lead to the formation of polycyclic condensation products (BRAUNS and BRAUNS [1960]).

The quotient of the gram-atoms of methoxyl divided by the gram-atoms of nitrogen of the lignin fractions isolated from decomposing rye straw decreases with time and amount of nitrogen added (FLAIG [1962]). (See Table 13.)

Table 13. Ratio of methoxyl groups to nitrogen in decomposing straw.

Days of decomposition	$\dfrac{\text{g-atoms OCH}_3}{\text{g-atoms N}}$ in the lignin fractions		
	N 0*	N 1†	N 2‡
70	13.2	6.3	4.0
120	12.3	4.6	2.8
180	11.7	3.4	2.8
240	10.8	3.1	2.5

* N 0 = No nitrogen.
† N 1 = 0.5% nitrogen.
‡ N 2 = 1.0% nitrogen as NH_4NO_3 per dry weight of straw.

The value of the quotient can be used as a measure of the increasing alteration of the lignins during humification. Fractions giving the same quotient were comparable, as was the case for the infrared spectra of the lignin fractions 70 days/N 2 and 120 days/N 1 or 120 days/ N 2 and 240 days/N 1. Further analytical data for the different fractions isolated from humified rye straw are given in Figure 2.

5. *Spectroscopic Investigations*

The absorption spectra of the lignin fractions change with increasing periods of humification. The maximum in the region of 275 to 285 nm, (Figure 6) which is attributed to the benzene, ring substituted with oxygen, and the maximum between 300 and 350 nm, which is caused by the presence of chromophoric groups of the side chain (carbonyl groups, olefinic double bounds) (JONES [1949]), are reduced in the lignin fraction from rye straw decomposed for 410 days with added nitrogen (SCHOBINGER [1958]). The absorption curves become less characteristic with increasing time of humification and tend to become straight lines. The slope depends on the composition and gradually approaches the spectra of humic acids, which have been investigated more extensively by FRÖMEL [1938a, b; 1941].

The transformation of the lignin fractions during humification has also been investigated by infrared spectroscopy (FLAIG, SCHOBINGER, and DEUEL [1959], FLAIG [1958, 1964a], FLAIG and BEUTELSPACHER [in press], FARMER and MORRISON [1960]).

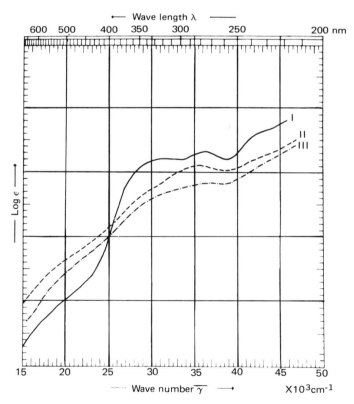

Figure 6. Ultraviolet absorption spectra of lignin fractions from rotted straw (SCHOBINGER [1958].)
I = lignin according to Björkman, fresh.
II = lignin according to Björkman, 410 days (+N).
III = lignin according to Björkman, 410 days.

The lignin fractions of fresh wheat and rye straw show the same bands between 5.8 and 7.75 μ. The method of isolation, whether with 96% alcohol according to BRAUNS [1939], with 90% dioxane according to BJÖRKMAN [1954, 1956], with sulfuric acid according to a modified method of Klason (RITTER, SEBORG, and MITCHELL [1932]), or with thioglycolic acid according to HOLMBERG [1934], does not change the pattern. The C = C valency vibrations of carboxyl groups at 5.8 μ are caused by the thioglycolic acid, if this acid is used for the isolation. Changes in the bands at 8 to 9 μ occur when sulfuric acid is used for isolation. These changes may be related to acid-catalyzed ring condensations similar to those which occur with coniferous lignin (BRAUNS and BRAUNS [1960]).

An increase of the intensity of the C=O band at 5.85 μ with increase of time of humification (Figure 7) may be due to an increase of carboxyl groups, because the consumption of alkali increases during titration (FLAIG, SCHOBINGER, and DEUEL [1959], BROADBENT [1954]). The number of carboxyl groups determined by chemical methods also increases (BROADBENT [1954]).

The absorption of the C=C valency vibrations at 6.25 and 6.65 μ may be caused chiefly by the aromatic parts of the lignin molecules and does not change very much with time. The increasing background absorption may be explained by the assumption that with time a greater variety of aromatic components are present, although the total quantity may remain

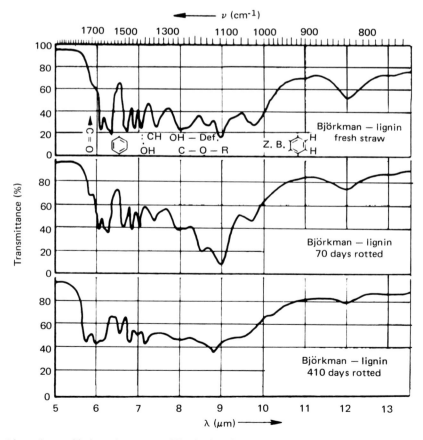

Figure 7.   Alterations of infrared spectra of lignin fractions during humification (SCHOBINGER [1958]).

the same. The steady decrease of the methoxyl content of the lignin fractions, the increase of the syringyl component, and the increase of nonhydrolyzable nitrogen support this explanation.

A mixture of model substances such as p-hydroxybenzaldehyde, vanillin, and syring-aldehyde, which approximately simulates the composition of the corresponding building blocks of the lignin, gives a spectrum with approximately the same arrangement of the bands as the lignin of the fresh straw isolated according to Brauns or Björkman. The spectrum of this mixture also shows an increased background absorption in comparison with the spectra of the single aldehydes (FLAIG and BEUTELSPACHER [in press]). The bands of the aldehyde groups are not in the lignin spectra.

The alterations of the C—H deformation vibrations at 6.85 and 7 μ may be connected with alterations in the side chains. The strongly marked bands at 7.95 μ and 8.95 to 9.10 μ are due to arylalkyl and other ether linkages based on investigations with model substances. The strong absorption of the ether bands at 8.95 μ in the case of the lignin fractions of fresh or rotted straw and the weak absorption for polymers of guaiacol may be caused by syringyl components (FLAIG [1964a]). Syringaldehyde as well as a mixture of the aldehydes, p-hydroxy-benzaldehyde, vanillin, syringaldehyde in the ratio 0.5:1:0.5 (which is nearly the same ratio of these monomers in straw lignin) also show strong absorption at 9.05 or 9.00 μ. The absorption at 6.1 to 6.20 μ increases with the increased nitrogen content of the lignin fractions during humification. In this region the N—H deformation and C=N valency vibrations occur. The latter may be due to heterocyclic compounds. The spectra of lignin fractions isolated with sulfuric acid from the same humified straw samples to which different amounts of nitrogen were added but which had the same nitrogen content were identical in the region from 5.8 to 8.8 μ (FLAIG [1964a]).

Polypeptides have a strong amide absorption band between 6.40 and 6.70 μ. Albumin from cow serum, for example, has a broad maximum from 6.60 to 6.70 μ. The intensity of the absorption at 6.65 μ is not changed during humification.

If a mixture of lignin and proteins was precipitated during the isolation of the lignin fractions, the absorption of this band should become stronger, because the nitrogen content of the lignin fractions increases with time. When lignin from fresh straw was treated with albumin nitrogen in amounts equal to the nitrogen in humified straw lignin, the 6.65 μ band was intensified.

A further proof that the isolated lignin fractions are not mixtures of lignin and proteins is available from the treatment of mixtures of nitrogen-containing lignin fractions with proteins such as albumin with 72% sulfuric acid.

The nitrogen content of reprecipitated lignin fractions is not altered because the added albumin is hydrolyzed by the 72% sulfuric acid. Furthermore, the hydrolysis of the lignin fractions with 6 N hydrochloric acid shows that the percentage of α-amino nitrogen decreases continuously with time, whereas nonhydrolyzable nitrogen increases with time. The latter is probably heterocyclic-bound.

The α-amino nitrogen, which is hydrolyzable only with 6 N hydrochloric acid, presumably consists of peptides that are difficult to hydrolyze and are linked to the lignin parts through the amino groups of the N-terminal amino acids. These problems require further elucidation.

So-called "ligno-protein" complexes have been isolated from soils by TINSLEY and ZIN [1954], JENKINSON and TINSLEY [1959, 1960], and GOULDEN and JENKINSON [1959] by extraction with aqueous sodium sulfite solution. An interface is formed by shaking the extracts with carbon tetrachloride, from which the ligno-protein is isolated (modified Sevag technique). The carbon tetrachloride extracts contain 20% of the nitrogen of the soil extract and 57% of the nitrogen of an extract of wheat straw composted for 480 days.

The nonhydrolyzable part of the ligno-protein from straw compost yielded an infrared spectrum that was almost identical with the infrared spectrum of straw lignin. The spectrum of the preparation isolated from a soil showed some differences.

6. *Degradation to Low Molecular Weight Components*

Up to this point the transformation of lignin that may occur within or on the edges of the intact molecules has been described. Another type of transformation that definitely occurs is the degradation into low molecular weight compounds. Various phenolic lignin degradation products are found (Figure 8), even after a short treatment of lignin with acids or with alkali (BRAUNS and BRAUNS [1960], SAALBACH [1958]). Several phenols that appear to be lignin-derived have been isolated from decomposing plant materials and from microbial cultures (especially of lignin-decomposing fungi) living on Brauns lignin, conifers, deciduous trees, and grasses (BÖRNER [1955, 1956, 1957], HENDERSON [1955], ISHIKAWA, SCHUBERT, and NORD [1963a], SEIFERT [1962]).

Some of these compounds, such as p-hydroxybenzaldehyde, vanillin, vanillic acid, p-hydroxycinnamic acid, and ferulic acid, could be isolated during 240 days of humification of rye straw (MAEDER [1960]). Obviously, these compounds were continuously released during the entire incubation period, because their resistance to oxidation or biological decomposition varies (SÖCHTIG [1961a, b]). Their formation can occur by oxidation and cleavage reactions on the side chains and by cleavage of the benzyl- and the arylmethyl ether. These lignin degradation products are important for the formation of humic acids or their precursors, in that some of them condense to dark-colored polymers in the presence of nitrogen-containing compounds

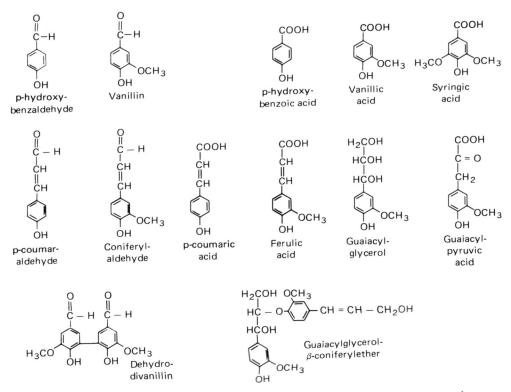

Figure 8. Identified lignin degradation products in soils and cultures of microorganisms.

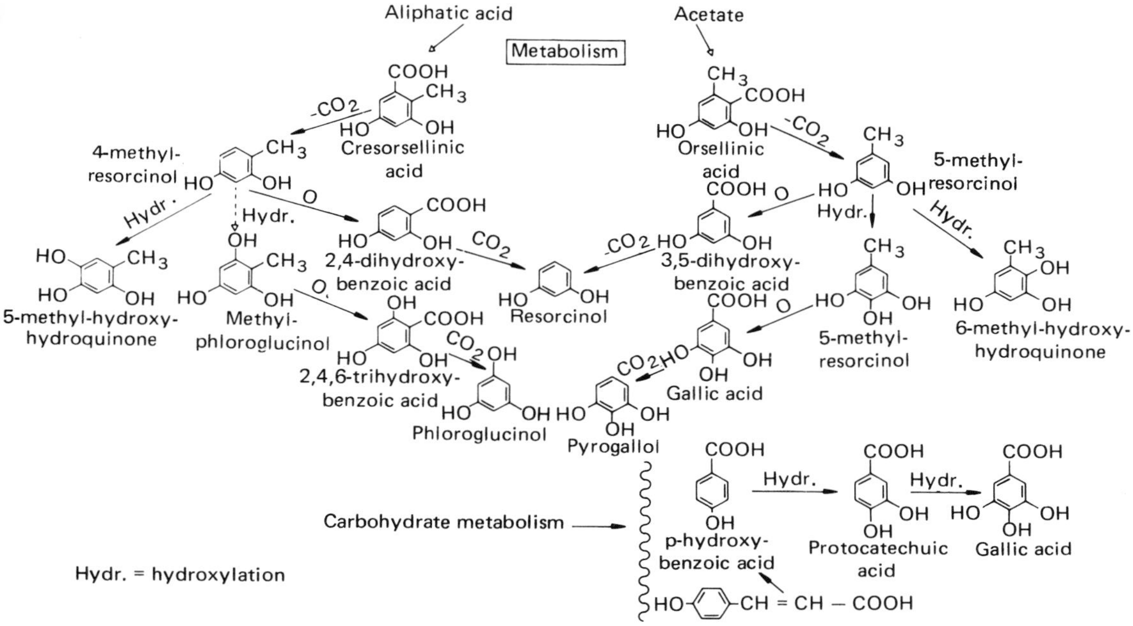

Figure 9. Formation and transformation of phenols by microorganisms (*Epicoccum nigrum*) (HAIDER and MARTIN [1967]).

derived from proteins. The condensation occurs in the presence of phenoloxidases in an oxidizing environment. The properties of the dark-colored polymers are similar to those of humic acids (FLAIG and HAIDER [1961a]). Further attention is given to these reactions in a later section. Some of the lignin degradation products have been identified in soils. WHITEHEAD [1964] identified p-hydroxybenzoic acid, vanillic acid, p-hydroxycinnamic acid, and ferulic acid in concentrations of around 10^{-5} mol in four different soil solutions obtained by pretreatment with calcium oxide in aqueous solution and followed by filtration and acidification. The phenols were recovered by extraction of the filtrates with ether. Previously, only p-hydroxybenzoic acid (WALTERS [1917]) and vanillin (SHOREY [1913]) had been isolated.

III. Microbial Synthesized Phenols as Units for the Formation of Humic Acids

Especially by the work of BURGES, HURST, and WALKDEN [1963, 1964] or MORRISON [1963] and FARMER and MORRISON [1964], it is known that besides the phenolic derivatives, which can be derived from lignin, phenols derived from flavonoids are also important initial materials for the formation of humic substances. These authors isolated, by reductive or oxidative cleavage of humic acids, compounds that belong to 1,3-di or 1,3,5-triphenols and that cannot be derived from lignin or its degradation products.

Recently, MARTIN, RICHARDS, and HAIDER [1967] and HAIDER and MARTIN [1967] have demonstrated the participation of phenols of the resorcinol type in the formation of humic substances that are formed in the culture of microorganisms such as *Epicoccum nigrum* (see Figure 9). This fungus forms dark-colored substances with properties comparable with those of the humic acids. Different resorcinol derivatives could be identified after reductive cleavage of the microbial synthesized humic acids. Other phenolic intermediates could be isolated from the culture media.

Orsellinic acid was identified. This is formed through the acetate metabolism of the organism. 3,5-Dihydroxytoluene (Orcinol) is formed by decarboxylation. By oxidation 3,5-dihydroxybenzoic acid is formed. Decarboxylation forms resorcinol, which is seldom found. 3,5-Dihydroxytoluene, 6-methylhydroxyhydroquinone (2,3,5-trihydroxytoluene), 4-methylpyrogallol, and gallic acid are formed by hydroxylation. Presumably pyrogallol is formed by the decarboxylation of gallic acid.

Cresorsellinic acid, which could also be identified, may be formed through the aliphatic acid metabolism. By reactions similar to those in the case of orsellinic acid, 2,4-dihydroxytoluene, 2,4-dihydroxybenzoic acid, and also resorcinol are formed. The hydroxylation of 2,4-dihydroxytoluene leads to 2,4,6-trihydroxytoluene and 5-methylhydroxyhydroquinone (2,4,5-trihydroxytoluene).

In the culture solution of the fungus other substances such as p-hydroxybenzoic-, protocatechuic-, gallic-, and p-hydroxycinnamic acids could be identified. These acids are proposed to be formed through the carbohydrate metabolism by the reaction of phosphoenolpyruvic acid with erythrose-4-phosphate to shikimic acid and then by some further metabolic pathways.

The question arises to what extent the phenolcarboxylic acids isolated from decomposing plant materials or from soils are derived from lignin-degradation products or from microbial-synthesized compounds.

The 5- and 6-methylhydroxyhydroquinones (2,4,5- and 2,3,5-trihydroxytoluene), as well as pyrogallol and its methyl and carboxyl derivatives, are responsible for the formation of the higher molecular weight humic acid such as substances in the culture media of *Epicoccum nigrum*, because resorcinol derivatives do not react with amino acids under these conditions and do not form this type of nitrogenous compound.

In this connection it must be mentioned that hydroxyhydroquinone is formed after the enzymatic oxidation of different lignin degradation products in microbial cultures to protocatechuic acid and by further oxidation of this acid in the presence of phenol oxidazes (FLAIG and HAIDER [1961a], HAIDER, LIM, and FLAIG [1964]). These derivatives of polyphenols can be oxidized to quinones in the culture media at pH values of 6 to 8. This is not the case with derivatives of resorcinol and phloroglucinol. This means that only compounds that are able to form quinones would be important for the formation of nitrogenous humic substances.

IV. Isolation of Humic Substances

The substances that are formed from plant material during humification consist of a mixture of various compounds. Because these compounds have very similar properties, the extraction of substances with uniform compositions is very difficult. The extraction is accompanied by more or less large alterations, such as hydrolytic degradation or condensation reactions, depending upon the procedure. These alterations especially accompany colored substances, such as fulvic acids, hymatomelanic acids, and humic acids, which are formed during humification. Therefore, the separation procedures lead to very different results, according to the physicochemical properties.

A further difficulty in extracting substances is the fact that the organic substances in the soil are bound by bivalent and trivalent cations and possibly by hydrogen-bridge linkages of phenolic or aliphatic hydroxyl groups or of carboxylic groups, and also, to a lesser extent, of amino groups with silicic clay minerals. Polar solvents with high dielectric constants, which increase the dispersity of humic substances or which increase the solubility by disrupting the hydrogen bonds of the fixed metallic cations and immobilizing them, are most efficient extractants causing little chemical alteration (WHITEHEAD and TINSLEY [1964]). The solvents used should be selected for the substances desired, so that purification and separation in substantially uniform components without chemical alterations are possible.

Difficulties in the separation of humic acids from inorganic soil constituents led to the investigation of humic substances of B horizons of podzolic soils; the extraction of these is less difficult than from B horizons of other soils (BURGES and LATTER [1960], COFFIN and DeLONG [1960], DUBACH, MEHTA, and DEUEL [1961], JACQUIN [1960], MARTIN and REEVE [1955], MARTIN, DUBACH, MEHTA, and DEUEL [1963], MEHTA, DUBACH, and DEUEL [1963a], RUCHTI [1962], SCHNITZER, WRIGHT, and DEJARDINS [1958], SOWDEN and DEUEL [1961], STEELINK, BERRY, HO, and NORDBY [1960], WRIGHT and SCHNITZER [1960], MARTIN and REEVE [1957a, b]). Humic acids are easily dissolved only as salts of monovalent cations, such as sodium, potassium, and ammonium. The hydrogen, calcium, ferrous, or aluminum salts, and salts of heavy metals are completely insoluble. Therefore, the efficiency of the applied extraction procedures depends on an increase of the specific solubility of one or all components for maximum separation from the soil minerals, which is often by specific fraction procedures.

Many extractants have been used for separation of organic soil substances (see SCHEFFER and ULRICH [1960], TINSLEY and SALAM [1961], DUBACH and MEHTA [1963]). The solvents can be divided according to their chemical and physical efficiency into several groups: (1) acids, (2) strong alkali, (3) weak alkali, (4) complexing agents, and (5) organic solvents.

A pretreatment with acids is applied for desorption of humic substances that are fixed on sesquioxides, free silicic acid, or clay minerals, and for removal of carbonates or exchangeable bases. A pretreatment of soils with hydrochloric acid and/or a mixture of hydrochloric acid with hydrofluoric acid and followed by leaching with water until complete decalcification leads to increased solubility in alkalis and therefore increased yield by extraction (BREMNER and HARADA [1959], CHOUDRI and STEVENSON [1957], DUBACH, MEHTA, and DEUEL [1963], POSNER [1966]).

SCHNITZER and WRIGHT [1956, 1957] established, by extraction of podzolic soils with diluted inorganic acids such as hydrofluoric or hydrochloric acid or a mixture of them in 0.5% concentration, that the acids extracted only 2% of the total carbon content from the A_0 horizon and up to 96% of the carbon content from the B horizon of a podzolic soil. Higher concentrations of acids were less effective. Furthermore, a correlation exists between the determined carbon of the organic substance extracted with 0.5% hydrochloric acid and the content of iron and aluminum. The maxima were in the extracts of the B horizon. These results confirm the investigations of PONOMAREVA [1947], who found that large quantities of fulvic acid fractions are dissolved by cold, diluted hydrochloric acid from the podzol B horizons of Russian tundra soils, but that very small quantities are dissolved from chernozem and forest soils. PARSON and TINSLEY [1960] dissolved 23% of the organic material from a calcareous meadow soil by refluxing with formic acid and precipitation with di-isopropylether. It, therefore, cannot be denied that chemical alterations, such as hydrolysis of carbohydrates and proteins, occur.

Since the first isolation of humic acids from a peat soil with sodium hydroxide by ACHARD [1786], aqueous solutions of alkali hydroxides have been used most often, because these solutions possess a maximum percentage extraction. Later, objections arose from their application as follows:

(1) Dissolution of silicic acid, which leads to impurification of the humic substances.
(2) Dissolution of not yet degraded plant material from fresh tissues, such as hemicellulose and lignin (SCHLICHTING [1953a, b], SCHEFFER, ULRICH, MEYER, and OEHMICHEN [1956], SPRINGER and KLEE [1958]).
(3) Chemical alterations, such as hydrolysis of polymers, amino acids, and sugars, or condensation reactions between nitrogenous compounds, phenols, or perhaps aldehydes, which might lead to the formation of humic acids.

As another objection to this method, the substances extractable with alkali might become oxidized by air in alkaline solution (DUBACH, MEHTA, and DEUEL [1963], MEHTA, DUBACH, and DEUEL [1963a], SCHEFFER and ULRICH [1960]).

MULDER [1862], BERTHELOT and ANDRÈ [1892], SCHREINER and SHOREY [1910], GORTNER [1916], ODEN [1919], HOBSON and PAGE [1932], FORSYTH [1947a], BREMNER [1949, 1955a], LYNCH, WRIGHT, and ONLEY [1957], and MORRISON [1958] used, after treating the soil with acid or directly, a 2% or 0.5 N solution of sodium hydroxide. SPRINGER [1934] used a hot 2.5 N solution of sodium hydroxide for extraction of "brown" humic acids whereas SPRINGER [1938] and SPRINGER and KLEE [1958] used 0.125 N sodium hydroxide to extract "gray" humic acids from soil humus, while SHMUK [1930] and later TYURIN [1940], KONONOVA [1947], KONONOVA and BELCHIKOVA [1956], KONONOVA and ALEXANDROVA [1956a] extracted with 0.1 N sodium hydroxide solution for extraction.

SIMON [1929] dissolved insoluble salts by extraction with neutral alkali salts such as sodium fluoride or sodium oxalate, which precipitate calcium ions. BREMNER and LEES [1949] found, in detailed investigations with 80 different solutions of sodium salts, that 0.1 N sodium pyrophosphate solution was the most effective. They established that the efficiency of the extraction with sodium salts of inorganic acids decreases as follows: pyrophosphate > fluoride > hexametaphosphate > orthophosphate > borate > chloride > bromide > iodide. For the sodium salts of organic acids the order is the following: oxalate > citrate > tartrate > malate > salicylate > benzoate > succinate > p-hydroxybenzoate > acetate.

TINSLEY and SALAM [1961] investigated the extraction of calcareous meadow soils. Their results agree with other authors that sodium pyrophosphate, oxalate, and citrate are the most effective. However, they found, too, that sodium pyrophosphate and sulfite gave nearly the same results in 0.25 M solution at pH values from 8.0 to 8.2.

SCHNITZER, WRIGHT, and DESJARDINS [1958] compared the efficiency of 15 different solvents for the separation of organic substances from two horizons of a podzol profile. As is shown in Table 14, only sodium hydroxide extracted reasonable amounts from an A_0 horizon. From the B_{21} horizon 96% of the total carbon content was extracted with sodium

Table 14. Effectiveness of various extractants for organic matter from two horizons of a podzol profile (according to SCHNITZER, WRIGHT, and DESJARDINS [1958]).

Extracting solution	Concentra- tion	pH before extraction	pH after extraction of		Percentage of carbon extracted from	
			A_0 horizon	B_{21} horizon	A_0 horizon	B_{21} horizon
$Na_4P_2O_7 \cdot 10 H_2O$	0.1 M	9.8	6.5	8.9	6.1	91.7
$Na_4P_2O_7 \cdot 10 H_2O$	0.1 M	7.0	5.2	6.2	5.5	82.6
$Na_3PO_4 \cdot 12 H_2O$	0.1 M	12.0	7.3	9.4	9.1	93.6
$(NaPO_3)_6$	1% (W/V)	6.6	4.5	6.0	1.2	37.0
$Na_2B_4O_7 \cdot 10 H_2O$	0.1 M	9.2	7.7	8.1	7.7	80.8
NaF	0.5 M	8.2	6.4	10.4	6.6	88.4
NaF	0.5 M	7.0	6.4	10.3	4.4	88.7
NaCl	0.5 M	5.5	2.9	3.9	2.5	2.2
NaBr	0.5 M	6.0	2.9	3.9	T*	2.7
NaI	0.5 M	8.8	3.1	4.0	T*	3.2
Na_2CO_3	0.5 M	11.2	9.5	9.9	8.3	92.3
NaOH	0.5 M	13.1	12.6	13.1	24.3	96.3
HCl	0.5 %(V/V)	1.4	1.7	2.9	T*	11.2
HF	0.5 %(V/V)	2.2	2.1	4.0	T*	83.1
$EDTA-Na_2$	0.2 M	4.2	3.7	4.2	T*	97.0

* T = traces.

hydroxide, 94% with sodium phosphate, 92% with sodium pyrophosphate, 88% with sodium fluoride, and 83% with hydrofluoric acid. According to ALEXANDROVA [1960], sodium pyrophosphate forms complexes with calcium ions and is therefore especially suitable for extraction of soils with a high calcium content (*e.g.*, serosems). A disadvantage of this extractant is its additional complex formations with water-soluble compounds of aluminum or iron.

GÄTKE [1961] extracted at the boiling point chernozem and different degraded brown-earths with 7.5% sodium carbonate solution. He investigated the effectiveness of the extraction by absorption measurements in the visible range of light. On account of the small deviation of the colorimetric measured values of the "Farbtonwert" (Welte factor: $Q_{4/6}$) and the "Farbwert" (color intensity), which means the maximum extinction of the extracted humic acid solution in relation to the original humus concentration, which show indirect proportionality from a so defined hyperbolic function, he assumes that in contrast to a mild extraction with 0.01 N sodium hydroxide, no essential impurities occur in the extract from soluble decomposition products.

The examples show that the amount of extracted substances increases with increasing pH value of the extractant. Furthermore, it could be established that neutral solutions of pyrophosphate and fluoride extract more organic substances than sulfate, chloride, and ortho-phosphate at comparable pH values. The yields of extraction of all reagents are increased by raising the extraction temperature.

To avoid the undesired fixation of the organic substance on inorganic soil components, reagents for complex formation such as ethylendiamine tetra-acetic acid (EDTA), acetylacetone, and others were used (MARTIN and REEVE [1955, 1957a, 1958]). SCHNITZER, WRIGHT, and DESJARDINS [1958] compared the extraction ability of EDTA-Na$_2$ with different solutions. They demonstrated that 97% of the total carbon content was extracted at a pH value of 4.2 from a podzol B horizon, but only traces from the A$_0$ horizon under the same conditions (Table 14). The disadvantage of this extraction method is the impossibility of determining the carbon or nitrogen content of the extracted material.

DEUEL and DUBACH [1958b] and DUBACH, MEHTA, and DEUEL [1961, 1963] confirmed that about 90% of the organic substances can be extracted from the B horizon of a podzol by EDTA at pH value 7. They succeeded in isolating about 40% of a humic acid fraction, which is precipitated by a cation exchanger; about 50% as fulvic acids, precipitable by barium; and 6% as fulvic acids, preciptiated by lead salts. The extractants, such as acetylacetone and 8-hydroxyquinoline and their metal complexes, can easily be removed with organic solvents (COFFIN and DELONG [1960], MARTIN and REEVE [1955, 1957, a, b]). Some of the ionic extractants (EDTA, sodium pyrophosphate, perchloric acid, hydrochloric acid, and a mixture of hydrochloric/hydrofluoric acid) can be removed by precipitation (SOWDEN and DEUEL [1961]) or by desalting with ion exchangers from the solution; thereby, a remarkable loss of humic substances often occurs (DUBACH, MEHTA, and DEUEL [1961, 1963], PROKH [1961], HORI and OKUDA [1961]).

The pretreatment of soils with acid or with complex forming agents has an advantage in that the humic substances become soluble in organic solvents also (acetone, dimethylformamide, pyridine, dioxane, tetrahydrofurane), especially when these are mixed with water (DUBACH, MEHTA, and DEUEL [1963], SCHEFFER and ULRICH [1960], BROMFIELD, COULSON, and DAVIES [1959], BURKAT [1959], FORSYTH [1947a], KLAMROTH [1954], KOSAKA, HONDA, and IZEKI [1961], ZETZSCHE and REINHART [1937]).

WHITEHEAD and TINSLEY [1964] investigated the suitability of dimethylformamide as an extractant of organic substances. They found that dimethylformamide dissolves a small amount of organic soil substance, but that 55 to 65% of the total organic substance of soils and 80% from a podzol B horizon are dissolved after the addition of a 0.3 to 0.8 molar solution of hydrofluoric or hydrofluoric-boron acid. Dimethylformamide has a high dipole moment in comparison with other solvents and a strong tendency to form hydrogen-bridge linkages with compounds containing hydrogen atoms in resonance. Its disadvantage for extraction purposes is its nitrogen content, since it is not possible to follow the extraction by determination of nitrogenous compounds.

Because the selection of an extractant has a decisive influence on the result of fractionation, WIESEMÜLLER [1965] compared different extraction procedures reported in the literature in the case of chernozem. The results (Figure 10) show that the use of strong and hot alkaline or acidic solutions causes degradation.

V. Fractionation of Humic Substances

The exhaustive extraction of the organic soil constituents is usually followed by a treatment with acid to separate the extracts in the acid-soluble fraction (fulvic acids) and in the acid-precipitable fraction (humic acids). The problem of separation of the acid-precipitable fraction is considered to be relatively simple in the case of humic acids. The purification of

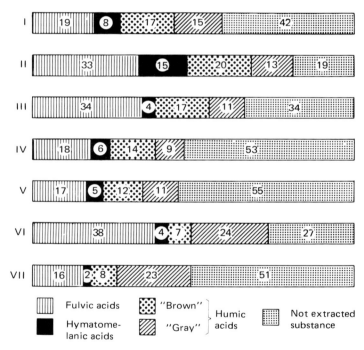

Figure 10. Comparison of different extractants (data in percentage of total C) (WIESEMÜLLER [1965]).

I = 2% HCl cold, 1% NaOH, cold: NEHRING [1955], TYURIN [1937].
II = 5% HCl, 70°C, 0.5% NaOH, boiling: ⎫
III = 5% HCl, 70°C, 1% NaOH, cold: ⎬ SPRINGER [1938].
IV = 0.1 M Na₄P₂O₇, 0.1 N NaOH, 16 hours, cold: KONONOVA and BELCHIKOVA [1961].
 V = 0.1 N Na₄P₂O₇, 0.1 N NaOH, 16 hours, cold: ⎫
VI = 0.1 N Na₄P₂O₇, 8 hours, boiling: ⎬ WELTE [1956].
VII = 0.1 N Na₄P₂O₇, 5% hydrazinehydrate, 16 hours, cold: ⎭

fulvic acids, which are found in the acid solutions in a smaller amount, is considered more difficult. Usually humic acids are purified by repeated precipitation, dialysis in acid solution, and electrodialysis (COFFIN and DeLONG [1960], DUBACH, MEHTA, and DEUEL [1963], EVANS [1959], MARTIN and REEVE [1955], SCHEFFER and ULRICH [1960]).

The difficulty in separating uniform fractions is because many experiments have been made in different ways using the classic methods of precipitation at different pH values, by organic solvents by polyvalent nitrogen bases, by formation of precipitates with salts of heavy metals or salts of alkaline earths, by paper chromatography, by column chromatography on cellulose starch, silica gel, or aluminium oxide, by anion exchangers, by cellulose derivatives, by electrophoreses, or by gel filtration.

Principally, two different methods were followed for further fractionation of the organic parts of soil extracts:

(1) Separation of low molecular weight substances, fulvic acids, by-products of carbohydrate, protein and lignin degradation from the fraction of humic acids extracted with sodium hydroxide for operation of purification.

(2) Experiments for specific separation of the main fractions (fulvic and humic acids) by means of different methods of physicochemical and quantitative analyses.

1. *Isolation and Removal of Low Molecular Weight Substances and Fulvic Acids*

As was previously mentioned, carbohydrates, proteins, and lignin are dissolved by extraction with sodium hydroxide. Therefore, many attempts have been made with relatively mild methods to separate these accompanying substances. Carbohydrates and proteins have been removed by hydrolysis with acids (FORSYTH [1947b], SCHEFFER and KICKUTH [1961a, b], or by enzymatic degradation (HOBSON and PAGE [1932], JENKINSON and TINSLEY [1960]). Only from the B horizon of podzols has anyone succeeded in isolating humic fractions free of carbohydrates and proteins, by separation with ion-exchange resins (RUCHTI [1962], WRIGHT and SCHNITZER [1960]), by precipitation with acids (JACQUIN [1960]), by precipitation with polyvalent cations (DUBACH, MEHTA, and DEUEL [1961], SOWDEN and DEUEL [1961]), or by precipitation with quaternary ammonium bases (COFFIN and DeLONG [1960]).

FORSYTH [1947a] isolated some of the carbohydrates in the fulvic acid fractions of podzolic soils by fractionation on charcoal. The carbohydrate content of fulvic acids can be reduced by gel filtration (ROULET, MEHTA, DUBACH, and DEUEL [1963]) from 25% to below 5%. Furthermore, the nitrogen content has been reduced by 50% with this method (RUCHTI [1962]).

Similar investigations have been made with different soils by column chromatography on DEAE-cellulose and SM-cellulose (ROULET, MEHTA, DUBACH, and DEUEL [1963]). From a large number of soils SCHARPENSEEL and KRAUSSE [1962] obtained "gray" and "brown" humic acid fractions with a nitrogen content ranging from 0.8 to 3%.

Because they have similar properties, it is especially difficult to remove aromatic constituents, which may be derived from lignin, tannins, or metabolic products of microorganisms.

FORSYTH [1947a] separated the fulvic acids by adsorption on charcoal; this has been done in a similar way by SCHLICHTING [1953a, b], and STEVENSON [1960]. For this purpose DUBACH, MEHTA, and DEUEL [1961] used DEAE-cellulose.

Another procedure for separating fulvic acids is the precipitation with barium, copper, or iron salts (DUBACH, MEHTA, and DEUEL [1961, 1963], KONONOVA [1958], MÜCKE and OBENAUS [1961]). Here it is assumed that solvents such as pyrophosphoric acid and hydrofluoric acid are not precipitated at the same time, because they may form insoluble salts with the above-mentioned cations. After dialysis the heavy metals can be removed with hydrogen sulfide or with strongly acidic cation exchangers (DUBACH, MEHTA, and DEUEL [1963]). In the latter case small losses occur.

2. *Chromatographic Methods*

Two different procedures are used for chromatographic separation of extracts of soil organic matter.

(1) Partition chromatography (paper or thin-layer chromatography).
(2) Adsorption chromatography (column chromatography).

HOCK [1937] chromatographed extracts of humic substances with a column of aluminum oxide and studied the luminescence of the chromatograms in ultraviolet light. He observed differences in the luminescence behavior of the different extracts from various soils. FORSYTH [1950] separated humic substances with active coal into four fractions: A colorless fraction (A), a fraction (B) that is unstable and contains phenolic glucosides, a fraction (C) with polysaccharides, and a fraction (D) that is rich in nitrogen. SCHLICHTING [1953a, b] repeated this work and could not confirm that the fraction (C) contained phenolic glucosides. However, he could separate the fraction (B), of which one part is soluble in ether and another is precipitable. The ether-soluble fraction was not altered in the presence of air after drying, but became insoluble with time and was therefore considered to be a mixture of precursors of humic acids. DROSDOVA [1955] obtained a low molecular weight acid by adsorption of humic substances on coal after elution with sodium hydroxide; this acid had properties similar to the fulvic acids. BROMFIELD, COULSON, and DAVIES [1959] demonstrated by measurements of optical density that different fractions that have similar properties could be isolated if humic acids were eluted from a cellite column by acetone-water or acetone-sodium pyrophosphate solutions. MEINSCHEIN and KENNY [1957] isolated a group of substances by extraction of soils with benzene-methanol and column chromatography with silica gel. They identified these by infrared analysis and mass spectrometry after hydration of essential constituents of waxes, such as aliphatic acids, normal primary aliphatic alcohols, and different pentacyclic and hexacyclic compounds, which might be triterpenes. KONONOVA, BELCHIKOVA, and NIKIFOROV [1958] chromatographed fulvic acids and humic acids on starch and activated aluminum oxide. The experiments show an heterogenity of the material (brown, partly fluorescing bands) and some analogy in chromatographic behavior of both fractions. A complete separation was not possible, because several components seemed to react with the adsorbents used.

WRIGHT, SCHNITZER, and LEVICK [1958] could find, after hydrolysis, only traces or small amounts of glucose, galactose, arabinose, and xylose in extracts with hydrofluoric or hydrochloric acid. In addition, traces of rhamnose could be observed in the hydrolyzed extract of hydrofluoric acid. It could be demonstrated by semiquantitative determination that these parts were less than 1% of the total organic substances. Synthetic humic acids from hydroquinone were separated into eight components by column chromatography by SCHEFFER, ZIECHMANN, and SCHOLZ [1958c]. The electronic spectra of the substances were similar to those of humic acids. They were very uncharacteristic and showed only small differences, but a maximum of absorption at 250 nm. In comparison, the infrared spectra of the fractions showed no differences. COULSON, DAVIES, and KHAN [1959] were unsuccessful in attempts to fractionate humic substances by chromatography with ion exchangers. Separations of extracts with sodium hydroxide from muck soils by thin-layer chromatography were carried out by BLISLE [1959]. He found that the best separation in several zones occurred on plates that were coated with aluminum oxide and by use of a mixture of 0.25 M sodium diphosphate, 1 M sodium fluoride, and 1 N ammonia (7:3:2).

SCHARPENSEEL [1960] made extensive studies on the separation of humic substances, and degraded humic acids and different model substances by paper chromatography. Preparations of humic acids from chernozem and podzol could be separated into two zones by means of quaternary ammonium bases such as hyamine hydroxide in organic solvents. Scharpenseel's results point to a fundamental identity of the chemical constitution, on one hand, of "gray" and, the other hand, of "brown" humic acids. Pyrocatechol, pyrogallol, and small amounts of protocatechuic and isophthalic acid could be isolated in degradation products of humic acids which were formed by alkaline fusion, alkaline nitrobenzene oxidation, or hydrolysis with acids, following extraction with butanol and neutralization with quaternary ammonium bases.

KONONOVA and BELCHIKOVA [1960] succeeded in separating fulvic acids and humic acids from chernozems and podzolic soils in two zones by fractionation with partition paper chromatography. The authors supposed that the fulvic acids were precursors or degradation products of humic acids, because after extraction, the carbon and hydrogen content and the optical density of the substances, which pass the column in front, are the same as those of fulvic acids. Furthermore, it was found that the portions of fulvic acids in the humus fractions were larger in the case of podzolic soils than in chernozems (KONONOVA and TITOVA [1961]). SOWDEN and DEUEL [1961] obtained single fractions after fractionated precipitation and by separation with a column of cellulose, which show a different behavior during paper chromatography. Their color changed from yellow to brown in solution or in dry condition during aeration.

RUCHTI [1962] treated hydrochloric acid extracts with ether. After treatment with water and percolation of the water-soluble portion of the ether extract through a dowex-50 column, he obtained a nitrogen-free fraction. This could be separated by paper chromatography in three zones, which were made visible in ultraviolet light. They were noted as low molecular weight substances of red, brown, and yellow color. KUMADA and SATO [1962] chromatographed P-type humic acids from a podzolic soil on a cellulose column with ethanol/0.1 N sodium hydroxide (2:1) and obtained a brown zone, which could easily be eluted, and a green zone, which remained in the upper part of the column. The elution with 0.1 N sodium hydroxide gives a green solution. By acidification with hydrochloric acid red-brown precipitates are formed with properties similar to those of humic acids.

3. Electrophoresis

Humic acids have electrical charges (ODEN [1922]). Therefore, they have different kinetic potentials (zeta-potential). Because of their different rates of migration in solutions of electrolytes, they can be separated in fractions in an electric field. This electrochromatographic effect is the fundamental principle of electrophoresis, which is used more and more for fractionation and characterization of humic substances.

BEUTELSPACHER [1952] was the first to measure the electrophoretic migration velocities for the determination of the zeta-potentials of different electrodialyzed humic acids in solution. He found that humic acids extracted from chernozem with sodium fluoride (2%) and sodium carbonate (1%) have nearly the same zeta-potential of -21.6 and -22.7 mvolt at pH values between 2.5 and 3.2. Nitrogenous hydroquinone humic acids have a higher zeta-potential of -33 mvolt at the same pH values.

Paper electrophoresis was introduced by ROBINSON, MARTIN, and PAGE [1951] for investigation of humic acids. The results of BEUTELSPACHER [1952]—that humic acids are colloid anions and migrate as charge carriers to the anode—were confirmed by THIELE and KETTNER [1953] and WELTE [1953]. Charge effects that are similar to the proteins are not observed in the case of humic acids. According to KLAMROTH [1954] and ZIECHMANN [1954], the best results for separation are obtained at pH values of 8.6. SCHEFFER, ZIECHMANN, and SCHLÜTER [1955] examined the influence of the electric field on humic acids for the determination of experimental conditions. They established that a different behavior could not be observed with different buffer solutions at the same pH values. The migration was influenced to a larger extent by the concentration of the humic acids. The higher the concentration, the faster and easier precipitation occurred, especially with "gray" humic acids, but overlapping was observed. Furthermore, it was demonstrated that differences of 5 to 10 volt/cm were without influence on the duration of separation. Influences of paper capillarity on separation can be excluded sufficiently by high-voltage electrophoresis (50 volt/cm at low amperage). From their electrophoretic investigations, the authors believe that there are no principal differences between natural and synthetic humic acids. Three fractions are found, but the transitions of the fraction are often indefinite.

FLAIG, SCHEFFER, and KLAMROTH [1955] detected two migrating components by paper electrophoresis of humic acids of a chernozem; one had the properties of "gray" and the other the properties of "brown" humic acids. COULSON, DAVIES, and KHAŃ [1959] investigated the electrophoretic behavior of humic acids extracted with pyrophosphate from different peat bogs and some preparations of lignin. They found similarities in the behavior of the preparations of lignin and humic acids. The humic acids could be separated in alkaline buffer solutions in two or three fractions by low-voltage electrophoresis: A gray-colored fraction remains at the starting point; a brown-colored is more mobile, whereas the third consists almost completely of a low molecular weight portion and migrates in front of the electropherogram. After electrophoretic fractionation of humic acids, SCHEFFER, ZIECHMANN, and SCHLÜTER [1959] recorded the ultraviolet spectra of the separated substances and demonstrated that the extinctions of the fractions were different in slope; however, no typical maxima could be observed.

Sixteen different samples of soils and peats were investigated by HAYASHI and NAGAI [1960]. Aluminum oxide was added to a solution of humic acids in 2% sodium hydroxide. The unabsorbed portion of humic acids was separated into two fractions (I and II) by paper electrophoresis. With a 5% sodium hydroxide solution a fraction (III) was extracted from the aluminium oxide. Fraction IV was determined by subtracting the amount remaining on the aluminum oxide. The authors found a dependence of the electrophoretic distribution rate on the kind of extraction reagent and on the depth of the soil. The fractions I and II are found in larger amounts in the upper soil layers, whereas the fractions III and IV are from much deeper horizons. KAURICHEW, FEDOROV, and SHNABEL [1960] used continuous electrophoresis for separation of humic acids. The individual fractions of humic and fulvic acids were described by $Q_{4/6}$ coefficients of WELTE [1956]. According to the author's results, the fractions with a high-color coefficient are the most mobile. Primarily two fractions were obtained: a gray one and a brown one. The brown zone showed a white fluorescence in ultraviolet light. A comparison of humic acids extracted with 0.1 N sodium hydroxide from podzols, sod-podzols, chernozems, red earths, and chestnut solonez showed that the humic acids extracted from red earths have ultraviolet light fluorescing fractions, but that chernozems do not yield these fluorescing fractions.

According to their fluorescence behavior, all other soil types are in between these extremes. The strongest fluorescence was found in fulvic acids. Investigations with paper electrophoresis of extracts of chernozems and podzolic soils by SCHARPENSEEL [1960] through the use of sodium veronal buffer (pH = 9.6) and phosphate buffer (pH = 8.6) at 110 v, two zones were separated, which were named "gray" and "brown" humic acids. BURGES [1960] also obtained a separation of humic acids in two components by paper electrophoresis. Observation in ultraviolet light showed a weak yellow fluorescence at the side of the anode and, in addition, two marked bands. All three fractions had a similar elemental composition, whereas the infrared spectra showed no differences. The author concluded from these results that a separation occurred only according to particle size and not according to different chemical composition. KONONOVA and TITOVA [1961] separated humic acids of different soil types in three fractions (the substance at the starting point was included). They established that the ratio of the fractions of the different acids varied with the soil type.

WALDRON and MORTENSEN [1961] obtained a different fractionation of humic substances by means of continuous paper electrophoresis by the use of 0.022 M borate buffer at pH = 9. The humic substances were extracted from soils by sodium hydroxide, pyrophosphate, and hydrochloric acid. Waldron and Mortensen investigated the extraction of total nitrogen and established losses of nitrogen by desalting. The pyrophosphate extracts led to a separation into two components. The sodium hydroxide extracts divided into five components, whereas the extracts with hydrochloric acids had two main and several complementary components. The distribution of the total nitrogen, α-NH_2-nitrogen, and amino sugars was essentially in

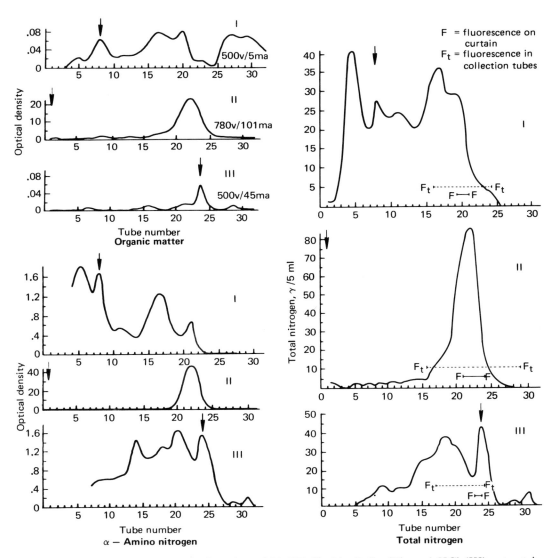

Figure 11. Electrophoretic fractionation of NaOH (I), $Na_4P_2O_7$ (II), and HCl (III) extract in borate buffer (I and II) and citrate buffer (III), according to WALDRON and MORTENSEN [1961].

agreement with the distribution of the organic soil substance. The brown-colored extract contained a fluorescent component.

POSPISIL [1962a, b] demonstrated, by densitometric measurements after electrophoresis of humic acids, that a clear separation into three fractions is obtained—one substance at the starting point and one or two substances which migrate. Samples from different origins showed different densitograms.

WIESEMÜLLER [1965] fractionated humic acids by paper electrophoresis with field intensities of about 10 v/cm and with veronal buffer at pH = 8 and investigated their spectroscopic behavior by absorption and fluorescence in ultraviolet light. They found that the fractions migrate more slowly in the electric field with increasing intensity of color and a decline in the color curve. "Gray" humic acids migrate slowly, and the alcohol-insoluble "brown" humic acids migrate still less. The hymatomelanic and fulvic acids are found more often in the direction of the anode. The fulvic acids show a distinct separation into a portion that moves faster and has a white fluorescence in ultraviolet light, whereas another portion has a yellow color visible in daylight. But like POSPISIL [1962a, b], Wiesemüller finds portions in nearly all fractions that fluoresce in ultraviolet light, when phosphate buffers are used. Furthermore, transitions of fractions can be established between hymatomelanic acids and "brown" humic acids or also between "gray" and "brown" humic acids by the use of this method.

ZIECHMANN [1967] found a separation of three fractions by low-voltage electrophoresis, whereby two fractions were mobile phases. The fraction that did not migrate consisted of humic acids with few polar groups. An increase of field intensity led to an increase of the portion that remained at the starting point and to a decrease of the migrating amount. The substances that accumulated in the front of the electropherogram consisted of precursors of humic acids, which migrate in the electric field.

DUCHAUFOUR and JACQUIN [1963] controlled the extraction and fractionation of fractions of humic substances by electrophoresis.

4. *Gel Filtration*

Different types of gels, such as synthetic cross-linked polydextranes (Sephadex), or starch are used to fractionate mixtures of molecules such as humic substances according to their different particle size (approximate molecular weight). The range of size of molecules fractionated depends on the degree of cross-linking in the gel, which in turn determines the size of pores in the gel particle. The gel is used in columns as in elution chromatography.

The fractionation by gel filtration is based on the fact that those molecules whose particle size is larger than the pores of the hydrated gel go faster through the column by elution with a solvent than those molecules that are small enough to diffuse inside the gel spheres. The exclusion limit is defined as the upper limit of particle size, which can be fractionated effectively. Therefore, at first, the larger molecules appear as bands during elution of the gel column; these are followed by one or several bands of smaller molecules. But this sequence is influenced by the charge of the gel matrix. Molecules with opposite charges are sorbed by the matrix, and like-charged molecules are repelled.

Often an automatically registering colorimeter for the determination of the distribution rate of the humic substances and an automatically registering apparatus for the measurement of conductivity to follow the desalting of the different fractions is combined with the column. This is connected with a fraction collector (*e.g.*, SÖCHTIG [1966]).

FERRARI and DELL'AGNOLA [1963] have separated humic substances extracted with 0.1 M sodium fluoride in three fractions by means of Sephadex. Two of the fractions had properties similar to those substances, which were prepared conventionally by fractionated precipitation. The electrophoretic behavior was different from these substances, which were isolated in the usual way. The authors concluded that the gel filtrated substances were more homogenous in their composition and less degraded (comp. DELL'AGNOLA, MAGGIONI, and FERRARI [1965a, b].

MEHTA, DUBACH, and DEUEL [1963a] found through investigations of molecular weight distribution by means of Sephadex gel filtration that an elution solvent with a higher pH value and sufficient salt concentration must be used to avoid an influence on the exclusion limit by interaction of the gel matrix with the acidic groups of the substances separated. Therefore, they used a glycine/sodium hydroxide buffer at pH 10, because the method is also useful for the determination of changes of the molecular weight. Experiments were made to determine alterations of humic acids by drying or treatment with acid or alkali. It could be demonstrated that the average molecular weight was not changed after drying at 100°C and after storage in 1 N hydrochloric acid at room temperature for one month. The average molecular weight was increased by boiling with acid and decreased by treatment with alkali at room temperature.

By gel filtration, the content of carbohydrates of fulvic acids from humus carbonate soils, rendzina, and brown earths, was lowered from 25 to 5% (ROULET, MEHTA, DUBACH, and DEUEL [1963]). The nitrogen content also decreased from 2.1 to 1.7% whereas it was decreased by ionic exchange to 1.3%. Because the ratio of the carbohydrates and nitrogenous compounds changed very much during gel filtration, the authors concluded that these compounds were only weakly attached to the humic substances. POSNER [1963] also investigated extracts of humic acids by Sephadex gel filtration. The humic acids were isolated with 0.4 M sodium carbonate/sodium bicarbonate (2:1). After elution with a mixture of water and ammonium chloride, two fractions were obtained. DELL'AGNOLA, FERRARI, and MAGGIONI [1964] determined the distribution of molecular weights of an extract of humic acids from a soil by the use of different gel types. They found 12% of the total carbon content in a range of molecular weights between 0 and 4000, 3% between 4000 and 9000, 23% between 9000 and 100,000. 62% were in range between 100,000 and 200,000.

A comparison of the results of gel filtration with different extractants such as 0.5 N sodium hydroxide, sodium carbonate/sodium bicarbonate, sodium pyrophosphate, or sodium fluoride demonstrated that three fractions were obtained by extraction with sodium hydroxide, whereas all the other extractants separated only in two fractions. DELL'AGNOLA and MAGGIONI [1965a] established that some other bands occur with water being used as an extractant. Furthermore, a shift to higher molecular weights occurs when the ionic strength increases and the pH value decreases. They concluded from the results of their investigations that the fractionation of humic acids agrees with the molecular size of the compounds at pH values larger than 10 and ionic strengths smaller than 0.06. They established also, by means of gel filtration with Sephadex, that the low molecular weight substances of soil humic substances obtained by gel filtration polymerize when they are desalted by dialysis or dissolved in sodium tetraborate (DELL'AGNOLA and MAGGIONI [1965b]).

KLEINHEMPEL and HIEKE [1965] investigated with Sephadex humic acids which were dissolved in sodium hydroxide. They found that slowly diffusing portions had a higher color quotient. Infrared-spectroscopic investigations demonstrated that the fractions with a higher color quotient had a higher content of carboxyl groups, arylether groups, etc. Finally, it was shown that humic acids synthesized by fungi agreed in their behavior during filtration with soil humic acids.

SÖCHTIG [1965] filtrated through Sephadex G-25, alkaline extracts of humic acids, which were isolated from chernozem pretreated with hydrochloric acid. By elution with three different solvents two zones were obtained with water, only one with 0.1 N sodium hydroxide and with a mixture of 0.1 N sodium hydroxide/0.1 N sodium sulfate. By gel filtration of an extract of a podzol B horizon with sodium hydroxide, three zones were obtained by elution

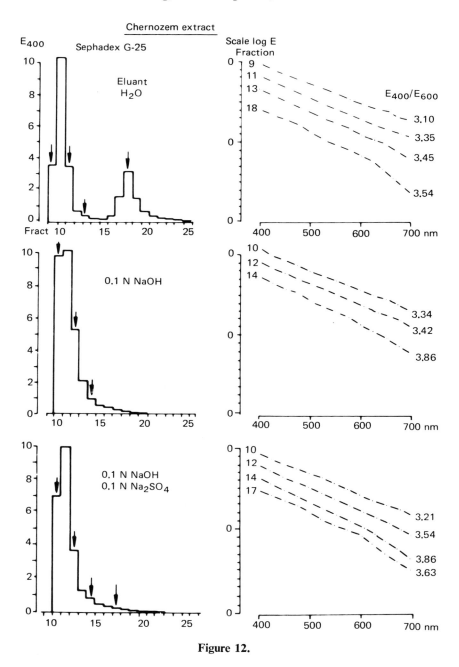

Figure 12.

with water: a strong and a weak zone by elution with 0.1 N sodium hydroxide and two zones by elution with a mixture of sodium hydroxide and sodium sulfate (Figure 12).

The spectra of the fractions 9 to 14 of the chernozem extract show nearly the same properties in all cases. However, small differences occur in the spectrum of fraction 18 and to a lesser amount in fraction 17 at 620 nm. In the case of the podzol extract the spectra of the fractions 9 to 12 were weakly differentiated, as were those of fractions 17 and 18 of the chernozem

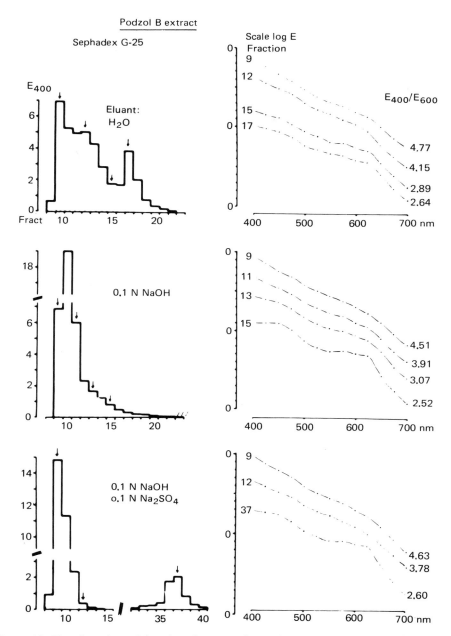

Figure 12. Fractionation of humic substances from chernozem and a podzol B horizon with Sephadex G-25 after elution with different solvents (Söchtig [1965]).

extract. The spectra of the fractions 15, 17, and 37 are remarkably different and have distinct maxima. A comparison of the elution diagrams of 15 different samples of chernozems or podzolic soils demonstrated a nearly complete agreement of the distribution rate of the organic materials in the different soil types.

In another paper Söchtig [1966] investigated fractionation using different elution solvents such as chlorides, phosphates, sodium tetraborate, sodium hydroxide, and hydrochloric and

sulfuric acids. He found (Figure 12) that the elution from the Sephadex column was independent of the molarity of all of the above compounds, with exception of sodium hydroxide and
sodium tetraborate. In the case of sodium hydroxide a slight shift, and in the case of sodium
tetraborate a strong shift, occurred to low values of percentage of total volume with increasing
molarity. According to the author's opinion, the inhibition of elution by sodium hydroxide
was caused by the interaction of ionizing groups, whereas the borate ion formed complex
compounds with the saccharides of the gel.

The investigations on the fractionation of humic substances with Sephadex gels indicate
that solutions of humic substances become almost completely desalted (for example, OBENAUS
and NEUMAN [1965]) and that fractionation occurs according to chemical composition. But
the results of the experiments on the effectiveness of separation need further confirmation by
other methods of particle weight determination.

5. Contribution to the Problem of Fractionation of Humic Acids

A review of the main methods of fractionation indicates that all so-called humic substances prepared by conventional methods very seldom are chemically uniform. This is shown
by attempts of many authors who have tried to obtain further fractionation of humic
acids or fulvic acids. At first FORSYTH [1947a] succeeded in a subfractionation of a fulvic acid
in several parts. DUBACH, MEHTA, and DEUEL [1963] fractionated fulvic acids in an extensive
work without getting a uniform fraction. By fractionated precipitation with the aid of electrolytes, FLAIG, SCHEFFER, and KLAMROTH [1955] obtained a further separation of humic acids

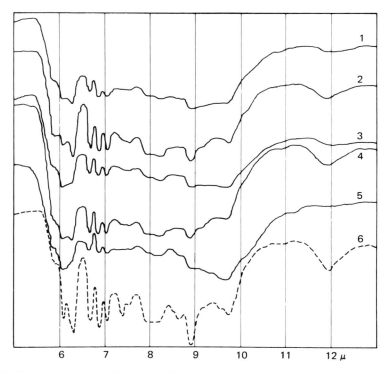

Figure 13. Infrared spectra of humic acid fractions from decomposed rye straw prepared with
different solvents. 1 = total extract; 2 = alcohol-soluble fraction; 3 = alcohol-insoluble fraction; 4
= acetone-soluble fraction from (3); 5 = acetone-insoluble fraction from (3); 6 = Björkman lignin
from straw.

in "brown" and "gray" humic acids, but uniform fractions were not obtained. The following investigations show how much different preparations of humic acids, fulvic acids, hymatomelanic acids, etc., may vary.

FLAIG and TROJANOWSKI (unpublished data) isolated a fraction defined as "humic acids" with a mixture of 0.1 N sodium hydroxide and 0.2 N sodium fluoride from decomposed rye straw, which was previously extracted with ether and water. After the usual purification by repeated precipitation, a product was obtained from which a fraction with 95 % alcohol was extracted. The yield of this fraction decreased with increasing time of decomposition. This fraction was very similar in elemental composition and light absorption properties to the original lignin, whereas the residue of the extraction became more and more like humic acids isolated from soils. By treatment of the alcohol-insoluble residue with aqueous acetone, a fraction was obtained whose properties are more like lignin than those of the part remaining insoluble in acetone (SALFELD [1964]).

A comparison of the infrared spectrum of Björkman lignin (Figure 13) with the alcohol = and acetone-soluble fractions shows that from the humic acids fraction a relatively large portion can be isolated whose spectrum is similar to unchanged lignin.

Most of the absorption maxima of the soluble fractions are unchanged. Only in the range of NH or C=N vibrations at 6.1 to 6.2 μ (1640 to 1610 cm^{-1}) is an increase of absorption seen. The residues seem to have been formed by condensation of lignin decomposition products with nitrogenous compounds.

The fractions insoluble in alcohol and acetone show larger differences in their infrared spectra than the spectrum of lignin. Besides the changes of absorption in the range of 9 to 10 μ (1110 to 1000 cm^{-1}), an increase of the background absorption can be observed.

The changes in the range of the NH or C=N vibrations between 6.1–6.2 μ (1640 to 1610 cm^{-1}) are larger in the spectra of the residues than in the spectra of the fractions soluble in organic solvents.

The elementary analysis and absorption spectra of the insoluble residue are somewhat similar to those of the humic acids.

SALFELD [1964] demonstrated that brown humic acids from a chernozem could be subfractionated by extraction with aqueous organic solvents dependent upon the water content (Table 15). Figure 14 shows the infrared spectra of such a fractionation series that was obtained by successive extraction with tetrahydrofurane mixed with 5, 10, and 25% water, and also the spectra of the residue and the initial material. The spectra indicate that fractions of different properties are dissolved, depending on the water content of the tetrahydrofurane.

The infrared spectra show a relative decrease of the carbonyl absorption at 5.9 μ (1695

Table 15. Elemental composition of fractions of "brown" humic acids from chernozem isolated with aqueous tetrahydrofurane according to SALFELD [1964].

	Ash, %	Carbon*	Oxygen*	Nitrogen*	OCH$_3$	log† 400–600 nm
1. Fraction 5 % H$_2$O/thf‡	1.29	58.81	33.20	2.79	3.00	0.82
2. Fraction 25 % H$_2$O/thf	3.72	55.65	33.02	5.89	2.08	0.68
3. Residue	6.73	53.17	34.59	5.50	1.49	0.58
4. Initial material	3.18	56.04	33.64	4.91	2.13	0.66

* in percent ash-free substance (values for hydrogen no large differences).
† measured in 0.1 N NaOH.
‡ thf = tetrahydrofurane.

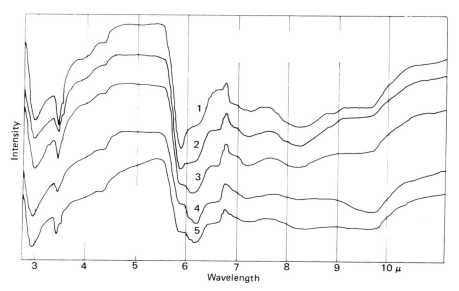

Figure 14. Infrared spectra of the subfractions from brown humic acids isolated from chernozem and extracted with tetrahydrofurane of different water content (SALFELD [1964]).
1 = 5% H₂O (39% soluble of initial material); 2 = 10% H₂O (10% soluble of initial material); 3 = 25% H₂O (24% soluble of initial material); 4 = residue (25% insoluble); 5 = initial material.

cm⁻¹), followed by a reduction of the hydroxyl vibration, which means a decrease of the free carboxylic groups. On the other hand, the C=C vibration at 6.2 μ (1610 cm⁻¹) tends to strengthen and may be attributed to increasing aromatic and olefinic double bonds. The absorption of the carboxylic group at 7.2 μ (1390 cm⁻¹) decreases with increasing water content, similar to the decrease of the carbonyl function.

The results of infrared analysis are in a good agreement with the elemental composition, indicating that the fractions isolated by extraction increase in oxygen and nitrogen content but decrease in methoxyl groups.

Measurements of the electronic spectra in the visible and ultraviolet range reveal a decrease in the slope of the nearly linear absorption curve with increasing water content.

Similar fractionation series were also obtained by KYUMA [1964] by successive precipitation with alcohols in alkaline solutions.

These examples demonstrate that the chemical and physical properties of the humic acids depend very much on the methods of isolation. In spite of this, all isolates are called "humic acids." Therefore, it is necessary that humic acids be isolated by standardized methods so that various workers will be able to compare their results (FLAIG [1964b], KONONOVA [1964].)

VI. Chemical Composition of Humic Acids

1. Cleavage by Oxidation and Reduction

For investigations on the composition of humic acids by oxidation procedures, different authors have often used very different preparations. Therefore, it is surprising that the isolated oxidation products of humic acids or the structural units found have consisted of chemically similar compounds. The yields of defined oxidation products would not be expected to be high, because the oxidation of very carefully isolated coniferous lignin with nitrobenzene in alkaline solution seldom yields more than 25% vanillin (BRAUNS and BRAUNS [1960]), although this lignin is almost a pure polymer of coniferyl alcohol. Therefore, a discussion is in order on the dependence of the yield of vanillin and other lignin degradation products upon the pretreatment of the lignin preparations or on the presence of other plant constituents. When lignin is isolated with alkali or with sulfuric acid by the method of Klason, the yield decreases to 13.87% and 1.5% vanillin, respectively, calculated on the basis of the total lignin (ODINTSOV and KREITBERG [1953]).

ROADHOUSE and MACDOUGALL [1956], working with wheat straw lignin, obtained yields of 30% syringaldehyde, 24.4% vanillin, and 4.4% p-hydroxybenzaldehyde upon the oxidation with nitrobenzene in 2 N sodium hydroxide. These high values are attributable to the fact that the authors made corrections for the loss of aldehydes, which could be formed during the oxidation in the presence of plant material. The aldehydes were separated chromatographically and the quantities were determined spectrometrically.

PEPPER, MANOLIPOULO, and BURTON [1962] suggested that gas chromatography was a more exact method for the determination of the aldehydes which are formed upon oxidation with nitrobenzene. They obtained values of pre-extracted straw or lignin isolated from it (Table 16).

Table 16.   Composition of wheat straw and its lignin.

	Percent of the original lignin		
	Vanillin	Syringaldehyde	p-Hydroxybenzaldehyde
Wheat straw	9.3	8.0	1.3
Lignin	7.5	7.7	1.3

HIGUCHI [1953] oxidized a sample of pre-extracted bamboo (Phyllostachis edulis), which contained 26.1% Klason lignin with nitrobenzene in 2 N sodium hydroxide. Syringaldehyde (2.5%), vanillin (8.5%), and p-hydroxybenzaldehyde (5.5%) were also obtained. These percentages are based on the original Klason lignin content and are in a molecular ratio of 2.4:1.9:1. Syringic- vanillic, and p-hydroxybenzoic acid were also found. In a similar test with spruce (*Picea mariana*) the yield of vanillin decreased more than 40% following a 65% decomposition of the fresh wood by brown-red fungi.

LEOPOLD [1950] and LEOPOLD and MALMSTROM [1951] investigated a large number of model substances in an attempt to elucidate the mechanism of the reaction yielding vanillin during the alkaline oxidation of lignin with nitrobenzene. Ferulic acid was transformed to vanillin, with a yield of 86%, and to vanillic acid, with a yield of 4%, while 5-carboxy vanillin gave a yield of only 2%. The carboxyl group in 5-carboxy-vanillin may be formed by the oxidation of the linkage of two monomers through the side chain and the ring or by the cleavage of one ring of a diphenyl linkage.

$$HOOC \overset{\displaystyle OH}{\underset{\displaystyle CHO}{\bigcirc}} OCH_3$$

5-Carboxy-vanillin
(See the formula for lignin, Figure 3, rings 2 and 3 or rings 12 and 13a.)

Syringaldehyde, vanillin, and p-hydroxybenzaldehyde are recovered quantitatively after treatment with nitrobenzene in alkaline medium at 160°C, while considerable losses occur

Table 17. Oxidative degradation of humic acids with nitrobenzene in alkaline medium according to MORRISON [1963].

Humic acids	Carbon — Percent of the sample	Carbon — Percent of the soil	Methoxyl	Carbon determined as aldehydes and acids in percent of total carbon	Carbon determined as oxidation product (calculated for vanillic acid = 1)							
					S	V	HB	Sa	Va	HBa	Fa	Ca
Wood soil	48.1	20.9	1.37	1.88	10.1 (0.4)	26.9 (1.2)	11.2 (0.45)	23.9 (0.95)	25.0 (1)	Trace	—	—
Alpine humus podzol H (0–2.5 cm)	47.7	9.9	1.00	0.22	15.6 (0.73)	30.3 (1.44)	19.6 (0.92)	13.1 (0.61)	21.3 (1)	Trace	—	—
Garden soil (10–20 cm)	40.0	10.2	1.64	1.48	18.1 (0.86)	23.9 (1.13)	14.5 (0.63)	22.4 (1.07)	21.0 (1)	Trace	—	—
Phragmites Peat	52.3	25.8	3.34	7.89	31.2 (4.5)	23.8 (3.1)	20.0 (2.6)	10.7 (1.4)	7.6 (1)	6.3	—	—
Straw	41.4		2.87	8.34	29.3 (7.2)	30.5 (8.1)	16.1 (4.01)	7.8 (2.05)	3.8 (1)	0.8 (0.18)	5.9 (1.5)	5.7 (1.5)

S   = syringaldehyde [13]
V   = vanillin [11]
HB = p-hydroxybenzaldehyde [10]
Sa = syringic acid [9]
Va = vanillic acid [8]
HBa = p-hydroxybenzoic acid [5]
Fa = ferulic acid [2]
Ca = p-coumaric acid [1]

from the addition of older plant material and even greater losses from the addition of younger plant material (see ROADHOUSE and MACDOUGALL [1956]).

The oxidation of lignin carefully isolated with potassium permanganate in alkaline solution gives a low yield of benzenecarboxylic acids, namely, about 5% (READ and PURVES [1952], BRAUNS [1952]). The yields of benzenepolycarboxylic acids can be increased to over 20% (CHUDAKOV, SUKANOVSKII, and AKIMOVA [1959]) by treatment with acid or especially with alkali at increased temperatures. The high yields may be explained on the basis of a condensation of rings to a conidendrin-type structure, ring closure, and dehydrogenation (BRAUNS and BRAUNS [1960]).

Condensed conidendrin structure

Not only are polymerization and condensation reactions possible on the altered lignin or its degradation products during the formation of humic acids, but nitrogen-containing compounds can also participate in these reactions. For instance, methylated lignin does not react with ammonia during shaking in an atmosphere of oxygen, whereas demethylated lignin absorbs 4.8% nitrogen in organic linkage (BENNET [1949]; BROADBENT, BURGE, and NAKASHIMA [1960]). The authors conclude that the oxidation and the fixation of nitrogen occur through the phenolic hydroxyl groups.

These preliminary remarks on the degradation reactions of the humic acids are made so that results found under very different oxidizing conditions can be properly evaluated. A low yield of phenolic compounds does not prove that only a few phenolic building blocks exist in the humic acids, while a high yield must be considered as proof, in every case, that they do exist.

One should also be cognizant that the procedures for extracting humic acids may remove partly decomposed lignin or other plant phenolic constituents simultaneously. At present it is not known to what extent humic acids as such are isolated, or whether other organic soil constituents are included in these fractions.

In an extensive and very detailed paper MORRISON [1963] reports on studies on oxidation of plant material with nitrobenzene in 2 N sodium hydroxide. Earlier [1958] he had reported some experiments on the same subject. Initial materials of humic substances such as wood or leaves, as well as peat, soils, and humic acids, were investigated. Not only did the author isolate the polyphenols by paper chromatography and spectrographic analyses of the quantities obtained, but he also determined the conditions giving the best reproducibility of the results.

The largest amount of carbon in the form of phenolic compounds (15.68% of the total carbon) was obtained from milled, lignified shoots of bamboo. In the case of woods, yields ranging from 9.3 to 12.77% of total carbon were obtained as carbon of phenols, aromatic aldehydes, and carbonic acids; leaves contained 2.6 to 4.2% of phenolic carbon; soils (upper layers), 0.3 to 7%; peat, 1.65 to 3.67%; and preparations of humic acids from soils, which were isolated by different procedures, from 0.24 to 1.88% (Table 17).

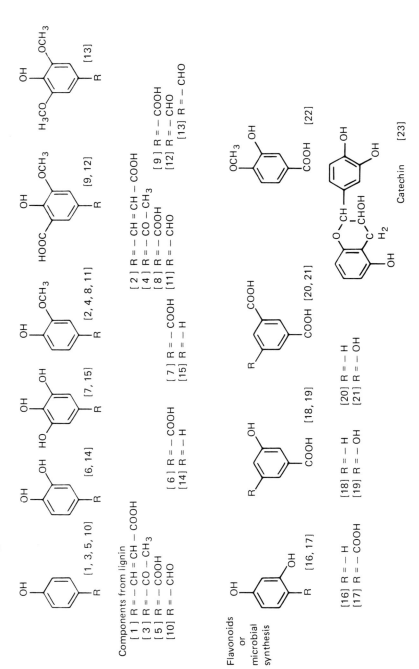

Degradation products of humic acids

Oxidative degradation

[1, 3, 5, 10]
[6, 14]
[7, 15]
[2, 4, 8, 11]
[9, 12]
[13]

Components from lignin

[1] R = – CH = CH – COOH
[3] R = – CO – CH₃
[5] R = – COOH
[10] R = – CHO

[2] R = – CH = CH – COOH
[4] R = – CO – CH₃
[8] R = – COOH
[11] R = – CHO

[9] R = – COOH
[12] R = – CHO

[13] R = – CHO

[6] R = – COOH
[14] R = – H

[7] R = – COOH
[15] R = – H

Flavonoids
or
microbial
synthesis

[16, 17]
[18, 19]
[20, 21]
[22]

[16] R = – H
[17] R = – COOH

[18] R = – H
[19] R = – OH

[20] R = – H
[21] R = – OH

Catechin   [23]

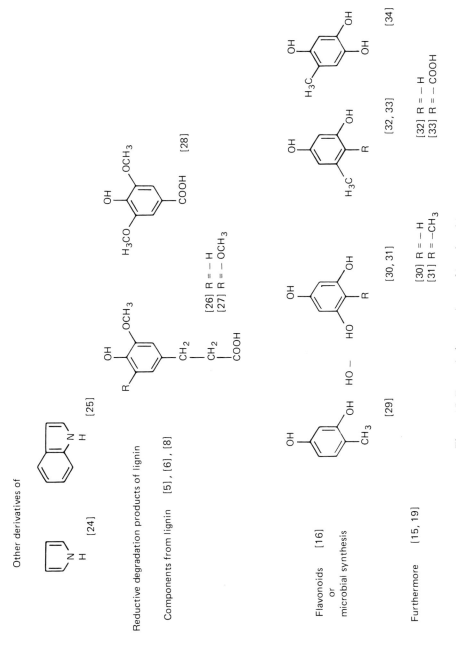

Figure 15. Degradation products of humic acids.

The humic acids were obtained by extraction with 0.5 N sodium hydroxide followed by precipitation, centrifugation, freezing, thawing, filtration, and drying in the air. A total carbon content of 7.89% for the phenolic compounds of methylated peat humic acids was noted, but only 3.21% of this represented identified phenolaldehydes and their acids.

The ethanol-benzene extracts of samples of humic acids, which contain 1.88% of the total carbon as phenolaldehydes and carboxylic acids, gave a value of 2.27%. In the case of humic acids isolated from phragmites peat, the value of the carbon for the isolated phenolic compounds was 7.89%. The value for the fractions of the peat that were soluble in ethanol-benzene was 10.19%.

The increased quantity of phenolic compounds from the extracts with ethanol-benzene supports the conclusion that these extracts are more similar in their composition to the lignin than to humic acids.

Figure 15 shows a summary of degradation products of humic acids that have been isolated by different degradation procedures.

p-Coumaric [1] and ferulic [2] acids could be identified in traces only, while p-hydroxy-acetophenone [3] and acetovanillone [4] were occasionally found in small amounts in soil humic substances. All of the phenolic compounds noted in Table 17, as well as p-hydroxy-acetophenone [3] from some plants, were found in beech woodmeal, larch woodmeal, bamboo stems, phragmites peat, straw, beech leaves, larch leaves, sitka spruce leaves, and spagnum. From peat soil MORITA [1962] isolated by alkaline oxidation with mercuric chloride in addition to compounds [4], [5], [8], [9], [10], and [11], acetosyringone, and 5-carboxy-vanillin [12] in yields as high as 1%. Calculated on the basis of vanillic acid = 1, the relative amount of the isolated substances varied with the type of plant cover.

SCHARPENSEEL [1960] made a qualitative determination of the oxidation products with nitrobenzene by means of paper chromatography. Twenty-five compounds were found. In addition to different phenolic substances, nitrogen-containing heterocyclic compounds [25] were noted.

Other degradation experiments in alkaline medium at temperatures between 170 and 250°C in the presence or absence of oxidation agents (copper oxide) have given different phenolic compounds in yields of about 5% (GREENE and STEELINK [1962], JAKAB, DUBACH, MEHTA, and DEUEL [1963], STEELINK, BERRY, HO, and NORDBY [1960]). Phenols, phenol-carboxylic acids, and some phenolaldehydes were found. SHMUK [1930] refers to nitrogen-containing heterocyclic compounds, such as pyrrol derivatives [24], which are obtained by the fusion of the humic acids with alkali. Only small amounts of organic compounds and aliphatic acids from which few conclusions can be drawn with respect to the constitution of humic acids are obtained by the use of such other oxidation agents as potassium permanganate (WRIGHT and SCHNITZER [1959]), hydrogen peroxide (MEHTA, DUBACH, and DEUEL [1963a], SAVAGE and STEVENSON [1961]), or chlorodioxide (ESH and GUHA-SIRCA [1940], SCHMIDT and ATTERER [1927]). KUMADA, SUZUKI, and AIZAWA [1961] found benzenepolycarboxylic acids by the oxidation of humic acids from alkaline soils with potassium permanganate. By following decarboxylation with copper-quinoline, traces of anthraquinone, benzene, naphthalene derivatives, and pyridine were noted. Tetrabromoquinone is obtained by oxidation with bromine (FEUSTEL and BYERS [1936]). It is formed in larger yield from lignin.

SCHNITZER and DESJARDINS [1964] extracted the B_h horizon of a podzol with 0.5 N sodium hydroxide under nitrogen. The extracted organic material was passed through a column of exchange resin, amberlite IR-120. The eluate was evaporated in a steam bath and the residue was dried in an oven. The portion of this material that was soluble in 1% sodium hydroxide was oxidized with potassium permanganate (5 to 1). The residue of the ether and ethylacetate extraction of the oxidation products was esterified with diazomethane. By means of gas chromatography 0.3% o-phthalic, 0.05% trimellitic, 0.05% pyromellitic, 0.05% pre-phenic, 0.06% mellophanic, and 0.08% benzenepentacarboxylic acid were found (calculated from ash-free original organic substance). The identity of these acids was ascertained by further physicochemical measurements (recent summary SCHNITZER and KAHN [1972]). Additional oxidation experiments have been made with nitric acid (JAKAB, DUBACH, MEHTA, and DEUEL [1962], SCHEFFER and KICKUTH [1961a, b], SCHNITZER and WRIGHT [1960]). A hydrolysis occurs in the course of this treatment. Some time ago FUCHS [1928] and FUCHS and STENGEL [1930] showed that a strong degradation occurs which leads to the formation of large amounts of carbon dioxide, sometimes as much as 60%, and to products that are soluble in organic solvents. Fuchs prepared acetone-soluble "nitro-humic acids" in this way.

Up to 20% of ether-soluble products are also formed by this type of oxidation. Among the reaction products are phthalic, isophthalic, terephthalic, and tetra- and pentabenzene-carboxylic acids. Nitro compounds are also formed: p-nitrophenol; 3,4-dinitrophenol; picric acid; p-nitrocresol; o-, m-, p-nitrobenzoic acid; (3-nitro-4-hydroxy-benzoic acid, 3,5-dinitro-4-hydroxy-benzoic acid, 3,5-dihydroxy-4-nitrobenzoic acid;) 2-nitrovanillic acid; 2-nitro- and 2-hydroxy-3-nitrobenzaldehyde; 4-nitro-p-phthalic acid; 2-nitro-3,6-dihydroxyterephthalic acid; and others.

HAYASHI and NAGAI [1961] found 2- and 4-hydroxybenzoic acid [5], 2,4- [17], 3,5-di-hydroxybenzoic acid [19], and 5-hydroxyisophthalic acid [21] as additional products. The yield of the three different benzenedicarboxylic acids was 18% when vanadiumpentoxide (V_2O_5) was added as a catalyst (SCHEFFER and KICKUTH [1961a, b]).

It was concluded that structure existed in which the rings were connected through carbon linkages or contained side chains. As was previously mentioned, the yield of benzenecarboxylic acids increases when lignin is pretreated with acids. Therefore, it is not known whether the isolated aromatic acids were in the original humic acids or were formed by the treatment with acids.

WRIGHT and SCHNITZER [1961] oxidized fractions of humic acids isolated from the A_o and B_h horizons of a podzol in an air stream at 170°C for 500 hours. The composition of the products formed and the distribution of the functional groups in comparison with the initial materials were studied by the methods of MAZUMDAR, CHAKRABARTTRY, and LAHIRI [1957] and CHAKRABARTTRY, MAZUMDAR, ROY, and LAHIRI [1960]. It was estimated that nearly half of the carbon atoms and about 21 to 36% of the hydrogen atoms belonged to the aromatic parts of the molecules.

Small amounts of benzenecarboxylic acids [5], [6], [8], or their aldehydes (JACQUIN [1960], JAKAB, DUBACH, MEHTA, and DEUEL [1962]) are found after hydrolysis with water or hydro-chloric acid, sulfuric acid, or perchloric acid. Catechin [23] has also been identified (COULSON, DAVIES, and LEWIS [1960a, b]).

Reductive degradation of humic acids has been tried in different ways. Hydrocarbons partially of hydro-aromatic nature have been found after reduction with hydrogen iodide and phosphorus (ELLER [1925a]). Cyclopentanols are obtained by hydrogenation, with molybdenum sulfide as a catalyst (GANZ [1944]), while phenols are formed by hydrogenation under pressure (KUKHARENKO and SAVELEV [1951, 1952]). Phenols and phenolcarboxylic acids were also obtained by hydrogenation of humic acids from brown coals with sodium in liquid ammonia (KUKHARENKO and KREDENSKAJA [1956]), but their constitution was not established.

Hydrogenation under pressure and temperature up to 350°C in the presence of catalysts produces large amounts of distillable oils (FELBECK [1965], GOTTLIEB and HENDRICKS [1945], MURPHY and MOORE [1960]). More than 50 different compounds were noted by gas chromato-graphic separation, but the chemical constitution of most was not established.

The most successful method for the elucidation of the structural units of humic acids appears to be reduction with sodium amalgam in an oxygen-free atmosphere at 100 to 110°C (BURGES, HURST, and WALKDEN [1963]). These investigators identified the following phenolic units: p-hydroxybenzoic acid [5], protocatechuic acid [6], vanillic acid [8], guaiacylpropionic acid [26], syringylpropionic acid [27], and syringic acid [28], which could be derived from lignin units; resorcinol [16], 2,4-dihydroxytoluene [29], phloroglucin [30], and methylphloroglucin [31], which could be degradation products of flavonoide; and 3,5-dihydroxybenzoic acid [19] and pyrogallol [15]. The quantities of the different phenolic compounds separated by thin-layer chromatography varied, depending on the origin of the humic acids. It was concluded that the composition of the phenols depended mainly on the plant formation, from which the humus was formed.

This procedure has yielded the greatest percentage of phenols yet reported, namely as high as 30 to 35% after repeated reductions. Additional phenols were obtained after fusion of the remaining residue with potassium hydroxide at 280°C.

Reductive degradation of humic acids synthesized by *Epicoccum nigrum* with Na-amalgam yielded ether-extractable phenols equivalent to 15 to 20% of the acid = hydrolyzed humic acid. Chromatographic analysis separated 14 phenols. The most prominent were orcinol [32], orsellinic acid [33], 4-methyl-resorcinol [29], 2,4-[17], 3,5-dihydroxybenzoic acid [19], and 2,4,5-trihydroxytoluene [34] (MARTIN, RICHARDS, and HAIDER [1967]).

The studies on the degradation of humic acids lead to the conclusion that they consist of a large number of phenolic units, which are linked together in different ways. The degradation products suggest that the linkage occurs through carbon atoms of two rings or through rings with side chains. According to BURGES, HURST, and WALKDEN [1963] there may also be ether linkages.

The degradation studies do not elucidate how nitrogen participates in the linkages between the phenolic compounds and the nitrogen-containing compounds. A part of the nitrogen is presumably heterocyclic.

During the degradation of humic acids to aromatic compounds it is possible that the acids or the alkalis cause condensation reactions, as in the case of lignin, which diminishes the yield of phenolic compounds. The relatively high amount of benzenecarboxylic acids sometimes obtained suggests that condensation reactions do occur during the degradation of humic acids.

The 1,2- and 1,2,3-phenols are obtained by the degradation reactions of phenols which are partially methylated. The methylated phenols could be derived from unaltered parts of lignin molecules or from its degradation products. It is not known whether the 1,2- or 1,2,3-phenols were indeed in the humic acids or if they were formed by the treatment with alkali by demethylation, as suggested by the work of KRATZL and RISNYOVSKY [1964] and KRATZL, RISNYOVSKI-SCHÄFER, CLAUS, and WITTMANN [1966]. Presumably some of these phenols will be present in the humic acids, because the partially methylated degradation products are not altered by some of the chemical reactions, such as the alkaline oxidation with nitrobenzene. It is also possible that they are derived from flavone components. According to the investigations of BURGES, HURST, and WALKDEN [1963], the 1,3- or 1,3,5-phenols, presumably derived from flavone components or synthesized by microorganisms, constitute a considerable portion of the isolated phenolic degradation products.

2. Functional Groups Containing Oxygen

The reactions of all groups containing oxygen in humic acids has not been completely determined. Many of these groups are acidic. Consequently, their reactivity may be associated with their acidic properties. The difficulties in the functional analysis of the oxygen in the humic acids is associated with their properties as polymeric compounds, because these do not react completely with the reagents and the reagents can be adsorbed.

The treatments with acids and alkali exclude many reactions, especially at higher temperatures, because the humic acids are thereby altered. It is also possible that oxygen-containing groups of different kinds are present in the humic acids that do not react in the same manner with the added reagents.

Carboxylic (BROADBENT and BRADFORD [1952]) or phenolic hydroxyl groups (FLAIG, SCHEFFER, and KLAMROTH [1955], LEWIS and BROADBENT [1961], MEYER [1962], and others) are responsible for the acidity of the humic substances. It is not possible to distinguish with certainty between the carboxylic and phenolic hydroxyl groups, because the acidities of these groups overlap. The acidity of the hydroxyl groups of 2,4-dihydroxy-p-benzoquinone is stronger than that of acetic acid (FLAIG, SCHEFFER, and KLAMROTH [1955]). DUBACH and MEHTA [1963] mention in a review paper that those methods that measure pK and are based on titration in aqueous and nonaqueous solvents do not truly differentiate between these two groups (VAN DIJK [1959, 1960], MEYER [1962], POMMER and BREGER [1960a, b], WRIGHT and SCHNITZER [1960]). The same is true for reactions with calcium acetate (STADNIKOV [1937], WRIGHT and SCHNITZER [1960]), iodide-iodate (WRIGHT and SCHNITZER [1960]), barium hydroxide (STADNIKOV [1937], WRIGHT and SCHNITZER [1960]), and methylation and saponification (BROADBENT and BRADFORD [1952], FUCHS and HORN [1930], KONONOVA [1958]). Furthermore, a variation in pK values exists within each group, which is due to the polybasicity of the materials (VAN DIJK [1960]). This fact has not been given sufficient consideration in investigations of the complex- or chelate-forming properties of the humic substances (VAN DIJK [1959]), especially when the titrations were made in the presence of metals (BECKWITH [1959], KHANNA and STEVENSON [1962], MARTIN and REEVE [1958]).

The presence of carboxyl groups has been established by different authors (FARMER and MORRISON [1960], KUMADA and AIZAWA [1958], MARTIN, DUBACH, MEHTA, and DEUEL [1963], SCHNITZER, SHEARER, and WRIGHT [1959], SCHNITZER and DESJARDINS [1965]) by the shift of the maxima of the carbonyl bands of the undissociated carboxyl group from 5.7 to 5.9 μ to that of the carboxylate group at 6.1 to 6.3 μ.

This has been confirmed by the disappearance of the carboxyl band in the infrared spectrum after reduction with diborate. The presence of carboxyl groups is indicated by an increase in the hydroxyl groups and a decrease of the oxygen content of the reduced material. This method agrees with the titration in sodium hydroxide to a pH value of 8.5. These results seem to exclude possible contributions of hydroxyquinones and enols to the acidity of humic acids. The total acidity, determined by the reaction with barium hydroxide, is in agreement with the amount of the active hydrogen, determined with diborane (MARTIN, DUBACH, MEHTA, and DEUEL [1963]).

The total hydroxyl content (phenolic and aliphatic) has been determined by acetylation, followed by titrimetric determination of the excess of acetic anhydride (MEYER [1962], WRIGHT and SCHNITZER [1960], MARTIN and JAY [1962]). Values obtained by deacetylation of the acetylated products are much smaller (FORSYTH [1947b], MARTIN, DUBACH, MEHTA, and DEUEL [1963], MEYER [1962]). Dinitrofluorobenzene gives very low values for hydroxyl groups (FLAIG, SCHEFFER, and KLAMROTH [1955]).

A method for the determination of the aliphatic hydroxyl groups (WRIGHT and SCHITNZER [1960]) is based on the subtraction of the values for phenolic hydroxyl groups from total hydroxyl groups. The values for the phenolic groups are obtained by subtraction of the acidic groups by the calcium acetate method from those obtained upon titration with barium hydroxide. The values obtained for total acidity with barium hydroxide agree with the values for active hydrogen determined with diborane, and indicate the absence of aliphatic hydroxyl groups (MARTIN, DUBACH, MEHTA, and DEUEL [1963]).

In a study of SCHNITZER and GUPTA [1965] the carboxyl groups were determined by the calcium acetate method, as well as by decarboxylation; the "total acidity" was determined with the barium hydroxide method, as well as by discontinuous titration. The methods were tested with numerous model substances, such as benzene and phenolcarboxylic acids. After application of further chemical and infrared spectroscopic methods SCHNITZER [1965] gave analytical data (shown in Table 18) for fractions of humic and fulvic acids of podzol.

Table 18. Analytical characteristics of a humic and a fulvic acid according to SCHNITZER [1965].

Element	Elementary composition	
	Humic acid, %	Fulvic acid, %
Carbon	56.72	50.92
Hydrogen	5.21	3.34
Nitrogen	2.37	0.74
Sulfur	0.35	0.26
Oxygen (by difference)	35.35	44.74
Oxygen-containing functional groups (m-equivalent per gram of dry ash-free material)		
Total acidity	5.7	12.4
Carboxyl	1.5	9.1
Total hydroxyl	6.9	6.9
Phenolic hydroxyl	4.2	3.3
Alcoholic hydroxyl	2.7	3.6
Carbonyl	0.9	3.1
Molecular weight*	1,684	669

* The average molecular weights were determined by SCHNITZER and DESJARDINS [1962] by lowering of the freezing point in sulfolane.

It is known that the molecular weights of high molecular weight substances determined by the sulfolane method are very often much lower than the weights determined by osmometry or by ultracentrifuge. Other authors investigated the distribution of the functions of oxygen to characterize humic acids from different origins (FISCHER, SCHLUNGBAUM, and KADNER [1964], FISCHER, SCHLUNGBAUM, and BELAU [1964], FISCHER, SCHLUNGBAUM, BELAU, and PREU [1965], KASATOCHKIN, KONONOVA, LARINA, and EGOROVA [1964], ZIECHMANN [1964], KRÖGER, DARSOW, and FÜHR [1965], KRÖGER [1966], SCHNITZER and SKINNER [1966]).

Values for carbonyl groups, determined by means of hydroxylamine (FLAIG, SCHEFFER, and KLAMROTH [1955], MEYER [1962], WRIGHT and SCHNITZER [1960], SCHNITZER and SKINNER [1965]), 2,4-dinitrophenylhydrazine (FLAIG, SCHEFFER, and KLAMROTH [1955], SCHNITZER and SKINNER [1965]), phenylhydrazine (SCHNITZER and SKINNER [1965]), and sodium boron-hydride (NaBH₄) (MARTIN, DUBACH, MEHTA, and DEUEL [1963]) range between 1 to 3 meq/g. With these reagents, interfering reactions occur in the presence of quinones, and therefore exact values are not obtained (FARMER and MORRISON [1960]). Reductive acetylation gave no indication of the presence of quinones (FARMER and MORRISON [1960], SCHNITZER, SHEARER, and WRIGHT [1959], SCHNITZER and SKINNER [1965]).

BAILEY, BRIGGS, LAWSON, SCRUTON, and WARDS [1965] explain absorptions in infrared spectra of brown-coal humic acids by hydrogen bridge linkage between the hydrogen of the phenolic hydroxyl groups and the carbonyl groups of quinones.

The content of methoxyl groups of humic acids from soils ranges up to about 2%. It decreases with the age of the humic acids. Fractions of the humic acids of composted plant material usually have a higher methoxyl content than those from soils or brown coals. The content is also dependent upon the soil type (TISHENKO and RYDALEVSKAYA [1936], NATKINA [1940], NEHRING [1955]).

The major portion of oxygen cannot be assigned to the different functional groups according to the investigations on the chemical composition of humic substances such as humic and fulvic acids (MARTIN, DUBACH, MEHTA, and DEUEL [1963], DUBACH, MEHTA, JAKAB, and ROULET [1964], MEYER [1962], SCHNITZER [1965], WRIGHT and SCHNITZER [1959, 1960]).

Another possible position of the oxygen is in ether linkages. MARTIN, DUBACH, MEHTA, and DEUEL [1963] concluded, by calculation, that the portion of oxygen that is not present as carboxyl, hydroxyl, or carbonyl oxygen in the molecule of the humic acids is of such quantity that up to four ether linkages in one aromatic ring of the molecule of the humic acids would have to be present to account for it in this form. This was considered very unlikely. They suggested that extremely stable quinones or lactones might be present, because even carefully reduced humic substances show a broad carbonyl band in the infrared spectrum, which cannot be assigned to carboxyl groups.

These authors also call attention to the interesting fact that with increasing average molecular weight, the number of carboxyl groups decreases and the number of hydroxyl groups increases. This fact is in agreement with the observation that a decarboxylation occurs during the dehydrogenating polymerization of phenolacrylic acids to humic acid type substances (HAIDER, LIM, and FLAIG [1964]).

(a) Carbohydrates in Humic Acids

Some humic acids contain carbohydrates, which may participate in the oxygen content. They are usually regarded as impurities. SCHEFFER and KICKUTH [1961a, b] hydrolyzed the carbohydrates of a humic acid from an oligotrophic moor and by extraction with copper oxide-ethylendiamine, found 16% free cellulose, 9% bound cellulose, and 16% bound low molecular weight sugars. The hydrolysates did not contain amino acids in these investigations.

LYNCH, WRIGHT, and OLNEY [1957] found 0.55% sugar in humic acids from a fallow soil and 1.96% in a cultivated soil. By means of paper chromatography, the presence of glucose, galactose, arabinose, and xylose (but only in very small amounts) were identified in hydrolysates of humic acids (COULSON, DAVIES, and KHAN [1959], WRIGHT, SCHNITZER and LEVICK [1958]).

The sugar content of humic acids may arise from coprecipitation of altered carbohydrates during isolation. Furthermore, the possibility exists that sugars may form copolymers during dehydrogenative polymerization of lignin degradation products, as is described by FREUDENBERG and HARKIN [1960] in the case of coniferyl alcohol (Figure 16).

The possible participation of sugars or their degradation products in the formation of humic acids is discussed later on.

3. *Functions of Nitrogen (Hydrolysis)*

The humic acids isolated from soils have nitrogen contents ranging from 1 to 5%. Some authors are of the opinion that proteins or nitrogen-containing compounds derived from proteins are dissolved during the extraction of humic acids with alkaline or neutral solutions. Therefore, many attempts have been made to decrease the nitrogen content of humic fractions. The separation of a mixture of humic substances and of proteins is very difficult to obtain because the solubilities of the two components of the mixture are similar (MEHTA, DUBACH, and DEUEL [1963a]). The hydrolysis methods are discussed in detail later. Enzymatic degrada-

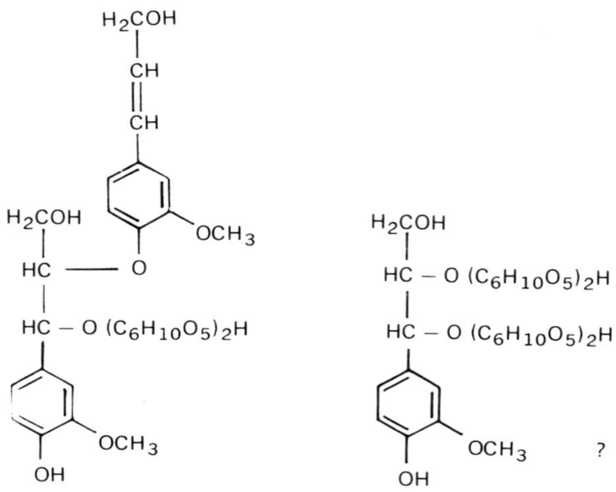

Figure 16. Fixation of sugars during polymerization of coniferyl alcohol (FREUDENBERG and HARKIN [1960]).

tion (HOBSON and PAGE [1932], JENKINSON and TINSLEY [1960], SCHARPENSEEL and KRAUSSE [1962], HAIDER and MARTIN [1967]) has also been investigated.

The nitrogen content has been lowered by 50 % in some fractions by the use of gel filtration with Sephadex (RUCHTI [1962]). It has also been possible to lower the nitrogen of the fulvic acid of a humus-carbonate soil by the use of column chromatography with DEAE-cellulose (anion exchange) and SM-cellulose (strong acidic cation exchanger) (ROULET, MEHTA, DUBACH, and DEUEL [1963]). Humic substances prepared by ether extraction of a hydrochloric acid extract could be freed from nitrogen by treatment with a cation exchanger (RUCHTI [1962]).

The most detailed investigations have been made using hydrolysis with 6 N hydrochloric acid for several hours. Nearly half of the nitrogen can be found in the hydrolysate. The values

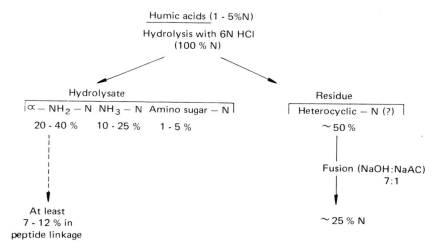

Figure 17. Distribution of nitrogen in humic acids in percent.

for the α-amino nitrogen in the hydrolysates of the humic acids from soils average 30 to 40 % of the hydrolyzable nitrogen, and increase to 60 % for humic acids produced by fungi (*Spicaria elegans*) (LAATSCH, HOOPS, and BIENECK [1952], BREMNER [1952]). Even at the beginning of this century a part of the hydrolyzable nitrogen had been identified as amino acids (SUZUKI [1906–1908a, b, c], ROBINSON [1911], SHMUK [1924]).

In recent years several papers have been published on the determination of amino acids in humic substances (BREMNER [1954, 1955b], SOWDEN and PARKER [1953], STEVENSON, MARKS, VARNER, and MARTIN [1952], PAVEL, KOLOUSEK, and SMATLAK [1954], HAYASHI and NAGAI [1956], SCHARPENSEEL and KRAUSSE [1962, 1963], BREMNER, FLAIG, and KÜSTER [1955], KONONOVA and ALEXANDROVA [1956a, b], OKUDA and HORI [1954, 1955], SOWDEN [1955, 1957], YOUNG and MORTENSEN [1958], CARLES and DECAU [1960a, b]). BREMNER [1955a] and YOUNG and MORTENSEN [1958] found that 3 to 10 % of the hydrolyzable nitrogen (1 to 5 % of total nitrogen) came from amino sugars. Degradation products of nucleic acids have been identified by ANDERSON [1958]. The amount of amino acids differ, depending upon the origin of the humic acids. No significant relationship between the content of amino acids and the soil type could be established, even in the case of samples of humic acids from 27 very different well-defined soil types.

The experiments show that a qualitatively analogous composition exists. Usually the amino acids, glycine, alanine, leucine + isoleucine, and lysine are found in greatest abundance, whereas the sulfur-containing amino acids are very rare (SCHARPENSEEL and KRAUSSE [1962]). Tryptophan is not found after either acidic or alkaline hydrolysis (BREMNER [1955b]). After separation of a humic acid into a nitrogen-rich fraction ("brown" humic acid) and a low-nitrogen fraction ("gray" humic acid) by precipitation with sodium chloride and hydrolysis of the two fractions, it was found that the "brown" humic acid had the higher α-amino nitrogen content (BREMNER [1955c], SCHARPENSEEL and KRAUSSE [1962]). The molar amounts of amino acids averaged three times higher in the "brown" humic acids than in the "gray" humic acids (Table 19).

Nearly all the amino acids noted are constituents of proteins of plants or microbes. Some are transformation products of amino acids. KUO and BARTHOLOMEW [1966] established by experiments with ^{15}N that the α-amino nitrogen is derived mostly from microbial proteins. A treatment with hydrofluoric acid before the hydrolysis increased the quantity of amino acids recovered by 10 % (SCHARPENSEEL and KRAUSSE [1962]).

SCHARPENSEEL and KRAUSSE [1962] concluded from studies of the hydrolysis of uniformly labeled humic acids with papain that approximately 33 % of the amino acids are in peptide linkage. These investigations with labeled humic acids also demonstrated that the main part of the hydrolyzable portions were indeed a part of the humic acids. The content of amino acids found in the hydrolysates from humic acids is influenced by the method of isolation; therefore, a comparison of the results of different authors is often difficult.

Further details about the occurrence of proteins in humic substances is mentioned later.

The finding that a part of the nitrogen in the humic acids is present as peptides is in agreement with the results of studies on the utilization of the nitrogen by microorganisms (FLAIG and SCHMIDT [1957]). The amounts utilized by the microorganisms are of the same order as those determined as α-amino nitrogen by chemical procedures.

More than half of the nitrogen remaining in the residue after hydrolysis can be recovered as ammonia by fusion in a mixture of potassium hydroxide and sodium acetate (7:1) in an iron crucible at 270°C for 7 minutes. On the basis of model experiments with N-phenylglycine, it is assumed that this part of nitrogen consists partly of nonhydrolyzable amino acids (FLAIG and BREYHAN [1956]).

Upon acidification of the aqueous solutions of the alkali fusion, followed by extraction

Table 19. Relative abundance of amino acids in the hydrolysates of humic acids after hydrolysis with 5 N HCl (24 hours at 110°C in the presence of stannous-II-chloride), nitrogen content of the dialyzed humic acids (according to SCHARPENSEEL and KRAUSSE [1962]).

	Chernozem Hungary (b.h.a.)*	Chernozem Hungary (g.h.a.)†	Acid brown-earth (Buntsandstein) (b.h.a.)	Acid brown-earth (Buntsandstein) (g.h.a.)	Para-brown-earth on loess (woods) (b.h.a.)	Para-brown-earth on loess (woods) (g.h.a.)	Podzol Bh (b.h.a.)	Podzol Bh (g.h.a.)	Pseudogley (Rhine) (b.g.a.)	Pseudogley (Rhine) (g.h.a.)
Cysteine	336	277	271	—	315	—	525	—	951	336
Methioninesulphoxide	331	53	475	25	174	—	333	—	—	195
Hydroxyproline	—	—	—	—	—	—	ca. 293	—	—	—
Aspartic acid	766	153	6666	230	> 8774	326	> 1359	—	2651	1896
Threonine	1180	135	5185	588	—	1024	804	> 267	2753	1635
Serine (glutamine, asparagine)	755	260	5876	972	3143	830	886	—	4668	674
Glutamic acid (citrulline)	637	466	7085	218	6517	1386	—	145	4961	—
Proline	—	—	—	—	—	—	—	—	—	—
Glycine	1147	539	6975	981	4132	1440	228	143	1886	2076
Alanine (amino adipic acid)	2415	286	8069	1784	3780	1188	668	182	2267	2377
α-Aminobutyric acid	—	—	—	—	—	—	—	—	—	—
Valine	1365	120	2187	119	2351	1474	449	482	5709	1258
ε-Diaminopimelic acid	—	—	687	—	—	1193	206	191	—	—
Methionine	1166	> 90	960	59	> 1140	1077	251	133	> 1008	> 408
Isoleucine	1083	354	4222	678	> 5388	2193	462	> 600	1147	516
Leucine	917	460	3679	1031	—	2003	776	—	4291	1971
Tyrosine	—	145	1762	> 2169	> 2483	2034	238	—	> 597	499
Phenylalanine	—	320	2963	—	498	1340	202	841	406	1190
γ-Aminobutyric acid	—	—	1363	—	—	936	111	—	—	—
NH₃, ethanolamine	748	676	4800	560	2620	1144	545	519	1798	—
Lysine	819	471	10,007	1940	2844	892	931	1218	4519	180
Histidine	699	374	4460	773	872	198	341	687	444	540
Methylhistidine	314	—	1053	253	387	550	139	293	—	335
Arginine	535	213	2565	1170	1257	198	438	663	1998	395
Ornithine	249	—	—	395	723	—	135	178	589	—
Total	15,282	5392	81,310	13,943	47,398	21,626	10,377	6542	44,733	16,481
Percentage of amino acids	1.53	0.54	8.13	1.39	4.74	2.16	1.04	0.65	4.47	1.65
Total nitrogen (Kjeldahl) of humic acids	2.04	1.08	4.53	3.08	3.73	0.69	0.92	0.7	3.30	1.57
Gray humic acids, %	31.5	68.5	74.5	25.5	95.6	4.4	71.9			
Brown humic acids, %								28.1	86.0	14.0

* b.h.a. = brown humic acid fraction.

† g.h.a. = gray humic acid fraction.

with ether and separation by paper chromatography, compounds were noted that gave positive color reactions for indoles (dimethylaminobenzaldehyde in 1 N hydrochloric acid) (FLAIG and BREYHAN [1956]). Earlier, SHMUK [1924] identified compounds with heterocyclic-bound nitrogen such as pyrrol [24], indole [25], and skatole after alkaline fusion of humic acids from chernozem. The percentage of nitrogen in the residue of the fusion (potassium hydroxide and sodium acetate) of humic acids from rotted wheat straw according to SCHOBINGER [1958] is much higher in nonhydrolyzed than in previously hydrolyzed humic acids.

This result may be due to the fact that tyrosine and especially tryptophane are more sensitive to hydrolysis with strong mineral acids and may be altered. Nitrogen-free fractions of lignin do not give reactions for indole compounds after fusion and acidification of the alkaline solution, so that the formation of indole-type compounds is not possible during the fusion. This objection of STEELINK [1959] to this identification is not valid. However, in the case of nitrogen-containing lignin fractions from rotted straw, indole compounds can be detected by the indicated procedure.

The hydrolysis experiments do not lead to conclusive evidence concerning the mechanism of the reactions involved in the synthesis of the nitrogen-containing humic substances.

VII. Model Investigations on the Formation of Dark-Colored Substances in Humus with Naturally Occurring Phenolic Compounds

Various investigators have attempted to gather information about the constitution of humic acids by experiments with model compounds. The three following facts:

(1) that the phenolic plant constituents are normally utilized more slowly by microorganisms than carbohydrates and their transformation products;
(2) that microorganisms synthesize phenols or quinones in their metabolism;
(3) that the isolation of phenolic lignin degradation products is possible from soils, as degradation products of the humic acids or their precursors

have led to investigations on the relation of the chemical properties of phenolic compounds to the formation of humic substances. The reactions of phenolic compounds which lead to dark-colored humus-like compounds have been studied most intensively.

Because the humic acids isolated according to the usual methods—by extraction with alkaline or nearly neutral solutions and precipitation with acids—contain nitrogen in nearly all cases, many investigations have dealt with the participation of nitrogen-containing compounds in the formation of humic acids. Reactions of the sugars and their degradation products with nitrogenous compounds have also been investigated for possible connections with humic acid synthesis.

A prime difficulty in the elucidation of the chemical constitution of humic acids is that they are high molecular weight compounds and cannot be split by hydrolysis or oxidative degradation into definite intermediate compounds, as can be done with proteins, polysaccharides, or caoutchouc, from which a clue to their constitution can be drawn. The humic acids according to physical investigations are polymers. The building blocks are connected three-dimensionally. The constitution of other three-dimensional polymers, such as condensates of urea-formaldehyde, likewise cannot be fully elucidated by degradation reactions.

It is important to establish the chemical nature of the bonds between different monomeric building blocks of the humic acids and the reactions that lead to their formation. Because the various constituents of the humus decompose continuously but at different rates, the reactions involved in decomposition should be elucidated. Many properties of soils as media for plant growth are directly related to the "dynamics" of the humic substances. These reactions are dependent upon the existing climatic conditions, which affect the physical and chemical properties of the different inorganic soil constituents, and on the initial soil materials including plant cover. It cannot be expected that a uniform humic acid with the same chemical constitution would be formed under various conditions of soil formation. Proteins are also formed from monomers of various chemical constitutions, but the reactive groups are the same, which lead, through amide linkages, to linear- or spheromacromolecules. Other reactive groups of some amino acids offer additional possibilities for linkages within the molecule or between several molecules, as for example, through the S-S linkages of the cysteine or through secondary valences or hydroxyl, amino, or carboxyl groups. This fact is also important for physical behavior of humic acids.

In the following discussion the reactions of phenols and carbohydrates are mentioned insofar as they may contribute to the formation of humic acids in nature and agree with the investigations of natural humic acids.

1. *Alterations of Lignin*

Lignin is the phenolic compound that occurs in large amounts in plants. The methoxyl and nitrogen content of lignin differs according to plant groups and varies from 5 to 20% (see Table 20).

Table 20. Methoxyl and nitrogen content of different lignins in percent (BONDY and MEYER [1948]).

	OCH$_3$, %	N, %
Hardwood groups	20.0	0
Softwood groups	15.0	0.2–0.3
Graminees	10.0	1.2–1.6
Leguminoses	5.0	2.9–3.4

As was previously indicated, during decomposition the methoxyl content of lignin fractions decreases while the nitrogen and the oxygen contents increase. The products formed are dark-colored and have physicochemical properties (spectra, solubility, cation exchange capacity) that are comparable to those of humic acids (FLAIG, SCHOBINGER, and DEUEL [1959], FLAIG [1959]). The products formed from lignin decomposition have a methoxyl content that is higher than that of humic acids from manures or composts (NEHRING and SCHIEMANN [1952a, b]).

If one isolates the lignin fractions from rotted plant material (for instance straw) in the usual way (with sulfuric acid), the content of methoxyl groups decreases and the nitrogen content increases concurrently with the time of rotting (BROADBENT [1954], BARTLETT and NORMAN [1938], BARTLETT [1939], FLAIG, SCHOBINGER, and DEUEL [1959], RITTER, SEBORG, and MITCHELL [1932], FLAIG [1960b], WAKSMAN and SMITH [1934], STÖCKLI [1952], NEHRING and SCHIEMANN [1952a, b]). This suggests that the cleavage of the methoxyl group precedes the introduction of nitrogenous groups (FLAIG [1960b]) (Figure 18).

The degradation of lignin is caused by enzymes of special species of microorganisms (comp. SCHANEL [1962], TROJANOWSKI, LEONOWICZ, and HAMPEL [1966], TROJANOWSKI, LEONOWICZ, and WOJTAS [1967]). Addition of nitrogenous salts (1 N against 0 N) accelerates degradation. The fixation of nitrogen compounds after alteration of substituent groups in the lignin preserves its high molecular structure and renders the nitrogen less accessible to microorganisms.

One possibility exists that humic acids are formed by the reaction of nitrogen-containing

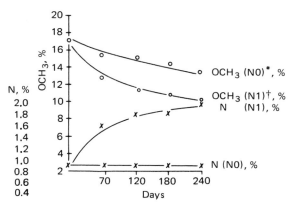

Figure 18. Alterations of the content of nitrogen and methoxyl of the lignin fractions of straw during humification.
*0 N = without additional nitrogen source.
†1 N = with 1 % N as NH_4NO_3 per dry weight of straw.

compounds such as proteins, amino acids, or their degradation products with a lignin that is still more or less polymeric but altered by oxidation and demethylation.

According to other investigations (WAKSMAN and HUTCHINGS [1936], GOTTLIEB, DAY, and PELCZAR [1950], PELCZAR, GOTTLIEB, and DAY [1950], FISCHER [1953], WEHRMANN [1955]), lignin can serve as a carbon source for white rot fungi and other soil microorganisms.

(a) Biochemical Degradation of "Synthetic Lignins"

During the degradation of lignin by the attack of special groups of microorganisms, the cleavage of the rings seems to be an important step for the degradation of lignin into smaller units (HAIDER [1967]).

With models of lignin synthesized by dehydrogenative polymerization of variously labeled lignin monomers by Freudenberg's (FREUDENBERG [1956, 1962]) mushroom phenol oxidase method it was determined which of the carbon atoms of the lignin building blocks is split off first by biochemical decomposition. Under the same conditions, less activity per carbon is split off from the methoxyl group (I) or from either of the three carbon atoms in the side chain (IIa through IIc) than from the polymers of the lignin building blocks labeled in the ring (III) (HAIDER and GRABBE [1967]) (Figure 19). This means that the cleavage of the ring is an essential reaction for the degradation of the lignin. In this way the high molecular, originally

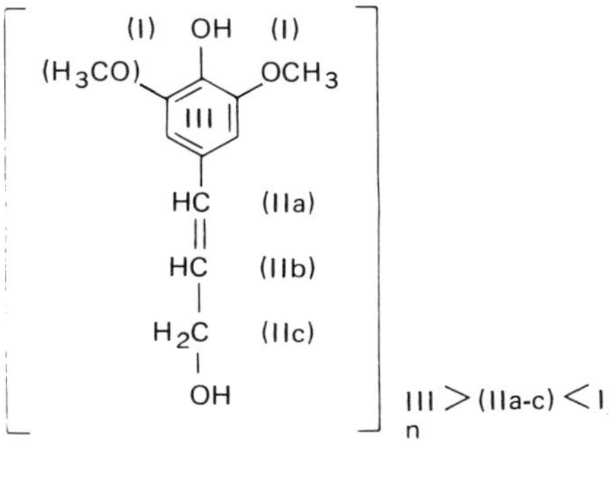

Figure 19. Rate of release of ^{14}C during microbial de- composition of "synthetic lignins."

three-dimensional, structure of the lignin is more or less disrupted, and degradation products with lower molecular weight are formed.

2. *Lignin Degradation Products*

Phenolic lignin degradation products have been identified in extracts of soils and decomposing plant material which contain side chains with *three* (C_6—C_3 compounds) or side chains with *one* carbon atom (C_6—C_1 compounds). Therefore, the reactions of these compounds in relation to the formation of humic substances are interesting.

(a) C_6—C_3 Compounds

Different substituted p-hydroxyphenylacrylic acids polymerize to polymers during oxidation in the presence of phenol oxidases from white rot fungi. During polymerization, carbon dioxide is released (HAIDER, LIM, and FLAIG [1964]).

$$R_1 = R_2 = H \text{ (p-hydroxycinnamic acid)}$$
$$R_1 = H; R_2 = OH \text{ (caffeic acid)}$$
$$R_1 = H; R_2 = OCH_3 \text{ (ferulic acid)}$$
$$R_1 = R_1 = OCH_3 \text{ (sinapic acid)}$$

It was shown by the carbon-14 technique that 50 to 60% of carbon atom 1 is released as carbon-14 dioxide. The specific activity of the polymers formed was approximately the same as for the initial products when the other carbon atoms of the side chain or the carbon atom of the methoxyl groups were labeled. The elimination of carbon dioxide cannot be caused by a direct effect of the phenoloxidases. Quinonoid intermediate products are formed at the beginning of the polymerization, whereby the linkage of the carboxyl group becomes unstable and is split off as carbon dioxide. The formation of quinonoid intermediate products can be deduced from the fact that no polymers are formed from 3,4-dimethoxy-cinnamic acid, because the phenolic hydroxyl group in p-position to the side chain is methylated and cannot be transformed into an intermediary quinone (LIM [1965]). These investigations constitute an important contribution to our understanding of the transformations of the substances during the course of humification, as it was shown that the lignin degradation products may not be completely decomposed to smaller fragments but recombine to form other polymers.

The polymers formed from the labeled compounds were added to cultures of white rot fungi to study their decomposition. The specific activities of the carbon dioxide released and that of the mycelium gave an indication of the extent of utilization of the different carbon atoms (see Figure 20).

It was found that the carbon atom of the methoxyl group of the polymers is readily utilized by the microorganisms. The demethylation occurs faster than the degradation of the portion of the side chain, which remains after the polymerization. The slight utilization of carbon atom 2 may be due to the fact that this atom is frequently involved in the linkage reactions of the acrylic acid side chain (ERDTMAN and WACHTMEISTER [1957], MÜLLER, MAYER, SPANAGEL, and SCHEFFLER [1961]).

Vanillin and vanillic acid were identified among the degradation products of the polymers

Polymerized from	% radioactivity in evolved CO_2	% radioactivity in mycellium
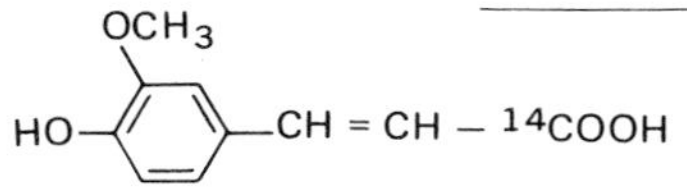 OCH_3 HO—⟨ring⟩—CH = CH — $^{14}COOH$	8.7	35
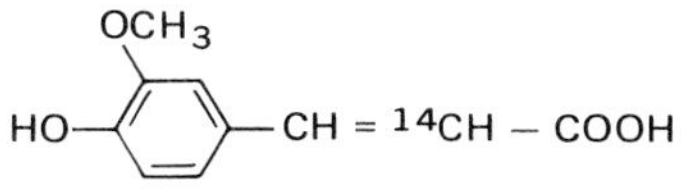 OCH_3 HO—⟨ring⟩—CH = ^{14}CH — COOH	4.5	18
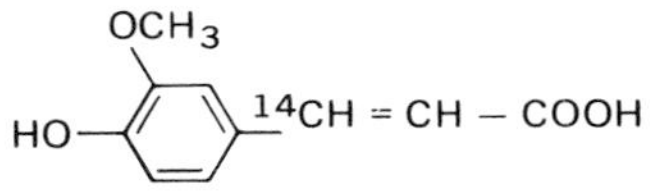 OCH_3 HO—⟨ring⟩— ^{14}CH = CH — COOH	11.2	34
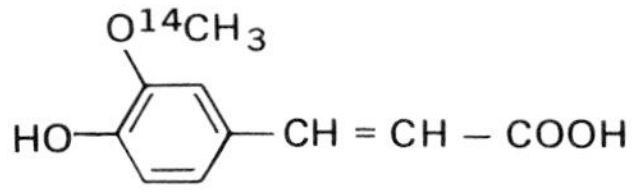 $O^{14}CH_3$ HO—⟨ring⟩—CH = CH — COOH	22.0	30

Figure 20. Degradation of differentially labeled groups of polymers of ferulic acid by *Pleurotus ostreatus* in shake cultures (HAIDER [1966]).

of ferulic acid (HAIDER, LIM, and FLAIG [1964]). Similar experiments were made with labeled coniferyl alcohol (HAIDER [1966]).

Styrene compounds were formed by microbial decarboxylation of phenylacrylic acids. According to FINKLE [1965], it is possible that these compounds react to form ethylene polymers. Ethylene polymerization during the formation of coniferous lignin from coniferyl alcohol has been assumed (ADLER [1957]).

Only aliphatic ethers were formed from o-hydroxystyrene. TANAKA, YAMASHITA, and YOKOYAMA [1957] established by means of infrared and ultraviolet spectra that the polymerization of hydroxystyrene occurs primarily through the ethylene group, but in contrast to the unsubstituted styrene, only carbon atom 2 of the vinyl group reacts.

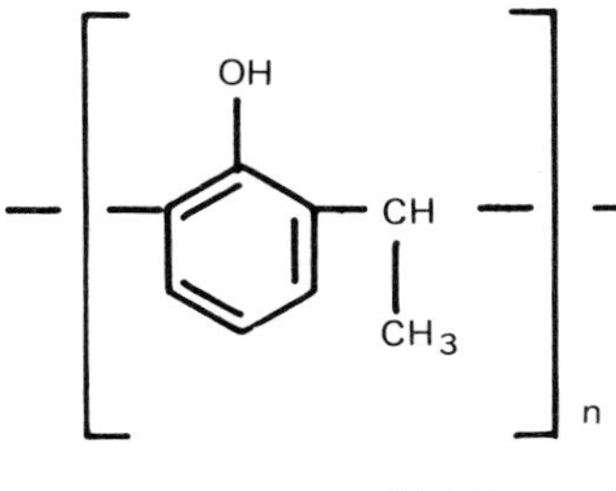

$$\left[\; \underset{CH_3}{\underset{|}{\overset{OH}{\underset{}{\bigcirc}}}}\!\!-CH- \;\right]_n$$

CASSIDY [1949], UPDEGRAFF and CASSIDY [1949], and STERN, ENGLISH, and CASSIDY [1957] prepared polymers of vinylhydroquinone as models for humic acids or as "electron exchangers." Based on a steeper slope of the curve of the potentiometric titration in acidic aqueous solution, it was concluded that these polymers have a stronger tendency to form semiquinones than the monomers (MICHAELIS [1933]). In a later paper GIZA, KUN, and CASSIDY [1962] investigated the possibilities of linkage of the polymers through the ring by use of different p-benzoquinones as model substances.

(b) C_6—C_1 Compounds

Phenolic compounds as lignin degradation products with a side chain of *one* carbon atom can be formed by direct decomposition of the lignin or by the action of white rot fungi on phenolacrylic acids such as ferulic acid [1], p-hydroxycinnamic acid, and sinapic acid (HAIDER, LIM, and FLAIG [1962], ISHIKAWA, SCHUBERT, and NORD [1963a, b]). The degradation of the side chain can occur at the double bond. Through this reaction aldehydes are formed, which are oxidized in the culture solutions to the corresponding acids, namely vanillic acid, p-hydroxybenzoic acid, and syringic acid. Presumably, glyoxylic or oxalic acid is formed as a C_2 fragment of the side chain.

The formation of vanillin and vanillic acid during the enzymatic oxidation of guaiacyl-glycerol [2] and guaiacylpyruvic acid [3] (Figure 21) has also been established (ISHIKAWA, SCHUBERT, and NORD [1963a, b]).

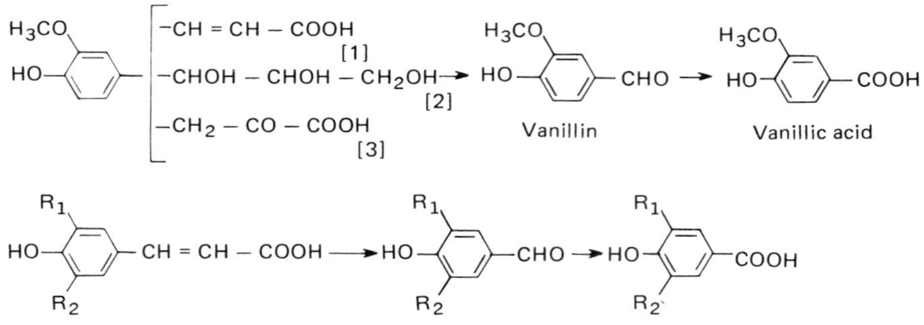

Figure 21. Formation of lignin degradation products.
$R_1 = R_2 = H$ p-hydroxycinnamic acid $R_1 = R_2 = H$ p-hydroxybenzoic acid
$R_1 = R_2 = OCH_3$ sinapic acid $R_1 = R_2 = OCH_3$ syringic acid

Vanillic acid and syringic acid also polymerize in the presence of phenoloxidases, but the polymers are only faintly colored. Working with carbon-14-labeled compounds, HAIDER, LIM, and FLAIG [1962, 1964] established that, depending upon substitution, up to 60% of the carboxyl groups could be released in the form of carbon dioxide. The polymers contained very little nitrogen. The fungi made good growth in the shake cultures, indicating that they could utilize phenolcarboxylic acid polymers as a carbon source. This was not true with the polymers of other phenolcarboxylic acids, such as protocatechuic acid.

(c) Differences in Biochemical Degradation of Polymers of Lignin Degradation Products

The biochemical decomposition of the polymers of lignin degradation products has been investigated. Some results of the experiments for the study of the decomposition of the polymers of the lignin degradation products by white rot fungi, such as *Pleurotus ostreatus*, are demonstrated by HAIDER [1966] in the case of vanillic, syringic, and ferulic acids. For comparison, the number of the polymerized coniferyl alcohol as a model substance for coniferous lignin are also given (Figure 22).

The values reveal the percentage of the split off, labeled groups as [14]carbon dioxide. In brackets the values of the percentage of the activity, which was in the mycelium of the fungus, are given. One must consider both sets of data to evaluate the turnover of the labeled carbon atoms, because not all carbon assimilated by the fungi is transformed into carbon dioxide.

Figure 22 demonstrates that a large quantity of the activity is derived from the carboxyl

Polymer from :	$-^{14}COOH$	$-O^{14}CH_3$	$-^{14}CH=CH-$	$-CH=^{14}CH-$	$-^{14}CH_2OH$
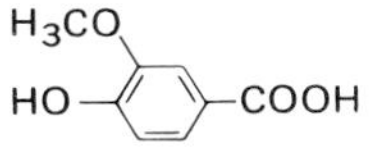 H_3CO, HO—⟨⟩—$COOH$ Vanillic acid	22	12[15]			
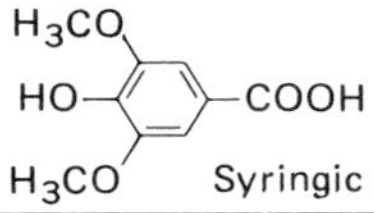 H_3CO, HO—⟨⟩—$COOH$, H_3CO Syringic acid	45	13 [12]			
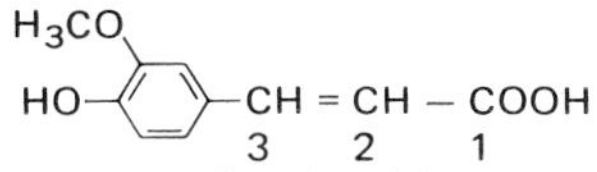 H_3CO, HO—⟨⟩—$CH=CH-COOH$ 3 2 1 Ferulic acid	$1-^{14}C$ 14 [23]	16 [58]	$3-^{14}C$ 7 [40]	$2-^{14}C$ 3 [35]	
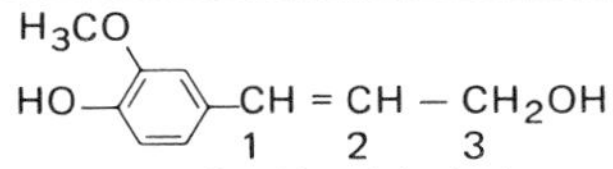 H_3CO, HO—⟨⟩—$CH=CH-CH_2OH$ 1 2 3 Coniferylalcohol		4.5 [5]	$1-^{14}C$ 3.5 [2,8]	$2-^{14}C$ 2.6 [3,6]	$3-^{14}C$ 4.4 [3,4]

Figure 22. Cleavage of different ^{14}C-labeled groups of polymers of lignin degradation products and of polymers of coniferyl alcohol ("synthetic lignin") as ^{14}carbon dioxide and the fixation of the activity in the mycelium (in parentheses) by the action of *Pleurotus ostreatus*, measured after eight days (HAIDER [1966]).

groups. From ferulic acid polymers presumably a cleavage product is utilized by the fungus in its metabolism, which consists of the carbon atoms 1 and 2 of the side chain. This may be supposed from the high activity in the mycelium when the carbon atom 2 is labeled. But also, the carbon atom 3 is found in a larger amount in the mycelium. And the cleavage of the carbon atom of the methoxyl groups occurs to a larger amount. This is demonstrated by the values for the labeled carbon dioxide and for the activity in the mycelium.

In contrast to the high rate of turnover of the polymers of the carboxylic acids, a small cleavage of activity in CO_2 and in the mycelium is observed in the case of the polymers or the copolymers of the lignin building blocks (HAIDER [1966]). Not only are the numbers for carbon dioxide relatively low, but also the activity in the mycelium is small.

The lignin-like polymers seem to be more resistant to microbial decomposition than the polymers of lignin degradation products. For this reason the lignin degradation products and their polymers may not be accumulated in larger amounts in soils. It is not yet known to what extent these polymers contribute to humic acids.

In Figure 22, only the degrees of degradation of the polymers of the guaiacyl or syringyl type have been given. These polymers have no nitrogen in their molecule. There is a difference between these and polymers that are formed from 1,2-dihydroxy- or 1,2,3-trihydroxybenzene carboxylic such as protocatechuic or gallic acid, and phenylacrylic acids such as caffeic or p-coumaric acid as monomers. These polymers contain nitrogen in their molecules and are more stable against microbial attack, as has also been demonstrated with natural nitrogenous humic acids (FLAIG and SCHMIDT [1957]).

The possible formation of mono- and polyphenol carboxylic acids could occur by methylether cleavage or by demethylation. In addition, protocatechuic acid polymerizes in an oxidative medium in the presence of phenoloxidases with a loss of up to 80% of its carboxyl groups. The polymers have a variable nitrogen content. Nitrogen-containing polymers from caffeic acid are also formed under these conditions.

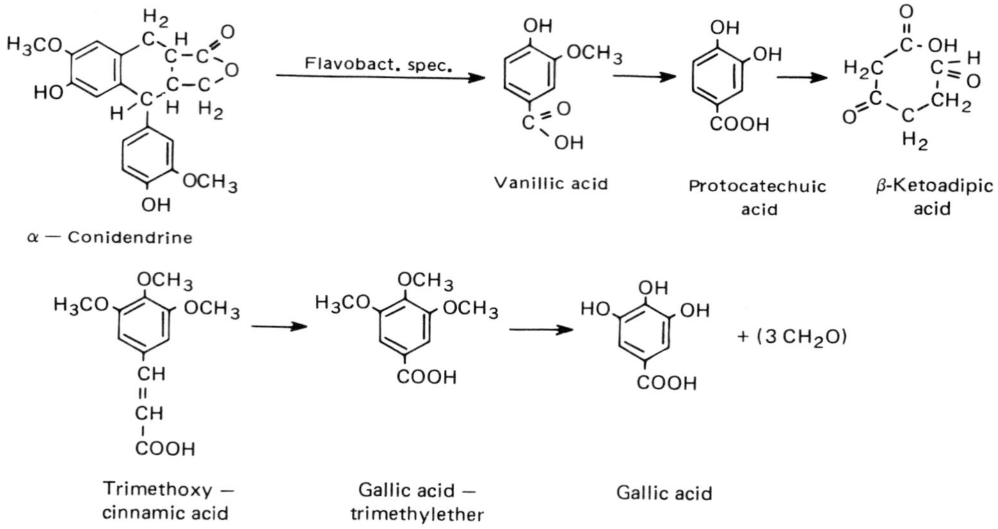

Figure 23. Methylether cleavage of lignin degradation products.

3. Reactions of the Ether Groups

ISHIKAWA, SCHUBERT, and NORD [1963a, b] showed that in microbial cultures guaiacylglycerol [2] may be formed by benzylether cleavage of guaiacylglycerol-β-coniferylether [1] (Figure 23).

(a) Benzylether cleavage

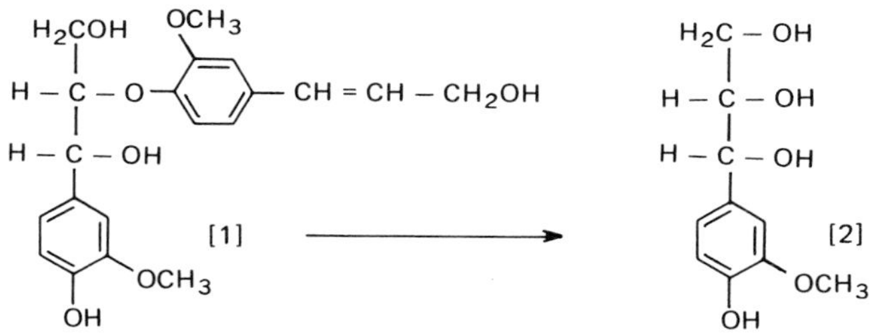

(b) Methylether cleavage

As has been previously mentioned, a decrease in the methoxyl content of the lignin fractions takes place during humification. It was also noted that the methoxyl content of C_6—C_3 compounds decreases when they are added to cultures of white rot fungi. Several investigations with model substances have been made in order to elucidate the mechanism of demethylation. KONETZKA, WOODINGS, and STOVE [1957] believe that the microbial degradation of a lignin, α-conidendrine, occurs as indicated in Figure 23. According to the investigations of HENDERSON [1957], the side chains of phenolacrylic acids are oxidized to a one-carbon carboxyl group before demethylation takes place.

The demethylation of additional lignin degradation products with carbon-14-labeled methoxyl groups has been investigated by HAIDER, LIM, and FLAIG [1962, 1964], as well as by ISHIKAWA, SCHUBERT, and NORD [1963a, b].

Some examples of demethylation of methoxylated lignin degradation products by the attack of white rot fungi are shown in Figure 24.

HENDERSON [1957] describes, in addition to the demethylation of vanillic acids to proto-

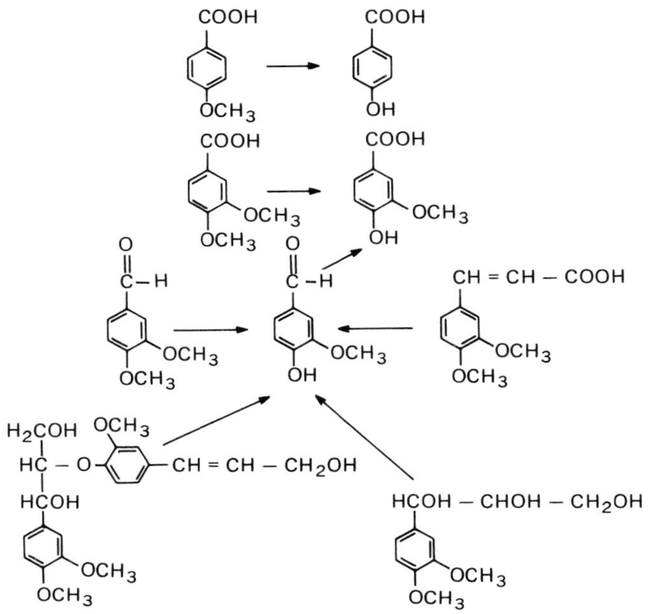

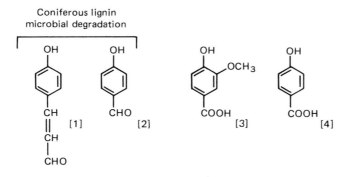

Figure 24. Demethylation of methoxylated phenols by white rot fungi.

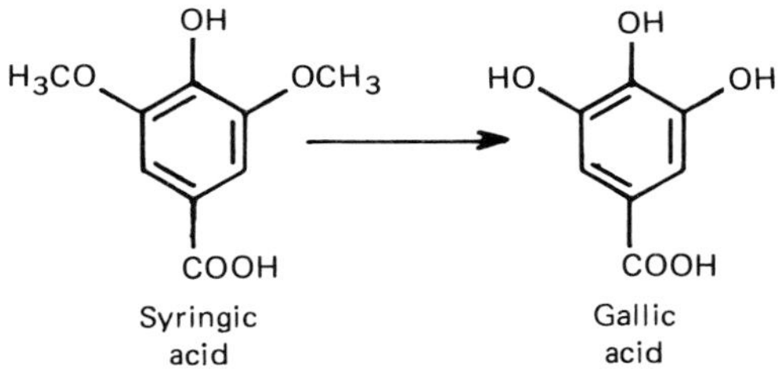

Figure 24a. Demethylation of Syringic acid.

Figure 24b. Compounds formed during the demethoxylation of lignin degradation products.

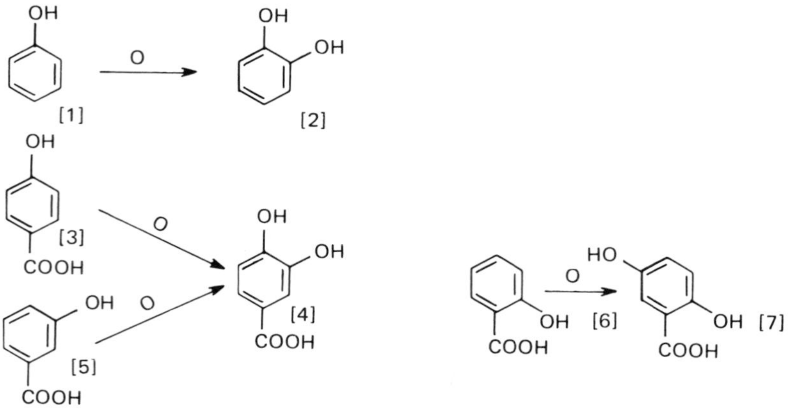

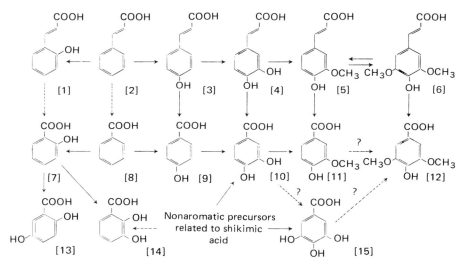

Figure 24c. The hydroxylation of phenols.

catechuic acid, the enzymatic demethylation of syringic acid to gallic acid (Figure 24a). Through these reactions polyphenolcarboxylic acids are formed. They are very reactive and easily oxidized.

(c) Demethoxylation

Through side chain oxidation and demethoxylation, hydroxybenzoic acid is formed from the lignin degradation products, ferulic acid and vanillin.

Figure 25. Interrelations of acids of the cinnamic and benzoic series in higher plants. Dotted arrows indicate possible routes (EL BASYOUNI *et al.* [1964]).

[1] m-hydroxycinnamic acid
[2] cinnamic acid
[3] p-coumaric acid
[4] caffeic acid
[5] ferulic acid
[6] sinapic acid
[7] salicylic acid

[8] benzoic acid
[9] p-hydroxybenzoic acid
[10] protocatechuic acid
[11] vanillic acid
[12] syringic acid
[13] hydroquinone carbonic acid
[14] o-protocatechuic acid
[15] gallic acid

SEIFERT [1962] found p-coumaric aldehyde to be the sole lignin degradation product [1] during the decomposition of pine wood by *Coniophora cerebella* (Figure 24b). ISHIKAWA, SCHUBERT, and NORD [1963a, b] also found p-hydroxybenzaldehyde [2]. These reactions point to a demethoxylation of lignin degradation products. The demethoxylation of vanillic acid [3] to p-hydroxybenzoic acid [4] inside of the plant has been demonstrated by EVEN-HAIM [1966] and EL BASYOUNI, SAID, CHEN, IBRAHIN, NEISH, and TOWERS [1964] by means of labeled compounds.

4. *Oxidation Reactions*

(a) Hydroxylation

A second hydroxyl group is introduced into monophenols by "mixed function oxidases" (MASON [1957]) (Figure 24c). By this type of reaction, phenol [1] is transformed into pyrocatechol [2] and p-hydroxybenzoic acid [3] into protocatechuic acid [4] (EVANS [1947]). The p-hydroxybenzoic acid may also be formed by demethylation of vanillic acid or from ferulic acid or by partial decomposition of p-coumaric acid (EL BASYOUNI, SAID, CHEN, IBRAHIM, NEISH, and TOWERS [1964]). Protocatechuic acid is produced from m-hydroxybenzoic acid [5] (EVANS [1947]), and hydroquinone carboxylic acid [7] (WALKER and EVANS [1952]) from salicylic acid [6].

Reaction sequences of phenols have been worked out by EL BASYOUNI, SAID, CHEN, IBRAHIM, NEISH, and TOWERS [1964]. It illustrates the transformations of phenylacrylic and benzoic acids in higher plants (Figure 25). The technique used was to place small sections of

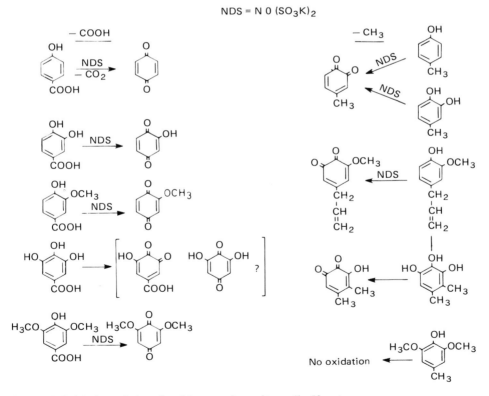

Figure 26. Oxidation of phenols with potassium nitrosodisulfonate.

leaves of different plants in aqueous solutions of the acids for 24 hours and to isolate and identify the reaction products.

These reactions may also play a role during humification, insofar as the enzymes remain active after the death of plants. As has been previously indicated, some of these reactions occur in microbial cultures.

(b) Formation of Quinones

Quinones are formed by further oxidation of the lignin degradation products. To illustrate this type of reaction, the action of potassium nitrosodisulfonate (NDS $= \mathrm{N\,O\,(SO_3K)_2}$) may be mentioned. This mild oxidizing reagent is comparable in its effect on phenols to that of tyrosinase.

According to MASON [1957], tyrosinase is a mixed function oxidase that has a twofold activity. The first is the introduction of a second hydroxyl group in o-position to one which is already present; the second is the dehydrogenation of 1,2-diphenols to o-quinones.

Upon oxidation of phenols substituted with a carboxyl group or an aliphatic side chain in which the carbon atom near the ring has an OH group, carbon dioxide is released and p-benzoquinones are formed (see the left side of Figure 26) (TEUBER and JELLINEK [1952], TEUBER and RAU [1953], FLAIG [1960a], ADLER and LUNQUIST [1961]).

The introduction of an additional OH group to phenols substituted with an aliphatic or unsaturated group leads to the formation of o-benzoquinones (see the right side of Figure 26) (HORNER and DÜRCKHEIMER [1959]). If both o-positions of the hydroxyl group are occupied by methylether groups, no oxidation occurs.

The substances shown in Figure 27 (FLAIG, and HAIDER [1961a]) have been noted in the course of investigations of the transformations of lignin degradation products, such as vanillin or vanillic acid, in cultures of lignin decomposing fungi.

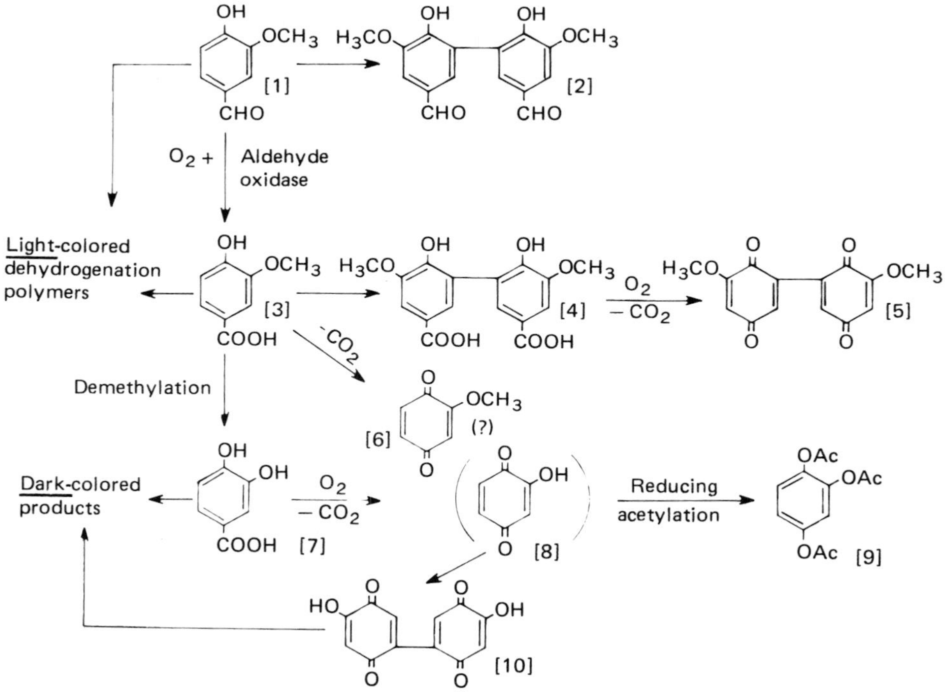

Figure 27. Formation of quinones from lignin degradation products in cultures of microorganisms.

Vanillin [1] is transformed to dehydrodivanillin [2] by enzymatic dehydrogenation. Vanillin is oxidized to vanillic acid [3] by an aldehyde oxidase (Figure 27). Vanillic acid polymerizes to dehydrodivanillic acid [4]. By further oxidation and decarboxylation 3,3'-dimethyl-diphenyl-diquinone-2,5,2',5' [5] is formed from the dimeric acid. Methoxy-p-benzoquinone [6] was found by SUNDMAN and HARO [1966] by microbial degradation of α-conidendrin, which is decomposed to vanillic acid. Vanillic acid is transformed to protocatechuic acid [7] by demethylation. Protocatechuic acid can be further oxidized to hydroxy-p-benzoquinone [8] (FLAIG and SALFELD [1958]). The hydroxyquinone could not be isolated as such from the culture solutions of microorganisms, but was identified as hydroquinone triacetate [9] by reductive acetylation (FLAIG and HAIDER [1961a]). Hydroxy-p-benzoquinone [8] is very reactive. In further studies, 4,4'-dihydroxy-diphenyl-diquinone-2,5,2',5' [10] was formed at pH 5 to 6. Vanillin and vanillic acid yield light-colored dehydrogenation polymers, and protocatechuic acid yields dark-colored polymers.

FORSYTH and QUESNEL [1957] and FORSYTH, QUESNEL, and ROBERTS [1960] found three different tetrahydroxydiphenyls during the enzymatic oxidation of pyrocatechol [11] (Figure 28).

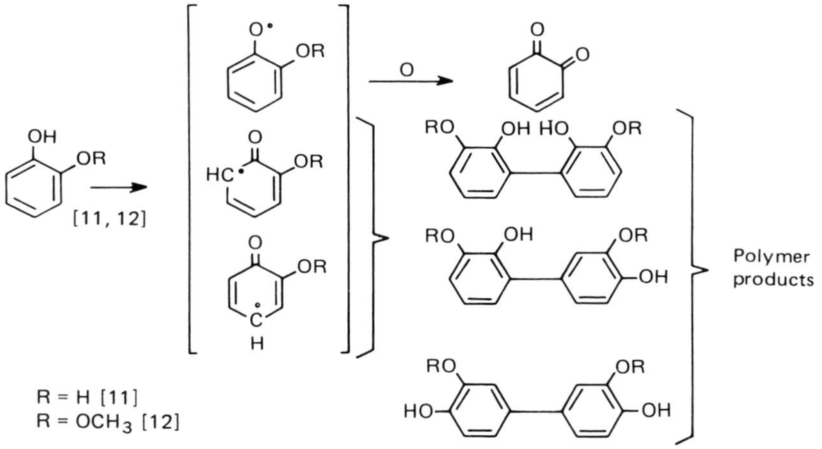

Figure 28. Dimerization of pyrocatechol and guaiacol.

The three corresponding isomeric dihydroxydimethoxy diphenyls were found with guaiacol [12] (Figure 29). The linkage of the rings occurs in this case in the o- or p-position to the OH groups. During these investigations the observation was made that in the formation of the ring systems, in both cases, the o-position to the hydroxyl group is preferred, while in the presence of ferric salts all three dimers are formed in about the same amounts. The three corresponding diphenoquinones have been identified as a further step in the oxidation (WALTER [1960], FLAIG [1964a]).

The dimerization of pyrocatechol, as well as that of guaiacol, occurs through radicals. In both cases polymeric products are formed from the dimers by further oxidation.

FORSYTH, QUESNEL, and ROBERTS [1960] found, as a further oxidation product, di-o-phenylenedioxide-quinone-2,3, during enzymatic oxidation of pyrocatechol (Figure 29a). The isolation of this compound is unusual, as the rings are linked through ether bridges. Connections of the rings through ether linkages had previously not been established, even though the oxidation products of guaiacol had been carefully investigated.

From the studies of BURTON and HOPKINS [1952], as well as MAJIMA and TAKAJAMA [1920],

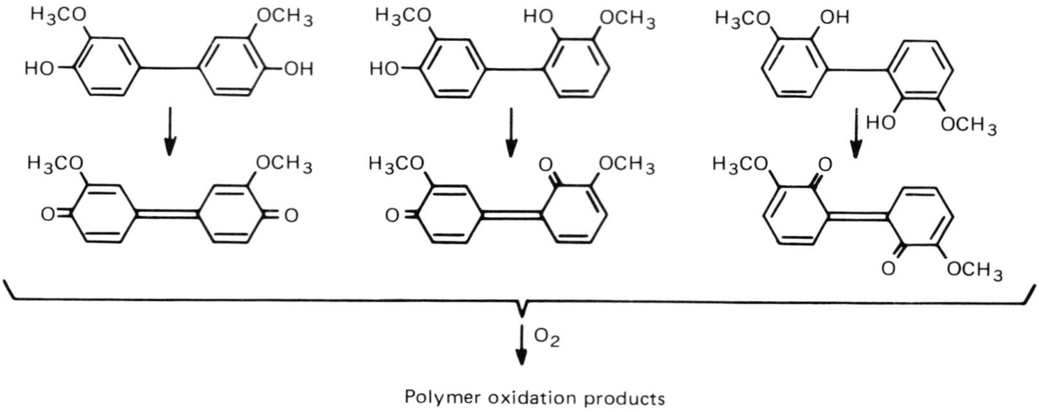

Figure 29. Oxidation of the dimers of guaiacol to diphenoquinones.

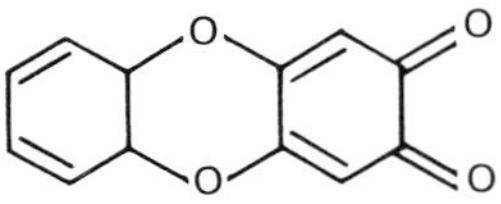

Figure 29a. Di-o-phenylenedioxide-quinone-2, 3.

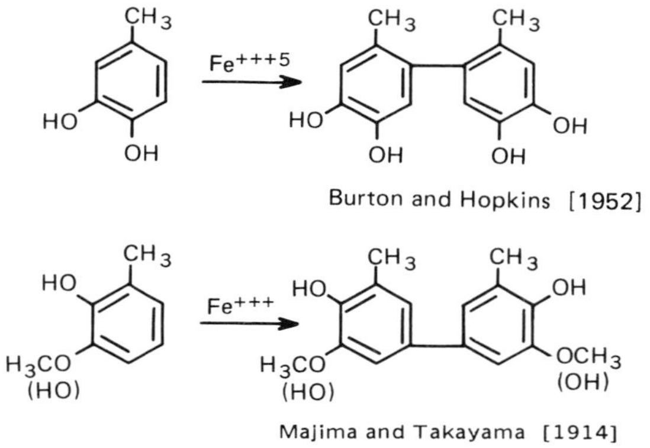

Burton and Hopkins [1952]

Majima and Takayama [1914]

Figure 29b. Dimerization of methyl-substituted pyrocatechols.

examples of the dimerization of methyl-substituted pyrocatechols or their derivatives may be noted. The linkage of the rings in the presence of ferric ions always occurs in the p-position to a hydroxyl group (Figure 29b).

(c) Cleavage of the Aromatic Rings

An important reaction of the oxidation of phenols is the cleavage of the aromatic ring which has been investigated most intensively with protocatechuic acid. The ring cleavage is effected by oxygen transferases (MASON [1957]), which transfer a molecule of oxygen directly to the substrate. Investigations with the oxygen isotope ^{18}O have shown that only the oxygen molecule participates in this reaction, and that the oxygen of water is not involved (HAYAISHI, KATAGIRI, and ROTHBERG [1955], MASON, FOWLKS, and PETERSON [1955]).

The enzymatic ring cleavage of phenols and phenolcarboxylic acids has been intensively investigated (STANIER [1947], KILBY [1948, 1951], GROSS, GAFFORD, and TATUM [1956], STANIER and INGRAHAM [1954], EVANS, SMITH, LINSTEAD, and ELVIDGE [1951]). These substances have been added to cultures of soil bacteria such as *Pseudomonas fluorescence* and species of *Vibrio, Fluorobacteria*, and *Neurospora*.

Recently, FLAIG and HAIDER [1961a] and HAIDER, LIM, and FLAIG [1962] isolated from *Polystictus versicolor* an enzyme system that also cleaves the ring of protocatechuic acid. By labeling the carboxyl group of the protocatechuic acid with carbon-14, these authors showed that the reaction does not occur through o-quinonecarboxylic acid as an intermediate product (Figure 30).

By cleavage of the ring muconic acid is formed from pyrocatechol and β-carboxymuconic acid from protocatechuic acid. Both acids are transformed by the addition of water and, in the case of the oxidation of protocatechuic acid, by elimination of CO_2 into β-ketoadipic acid. The elimination of carbon dioxide is catalyzed by a separate enzyme.

The lactone of carboxymuconic acid can be isolated as an intermediate product during cleavage of the ring of protocatechuic acid in cultures of *Neurospora* (GROSS, GAFFORD, and TATUM [1956]). The protocatechuic acid oxidase, which can be isolated from *Pseudomonas fluorescence*, is very specific and does not catalyze the oxidation of pyrocatechol, hydroxy-p-benzoquinone, 2-, 3-, 4-, 2,3-, or 2,4-hydroxybenzoic acid (CAIN [1961]). FUKUZUMI, MINAMI, and SHIBAMOTO [1959] isolated other ketonic acids such as aceto acetic acid and α-ketoglutaric acid from cultures of white rot fungi. β-ketoadipic acid and possibly other ketonic acids, which may be produced, can serve as carbon sources for the microorganisms.

FLAIG and HAIDER [1961b] found that different phenolic lignin degradation products are utilized by white rot fungi. These investigations have not shown whether all lignin degradation products of a phenolic nature must first be transformed to protocatechuic acid or if other mechanisms lead to cleavage of the ring and utilization of the aliphatic compounds that are produced.

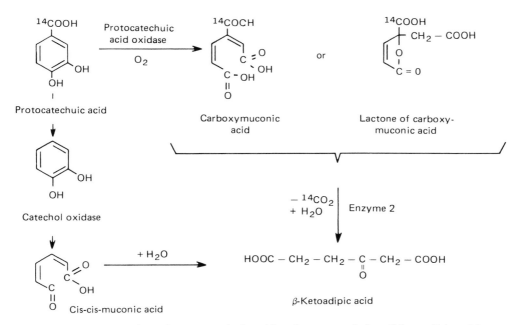

Figure 30. Degradation of protocatechuic acid and pyrocatechol to β-ketoadipic acid.

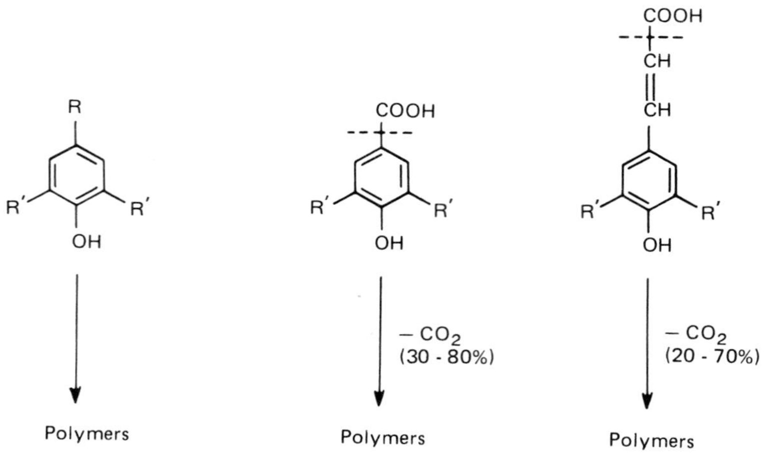

R = CH₂OH; CHO; CH = CH — CH₂OH

R = CH_2OH; CHO; $CH = CH - CH_2OH$

R′ = OH; H; OCH_3

Figure 31. Reactions that are affected by phenol oxidases (HAIDER [1966]).

5. Reactions Affected by Phenoloxidases or Other Enzymes

HAIDER [1966] summarized the reactions of lignin degradation products in the presence of lignin-decomposing fungi or by phenoloxidases (see Figure 31).

In the presence of phenoloxydases produced by lignin-decomposing fungi lignin degradation products are polymerized. During the reaction, 30 to 80% of the carboxylic acid carbon of phenol carboxylic acids and 20 to 70% of the carboxylic acid carbon of phenol acrylic acids is released as carbon dioxide.

The cleavage of the carboxyl group is an indirect effect of the enzyme action. Quinonoid intermediate compounds are formed by the oxidation. The carboxyl group thus becomes unstable and is spontaneously decomposed to give carbon dioxide.

Other reactions occur during the degradation of lignin degradation products in cultures of lignin-decomposing fungi, which are not catalyzed by phenol oxidases. The side chain C_6—C_3 compounds may be reduced by two carbon atoms (Figure 32), whereby a phenol-aldehyde is produced which may then be oxidized to the corresponding carboxylic acid. FARMER, HENDERSON, and RUSSELL [1959] describe a reduction of phenolaldehydes to alcohols.

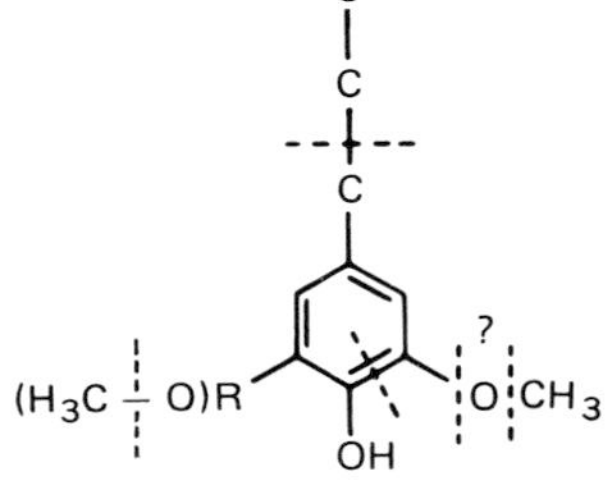

R = H or OCH_3

Figure 32. Degradation reactions that are not affected by phenol oxidases (HAIDER [1966]).

Another reaction is the cleavage of the phenolethers or the cleavage of methoxyl groups. A cleavage of the phenol rings between hydroxyl groups is affected by other enzymes. Thereby β-ketoadipic acid (FLAIG and HAIDER [1961a]) or aceto acetic acid and α-ketoglutaric acid are formed (FUKUZUMI, MINAMI, and SHIBAMOTO [1959]). These acids can be metabolized by fungi or other soil microorganisms.

6. *Possible Coupling Reactions Which May Lead to Formation of Polymers*

(a) Lignin Degradation Products

Coupling possibilities of monomers with a three-carbon side chain depend on the nature of the side chain. With phenolacrylic acids, coupling may occur through the carbon atoms of the side chains, through carbon atoms of two rings, or through phenol ether linkages in which the unsaturated side chain participates. Additional possibilities of bonds through ether linkages are possible with phenylpropenyl alcohol or phenylglycerol derivatives. Possibilities for polymerization of coniferyl alcohol have been discussed by ADLER [1961] and FREUDEN-BERG [1962, 1964]. The results of these investigations constitute the basis for the structural scheme of coniferous lignin given in Figure 3.

FREUDENBERG [1959] made the interesting observation during his investigations of dehydrogenation polymerization of lignin building blocks that sinapyl alcohol alone does not polymerize in the presence of oxygen transferring enzymes, but is built into the dehydrogenation polymers from a mixture with coniferyl alcohol.

The number and the position of the hydroxyl and methoxyl groups influence the coupling reactions of the monomers during polymerization of lignin degradation products formed during biological degradation of lignin (comp. Figure 8). Coupling reactions involving the side chain may occur with units possessing a three-carbon side chain. Lignin degradation products having only *one* carbon atom in a side chain polymerize mainly through the ring. These findings are important for a conception of the formation of humic acids and also in connection with the possibilities for the reactions of nitrogen-containing compounds during the polymerization. In addition, the reactions of those phenols that do not have hydroxyl groups in the o-position and that are derived from nonlignin phenols of plant or microbial origin must be included.

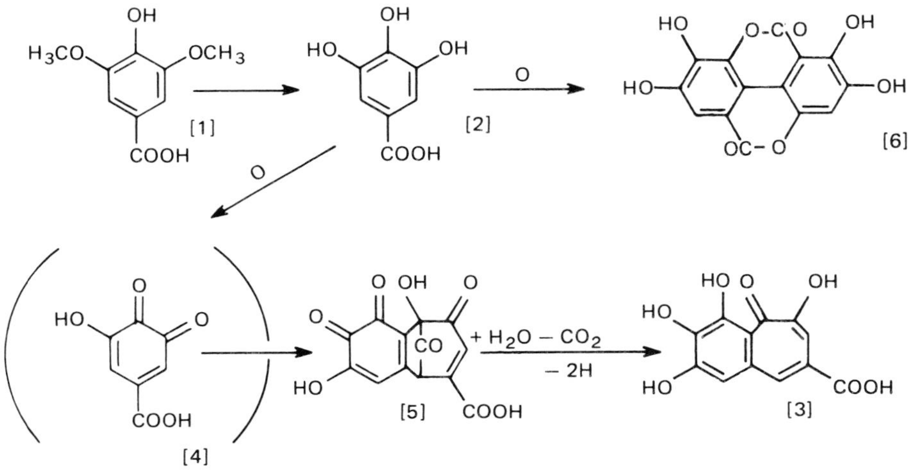

Figure 33. Possibilities of reactions of gallic acid.

As has been previously mentioned, carboxylic acids are formed by the splitting of methyl-ethers, which have three vicinal hydroxyl groups. These dimerize under the influence of phenoloxidases.

For example, syringic acid [1] is demethylated to gallic acid [2] (HENDERSON [1957]) (Figure 33). Purpurogallin-β-carboxylic acid [3] is formed from gallic acid (HERRMANN [1954], STEINMETZ [1956]). The mechanism of this reaction involves the formation of a 3-hydroxy-p-benzoquinone derivative [4], and then an adduct [5] (SALFELD [1957, 1960], SALFELD and BAUME [1960, 1964]) is formed by 1,3-addition. Pyrogallol is also transformed through an intermediary 3-hydroxy-o-benzoquinone and then through a further intermediate product to purpurogallin. Pyrogallol and gallic acid form dark-colored, humic acid-like products during the enzymatic oxidation (STEINMETZ [1956]). In the presence of phenoloxidases ellagic acid [6] is formed from the methylester of gallic acid by elimination of methanol (STEINMETZ [1956]).

3-Methylgallic acid [7], which may be considered as a partially demethylated product of syringic acid, is transformed to a slight extent to 4,4'-dimethoxy-2, 3,2'-, 3'-tetrahydroxy-diphenyldicarboxylic acid-6,6' [8] during oxidation in weak alkaline solution (Figure 34).

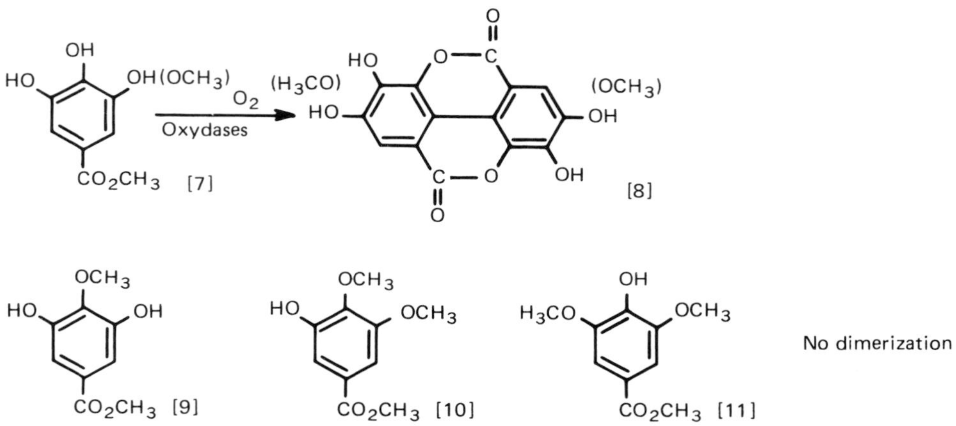

Figure 34. Dimerization of gallic acid derivatives.

The esters of the gallic acids methylated in 4- [9] or 3,4-positions [10] and syringic acid [11] itself do not dimerize to form diphenyldicarboxylic acids (HATHWAY [1958]).

HATHWAY [1957] is of the opinion that dimerization occurs only in the case of two ortho hydroxyl groups, through an o-quinonoid intermediate compound. The 2,3,4,2',3',4'-hexa-hydroxydiphenic acid formed from gallic acid is found in nature in the ellagtannins. Gallic acid or hexahydroxydiphenic acid is esterified with glucose in these tannin derivatives (SCHMIDT and LADEMANN [1951], SCHMIDT and SCHMIDT [1952]).

No diphenylcarboxylic acids have been found in the course of the enzymatic oxidation of protocatechuic acid [12]. It is known, however, that protocatehuic acid [12] reacts in dilute potassium carbonate solutions in the presence of the oxygen of the air to form catellagic acid [13] (SCHIFF [1882]). Upon further oxidation of protocatechuic acid dimers can be isolated (see Figure 27).

(α) *Mechanism of Decarboxylation and Polymerization.* The mechanisms of oxidative decarboxylation and polymerization go through radicals. So three mesomeric structures can be formulated for the semiquinones. It is noteworthy that no compounds that can be derived from the oxygen radical have been detected during humification.

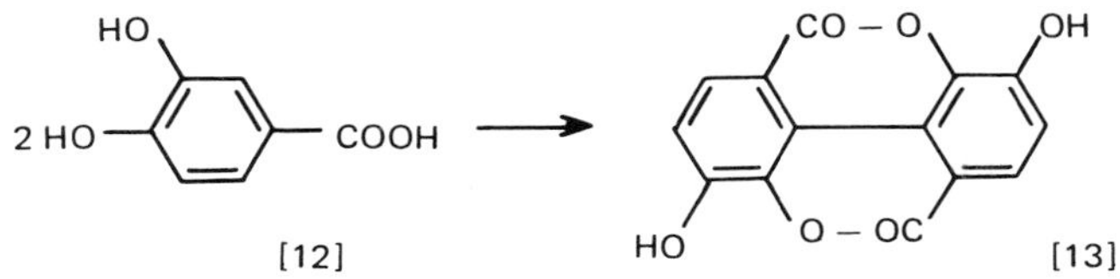

[12]   [13]

Semiquinonoid carboxylic acids are not stable. They decarboxylate to quinones with the addition of oxygen. The rate of decarboxylation depends upon the derivative of the compound. The formation of p-benzoquinone radicals in the course of the reduction of the quinone is illustrated in Figure 35. The mechanism of the formation of radicals of o-benzoquinones or 1,2-diphenols is the same.

Lignin degradation products with side chains of three carbon atoms such as ferulic acid also form radicals in an oxidizing medium. The decarboxylation is not a direct effect of the phenoloxidase but must be considered a more or less secondary reaction of the formation of quinone methides, which renders the carboxyl groups unstable.

If the hydroxyl group in p-position is etherified, the formation of a semiquinone is not possible and no polymerization can occur. The shortening of the side chain with three carbon

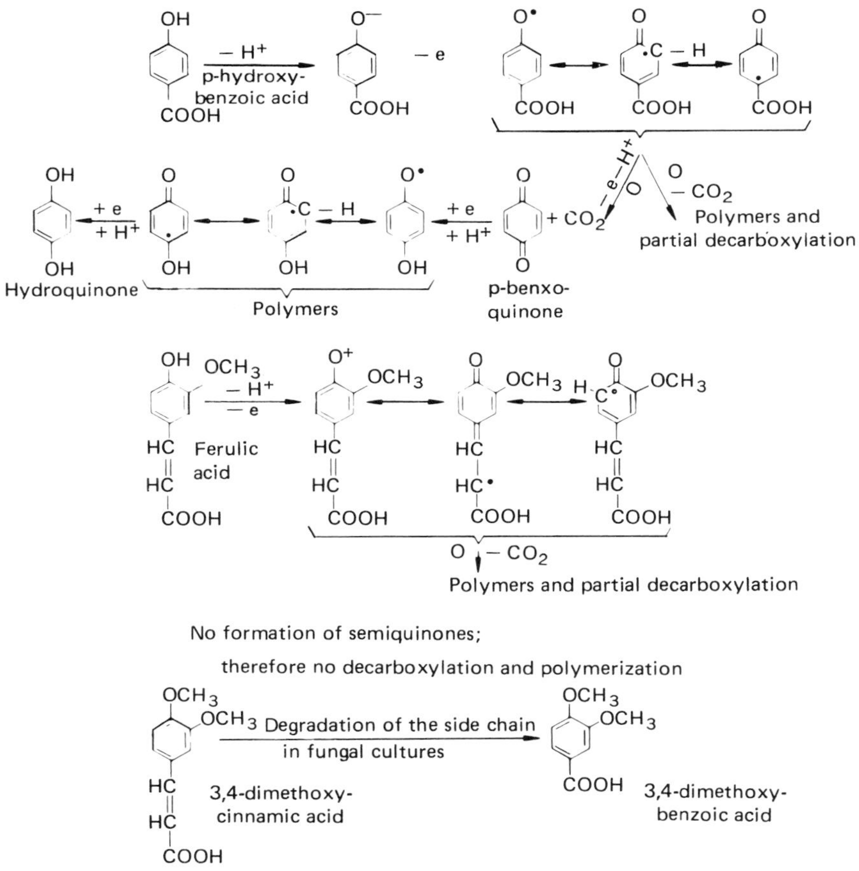

Figure 35. Formation of semiquinones (mechanism of oxidative decarboxylation and polymerization).

atoms in the case of 3,4-dimethoxy-cinnamic acid is effected by enzymes other than phenol-oxidases in the fungal cultures.

(b) Nonlignin Phenolic Plant Constituents

Compounds are obtained by the oxidative and reductive degradation of humic acids which could be derived from flavone components or microbial synthesis (Figure 15). These differ from the degradation products of lignin, in that they have hydroxyl groups in the m-position, such as in catechin [10] (Figure 36). This type of polyphenol cannot be dehydrogenated to quinones, as with lignin degradation products. Several investigations have demonstrated that such phenols react by oxidative coupling with equimolar mixtures of pyrocatechol derivatives to form dimeric or polymeric products. For example, 3,5-dimethylpyrocatechol [1] reacts with 2,4-dimethylresorcinol [2] to form 2,3,3',4'-tetrahydroxy-3,5-2',6'-tetramethyl-diphenyl [3] (BAKER and MILES [1955]) (Figure 36). An equimolar mixture of 4-methylpyrocatechol [4] and phloroglucinol [5] or certain other derivatives produces colored polymers by oxidative coupling. The absorption spectra of these compounds are very similar to those of oxidized tannin structural units such as catechine [10] (HATHWAY and SEKINS [1957]).

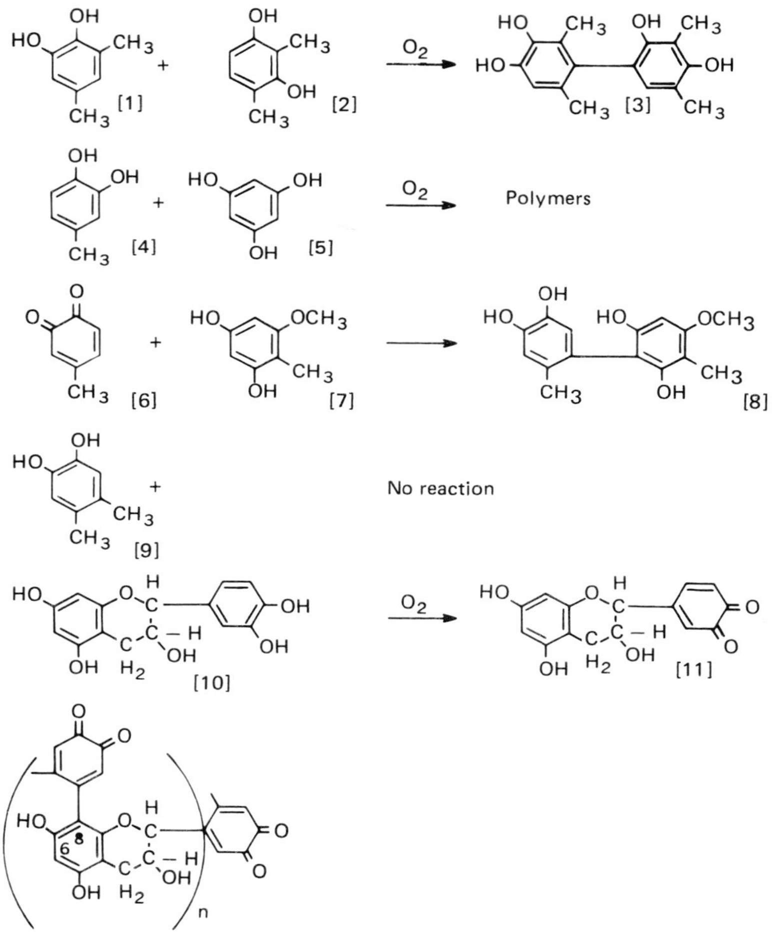

Figure 36. Oxidative coupling of polyphenols.

4-methyl-o-benzoquinone [6] reacts with 5-methoxy-4-methylresorcinol to give a corresponding diphenyl derivative [8].

No oxidative coupling occurs when pyrocatechol is substituted in the 4 and 5 positions with methyl groups [9]. For these reasons the authors conclude that oxidative polymerization of catechine [10] occurs by "head to tail" addition through intermediate quinones [11]. Based on investigations of the addition of amino compounds to phloroglucinol [5] in a neutral oxidizing medium STEIN and TENDELOO [1956a, b; 1958; 1959] concluded that the first reaction step is the formation of a radical.

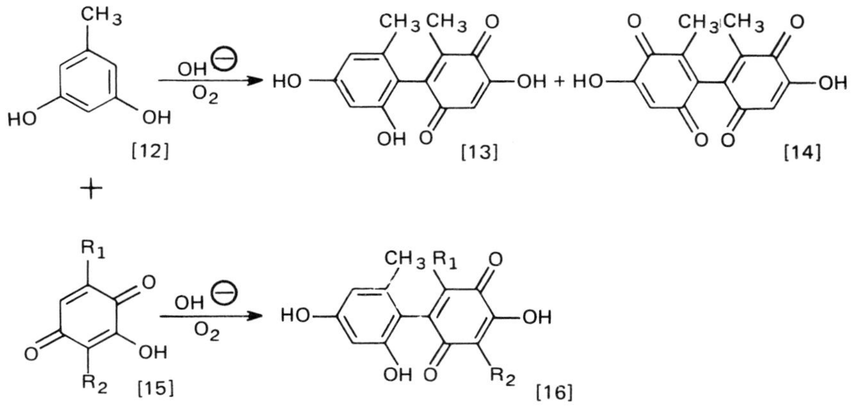

Figure 36a. Autoxidation of orcinol.

$$R_1 = R_2 = H$$
$$R_1 = H; R_2 = CH_3$$
$$R_1 = CH_3; R_2 = H$$
$$R_1 = R_2 = CH_3$$

The autoxidation of orcinol [12] (HENRICH, SCHMIDT, and ROSSTEUTSCHER [1915]) leads, in aqueous solution, to a mixture of quinones [13, 14] that have been isolated by MUSSO [1958] (Figure 36a). Orcinol [12] and resorcinol derivatives, which are substituted in 2- and in 2,4-positions with methyl groups, react in aqueous alkaline solution with hydroxy-p-benzoquinone [15] or with methyl-substituted derivatives to form the corresponding dihydroxyarylhydroxy-p-benzoquinone [16] (MUSSO, GIZYCKI, ZAHORSKY, and BORMANN [1964]).

Investigations of the oxidative coupling of 1,3- and 1,3,5-phenols show that natural phenolic compounds other than lignin, and indeed those which are formed through plant and microbial metabolism, can participate in the formation of dark-colored polymeric products that are similar to humic acids in their properties.

VIII. Model Experiments with "Synthetic" Humic Acids

1. *Alkaline Oxidation of Phenols*

ELLER [1921, 1923, 1925a, b], ELLER and KOCH [1920], and ELLER, MEYER, and SAENGER 1923] were the first to study the dark-colored reaction products that are formed by the alkaline oxidation of phenols, such as hydroquinone, pyrocatechol, pyrogallol, and p-benzoquinone for comparison with the humic acids isolated from the soil. They established that reaction products formed by chlorination (ELLER, HERDICKERHOFF, and SAENGER [1923]) also have the same properties. The formula proposed by ELLER [1923] corresponds to a polymeric hydroxy-p-benzoquinone. The alkaline oxidation of hydroquinone in relation to "synthetic" humic acids has also been investigated by other authors (ERDTMAN [1933a, b, c, d, e, f; 1935; 1955; 1962], ERDTMAN and GRANATH [1954], ERDTMAN and STJERNSTRÖM [1959], ERDTMAN, GRANATH, and SCHULTZ [1954], FLAIG [1950; 1955b; 1956; 1959; 1960a, b; 1964a], FLAIG and SALFELD [1958, 1959, 1960a], FLAIG, PLOETZ, and KÜLLMER]1955], PLOETZ [1955], SALFELD [1961], SCHEFFER and ZIECHMANN [1958], SCHEFFER, WELTE, and ZIECHMANN [1955], SCHEFFER, ZIECHMANN, and SCHOLZ [1958], WELTE, SCHATZ, and ZIECHMANN [1954], ZIECHMANN [1960a, b; 1961], ZIECHMANN and KICKUTH [1961]). Oxidation in air and the use of oxidizing agents such as sodium persulfate have been studied. The reaction solution at first becomes dark red and then changes to dark brown and black, depending upon the concentration of the alkali and on the constitution of the phenolic compounds. The solution again turns red-brown, if an excess of oxygen is used. The largest yields of "synthetic" humic acids, the products precipitated with acid, are obtained at pH values between 8 and 10. The purification of the "synthetic" humic acids is accomplished by repeated alternate dissolution and reprecipitation with alkali and acids, followed by dialysis and electrodialysis. Freeze-dried phenol humic acids have a dark brown color.

The largest yield from the oxidation of hydroquinone is obtained by slow addition of sodium hydroxide to the solution of hydroquinone and by not exceeding a pH value of 10 during the oxidation. A yield of about 40% (FLAIG [1950]) is obtained by an uptake of 2,5 atoms of oxygen per mol hydroquinone. Traces of copper, iron, or manganese ions, as well as the addition of dispersing reagents, accelerate the uptake of oxygen.

Carbon dioxide, which must originate from ring cleavage, is formed upon acidification of the dark-colored solutions. Acetic acid and oxalic acid (WELTE, SCHATZ, and ZIECHMANN [1954]) could be isolated from the reaction solutions.

In an attempt to approach natural conditions, investigations on the formation of "synthetic" humic acids from hydroquinone were made in acidic medium in the presence of silicic acid as catalysts (ZIECHMANN [1960a, b]). The reactions of hydroquinone in acidic solution were observed for several days instead of hours. The yields of "synthetic" humic acids were very low.

The aerobic oxidation of hydroquinone in the presence of silica gels in dilute solutions at pH ranges between 4.5 and 7.0 occurs more slowly than in an alkaline medium (SALFELD [1961]). An accelerating effect of different preparations of silica gel may have been related to their iron content, because ferric ions accelerate the oxidation and iron compounds that occur in the soil have a comparable effect (SCHEFFER, MEYER, NIEDERBUDDE [1959]). The catalytic effect of the silica gel decreases when they are purified (SALFELD [1961]). Another possible explanation for the effect of the silicic acid may be an acceleration of the reaction of hydroquinone with oxygen on its surface.

ZIECHMANN [1961] succeeded in isolating a number of substances by extraction of the oxidation reaction mixtures containing silicic acid and o-amino phenol with ether and hot xylene. One of the substances was 3-hydroxyphenoxazone. He concluded that 2,5-dihydroxy-p-benzoquinone is formed. This compound is relatively stable against the oxygen of the air in dilute sodium hydroxide solution and does not form dark-colored polymers, as does hydroquinone or p-benzoquinone.

The oxidation of hydroquinone is very much dependent upon the pH of the reaction solution. From oxygen utilization studies it was found that in the first phase of the reaction, oxygen uptake is proportional to the square of the concentration of OH ions (REINDERS and DINGEMANS [1934], JAMES and WEISSBERGER [1938], JAMES, SNELL, and WEISSBERGER [1938]). The subsequent reactions of the preliminary oxidation products are also dependent upon the pH value. REINDERS and DINGEMANS [1934] found p-benzoquinone as the main product of the oxidation at a pH value of 6.9 and humic acids at a pH value of 7.9. The formation of p-benzoquinone from hydroquinone goes through a semiquinonoid radical in two steps in alkaline solution (MICHAELIS [1951]). The concentration of the semiquinonoid radical is very low during the oxidation of hydroquinone. Its formation was established by the corresponding electron-spin resonance spectrum (VENKATARAMAN and FRAENKEL [1955], BLOIS [1955]). The initial yellow color during the autoxidation of hydroquinone in weak alkaline solution may be caused by the semiquinonoid radical, since the spectrum of this solution shows the same double band at 400 to 450 nm (FLAIG and SALFELD [1960b]) as the p-benzosemiquinone anion (EIGEN and MATTHIES [1961]) formed from hydroquinone and p-benzoquinone.

EIGEN and MATTHIES [1961], and DIEBLER, EIGEN, and MATTHIES [1961] found additional products. The reaction solution becomes red upon further oxidation of hydroquinone. The broad band of the anion of hydroxy-p-benzoquinone at 488 nm occurs in the spectrum. The spectrum quickly changes, however, because additional compounds are formed during the oxidation.

JAMES, SNELL, and WEISSBERGER [1938] demonstrated that besides p-benzoquinone, hydrogen peroxide is formed simultaneously during the alkaline oxidation of hydroquinone. SCHEFFER, ZIECHMANN, and SCHMIDT [1958] erroneously believed that p-benzoquinone could not be an intermediate product during the formation of humic acids by alkaline oxidation of hydroquinone. Therefore, they concluded that a red color could not be formed at the beginning of the alkaline oxidation of quinone. Although p-benzoquinone remains yellow for a long time in weak alkaline solution (0.1 N sodium acetate), when hydrogen peroxide is added, the solution suddenly becomes red. Hydroxy-p-benzoquinone (Figure 37) can be isolated from the reaction mixture and identified spectrographically (FLAIG and SALFELD 1960b]).

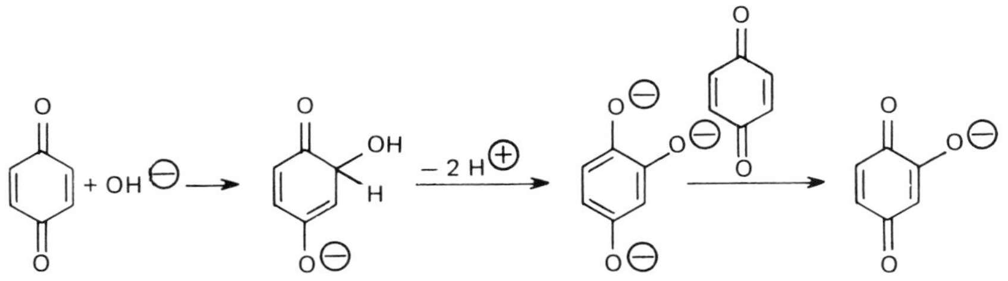

Figure 37. Formation of hydroxy-p-benzoquinone.

However, the formation of hydroxy-p-benzoquinone from p-benzoquinone also occurs at higher pH values in the absence of hydrogen peroxide through hydroxyl ion reactions. The reaction is dependent upon the concentration of hydroxyl groups (EIGEN and MATTHIES [1961]).

These investigations demonstrate that the first monomeric reaction products during the alkaline oxidation of hydroquinone, which lead to the formation of "synthetic" humic acids, are the semiquinonoid radicals p-benzoquinone and hydroxy-p-benzoquinone. The amounts formed depend upon the concentration of hydrogen ions.

It was of further interest to determine what other compounds could possibly cause the red color of the solution during the alkaline oxidation of hydroquinone (FLAIG and SALFELD [1960a]).

When hydroquinone is present in excess, a red color appears in acidic to neutral solutions at the beginning of the oxidation by the formation of a compound [1] related to p-benzoquinone and hydroquinone (Figure 38).

Hydroxy-p-benzoquinone [2] is formed in a weak alkaline solution and can be identified

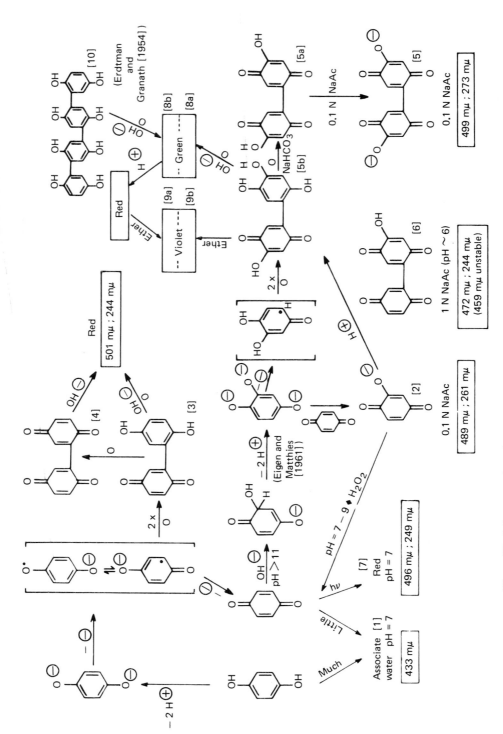

Figure 38. Substances that give a red color during the oxidation of hydroquinone.

spectroscopically after separation from the other components, hydroquinone and p-benzoquinone, which are still in the solution (FLAIG and SALFELD [1960a]). All hydroxy-p-benzoquinones, which are not substituted in the o-position to the OH group, give a red color as anions in alkaline solution (FLAIG and SALFELD [1958]).

Dimerization can also occur during these oxidation reactions (FLAIG and SALFELD [1960a]). ERDTMAN and GRANATH [1954] observed that dihydroquinone is formed in small amounts during the formation of the dark-colored products in alkaline solution from p-benzoquinone in the absence of oxygen.

During oxidation of hydroquinone, it is possible that the intramolecular quinhydrone of diphenyldiquinone-2,5,2′,5′ is formed. The intramolecular quinhydrone [3] autoxidizes in alkaline solution to a red-colored compound of which the spectrum has the same maximum as an alkaline solution of diphenyldiquinone-2,5,2′,5′ [4].

The compound 4,4′-dihydroxy-diphenyl-diquinone-2,5,2′5′ (5a or 5b, respectively) may also give a red color in alkaline solution. This diquinone is formed by dimerization of hydroxy-p-benzoquinone [2] in acidic solution through the intramolecular quinhydrone. Using paper chromatography the dimers [5a] of hydroxy-p-benzoquinone [2] and of other compounds were identified as oxidation products in neutral solution. Dark-colored "synthetic" humic acids are the main products formed in alkaline solution. It is also possible that the previously mentioned 4,4′ dihydroxy-diphenyl-diquinone-2,5,2′,5′ [5a] is formed as an intermediate product during the oxidation of protocatechuic acid to dark brown polymers in the presence of phenolases (FLAIG and HAIDER [1961a]).

A further possibility of oxidative dimerization of hydroquinone under alkaline conditions is the formation of 4-hydroxy-diphenyl-diquinone-2, 5,2′,5′ [6]. This compound is also soluble and gives a red color in dilute alkaline solutions. The spectrum changes faster than in the case of compounds [2], [3, 4], [5], and [7], indicating a different reaction behavior.

The alteration of aqueous solutions of p-benzoquinone to red-brown substances is greatly accelerated by exposure to light. For this reason the red-colored products [7] isolated from such solutions have been more precisely investigated. The spectra of the isolated products are similar to those of the oxidation products of the intramolecular quinhydrone [5b] of 4,4′-dihydroxy-diphenyl-diquinone-2,5,2′,5′ and its quinone [5] in alkaline solution.

The red color at the beginning of the oxidation in solutions of hydroquinone may be related to the formation of hydroxy-p-benzoquinone [2] and eventually other hydroquinones as subsequent products. In the first phase of the oxidation the absorption spectrum has, in addition to the maximum of hydroxy-p-benzoquinone at 488 nm, a double band at 407 and 429 nm, which is due to the semi-p-benzoquinonoid radical, according to EIGEN and MATTHIES [1961].

Compounds giving other colors than red are produced during the oxidation of dimeric products, and these substances have also been investigated. The intramolecular quinhydrone [5b] of 4,4′-dihydroxy-diquinone-2,5,2′,5′ produces a green color [8a] in alkaline sodium solutions, which becomes red upon acidification. A compound with a violet color [9a] can be extracted from this red solution. ERDTMAN, GRANATH, and SCHULTZ [1954] have also observed a green color [8b] during the oxidation of tetrahydroquinone [10] in alkaline solution. It is not certain whether the green colors [8a] and [8b] can be ascribed to compounds of the same constitution. Further investigations are necessary to elucidate this point.

The intramolecular quinhydrone of 4,4′-dihydroxy-diphenyl-diquinone-2,5,2′,5′ is soluble in ether and gives a violet color [9b]. It is not known if this color [9b] is produced by the same compound as that of [9a].

The dimerization of hydroquinone occurs during the course of the oxidation through the semi-p-benzoquinonoid radical, which can be formulated in two mesomeric forms (Figure 38a).

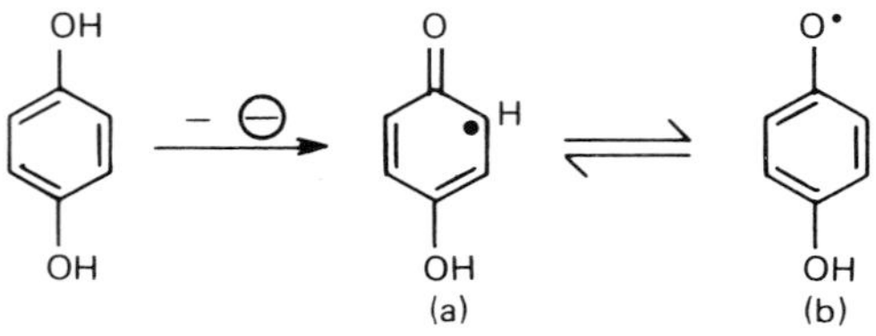

Figure 38a. Mesomeric forms of hydroquinone.

The formation of diphenylethers can also be expected, but diphenylethers have not yet been found after the alkaline oxidation of hydroquinone.

GÜNDEL and PUMMERER [1937] demonstrated that 4,4'-dihydroxy-diphenylether is easily split into two molecules of p-benzoquinone by ferric salts (Figure 38b).

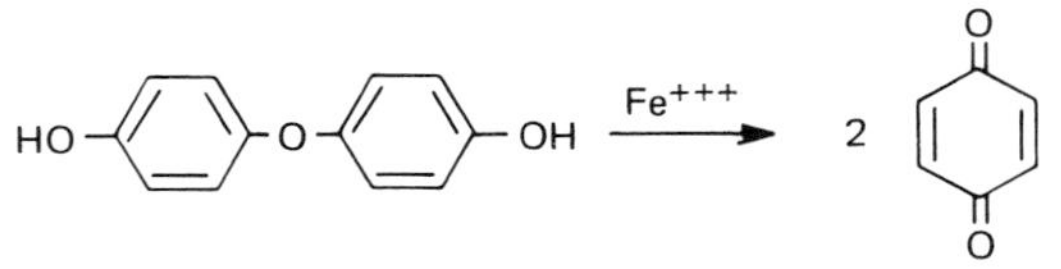

Figure 38b. Cleavage 4,4' dihydroxydiphenylether.

The trimerization of p-benzoquinone in the presence of pyridine in ether is described by DIELS and KASSEBART [1937]. The same compound was also found during the reaction of sodium phenolate and p-benzoquinone in dimethylformamide and dimethylsulfoxide (MUSSO, GIZYCKI, ZAHORSKY, and BORMANN [1964]), whereas only polymers and hydroquinone were identified in benzene and dioxane.

Presumably the formation of the trimer occurs partially through formula (b) (Figure 38c).

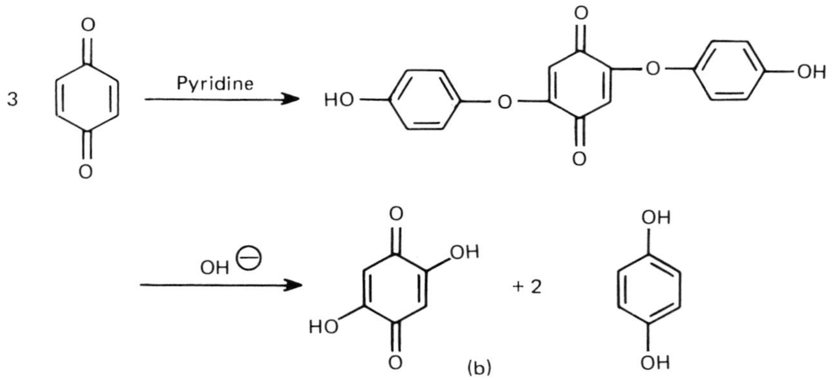

Figure 38c. Trimerization of p-benzoquinone.

Two molecules of hydroquinone and 2,5-dihydroxy-p-benzoquinone are formed by alkaline hydrolysis. Polymers connected through ether linkages can hardly be expected during the formation of "synthetic" humic acids by alkaline oxidation of hydroquinone, because these substances are less stable in alkali.

When p-benzoquinone [11] is allowed to stand for nine days in a neutral aqueous solution of methanol, [4-hydroxy-phenoxy]-p-benzoquinone [12] is formed with a yield of only 1%

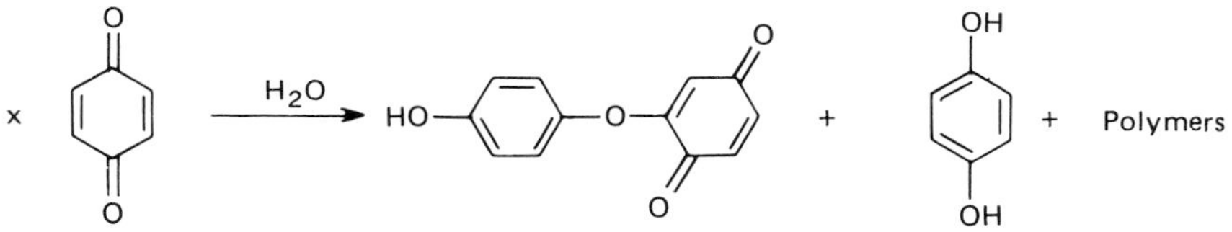

Figure 38d. Dimerization of p-benzoquinone in water.

(Figure 38d). Hydroquinone and polymers are, of course, also formed (MUSSO, GIZYCKI, ZAHORSZKY, and BORMANN [1964]).

Monophenols substituted with tert. butyl groups in 2,4,6-positions react with oxygen to give quinonoid peroxides, which react with phenols to form quinolethers (MÜLLER, LEY, and SCHLECHTE [1957]).

MUSSO [1963] summarized the oxidation reactions of phenols under highly variable conditions, many of which could not be involved in the formation of humic substances.

In connection with the radical mechanism involved in the formation of polymeric phenols or quinones, it is of interest to note that the presence of free radicals has been established in the humic acids (REX [1960]).

STEELINK and TOLLIN [1962] and STEELINK [1964] presumed, by investigations with electron paramagnetic resonance spectrometry, that the humic acids contain semiquinone radicals, together with some quinhydrones, which may possibly be derived from o- or p-benzo-quinones. The radicals are stable against chemical attack and increase in the following series: lignin→degraded lignin→fulvic acids→humic acids. These results agree with the conceptions presented concerning the chemical processes occurring during the transformation of lignin or its degradation products to humic acids. The content of radicals in humic substances was found to be 10^{18} per gram of humic acids. PLOETZ [1955] discussed some of the problems of radical structure of humic acids using 1,4-diquinoylbenzene as a model substance.

Some reactions of the substituted p-benzoquinones are important, because products with aliphatic side chains are formed during the degradation of lignin. These are of interest in connection with reactions of nonlignin phenolic plant constituents and phenols, synthesized by microorganisms. Reactions of thymohydroquinone have been investigated as models.

In alkaline solution with pH values between 8 and 9, thymohydroquinone [13] is transformed to dark-colored products by shaking with oxygen. Similar reactions take place with hydroquinone. The oxidation products of thymohydroquinone are ether-soluble, whereas those of hydroquinone are not. Furthermore, the two isomeric hydroxy-thymohydroquinones formed react differently depending upon the position of the hydroxyl groups (Figure 39).

3-Hydroxy-thymoquinone [14] is oxidized to 3,6-dihydroxy-thymoquinone [15]. 6-Hydroxy-thymoquinone [16] dimerizes to the corresponding diquinone [17]. Free rotation of both quinone rings is impossible in the latter, as is indicated by the spectrum in ultraviolet and in visible light (FLAIG and SALFELD [1958]). 2,5-Ditert. butyl-p-benzoquinone [19] precipitates during oxidation of its hydroquinone [18] in alkaline aqueous solution (FLAIG and SCHULZE [1952]). With this type of substitution, neither the introduction of a hydroxyl group nor a dimerization takes place.

Seneca 4 B.C. - 69 A.D. said:
To err is human

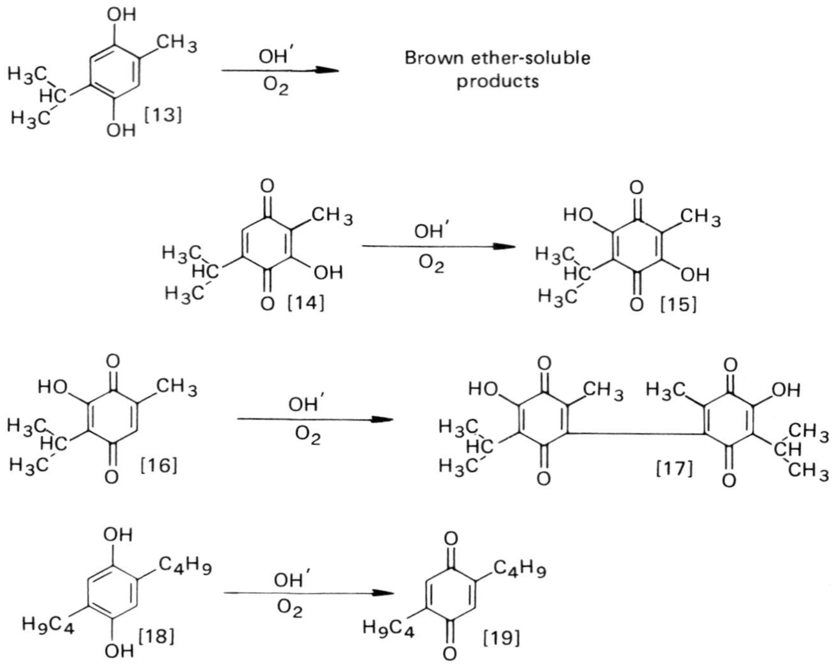

Figure 39. Oxidation of thymohydroquinone.

2. *Evaluation of Experiments* in vitro *for Reactions Occurring in Nature*

During the enzymatic oxidation of ligneous materials (see Figure 26), p-benzoquinone or methoxy-p-benzoquinone may be formed from p-hydroxybenzoic or vanillic acid. Although p-hydroxybenzoic and vanillic acid have been isolated from soil-compost mixtures, the two quinones, as indicated above, are very reactive compounds, and it is rather unlikely that they could persist long in soils. Experiments concerned with the microbial transformation of vanillic acid or protocatechuic acid, probably formed from vanillic acid by demethylation, show that formation of hydroquinone or p-benzoquinone from p-hydroxybenzoic acid occurs in microbial cultures, especially because hydroxyhydroquinone or hydroxy-p-benzoquinone have been identified as oxidation products of hydroquinone. The formation of hydroxy-p-benzoquinone after the oxidation of protocatechuic acid has also been demonstrated in fungal cultures through the triacetate of hydroxyhydroquinone.

The experiments on the formation of dark-colored, humus-like substances—"synthetic" humic acids—which are formed during the alkaline oxidation of hydroquinone, p-benzoquinone, or their hydroxyl derivatives, provide an interesting contribution to the problem of humic acid synthesis in nature (FLAIG, SALFELD, and HAIDER [1963]). ELLER (1921) did not know during his investigations in the 1920s that various phenols that he used for his experiments were formed during oxidation of lignin degradation products and, therefore, could take part in the formation of humic acids in nature. At this time various investigators (FISCHER and SCHRADER [1932]) believed that phenolic plant constituents, such as lignin, played a role in the formation of humic acids, but they did not know the manner in which the dark-colored polymers were formed. These mechanisms have been recently discovered.

The oxidation experiments with hydroquinone *in vitro* have been made at alkaline pH levels between 8 and 10. Such high pH values are not often obtained in soils, except in localized areas in soil materials where ammonia is being formed from nitrogenous organic materials.

As is well known, a reduction of the concentration of hydrogen ions increases the rate of the reactions. Furthermore, it is known that quinonoid compounds are formed by enzymatic dehydrogenation in weakly acidic media and that polymerization occurs. The substitution of quinones with aliphatic or aromatic substances lowers their redox potential. Therefore, the released quinone can again oxidize the reaction products that then take part in the polymerization processes. This is especially true in the presence of enzymes that accelerate the dehydrogenation. In this way polymeric phenols are formed *in vitro* or in plants. For example, different dimers are first formed during dehydrogenation and by the reaction of the released initial compounds or other dimers the three-dimensional connected coniferous lignin is formed (FREUDENBERG [1962]). In a similar way, the formation of humic acids from phenolic degradation products of lignin may be envisioned by polymerization by dehydrogenation. The degradation products may be demethylated and/or partially degraded in the side chains.They may be of high or low molecular weight. Other phenolic substances synthesized by plants or microorganisms take part in these reactions.

There is, however, an essential difference between formation of lignin on the one hand and of humic acids on the other. At most, only three monomers take part in the formation of lignin (see Figure 3), but numerous phenolic compounds take part in the formation of humic substances. An additional difference is the fact that organic nitrogen compounds that are formed by plants and especially by microbial metabolism take part in the reactions leading to the formation of humic acids. The relative participation of the altered lignin molecule, of the numerous lignin degradation products, and of the microbial synthesized phenols has not yet been determined. The importance of these phenolic products may vary with soil conditions.

IX. Participation of Nitrogenous Compounds in the Formation of Humic Substances

In cultures of microorganisms the formation of dark-colored nitrogenous humic-like substances is also observed (BREMNER, FLAIG, and KÜSTER [1955], KÜSTER [1952, 1956], LAATSCH, HOOPS, and BIENEK [1952], VON PLOTHO [1950, 1951], SCHEFFER, VON PLOTHO, and WELTE [1950], and others). KÜSTER [1952, 1953, 1958] found that the formation of these substances in microbial cultures may occur in relation to the activity of phenol oxidases. This indicates that this type of humic acid would also have an aromatic structure (FLAIG [1958]). Various phenolic compounds are formed as typical products of the metabolism of certain species of microorganisms (see ANSLOW and RAISTRICK [1938a, b, c], QUILICO [1951], ANSLOW, BREEN, and RAISTRICK [1940], and HAIDER and MARTIN [1967]).

The content of hydrolyzable α-amino nitrogen of the lignin fractions decreases during the humification of plant material, but the percentage to total nitrogen increases. This can be explained in two ways. Either nitrogen as amino acids or peptides continues to condense with the lignin during humification, or the peptide chains are increasingly decomposed by the microorganisms. There is the possibility that both of these latter nitrogen reactions occur simu taneously.

It has been noted that organic compounds extracted from composted straw, pre-extracted with ether, had a higher nitrogen content when nitrogen salts were applied to the straw. However, the methoxyl content was higher in the composted straw without added nitrogen (FLAIG [1962]).

Lignin degradation products can always be isolated during the course of humification. Therefore, the assumption may be made that condensation products of lignin degradation units and organic nitrogen compounds from proteins are present among the nitrogen-containing, water-soluble organic compounds. Furthermore it can be shown that the formation of phenols or quinones is possible during oxidative degradation or by enzymatic dehydrogenation of lignin. In addition, it is known that dark-colored polymers are formed during the oxidation of 1,2-di- or 1,2,3-triphenols, together with amino acids in the presence of phenoloxidases (STEINMETZ [1956], SWABY [1956, 1958, 1962], FLAIG [1956], or other oxidants (LADD and BUTLER [1966]).

   Investigations on the reactions occurring between amino acids and lignin degradation products in the presence of phenoloxidases showed that, besides the usual oxygen consumption during the addition of amino acids on quinones, the amino acids may be deaminated and decarboxylated (Figure 40). These oxidative reactions are influenced by chemical structure,

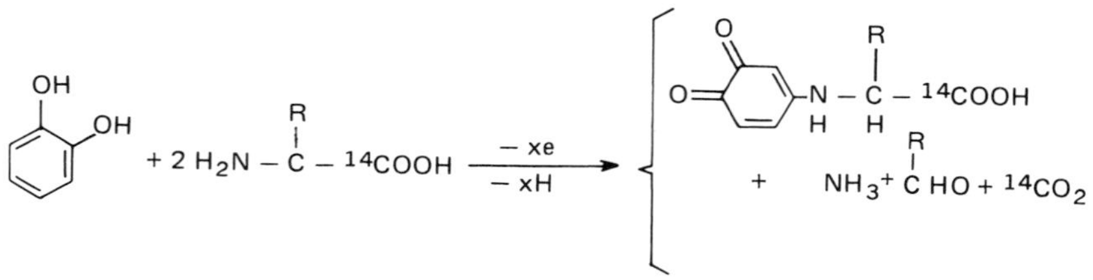

Figure 40. Addition of amino acids on 1,2-diphenols during oxidation.

concentration, and pH. Phenols that contain methylated hydroxyl groups such as guaiacol, vanillic acid, ferulic acid, and syringic acid show very little additional uptake of oxygen in analogous experiments. The investigations were made with labeled amino acids to elucidate the mechanism of the reaction (HAIDER, FREDERICK, and FLAIG [1965]).

1. *Nucleophilic Addition of Amino Acids on Phenols and the Deamination of Amino Acids*

(a) Conditions of the Formation of Nitrogen-Free Polymers

   The first step in the oxidation of phenols is the formation of a semiquinonoid phase, by which polymers are formed and added nitrogenous compounds are fixed. In Figure 41 some examples of important reactions of the semiquinonoid radicals are shown.

   During the oxidation of guaiacol two radicals are stabilized by the formation of three diphenyl derivatives. The identified dihydroxy-dimethoxy-diphenyl isomers show that only two mesomeric structures formulated as carbon radicals take part in the reaction.

   Diphenylethers, which might be derived from the oxygen radical, have never been found. In a summary of the oxidation of phenols MUSSO [1963] reported also that diphenylethers are not formed in the presence of phenol-oxidizing enzymes (see TODD [1962], LINDGREEN [1960a, b]).

   A nitrogen-free polymer with a weak red color is formed during the oxidation of guaiacol in the presence of mushroom phenoloxidase. STEELINK [1964] found an increase of radicals in the lignin fractions during the humification by means of electron spin resonance. This means that there exists some resonance structure in the polymers. In spite of this, neither the lignin fractions nor the polymers of guaiacol add nitrogenous compounds without further oxidation.

   Vanillic acid dimerizes also by oxidation in the presence of phenoloxidases. The acid and its dimer are oxidatively decarboxylated to the corresponding quinones.

   From the five possible dimerization products of ferulic acid, only two are known (MÜLLER, MAYER, SPANAGEL, and SCHEFFLER [1961], ERDTMAN [1935], CARTWRIGHT and HAWORTH [1944], RICHTZENHAIN [1949]). In this case the side chain is also included in the formation of one dimer. The polymers of ferulic acid have no nitrogen in their molecules.

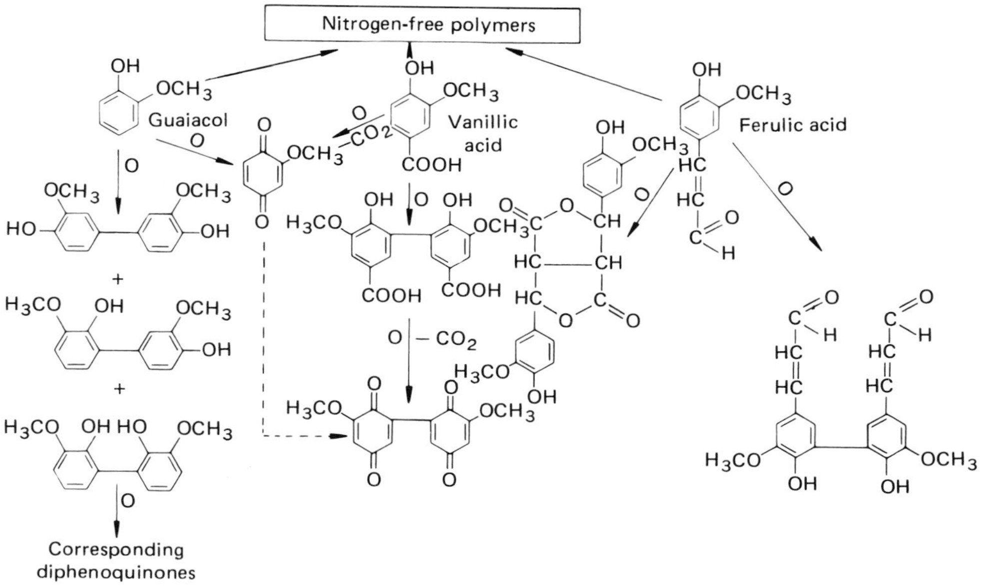

Figure 41. Formation of nitrogen-free polymers by oxidation of partially methylated o-diphenols in the presence of phenoloxidases and identified dimers.

As a rule it can be established that the lignin degradation products of the guaiacyl type form nitrogen-free polymers and do not form dark-colored products comparable with humic substances. Furthermore, no oxidative deamination of amino acids occurs.

These results agree with the findings that lignin adds increasing amounts of nitrogenous compounds with decreasing methoxyl content during humification and, furthermore, that the nitrogen content of soil humic acids is generally higher than the methoxyl content.

(b) Possibilities for the Formation of Nitrogenous Polymers

When nitrogenous compounds derived from proteins are added to the enzymatic oxidation of lignin degradation products, an increased uptake of oxygen and also a partial deamination of the amino acids occur. Ammonia, carbon dioxide, and carboxyl compounds are formed during this reaction at pH levels higher than 6.

(α) *Catechol and Hydroxy-hydroquinone Derivatives.* For the nucleophilic addition of one amino acid on the substituted quinones, one oxygen atom is necessary. When two amino acids are added, as in the case of p-benzoquinone, two oxygen atoms are needed.

In some cases more oxygen is combined in a reaction mixture of one phenol and one amino acid in the presence of a phenoloxidase at pH 6 to 8 than in a control without the addition of an amino acid. The additional uptake of oxygen is in some cases positively related to the quantity of the released ammonia, which is formed by the deamination of amino acids.

p-Benzoquinone absorbs an additional amount of oxygen, but does not deaminate the amino acids. Deamination is observed in small amounts in the presence of hydroxy-hydroquinone, or to a smaller amount in the presence of pyrogallol or gallic acid. The extent of the deamination of an amino acid depends upon the chemical constitution of the phenols and of the amino acids. At pH levels higher than 8, hydroquinone, hydroxy-hydroquinone, and gallic acid or pyrogallol form dark-colored nitrogenous polymers during oxidation in the presence of amino acids; this is not the case at pH values below 8.

The addition of catechol, protocatechuic acid, and caffeic acid to fungal cultures has a remarkable effect on the additional oxygen uptake and on the deamination of amino acids, whereby ammonia and carbon dioxide are formed (Figure 42). The ratio of the two reactions,

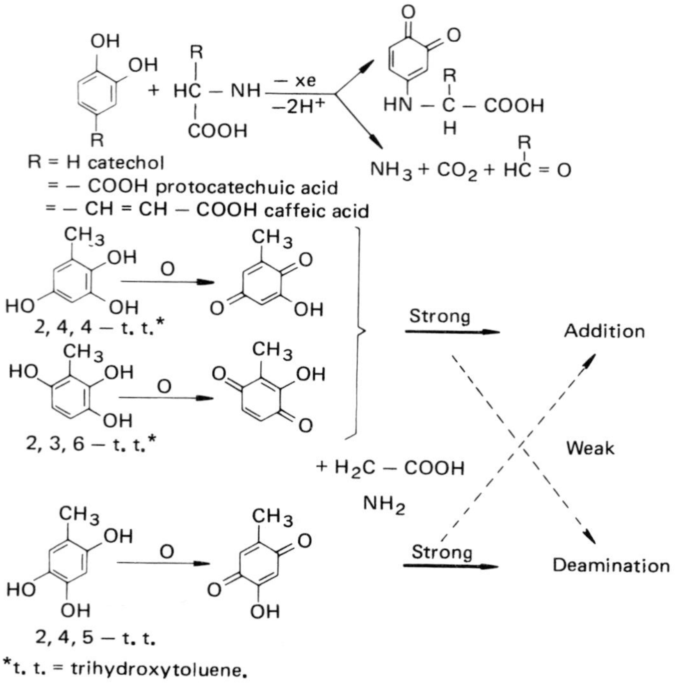

Figure 42. Addition and oxidative deamination of amino acids by o-diphenols during oxidation in the presence of phenoloxidases.

the addition of an amino acid and its deamination, depends upon the constitution of the added phenol. The presence of 2,3,5- and 2,3,6-trihydroxytoluene effects a strong addition reaction and a weak deamination, while in the presence of 2,4,5-trihydroxytoluene the deamination occurs stronger than the addition of an amino acid (HAIDER and MARTIN [1967]). Two of these compounds 2,4,5- and 2,3,5-trihydroxytoluene have been mentioned above as phenols that are formed in the culture media of *Epicoccum nigrum*.

    (β) *Influence of the Constitution of the Amino Acids.* The additional oxygen uptake and the release of ammonia during the oxidation of 1,2-diphenols depend also upon the chemical constitution of the amino acids.

    The amino acids have been separated into three groups. Average numbers are mentioned in Figure 43 because the influence of the three applied phenols, catechol, protocatechuic acid, and caffeic acid, is less than the influence of the amino acids on the values of oxygen uptake or ammonia release.

    The results with hydroquinone are also recorded for comparison. In this case, much oxygen is consumed, but no measurable amount of ammonia could be found.

    The CO_2 released by oxidizing catechol in the presence of amino acids is higher than the NH_3 released. Also, the content of nitrogen in the HCl precipitable reaction products is higher than the content of radioactive carbon.

    In previous experiments it could be shown that fixation of ammonia is not the reason

Amino acids	Catechol, protocatechuic acid, caffeic acid		Hydroquinone	
	μ atoms oxygen	μ atoms $NH_3 - N$	μ atoms oxygen	μ atoms $NH_3 - N$
Glycine, serine, threonine, tryptophan	29	18	20	no measurable amount
Glutamic acid, lysine, leucine, methionine, alanine, valine	15	10	18	
Aspartic acid	8	5	5	

Figure 43. Oxygen uptake in μ atoms and released ammonia in μmoles during the oxidation of 20 μmoles of catechol in the presence of 100 μmoles of various amino acids (according to HAIDER, FREDERICK, and FLAIG [1965]).

for the higher CO_2 release nor for the higher nitrogen content, because ammonia is fixed to a very limited extent during oxidation of catechol or hydroquinone below pH 8.5. Catechol was oxidized in the presence of increasing amounts of amino acids. Figure 44 shows the curves for the oxygen uptake in μ atoms, the CO_2 and ammonia release in μmoles by oxidizing 50 μmoles of catechol in the presence of amounts of glycine ranging from 10 to 1000 μmole.

Low amino acid concentrations release more CO_2 from the carboxyl groups of the amino acids than the release of ammonia from the amino groups. If the amino acid content is increased, the difference between these two values becomes smaller and finally approaches zero. It seems that the reaction of catechol during oxidation in the presence of amino acids follows different paths, depending on the concentration of the two reactants. In the range of low amino acid contents side reactions take place presumably by condensation reactions of an amino phenol that is formed as an intermediate by the secondary oxidation of the added amino acid.

If the amino acid content exceeds the content of catechol, the nitrogen fixation in the reaction products occurs mainly through the nucleophilic addition of the amino acid on the corresponding o-quinone.

The relatively low, additional oxygen uptake in the case of high concentrations of amino

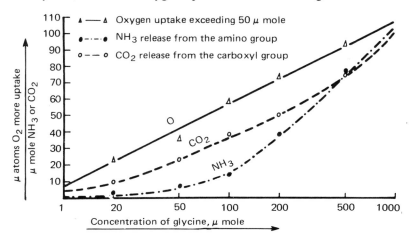

Figure 44. Dependence of oxygen uptake, CO_2, and ammonia release upon concentrations of amino acids (according to HAIDER, FREDERICK, and FLAIG [1965]).

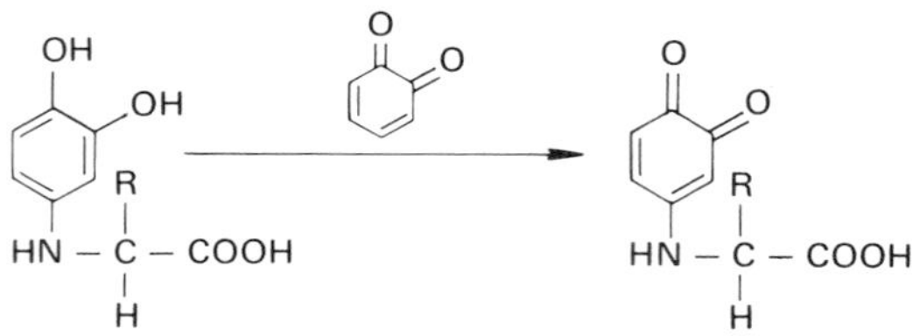

Figure 44a. Oxidation of phenolic amino acid addition products by o-benzoquinone.

acids may be explained by the presumption of MASON [1955] that the phenolic addition products are not oxidized by the action of the phenoloxidase, but by the o-benzoquinone from the applied phenol (Figure 44a). This can occur because the redox potential of the addition product is lower than that of the original quinone.

(γ) *Experiments with Ammonia.* Ammonia is formed by deamination of amino acids. Therefore, it was necessary to establish any part it might have in polymerization. With further experiments it could be demonstrated that ammonia is not added during the oxidation of the phenols in the presence of phenoloxidases. No noticeable additional oxygen uptake was found at pH values between 6 and 8. Even hydroxy-hydroquinone does not add ammonia at these pH values. Therefore, the total nitrogen in the addition products of the reaction between the phenols and the amino acids is derived from the amino acids.

(c) Properties of Substances Precipitable with Acids

Precipitates can be isolated by addition of mineral acids to the reaction solution. The precipitates are the main part of the addition products.

HAIDER, FREDERICK, and FLAIG [1965] reacted glycine with catechol in the presence of mushroom phenoloxidase. The molal ratio of glycine to catechol was 5 to 1 and optimum

Amino acid	Percent of applied activity in CO_2	Percent of applied N found as NH_3 in sol.	Percent of applied activity in precipitate	Percent of applied N in precipitate
NH_2 | $HC - {}^{14}COOH$	31	20	14	18
NH_2 | $H{}^{14}C - COOH$	1	21	15	18
NH_2 | $(CH_3)_2 - CH - CH - {}^{14}COOH$	14	10	8	10
NH_2 | $HOOC - (CH_2)_2 - CH - {}^{14}COOH$	21	16	6	10
NH_2 | $HOOC - ({}^{14}CH_2)_2 - CH - COOH$	0.2	16	8	10

Figure 45. Oxidative deamination of differently labeled amino acids during the oxidation of catechol in the presence of mushroom phenoloxidase (pH = 7). Distribution for the ^{14}carbon and the nitrogen in different fractions (HAIDER, FREDERICK, and FLAIG [1965]).

conditions for the action of the phenoloxidase were used in the investigation. Addition products were formed that were precipitated by mineral acids. The precipitate contained 12% of the original glycine nitrogen in the catechol reaction mixture, and 18% of the original nitrogen was released as ammonia.

Some properties of the precipitates were comparable to those of humic acids. The catechol-fixed glycine could not be hydrolyzed with 6 N HCl. After electrophoresis or paper chromatography of the hydrolysates, some spots are observed, which fluoresce in ultraviolet light.Only very weak colored spots are observed after spraying with ninhydrin. This means that little free glycine was present and that not all nonhydrolyzable nitrogen of the humic acids is heterocyclic-bound. The linkage can also occur in the form of secondary amines, in which one hydrogen atom is substituted by a phenol or perhaps a quinone ring and the other by an acid-containing group.

(d) Distribution of the Carbon Skeleton and Nitrogen as Constituents of the Amino Acids in the Reaction Products and in the Solution

Some further experiments have been made to elucidate how the amino acids are transformed and where the parts of the transformed amino acids remain during the addition and during the deamination. For this purpose different ^{14}C-labeled amino acids have been used (HAIDER, FREDERICK, and FLAIG [1965]).

When the oxygen uptake was complete, the reaction solution was adjusted to pH 1. The released carbon dioxide was absorbed in sodium hydroxide.

The larger amount of released $^{14}CO_2$ in the case of the carboxyl-labeled amino acids in comparison to those that are labeled on another carbon atom (Figure 45) is a measure of the deamination, because the activity in the precipitates formed in the presence of the two differently labeled amino acids is the same. Furthermore, this result demonstrates that the amino acids are added as a whole in the oxidation product without a loss of carbon dioxide.

In all cases the quantity of the released carbon dioxide in the carboxyl-labeled amino acids was larger than the amount of the ammonia nitrogen in the solution. Furthermore, the nitrogen content in the precipitates was larger than the content of labeled carbon. These data indicate that some nitrogen in the precipitates is not bound in the form of an amino acid. Further condensation must have taken place, by which nitrogen is bound in another way.

2. *Proposed Mechanism of the Nucleophilic Addition and the Deamination of Amino Acids in the Presence of Phenols*

The mechanism of the reaction of o-diphenols with amino acids may be explained by reference to the work of TRAUTNER and ROBERTS [1950].

The o-diphenol is oxidized to the o-quinone, which combines with an amino acid by nucleophilic addition (Figure 46). The phenolic addition product is oxidized to the corresponding quinone. A second amino acid forms a Schiff base. The double bond is rearranged and a hydrolysis occurs. Carbon dioxide and a carbonyl compound are split off and an o-aminophenol is formed. By hydrolysis of the amino group ammonia is split off. The resulting o-diphenol is oxidized to the corresponding o-quinone by the o-quinone of the applied phenol. A further amino acid is added again, whereby the deamination reaction continues.

The scheme shows that one molecule of the addition product formed by one amino acid with one o-quinone deaminates several amino acids with a consumption of one atom of oxygen per amino acid. Before this, it had been demonstrated that in some cases a larger release of carbon dioxide than ammonia takes place, and it has been shown that further reactions take place to bind nitrogen. An explanation for these results may be a reaction of the aminophenol with o-quinones, which leads to the formation of phenoxazines or phenoxazones.

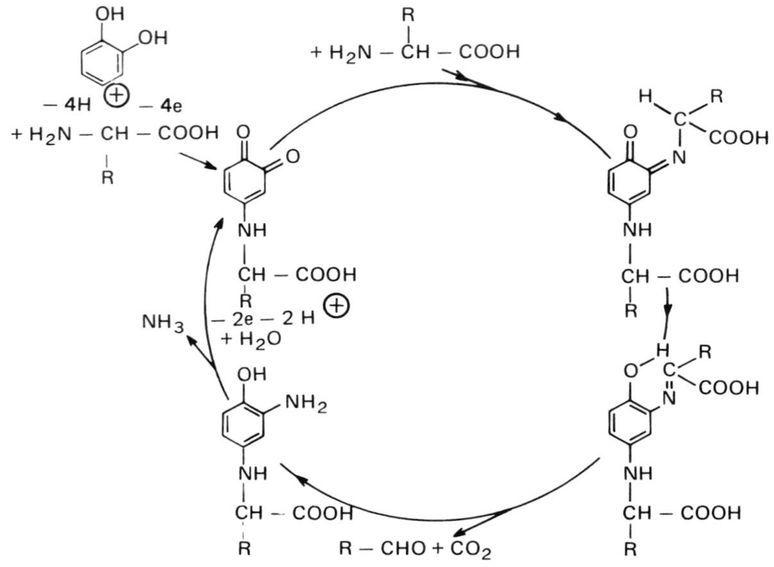

Figure 46. Reaction scheme of o-quinone with amino acids, according to TRAUTNER and ROBERTS [1950].

The formation of phenoxazone derivatives at low concentrations of amino acids has been observed in other studies (BUTENANDT, BIECKERT, and SCHÄFER [1960], LANTZ and MICHEL [1961], BEEKEN *et al.* [1961]).

The catalytic effect of quinones in the deamination of amino acids has been frequently investigated (LANGENBECK, SCHALLER, and ARMBERG [1942], LANGENBECK, TARNAN, KERTEL, and HIRSCH [1948], FLAIG and REINEMUND [1955], REINEMUND and FLAIG [1955]).

3. *Nucleophilic Addition of Peptides and Proteins by Phenols and Their Deamination*

In further work, reactions occurring with phenols such as pyrocatechol, caffeic acid, hydroxyhydroquinone, and hydroquinone, with peptides such as glycyl-alanine, glycyl-valyl-leucine, glycyl-valyl-isoleucine, glycyl-methionyl-isoleucin, leucyl-glycine, and tryptophanyl-alanine were studied (HAIDER, FREDERICK, and FLAIG [1965]) (Figure 47).

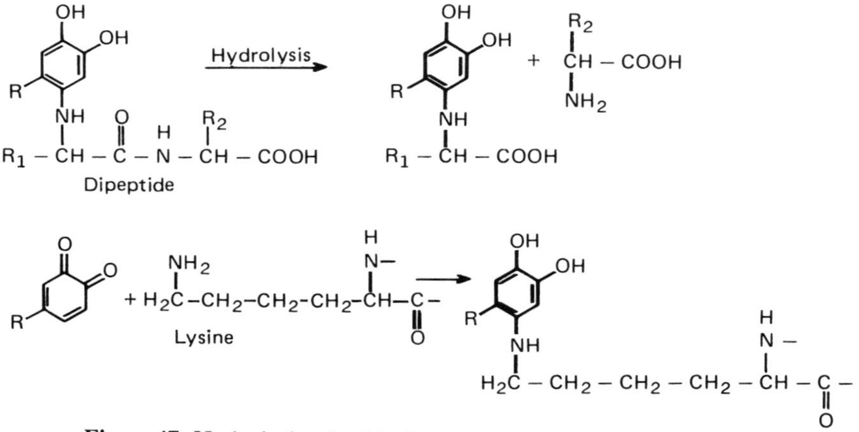

Figure 47. Hydrolysis of oxidation products of phenols and peptides.

In all these investigations it was established that the nature of the peptide chain did not appreciably influence the activity of the free amino groups.

Experiments with di- or tripeptides showed that both nucleophilic addition of a peptide by a quinone and deamination occurred. When the addition products had been hydrolyzed, all the amino acids were identified except the nitrogen terminal, which is bound to the oxidized phenols.

Furthermore, experiments have been made with serum albumin. According to THOMPSON [1958], this protein has a molecular weight of 65,000 and has a nitrogen terminal amino acid, aspartic acid. According to WOLD [1961], 14 units of lysine are in the molecule.

Serum albumin also formed addition products with various diphenols. The dinitrophenyl protein, prepared by the method of SANGER and THOMPSON [1953], was hydrolyzed. The content of the nitrogen terminal amino acid of the original serum albumin the aspartic acid was greatly decreased. Furthermore, it was found that after the addition, the amount of lysine in the addition product after hydrolysis was smaller in comparison to the amount in the original protein. Therefore, the ϵ-amino groups also take part in the reaction with the oxidized phenols. The course of the proteolysis of the addition product by nagarse (HAGIHARA [1954]), a protease of *Bacillus subtilis*, was not very much different from that of the protein alone.

4. Autoxidation of Phenol Derivatives in the Presence of Ammonia at pH Values Higher than 8

(a) 1,2- or 1,4- Diphenols

In the preceding sections the addition reactions in oxidizing medium have been reported, which were catalyzed by phenoloxidases and which occur in a range of pH values between 6 and 8. At higher pH values the phenols with at least two hydroxyl groups in o- or p-position react with ammonia to give dark-colored nitrogenous polymers.

The linkage of the nitrogen in phenol polymers has been investigated with model substances. The properties of these dark-colored substances have been linked to those of humic acids isolated from soils. The amount of oxygen utilized in the reactions is as high as four atoms per molecule of hydroquinone, and yields are about 80% (FLAIG [1950]). Up to 6% nitrogen is present in the products precipitable with acid after distillation with magnesium oxide.

The release of carbon dioxide upon acidification is less than in the case of oxidation products formed in the absence of ammonia. The uptake of oxygen in the presence or absence of ammonia is accelerated by copper or ferric ions and by the addition of a dispersing agent. In comparative experiments of the alkaline oxidation of hydroquinone in the absence or presence of ammonia, it appeared that with ammonia a certain stabilization of the synthetic products against oxidative attack occurs (FLAIG [1950]).

Two of the hydrogen atoms of ammonia seem to participate in the reactions: dimethyl-

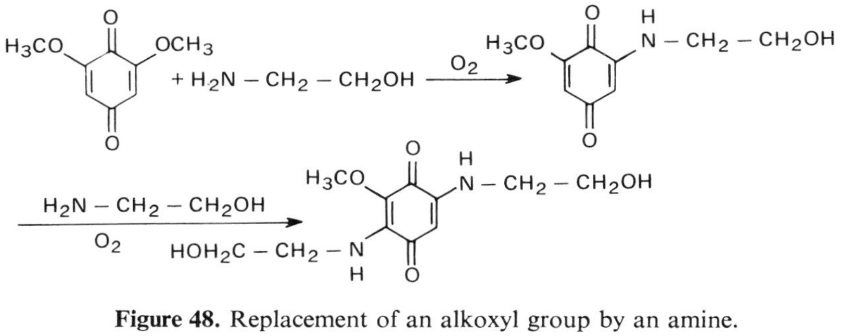

Figure 48. Replacement of an alkoxyl group by an amine.

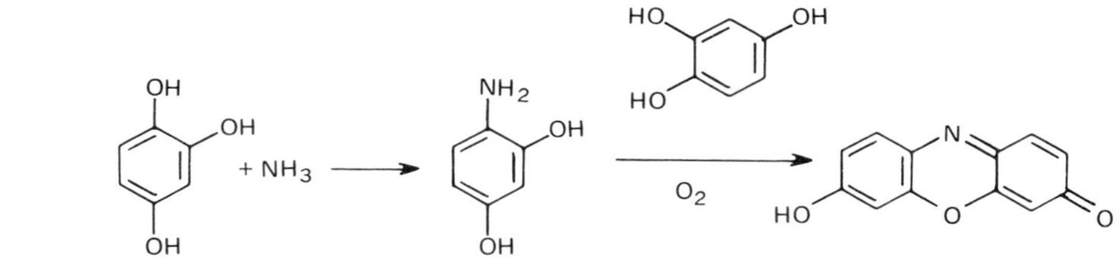

Figure 49. Formation of phenoxazone derivatives.

amine reacts with p-benzoquinone to *bis*-dimethylamino-quinone (BALTZLY and LORZ [1948]), whereas monomethylamine leads to resinous products. No high molecular weight nitrogen-containing products are formed with trimethylamine.

Besides addition, the possibility exists that a hydroxyl or alkoxyl group is replaced by an amino group. The reaction of ethanolamine with 2,6-dimethoxy-p-benzoquinone is an example of this (Figure 48). The second methoxyl group remains after the introduction of a second ethanolamine group (HUISMAN [1950]).

In the absence of air and in the presence of ammonia, a hydroxyl group can be replaced

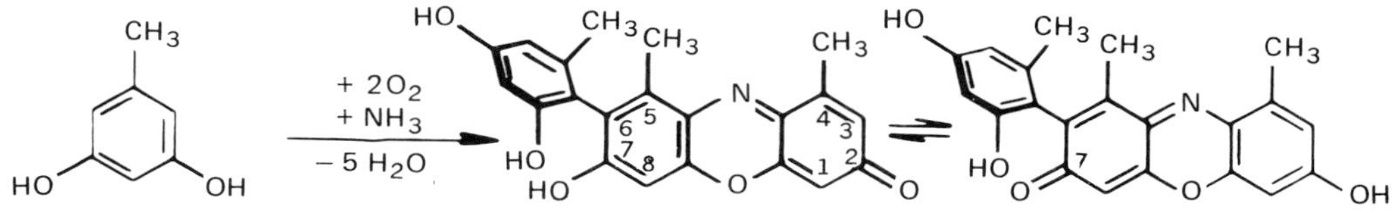

Figure 50. Reaction of resorcinol derivatives with ammonia in the presence of oxygen to form phenoxazones.

in hydroxyhydroquinone by an amino group (LANTZ and MICHEL [1961]). The amino-resorcinol thus formed reacts in the presence of oxygen with a further hydroxyhydroquinone to produce 7-hydroxyphenoxazone (Figure 49).

(b) 1,3-Diphenols

It has already been mentioned that 3,5-dihydroxytoluene (orcinol) was found in the culture media of *Epicoccum nigrum* in the humic acids after reductive cleavage. These substances react with ammonia in an oxidizing medium, but only at higher pH values than 8. MUSSO

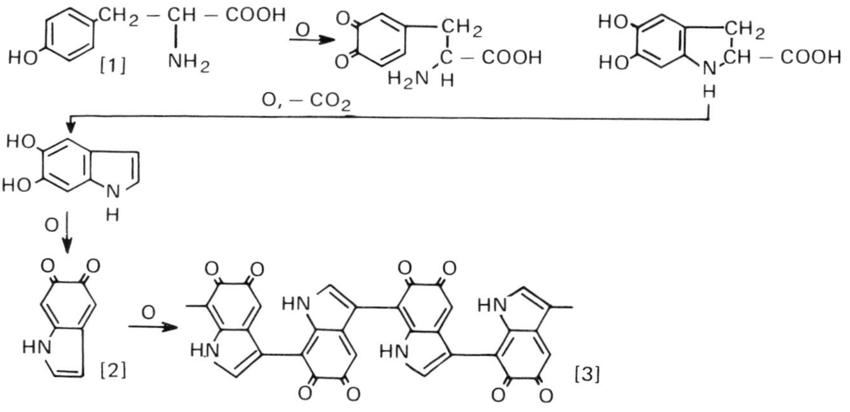

Figure 51. Formation of melanin.

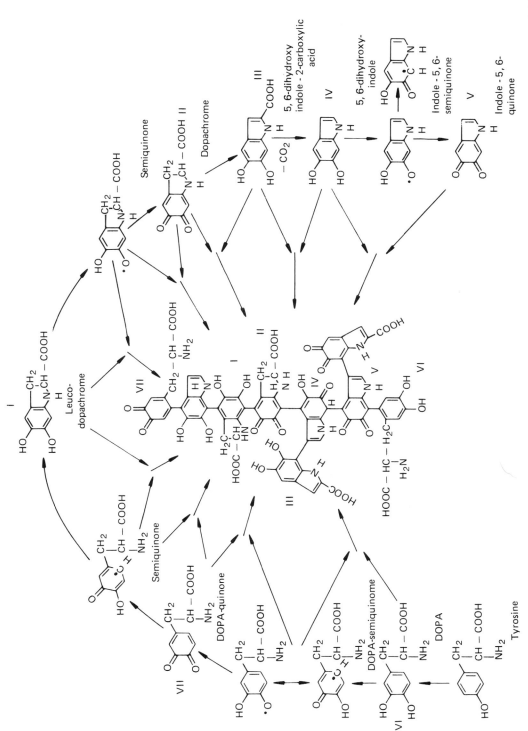

Figure 52. Hypothetical scheme of melanin synthesis assuming the participation of semiquinone free radicals (RAPER [1928], MASON [1948], NICOLAUS [1962], and BLOIS [1965]).

[1963] and MUSSO, GIZYCKI, KRÄMER, and DÖPP [1965] investigated these reactions to elucidate the constitution of the orcein dyestuff (Figure 50). For instance, α-hydroxy-orcein is a tautomeric mixture of the derivatives from phenoxazone-2 and phenoxazone-7, which are also formed through a quinonoid intermediate by the addition of ammonia on orcinol.

5. *Possibilities of Participation of Heterocyclic Nitrogen Compounds in Composition of Humic Acids*

A further possibility of the formation of dark-colored humic substances is the participation of aromatic amino acids from the protein of the microorganisms. For example, the oxidation of tyrosine [1] could lead through dopaquinone [2] (RAPER [1927]) to dark-colored polymers [3] (CROMARTIE and HARLEY-MASON [1953]).

The proposed random polymerization of tyrosine to form melanin (Figure 51) (BLOIS [1965], MASON [1948], NICOLAUS [1962], RAPER [1928]) can be used for the explanation of the formation of humic acids. Both are dark-colored, higher molecular substances with spherical shapes. The polymerization occurs in each case through quinonoid intermediates, such as semiquinone free radicals.

The random polymerization to melanin starts from tyrosine (Figure 52); all intermediates participate in the formation of the polymer. Similar but much more complicated processes occur during the formation of humic substances, since not one but several phenolic components and various nitrogenous compounds take part in the formation of the polymers. The formation of dark-colored substances in structurally intact dead plant tissues at the beginning of humification could be explained in a similar way (KONONOVA [1961]).

Some other possibilities exist for the formation of heterocyclic nitrogen compounds, which may participate in formation of humic substances.

As has been mentioned before, phenoxazines may be formed by condensation of o-aminophenols with catechol derivatives under oxidizing conditions; in the case of hydroxy-hydroquinones, phenoxazones are formed. Further reactions of such type under natural conditions are described, for instance, by BUTENANDT, BICKERT, and SCHÄFER [1960].

HORNER and STURM [1957] have demonstrated that 4-amino catechol reacts with itself to a phenazine derivative (Figure 52a), and 3-amino catechol condenses to polymers (Figure 52b).

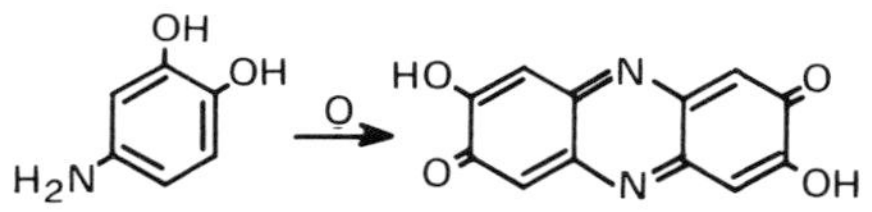

Figure 52a.

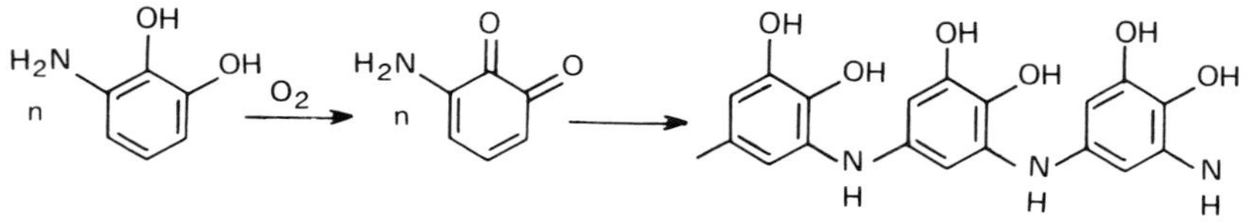

Figure 52b.

Heterocyclic condensation products are also formed by condensation reaction between quinones and amino acids which contain sulfur in the molecule.

KUHN and BEINERT [1944] described the condensation of 2,5-dimethyl-p-benzoquinone with the ester of cysteine to form a dark-colored heterocyclic compound, which contains nitrogen and sulfur in one ring (Figure 52c). The reactions of sulfur-containing amino acids during the

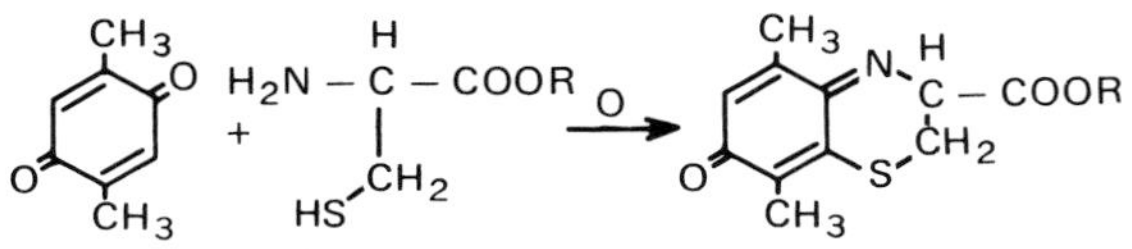

Figure 52c.

formation of humic substances and their liberation from these substances have been investigated by SCHARPENSEEL and KRAUSSE [1963].

Nitrogen ring compounds such as pyridine derivatives are formed after the oxidative cleavage of catechol or protocatechuic acid through the half-aldehyde of carboxymuconic acid and reaction with ammonia (TRIPPETT, DAGLEY, and STOPHER [1960], DAGLEY, EVANS, and RIBBONS [1960]) (Figure 52d).

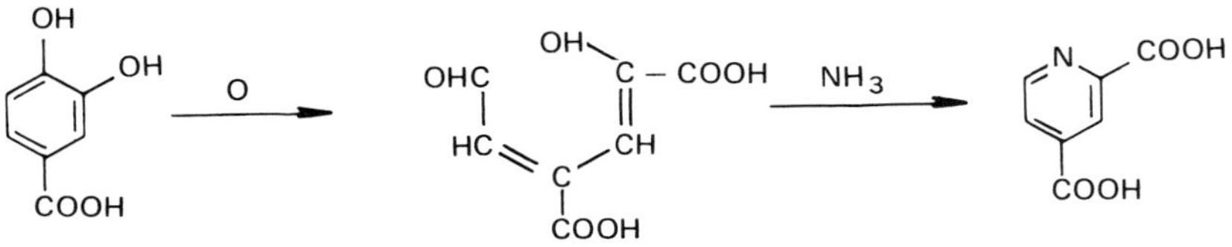

Figure 52d.

Also, the formation of brown polymers is described during the reaction of phloroglucinol with amino acids in the presence of potassium ferricyanide (Figure 52e).

Figure 52e.

Phloroglucinol cannot be oxidized to a quinone and therefore no nucleophilic addition of amino acids can occur. STEIN and TENDELOO [1955; 1956a, b; 1958; 1959] suggested that the first step of the reaction is the formation of a free radical.

X. Possible Participation of Carbohydrates in the Formation of Humic Acids

During humification of plant material, sugars are present in the plants as such or are formed by the degradation of the high molecular weight carbohydrates. When glycine and glucose are heated in a small amount of water in a water bath, dark-colored nitrogen-containing condensation products precipitate following formation of a yellow and later a brown color (MAILLARD [1912, 1913]). These colored condensation products are named "melanoidins."

Extensive studies of this reaction in regard to the formation of humic acids were made by Enders and coworkers (ENDERS and THEIS [1938a, b, c], ENDERS [1942a, b, 1943], ENDERS, SIGURDSON [1948], ENDERS, TSCHAPEK, and GLAWE [1948]). DANCHY and PIGMAN [1951], HODGE [1953], DROSDOVA [1959], and ELLIS [1959] have written summary papers on this subject, and MANSKAJA and DROSDOVA [1964] have surveyed the question involved in relation to melanoidins as humic substances. Many experiments have shown that the amount of the brown products formed is dependent on the chemical constitution of the amino acids and of the sugars (KRETOVICH and TOKAREVA [1948], BHATIA, KAPUR, BHATIA, and GIRDHARI [1957],

MARKUZE [1960], and others). The brown color occurs only after the addition of reducing sugars. Furthermore, the dark color appears faster in the presence of amino acids with an indole or phenol ring (such as tryptophane or tyrosine), whereas cysteine delays the reactions (LÜERS [1953], MENGER [1953]).

At the beginning of the reaction, a nitrogen glucoside or a Schiff base is formed between the carbonyl group of the sugar and the amino group of the amino acid (FRANKEL and KATCHALSKY [1941], HAUGAARD, TUMERMAN, and SILVESTRI [1951], MENGER [1953], HODGE and RIST [1953]). The nitrogen glucosides are transformed into isomeric dihydroxyaminoketoses of the iso-glucosamine type. This reaction, known as an Amadori rearrangement (Figure 53),

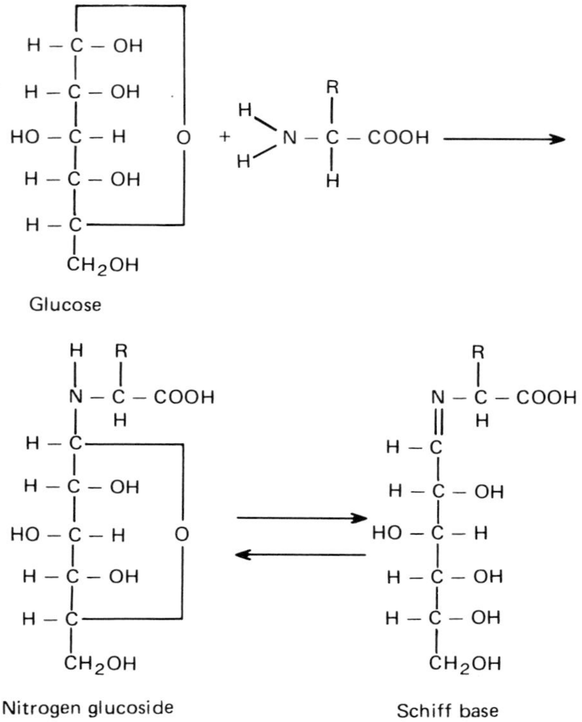

$$
\begin{array}{l}
\text{H} - \text{C} - \text{OH} \\
\text{H} - \text{C} - \text{OH} \\
\text{HO} - \text{C} - \text{H} \qquad \text{O} \\
\text{H} - \text{C} - \text{OH} \\
\text{H} - \text{C} \\
\quad\ \ \text{CH}_2\text{OH}
\end{array}
\qquad + \qquad
\begin{array}{c}
\text{H} \\
\diagdown \quad\ \ \text{R} \\
\quad\ \ \text{N} - \text{C} - \text{COOH} \\
\diagup \qquad | \\
\text{H} \qquad\ \ \text{H}
\end{array}
\longrightarrow
$$

Glucose

$$
\begin{array}{l}
\quad\ \ \text{H} \quad\ \text{R} \\
\quad\ \ \text{N} - \text{C} - \text{COOH} \\
\qquad\quad \text{H} \\
\text{H} - \text{C} \\
\text{H} - \text{C} - \text{OH} \\
\text{HO} - \text{C} - \text{H} \qquad \text{O} \\
\text{H} - \text{C} - \text{OH} \\
\text{H} - \text{C} \\
\quad\ \ \text{CH}_2\text{OH}
\end{array}
\qquad \xrightarrow{\ \ \longrightarrow\ \ } \atop \xleftarrow{\ \ \longleftarrow\ \ }
\qquad
\begin{array}{l}
\qquad\quad \text{R} \\
\quad\ \ \text{N} - \text{C} - \text{COOH} \\
\qquad\ \| \quad \text{H} \\
\text{H} - \text{C} \\
\text{H} - \text{C} - \text{OH} \\
\text{HO} - \text{C} - \text{H} \\
\text{H} - \text{C} - \text{OH} \\
\text{H} - \text{C} - \text{OH} \\
\quad\ \ \text{CH}_2\text{OH}
\end{array}
$$

Nitrogen glucoside   Schiff base

Figure 53. An Amadori rearrangement.

has been intensively investigated (MICHEEL and SCHLEPPINGHOFF [1956], MICHEEL and BÜNING [1957]).

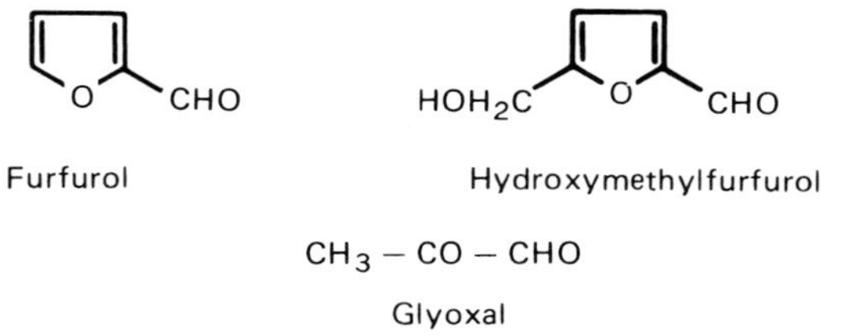

Furfurol

Hydroxymethylfurfurol

$$\text{CH}_3 - \text{CO} - \text{CHO}$$

Glyoxal

Furfurol and hydroxymethylfurfurol have been identified during the formation of melanoidins (KRETOVICH and TOKAREVA [1949], LÜERS [1953], LANGEL and DEKKER [1954], WOLFROM, SCHUETZ. and CAVALIERI [1948, 1949], GOTTSCHALK and PATRIDGE [1950], TÄUFEL and IWAINSKY [1952]).

Enders considered methylglyoxal as an intermediary product. This compound was identified in small amounts as the 2,4-dinitrophenylhydrazone after passing of air through the soil (ENDERS and SIGURDSSON [1942]). HOUGH, JONES, and RICHARDS [1952] found methylglyoxal chromatographically in the heated solution during the formation of the melanoidins. SCHUFFELEN and BOLT [1950] investigated the reaction of glycine and methylglyoxal in regard to similarity of the brown condensation product with humic acids from soil; they isolated 3% fulvic acid, 70 to 80% hymatomelanic acid, and 5 to 7% humic acid from the reaction mixture.

Melanoidins are formed from chitin and glucosamine. Their formation is accelerated by the addition of copper ions (MANSKAYA, DROSDOVA, and EMELYONOVA [1958]). Furfurol and hydroxymethylfurfurol are also formed.

Using ^{14}C-glucose SIMONART and MAYAUDON [1958b] concluded that this sugar participates in the formation of humic acids through the metabolism of microorganisms in the form of synthesized amino acids, which can be recovered by hydrolysis with 6 N hydrochloric acid, and not in the formation of the residual material, which is largely derived from the lignin. Further investigations are necessary to fully clarify unanswered questions concerning the participation of melanoidins in the formation of humic substances.

A further proposal (FLAIG [1958]) of an eventual participation of the sugars in the formation of humic substances is the formation from degradation products of sugars through shikimic acid as suggested by BROWN and NEISH [1955] or the condensation of acetyl units from acetyl-coenzyme-A to form aromatic compounds (GRIESEBACH [1957]).

C. Physical Properties

I. Absorption in the Ultraviolet and Visible Ranges of Light

The spectroscopic investigations of soil organic matter and its fractions in the ultraviolet and visible range of light have become more and more important for both the characterization and the determination of genetic differences and also for transitions between the types of humic substances. To elucidate the contradictions and the relationships of published results, some theoretical presumption of the method and some special spectroscopic properties of the fractions of soil organic matter such as color intensity and type of color are mentioned below

1. *Principles of Absorption Spectroscopy of Humic Acids*

The concentration used in the analysis were of about 0.01 to 0.05 mg/100 ml concentration in case of substances with unknown constitution (concentration of chemically defined compounds mol/liter). The absorption of humic substances in solution follows the Lambert-Beer law:

$$\text{absorption: } A = E = \log P_0/P; P = \epsilon \cdot c \cdot d$$

The *extinction coefficient K* is $K = \epsilon \cdot c$, if the length of light path d in the cell is 1 cm. The extinction coefficient K is therefore directly proportional to the concentration c (mg/100 ml), whereby the proportionality constant ϵ is specific for the investigated substance and is named *molar extinction coefficient* (molar absorptivity). Because the value of ϵ is dependent upon the wavelength, the determination of concentrations must be made at a defined wavelength (*e.g.*, by means of special filter or monochromator). Owing to the still-undetermined molar absorptivity of the humic substances, due to their unknown chemical constitution, the absolute concentration of humic acids in mol/liter cannot be determined. Therefore, the spectroscopic measurements of absorption of soil organic matter or its fractions are used only for the determination of the K values.

The light absorption of a solution of humic substances of constant concentration is measured in the visible and ultraviolet ranges and the logarithm of extinction (log K is plotted against the wavelength (log $K = (\lambda)_c$). In contrast to the higher molecular weight substances, these electronic spectra show no distinct maxima but only a monotonously rising, almost-straight line in the direction of shorter wavelengths. With the applicability of the Lambert-Beer law, whereby log ϵ and log c became additive (log K = log ϵ + log c), the log of extinction plotted against wavelength gives a family of curves, at various concentrations, that are parallel. The determination of relative concentrations by changes of absorption are possible. On the other hand, this indicates that the slope of the curves is independent of variation in the concentration, leading to the possibility of characterization of the color type of humic substances by determination of the slope of the absorption curve. This characterization is based on the differences of the molar absorptivity of these substances.

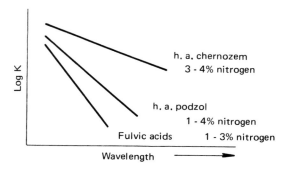

Figure 54. Spectra of humic acids (h.a.) from chernozem, podzol, and fulvic acids. The absorbance (log K) is depicted schematically as straight lines, although more exact measurement demonstrates more or less deviation from linearity.

Commonly, the spectra of humic acids of different genetic origin or methods of isolation can be distinguished by absorbance differences at any wavelength or by differences in the slope of the absorption curve.

The attempt to represent the spectroscopic results uniformly for comparison led to the definition of different key numbers, especially in the German literature. SIMON and SPEICHER-MANN [1938] introduced the color value to characterize spectra of humic acids; this value is the absorption at 619 nm of the sample divided by absorption at 619 nm of "standard" humic acids (e.g., Huminsäuren Merck) (E_{619} sample/E_{619} "standard" humic acids). These authors also introduced the color factor (E_{463}/E_{619}). This was also called the color quotient (E_{470}/E_{664}) by HOCK [1942] and WELTE [1956]. NEHRING [1955] introduced the color area value for the characterization of humic acids. He defined this value as the area below the absorption curve (log $E = f(\lambda)_c$) that summarizes an indication of the course of the integral absorption and the color intensity. Of the three key numbers, the color quotient ($Q_{4/6}$ value) is still mentioned most often in contemporary literature.

More recently, other authors such as KUMADA [1955]* and SALFELD [1965, 1968]* have characterized the slope of the color curve by the difference of logarithmic extinction of at least two different wavelengths (Δ log K = log K_{400} − log K_{600}).

2. Spectrophotometric Determination of Concentration of Soil Organic Matter Extracts or Its Fractions

The necessity of quantitative determination of fractions from soil organic matter for an

*Compare C.I.7., page 105.

explanation of soil fertility problems has led in recent times to the elaboration of different calculation procedures.

The procedure proposed by WELTE [1956] is based on the assumption that humic acids are a mixture of "brown" and "gray" humic acids. This distinction was first established in colorimetric measurements by SPRINGER [1938], who distinguished the two types of humic acids, the "brown" humic acids with a low absorption and a large slope of the absorption curve or the "gray" humic acids with a high absorption and a small slope.

WELTE [1956] determined the extinctions of different synthetic and natural humic acids at 472 nm (E_4) and 664 nm (E_6) and developed, from his mathematical considerations of the photometric analysis of systems of two components, a formula for calculation of the concentration as follows:

$$c = 79.9\ E_4 - 103\ E_6$$

This means the quantity of humic acids in 100 mg/l. in 0.5% sodium hydroxide solution. Welte concluded, by the constancy of his determined coefficients, that the organic constituents of soil extracts have the same optical properties, independent of their origin.

BOLDYREV and ALESHIN [1963] simplified Welte's formula by the assumption that humic acids from different soil types have the same absorption coefficients, not only at a wavelength of 400 nm but also from 260 to 400 nm. According to these authors the constant absorption coefficient at 400 nm is 0.157; they propose the following formula to compute the concentration:

$$c = \frac{E_{400} \cdot d}{0.157}$$

(c in mg/100 ml; d = width of the cuvette).

The quantitative spectrophotometric determination of humic acids according to PAVEL and ZAZVORKA [1965] starts from the assumption that soil humic acids are a mixture of polycomponents of analogous heteropolymers. Starting from the concept of the nearly linear dependence of the log K values from the wavelength, the authors assumed a decrease of the optical density with increasing wavelength according to an exponential curve ($E = A \cdot e^{-a \cdot \lambda}$), and they assumed a direct proportionality between the concentration of the humic acids and the area P, which is formed by the absorption curve and the λ axis. This leads to the equation

$$c = k \cdot P = k\ \frac{A}{a}$$

(A, a = constants of the exponential equation; k = proportionality factor).

In contrast to the equations for spectrophotometric determination of concentrations mentioned above, this formula would have the advantage of taking into consideration the total integral light absorption in the measured range of wavelength, and not just the absorption at one distinct wavelength. The coefficients A and a represent the form and the position of the spectroscopic curve in log K/λ diagram. This means that the larger the value A, the higher the spectrum of the humic acids is located in the diagram, whereas an increase of the slope of the absorption curve is connected by the value of a. The computation formula of Pavel and Zazvorka supposes a constancy of the quotient A:a (i.e., for a spectroscopic comparison of two different humic acids of the same concentrations, an increase of A should result in a decrease of a by the same factor, and vice versa).

ORLOV [1966] examined all the mentioned formulae for the spectrophotometric determination of concentration of humic acids with samples of different origin (Table 21). He found that neither the formula of WELTE [1956] nor that of BOLDYREV and ALESHIN [1963] is

Table 21. Results of the spectroscopic determination of concentration of humic acids by different methods ($d = 2$ cm), according to ORLOV [1966].

	E_{474}	$\dfrac{E_{474}}{E_{666}}$	Real concentration of humic acids, mg/l.	WELTE, BOLDYREV, and ALESHIN		PAVEL and ZAZVORKA	
Sod-podzolic soil, forest	0.290	3.98	40	15.7	15.6	10.0*	15.3†
Sod-podzolic soil, arable land	0.320	3.56	40	16.3	16.2	9.0	13.8
Sod-podzolic soil, meadow	0.340	3.10	40	15.9	17.5	7.5	11.5
Gray forest soil	0.824	3.30	40	−40.0	−39.8	20.5	31.4
Humus-carbonate soil	0.602	3.42	40	30.1	31.8	15.5	24.8
Deep chernozem, steppe	0.860	3.20	40	−41.1	−44.3	20.0	30.6
Deep chernozem, forest	0.832	3.20	40	−39.8	−41.7	19.0	29.1
Ordinary chernozem	0.850	3.24	40	−41.0	−42.0	20.0	30.6
Chernozem in southern area	0.897	3.24	40	43.2	47.8	21.5	32.9
Chernozem, Krim region	0.968	2.58	40	38.7	43.3	16.0	24.5
Solonets, meadow	0.556	3.53	40	28.2	31.5	15.5	23.8
Gray earth	0.610	1.8	40	13.9	30.0	6.5	9.9
Brown earth, forest highland	0.480	2.45	40	18.1	22.6	7.5	11.5
Red earth, forest	0.379	3.73	40	19.8	23.9	11.5	17.0
Light red earth	0.620	2.52	40	25.1	30.2	10.0	15.3
Brown earth on calcareous soil	0.460	2.50	40	17.8	22.9	7.5	11.5
Brown earth on shale	0.500	3.52	40	19.5	23.6	8.5	13.0
Mountain-meadow soil	0.690	2.92	40	30.7	32.8	14.0	21.4

The "WELTE, BOLDYREV, and ALESHIN" and "PAVEL and ZAZVORKA" columns fall under the heading *Concentration of humic acids according to the formulae of*.

* According to the nomogram of PAVEL and ZAZVORKA [1965].
† According to the same nomogram with a corrected value of the constant K.

suited for the determination of organic matter of most soil types. The formulae can only be used if the $Q_{4/6}$ values for the humic substances are high and the E_4 values are very high.

According to the investigation of Orlov, the formula of Pavel and Zazvorka deviates the most from real concentrations. According to the author's opinion, it is caused by the fact that the theoretical and the experimental curves of light absorption do not coincide. This leads to an incorrect determination of the proportionality factor. The method of Pavel and Zazvorka is in principle suited only for the analysis of such soils that are similar in genetic origin. Therefore, this method is not as universally valid as the others.

3. *Connections Between Chemical Composition and Light Absorption of Humic Acids*

Since the beginning of the work on humus, the color of some fractions of soil organic matter such as humic acids was of interest in different investigations and led to the definition of the term "humic acids" as dark-colored, amorphous, high molecular weight substances of soils; their classification with optical methods is also today the subject of many investigations. Primarily, the spectroscopic and physicochemical differences of the humic acids are influenced by the different genetic conditions in soils and are furthermore varied by the methods of extraction and purification.

These investigations did not result in a further differentiation, because the multitude of molecular constituents of humic acids, which absorb in the ultraviolet and visible range, such as phenolic compounds and their oxidation products, amino acids and their condensation products with phenols in oxidizing medium, and the formation of heterocyclic components, leads to uncharacteristic spectra of mixtures. These show no distinct differences in the absorption properties, if different types of humic acids are compared.

Lignin is a typical example for the absorption of single or repeated nucleus-substituted polymeric phenol derivatives. Its absorption has already been discussed. In connection with humic substances it was demonstrated that the absorption by lignin becomes more and more uncharacteristic with humification. In all cases the curve of absorption became more and more similar to that of humic acids.

(a) Contribution of Phenols and Their Transformation Products

The absorption of humic acid in the ultraviolet range may be caused by phenolic components and in the visible by chromophoric groups formed by oxidation. It could be demonstrated by oxidative or reductive cleavage of humic acids that these contain phenolic components. FLAIG and SALFELD [1958] investigated especially the changes of the optical properties of differently substituted o- and p-benzoquinones as model substances. They demonstrated that the absorption curves of these compounds have two or three significant maxima according to their substitution. The kind and the position of the substituents cause changes of the absorption properties or shifts of specific absorption bands. For instance, substitution with an aliphatic group such as a methyl group results in a shift in the second maximum of p-benzoquinone at 282 nm (in carbon tetrachloride) of 27 nm, in the case of a methoxyl group of 60 nm, and in the case of a hydroxyl group of 90 nm in direction of longer wavelengths (Figure 55).

Furthermore, it was established that a shift of the second maximum to longer wavelengths depends less on the chain length of the aliphatic group, and more on the kind of substitution in the ring with further alkyl, alkoxyl or hydroxyl groups in m- or p-positions, and especially in the o-position. Strongly effective chromophoric groups are phenyl as well as aromatic or aliphatic amino groups (amino acids).

The facts demonstrate that spectrophotometric investigations of organic soil extracts are not very suitable for the explanation of molecular differences. Because the complex chemical composition of the macromolecules is not yet elucidated, the recorded absorption curves must

	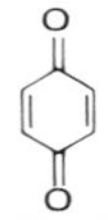 Second maximum in nm	Shift in nm	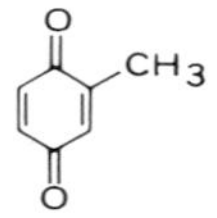 Second maximum m-, p-substitution in nm	Shift in nm	Second maximum o-substitution in nm	Shift in nm	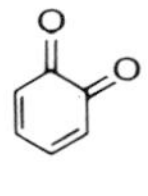 Second maximum in nm	Shift in nm
Unsubstituted	$282(CCl_4)$ $288(CHCl_3)$	—	$309(CCl_4)$	—	$309(CCl_4)$	—	$375(CCl_4)$	—
—CH₃	$309(CCl_4)$	27	$310(CCl_4)$	1	$332(CCl_4)$	23	$328(CCl_4)$ in 4-position	$7(CCl_4)$
—OCH₃	$351(CCl_4)$	69	$355(CCl_4)$	46	$374(CCl_4)$	65	$450(CCl_4)$ in 3-position	$75(CCl_4)$
—OH	$369(CHCl_3)$ $372(CCl_4)$	81 90	$381(CHCl_3)$	72	$396(CHCl_3)$	85	$505(CCl_4)$ in 3-position (3-hydroxy — 4,6-di-tert. butyl)	$116(CCl_4)$
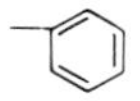 ⬡	$369(CCl_4)$	87						
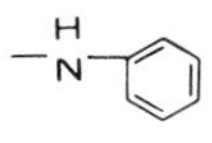 —N(H)—⬡	$550(CHCl_3)$	262 minus benzene ring-175						
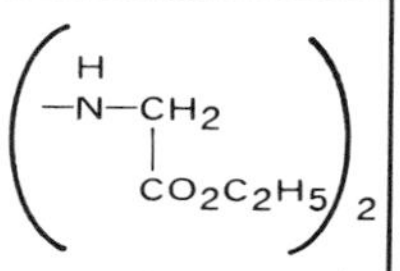 (—N(H)—CH₂ | CO₂C₂H₅)₂ p - position	$388(CHCl_3)$	100						

Figure 55. Shift of the second maximum of absorption curve of benzoquinones by substitution with different groups.

be considered to represent only mixed spectra of the specific absorptions of molecular constituents. This can be demonstrated by mixtures of known corresponding compounds.

(b) Nitrogen Content

The total nitrogen content plays no important role for the brown color of humic acids. Only hydrolyzable amino acids participate in the absorption, because these show no absorption in the visible range.

As a rule, the humic acids from chernozems, which are richer in nitrogen, have not only a strong absorption but also absorb in a range of longer wavelengths than the humic acids of podzols, which have a lower content of nitrogen (compare with Figure 54).

If there exists a relationship between the light absorption and the nitrogen content of humic acids, then one has to bear in mind that the nonhydrolyzable nitrogen, presumably the heterocyclic-bound nitrogen, contributes remarkably to the absorption by its bathochromic effect.

The heterocyclic ring system containing nitrogen and oxygen often has properties of dyestuffs. Their main absorption is found mostly in the range of longer wavelengths.

In spite of the higher nitrogen content of "brown" humic acids to that of "gray" humic acids, the absorption of gray humic acids is larger in the total range of wavelengths (Figure 56). This result may be due to the fact that brown humic acids have a higher part of hydrolyzable amino acids than gray humic acids. Although the content of nitrogen in the residues of hydrolysis of brown humic acids is smaller than that of brown humic acids themselves, the absorption of the residues is stronger.

These conclusions are supported by investigations of KLEIST and MÜCKE [1966a, b] with electron spin resonance. They assumed that the higher absorption of gray humic acids is related to a higher content of radicals, which may be caused by resonance systems of heterocyclic ring components. Also, the nonhydrolyzable nitrogen increases in the lignin fractions during humification (SCHOBINGER [1958]).

HARGITAI [1962] and ANDRZEJEWSKI [1964, 1965] demonstrated also possible relationships between nitrogen content and absorption of humic substances in the ultraviolet and visible ranges.

The character of light absorption of fulvic acids is published in several papers (KONONOVA [1956, 1958, 1961], ORLOV [1959, 1960], ORLOV and ZUB [1964], ORLOV and GRINDEL [1967]). From the absorption curve, it seems that the absorption is mainly influenced by oxidized phenols. Phenols were isolated from soil extracts (WHITEHEAD [1964], BRUCKERT, JACQUIN, and METCHE [1967]).

ORLOV and ZUB [1964] believe that the pH-dependent increase of the absorption of fulvic acids is affected by ketoenol-tautomerism, whereby the formation of polar auxochromic hydroxyl groups occurs. The nitrogen content of fulvic acids is in most cases lower (KONONOVA [1966]) and the content of carbonyl groups is higher in comparison to humic acids (SCHNITZER [1965]).

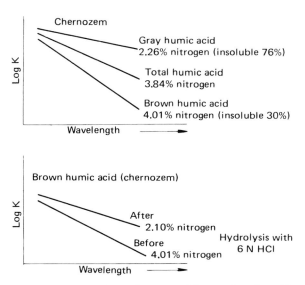

Figure 56. Schematic of the absorption of gray and brown humic acids of chernozem and the residue of hydrolysis of brown humic acids.

4. Influence of Hydrogen Ion Concentration

From the investigations of SPRINGER [1934] and HOCK [1938a, b] it is known that the absorption of humic substances in solution increases with decreasing hydrogen ion concentration. SALFELD [1965] studied the dependence of the absorption of humic acids, purified by ion exchange from different origins, on the pH (Figure 57). He found that the decrease of the hydrogen ion concentration by addition of increasing amounts of 0.1 N NaOH up to pH = 10 caused only a parallel shift of the spectra in the direction toward higher values of absorbance. A further decrease of the hydrogen ion concentration did not effect changes in the light absorption up to 500 nm. Above 500 nm a continuous increase of the absorption up to 620 nm occurred, which led to the formation of a weak shoulder and, furthermore, to a parallel shifting of the extinction values. A similar effect on the absorption spectrum was observed in the case of humic acids formed by autoxidation of pyrocatechol in sodium hydroxide solutions. Therefore, it seems improbable that the changes in absorption in the range of 620 nm are only caused by chlorophyll-like porphyrin systems.

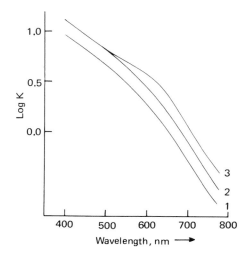

Figure 57. Dependence of absorption of humic acids from a podzol at different pH levels (1:pH = 3.75, 2:pH = 1.98, 3:pH = 11.80), according to SALFIELD [1965].

WIESEMÜLLER [1965] investigated brown and gray humic acids. He compared alkaline solutions of these humic acids with their hydrogen form, which were prepared with strong acid cation exchangers. The slope of the absorption spectra did not change. Therefore, he assumed no influence of the hydrogen ion concentration on the absorption properties of these two types of humic acids.

5. Transition of Fractions of Humic Substances

The chemical composition of fractions of humic substances, which are separated according to the usual methods, is largely unknown. Degradation reactions and investigations with model substances show that phenolic substances of monomer or polymer type and nitrogenous compounds derived from proteins participate mainly in the formation of humic substances.

Many authors conclude from the results of their experiments that there exists a genetic relationship between fulvic acids and humic acids, whereby the latter are formed by condensation or polymerization of lower molecular weight parts, such as constituents of fulvic acids (e.g., WELTE [1952], FREYTAG [1955, 1961a, b, 1962], SCHEFFER, WELTE, and ZIECHMANN [1955],

KONONOVA [1958], LAATSCH [1958], KOBO and TATSUKAWA [1959], and others). Humic acids may be degraded to fulvic acids after standing a longer time in solution. This is caused by the oxidative degradation by the oxygen of the air. It cannot yet be decided if the formation of high molecular weight substances of soil organic matter is caused by chemical linkages alone or, in addition, by strong van der Waals forces. The idea that humic acids are exclusively polymer homologs of fulvic acids is uncertain, because the chemical constitution of the "monomers" is not known.

Searches for transitions between the different fractions of humic substances were made, for instance, by FREYTAG [1961a, b, c]. He indicated that alterations of the type of humic fractions can be followed easier in the range of shorter wavelengths (e.g., $\lambda = 496$ nm) (FREYTAG [1955]), because here there is a better proportionality between absorption and concentration. The observed changes in absorption during time of humification of plant roots (potatoes and sheep's fescue) and of coalified glucose as a model substance may point to a possible transition from fulvic acids to hymatomelanic acids and finally to humic acids. This genetic order is revealed by a time-dependent raising of the absorption synchronously followed by a decrease of the slope, with distinct maxima for each of the transition stages, which more or less overlap each other.

6. Influence of Soil Constituents on Light Absorption

The inorganic materials with active surface are distributed differently in the soil types. The humification is influenced by the more or less strong adsorption of organic substances. Furthermore, the soil acts as a sorption column, and a "chromatographic" fractionation occurs when substances are transported by leaching in the deeper layers of soils. According to NEHRING [1955], this is the reason why the absorption increases with increasing soil depth.

HOFFMANN [1964] investigated the influence of the clay content in the upper soil horizons on the absorption of humic acids by means of regression analysis. An unexpected strong negative correlation ($r = -0.93$) was found between the logarithmic values of the clay content and the absorption (log K, S 53 filter). Hoffmann concluded from the evident decrease of the absorption with increasing clay content that humification of organic substances are inhibited by increasing clay content (compare ENSMINGER and GIESEKING [1939]). During his investigations he found that chlorophyll-like substances are transported more easily than other materials or are enriched in the deeper layers of the soil by degradation processes, which occur during the transport.

Furthermore, KUMADA and SATO [1962] and KUMADA [1965] indicate that their organic soil extract, the so-called P-humic acids, can be separated in two fractions, a green (Pg) and a red brown (Pb). According to these authors, the green fraction is caused by the presence of metal chelates with porphyrin systems. This assumption is supported by a specific absorption at 620 nm (KUMADA [1955]), where protochlorophyll has a maximum absorption. SIMONART, MAYAUDON, and BATISTIC [1959] demonstrated by fermentation experiments that 27% of the chlorophyll released during humification was found in the humic acids; most of the rest was in the insoluble humic substances, and only 8% was in fulvic acids.

7. Classification of Humic Substances by Means of Optical Methods

In spite of the fact that the absorption spectra of different humic acids are very similar in their slope, KUMADA [1965] concluded from his spectrometric investigations that extracts of humic acids from different soil types can be divided into three types, which he denotes as A-, B-, and Rp-types (Figure 58).

The A-type is characterized by those humic acids that have no specific absorption bands over the total range of wavelengths with lower values (about 0.6) of Δ log K (Δ log K = log

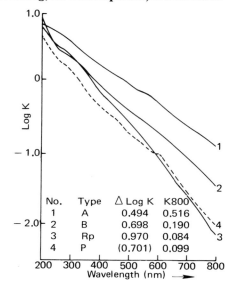

No.	Type	$\triangle$ Log K	K800
1	A	0.494	0.516
2	B	0.698	0.190
3	Rp	0.970	0.084
4	P	(0.701)	0.099

Figure 58. Absorption spectra of the A-, B-, Rp-, and P-types of humic acids (KUMADA [1965]).

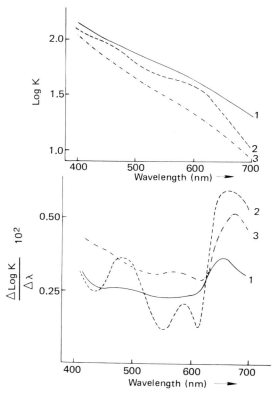

Figure 59. Electronic spectra and differential spectrograms of chernozem, podzol-B, black peat humic acids in 0.1 N NaOH (SALFELD [1965]).

$K_{400} - \log K_{600}$). This type of humic acid is found mainly in the A horizons of volcanic ash soils and in calcareous soils. The B-type shows a weak shoulder at 275 nm in the absorption spectrum and a higher $\Delta \log K$ value (0.6 to 0.8). Humic acids of this type are found in a variety of soils, for instance, in brown forest soils, paddy soils, etc. The Rp-type has the same shoulder in the spectrum as the B-type, but considerably higher $\Delta \log K$ values (0.8 to 1.1). Humic acids that can be isolated from peat, rotted gramineous plants, or stable manure belong to this type. These three types of humic acids might occur in the same soil horizon. Therefore, no correlation with genetic processes in soil seems to exist. As an exception, the so-called P-type of humic acids has three, more or less specific absorptions at 615, 570, and 450 nm. This type was found in podzolic soils and in B and C horizons of volcanic ash soils, brown forest soil, and red-yellow soils.

HOFFMAN [1964] checked the deviations of spectra of humic acids from the straight line, which were regarded by HOCK [1938a, b] as merely errors in measurement, by determination of the absorption of humic and fulvic acids. As an index he used the standard values of deviation

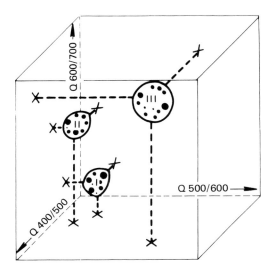

Figure 60. Three-dimensional model of the similarity of humic acids, by means of three extinction quotients, strongly schematized (SALFELD [1968]).

s of the measured absorption around the regression line, which was plotted through the absorption curve. The smallest deviation occurred in the case of humic acids of chernozem (s < 0.025 and raw humus ($s = 0.035$ to 0.045). The brown humic acids gave s values of 0.045 to 0.10.

The weak distinction between the type of humic acids by differences in absorption and slope of the absorption curves as key numbers led SALFELD [1965] to describe the spectra of humic acids as differential spectrograms, with the parameters $\Delta \log K/\Delta\lambda$ and the wavelength λ, whereby the differences were formed in the range of 20 nm. The plot shows significant changes of the absorption of humic acids in form of peak-like maxima and minima of the differential spectrograms (Figure 59).

In further work SALFELD [1968] tried to classify humic acids of different origin according to the similarity of absorption properties which were defined by three extinction quotients ($Q_{400/500}$, $Q_{500/600}$, and $Q_{600/700}$) as key numbers. He arranged the values according to the statistical method of SNEATH [1957].

He plotted these values in a cartesian system, with the three extinction quotients as para-

Table 22. Spectral assignment of the main infrared absorption bands of humic acids.

$\lambda(\mu)$	$\mu^*(\text{cm}^{-1})$	Assignment	Authors*
2.90–3.03	3450–3300	Hydrogen-bonded OH groups, free OH, intermolecular-bonded OH	(1), (3), (5), (6), (7)
3.25–3.30	3077–3030	Aromatic C–H stretching	(1), (2), (3)
3.39–3.50	2950–2850	Aliphatic C–H, C–H$_2$, C–H$_3$ stretching	(1), (3), (4), (5), (6), (7)
3.50–4.00	2850–2500	Carboxylate ion	(1), (3), (4), (5), (6), (7)
5.80–6.10	1725–1640	C=O stretching of carboxylic acids, cyclic and acyclic aldehydes and ketones, quinones	(1), (2), (3), (4), (5), (6), (7), (8),
6.10–6.31	1640–1585	C=C stretching vibration of double bonds in cyclic and acyclic compounds, benzene rings, substitution	(1), (2), (3), (4), (5), (6), (7), (8)
6.35	1575	Carboxyl ion (metallic derivatives of chelated carbonyl groups)	(2), (5)
6.50	1540	NO$_2$ vibrations of nitro groups (mainly in humic acids prepared by nitric acid oxidation from coals)	(2)
6.60	1515	C=C stretching vibration of benzene, pyridines, etc., benzene ring substitution, secondary amines	(3), (5)
6.80–7.05	1470–1420	Aliphatic C–H deformation	(3), (4), (6), (8)
7.20–7.50	1390–1332	Salts of carboxylic acids	(4), (5), (7)
7.80–8.80	1280–1137	C–O stretching of esters, ethers, and phenols	(1), (3), (4), (5), (6), (7), (8),
9.75	1025	Si–O–Si vibration of silicates	(6)

* (1) Brown [1955a.b]; (2) Ceh and Hadzi [1956]; (3) Kumada and Aizawa [1958]; (4) Orlov, Rozanova, and Matyukhina [1962]; (5) Scharpenseel, König, and Menthe [1964]; (6) Schnitzer[1965]; (7) Wagner and Stevenson [1965]; (8) Kononova [1966].

meters, and proposed the distances between the points as a measure of the degree of similarity (Figure 60). In this model the groups of humic acids with similar absorption properties occur as three-dimensionally arranged accumulations of points, which originate from the same soils.

II. Infrared Absorption of Humic Acids and Their Fractions

The infrared spectroscopy of humic acid preparations and their fractions is used by many authors for characterizing humic substances from different soil origins (KASATOCHKIN and ZILBERBRAND [1956], KASATOCHKIN, KONONOVA, and ZILBERBRAND [1958], KUMADA and AIZAWA [1958, 1959], ZIECHMANN [1958, 1959, 1964], SCHARPENSEEL and ALBERSMEYER [1960], ZIECHMANN and SCHOLZ [1960], ORLOV, ROZANOVA, and MATYUKHINA [1962], SCHARPENSEEL, KÖNIG, and MENTHE [1964], KONONOVA [1966], THENG and POSNER [1967], TOKUDOME and KANNO [1968], POSNER, THENG, and WAKE [1968]). It gives some information about the modification by chemical treatments (FARMER and MORRISON [1960], ORLOV, ROZANOVA, and MATYUKHINA [1962], SCHNITZER [1965]), such as methylation, acetylation, esterification, saponification, and the formation of other derivatives (SCHNITZER and SKINNER [1965a, b], SCHNITZER and DESJARDINS [1965]). It is possible by this method to detect changes in the chemical structure of the investigated material during oxidation and pyrolysis (SCHNITZER and HOFFMANN [1964], SCHNITZER [1965]). Furthermore, metal-humate complexes, which may occur by organometallic interactions in soils, have been investigated (KASATOCHKIN, KONONOVA, and ZILBERBRAND [1958], SCHNITZER, SHEARER, and WRIGHT [1959], SCHNITZER and SKINNER [1963a, 1965], SCHNITZER [1965], LEVESQUE and SCHNITZER [1967b]).

Unfortunately, assignment of the specific absorption bands is limited by the fact that soil organic matter preparations represent in most cases mixtures of more or less complex molecules containing different types of linkages and functional groups. This leads to an overlapping of the absorption bands. The infrared absorption spectra of humic acids therefore show only some bands, which are characteristic for the chemical nature of the molecule, since the spectra indicate that absorption bands do not originate from identical structural features, but probably from similar groups in different molecular surroundings. The infrared absorption properties of these samples are also strongly influenced by different methods of sample preparation (*e.g.*, different methods of soil extraction and subsequent fractionation). In cases where pellet techniques are used, the time of grinding and evacuation are also influential factors (THENG, WAKE, and POSNER [1966]).

1. *Assignment of the Main Absorption Bands in Humic Acid Spectra*

There are some difficulties in assigning the infrared absorption bands of humic acid preparations. Some main absorption regions are found which appear in nearly all soil organic matter preparations, with some differences only in intensity or specific wavelengths of the absorption (see Table 22).

In order to give a general assignment of the main infrared absorption bands, humic acids are chosen from chernozems (FLAIG, SCHEFFER, and KLAMROTH [1955]) for comparison with the results published in the literature. This sample was prepared by 0.5% hydrochloric acid pretreatment and extracted with 0.5% sodium hydroxide. After centrifugation and several reprecipitations and dialyses, these raw humic acids were subfractionated with 2N NaCl at pH7. After centrifugation of this sample, a gray humic acid fraction (G fraction) was precipitated, while a dark-colored brown humic acid fraction (B fraction) remained in solution. After acid treatment and dialysis, the G fraction showed an ash content of 10%, while the B fraction had only 3.5% of inorganic constituents.

A comparison of the infrared spectra of the raw humic acids (Figure 61 (1)), the gray humic acid fraction (Figure 61 (2)), and the brown humic acid fraction (Figure 61 (3)) reveals the main differences in intensity of absorption bands in the region at 9.72 μ (1029 cm^{-1}), which is assigned to the Si—O—Si vibration frequencies of a complex silicate mineral component.

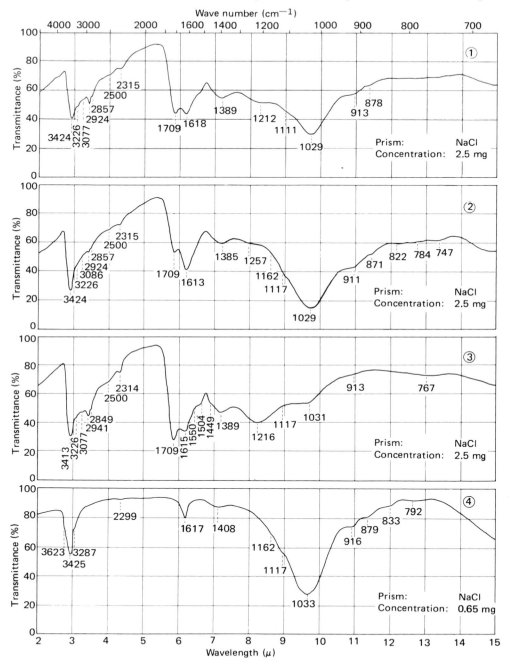

Figure 61. Infrared spectra of humic acids from chernozem. (1) Original; (2) gray humic acid fraction; (3) brown humic acid fraction; (4) standard montmorillonite.

The specific investigations of the gray humic acid fractions by infrared analysis and electron microscopy gave evidence of the presence of a smectite mineral of the montmorillonite type, which is shown in case of standard montmorillonite (Figure 61 (4)). This clay mineral shows medium absorption bands at 2.92 μ (3425 cm^{-1}) due to lattice hydroxyl and interlayer water vibrations, free water at 6.20 μ (1613 cm^{-1}), and the strong vibrations of the silicate anion at 9.68 μ (1033 cm^{-1}), with some minor inflections at 8.60 μ (1162 cm^{-1}), 8.95 μ (1117 cm^{-1}), 10.92 μ (916 cm^{-1}), 11.38 μ (879 cm^{-1}), and 12.00 μ (833 cm^{-1}), which partly arise from the vibrations of the hydrargillite interlayers.

The nearly ash-free brown humic acid fraction (Figure 61 (3)) shows a strong absorption in the range of shorter wavelengths, 2.93 μ (3413 cm^{-1}), which is assigned to hydrogen-bonded hydroxyl groups. Some medium absorption intensities are assigned to intermolecular-bonded hydroxyl groups at 3.10 μ (3226 cm^{-1}) and for aromatic C—H stretching vibrations at 3.25 μ (3077 cm^{-1}). The maxima of absorption at 3.42 μ (2941 cm^{-1}) and 3.51 μ (2849 cm^{-1}) are defined by stretching vibrations of aliphatic CH, CH$_2$, and CH$_3$ groups of side chains of aromatic nuclei. The broad band of low intensity at 4.0 μ (2500 cm^{-1}) may be assigned to the O—H stretching vibrations of bonded hydroxyl in carboxylic acids, while the low intensity at 4.32 μ (2314 cm^{-1}) is possibly due to minor contents of charged amine derivatives or amino acids.

The most characteristic feature of humic acid spectra is given by the strong but relatively broad absorptions of the C=O stretching vibrations of several functional groups, which may be present in preparations of humic substances such as carboxylic acids, cyclic and acyclic aldehydes, and ketones at 5.85 μ (1709 cm^{-1}) and of the C=C stretching vibrations of double bonds in aliphatic and aromatic compounds at 6.19 μ (1615 cm^{-1}), also pointing to different types of benzene ring substitution. A comparison of the differences of the C=O vibration absorption of the samples shows the influence of the fractionation. In the gray humic acid fraction a decrease of the intensity of the carbonyl absorption occurs, while the brown humic acids are enriched in functional groups containing carbonyl or carboxyl. On the other hand, the intensity of the C=C stretching vibration band at 6.19 μ (1615 cm^{-1}) is nearly unchanged.

In the direction of longer wavelengths there are some minor inflections at 6.45 μ (1550 cm^{-1}) and 6.65 μ (1504 cm^{-1}) which may develop from C=C, as well as from C=N stretching vibrations of possible nitrogen-containing heterocycles. The small shoulder at 6.90 μ (1449 cm^{-1}) is generally assigned to C—H deformation vibrations of methyl or methylene groups. The broad band of medium intensity with an absorption maximum at 7.20 μ (1389 cm^{-1}) is derived from salts of carboxylic acid groups, while the band at 8.22 μ (1216 cm^{-1}) is due to C—O stretching vibrations of esters, ethers, and phenols. The shoulder at 9.70 μ (1031 cm^{-1}) points to minor contents of the above-mentioned silicate mineral. In the range of wavelengths above 10.0 μ (1000 cm^{-1}) there is no absorption band, which might be assigned to vibration frequencies of further functional groups. This is in agreement with the results of KUMADA and AIZAWA [1958], who studied the infrared absorption properties of humic substances from various soil origins.

2. Changes of Infrared Absorption of Humic Acids by Formation of Derivatives

In order to get some knowledge about the variety of functional groups in humic acids several authors tried to combine chemical and infrared studies (FARMER and MORRISON [1960], WAGNER and STEVENSON [1965], and SCHNITZER [1965]). The first attempts in this direction were made by CEH and HADZI [1956] and BROOKS, DURIE, and STERNHELL [1958], using humic acid samples derived from coal.

Humic acids that had been fractionated by solvent extraction from phragmites peat were investigated by FARMER and MORRISON [1960]. They studied the changes in the infrared absorption following methylation and reduction in order to get information on the acidic unsaturated structures. They found, agreeing with FORSYTH [1947a], that about 40% of the humic acids were converted by alkaline hydrolysis to water-soluble products, presumably sugars and amino acids. This was shown by a marked drop in absorption near $9.50\,\mu$ ($1052\,cm^{-1}$), ascribed to a loss of carbohydrates. A drop near $6.05\,\mu$ ($1653\,cm^{-1}$) and $6.50\,\mu$ ($1538\,cm^{-1}$) was ascribed to the loss of peptides. According to their results, the C=O absorption band at $5.85\,\mu$ ($1709\,cm^{-1}$) must originate principally from carboxylic acids, since this band is absent in the spectrum of sodium salts of both humic acids and alkaline hydrolyzed humic acids. By treatment with sodium-borohydride ($NaBH_4$), a reagent that reduces most aldehydes and ketones but not carboxyl groups, the intensity of this band was found to be reduced to only 10%. The authors assume that the changes are a contribution of strongly hydrogen-bonded α, β-unsaturated carbonyl groups.

BAILEY et al. [1965] explained that the changes of absorption in the infrared spectra of brown-coal humic acids in this region are due to hydrogen bridge linkages between the hydrogen of phenolic hydroxyl groups and the carbonyl groups of quinones.

Similar relationships were found by FLAIG and SALFELD [1959], who made an extensive infrared investigation of the relationship between the relative intensity of the C=O and C=C stretching bands and the substitution. In the case of monomethoxy-p-benzoquinones with different numbers of methyl groups substituted in various positions, there is a dependency of relative intensity of these absorption bands on the position of one or several methyl groups. They found that in the case of those quinones with methoxyl groups adjacent to a hydrogen atom, a decrease resulted in the intensity of the absorption band in the longer wavelengths region (C=C), while those with a methyl group adjacent to the methoxyl group produced a decrease in intensity in the range of shorter wavelengths (C=O). In addition to this effect, a shift of the C=C bands occurred in the direction of shorter wavelengths, with increasing substitution with methyl groups.

An alternative explanation for the decrease in absorption near $6.20\,\mu$ ($1613\,cm^{-1}$) is given by FARMER and MORRISON [1960] as an additional reaction of diazomethane to the double bond in β-unsaturated esters, which is a known reaction in organic synthesis (EISTERT [1948]).

FLAIG, SALFELD, and RABIEN [1963] found that some substituted hydroxy-p-benzo-

*R = aliphatic rests

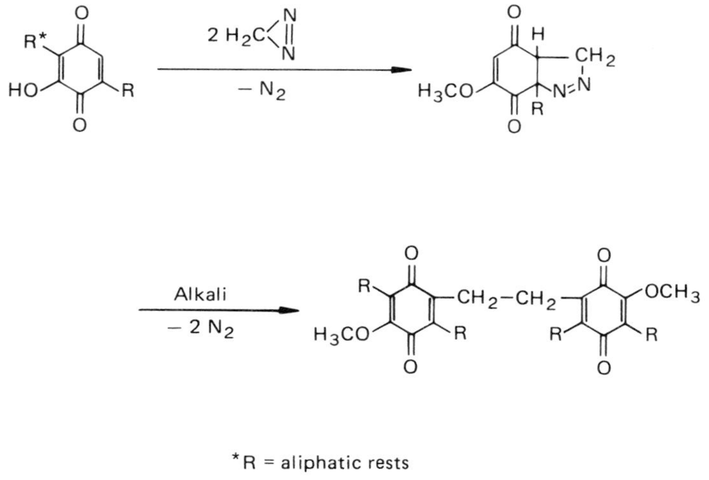

Figure 62. Reaction of substituted hydroxy-p-benzoquinones with diazomethane and treatment with alkali.

quinones react not only to form methoxyquinones but also in different ways with diazomethane. In some cases substituted hydroxybenzoquinones react to form diketo-tetrahydro-indazoles. These are transformed in weak alkaline solution to 1,2-di (benzoquinone-2′,5′)-ethane by splitting off nitrogen (Figure 62).

Therefore, it may be assumed that the decrease in intensity of the C=O stretching band is not always caused by methylation of hydroxyl groups of quinones, with diazomethane, but also by addition reactions, which can shift this band toward shorter wavelengths.

For the determination of the number of the arrangements of functional groups WAGNER and STEVENSON [1965] subjected humic acids (Figure 63 (a)) from brunizem soil to methylation,

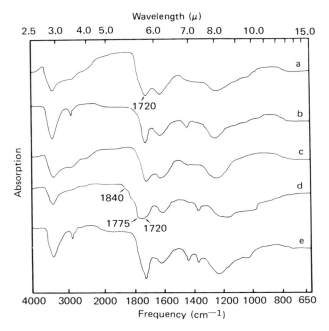

Figure 63. Infrared spectra of humic acids and derivatives (WAGNER and STEVENSON [1965]).
(a) Parent humic acid; (b) methylated with diazomethane; (c) product of (b) saponified and reacidified; (d) product of (c) refluxed in acetic anhydride for 24 hours; (e) product of (b) refluxed in acetic anhydride for 24 hours.

saponification and acetylation. By treatment with diazomethane (Figure 63, (b)), they found an increase in intensity of the C=O stretching vibrations at 5.80 μ (1720 cm^{-1}) and a sharpening of the C—O stretching vibration at 8.0 μ (1250 cm^{-1}) as a result of ester formation. The addition of the methyl groups, furthermore, caused the introduction of bands in the region C—H stretching vibrations at 3.42 μ (2920 cm^{-1}) and of hydrogen-bonding vibrations at 6.94 (1440 cm^{-1}).

After saponification of the methylated humic acid by heating with 1 N NaOH for two hours and reacidification, a demethylated product (Figure 63 (c)) was isolated, whose characteristic features are similar to those of the initial material. Treatment of this sample by a 24-hour reflux with acetic anhydride (Figure 63 (d)) led to the appearance of at least four new bands in the region of the C=O absorption vibrations, with approximate bands at 5.45 μ (1840 cm^{-1}), 5.50 μ (1815 cm^{-1}), 5.63 μ (1775 cm^{-1}), and 5.75 μ (1740 cm^{-1}). According to WOOD, MOSCHEPEDIS, and DEN HERTOG [1961], the new bands might be explained as C=O vibrations of phenol acetates and anhydrides. After reacting the unsaponified but diazomethane-treated sample for 24 hours with acetic anhydride (Figure 63 (e)), the spectra showed a strong band at 7.33 μ (1365 cm^{-1}) assigned to the C—CH$_3$ vibration of the added acetyl groups and several faint bands as shoulders in the region of the carbonyl vibration at about 5.8 μ (1720 cm^{-1}). Because they found no bands typical for phenol acetate in the region of the C—O vibration at 8.35 μ (1200 cm^{-1}), they concluded that there was no direct evidence for the presence of unmethylatable but acetylatable C=O groups as in quinones. The titration of the original humic acids to pH 9 with dilute sodium hydroxide resulted in almost complete loss of absorption of the strong C=O absorption band at 5.8 μ (1720 cm^{-1}), indicating that this vibration is due almost entirely to the occurrence of carboxyl groups.

SCHNITZER [1965] studied the modification of functional groups by formation of derivatives and selective blocking of the functional groups in fulvic acids from a podzolic soil. As a result of this and of some other investigations, SCHNITZER and GUPTA [1964] and SCHNITZER and SKINNER [1965a, b] came to the conclusion that the most striking difference between the spectra of the humic acids and fulvic acids is the intensity of the C=O band at 5.80 μ (1785 cm^{-1}). It is very strong in fulvic acids because of a higher content of carboxyl groups; for humic acids it is only a shoulder. According to SCHNITZER [1965] the similarity between the infrared spectra of fulvic acids and humic acids indicates essentially similar chemical structures, which differ mainly in the number of carboxyl groups.

By acetylation of fulvic acids (Figure 64 (1)), a product was formed (Figure 64 (2)) that showed some decrease of the hydroxyl content at 2.94 μ (3400 cm^{-1}), and no change in the region of aliphatic absorption at 3.43 μ (2920 cm^{-1}) and 3.50 μ (2860 cm^{-1}). Due to the formation of phenolic and alcoholic acetates, a shoulder at 5.63 μ (1775 cm^{-1}) appeared, which was in the region of the C=O absorption of anhydrides. A new band at 7.29 μ (1375 cm^{-1}) as C—H deformation vibration of C—CH$_3$ groups and a strengthening of the absorption near 8.33 μ (1200 cm^{-1}) were observed.

Methylation (Figure 64 (3)) caused a slight decrease in the range of the hydroxyl vibrations

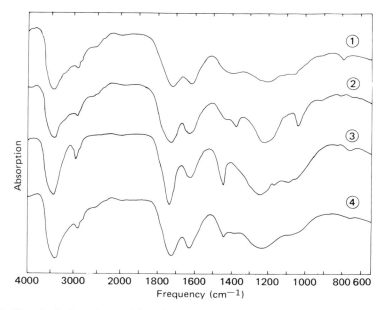

Figure 64. Chemical alterations of functional groups in fulvic acids and their effects on infrared spectra (according to SCHNITZER [1965a]). (1) Untreated fulvic acids; (2) twice-acetylated fulvic acids; (3) twice-methylated fulvic acids; (4) esterified fulvic acid.

at 2.94 μ (3400 cm^{-1}) and an increased aliphatic C—H absorption at 3.38 μ (2960 cm^{-1}), 3.51 μ (2850 cm^{-1}), and 6.92 μ (1445 cm^{-1}). Marked changes were an increased absorption of the C=O carbonyl at 5,80 μ (1720 cm^{-1}) and a decrease of absorption in the region of the C=C vibrations at about 6.20 μ (1600 cm^{-1}). This might be caused by:

(a) a shift of the absorption of hydrogen-bonded carbonyl groups from 6.12 μ (1630 cm^{-1}) to 5.82 μ (1720 cm^{-1}) after methylation; or

(b) C=O groups that absorb originally at 6.14 μ (1630 cm^{-1}) and now after methylation and esterification, about 5.82 μ (1720 cm^{-1}).

Esterification of the original fulvic acids (Figure 64 (4)) increased the intensity of the C=O band at 5.80 μ (1725 cm^{-1}) and reduced absorption at 6.13 μ (1630 cm^{-1}), so that a participation of symmetric and asymmetric stretching modes of the carboxylic acid groups at 6.13 μ (1680 cm^{-1}) seems to be a possible explanation of the effects of methylation.

In order to distinguish the C=C vibrations of double bonds of the benzene ring at about 6.25 μ (1600 cm^{-1}), ORLOV, ROZANOVA and MATYUKHINA [1962] studied the effects of chemical alterations on infrared spectra of humic acids from deep chernozem (on the steppes of Russia) after treatment with various oxidants. They found that the intensity of absorption in this region decreases somewhat after cold treatment with a dilute solution of potassium permanganate. This result points to the presence of double bonds of the olefinic type. Because the carbonyl absorption is absent in permanganate-treated samples, the authors assume that the band with a maximum at 6.20 μ (1613 cm^{-1}) is a complex absorption, where the C=C absorption and that of the ionized carboxyl group coincide. These results agree with those by treatment with peroxide (3 and 6%), where a strong absorption at 6.12 μ (1630 cm^{-1}) occurs, while the carboxyl vibrations at 5.90 μ (1695 cm^{-1}) disappear completely by increased oxidation. Similar effects were found by oxidation with 2% nitric acid.

3. *Comparative Investigations of Infrared Absorption of Humic Acids from Various Soil Types*

The first comparative investigations of the differences in infrared absorption of a great number of soil humic acids from various soil types were made by KUMADA and AIZAWA [1958]. The humic acids were extracted by a hot 0.5% sodium hydroxide solution from the following soils: muck soil, calcareous soil, alluvial soil, peat soil, and podzol soil. From their infrared patterns, the authors classified these humic acids into four types: AA_1, AA_2, B, Rp, and P. This corresponds to their absorption properties in the visible and ultraviolet ranges of light (see page 106) and to the soil type.

The main differences of infrared absorption are given in the case of some selected samples in Figure 65 and Table 23. According to these results humic acids, which belong to the A-type and are subdivided into AA_1 (isolated from volcanic ash or muck soils) and AA_2 (from calcareous soils), show relatively strong aromatic and aliphatic stretching vibrations at 3.25 μ (3077 cm^{-1}) and 3.4 μ (2941 cm^{-1}), respectively.

The same high values of optical density are found in the region of the C=O stretching vibrations at 5.9 μ (1695 cm^{-1}) and of C=C stretching vibrations at 6.2 μ (1613 cm^{-1}). The other C=C stretching vibrations at 6.1 μ (1639 cm^{-1}) and 6.2 μ (1613 cm^{-1}), which are found in the spectra of B- and Rp-types, are absent in AA_1, and are only a shoulder in AA_2.

In the region of the aliphatic C—H vibrations at 6.8 to 7.3 μ (1470 to 1370 cm^{-1}) are two broad bands. There is one broad band at 7.8 to 8.3 μ (1281 to 1205 cm^{-1}), where the vibrations of the oxygen-containing groups are found. Characteristic to all investigated samples is the complete failure of any distinct absorption band beyond 10 μ (1000 cm^{-1}).

The pattern of the B- and Rp-types in the region below 7.5 μ (1333 cm^{-1}) is somewhat analogous to that of lignin, supporting the hypothesis that humic acids belonging to this type originate from lignin. The infrared absorption properties of the P-type humic acids are similar

Table 23. Optical densities of the main absorption bands in the infrared spectra of soil humic acids, according to KUMADA and AIZAWA [1958].

Type	Soil type	Humic acids	$3.25\ \mu$	$3.4\ \mu$	$\dfrac{D_{3.25\mu}}{D_{3.4\mu}}$	$3.8\ \mu$	$5.9\ \mu$	$6.1\ \mu$	$6.2\ \mu$	$\dfrac{D_{5.9\mu}}{D_{6.2\mu}}$	$6.6\ \mu$	8.0 $8.3\ \mu$	$9.8\ \mu$
AA$_1$	Muck soil	Tanemori (M)	0.542	0.497	1.09	0.325	1.124	—	0.938	0.14	—	0.889	0.274
AA$_2$	Calcareous soil	Jagaru	0.365	0.398	0.92	0.245	0.383	—	0.755	1.11	—	0.720	0.364
B	Alluvial soil	Busshozan	0.345	0.379	0.91	0.151	0.717	0.862	0.851	0.84	0.539	0.696	0.411
Rp	Peat soil	Tanemori (P6)	0.437	0.409	1.07	0.196	0.753	0.745	0.907	0.83	0.489	0.778	0.394
P	Podzol soil	Kinpo	0.369	0.358	1.03	0.160	0.779	—	0.928	0.84	0.503*	0.634	0.436

(Columns $3.25\ \mu$ through $9.8\ \mu$ fall under the heading: Optical density (D))

* The 6.5 μ band.

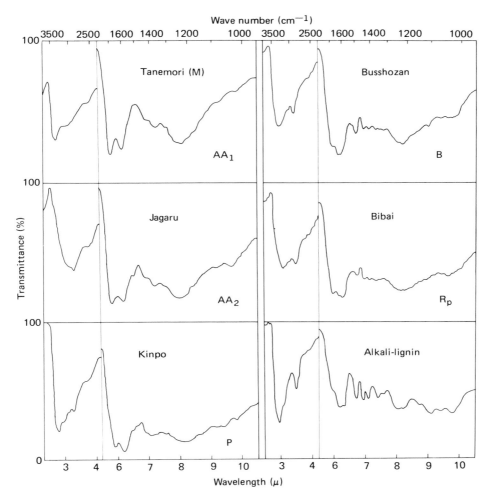

Figure 65. Infrared spectra of soil and peat humic acids and alkali lignin (according to KUMADA and AIZAWA [1958]).

in shape to that of A-type, but the intensities at 3.25 μ (3077 cm^{-1}), 3.4 μ (2941 cm^{-1}), and 3.8 μ (2631 cm^{-1}) are relatively weaker, and there can be found a shoulder-like absorption band near 6.5 μ (1538 cm^{-1}). Otherwise, the pattern shows a fairly strong absorption in the region of the C—O stretching vibrations of esters, ethers, and phenols at about 8.5 to 9.7 μ (1176 to 1031 cm^{-1}), as in case of B- and Rp-types. The authors conclude that the most significant change in absorption bands of soil humic acids in the course of humification is the increase of intensity of the bands of oxygen-containing groups, accompanied by an increase of the bands of aromatic groups and by a decrease of bands of the aliphatic side chains of the humic acids.

ORLOV, ROZANOVA, and MATYUKHINA [1962] studied the infrared absorption of humic acids which were isolated by alkali extraction from the decalcified humus-accumulation horizon (A$_1$) of the major soil groups, such as sod-podzolic soil, gray forest soil, deep chernozem (steppe), ordinary chernozem, humus calcareous soil, meadow bottom solonetz, and red-colored soil. By comparison of the infrared spectra, they found that humic acids from various soils are characterized by a common structural pattern, which is determined by the occurrence of the benzene ring vibrations with a high degree of substitution. These characteristic properties are most pronounced in humic acids from chernozem (compare KASATOCHKIN, KONONOVA, and ZILBERBRAND [1958]).

All the samples absorb in the region of the carbonyl vibrations at about 5.8 μ (1725 cm^{-1}), indicating the occurrence of carboxylic acid groups or of carbonyl groups in ketones or aldehydes. Only in case of the humic acids from sod-podzolic soil was a splitting of a carbonyl absorption band noted, with a maximum intensity at 5.85 μ (1709 cm^{-1}) and 5.93 μ (1686 cm^{-1}). This points to the presence of carbonyl functions in different combinations, possibly caused by a simultaneous occurrence of aliphatic and aromatic carboxyl groups or of quinones. Vibrations of terminal CH_3 and CH_2 groups were found only in samples extracted from forest soils with a relatively reduced content of substances of aromatic structure.

Humic acids prepared by alkaline extraction and dialysis from various soils were studied in a comparative infrared analysis by SCHARPENSEEL, KÖNIG, and MENTHE [1964]. According to their results, the spectra of the preparations from pseudogley, podzolic soil, brown earth, podzol, and chernozem show no representative differences but show distinct changes of the relative intensities of some absorption bands. Otherwise, gray and brown humic acid fractions reveal no marked differences. The gray humic acids are somewhat poorer in absorption bands, but the bands are not easily detectable.

TOKUDOME and KANNO [1968] studied the nature of the humus of humic allophane soils, altered humic allophane soils, red-yellow soils, and peat soils in Japan by elementary analysis, X-ray diffraction, and infrared and light absorption spectroscopy. They found that the humic acids from altered humic allophane soils, which are characterized by a relatively high C_h/C_f ratio of 3.14 (C_h/C_f = ratio of the amounts of humic and fulvic acids expressed as percent of the total organic carbon), and are distinguished from humic allophane soils by the predominance of aluminum-vermiculite and kaolinite minerals in their clay fractions, show a high degree of aromaticity according to the infrared spectrograms.

In contrast, the samples from red-yellow soils with low C_h/C_f ratios (0.4 to 0.8) are characterized mainly by aliphatic groups. Japanese peat soils whose humus of the uppermost layer had a C_h/C_f ratio of 1.2 to 2.8 and whose layers directly beneath had C_h/C_f ratios of 2.9 to 5.8, were investigated by infrared analysis. In all cases the humic acids had a relatively low degree of aromaticity and therefore they resemble the red-yellow soils.

From all the above comparative infrared spectroscopic studies of effects of chemical treatment of fractions of humic substances and the results of infrared analysis of humic acids isolated from various soils, including those from coals (BROWN [1955a, b], CEH and HADZI [1956]), it may be concluded that these humic substances are similar in general structure. The most characteristic features are the aromatic nucleus, probably containing heteromoieties, and the carboxyl groups. There is some evidence that the carboxyl group is conjugated or directly bound to the aromatic nucleus, because the absorption of terminal CH_3 or CO_2 groups is found to be very small. Furthermore, the investigations point to the presence of different types of carbonyl functions. They are influenced by different combinations with other functional groups. Generally, the main differences in infrared absorption may be caused by different degrees of condensation during formation of humic substances.

There are some hints from infrared absorption studies that the basic principle of structure of humic acids is influenced by the soil type in connection with the regional environmental conditions. Probably, infrared spectroscopic investigations of humic acids will become more effective for the elucidation of the structure of humic acids if more detailed fractionation precedes the application of further physicochemical methods.

III. Characterization of Humic Acids and Their Fractions by Thermoanalytic Methods

While in coal research the investigation of organic substances by thermal decomposition during pyrolysis plays an important role for the characterization of coals and coalification (see VAN KREVELEN, HEERDEN, HUNTJES [1951], CHERMIN and VAN KREVELEN [1957], LUTHER [1960], LUTHER, ABEL, KETT, TRAENCKNER [1960], WELTNER [1961], KRÖGER and BRÜCKNER [1961a, b], ABEL and LUTHER [1962], ABEL [1965], LUTHER, EISENHUT, and ABEL [1966]), thermoanalytic methods for the characterization for humic acids and their fractionation products have been used by some authors (STADNIKOFF, SSYSKOFF, and USCHAKOVA [1936], BALINT and TREIBER [1959], VALEK [1962], TURNER and SCHNITZER [1962], NETZSCH-LEHNER and KLEE [1962], SCHNITZER and HOFFMAN [1964, 1965, 1966, 1967], BELKEVIC and GAJDUK [1964], SCHARPENSEEL, KÖNIG, and MENTHE [1964], NAUCKE [1967, 1968]). That was because until recent years there was only a small amount of information about the reaction kinetics and mechanism of thermochemical processes, and on the other hand, suitable supplementing methods such as gas chromatography and mass spectrometry for comparative investigation, were still being developed for application to thermoanalytic methods. Therefore, the interpretation of the results of investigations are mainly limited to a qualitative description of the thermal degradation curve. SCHARPENSEEL, KÖNIG, and MENTHE [1964], for example, found by investigations of the characterization of several types of humus, which were made in order to find analytical differences of humic substances from soils with the aid of differential thermal analysis (DTA), that making a distinction between gray and brown humic acids by typical exothermic peaks is nearly impossible. Thus, the elucidation of pedogenic differences of humic acids from different origin for determination of genetic types with respect to thermoanalytic data is difficult and questionable. These difficulties are reduced by application of thermogravimetric methods to the investigation of the pyrolytic degradation reactions of organic soil substances. Therefore, the following discussion concerns only this subject.

1. *Principles of the Thermogravimetric Method*

While by differential thermal analysis (DTA) it is possible to determine only phase transformations of the samples by heat of reactions, the thermogravimetric method (TG) enables researchers to detect and record quantitatively the alterations of the sample weights continuously during the course of heating in pyrolysis, with the aid of a thermobalance. Because of the possible overlapping of the decomposition temperatures during the investigation of pyrolysis of polycomponent systems, today the differential thermogravimetric method (DTG) is generally applied. This method permits the direct recording of the first derivative of the thermogravimetric curves. DTG curves may also be derived from thermogravimetric curves by plotting the weight differences per time unit versus temperature.

VAN KREVELEN, HEERDEN, and HUNTJES [1951] investigated in detail the kinetics of pyrolysis. They found that the determination of the weight loss during the constant rate of heating is more suited for the characterization of different samples than the determination of weight loss at constant temperature per time unit.

With the assumption that the reaction of pyrolysis of coal substances and coal precursors is a reaction of the first order, they demonstrated that if degasification maxima occur, the maximum temperature and the half-width value of the degasification peak is characteristic for the substances investigated. Beyond this, they developed a mathematical theory for the calculation of the reaction velocity constant and the heat of activation from the degasification curves, resulting from a constant rate of heating.

The kinetic theory of the thermal decomposition of organic substances extracted from soils was proved and extended by TURNER and SCHNITZER [1962] by application of preparations from podzol. They found that the energy of activation of the main decomposition reaction of humic substances derived from the A_0 horizon of a podzol is in the same order as that found by VAN KREVELEN, HEERDEN, and HUNTJENS [1951] in brown coals.

The energy of activation of the main decomposition reaction of fulvic acids isolated from the B_h horizon of a podzol is also in agreement with the results published by the above-mentioned authors for bitumen coal of low quality.

2. *Bonding Energy and Thermostability of Functional Groups*

By thermoanalytic investigations of model substances (VAN KREVELEN, HEERDEN, and HUNTJES [1951]), it follows that in the case of increasing substitution of hydrogen atoms by aromatic rings in aliphatic linear polymers, the value of the main decomposition temperature and the energies of activation decreases, while the half-width value of the decomposition peaks remains nearly constant. In contrast, a polycondensate of cyclohexanone shows similarly to cellulose a larger value of the half-width value of the main decomposition maximum. Dibenzanthrone as a model substance of a completely condensed molecule is very thermo-resistant and results in only a small half-width value of the main decomposition peak. From these facts it may be gathered that the thermostability is strongly decreased by the presence of oxygen linkages. The thermal degradation will occur in narrow ranges of temperature. The number of peaks will depend on the uniformity and simplicity of the molecular types of bonds and functional groups. Therefore, the thermal decomposition of a nonuniform substance will depend on the relative content of components, which occur in the molecule. The complexity of decomposition reactions will increase, corresponding to the number of linkage types and

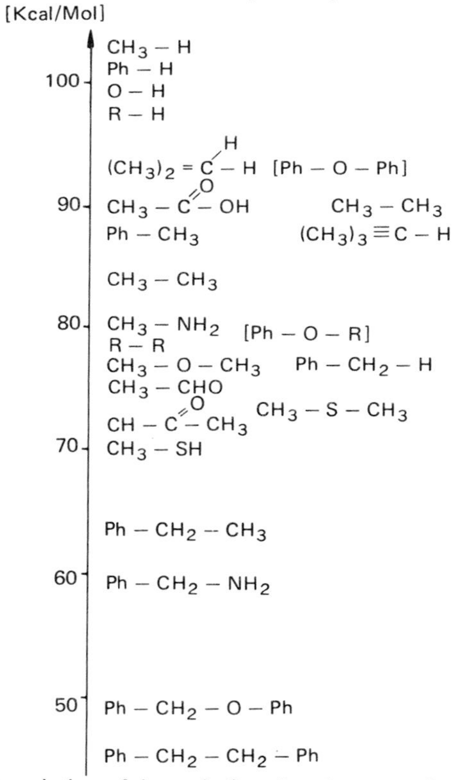

Figure 66. Energies of dissociation of the main functional groups of soil organic matter substances.

the variations of the different radicals and atom groups, so that decomposition reactions will overlap each other (compare PAULIK and WELTNER [1958]).

However, apart from the knowledge of a structural formula for humic acids, it is possible by investigation of the thermal decomposition to make some general remarks about the type of linkages of carbon, hydrogen, and oxygen atoms, which occur in the macromolecule. This should allow evaluation of the energies of dissociation of components of the complex molecule.

According to a comparison of the energies of dissociation (Figure 66) of the different types of —C—C—, —C—H, —C—OH, and C—O— linkages, as well as O— and N— linkages, which may occur in the complex heterocyclic structure of humic acids (compare LUTHER [1960]), it may be supposed that heteroatoms will decrease the thermostability of aliphatics. On the other hand, the cleavage of hydrogen atoms from aromatic rings and methane will answer for the highest thermal energies, which should appear only at high temperature ranges.

From these kinetic data an overlapping of several degasification ranges during the pyrolysis of humic acid is expected. Because of changes in relative amounts of the residual functional groups dependent on the experimental conditions of extraction, difficulties in the reproducibility of the thermal degradation reaction of these substances may be normal. A further difficulty of the correct interpretation of the thermal decomposition is given by possibly occurring autocatalytic influences of cleaved radicals, from which thermal polymerizations may result, so that intermediates will be omitted. Therefore, the assignment of maxima in the degasification curves in relation to the decomposition of distinct structural components is troublesome. The investigation of the thermal decomposition of soil organic substances will therefore be reduced to the problem of the suitable application of comparative chemical and physicochemical methods to the investigation of the residues of heating and the released degasification products. In spite of this, it should be possible to study quantitative relationships

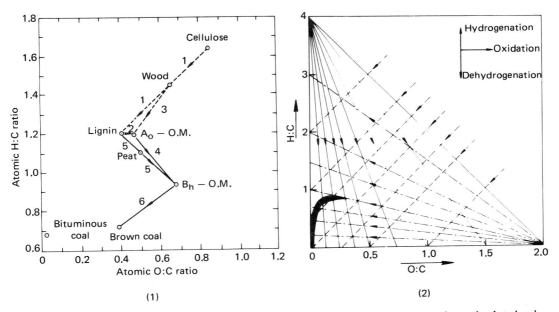

Figure 67. (1) Atomic H:C versus O:C diagrams for lignin, humic compounds, and related substances. According to SCHNITZER and HOFFMANN [1965]. (2) Atomic H:C versus O:C diagram. According to VAN KREVELEN [1950]. ———— demethylation –.–.–. decarboxylation ————— dehydration.

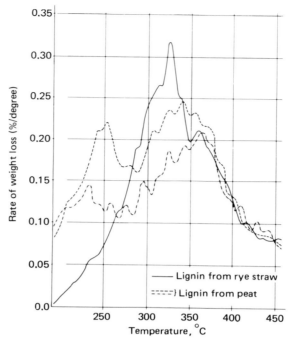

Figure 68. Differential thermogravimetric curves of lignin from several sources, according to ABEL [1965].

between the course of the degasification curves and the amount of distinct components by application of thermogravimetric methods.

3. *Genetic Relations between Lignin, Humic Acids, and Different Products of Coalification*

According to VAN KREVELEN [1950], the investigation of the reactions occurring during the course of the thermal degradation of organic substances is made by plots of the atomic ratios H:C versus O:H, obtained from elementary analysis of the residues of the degradation at the corresponding temperatures. This so-called graphical-statistical method distinguishes between decarboxylation, demethylation, dehydration, dehydrogenation, and oxidation, which occur during the thermal decomposition of organic substances in pyrolysis (Figure 67(2)).

SCHNITZER and HOFFMANN [1965] used for comparison the H:C versus O:C ratios of several untreated organic soil extracts and some related substances of the coalification series to establish genetic relationships between these groups of substances, using the methods of VAN KREVELEN [1950]. Figure 67(1) shows that lignin might be formed by dehydration from cellulose (1),* while humic acids of the A_0 extracts of podzol could be derived primarily from lignin (2) by oxidation and formation of carboxyl groups (in agreement with the results of the investigations made by FLAIG, SCHOBINGER, and DEUEL [1959] by demethylation), but also directly from wood (3) by dehydration. The formation of fulvic acids derived from extracts of the B_h horizon of a podzol might be possible by demethylation and oxidation of the A_0 material (4). In a similar way, the fulvic acids might be directly formed from lignin and peat intermediates (5). The alteration of the B_h material to brown coal includes a dehydration (6), which can be confirmed by heating to temperatures above 500°C. Since the structure and pyrolytic reactions of lignins are quite well known, hints of the nature of humic acid structural

---

* The numbers between brackets refer to the pathway of the assumed alterations.

units can be obtained by comparing the pyrolytic reactions of these two groups of substances.

4. *Thermal Degradation Properties of Lignin*

PAULIK and WELTNER [1958] first published derivative thermogravimetric curves obtained from lignin. They investigated straw lignin between 150 and 450°C. They found low peaks from 200 to 290°C, with the main decomposition peak appearing at 350°C. They explained the exothermal reaction occurring in the temperature range of the main decomposition as a formation of a more thermoresistant, possibly strongly condensed aromatic structure. ABEL [1965] investigated, by thermogravimetric analysis, the degradation of rye straw and peat samples from various origins. He found that the thermal degradation is spread over a wide range of temperatures from 250 to 400°C. Lignin from rye straw gave distinctly recognizable peaks at 285, 315, 330, and 360°C, as well as shoulders of further single reactions at 250 and 390°C (Figure 68).

DTG curves of lignins from peat are most differentiated between 230 and 360°C. Furthermore, they show certain similarities in the properties of decomposition.

URBANSKI, KUCZINKSKI, HOFMAN, URBANIK, and WITANOVSKI [1959] studied, by infrared spectroscopy, the residues of the thermal decomposition of lignin. They found that phenolic hydroxyl groups are split off between 205 and 275°C, while the cleavage of phenolic hydroxyls and aromatic ethers occurs between 245 and 375°C. The carbonyls gave a band at 325°C. Alcoholic hydroxyl and aliphatic ether groups are decomposed above 350°C, a temperature range where the condensation of aromatic groups is predominant.

NAUCKE [1967, 1968] published a comparison of the results of experiments on the degasification of different substances present in peat. According to the temperature ranges of the main decomposition, degasification reaction follows the order:

150–225°C	water-soluble moieties, hydroxycarboxylic acids, and sugars
225–260	sugars and hemicellulose
260–290	hemicellulose
290–315	cellulose
315–350	lignins (humic acids)
350–390	humic acids (bitumens)
390–450	bitumens (humic acids)

5. *Characterization of Humic Acids and Their Fractions by Differential Thermogravimetric Analysis (DTG)*

Although humic acids and lignin might possibly contain similar structures, humic acids were more thermoresistant in thermogravimetrical investigations. PAULIK and WELTNER [1958] found by pyrolysis of humic acids from peat that the decomposition starts at lower temperature ranges, where the changes probably arise from reactions such as dehydration and decarboxylation. The cleavage of terminal groups accompanied by the cleavage of acetyl groups and demethylation of methoxyls with formation and release of acetic acid, methanol, and methane also starts in the lower temperature ranges. Accordingly, decompositions of organic structures take place by stepwise release of phenols and their homologues and greater amounts of H_2O, CO_2, and CO.

The DTG curves (Figure 69) show main decomposition reactions at 340 to 370°C and 400 to 420°C. The temperature range above 450°C is assigned to the decomposition reactions of aromatic condensation products with lower contents of oxygen but proportionally higher contents of carbon and hydrogen.

According to the thermal decomposition curves, the differences in composition result

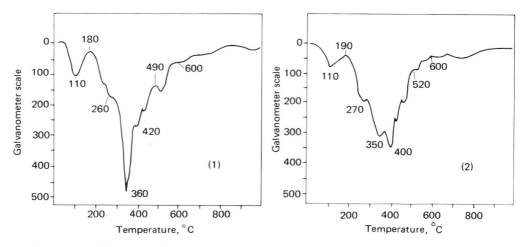

Figure 69. DTG curves of humic acids from peat, according to PAULIK and WELTNER [1958].

from different degrees of peatification, which is higher in case of sample 2 (compare Figure 69). There is a broad analogy in the relative position of the temperature values of maximum degasification, which points to a general similarity in the composition of the main components of the samples of humic acids.

SCHNITZER and coworkers published some thermogravimetric results on the decomposi-

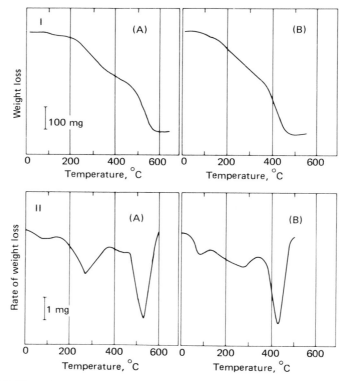

Figure 70. TG (I) and DTG (II) curves of (A) organic matter (humic acids) and (B) B_h organic matter (fulvic acids) of a podzol. According to SCHNITZER and HOFFMANN [1961].

tion of organic soil extracts derived from the A_0 and B_h horizons of a podzol (TURNER and SCHNITZER [1962], SCHNITZER [1965], SCHNITZER and SKINNER [1963, 1964], SCHNITZER and HOFFMANN [1965, 1966]). They found very similar curves from the thermal degradation of both the samples (Figure 70).

In the material from the A_0 horizon (humic acids) two characteristic peaks of the main deomposition reaction at 280 and 540°C could be determined, while in the material of the B_h horizon (fulvic acids) only a weak thermal reaction at 270°C was obtained. The main decomposition reaction reached a maximum at 420°C. By chemical and spectroscopic analyses of the residues of thermal decomposition taken at different temperatures, Schnitzer and his coworkers established that the decarboxylation of the A_0 sample started above 150°C, and seemed to be completed at 250°C. The decomposition of phenolic hydroxyl groups started at 150°C and reached a maximum at 200°C. Both functional groups were completely released at 400°C. The sample from the B_h material was more resistant to decarboxylation and dehydroxylation. Therefore, the cleavage of carboxyl groups started at temperatures above 250°C and was completed at 450°C. The elimination of hydroxyl groups increased up to 300°C and rapidly decreased to the final temperature of 450°C. According to the results from investigations on the oxidative degradation of these substances (SCHNITZER and WRIGHT [1960], SCHNITZER and DESJARDINS [1964]), the A_0 material contains both aliphatic and aromatic groups, and the B_h material is predominantly aromatic. The differences in the decomposition rates are related to the different states of aromaticity. This might explain the more pronounced thermostability of the sample derived from the B_h horizon.

From stepwise examination of the residues of decomposition by elemental analysis, and by plotting cleaved amounts of carbon, hydrogen, and oxygen against the loss of weight, it can be demonstrated that a very similar course of the degradation reaction occurred for both of the A and B organic materials. At lower temperature ranges the main reaction was dehydrogenation.

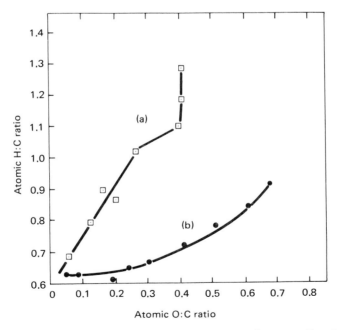

Figure 71. Atomic H:C versus O:C diagrams for (a) A_0 organic matter (humic acids) and (b) B_h organic matter (fulvic acids). According to SCHNITZER and HOFFMANN [1964].

Graphs of the atomic H:C versus O:C ratios according to the graphical-statical methods of VAN KREVELEN [1950] (Figure 71) show dehydrogenation as the main reaction for the A_0 material when heated to 200°C, which is followed by a combined reaction of dehydrogenation and decarboxylation to 250°C, followed by a constant dehydrogenation with higher temperatures. The material of the B_h horizon shows a constant dehydration up to 400°C, which was followed by a cleavage of oxygen-containing groups. It is remarkable that the residues of pyrolysis of both substances, after a complete loss of oxygen at 540°C, are nearly identical in their carbon and hydrogen content.

Degasification of peat humic acids (T 70) and commercially available humic acids (Fa. Roth) by ABEL [1965] (compare NAUCKE [1967]) gives distinct maxima at 295 to 305, 315 to 325, 345 to 350, 370 to 375, 390 to 395, and 410 to 420°C (Figure 72).

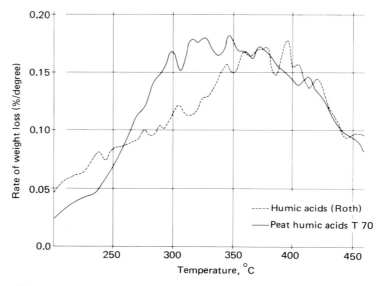

Figure 72. DTG curves of humic acids, according to ABEL [1965].

By plotting the amounts of the volatile products in the range between 270 and 320°C against the content of methoxyl, Abel found a linear correlation. By extrapolation to the origin, it is evident that other groups release gaseous products in this temperature range. Here, in general, a cleavage of carboxyl, carbonyl, and phenolic hydroxyl groups takes place which releases H_2O, CO_2, and CO. This is in agreement with the results from the thermal decomposition of humic acids from brown coal and peat, studied by STADNIKOFF, SSYSKOFF, and USCHAKOVA [1936], as well as by UBALDINI [1937]. They established that in the temperature range between 250 and 300°C, methoxyl groups are cleaved from the molecule, more pronounced, than in the adjacent temperature ranges. Furthermore, it could be established that decarboxylation is a constant background reaction of the thermal decomposition.

IV. Oxidation-Reduction Potentials and Polarographic Properties of Humic Acids and Their Fractions

Oxidation-reduction potentials are assumed to be important parameters for the investigation of soil fertility (DIRKS [1940]). Despite the knowledge that important functional groups, and presumed essential units of humic acids, include both oxidizible and reducible components, there are few attempts to apply redox titration or polarographic methods to the characterization of soil humic substances.

1. *Principles of the Method*

Oxidation-reduction potentials are defined as difference in voltage between a reference electrode (*e.g.*, SCE = saturated calomel electrode) and a measuring electrode (*e.g.*, platinum). In general, the oxidation-reduction potential of a system is referred to as the normal hydrogen electrode with a potential of zero at all temperatures. The measured potential difference is characteristic for a so-called redox system, which consists of a mixture of a substance in oxidized or reduced state.

In polarography, redox potentials are measured in a cell in which one electrode is a dropping mercury electrode or other microelectrode by application of a continuously increasing voltage to the system and recording the resulting current. The current then rises rapidly when the oxidation-reduction potential of the redox system is reached. By plotting the current against the corresponding potentials, one obtains a curve which, under suitable conditions, indicates the nature and concentration of the reacting materials present and gives information about the charge transfer processes of the substances involved in the oxidation-reduction processes.

The dependency of the oxidation-reduction potential on concentration of the oxidized [ox.]* and the reduced form [red.] of the redox system at equilibrium is described in accordance with the law of mass action by the Nernst equation:

$$E = E_0 + \frac{RT}{nF} \ln \frac{[\text{ox.}]}{[\text{red.}]} \tag{1}$$

E_0 = redox normal potential, which means potential of a redox system at equal concentration of oxidant and reductant ([red.] = [ox.]) at pH = 0 against normal hydrogen electrode.

R = gas constant.

T = absolute temperature ($°K$).

n = number of electrons participating in the redox reaction.

F = Faraday constant.

Because oxidation-reduction potentials of organic redox systems are generally pH-dependent, the pH value must be given for comparisons of different systems. Therefore, equation (1) is transformed into:

$$E = E_0 - \frac{RT \cdot m \cdot \text{pH}}{n \cdot F} + \frac{R \cdot T}{n \cdot F} \cdot \ln \frac{[\text{ox.}]}{[\text{red.}]} \tag{2}$$

m = number of protons participating in the oxidation-reduction process.

2. *Oxidation-Reduction Properties of Humic Acids*

Besides the oxidation-reduction potential of a redox system, its "loading" and its normal potential are of interest. These two values may be determined by redox titration.

FLAIG and BEUTELSPACHER [1954a] determined by redox titration (Pt/SCE) the *rH* values†

* The brackets indicate concentrations.

† In measurements of potential differences of redox electrodes against the standard hydrogen electrode, the electromotive forces of this system depend only on the hydrogen pressure at both electrodes according to the equation:

$$-\log \text{pH}_2 = rH = \frac{E + 59 \cdot \text{pH}}{29.5} \text{ (mV; 25°C)}$$

In this way the point of neutralization of a redox system is found at *rH* = 27.2. Strong reductive agents are characterized by *rH* = 0, while *rH* values above 27 are characteristic for oxidants.

Table 24. *rH* measurements by redox titration of natural, biological, and synthetic humic acids, according to FLAIG and BEUTELSPACHER [1954a].

	pH	Before aeration	After with N_2	pH	Before aeration	After with N_2
Humic acids from soils						
Chernozem (Schöppenstedt)						
Extracted with 2% NaF sol.	3.2	—	21.7	8.2	—	26.7
Extracted with 1% NaOH sol.	2.5	—	20.4	7.8	—	26.1
Extracted with 1% Na_2CO_3 sol.	2.6	26.5	22.4	8.1	27.9	27.3
Black peat 1% Na_2CO_3 sol.	2.4	27.1	13.3	8.5	28.4	28.3
Black peat, hymatomelanic acids	3.2	—	19.3	8.5	—	22.9
Synthetic humic acids						
Hydroquinone	4.8	—	19.4	7.9	—	19.2
Pyrocatechol	3.2	—	21.1	8.1	—	21.6
Microbial synthetic humic acids						
Actinomycetes, 13 months old	2.5	28.5	23.0	8.7	28.2	29.5
Actinomycetes, 9 ,, ,,	4.1	29.6	20.1	8.2	27.3	27.5
Actinomycetes, 1.5 ,, ,,	3.5	—	20.9	8.4	—	20.4
Aspergillus, 2.5 ,, ,,	3.0	—	19.3	8.5	—	21.4

of some natural biological and synthetic humic acids. In this way it was established that in acidic medium under nitrogen all kinds of humic acids show reductive properties, while in air the *rH* values of natural humic acids are found near the point of neutralization of the redox system (Table 24).

EBERTOVA [1959] studied the methods of the measurement of oxidation-reduction potentials of humate solutions from soil and compost by application of several types of platinum electrodes. Small platinum cylinders were best. The different methods of cleaning the electrodes are discussed. The potentials in the humate solutions were stable after 21 minutes with an accuracy of ± 3.5 mV and in the compost with ± 10.8 mV after 18 minutes.

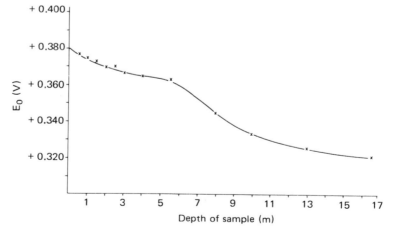

Figure 73. The normal redox potential at pH 7.0 of humic acids extracted from a tropical peat deposit, according to VISSER [1964].

VISSER [1964] determined the redox potentials of humic acids after reduction with Pd/H_2 by potentiometric titration with potassium ferricyanide. In the case of humic acids from tropical sphagnum peat, Visser found that the oxidation-reduction potentials measured at pH 7 decreased with increasing humification. Depending on the soil depth at which the samples are taken, the redox normal potential gave values between $+0.32$ and $+0.38$ (Figure 73).

Attempts to estimate oxidation-reduction potentials of humic acid fractions from soils by polarography failed, because the isolated organic components were not electroreducible at the dropping mercury electrode (KOLTHOFF and LINGANE [1952]).

TOKUOKA and RUZICKA [1934] found no reduction of humic acids but obtained a definite polarographic wave at -1.44 V against standard calomel electrode by application to a hymatomelanic acid fraction, with lithium hydroxide as supporting electrolyte. The half-wave potential was found to be dependent on concentration and shifted with increasing concentration to more positive values.

LUCENA-CONDE and GONZALEZ-CRESPO [1960] found that three of their isolated fractions of soil organic matter, with the addition of perchloric acid, lithium chlorate, and lithium hydroxide as supporting electrolytes, are not electroreducible at the dropping mercury electrode.

LINDBECK and YOUNG [1966] found a small reduction wave at -0.45 V against a mercury electrode in dimethylsulfoxide (DMSO) solutions of the natural humic acids. This was assigned to an easily reducible functional group, (e.g., a quinone or some other system of highly conjugated carbonyl groups). Partial cleavage of the humic substances from soils by nitric acid oxidation revealed products which show, in nonaqueous solution of dimethylsulfoxide, a reductive polarographic wave at -0.34 V, which is adjacent to the wave of nitrobenzoic acid (Figure 74).

Polarographic methods were also applied to the study of surface activities of humic acids, because these substances suppress polarographic maxima. To compare the surface activities of different substances, LUCENA-CONDE and GONZALEZ-CRESPO [1960] introduced a so-called semi-suppression coefficient, which is defined as the concentration of substance that is necessary to result in a 50% diminution of a polarographic maximum.

According to ORLOV and EROSICEVA [1967], a correlation between the height of the oxygen maximum (H) and the humic acid concentration (C) is found by the following equation:

$$H = a \cdot 10^{C \cdot N} \qquad (3)$$

where a and N are constants.

In the same publication Orlov reported on the use of polarography to study the complexing of copper, nickel, and cobalt by humic acids (compare p. 163).

Summarizing the results of the polarographic investigations, the redox properties of humic acids may be established. Since humic acids are not reducible at the dropping mercury electrode, the application of polarography in humic acid research is suited only for the investigation of "secondary effects," such as surface activity, complexing with metallic cations, or determination of carbonyl contents (SCHNITZER and SKINNER [1966b]). However, polarography is a useful tool for the investigation of kinetics during the formation of synthetic humic acids from phenol- or quinone-type precursors.

V. Investigations of Humic Acids by X-Ray Diffraction

The X-ray diffraction of crystalline substances, in which the atoms in molecular groups are regularly oriented in planes, with three-dimensional repetition forming so-called space lattices, is an important method of nondestructive analysis. It is possible to determine the degree of crystallization of unknown material by this method. Crystalline samples reflect

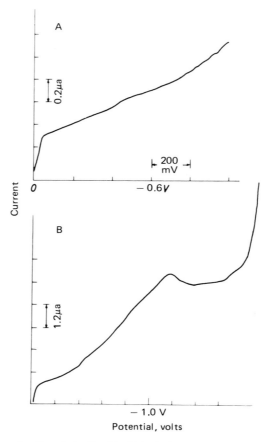

Figure 74. Polarograms in dimethylsulfoxide (DMSO) with 0.5 M LiCl supporting electrolyte of (A) humic acids directly soluble in DMSO and (B) nitric acid oxidation products of soil humic acids. According to LINDBECK and YOUNG [1966].

reinforced X-ray bands from repeating parallel planes. Amorphous materials reflect broad halos of scattered X-rays.

1. *X-Ray Diffraction Properties of Humic Acids*

SEDLETZKY and BRUNOWSKY [1935] made Debye-Scherrer X-ray diagrams-diffractograms of various humic acids. They observed that the humic acids samples derived from peat and chernozems show interference lines, which were assumed to be identical with some important diffraction bands of graphite. According to these results, they established that humic acids are crystalline substances. MEYER [1938] also observed X-ray diffraction lines from chernozem samples, which coincided with reflections of the graphite lattice. In contrast to these findings JUNG [1938] found that sodium hydroxide and sodium fluoride extracts from high-moor peat gave no well-defined X-ray reflections. SAEKI [1939] and PALLMANN [1942] confirmed the results of SEDLETZKY and BRUNOWSKY [1935], but in the X-ray diagrams of their preparations only small, weak, and broad diffraction lines were detected. Sedletzky and Pallmann proposed that X-ray–amorphous lignin alters to humic intermediates and finally to crystalline graphite. Comparative investigations of JODL [1941a, b, c, d] with synthetic humic acids derived from phenol and several natural humic acids revealed a good agreement of the X-ray spectra from each group.

From the similarity of the X-ray diffraction patterns of humic acids with those of graphite, the graphite-like orthohexagonal lattice distances have been assumed, namely, $d_{002} = 3.4 - 4.0$ Å and $d_{110} = 2.1 - 2.2$ Å. Later, KASATOTCHKIN [1951], KASATOTCHKIN and ZILBERBRAND [1956], and others defined the additionally occurring interferences in humic acids spectra at $d = 4.7 - 5.5$ Å as γ-bands. KASATOTCHKIN, KONONOVA, LARINA, and EGEROVA [1964] compared X-ray diffractograms of humic acids and fulvic acids from different soil types. They suggested that aromatic nuclei of humic acids molecules are characterized by reflection bands at d_{002} and d_{110}. The simultaneous occurrence of the interferences at d_{002} and of the so-called γ-bands ($d = 4.7$ Å) led the authors to assume that two types of intermolecular arrangements occur:

(1) those in which the lattice layers are formed by planar aromatic nuclei with the charge field of the condensed aromatic arrangements, and

(2) those that are formed by contact of radicals in the side-chain with hydrogen bridge linkages.

In comparison to humic acids from chernozem, the layers of carbon in the molecules of humic acids from sod-podzols show a lesser degree of structural organization because of more substitution of the aromatic hydrogen atoms of radicals in the side-chain. In the X-ray diffraction pattern of fulvic acids from sod-podzolic soil the γ-bands are more pronounced than the band of the basic reflection line at d_{002}.

KODAMA and SCHNITZER [1967] investigated in detail the X-ray diffraction properties of humic acids derived from podzol. They observed diffuse X-ray patterns, with a maximum at 4 Å. By Fourier analysis four maxima are calculated at 1.6; 2.9; 4.2, and 5.2 Å, which are similar to the main reflections of carbon black. From these results the authors assume that the carbon skeleton of the fulvic acids consists of a broken network of condensed aromatic rings with a certain amount of disoriented aliphatic or alicyclic structural units, with a molecular weight of 670 and a density of 1.61 g/cm^3. They calculated a particle volume of 690 $Å^3$.

VI. The Complexing Properties of Humic Acids with Inorganic Cations

1. *Introduction*

Organic soil substances are able to form complex linkages of various kinds with metals by ion exchange, adsorption on surfaces, and formation of chelates. Metal-organic colloids in soils are found as flocculates, aggregates, and various mixtures. Humic substances are important in soil genesis and in the formation of soil structure particles. Cation exchange reactions are important in plant nutrition. Consequently many publications deal with the problems of interaction between inorganic and organic soil constituents, especially between polyvalent cations and humic and fulvic acid fractions.

For the investigation of the mechanism of these interactions and the properties of the reaction products, several physicochemical methods have been used. Ion-exchange chromatography has been used by BROADBENT [1957], BROADBENT and OTT [1957] BERES and KIRALY [1957], KAWAGUCHI and KYUMA [1959], KAURICHEV and NOZDRUNOVA [1960], LEWIS and BROADBENT [1961a, b]; electrophoresis by HAYASHI and NAGAI [1956], MISTERSKI and LOGINOV [1959], KONONOVA and TITOVA [1961], TITOVA [1962], DROSDOVA, KRAVTSOVA, and TOBELKO [1962], POSPISIL [1962a, b], and JUSTE [1966]; potentiometric titrations by CHATTERJEE and BOSE [1952], ZADMARD [1939], BECKWITH [1955], THIELE and KETTNER [1953], HALLA and RUSTON [1955], MARTIN and REEVE [1958], VAN DIJK [1959, 1967], KAWAGUCHI and KYUMA

[1959], LEWIS and BROADBENT [1961a, b], POMMER and BREGER [1960a, b], SCHNITZER and DESJARDINS [1962], KHANNA and STEVENSON [1962], POSPISIL [1962b], WRIGHT and SCHNITZER [1963], SCHNITZER and SKINNER [1963a, b, 1964, 1966a], POSNER [1964], SCHNITZER and GUPTA [1965], POSNER [1966]; conductometric titrations by HALLA and RUSTON [1955], CHATTERJEE and BOSE [1952], ROY [1957], MUKHERJEE and LAHIRI [1958], PAVEL [1959], VAN DIJK [1960], PIRET, WHITE, WALTHER, and MADDEN [1960], SMITH and LORIMER [1964]; high-frequency titrations by PAVEL [1960], SCHNITZER, DESJARDINS [1962]; electron spectroscopy by HAYASHI and NAGAI [1956b], BROADBENT and OTT [1957], MANSKAYA, DROSDOVA, and EMELJANOVA [1958], PAVEL [1959], POSPISIL [1962a, b], KLEIST [1967]; infrared spectroscopy by DEMUMBRUM and JACKSON [1956], KASATOTCHKIN, KONONOVA, and ZILBERBRAND [1958], SCHNITZER, SHEARER, and WRIGHT [1959], MORTENSEN [1960], ORLOV and NESTERENKO]1960], KHANNA and STEVENSON [1962], WRIGHT and SCHNITZER [1963], SCHNITZER and SKINNER [1965a], JUSTE [1966], LEVESQUE and SCHNITZER [1967b], SCHNITZER and HOFFMANN [1967]; electron paramagnetic resonance by KLEIST [1965, 1967], KLEIST and MÜCKE [1966a, b]; and thermoanalytic methods by SCHNITZER and SKINNER [1964], SCHNITZER and HOFFMANN [1967], LEVESQUE and SCHNITZER [1967b].

According to the different plausible types of linkages between functional groups of humic acids and the inorganic soil constituents, three main types of metal-organic derivatives can be distinguished (compare SCHATZ, SCHATZ, SCHALSCHA, and MARTIN [1964], MORTENSEN [1963], ALEXANDROVA [1967]).

(1) The ionic (heteropolar) type, with participation of carboxyl and phenolic hydroxy-groups, leads to the corresponding humates and fulvates by the known reaction of salt formation.

(2) The semipolar type, formed by coordination linkages with participation of amino-, imino-, keto-, and thioether groups, leads to complex compounds of the chelate type.

(3) A type that is formed by polarization effects and hydrogen bridge linkages with special participation of terminal functional groups, forming compounds of the adsorption type.

The humic acids contain polyfunctional groups. Each group is affected differently by any given set of conditions. Consequently, it is difficult to measure and predict the reactivity of any one group unless it is known to be affected specifically by the conditions in the reaction medium.

2. Ion-Exchange Properties of Humic Acids

(a) General Remarks

In contrast to clay minerals, soil organic matter has no well-defined capacity to bind reversibly exchangeable cations. BARTELETT and NORMAN [1938] reported that the complexing of cations is influenced by the charge of the exchanging cations. According to their results, more barium than potassium is bound at the same hydrogen ion concentration. Generally, the intensity of exchangers follows the lyotropic order. That means that the exchange mechanism depends on the charge and hydrated size of the exchanging ion.

The hydrogen and hydroxyl ions are most strongly fixed; these are followed by the polyvalent cations; then the monovalent ions are arranged according to decreasing crystal lattice radii. In the exchange mechanism, ions exchange equivalent quantities of other ions. The equilibrium state depends on concentrations and activities of all ions in the exchange system. The exchange capacity depends on the amount and nature of the surface of the solid phase. Adsorption also occurs with neutral molecules through van der Waals forces. The selectivity is increased with increasing charges of the ions. The rule of Schulze and Hardy is valid for the mechanism of ion exchange (compare ZADMARD [1939]).

The ion-exchange capacity of polyfunctional exchangers, such as humic acids, is strongly influenced by the hydrogen ion concentration of the exchange solution (RYDALEVSKAYA and TISHCHENKO [1944], BROADBENT and BRADFORD [1952]). The cation exchange capacity increases with increasing pH values. This arises from the increasing dissociation of phenols and carboxylic acids at higher pH levels. LEWIS and BROADBENT [1961a] demonstrated by experiments with model substances that there is a limiting pH value, above which the adsorption of cations occurs, with hydrogen ions being displaced into solution. By the low increase of the hydrogen ion concentration of metal ions in solution, a back reaction takes place, and the adsorption ceases regardless of the amount of exchanger present. In the use of adsorption of copper and uranyl cations by carboxyl groups it was found that the limiting pH value of each ion is independent of the activity of the exchange sites of the exchangers, while the limiting pH value for barium is lowered proportionally to the increase of the exchange acidity. Phenols show a decrease of the limiting pH value, with increasing site acidity for all three cations.

Differences of the exchange properties of the humic acids may be expected. These variations arise from genetic, soil-forming, and laboratory-extraction sources.

TYURIN [1937], PALLMANN [1938], ZADMARD [1939], NATKINA [1940], KONONOVA [1943], and others find values of the exchange capacity of soil organic substances in the range between 150 and 450 meq/100 g. The values depend on the origin of the fractions. According to their results, the values of exchange capacity of humic acids isolated from peat are about 150 meq/100 g, while those isolated from black soil show the highest values in a range between 350 to 500 meq/100 g. Since humic acids may differ much from soil to soil, it is often difficult to compare results obtained by various workers on cation exchange (FLAIG, SÖCHTIG, and BEUTELSPACHER [1963]).

Starting from the assumption that chemical reactions occurring in the course of the titration of high molecular substances may be explained as ion-exchange reactions, POSNER [1966]

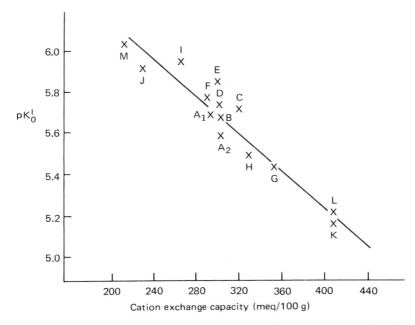

Figure 75. Relationship between cation exchange capacity and average dissociation constants (pK_0^I) of humic acids from red-brown soil obtained by various extractants and conditions. According to POSNER [1966]. (For an explanation of characters, see Table 25.)

investigated the cation exchange properties of humic acids, which are prepared by different extractants from a red-brown soil under different conditions. To determine the pK_0^I values by extrapolation, POSNER [1964] titrated 5 to 7 mg humic acids in a total volume of 22.5 ml with potassium hydroxide in a solution of different ionic strength (0.0023 up to 0.1 N KCl). From these data the values of the cation exchange capacity were determined by taking the end-point at the titration volume, where the changes in hydrogen ion concentration with added base reached a maximum (pH 7 to 7.6). Posner found in this way that the values of exchange capacity over the range of used ionic strengths show no systematic variation. A comparison of the results according to Figure 75 shows a negative correlation between the pH values and the values of cation exchange capacity.

The yield and the analytic data on humic acids obtained by various extractants and conditions shown in Table 25 demonstrate, with some exceptions, that in spite of different

Table 25. Yield and analytical data of humic acids from red-brown soil obtained by various extractants and conditions. According to POSNER [1966].

	Pre-treatment	Extractant	Conditions		Percent content in humic acids			Yield g/kg soil	Cation exchange capacity meq/100 g
			N_2	°C	Fe_2O_3	SiO_2	Al_2O_3		
A_1	—	NaOH	—	20	0.47	0.96	0.44	6.9	297
A_2	—	NaOH	—	20	0.63	0.81	0.51	6.0	305
B	—	NaOH	—	20	0.49	0.62	0.32	6.0	301
C	—	NaOH	—	60	0.061	0.026	0.057	10.8	320
D	—	NaOH	+	60	0.047	0.028	0.038	10.2	301
E	—	$Na_2CO_3/NaHCO_3$	+	20	0.46	0.21	0.11	2.7	300
F	—	$Na_2CO_3/NaHCO_3$	—	60	0.39	0.066	0.051	5.1	292
G	—	$Na_2CO_3/NaHCO_3$	+	60	0.50	0.11	0.055	4.9	353
H	HCl	NaOH	+	20	0.39	0.039	0.11	11.5	331
I	HCl	NaOH	+	60	0.056	0.05	0.061	12.0	269
J	HCl/HF	0.5N NaOH	+	20	1.11	0.26	0.10	9.8	231
K	HCl	$Na_2CO_3/NaHCO_3$	+	20	0.71	0.15	0.11	3.5	407
L	—	$Na_4P_2O_7$	+	20	0.086	0.064	0.045	4.0	409
M	$Na_4P_2O_7$	NaOH	+	20	0.24	0.10	0.087	4.6	217

yields of extraction, substances of relatively similar values of exchange capacity are obtained. Remarkable differences are found only with potassium hydroxide extracts of soils, which were pretreated by a mixture of hydrochloric and hydrofluoric acid.

Humic acids (J) had the lowest exchange capacity, possibly due to the high iron content of this sample. On the contrary, pyrophosphate extraction gave a material with the highest exchange capacity and the lowest pK_0^I value (L). This was similar to the humic acids (K) pretreated with 0.1 N HCl and having subsequent $Na_2CO_3/Na HCO_3$ extraction. The post-treatment of the pyrophosphate extract with sodium hydroxide gave humic acids with the highest pK_0^I values and the lowest exchange capacity (M). From the results it was concluded that a fractionation of humic acids according to the different pK_0^I values might be possible. Corresponding to the assumption that the extracts with the lowest pK_0^I values show the strongest links to cations typically occurring in soils, these properties depend on the different degradations of the original metal-organic complexes by the different treatments.

(b) The Nature and Properties of the Exchange Sites of Humic Acids

The characterization of the exchange sites, which are responsible for the cation exchange properties of organic soil substances, has been the object of many investigations leading to different results.

One of the first assumptions, that the active sites are of phenolic character, was based on the observation that there is a direct relationship between the contents of lignin-like materials of the soil organic fraction and the cation exchange capacity. The phenolic hydrogen atoms are the most acidic sites in lignin. GILLAM [1940] concluded from his results that the active groups must be carboxyls because active sites with lower acidity than pK = 6 are blocked by dimethyl-sulfate. Only a low decrease of the exchange capacity for calcium and barium was obtained. Investigations of the exchange properties of soil organic matter by BROADBENT and BRADFORD [1952] by selective blocking of the exchange sites with dimethylsulfate and diazomethane indicated that the exchange capacity was inversely proportional to the content of methoxyl groups. By treatment with dimethylsulfate, the exchange capacity was reduced. From the observation that an increase in the methoxyl content affected a linear decrease in exchange capacity, it may be concluded that carboxyl groups as well as phenolic hydroxyl groups participate in the exchange mechanism. HIMES and BARBER [1957] were able to eliminate the whole exchange capacity of a sandy loam for zinc by treatment with dimethylsulfate. They concluded from this result a greater importance of the phenolic hydroxyl groups for the exchange properties of the soil organic substances with which they were working. In contrast BROADBENT [1957] established by exchange chromatography that carboxyl groups are the predominant part of the exchange mechanism of soil-organic substances in the case of calcium and copper fixation.

Exchange reactions of barium, calcium, and uranyl cations with model compounds in solid phase (LEWIS and BROADBENT [1961a]) containing carboxyl and phenolic groups, such as different substituted hydroxynaphthoic acids, indicated that phenolic groups linked copper in the divalent ionic state. Substances with phenolic hydroxyl groups in o-position to the carboxyl groups bound the monovalent form of copper [Cu (OH)$^{+}$] in phenolic and carboxyl groups. This was attributed to a steric hindrance of a second ligand molecule of the complex.

Investigations of SCHNITZER and SKINNER [1965a] on the role of carboxyl and hydroxyl groups in metal fixation by organic soil substances by specific blocking of the functional groups indicate two types of reactions:

(1) a main reaction, with the simultaneous participation of carboxyl and phenolic groups, and

(2) a subsidiary reaction, involving only carboxyl groups.

Although these authors find no evidence for reactions of the carbonyl groups, according to their opinion these reactions cannot be excluded from the interaction mechanism between humic acids and metals.

(c) Ion Exchange and Metal Retention of Humic Acids

The ion exchange between humic acids and alkaline earth and alkali metals was first investigated in detail by ZADMARD [1939]. He found that potassium of the potassium humates was easily exchanged by alkaline earth and alkali metals and that the amount of potassium exchanged was directly proportional to the diameter of the adsorbed ions.

If the equivalent ratios of the ions in solution versus ion concentration of potassium humate are plotted (Figure 76), it can be shown* (compare DEUEL and HUTSCHENECKER [1955], SCHEFFER-ULRICH [1960]) that the lyotropic order of ions is followed.

---

* In case of equivalence of the ratio between humate and solution, the adsorbed and released cations are equally linked by the humate anion and the values are found on the diagonal. Curves beneath the diagonal indicate a stronger linkage of the released cation, while curves above the diagonal show a stronger linkage of the adsorbed cations.

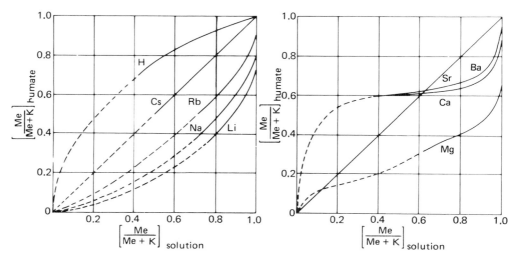

Figure 76. Cation exchange equilibriums in the systems of alkaline earth and alkali chlorides, hydrochloric acid, and potassium humate. According to Zadmard [1939].

According to the decreasing hydration of the monovalent cations in the order: Li > Na > K > Rb > Cs of the divalent cations: > Mg > Ca > Sr > Ba, the binding energy, and therefore, the intensity of adsorption is increased.

The exchange capacity of the calcium humate was about 30% higher than that of the potassium humate. However, in the case of calcium humate no linear correlations between the mean values of exchange and the ionic radii of the alkaline earths and alkali metals are found. The exchange values of the alkali ions with calcium ions of the calcium humate are distinctly lower than those of the potassium ions of the potassium humate. In contrast, the alkaline earth ions show a stronger adsorption with calcium humate than with potassium humate.

The comparable strong linkage of the alkaline earth metal ions in contrast to alkali metals with humic acids was observed by Schachtschabel [1940]. He found in a solution of equivalent amounts of Ca^{++} and NH_4^+ a selective sorption of these cations in a ratio of 92:8.

Broadbent [1957] used preparations of organic soil substances in the acidic state as ion exchangers. To investigate the ionic binding properties of copper and calcium, he placed a thin layer of the organic substance saturated with these ions on a chromatographic column. The experiments were carried out by elution of the columns with hydrochloric acid of increasing concentration, and the eluates were analyzed quantitively for the released cation. A plot of the concentration of the cation versus the volume of the elution showed four peaks for copper and two for calcium. This leads to the assumption that, in the copper fixation, some functional groups participate which show no interaction with the calcium ion.

A comparative determination of the exchange capacity of soil organic substances for copper and barium by ion-exchange chromatography was carried out by Lewis and Broadbent [1961a] with two methods. In the first case the sample was repeatedly treated with a 0.01 molar solution of ions at pH = 5.5 to 6.0, until treatments showed no further significant changes in hydrogen ion concentration.

In the other experiment, mixtures of organic soil substances with sand were saturated by ion-exchange chromatography with the copper and barium ions. Extraction was carried out by 0.01 N acetic acid and subsequent extraction with 0.01 N nitric acid. In former model investigations of the pH-dependent reaction of ion exchange of the carboxyl- and phenyl-hydroxyl-containing organic cation exchangers such as amberlite IRC-50 and amberlite

IRC-120 (LEWIS and BROADBENT [1961a]), it was demonstrated that 0.01 N acetic acid removes barium and copper from the phenolic hydroxyl and the carboxyl groups if the pH is 2.85. From the fact that the acetic acid removes the main amounts of both ions from the organic sand mixtures, it was concluded that the exchange capacity of these samples is associated with phenolic and carboxylic functional groups.

According to LEWIS and BROADBENT [1961a], phenolic groups of model substances with a 0.1 M copper solution at pH 6.0 fixed no copper, at pH 5.5, 437 meq of copper was fixed, and at pH 4.6, 294 meq was fixed. It was concluded that an interaction with the phenolic hydroxyl groups is responsible for the resulting exchange. The methylation of the phenolic hydroxyl groups with diazomethane and dimethylsulfate gave the same methoxyl content with either reagent. Samples methylated with diazomethane adsorbed no copper, while samples treated with dimethylsulfate showed a reduction of the exchange capacity of about 25%. Apparently these model systems contained phenolic hydroxyl groups and carboxyl groups of various degrees of acidity. Barium is found to be fixed only by the more acidic phenolic hydroxyl groups and other more acidic exchange sites. Copper is fixed as the divalent ion (Cu^{++}) and in the monovalent form as ($Cu(OH)^+$) by an interaction with all functional groups.

MANSKAYA, DROSDOVA, and EMELYANOVA [1958] found a maximum of copper fixation by humic acids at pH 2.5 to 3.5, while fulvic acids fixed the most copper from exchange solutions at pH values near 6 (Figure 77).

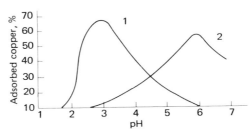

Figure 77. Formation of copper humates and fulvates at different pH values of the medium. 1—humic acids; 2—fulvic acids. According to MANSKAYA, DROSDOVA, and EMELYANOVA [1958].

The reaction between humic acids and/or their soluble alkaline humates with hydroxides of iron and aluminum also follows an exchange reaction. As DEB [1949a, b] demonstrated, a simultaneous reaction of coagulation and peptization occurs if extracts of organic soil substances are mixed with iron oxides. He found a variation of the peptized amounts of iron oxides, depending on the kind of the organic soil substances, the hydrogen-ion concentration, and the concentration of the sols. Generally, the organic substances are completely precipitated in mixtures of 7 parts of humic acids and 100 parts of iron oxide.

However, in investigations of the interaction of the organic soil extracts with iron and aluminum hydroxides, ALEXANDROVA [1954] found that the formation of metal-organic compounds of humic acids depends largely on the hydration state of the sesquioxides. She found that these compounds are formed in a wide range of pH values, from acidic to weak alkaline solutions by a reaction between the gels of humic acids of different states of hydration with the sols of the sesquioxides, and by reaction of humic acids gels with freshly precipitated gels of sesquioxides. She found no reaction between humic acids and sols of humates with aged and strongly dehydrated or with crystalline forms of sesquioxide sols. Furthermore, ALEXANDROVA [1967] demonstrated that some undefined functional groups are not affected in the reaction of the organic substances with metal hydroxides, so that a small residual exchange capacity remains.

In contrast, SCHNITZER and SKINNER [1963b] demonstrated the formation of stable complex compounds with iron and aluminum compounds and extracts from podzolic soils by application of goethite (FeOOH), gibbsite (AlOOH), and iron- or aluminum-saturated ion exchangers and soils, respectively. An increase of the pH value from 3 to 7 resulted in a decrease of the mobility of the ions from goethite and gibbsite, but increased the ionic mobility from the soils. Similar investigations with freshly precipitated iron and aluminum hydroxides in the range between pH 3 and 5 (LEVASHKEVICH [1966]) resulted in a strong adsorption of the humic acids by aluminum hydroxides with decolorization of the solution, while the adsorption by precipitated iron hydroxides occurs slowly. Levashkevich also demonstrated that humic acids from chernozems have the strongest binding power of cations in an amount 67 to 84 mg Fe/100 g on humic acid in contact with Fe_2O_3 and 100 to 114 mg Al/100 g humic acid. Organic soil extracts from the A_1 horizon of a sod-podzolic soil fixed 59 to 72 mg Fe/100 g humic acid from Fe_2O_3 and 87 to 96 mg Al/100 g humic acid in contact with Al_2O_3, fulvic acids from the B_h horizon from a sandy podzol 20 to 32 mg Fe/100 g humic acid in contact with Fe_2O_3, and 47 to 59 mg Al/100 g humic acid in contact with Al_2O_3. According to the author's opinion the metal hydroxides are converted by elimination of hydroxyl ions to the monovalent $(R(OH)^+{}_2)$ and divalent $(R(OH)^{++})$ ionic states, which may react with the anionic parts of the soil organic matter. The estimation of the relative complexing stability of the metal-organic complexes was measured by the solubility in sulfuric acid of increasing normality. Complexes with aluminum are less soluble than those of the similar iron compounds.

By application of the ion-exchange–equilibrium method RANDHAWA and BROADBENT [1965a, b] determined the pH-dependent stability constants of soluble metal-organic complexes of humic acids with zinc. This method was also used by HIMES and BARBER [1957] for investigation of the ^{65}Zn fixation by soils. By treatment of the soils with peroxide, HIMES and BARBER demonstrated that this reaction was significantly associated with the chelating ability of the organic soil substances. According to the equilibrium reaction of the metal ions (M) with the chelating agent (Ch) during the formation of the complexes (MCh_x):

$$M + x \cdot Ch = MCh_x,$$

with respect to the law of mass action the numerical value of the equilibrium constant (k) is defined as:

$$k = [MCh_x]/([M] \cdot [CH]^x),$$

where x means the amount of the chelating molecules in the complex. The unknowns x and Ch of this equation are found by application of the ion-exchange–equilibrium method according to the equation established by MARTELL and CALVIN [1952]:

$$\log \lambda_0/\lambda - 1 = \log k + X \cdot \log [Ch]$$

by plots of $\lambda_0/\lambda - 1$ versus $\log [CH]$. Here λ_0 is the distribution constant of the exchange reaction with the ion exchanger (R) in the absence of a chelating agent $(\lambda_0 = MR/[M])$, and λ is the distribution constant for exchange in the presence of the chelating agent $(\lambda = MR/[MChx])$. From the different values of x of the chelating molecules in the complex obtained at different hydrogen ion concentrations $(x = 1.25$ at pH 3.6; $x = 1.59$ at pH 5.6; and $x = 1.70$ at pH 7.0), Randhawa and Broadbent assumed a participation of monovalent and divalent form of ions in the reaction. At low pH values the amount of the monovalent form is predominant (75% at pH 3.6), which is lowered with increasing pH values in favor of the divalent form (30% at pH 7.0). According to this, the pH-dependent stability constants of the complex compounds between humic acids and zinc are found to be 4.42 at pH 3.6, 6.13 at pH 5.6, and 6.80 at pH 7. From these results it was concluded that the concentration of the ligands available for the complex formation is increased with decreasing hydrogen ion concentration of the exchange solution.

SCHNITZER and SKINNER [1966a] investigated in a similar way the pH-dependent stability from the B_h horizon of a podzol and the metal ions Cu^{++}, Fe^{++}, and Zn^{++} at pH 3.5 and 5.0, and in a further publication (SCHNITZER and SKINNER [1967]) the complexing properties with the metal ions Pb^{++}, Ni^{++}, Mn^{++}, Co^{++}, Ca^{++}, and Mg^{++}. Table 26 gives the moles

Table 26. Stability constants of fulvic acid complexes with divalent cations. According to SCHNITZER and SKINNER (1966(a), 1967).

Metal	pH 3.5		pH 5.0	
	x	$\log k$	x	$\log k$
Mg^{++}	0.53	1.23	0.79	2.09
Mn^{++}	0.55	1.47	1.10	3.78
Zn^{++}	0.58	1.73	0.56	2.34
Ca^{++}	0.83	2.04	0.90	2.92
Co^{++}	0.70	2.20	1.00	3.69
Pb^{++}	0.75	3.09	1.50	6.13
Ni^{++}	1.00	3.47	1.00	4.14
Fe^{++}	1.25	5.06	1.30	5.77
Cu^{++}	1.50	5.78	2.00	8.69

of fulvic acids (x), which combine with one mole of metal, and the logarithm of the stability constants ($\log k$). The stability at pH 3.5 decreases in the following order: $Cu^{++} > Fe^{++} > Ni^{++} > Pb^{++} > Co^{++} > Ca^{++} > Zn^{++} > Mn^{++} > Mg^{++}$. At pH 5 the order is: $Cu^{++} > Pb^{++} > Ni^{++} > Mn^{++} > Co^{++} > Ca^{++} > Zn^{++} > Mg^{++}$.

These results show significant deviations from the results published by IRVING and WILLIAMS [1948], who found the order of stability of divalent cation complexes independent of kind of complexing agent as follows: Pd > Cu > Ni > Co > Zn > Ca > Fe > Mn > Mg. This is in agreement with the results of investigations of BECKWITH [1959] and KHANNA and STEVENSON [1962]. The stability of the fulvic acid complexes with the divalent forms of iron (Fe^{++}) is remarkably high; this has not yet been explained. The $\log k$ values of the complexes between fulvic acids and zinc show, in contrast to the results of RANDHAWA and BROADBENT [1965a, b], only a slight influence of the hydrogen ion concentration of the exchange solution. Numeric calculations show that zinc complexes more with fulvic acids than other cations. From the fact that the stability of complexes between lower amounts of functional groups of soil organic matter (e.g., in Cu^{++} and Fe^{++} fixation) is stronger than in fulvic acid complexes with zinc, it was concluded that not the number of linkages, but the binding strength influences the values of the stability constants. No direct information about the mechanisms of interaction between metals and soil organic matter can be drawn from the stability constants.

3. Coagulation of Humic Acids by Electrolytes

In the investigation of the chemical interaction between metal ions and humic acids' difficulties arise because of the reduced solubility of these compounds with increasing hydrogen ion concentration, which is in most cases accompanied by a simultaneous precipitation of less-soluble hydroxycomplexes of polyvalent cations. If no chemical interaction between humic acids and metal ions occurs, the formation of mixed precipitates of humic acids and metal hydroxides is possible (ORLOV and NESTERENKO [1960]). Therefore, appropriate investigation of the coagulation on peptization properties during the pH-dependent interaction with electrolytes provides important information about the behavior of metal-organic colloids.

Like other organic colloids, humic acids may be precipitated from their solutions by strong electrolytes. As mixtures of differently composed high-polymer substances, they have no defined value of the coagulation threshold. SPRINGER [1938] observed that in certain types of soils the coloring properties of extracts of humic acids decreased with increasing concentration of alkali. By developing a range of conditions affecting coagulation by addition of electrolytes SPRINGER [1938] was the first to separate brown humic acids from gray humic acids. Later on FLAIG, SCHEFFER, and KLAMROTH [1955] used 2 N sodium chloride solutions to selectively precipitate humic acids.

If humic acids are precipitated, a control of the hydrogen ion concentration of the solution shows that the coagulation starts at certain pH values and stops at a certain hydrogen ion concentration. Application of higher concentration of acids precipitates not only humic acids but also fulvic acids. Repeated reprecipitation leads to the continuous removal of organic substances in solution.

BISLE [1959] investigated, by gravimetric analysis, the flocculation properties of humic acids from peat bog with hydrochloric acid. They were subfractionated by sodium hydroxide, sodium chloride, and sodium fluoride. He found that the precipitation of gray humic acids starts at pH 4 and is completed at pH 3.5, while the precipitation of brown humic acids occurs at higher hydrogen ion concentrations (pH = 2).

A detailed investigation of the coagulation properties of humic acids, which were carefully prepared by ammonium oxalate extraction, was carried out by MISTERSKI and LOGINOV [1959]. They measured, by colorimetry, the extinction of the amount remaining in solution. Plots, which show the relationship between the measured values of extinction and the concentration of the flocculates (in meq/l.), are defined as coagulation curves. They applied increasing amounts of coagulating ions to humic acids, which were adjusted to pH 7 by diluted potassium hydroxide solutions, and determined the changes in extinction at equal concentration of humic acids (1.25 g/l.). The resulting coagulation curve of the hydrochloric acid treatment (Figure 78) shows a decrease in extinction near zero, which is explained by a change in color of the solution after acidification.

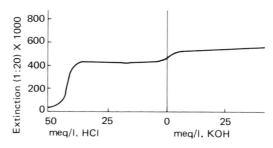

Figure 78. Coagulation and peptization of humic acids. According to MISTERSKI and LOGINOV [1959].

A stronger decrease in extinction was found after addition of 40 meq/l. HCl (pH = 2.3), which was caused by the precipitation of the soluble humic acids. The continuous change of extinction in the whole range of coagulation of the humic acids points to polymolecularity and to transitions between the fractions. Attempts to precipitate potassium humates with the chlorides of calcium, iron (III), and aluminum show that the metallic cation leads to a much faster coagulation of the humic acids than addition of hydrogen ions (rule of Schulze–Hardy).

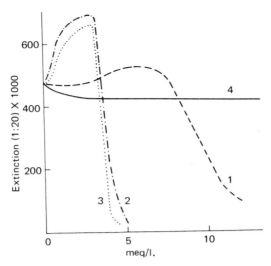

Figure 79. Coagulation of potassium humate by Ca^{2+} (1), Fe^{3+} (2), Al^{3+} (3), and H^+ (4). According to MISTERSKI and LOGINOV [1959].

The precipitation takes place by the addition of small amounts of iron and aluminum ions. On the contrary, the coagulation curve of the calcium treatment deviates distinctly from the iron and aluminum curves, pointing to another mechanism of interaction (Figure 79).

The pH of the pretreated K humate does not affect its subsequent flocculation by calcium, but pH of the pretreated K humates does affect the effectiveness of iron and aluminum in coagulating the humates (Figure 80).

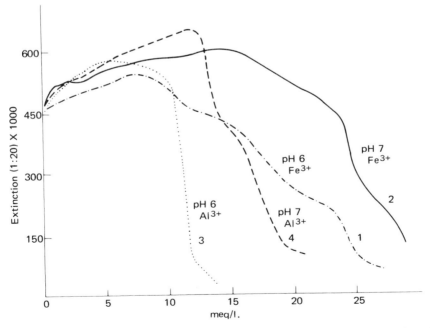

Figure 80. Coagulation of humates by iron and aluminum chloride with KOH adjusted to pH 6 (1.3) and pH 7 (2.4). According to MISTERSKI and LOGINOV [1959].

Attempts to precipitate the humates by addition of calcium chloride in the presence of the alkali salts of nitrate, chloride, sulfate, and phosphate showed that the coagulation of the humates is retarded by the phosphates and to a lesser amount by the sulfates. The retardation of coagulation by phosphates may be effected by the formation of nonionizable salts of the active polyvalent ions, as well as by the changes of the hydrogen ion concentration, because of the buffering action of the phosphate solution. Supplementary peptization experiments showed that the peptization of the flocculated molecules is decreased in the order: phosphate > sulfate > chloride > nitrate.

By flocculation of humic substances from a podzolic soil by iron (Fe^{2+} and Fe^{3+}) and nickel MARTIN [1960] demonstrated that a maximum of precipitation occurs at pH 2 and 3 at a C/Fe^{3+} ratio* of 3 and 6. By decreasing the hydrogen ion concentration to levels above pH 3, the humic acids were peptized.

Furthermore, Martin demonstrated that ionic iron is precipitated at pH 5 without co-precipitation of humic acids. Flocculation attempts with aluminum trichloride ($C/Al^{3+} = 12$) gave a maximum precipitation between pH 3 and 4, while the precipitation of divalent iron ($C/Fe^{2+} = 1$) is completed at pH values between 6 and 7. At a C/Fe^{2+} ratio > 1, no complete flocculation of the humic acids was obtained. Nickel ($C/Ni^{2+} = 3$) did not precipitate the humic substances in the range between pH 1 and 8. Therefore, the critical hydrogen ion concentration in the case of the application of trivalent iron is reached at pH 2.8, and for divalent iron at values from pH 6.8 to 7.8. In the case of aluminum, the limiting pH value was found to be pH 4.0, and for nickel, pH 8.9. In mixtures of the humates with metal solutions with respect to the type of cation used, different types of changes of the hydrogen ion concentration were observed. These were negatively directed below the limiting pH value, and positively above these values.

Flocculation experiments with hydrochloric acid carried out to determine the pH values of the beginning coagulation of humic acids by aid of light-scattering measurements were published by ORLOV and EROSICEVA [1967]. They found that the light scattering of humic acids is nearly constant in pH values between 8 and 3 and is increased in acidic media (Figure 81).

By addition of bases to the acidic solution, it was observed that the peptization of humic acids started at pH levels somewhat higher than the coagulation limit. A repeated precipitation by acidification of the same solution indicated a similar course of the curve but gave displace-

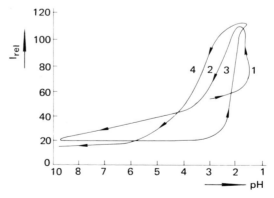

Figure 81. Hysteresis of coagulation and peptization of humic acids. 1 = coagulation by HCl; 2 = peptization by NaOH; 3 = recoagulation by HCl; 4 = repeptization by NaOH. According to ORLOV and EROSICEVA [1967].

* C/Me = ratio between percent carbon content and amount of added metals.

ments of the coagulation curve in the direction of lower pH values. From this fact the authors concluded that there was a dependency of the coagulation properties of humic acids upon ionic strength in relationship to the amount of salts formed during neutralization reactions. This seemed to cause the lowering of solubility of humic acids with high ash contents. Generally, investigations of the interaction between humic acids and different types of divalent and trivalent cations (Al^{3+}, Fe^{3+}, Cu^{2+}, Ni^{2+}, Zn^{2+}, Co^{2+}, and Mn^{2+}) indicated that trivalent ions cause incipient precipitations at low amounts of added electrolytes. The amount of metals which was bound to 1 g of humic acids was in the range between 0.7 to 20 mg/meq. The degree of complexing of the metals corresponds to the order of solubility of their hydroxides.

KONONOVA [1966] concluded from calcium chloride flocculation studies on humic acids isolated from different soil types that the ease of peptization of humic acids was dependent on soil origin and was inversely proportional to the ratio between the amount of aromatic and aliphatic molecular constituents. The decrease of the flocculation threshold conformed to the increase in the content of aromatic constituents. Furthermore, it was established that humic acids from strongly podzolized soils are most easily peptized and that the values of the coagulation threshold of humic acids are decreased in the following order: chernozem, chestnut soils, forest soils, red soils, and strong podzolic soil.

4. *Influence of Metal Complexing on the Electrophoretic Properties of Humic Acids and Their Fractions*

Besides the more general application of partition paper electrophoresis to problems of the analytical fractionation of organic polyelectrolytes, this method has also been applied to the investigation of the complexing properties of organic soil extracts by several workers. DROSDOVA and YEMEL'YANOVA [1960] indicated by paper electrophoresis that natural complexes of peat humic acids with copper move toward the anode, like humic acids which contain no copper, while cupric ions, which normally move toward the cathode, move to the anode with the humic complex. Complexing with copper was also demonstrated by counter electrophoresis. In the presence of large amounts of copper the rate of the movement to the anode was diminished. The same method was applied by DROSDOVA [1963] to study interactions between humic acids and the divalent nickel and cobalt cations, where similar results were obtained. KONONOVA and TITOVA [1961] used paper electrophoresis to study the complexing of humic acids from chernozem, sod-podzolic soil, and podzol, and of fulvic acids from red earths with iron. As an analytical test of the complex formation between iron and humic acids the pherograms were sprayed with a 4% solution of potassium ferricyanide in sulfuric acid, or the content in iron and carbon of the zonal eluated substances was determined by quantitative chemical analysis. Kononova and Titova found that these complexes are predominantly formed by low molecular fulvic acids and by humic acids of strongly podzolized soils. Mobile complexes of humic acids from chernozem and sod-podzolic soil are not observed. These results were confirmed from preparations of synthetic complex compounds between the pyrophosphate extracts of the above-mentioned soils and solutions of iron chloride. This demonstrated that iron and humic acids, as well as fulvic acids, form stable complexes in a wide range of pH values. The explanation of the complexing mechanism is deduced from the hydration properties of the structural constituents of the humic acid molecules. According to this, the aromatic parts of the molecules are hydrophobic, while peripheric parts of the humic acids molecules contain hydrophilic groups. This is why fulvic acids low in aromatic groups and high in hydrophilic groups show a dominant tendency to form mobile complexes with iron. Similar to this is the behavior of humic acids isolated from strong podzolized soils, which are similar in structure to fulvic acids. Humic acids from sod-podzolic soil and chernozem show a higher degree of aromaticity, which lowers the tendency to complex with heavy metals. This is in agreement with the results of KAWAGUCHI and KYUMA [1959], who found a direct relationship between the formation of chelates with iron and aluminum and the content of terminal groups in the humic substances.

TITOVA [1962] demonstrated by paper electrophoresis and the formation of synthetic complex compounds of humic acids from chernozem, red earth, and fulvic acids from sod-podzolic soils with iron that 1 g carbon of the fulvic acids is bound to 670 to 760 mg Fe_2O_3, while 1 g carbon of humic acids of the other mentioned soil types is bound to only 300 to 350 mg Fe_2O_3.

According to Figure 82 (I–III) the fulvic acids form predominantly mobile complexes, while metal complexes with humic acids of chernozems are immobile, and remain at the starting point of the pherogram. The mobility of humic acids from red earths is intermediate. Comparative experiments with extracts of sodium pyrophosphate with extracts of sodium fluoride showed a complexing of iron from inorganic soil constituents and a formation of metal organic soil extracts during the extraction only in the case of the pyrophosphate solutions. By application of sodium fluoride for extraction, no mobile organic complexes in chernozem, sod-podzolic soil, or in the upper horizons of red earths could be observed by electrophoresis.

The stability of the complex linkages of humic acids from different soil types with copper was investigated by POSPISIL [1962a]. According to the densitograms of electrophoretic experiments (Figure 82 (1–5)), one fraction separates in the beginning, and one or two separate later. Humic substances with a C_h/C_f ratio > 1 show an immobile fraction, while those with dominant amounts of fulvic acids (C_h/C_f < 1) show a greater mobility of the organic metal complexes with copper. Spectroscopic investigations of the fractions in the visible range of light show that

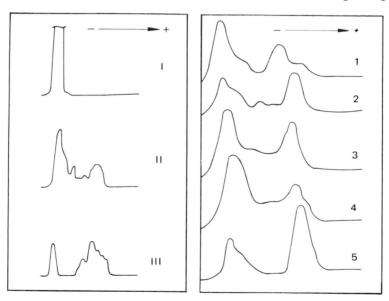

Figure 82. Densitograms of the electrophoretic movement of metal complexes of humic acids.
Left: iron complexes of humic acids from (I) chernozem; (II) red earth; and (III) of fulvic acids from sod-podzolic soil. According to TITOVA [1962].
Right: Copper complexes of humic acids from (1) chernozem [C_h/C_f* = 1.3]; (2) sod-podzolic soil [C_h/C_f = 0.7]; (3) podzol [C_h/C_f = 1.7]; (4) brown earth [C_h/C_f = 1.0]; and (5) brown forest soil [C_h/C_f = 0.8]. According to POSPISIL [1962a].

* C_h/C_f = the ratio between humic acids and fulvic acids.

in front of the electropherograms substances are moved with high color quotients, while those with low color quotients remain at the starting point. From this fact it is concluded that the fractions moving with copper are fulvic acids.

5. Investigation of Humic Acids by Potentiometric Titration

Through investigation of the potentiometric titration behavior, the properties of humic acids have been studied by several authors. In this way the acidic character of humic acids was confirmed. But also from the nature of the potentiometric titration curves the mechanism of the interaction between neutral salts and humic acids was discussed.

(a) Principles of the Titration Properties of Weak Organic Acids

The equilibrium of ionization of a polybasic acid is determined by n different ionization constants (each acid group is determined by one). If these constants differ only slightly and if the number of constants is very high, it is impossible to determine the constants separately. Furthermore, the degree of ionization is influenced by particle size and by salts with low molecular weights.

With the assumption that ionized and nonionized groups are evenly distributed in the polymer (no interaction) and that no linkage occurs between polyanions and counterions, KATCHALSKI and SPITNIK [1947] established the simplified Henderson–Hasselbach equation to calculate the titration curve:

$$pH = pK + n \cdot \log \alpha/(1-\alpha)$$

    α = degree of neutralization

    n = constant

    pK = dissociation constant under the condition of the experiment (increased with increasing concentration of neutral salts).

The dependency of the pH values on the degree of neutralization α, which is proportional to the amount of added base or acid, is graphically characterized by a so-called titration curve. The curve may be constructed directly from the experimental results by plotting the volume of added base or acid versus measured pH values. Plots of the degree of the dissociation α versus the pH values give titration curves with an S shape.

Titration curves of all weak acids show the same centrosymmetric course and differ only in their position in relation to the pH axis. If the added base has neutralized exactly one-half of the original amount of weak acids present, results are that $\alpha = 1/2$, and $\alpha/(1-\alpha) = 1$. The logarithm of this value ($\log \alpha/(1-\alpha)$) is then calculated to be zero, and the pH value of the solution is in this case equal to the pK value. At this point the hydrogen ion concentration is equal to the titration constant. Adjacent to this pH value the course of the titration curve shows the steepest slope.

(b) Potentiometric Titration Properties of Humic Acids

While SPRENGEL [1826] and ODEN [1919] assumed that humic acids are real organic acids, BAUMANN and GULLY [1910] were of the opinion that the acidity of humic acids results from the addition of salts, similar to a process of salt adsorption. Therefore Baumann and Gully believed that humic acids are not real acids, but colloidic complexes with a high adsorption capacity.

THIELE and KETTNER [1953] concluded from the fact that electrometrically titrated humic acids show no endpoint of titration that they are not real acids. ODEN [1919] titrated, potentiometrically, suspensions of peat with sodium hydroxide. He observed that the concentration of hydroxide ions calculated from the measured EMK values show at first a strong increase with the addition of sodium hydroxide. Therefore, he assumed that a combination of free hydroxyl groups with released hydrogen ions of the humic acids occurs to form water. ANDERSON and BYERS [1936] investigated, by potentiometric titration, electrodialized soil humic acids and humic acids from sucrose prepared by sulfuric acid treatment. They found that the base exchange capacity of the soil humic acids was much larger than that of the synthetic humic acids; but inflection points in the titration curves were not obtained. Similar results were found by FEUSTEL [1936], who studied the neutralization reaction of humic acids from peat by potentiometric titration with barium hydroxide. In diluted solutions he obtained inflections of the titration curves, but he did not obtain them at higher concentrations. ZADMARD [1939] found, by titration of humic acids with alkali hydroxides, a greater increase of pH values than by alkaline earth hydroxides. The determination of the exchange capacity of the bases at pH−7 gave the following order: barium hydroxide > calcium hydroxide > potassium hydroxide > lithium hydroxide > sodium hydroxide. GILLAM [1940] observed two distinct inflection points during the potentiometric titration of organic soil substances with sodium hydroxide. In contrast to the results of ZADMARD [1939], CHATTERJEE and BOSE [1952] found the reactivity of bases to be in the order: calcium hydroxide > barium hydroxide > sodium hydroxide. They assumed that Zadmard measured the hydrogen ion concentration of excess solution instead of total suspension, so that the values measured by Zadmard are supposed to be erroneous. According to their results the hydrogen ion activity of the humic acids is about 10^{-4} N, which shows that humic acids are weakly dissociated. The base exchange capacity depends on the kind of the base used for titration (cation effect). If humic acids react alone with bases, irregular cation effects occur, and potentiometric behavior does not follow the order of Hoffmann. In the presence of salts the titration properties of humic acids are regular. The authors believe that the neutralization reactions of humic acids are difficult to explain by the classic laws of the electrochemistry, but might be explained by secondary ion adsorption and the theory of the electric double layer.

HALLA and RUSTON [1955] investigated by conductive and potentiometric titration the anomalies of humic acids prepared from coals (Figure 83). From hysteresis of the titration curves they assumed an exchange reaction of the humic acids, if the base titrated solution is back-titrated with acids.

With respect to the hydration volumes and neglecting activity coefficients and the free

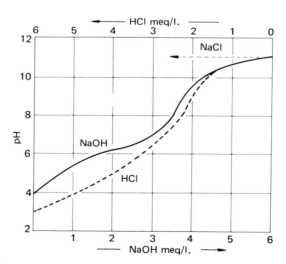

Figure 83. Potentiometric titration of humic acids from coal. According to HALLA and RUSTON [1955].

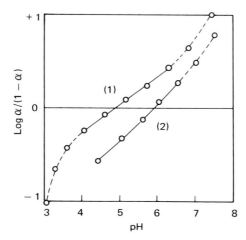

Figure 84. Relation between pH and log $\alpha/(1-\alpha)$ for humus from Beerwah soil. (1) Amberlite-treated suspension (C/Al = 15.8). (2) Original suspension (C/Al = 8.5). According to MARTIN and REEVE [1958].

hydration energies, they found values of the exchange constant K to be in the range between 6 and 10×10^{-4}. A comparison of the results of these authors with the values obtained by ZADMARD [1939] shows Halla and Ruston's values are five to seven times greater. These differences are not derived from different origin of the humic acids, but possibly from a slow alteration of the humic acids, effected by interaction with alkali metals, which changes the initial adsorption capacity.

SCHOBINGER [1958] titrated different types of hydroxycarboxylic acids in nonaqueous solvents such as ethylendiamide or dimethylformamide. He demonstrated the detection of weak acidic phenolic hydroxyl groups in the presence of carboxylic acids groups. In this case two distinct inflection points in the titration curves occur: a stronger one due to the reaction of carboxyl groups and a weaker one of phenolic hydroxyls. By comparative investigation of rotted lignins, he found an increase of the acidic groups in lignin molecules during the decomposition process.

To prove the controversial questions of the precipitation of metal organic products during the development of podzol soils MARTIN and REEVE [1958] investigated acetyl-acetone extracts of organic matter from podzol by potentiometric titration with 0.05 N sodium hydroxide in 0.1 N potassium chloride according to the method published by BECKWITH (1955). They established that, in absence of added metal ions, the shape of the titration curve and the average pK values are influenced by aluminum ions present in suspension (Figure 84). Plots of the pK values versus the C/Al ratio (C = carbon content) reveal linear correlations.

A characteristic effect of the titration of freshly prepared humic substances was the hysteresis during back-titration with acid. A comparative titration of amberlite showed that the forward- and back-titration nearly coincide (Figure 85).

The authors concluded that the irreversibility of the neutralization curve and the pH drop, as well as the yield of hydrogen ions at higher pH values, depends, possibly, on the aluminum atoms present in fresh preparations. They found a strong increase of the pK values with the content of aluminum, which are 3.8 to 4.0 without aluminum, and, therefore lower values than are normally associated with humic acids. Because aluminum ions are easily removed by ion exchangers, it is assumed that a large amount of aluminum is located on the surface of the organic molecules. Titration curves of freshly prepared humic acids with ions of aluminum,

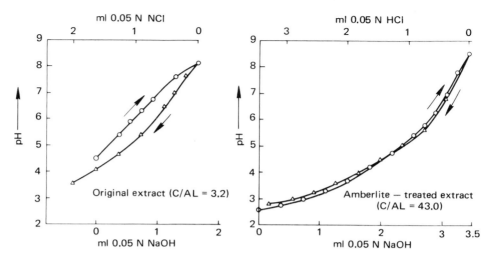

Figure 85. Effect of amberlite treatment on hysteresis of neutralization curves of Te Kompuru humus. According to MARTIN and REEVE [1958].

iron (Fe^{2+}), and of the divalent transition ions do not point to the presence of complexing sites in the molecules. Similar results were found by titration of amberlite-treated soil organic matter extracts with aluminum and copper. Therefore, it is assumed that organic substances are immobilized by electrostatic linkages with aluminum.

The interaction of cations with humic acids and the influence of metal ions on the titration curves of humic acids were studied in detail by alternating titration with different metal salts and bases by VAN DIJK [1959, 1960, 1962, 1967]. Decreases in pH during the addition of metal ions in small portions led VAN DIJK to conclude that protons were released from the weaker acidic groups by the metal ions. In this way copper, for example, may be linked by one valency bond to a free ionic group, while the other may be linked to a deprotonized phenolic

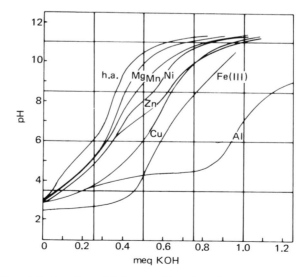

Figure 86. Potentiometric titration of 0.5 meq of humic acids in the absence and presence of 0.5 meq of metal ions. According to VAN DIJK [1967].

hydroxyl group. By this mechanism, two distant carboxyl groups and two hydroxyl groups may participate when brought in contact by thermal agitation. VAN DIJK established that the ease of replacement of the following ions by barium is in the order: manganese, cobalt, nickel, lead, copper, and trivalent iron. In the same order the tendency of the formation of hydroxyl complexes of the metal ions and the tendency of cleavage and formation of oxide hydrates are increased. From the pH-dependent displacement of the equivalence points van Dijk concluded that metal ions with the strongest complexing ability give the greatest pH drop during titration. Titration curves show that copper combines with one-half of the acidic groups of the humic acids. Figure 86 shows the titration curves of humic acids in the presence of various metal salts. The titration properties of aluminium are completely different from all the other metal ions. Here, as in the case of the inorganic aluminium salts, aluminum is transformed quantitatively to its hydroxide at pH 4.1.

Furthermore, it follows from the titration curves that all cations show a strongly different and pH-dependent influence on the course of the titration curve. At pH 3 the differences of the titration properties of the several applied cations are small, because at this pH value the ions are still in a dissociated state in solution. According to the values of the pH drop during the addition of inorganic salts at pH 5, no large differences were found for the weakly bound divalent ions such as barium, calcium, magnesium, manganese, cobalt, nickel, iron, and zinc. Lead, copper, and trivalent iron are more strongly bound. In spite of differences in nitrogen content and the ratio of weaker and stronger acidic groups present in the molecules, only small differences in the mechanism and the strength of bonds with the ions are observed between humic acids isolated from different soils. VAN DIJK thus concluded that the linkage between humic acids and ions of the transition metals is of covalent character.

The order of pH drop caused by the addition of metal ions of the first transition row was in agreement with the results of potentiometric titration carried out by KHANNA and STEVESON [1962]. They assumed that the pH drop originated from dislocation of the equilibrium of ionization of weakly acidic groups, and possibly also from the release of protons from functional groups, which are normally not ionized (alcohol groups, hydroxyl groups, and possibly amino groups).

SCHNITZER and SKINNER [1963a] determined by potentiometric titration the complexing properties of organic substances, isolated from alkaline extracts of the B_h horizon of a podzol soil with iron, calcium, magnesium, copper, and nickel cations.

A comparison of the titration curves of the metal salts with those of the equimolar metal organic complexes (Figure 87) shows that the formation of metal hydroxides is indicated by distinct inflection points: Fe^{3+} at pH 3.2, Al^{3+} at pH 4.5, Ni^{2+} at pH between 8.4 and 9.0, as well as Cu^{2+} between pH 5.4 and 5.6. From the fact that the titration curves of the equimolar mixtures of the organic substances with the metal salts are in more or less good agreement with the titration curve of the original sample, and because inflection points of precipitated hydroxides are not observed, the authors assume the existence of a complex formation of the metals with organic substances. Table 27 shows the release of protons during the titration of the cations in the presence and absence of the organic substances.

From the proposed molecular formula of these organic substances, $C_{20}H_{12}(COOH)_6$ $(OH)_5(CO)_2$, SCHNITZER and DESJARDINS [1962] assumed that at pH 3 one carboxylic acid group, at pH 6 five groups, and at pH 8 all six carboxylic acid groups were titrated. Additionally, they assumed a titration of two phenolic hydroxyl groups of the organic substances at pH 10. The determination of the release of protons during titration of metal complexes gives two protons for Fe^{3+} at pH 8. Therefore, it is concluded that in this pH range $[Fe(OH)_2]^+$ is combined with one negatively charged carboxyl group by an electrovalent linkage. Beyond pH 9 the complex seems to be destroyed by iron hydroxide formation and by release of three

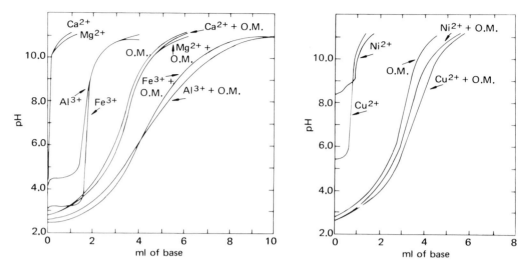

Figure 87. Potentiometric titration of 0.029 mmoles of metal ions in the absence and presence of 0.029 mmoles of organic matter (O.M.). According to SCHNITZER and SKINNER [1963a].

protons. In the case of Al^{3+} application, one proton is released at pH 3, corresponding to a linkage of $[Al(OH)]^{2+}$ with two carboxyl groups, while the release of two protons at pH 7 is obtained by a reaction between $[Al(OH)_2]^+$ and one carboxyl group. Three protons are released beyond pH 9 to form $Al(OH)^3$, and the organo-metalic complex is finally destroyed. For copper these authors find in agreement with results of KHANNA and STEVENSON [1962] a release of nearly one proton (0.8–1.1) in the range between pH 3 and 8, while the values for nickel at pH 9 are lower than 0.1, pointing to a remarkable stability of the complexes formed with this cation. With calcium and magnesium no release of protons was observed. According to these results it is assumed that the pH-dependent complexing is performed at molar ratios of 1:1 and 6:1. The solubility of the complexes in water decreases with increasing metal content.

pH effects during the potentiometric titration of metal organic complexes of humic acids and fulvic acids with iron and aluminium were confirmed by ALEXANDROVA (1967). The results of this investigation (Figure 88) show that the original acids (1) need lower amounts of alkali for neutralization than is necessary for titration of metal organic complexes (2, 3, 4).

Table 27. Potentiometric titrations of 0.0029 mmoles of metal ions in the absence and presence of 0.0029 mmoles (19.4 mg) of O.M. According to SCHNITZER and SKINNER [1963a].

pH	Number of protons released per mole of base					Number of protons released per mole of metal added			
	O.M.	Fe^{3+}	Al^{3+}	Ni^{2+}	Cu^{2+}	Fe^{3+}	Al^{3+}	Ni^{2+}	Cu^{2+}
3	1.2	—	—	—	—	2.0	1.0	0.5	0.8
4	3.2	2.7	—	—	—	1.8	1.3	0.4	0.9
5	4.4	2.9	2.4	—	—	1.6	1.4	0.4	0.8
6	5.2	2.9	2.7	—	1.3	1.8	1.5	0.4	0.8
7	5.7	3.0	2.8	—	1.5	1.9	2.1	0.4	0.9
8	6.2	3.1	2.9	—	1.6	2.5	2.9	0.4	1.1
9	6.7	3.2	3.2	1.6	1.6	3.1	3.7	0.7	1.3
10	7.6	3.7	3.7	2.0	1.7	3.8	4.7	1.1	1.6

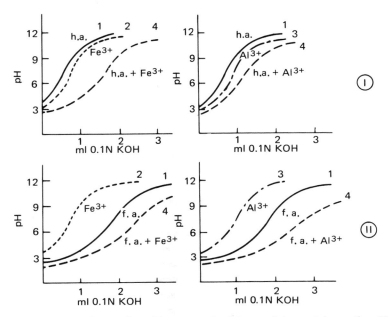

Figure 88. 1 in section I is humic acid untreated with metal ions. 1 in section II is untreated (control) fulvic acid. 2 is iron, 3 is aluminum sulfates, and 4 is iron and alumino-humus compounds. According to ALEXANDROVA [1967].

POMMER and BREGER [1960a] examined the contradictory results of the titration experiments of THIELE and KETTNER [1953] in a detailed publication. According to a new calculation of the results from the titrations carried out by Thiele and Kettner, they established that the humic acids are, in contrast to the opinion of these authors, weak acids and that the titration curves are not similar to adsorption isotherms (Figure 89).

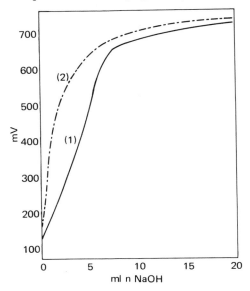

Figure 89. Potentiometric titration of humic acids (1) and phenol formaldehyde resin (2). Titration data from THIELE and KETTNER [1953]. According to POMMER and BREGER [1960a].

The titration curves of synthetic exchange resins, which conform with adsorption iso-therms, yielded linear functions according to Freundlich,* if the added amount of bases is plotted in a logarithmic scale versus the values of electrometric measurement (Figure 90). The

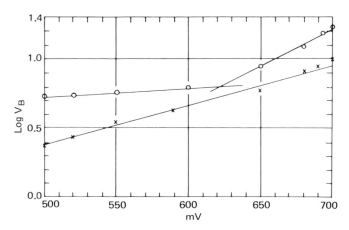

Figure 90. Plot of log V against millivolts for discontinuous titration of humic acids (O) and phenol formaldehyde resin (X). Titration data from THIELE and KETTNER [1953]. According to POMMER and BREGER [1960a].

titration of weak polanyions in the semilogarithmic diagram gives two independent linear functions, intersecting at the equivalence point. A prerequisite of this method is a discontinuous titration maintaining constant volume of solution to assure equilibrium conditions.

By transformation of the values obtained by Thiele and Kettner the equivalent weight of humic acids (Merck) was found to be 150, which was in agreement with the results obtained from their own titrations. In a further investigation POMMER and BREGER [1960a] determined the equivalent weight of humic acids from peat, obtaining values between 144 and 183. It was demonstrated that the titration is a strongly pH-dependent and time-dependent equilibrium reaction. Final equilibriums are obtained only after days. From the rise of the equivalent weights of the applied humic acids, a chemical transformation was assumed.

The method of discontinuous potentiometric titration of Pommer and Breger was also used by SCHNITZER and DESJARDINS [1962] to determine equivalent weights of the organic substances isolated from podzol. The A_0 material gave equivalent weights of 165; the B_h material gave 76. In contrast to the results of Pommer and Breger, the equivalent weights of the A_0 material decreased with time, while the B_h material changed little.

SCHNITZER and GUPTA [1965] determined the total acidity of humic acids from podzols in equilibrium with barium hydroxide by discontinuous potentiometric titration. The content in carboxyls was determined by ion exchange with calcium acetate and by decarboxylation with basic copper carbonate. From these results the authors calculated that the content of phenolic hydroxyl groups is equal to the difference between total acidity and carboxyl groups. But the authors found also, that the values of the phenolic hydroxyl groups of known organic compounds are not sufficiently determined by any of the mentioned methods.

POSNER [1964] investigated the influence of the ionic strength on the titration properties of humic acids to prove the polyelectrolyte behavior of humic acids by potentiometric titration

---

* The Freundlich adsorption isotherm can be represented by the equation $a = K \cdot c^n$, where a is the amount of solute adsorbed per weight of adsorbent from a solution of equilibrium concentration c; K and n are constants.

of humic acids from a red-brown soil in contact with various amounts of added salts. For this purpose Posner discussed in detail the application of the Henderson–Hasselbach equation, modified by KATCHALSKI and SPITNIK [1947]:

$$\text{pH} = \text{pK} + 1/n \cdot \log \alpha/(1-\alpha)$$

This equation is applicable to the titration curves of linear polyelectrolytes. MARTIN and REEVE [1958] and KHANNA and STEVENSON [1962] found, in contrast to VAN DIJK [1959], that the results of their investigations conform with this equation. With the presumption that all functional groups of the same molecule have the same intrinsic pK value and that the differences arise because of coulombic interactions between charges of the same molecule as the molecule dissociates, Posner established an equation for spherical molecules, which neglects the effect of polydispersion. As a result of these theoretical assumptions he found the plot of the difference between the pH value and $\log \alpha/(1-\alpha)$ versus α to be a straight line. The slope decreases with increasing ionic strength.

According to this, plots of the results from potentiometric titration obtained at different ionic strengths (Figure 91) show that the slopes of the curves are independent of ionic strength. The influence of the ionic strength effects only a parallel displacement of the curves.

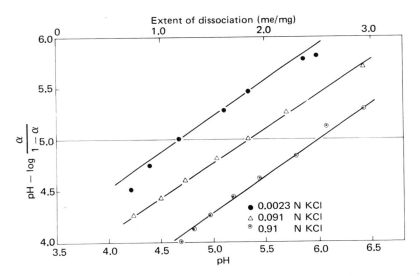

Figure 91. Relationship between pH and pH log $\alpha/(1-\alpha)$ and effect of ionic strengths on extent of dissociation. According to POSNER, 1964.

From these results it is assumed that the shape of the acid titration curve is not due to electrical interactions on the same molecule but rather to a distribution of the pK values not only within a molecule but also between molecules. From these results the authors conclude that the absence of expected influences of the ionic strength on the titration properties of humic acids is in contrast to the theory that humic acids behave like polyelectrolytes. Later it was possible to fractionate humic acids according to the magnitude of their pK values (POSNER [1966]). Furthermore, it was demonstrated that the shape of the titration curve is in agreement with the existence of a gaussian distribution of the pK values and that at lower ionic strength the pK values decrease linearly with the square root of the ionic strength. By extrapolation to zero ionic strength, an average pK value is obtained, which is corrected for effects of ionic strength (pK_o^I). The slope of this linear function depends on the charge of the ionizing groups and on the activity coefficients of the humate ion:

$$\log f = -0.5 \cdot z^2 \cdot \sqrt{\mu}$$

where z = charge
f = activity coefficient
μ = ionic strength

With reference to the approximation method given by TANFORD [1962] in case of proteins, Posner proposed a classification of the titration curve into three regions: region I from pH 3 through 7 assigned to the carboxyl groups, region II from pH 7 through 8 assigned to α-amino nitrogen, and region III from 8 through 12 to phenolic hydroxyls. Region II overlaps regions I and III.

POSNER [1964] also titrated humic acids in the presence of sodium chloride and lithium chloride, and found that the results conform with those of potassium chloride. The same behavior was found by application of calcium chloride and barium chloride with ionic strengths of 0.0023, 0.0096, 0.091, and 0.91. In solutions in which the ionic strength was 0.91 N of the above salts, the humic acids titrated like strong acids. This points to the fact that all protons from humic acids were released by application of divalent cations. At lower ionic strengths the titration curves were displaced toward higher pH values. Because the results obtained in the acid region are in agreement with the theoretical assumptions, it is supposed that the pK values are influenced only by the cation effect in the presence of the diluted solutions of divalent cations.

6. Conductometric Titration Properties of Humic Acids

The results of conductometric titration of humic acids published by different authors show a variation in the number of breaks in the titration curves (between one and four), whereby the polybasic character of the humic acids is assumed. In contrast to the results from

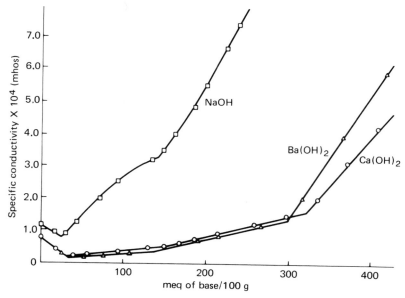

Figure 92. Conductometric titration curves of humic acids (SOL B) with bases. According to CHATTERJEE and BOSE [1952].

potentiometric titration, but in agreement with the results of ODEN [1919], CHATTERJEE and BOSE [1952] obtained three breaks in the titration curves of commercial humic acids (Merck) by conductometric titration (Figure 92), while in case of soil humic acids only one minimum of conductance was observed.

From the titration curves it appears that the neutralization of humic acids by bases follows the order $Ca(OH)_2 > Ba(OH)_2 > NaOH$. The pronounced influence of calcium ions on the neutralization reaction of humic acids is assigned to the great insolubility of calcium humate compared to barium humate. In contrast, by potentiometric titration of humic acids by application of the above bases in the presence of their chlorides it was established that barium humate is less soluble than calcium humate. According to the theory of the electric double-layer this may be explained by an increase of the total acidity in the presence of added salts, effecting a replacement of hydrogen ions bound to the double-layer by cations of the applied salts (secondary adsorption; compare MUKHERJEE [1920, 1922]).

HALLA and RUSTON [1955] investigated in detail the anomalies of conductance of electro-dialized humic acids prepared by nitric acid oxidation of coals (Figure 93). According to the

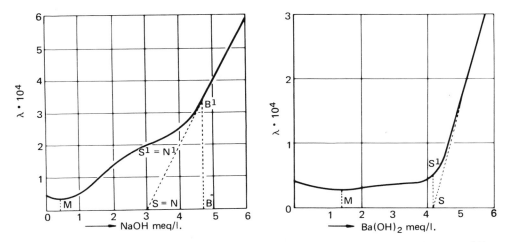

Figure 93. Conductometric titration of humic acids from coal. According to HALLA and RUSTON [1955].

theory of conductometric titration, the conductance of the solution depends on the number, the charge, and the mobility of the ions present, and therefore, several characteristic points of the conductometric titration curves can be described as follows:

(1) The minimum conductance at M occurs at the beginning because the hydroxonium ions in the added base are combined with the replaced, initially adsorbed hydrogen ions to form water and a cation-polyanion complex that is not highly dissociated.
(2) The points B (B'), at which the conductivity of the "free base" determines the further course of the titration curve with a constant slope.
(3) The intersecting point S of the linear backward extrapolation of the titration curve of the free base with the abscissa, which determines the unavailable bases; in case of alkali titration equivalent to the point of neutralization N (pH = 7).

These conditions are more diffuse by application of alkaline earth metal hydroxides. Here a more or less broad minimum occurs, and the intersecting point (S) with the abscissa is no more identical with the point of neutralization. It is displaced to lower pH values in the acidic range.

From this it is assumed that the humate is already precipitated before reaching the point of neutralization. The different kinds of ionic reactions, which determine the changes of conductivity in the course of titration, are explained by HALLA and RUSTON [1965] by the assumption that cations and especially alkali ions produce slow alterations of humic acids. Adsorption properties, caused by a cleavage of the hydrogen bridge linkage between carboxyl and hydroxyl groups of adjacent molecules are changed and the systems tend to disperse.

By conductometric titration with lithium hydroxide, potassium hydroxide, and barium hydroxide, ROY [1957] compared the titration properties of humic acids derived from peat, soils, and commercial humic acids (Merck). With the application of the commercial humic acids (Merck), two breaks of the conductometric titration curve occur, while soil humic acids reveal only one. The neutralization reaction of bases follows the order LiOH > Ba(OH)$_2$ > KOH, in agreement with the results from potentiometric titration. Because in each case a discrete dependence of the base exchange properties of the kind of applied bases was observed, Roy assumed a predominance of the "pH effect," as opposed to the "cation effect."

Similarly, VAN DIJK [1960] studied by conductometric titration the complexing reaction of bases with soil humic acids in the hydrogen state, which were prepared by cation exchange with amberlite IR 120 (Figure 94).

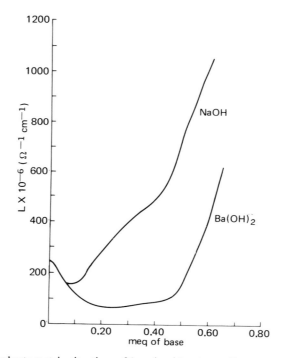

Figure 94. Conductometric titration of humi acids. According to VAN DIJK [1959].

A detailed comparative investigation of synthetic and natural humic acids revealed no essential differences in the titration properties. During conductometric titration of humic acids dissolved in dimethylformamide, sodium isopropylate in isopropyl alcohol/pyridine and in butylammonium hydroxide in isopropyl alcohol, no defined endpoint of titration was observed.

By conductometric titration of humic acids from peat, PIRET, WHITE, WALTHER, and MADDEN [1960] established discontinuities in the titration curves at humic acid concentrations in the range between 18 and 20 g/1. These are attributed to the formation of large and weakly charged association colloids. This is confirmed by measurements of the relative surface tension and the reduced viscosity, so that these authors describe this range of critical humic acids concentration as critical micellar concentration.

Sodium hydroxide extracts from the B_h horizon of a podzol, which were purified by amberlite IR 120 treatment, were studied by conductometric titration with sodium hydroxide and calcium hydroxide by SCHNITZER and SKINNER [1963a]. These authors assume the minimum of the sodium hydroxide titration to be the point of neutralization of two strongly acidic carboxyl hydrogen atoms (Figure 95).

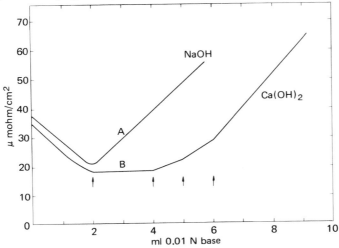

Figure 95. Conductometric titration of soil organic matter from podzol (B_h) with sodium hydroxide and calcium hydroxide. According to SCHNITZER and SKINNER [1963a].

The additional three breaks in the conductometric titration curve from the calcium hydroxide titration, in the range between 4 and 6 meq, is explained by a stepwise change of conductance of carboxyl groups as the acidity changes. The participation of phenolic groups in this reaction could not be established.

7. *Spectroscopic Investigations of the Complexing of Humic Acids and Their Fractions with Metals*

BROADBENT and OTT [1957] investigated, by *spectrophotometry* in the visible range of light, the complexing properties of the soluble amounts of hydrochloric acid precipitations of different organic soil extracts with copper. They found that a mixture of these substances shows a stronger absorption than comparable amounts of soluble organic substances, but less absorption than a comparable solution of copper sulfate above 535 nm (Figure 96).

For the investigation of the complexing properties, these authors applied the method developed by Job and described by VOSBURGH and COPPER [1941], which is used for the determination of the composition of soluble, colored complexes by variation of the cations. This technique depends on the fact that at a constant total concentration of the metal and the complexing agent the concentration of the complex is the highest, if the proportions of concentration of the components are the same as in the complex compound formed. If two colorless solutions form a colored complex and if the optical density of varying properties of the two components is determined, a maximum or minimum of optical density occurs at concentration ratios, when the stoichiometric composition is the same as in the complex

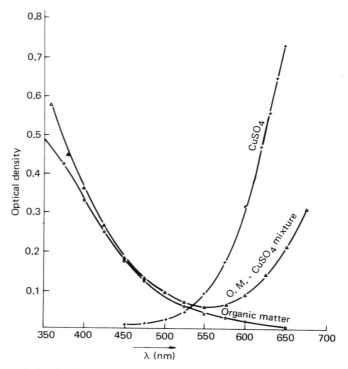

Figure 96. Spectral distribution curves of soil organic matter solution, copper sulfate, and a 1:1 mixture of the two. According to Broadbent and Ott [1957].

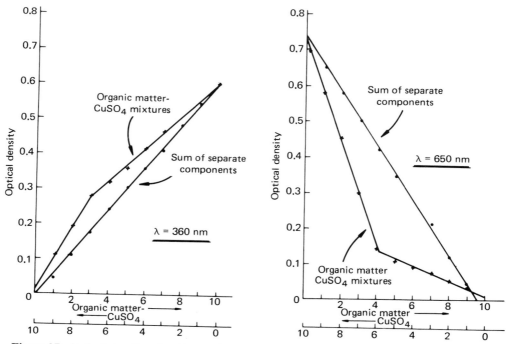

Figure 97. Optical density of soil organic matter–copper sulfate mixtures and of separate components at 360 and 650 nm. According to Broadbent and Ott [1957].

formed. If one or both of the reaction substances are colored, the differences between the observed optical density and that which would have resulted without complexing will show a maximum at the complex ratio. This method is not generally applicable to systems that form more than one type of these compounds. The results of such measurements at different wavelengths are demonstrated in Figure 97.

From the different ratios of 3:7 and 4:6 at 650 nm determined from the position of the maxima, BROADBENT and OTT [1957] assume the formation of more than one type of chelate during the complexing of the organic substances with copper.

PAVEL [1959] studied, in a similar manner, the complexing properties of fulvic acids from peat with copper, chromium, manganese, and cobalt. According to the results from difference-spectra obtained at different wavelengths, metal-organic mixtures of different composition are suggested. As revealed by spectrophotometric measurements, fulvic acids combine with copper by formation of at least one, with cobalt of at least three to four, and with chromium of at least four different types of chelate complexes.

The method of Job (compare VOSBURGH and COOPER [1941]) was also applied by SCHNITZER and SKINNER [1963a] for investigations of the complexing properties of alkali extracts from podzols. According to the Job plots of the complexes between the organic substances and copper, iron and aluminum at pH 3 (Figure 98(a)), a molar ratio of 1:1 is found. Copper and iron at pH 5 form 2:1 molar complexes (Figure 98(b)). Aluminum, with a complexing ratio of 1:1, shows no change at different hydrogen ion concentrations and therefore no pH dependency of the complex formation.

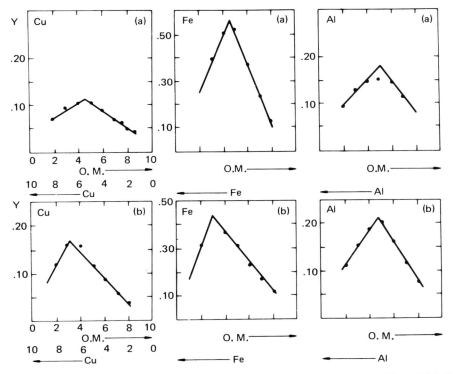

Figure 98. Job plots for Cu^{2+} plus organic matter (O.M.), $Fe^{3+} + O.M.$, $Al^{3+} + O.M.$ (a) at pH 3 and (b) at pH 5. Figures along the abscissa represent moles of each constituent. According to SCHNITZER and SKINNER [1963a].

By application of calcium and magnesium it was impossible to obtain Job plots, because the differences in the y-values (y is the difference between the optical density with complex formation and the optical density without complex formation) were too low to be determined at the different molar ratios of the organic substances and metals. *Infrared spectroscopic* investigations of the interaction between humic acids from chernozem and sod-podzolic soil with barium hydroxide were performed by KASATOTCHKIN, KONONOVA, and ZILBERBRAND [1958]. They found that the treatment of preparations with barium hydroxide leads to an overlapping of the bands occurring at 6.1 μ (1630 cm^{-1}) and 6.5 μ (1538 cm^{-1}) by a broad absorption at 6.35 μ (1575 cm^{-1}) (Figure 99). This absorption band is assigned to ionized carboxyl groups.

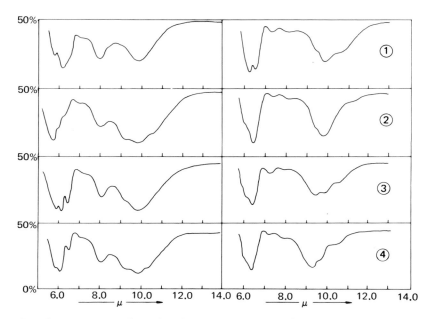

Figure 99. Infrared spectra of humic acids from chernozem (1), fulvic acids from chernozem (2), humic acids from derno-podzolic soil (3), and fulvic acids from derno-podzolic soil (4) before (left) and after treatment with Ba(OH)$_2$ (right). According to KASATOTCHKIN, KONONOVA and ZILBERBRAND [1958].

In all samples after treatment the absorption band at 8.0 μ (1250 cm^{-1}), which is assigned to the C—O vibration of C—OH groups in organic acids and phenols, disappeared during complex formation between organic substances and metals. Direct information about the kind of metal-organic linkage cannot be obtained from the infrared spectra.

By treatment of alkali extracts from podzol with increasing amounts of iron, SCHNITZER and SKINNER [1963a] determined, from infrared spectra, a decrease of the carboxyl absorption at 5.8 μ (1725 cm^{-1}), while the absorption of the carboxylate anion at 6.25 μ (1600 cm^{-1}) was increased (Figure 100).

Coincidentally, the intensity of the carboxylate absorption at 7.3 μ (1400 cm^{-1}) increased, while that at 8.32 μ (1200 cm^{-1}) decreased. After a molar ratio of 6:1 was reached, the bands at 5.8 μ (1725 cm^{-1}) and at 8.32 μ (1200 cm^{-1}) disappeared completely, while the intensity of the absorption band at 6.25 μ (1600 cm^{-1}) and 7.13 μ (1400 cm^{-1}) reached a maximum.* No

* Compare Table 22.

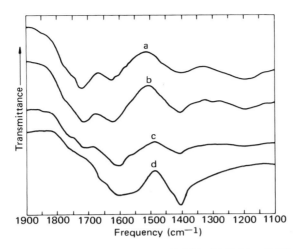

Figure 100. Infrared spectra of (a) organic matter (O.M.), (b) Fe^{3+} + O.M. mixture 1:1, (c) Fe^{3+} + O.M. mixture 3:1, (d) Fe^{3+} + O.M. mixture 6:1. According to Schnitzer and Skinner [1963a].

participation of phenolic hydroxyl groups in the interaction between soil organic matter and iron is detectable by infrared absorption. By use of aluminum instead of iron, similar results were obtained. From this fact a reaction between ionized carboxyl groups and partly hydroxylized iron and aluminium ions is assumed by participation of electrovalent linkages.

8. *Miscellaneous Methods Applied to the Investigation of Interactions between Humic Acids and Metals*

The possibilities for the application of *differential thermogravimetry* (DTG) for investigations of the complexing of soil organic substances with metals were demonstrated by Schnitzer and Skinner [1964] and Schnitzer and Hoffmann [1966]. These authors studied the properties of thermal decomposition of fulvic extracts from podzol. The thermal decomposition curve of this substance shows a maximum of main decomposition at 430°C, a weak maximum in the range between 250 and 270°C, as well as a low peak at 100°C assigned to the release of water.

By comparative investigations on the influence of metal linkages to the thermal decomposition properties of fulvic acids in relation to an increasing content in iron and aluminum, Schnitzer and Skinner [1964] demonstrated, in the case of iron, a displacement of the main decomposition reaction to lower temperatures at 270°C with increasing iron content and the complete loss of the original decomposition maximum at 250°C (Figure 101 (1–4)). DTG curves of the alumino-organic complexes show, at low metal contents, little change in the exothermal properties compared with the original curve, while with increasing aluminum content the decomposition curve is changed into a broad maximum in the range between 350 and 450°C, with a loss of the maximum at 250°C (Figure 101 (5–7)).

Because mixtures of organic substances with goethite and gibbsite in various ratios showed only slight decreases in temperatures of the main decomposition reactions, it was assumed that catalytic effects yield no changes in the thermal properties of organo-metal complexes.

Schnitzer and Hoffmann [1966] studied the influence of different mono- and divalent cations on the thermal decomposition properties of organic substances. In this way it was established that, in different types of alkali fulvates, the maximum in the range at 250°C remained nearly unchanged, while the main decomposition peak, originally at 420°C, was definitely decreased and partly displaced above 600°C in an unexplained manner. In this tem-

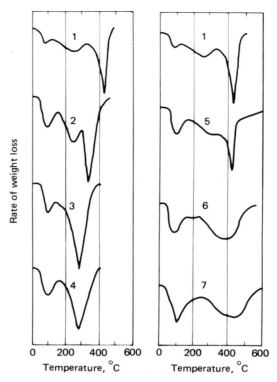

Figure 101. DTG curves of organic matter (O.M.) and of metal–O.M. complexes: (1) O.M.; (2) low Fe complex; (3) medium Fe complex; (4) high Fe complex; (5) low Al complex; (6) medium Al complex; (7) high Al complex. According to SCHNITZER and SKINNER [1964].

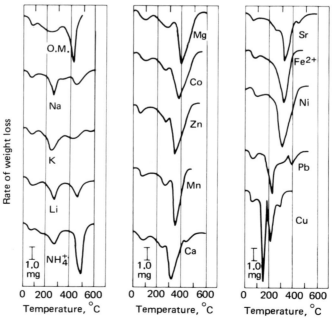

Figure 102. DTG curves of fulvic acids (O.M.) and their alkali salts and of their complexes with divalent metal ions. According to SCHNITZER and HOFFMANN [1966].

perature range exothermal reactions occur, caused by thermal decomposition of alkali carbonates formed during heating. Therefore, this range in Figure 102 for better comparison of the results of complexing of divalent and monovalent ions with organic substances is neglected.

The main decomposition temperatures of the various divalent metal complexes decreased in the order: Mg > Co > Zn > Mn > Ca = Sr = Fe > Ni > Pb > Cu. From this it may be assumed that the thermal stability constants are inversely related to the stability constants derived from the law of mass action. The stability constants were determined by ion-exchange equilibrium methods. The remarkably low, but strongly detailed, decomposition reaction of copper complexes at 200 and 300°C appears to be very significant, but no explanation of this was given by Schnitzer and Hoffmann.

ORLOV and EROSICEVA [1967] investigated the complexing properties of humic acids and metals by *polarographic* methods. Because humic acids show no polarographic waves, and results from the polarographic behavior of metal amines show no displacement of the half-wave potential of the metals, the complexing of metals with ammonium humate was quantitatively determined by measurements of the decrease of polarographic waves of metal amines. By stepwise additions of metal ions to solutions of ammonium humates, it could be established that a corresponding stepwise substitution of hydrogen ions of different functional groups occurred (Figure 103).

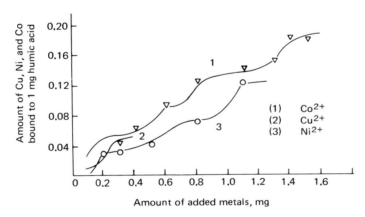

Figure 103. Interaction of metal amine complexes and humic acids. According to ORLOV and EROSECIVA [1967].

The parallel orientation of the wave points to the fact that the applied ions, Cu^{2+}, Ni^{2+}, and Co^{2+}, react in a very similar manner with humic acids. According to these results, cobalt forms four, nickel three, and copper more than one type of complex compounds.

By *electrolytic concentration* and quantitative analysis of the applied electrode materials by means of *emission spectroscopy*, KLEIST [1963] established a change of the original charge of metal ions in the presence of humic acids. He demonstrated that copper, iron, aluminum, and cobalt are transported by humic acids in the direction of the anode. These metals were detected by application of spectroanalytical methods after concentration. Sodium, magnesium, calcium, and strontium did not lose their ionic state, in agreement with the results of fractionation attempts in borate buffer (pH 8.6) by paper electrophoresis.

Investigations on the influence of metal complexing on the decrease in radical content of gray humic acids with originally high radical contents, by measurements of paramagnetic resonance (KLEIST [1967]) reveal that the resonance signal is strongly decreased by copper ions. From this effect, it is concluded that there is a pronounced reactivity of copper with humic acids. The influence of Fe^{3+}, Fe^{2+}, and Ni^{2+} is intermediate, while Co^{2+} and Al^{3+} show weaker effects. A slight influence on the resonance signal is found by application of Ag^+ and Pb^{2+}, while by application of Ba^{2+} and Sr^{2+}, the content of radicals in humic acids remains unchanged. These results show that paramagnetic ions produce a pronounced decrease of the resonance signals during the interaction with humic acids. Specific interactions of special functional groups, *e.g.*, semiquinones are not clearly detected, because certain mechanisms overlap each other by mesomeric stabilization of the radicals.

VII. Shape, Size, and Particle Weight of Humic Acids and Their Fractions

1. *Dialysis and Ultrafiltration*

Dialysis and ultrafiltration are the simplest methods of obtaining information about the colloidal nature of soil organic substances. Low molecular substances diffuse through the pores of suitable membranes, while colloidal particles cannot unless the particles are very small and the pores large. By application of membranes of appropiate pore size humic acids may be fractionated according to different particle sizes. They may be separated from low molecular admixtures by the use of membranes with smaller pores. For analytical determinations these methods are not important, because very often interactions between membranes and the filtered substances result in clogged pores.

2. *Determination of Particle Size by Ultramicroscopy and Electron Microscopy*

The well-known method for the determination of absolute particle sizes in colloidal suspensions is ultramicroscopy. Centers of light scattered by single particles of the colloidal system in a beam of light are observed from the side with a microscope. SVEN ODEN [1922] found the ultramicroscope to be unsuitable for the study of humic acids.

0.5μ

Figure 104. Electron microscopic photograph of natural humic acids from black soil at pH 10 to 11. According to FLAIG and BEUTELSPACHER [1954].

The observation of dispersed colloidal particles in the electron microscope depends on the scattering of electron rays by the electron shells of the observed object and not on account of their absorption. Because scattering of electron rays depends upon the number of atoms per unit volume and on the atomic number, difficulties arise in the sample preparation of humic acids for electron microscopic investigation, because the particles are very transparent to electron rays and therefore are not easily distinguished from the specimen supporting films. By shadowing with a heavy metal (*e.g.*, gold or platinum) in vacuum, it is possible to obtain contrast at the edges of the humic acid particles. Dilute humic acid solutions must be applied to the specimen supporting films and they must be carefully freeze-dried to obtain single particles rather than aggregates.

FLAIG and BEUTELSPACHER [1950, 1951, 1954] and BEUTELSPACHER [1955] studied the morphological features of humic acids by electron microscopy. They mounted peptized humic acids as little droplets on specimen supporting films and sublimed the solvent by freeze-drying in vacuum. For better contrast, the samples were shadow-casted with platinum. The results of the electron microscopic investigations of humic acids show that single particles may be obtained from humates and solutions in alkaline medium. That means that complete peptization was reached under the conditions (Figure 104).

According to the results of detailed electron microscopic investigations of humic acids samples from different origin, it was shown that:

(1) A remarkable number of particles (with diameters up to 100 Å) was beyond the upper limit (> 20 Å) of the electron microscope.

(2) The single particles occurred in the form of globular spheres, as shown by the oval shadows.

(3) All humic acid samples isolated from soils, which were purified by repeated precipitation, showed varying amounts of platy particles, pointing to admixtures of clay minerals, which were peptized during alkali extractions and were retained by humic acids during flocculation in acidic medium.

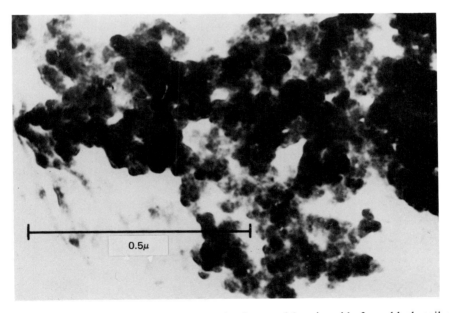

0.5μ

Figure 105. Electron microscopic photograph of natural humic acids from black soil at pH 3. According to FLAIG and BEUTELSPACHER [1954].

In the case of samples prepared by extraction from peat, the samples contain mostly small fibrous particles of plant residues. These admixtures to the humic acids are the chief reason why physicochemical results lead to different conclusions dependent on the kind of the impurities present.

Figure 105 is a picture of natural humic acids derived from a black soil. It was electro-dialized to pH 4 before sample preparation. Here the humic acids appear as cluster-like aggregates, and single particles are united to a larger globular structure due to coacervation.

3. Characterization of Humic Acids by Small-Angle X-ray Scattering

Typical interferences and reinforcements of X-rays by crystalline substances allow an assignment of characteristic diffraction bands for the purpose of quantitative and qualitative analyses of elements and compounds and enables, under certain conditions, the crystal structures to be determined. In contrast, unoriented structureless molecules (amorphous substances) show at a low angle of diffraction only a diffuse scattering of X-rays. From the maximum occurring at low scattering angles, information regarding particle shape, volumes, diameters, and under certain conditions, particle weights can be obtained.

(a) Principles of the Method

The principles of X-ray scattering of colloidal systems were established by KRATKY [1938] and GUINIER [1939] and can only be applied to diluted monodispersed solutions. The basic studies of Guinier showed that the diffuse scattering of X-rays of single corpuscular particles may be characterized in first approximation by a gaussian distribution curve according to the equation:

$$I = I_0 \cdot e^{-k \cdot R \cdot \theta^2} \tag{1}$$

where I_0 means the scattered intensity at zero diffraction, θ is one-half of the scattering angle, $k = \text{constant} = 16\pi^2/3^{\lambda 2}$, and $R = $ radius of gyration.

In actual practice the distance m of the detector above the primary beam is measured. This is related to the scattering angle by $2\theta = m/a$, where a is the distance from the sample to the detector. By transformation of equation (1) to logarithms, equation (2) is obtained:

$$\ln I_\theta = \ln I_0 - k \cdot R^2 \cdot \theta^2 \tag{2}$$

from which plots of the logarithm of the scattered intensity versus the squares of the scattering angles (θ^2) or the distances (m^2) give a linear function with the tangent $-k \cdot R^2$.

This diagram is known as Guinier curve. From the numeric value of the slope of the tangent it is possible to calculate the radii of gyration of the particles (R), and by linear extrapolation to the ordinate, the scattering intensity I_0 may be calculated, which is necessary for the determination of volumes. In practice there are in most cases deviations from the Guinier curve from linearity, but these are of no consequence for the calculation of the radii of gyration from the tangent (KRATKY [1960]). The particle volumes are calculated by the invariant $\bar{Q}$, which is the area under the $I \cdot m$ versus m curve, according to the equation:

$$V = \frac{I_0}{\bar{Q}} \cdot \frac{(\lambda \cdot a)^3}{4\pi} [\text{Å}] \tag{3}$$

(b) X-Ray Scattering of Humic Acids

In a recent publication WERSHAW, BURCAR, SUTULA, and WIGINTON [1967] studied the small-angle X-ray scattering of sodium humate solutions from a sandy soil. They were prepared

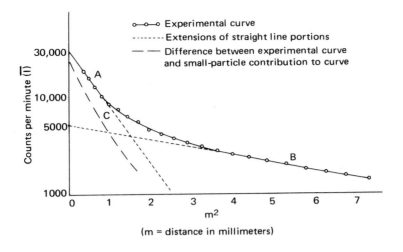

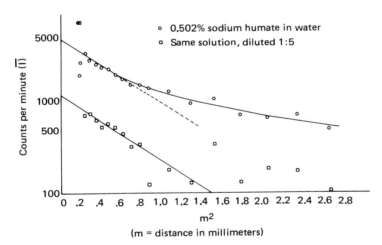

Figure 106. Guinier plots of humic acids and sodium humate as revealed by small-angle X-ray scattering. According to WERSHAW, BURCAR, SUTULA, and WIGINTON [1967].

by extractions with 0.1 N sodium hydroxide followed by gel filtration. The Guinier curve (Figure 106) of the measurements may be described by two intersecting tangents ($A + B$), separated by the concave-upward section. From the presumption that each tangent signifies differences in electron density, these authors assumed the presence of two different particle sizes. The calculation of the particle radii according to the tangent method gave a maximum radius of 110 Å.

Assuming a mean specific gravity of the particles of 1.6, these authors calculated molecular weights, which are in the range between 2.0×10^5 to 1.2×10^6. These particles are assumed to be hydrated.

With knowledge of the molecular weights and the density of the substances, the radii of gyration of the scattering particles (R_{min}) may be calculated for particles of globular shape. The ratio $R/R_{min} = f$ is called the "shape factor" and gives, similar to the friction quotient f from viscosity and ultracentrifuge measurements, the deviation from spheres. According to the measurements of WERSHAW, BURCAR, SUTULA, and WIGINTON [1967] f values are found between 1.3 and 2.5, which indicates more or less deviation from spheres. This is similar to the results of FLAIG and BEUTELSPACHER [1951] who used an electron microscope.

4. *Viscosimetric Properties of Humic Acids*

Because viscosimetric investigations of macromolecular substances give important information about particle shapes, and under certain conditions also about molecular weights, this method was applied to characterization of humic acids by several authors (BEUTELSPACHER [1952], FLAIG and BEUTELSPACHER [1954], RAJALAKSHMI, SIVARAJAN, and VOLD [1959], PIRET, WHITE, WALTHER, and MADDEN [1960], SMITH and LORIMER [1964], ORLOV and GORSKOVA [1965]). Special consideration was given to the polyelectrolytic properties of humic acids. Ionic strengths of added electrolytes have a pronounced effect on viscosimetric measurements (FLAIG and BEUTELSPACHER [1954], MUKHERJEE and LAHIRI [1958], RAJALAKSHMI, SIVARAJAN, and VOLD [1959], KUMADA and KAWAMURA [1965, 1968]).

(a) Relation between Particle Shape and Particle Weight of Colloidal Substances in Viscosimetry

Besides a dependency of the viscosity* of colloidal solutions on concentration, temperature, and solvent, there is also a dependency on particle shape. Spherical and linear colloidal particles have different viscosimetric properties. Measurements of viscosity of solutions of solvated nearly spherical shaped colloidal particles gives a direct proportionality between the specific viscosity η_{sp} and the concentration c (g/1.). Therefore, the specific viscosity of spherical colloids is independent of molecular weight and particle size. That means the ratio between specific viscosity and concentration, which is called reduced viscosity or viscosity number Z_η, is constant for this type of particle, giving always the same value for substances derived from any single homologous polymeric series:

$$\eta_{sp}/c = Z_\eta = K \tag{1}$$

In contrast, the viscosity of more or less rod-like linear colloids of equal concentration and derived from any single polymeric homologous series is dependent on the actual length of the molecules and increases with the degree of polymerization. In this case the specific viscosity depends upon molecular weight of the substances. According to the fundamental work of STAUDINGER [1930] this dependency is given by the equation:

$$\frac{\eta_{sp}}{c} = K_m \cdot M \tag{2}$$

Because the constant k_m depends insignificantly on the molecular weight, in actual practice the limiting viscosity number of intrinsic viscosity is used, which means the reduced viscosity extrapolated to zero concentration:

$$\lim_{c \to 0} \frac{\eta_{sp}}{c} = [\eta] \tag{3}$$

Furthermore, it was established that the limiting viscosity number of diluted solutions of

---

* The viscosity of a colloidal solution is determined by the relative viscosity η_r, which means the ratio between the viscosity of the solution η_1 and the viscosity of the pure solvent η_0:

$$\eta_r = \frac{\eta_1}{\eta_0} .$$

The ratio of the change in viscosity ($\eta_1 - \eta_0$) and the viscosity of the solvent is called the specific viscosity:

$$\eta_{sp} = \frac{\eta_1 - \eta_0}{\eta_0} = \eta_r - 1.$$

linear polymers depends on the morphology of the molecules due to interaction with the solvent. In this case the molecular weight is more generally described by the equation of Kuhn–Mark–Houwing (compare STUART [1953], p. 303):

$$[\eta] = k \cdot M^a \tag{4}$$

The exponent a, which is related to the morphology of the molecules, varies from 0.5 to 2.0 in case of unpenetrated coils and rigid rod-like molecules, respectively. The constants k and a may be determined from viscosity measurements, if the molecular weight is determined by one of the known absolute methods (e.g., osmometry, light scattering, or ultracentrifugation). An indirect statement about the particle shape of spherical colloids may be derived from the viscosity equation established by EINSTEIN [1906, 1911]; the relative viscosity is related to the blockaded volume (ρ) of the disperse medium of the system:

$$\eta_r = 1 + k \cdot \rho \tag{5}$$

With the presumption that spherical colloids are rigid globular particles and the diameter of the peptized particles is very large in comparison to the diameter of the solvent molecules, and the interaction between molecules may be neglected by dilution, Einstein calculated for the constant k, a value of 2.5. By transformation of this equation and substitution of the relative viscosity by the specific viscosity ($\eta_r - 1 = \eta_{sp}$) with consideration of the density ρ of the soluted substances, STAUDINGER and HUSEMANN [1935] determined the corrected limiting viscosity number for spherical colloids according to the equation:

$$\frac{\eta_{sp}}{c} \cdot \rho = k = 0.0025$$

With this equation and by comparison of the experimentally determined viscosity numbers with the theoretical values, it is possible to obtain information about the amount of solvent bound to the globular colloids during solvation, and the blockaded volume (ρ) $\left(\dfrac{0.0025c}{\eta_{sp}}\right)$ can be calculated. Because spherical colloids as lyophilic macromolecules are soluble only by solvation, the limiting viscosity number of the solutions is in most cases larger than the theoretical value. The intrinsic viscosity of suspensions of globular particles is between 0.05 and 5.0, higher than that of spherical colloids by the factor 25 under the same conditions.

(b) Viscosimetric Properties and Particle Shape of Humic Acids

FLAIG and BEUTELSPACHER [1954] investigated the viscosimetric properties of dialyzed humic acids and model substances prepared from hydroquinones by oxidation in alkaline media in the absence of and the presence of ammonia. For natural and synthetic humic acids they found a linear correlation between the concentration and the specific viscosity and in case of synthetic humic acids, an increase of intrinsic viscosity with increasing nitrogen content (Figure 107).

From these results it was concluded that the investigated natural and synthetic humic acids approximate in shape globular-shaped particles. This is in agreement with the results obtained by electron microscopic studies (FLAIG and BEUTELSPACHER [1950, 1951], BEUTEL-SPACHER [1955], FLAIG and BEUTELSPACHER [1954]). A comparison of the intrinsic viscosity of electrodialyzed humic acids obtained at various pH values with the theoretical value of 0.0025 (compare Table 28) gave wide deviations from the ideal shape of the spherical colloids.

The high values of blocked volume obtained at low pH values may be explained by differences of association at the starting coagulation. This gives coacervates with particles.

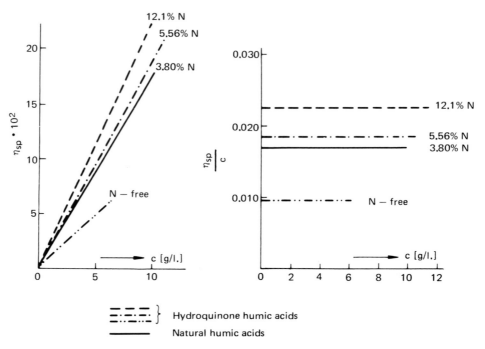

Figure 107. Effect of nitrogen content on the specific viscosity and the viscosity number of natural and synthetic humic acids. According to FLAIG and BEUTELSPACHER [1954].

Therefore, in the range of complete peptization of the particles at pH = 12, particle shapes deviate only slightly from theoretical values. Special investigations on the influence of hydration on viscosity properties by measurements of the temperature dependency of the viscosity of humic acids from sod-podzolic soil (3.1 % ash content) and strong chernozem (2 % ash content) were carried out by ORLOV and GORSKOVA [1965]. They established that the relative viscosity measured in a 0.1 N sodium hydroxide solution was more dependent on concentration in the case of humic acids from sod-podzolic soil than from chernozem. From this they assumed that humic acid particles derived from chernozem are more or less hydrated and more compact in structure. The decrease in viscosity with increasing temperature was explained by an increased dehydration of the humic acid particles. From the linear course of the viscosity curves, it was

Table 28. Effect of nitrogen content on the viscosity number and the volume ratio $(Z_\eta/Z_{\eta 0}) = 0.0025)$ of natural and synthetic humic acids. According to FLAIG and BEUTELSPACHER [1954].

	Z_η	Volume ratio
pH 3		
Hydroquinone humic acid, N-free	0.0097	3.9
,, ,, ,, , 5.56% N	0.0185	7.4
,, ,, ,, , 12.1% N	0.0228	9.0
Natural humic acid , 3.81% N	0.0172	6.9
pH 12		
Hydroquinone humic acid, N-free	0.0027	1.1
,, ,, ,, , 5.7% N	0.0027	1.1

concluded that the humic acid particles are spherical colloids, in agreement with the Einstein equation. By application of the Einstein equation and by extrapolation to the relative viscosity at 100°C, they calculated from the changes in viscosity $(\eta_r^{20°} - \eta_r^{100°})$ the amount of hydration water of the humic acids from sod-podzolic soil to be 2.5 g H_2O/g humic acids, and in case of chernozem it was 1.2 g H_2O/g humic acids.

(c) Influence of Electrolytes on the Viscosimetric Properties of Humic Acids

The influence of functional groups on the rheological properties of humic acids and their behavior as polyelectrolytes was investigated by several authors by comparative measurements of viscosity with and without the addition of salts. MUKHERJEE and LAHIRI [1958] found, with extracts obtained by alkali carbonate extraction from coal, a greatly reduced viscosity with increasing dilution. They concluded from this fact an increased dissociation of ionizing groups, with the restriction that smaller humic acid units may combine by association in solution to large chain-like polymer structures, which cause an increase of viscosity with increasing dilution by decoiling. The addition of electrolytes causes a decrease in the degree of dissociation of polymers, which is derived from viscosity curves and a slightly reduced viscosity with increasing dilution. Similar results were found by RAJALAKSHMI, SIVARAJAN, and VOLD [1959] by investigations on the influences of sodium chloride, which was added in various amounts, on the changes of viscosity of alkali humates from coal. They found that the added salts result in a strong decrease of the reduced viscosity and that curves of the reduced viscosity versus concentration in humate pass a maximum at low salt concentration and become linear at higher salt concentrations. This effect is explained by the assumption that the added salts decrease the electrostatic interaction of the particles by the decrease of the ζ-potential. At concentrations of sodium humate, which gives a sodium ion concentration larger than that resulting from added salts, the degree of dissociation is increased with decreasing concentration

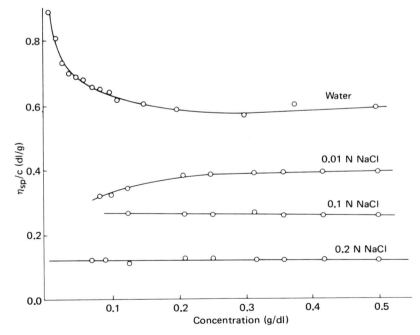

Figure 108. Dependence of reduced viscosity of sodium humate on concentration and on added salt. According to SMITH and LORIMER [1964].

in humate. This leads to an expansion of the coiled molecules, and increases the value of the reduced viscosity with increasing dilution. On the other hand, if the concentration in humate reaches values at which the sodium ion concentration is in excess of added salts, the dissociation of humate is reduced. In this case the density of the coiled molecules is increased, and the reduced viscosity decreases with decreasing concentration in humate. In the curves a maximum is obtained if the sodium ion concentration of the added electrolytes is equal to the sodium ion concentration formed by dissociation from humate. According to this, it is assumed that a limiting concentration in sodium chloride exists, at which humic acid polyelectrolytes are completely discharged. In the case of charged humic acid particles, the reduced viscosity is independent of concentration. This is confirmed by investigations of SMITH and LORIMER [1964] in the case of humic acids from peat. They demonstrated that the limiting concentration of their samples is reached at about 0.1 N sodium chloride (Figure 108).

PIRET, WHITE, WALTHER, and MADDEN [1960] obtained the best results in 0.26 N sodium chloride solutions of organic matter preparations derived from peat. For the intrinsic viscosity obtained by extrapolation to zero concentration of the 0.26 N sodium chloride solution it follows that the value of the blockaded volume is higher by a factor of 4.6 compared with the theoretical value of 0.0025. This leads to the assumption that the particles are strongly hydrated, forming elongated elliptic molecules, with the ratio of axes being 1:11. These results indicate that the polyelectrolytic behavior of humic acids is effected by terminal groups of phenolic hydroxyl and carboxyl type, which enables an increase hydration by formation of hydrogen bridge linkages with increasing ζ-potentials.

(d) Influence of Hydrogen Ion Concentration on the Viscosimetric Properties of Humic Acids

To investigate the influence of hydrogen ion concentration on the viscosimetric properties of humic acids, FLAIG and BEUTELSPACHER [1954] treated electrodialized suspensions of humic acids from black soil, which were adjusted to pH 13 by sodium hydroxide, with increasing amounts of hydrochloric acids at constant volumes and concentrations. They found that with increasing hydrogen ion concentration the viscosity was slightly increased up to the limiting value of pH 3, and was followed by a rapid increase with increasing hydrogen ion concentration due to the enlargement of the particle sizes by coagulation (Figure 109). Considering that in this experiment the concentration of free salts in solution was continuously increased and the

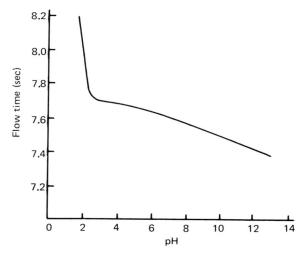

Figure 109. Influence of hydrogen ion concentration on the viscosimetric properties of humic acids. According to FLAIG and BEUTELSPACHER [1954].

counter-ions were blockaded by neutralization, this effect may be explained as a decrease of the electrostatic charges of the polyelectrolytes with increasing partial specific volumes.

KUMADA and KAWAMURA [1965] tried to find relationships between the types of humic acids determined by electronic spectra (compare p. 106) and the viscosimetric properties. They investigated the pH-dependent changes of viscosimetric properties of humic acids from various origins. These substances were characterized by chemical analysis and analysis of the functional groups. Viscosities were not found to be related to functional groups.

H-Amberlite generated 0.5% suspensions of humic acids were stepwise titrated with 0.5 N NaOH. The viscosity of these suspensions decreased to pH 5 to 6 and then increased (Figure 110).

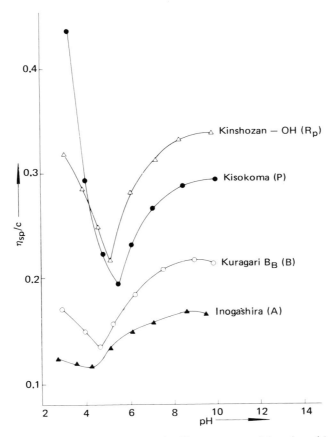

Figure 110. Change of reduced viscosity of different types of humic acids with hydrogen ion concentration. According to KUMADA and KAWAMURA [1968].

According to the differences in viscosimetric properties KUMADA [1955] distinguished three groups of humic acids, in agreement with the results of spectroscopic measurements:
(1) The slope of the descending and ascending parts of the viscosity curve is the lowest of all groups (A-type).
(2) The slope of the descending curve is similar to that of 1, but the slope of the ascending part shows larger values (B-type).
(3) A stronger inclination of the descending part is obtained, but the slope of the descending part of the curve gives similar values like those of 2 (Rp-type, P-type).

Although according to Kumada, a complete explanation is still impossible because certain transformations in the types of humic acids occur, it is assumed that the viscosimetric properties are a function of different degrees of dissociation of the carboxyl groups, which increases with the content in functional groups in the order: A < B < P < Rp. Similarly, the slope of the descending part of the viscosity curves decrease in direct proportion to the increase of the degree of dissociation in the following order: A = B < P < Rp. Because the increase in viscosity during the neutralization of weak polyelectrolytes is not only a function of the particle shape, but is also dependent upon several other factors such as particle weight, charge, density, coacervation, etc., it is difficult to decide whether humic acids exist as spherical or linear colloidal particles.

5. *Determination of Particle Weights of Humic Acids by Light Scattering*

For the determination of absolute values of molecular weights of high polymer substances, light scattering has been applied more often in recent years, because according to the theory of Rayleigh, there is a direct proportionality between the intensity of scattered light at the same concentration with the molecular weight of the particles in solution.

(a) Principles of the Method

Electron microscopic investigations of humic acid particles show that their radii of gyration are less than 1/40th of the wavelength of the light used in measuring molecular weights. Therefore, the equation established by DEBYE [1944] may be used:

$$\frac{k \cdot C}{R_\theta} = \frac{1}{M} + \frac{2B \cdot c}{R \cdot T} + \cdots \tag{1}$$

The measured value of the concentration-dependent molecular weight M is the molecular scattered light R_θ, which is the intensity of the rays at a unit distance of the primary light intensity equal to 1, dependent on the angle θ of observation and the wavelength λ (R = gas constant, T = absolute temperature, B and C = virial coefficients of osmotic pressure). The experimentally determined solvent-dependent constant k of unpolarized primary light for anisotropic particles is given by the equation:

$$K = \frac{4 \cdot \eta^2 \cdot n_1^2}{N_L \cdot \lambda^4} \cdot \left(\frac{dn}{dc}\right)^2 \tag{2}$$

(n_1 = refractive index of the solvent, dn/dc = refractive index increment of the solution, N_L = Loschmidt number, λ = wavelength of the primary light. Plots of the values $k \times C/R$ versus concentration result in a line with the slope $2B/R \cdot T$, intersecting the ordinates at the value of $1/M$, from which the average molecular weights may be calculated.

(b) Light-Scattering Properties of Humic Acids

Until now, investigations of light-scattering properties of soil organic matter have been carried out only by ORLOV and GORSKOVA [1965] with humic acids derived from the chernozem and a sod-podzolic soil. According to the results given in Figure 111 they found for both samples a linear correlation between the reduced reciprocal scattered light intensity and the concentration. By extrapolation nearly the same molecular weights resulted: 65.3×10^3 for humic acids from sod-podzolic soil and 66.2×10^3 for humic acids derived from chernozem. These values are very similar to those estimated from electron micrographs made by FLAIG and BEUTELSPACHER [1954]. Orlov points to the possibility that the light-scattering measurements may be influenced by admixtures of clay-minerals, which are normally associated with the humic acid samples.

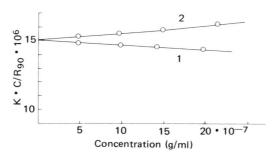

Figure 111. Correlation between reduced reciprocal scattered light intensity and concentration of humic acids from sod-podzolic soil (1) and chernozem (2). According to ORLOV and GORSKOVA [1968].

6. *Determination of Particle Shape, Particle Size, and Particle Weight of Humic Acids by Sedimentation Velocity and Diffusion*

By these methods information is gained about particle size (molecular weight) and particle size distribution (polymolecularity), particle shape, frictional coefficient, and charge effects of polyelectrolytes in solution (interactions), such as degree of association and dissociation.

(a) Principles of the Method

In free diffusion experiments with miscible solutions, a sharp boundary is formed between any two miscible phases. In the absence of external forces, the mobile polymers will migrate toward regions of lower concentrations to become uniformly distributed throughout the entire volume in the course of time. The driving force underlying the irrversible diffusion process is the thermodynamic free-energy gradient, which is determined by temperature, entropy, and composition of the solution.

In practice, the changes in concentrations during the transport processes may be detected, *e.g.*, by measurements of the changes of refractive indices. These are observable in an optical arrangement, according to PHILPOT-SVENSSON [1940] directly as refractive index gradients in the form of gaussian normal distribution curves.

The calculation of the diffusion coefficient D is based on Wiener's solution of Fick's law (FICK [1955], WIENER [1893]):

$$\frac{\partial n}{\partial c} = \frac{n_1 - n_0}{2\sqrt{\eta \cdot D \cdot t}} \cdot e^{-x^2/4 \, D \cdot} \tag{1}$$

where $\partial n / \partial c$ is the refractive index gradient, n_0 and n_1 are the refractive indices of the solvent and the solution, respectively, along the diffusion column.

If x is taken as zero, it follows from (1) that

$$D = \frac{(n_1 - n_0)^2}{4 \cdot t \cdot H_m^{\,2}} \tag{2}$$

where $(n_1 - n_0)$ is the area under the curve, which can be determined by integration. H_m is the maximum ordinate of the distribution curve at the point of inflection.

In sedimentation velocity experiments in ultracentrifuges the centrifugal force causes polymer particles in solution to obtain various terminal sedimentation velocities. These velocities increase with size, weight, or density, and decrease with the viscosity of the solvent medium and with deviations from the spherical shape. Also this leads to particle distributions that result in concentration gradients. Consequently, particle measurements of sedimentation properties of high molecular weight substances in solutions use the same optical arrangement as those used in diffusion experiments to record changes in concentration during centrifugation as moving boundaries of refractive index gradients in the form of gaussian distribution curves.

The most prominent application of the ultracentrifuge method is for the determination of sedimentation coefficients, which permits according to SVEDBERG [1940, 1960] with distinct assumptions, calculations of molecular weights (M_s), according to the equation:

$$M_s = \frac{f_s \cdot s_0 \cdot N_L}{(1 - V^* \cdot \rho_L)} \tag{1}$$

f_s = friction coefficient of sedimentation
s_0 = sedimentation coefficient
V^* = particle specific volume
ρ_L = density of the solvent
N_L = Loschmidt number

The friction coefficient in this equation depends on the kind of the sedimenting substances (mass, shape, and sizes) and the medium (density, viscosity, and interaction with the solvated colloidal particles). According to SVEDBERG and PEDERSON [1940, 1960], the friction coefficients of the sedimentation (f_s) and diffusion (f_D) are equal. Therefore, according to the Einstein equation, it follows that:

$$f_D = \frac{R \cdot T}{D_0 \cdot N_L} \tag{2}$$

R = gas constant
T = absolute temperature
D_0 = diffusion coefficient

By combining equations (1) and (2), the friction coefficients in the Svedberg equation (1) may be eliminated:

$$M_{s,D} = \frac{R \cdot T}{(1 - V^* \rho_L)} \cdot \frac{s_0}{D_0} \tag{3}$$

Thus, the determination of particle weights is possible by determination of the sedimentation coefficients in measurements of the sedimentation velocity and by the ultracentrifuge technique and by determination of the diffusion coefficients in measurements of the diffusion velocity of the particles in free diffusion experiments.

(b) Investigations of Humic Acids by Ultracentrifuge Measurements

The first determinations of particle weights by the ultracentrifuge method with humic acids of black soil and brown soil were carried out by BEUTELSPACHER [1950] by application of the method of light absorption. In this way it was established that the sedimentation velocity of the humates varies with the soil type. The results depended on the kind of peptizing medium and the methods of purification during preparation. In most cases, during the measurement, three groups of different particle weights were observed. The most prominent value (particle weight 3–5×10^4) settled relatively slowly, giving a diffuse gradient, while a relatively small, undetermined amount of particles with higher particle weights settled down with a higher sedimentation velocity. In spite of several hours of centrifugation, a weakly yellow-colored solution above the gradient remained visible. This third group of substances was assumed to be composed of humates with particle weights lower than 1×10^4.

Also, STEVENSON, VAN WINKLE, and MARTIN [1953] investigated the sedimentation properties of pyrophosphate extracts from Miami clay soil by light absorption. They concluded that the extracts contained polydispersed substances. They found that the sedimentation constants were independent of concentration. From the sedimentation constant $s_{20} = 2.8 \times 10^{-13}$, the diffusion constant $D_{20} = 3.9 \times 10^{-7}$ and the coefficient of buoyancy $(1 - V^* \cdot \rho_L) = 0.33$, the molecular weight $M_{s,D}$ of 5.3×10^4 was calculated, which was in the range of the values determined for the main fractions of substances investigated by BEUTELSPACHER [1950]. From the molecular weight, the friction coefficient was calculated to be $f = 6.25 \times 10^{16}$. Because spherical particles of the same sizes would have the theoretical friction coefficient $f_0 = 2.76 \times 10^{16}$, the friction ratio f/f_0, was 2.3. From these results it is concluded that the colloidal particles are nonspherical and/or solvated and that the humic acids particles are short, compact, and hydrated molecules of different particle sizes with unequal charge densities on the particle surface.

SCHEFFER, ZIECHMANN, and SCHLÜTER [1958], as well as SCHLÜTER [1959], investigated the sedimentation properties of synthetic humic acids from hydroquinone and electrophoretic fractionated extracts of humic acids from chernozem-type and various marsh soils in the gravity of the ultracentrifuge by light absorption and by refractrometry according to the method of Philpot–Svensson. From the occurrence of several concentration gradients and from cloudy centrifugates they assumed polymolecularity. During centrifugation of the solutions, which were in 0.1 N sodium hydroxide, and in mixtures of dioxan/water (1:1), they observed particle movement toward the center. These floating fractions occur in the case of negative values of the coefficient of buoyancy $(1 - V^* \cdot \rho_L)$. Differences in sedimentation and diffusion properties between synthetic and natural humic acids were not observed. Because polydispersity of the material made the exact determination of diffusion coefficients impossible, these authors propose to dispense with determinations of particle weights of humic acids but to use instead the standardized sedimentation coefficients as the relative measures of particle sizes. That means that the influence of the friction coefficients and the other functional values must be ignored. Particle size, particle shape, and solvent interactions are obtained only with the combined application of sedimentation and diffusion data.

PIRET, WHITE, WALTHER, and MADDEN [1960] studied the sedimentation properties of sodium hydroxide extracts from peat (ash content, 8.3 %) with the addition of 0.26 M sodium chloride in a concentration range between 0.73 and 7.3 g/l. with application of the refractrometric method of Philpot–Svensson. From the sedimentation constant of 1.83×10^{-13}, and the friction coefficients $f/f_0 = 1.6$ to 1.86 obtained by viscosimetric measurements, they calculated particle weights of 2.2 to 2.8×10^4, which are similar to those found by BEUTELSPACHER [1950] and STEVENSON, VAN WINKLE, and MARTIN (1953). From the change of *Schlieren* curves during sedimentation they also assumed polydispersity of their samples.

SCHNITZER and SKINNER [1968] investigated, by application of ultracentrifuge measurements and vapor pressure osmometry, the distribution of molecular weights of fulvic acid extracts from the B_h horizon of a podzol obtained by gel filtration. A fraction of fulvic acid that was 65 % excluded by a G-10 gel but 100 % retained by a G-50 gel, gave a sedimentation coefficient $s_{20} = 1.2 \times 10^{-3}$, a diffusion coefficient $D_{20} = 1.6 \times ^{-6}$ cm/sec, a partial specific volume $V^* = 0.69$, and a calculated molecular weight $M_{s,D} = 5893$, which was in good agreement with the corrected molecular weight of 5819 obtained by vapor osmometry. Generally, the molecular weights of fulvic acids obtained by gel filtration are 2 to 10 times higher than by application of other methods. Furthermore, it was found that the content of nitrogen decreased with decreasing molecular weight, while the total acidity increased.

(c) Influence of Hydrogen Ion Concentration and Ionic Strength on the Sedimentation and Diffusion Properties of Humic Acids

Because charge effects of dissociating high molecular weight substances influence the sedimentation and diffusion properties by hydration, FLAIG and BEUTELSPACHER [1968] investigated sedimentation and diffusion of electrodialyzed 0.1 N sodium hydroxide extracts from hydrochloric acid pretreated marsh soil over a range of hydrogen ion concentrations. From the asymmetry of the gaussian distribution curves of refractive index gradients during ultracentrifugation they assumed polydispersity of the applied samples. Besides that, they observed low amounts of very high molecular substances, which had settled to the bottom of the cells after a short time of sedimentation. Plots of sedimentation and diffusion constants versus the pH values revealed that the sedimentation constants increased with decreasing hydrogen ion concentration, while the diffusion constants decreased (Figure 112). These

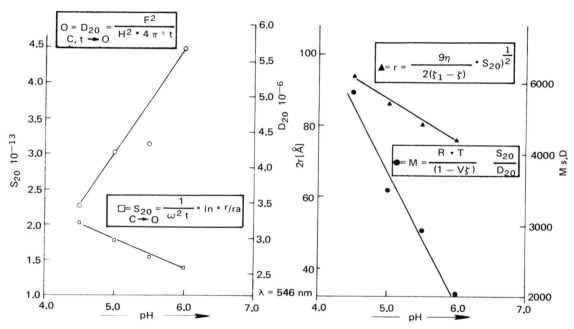

Figure 112. Effect of pH on particle weights and diameters of an electrodialized humic acid. According to FLAIG and BEUTELSPACHER [1968].

authors explained these effects by charge differences of humic acid particles, which lead to an increase of the partial specific volume (V^*) because of enhanced hydration with increasing hydrogen ion concentration favoring the formation of micellar colloids. According to this, the molecular weights calculated by the Svedberg equation are reduced from 4850 at pH 4.5 to 2030 at pH 6 (compare p. 180).

The same is true of the calculated particle diameters which decrease with increasing pH values from 94 Å at pH 4.5 to 76 Å at pH 6.0. The friction coefficients calculated were between 1.0 and 1.1, which was in agreement with the results of investigations of other authors (compare FLAIG and BEUTELSPACHER [1954]), confirming that the humic acids are spherical colloids with only small deviations from an ideal globular shape. These results are further supported by the evaluation of the diffusion measurements of SCHEELE and STEINKE (1936), who applied the methods developed by ÖHOLM (1904) by FLAIG and BEUTELSPACHER (1954).

Scheele and his coworkers found by investigation of the diffusion properties of sodium fluoride extracts from low rank coal (Kasseler brown), that the diffusion constants show small decreases with increasing pH values. Besides that, they observed at pH 8 a break in the curve, so that the pH-dependent system may be divided into two ranges (Figure 113).

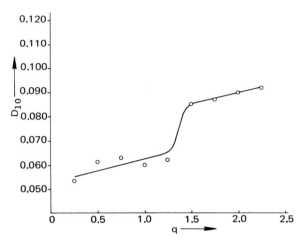

Figure 113. Influence of hydrogen ion concentration on diffusion properties of humic acids. According to SCHEELE and STEINKE [1936]; q = equivalents of NaOH.

By calculation of the mean diffusion constants at $D_{10} = 0.060$ (pH 8.0), and at $D_{10} = 0.088$ (pH 8.0), followed by the application of the Einstein equation $D\sqrt{M} = K$ ($K = 6.3$), it is found that the values $1/D^2$ are proportional to the molecular weight. The molecular weights show a 2:1 ratio. From this it is concluded that humic acids from this low rank coal in alkaline solutions have one-half of the molecular weights found in acidic solutions ($M_D = 11.2 \times 10^3$ at pH 8.0 compared with 5.15×10^3 at pH 8.0). According to this, the calculated mean particle diameters are 45.9 Å and 30.9 Å. This confirmed the opinion that humic acids are stable in alkaline solution as molecular colloids, while under acidic conditions particle aggregation and formation of micelles occur by coacervation.

Similar to the results of ultracentrifuge measurements on proteins, which indicate that addition of neutral salts reduces the primary charge on polyelectrolytes and causes electro-neutrality, which is important to unhindered sedimentation, FLAIG and BEUTELSPACHER [1968] investigated the influence of added sodium chloride on the sedimentation and diffusion properties of electrodialyzed 0.1 N sodium hydroxide extracts derived from hydrochloric acid pretreated marsh soils. A comparison of the results obtained at pH 4.5 and 5.0 with addition of 0.2 M sodium chloride with the values of the untreated samples (Table 29) shows that by addition of salts the sedimentation constant was increased from 1.77 to 4.46×10^{-13} at pH 5.0 and from 2.01 to 4.71×10^{-13} at pH 4.5, approximately in the ratio of 1:2.5, in each case.

The diffusion properties especially are influenced by the salt addition, leading to a nearly tenfold increase of the diffusion constant. From these results the particle weights calculated are 7.7×10^4 compared with 3.49×10^3 of the untreated sample at pH 5.0 and 6.04×10^4, instead of 4.85×10^3 at pH 4.5. According to this also the values of the particle diameters and the molecular friction coefficients are changed. The results show that humic acids act as polyelectrolytes, therefore resulting in strong differences in sedimentation and diffusion properties according to the charges of the particles in dependence on the hydrogen ion concentration and the ionic strength of the solution. It is assumed that the results obtained with addition of salts are most

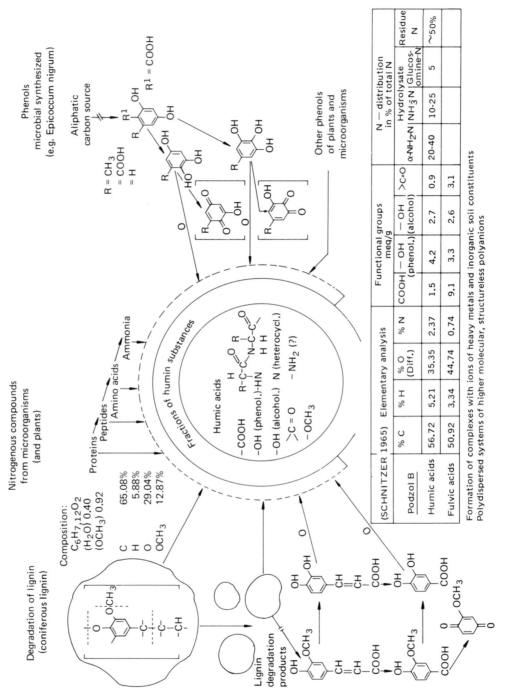

Figure 114. Scheme of formation of humic substances.

Table 29. Dependence of sedimentation (s_{20}) and diffusion constants (D_{20}), molecular weight ($M_{s,D}$), radii (r), and friction coefficients (f/f) of a fraction of electrodialyzed humic acids on hydrogen ion concentration and on added salts. According to FLAIG and BEUTELSPACHER [1968].

pH	$s_{20} \cdot 10^{-13}$	$D_{20} \cdot 10^{-6}$	$M_{(s,D)}$	r (Å)	f/f_0
6.0	1.38	5.66	2030	36	1.0
5.5	1.53	4.32	2950	40	1.1
5.0	1.77	4.21	3490	43	1.1
4.5	2.01	3.45	4850	47	1.1
5.0 In 0.2 M NaCl	4.46	0.48	77,000	69	1.5*
4.5 In 0.2 M NaCl	4.71	0.65	60,400	71	1.2*

* Corrected for the amount of salt added (A. G. Elias, *Ultrazentrifugen-Methoden.* Beckman Instr. GmbH., München, 2. Aufl., p. 126 (1961)).

probable, compared with the results of particle weight determination obtained by application of other methods (ORLOV and EROSICEVA [1967], FLAIG and BEUTELSPACHER [1954]).

D. Concluding Remarks

The characterization of humic substances by chemical and physical methods is still difficult, because they are a mixture of substances that are derived from different, original or transformed materials of organisms during humification. In Figure 114 the basic knowledge, which is elucidated by direct investigations of separate components of soil organic matter or by model experiments, is summarized.

One of the most important components is phenols, derived from lignin or formed by microbial synthesis. Other important components are the degradation products of proteins such as peptides, amino acids, or ammonia. Some other nitrogenous or nitrogen-free compounds such as sugars are found by chemical degradation of humic fractions.

The phenols formed by stepwise lignin degradation or synthesis by microorganisms, which are partly found in soils or occur by reductive or oxidative degradation of humic acids, react by random polymerization, mainly with nitrogenous compounds derived from proteins, when the former can be oxidized to quinones. This type of phenol also forms addition products with other phenols, containing hydroxyl groups in the meta position.

By these reactions humic substances are formed. They are separated into fractions by different methods. The usual method is the extraction by alkaline solvents followed by precipitation of the fractions of humic acids by mineral acids. The hymatomelanic acids are extracted from humic acids with alcohol.

The fulvic acids remain in the acidic solution. The insoluble fractions of soil organic matter are named humins; these remain in the inorganic part of the soil. The separation of soil organic matter is conventional and the quantities of the fractions depend upon the method. A standardization of extraction and separation methods would permit a comparison of the results of different investigators.

The elemental composition of the fractions of humic acids show a higher carbon and nitrogen content but a lower oxygen content than the fulvic acids. By chemical investigations some differences of the content of functional groups are determined. The nitrogen is bound in different organic linkages. Model reactions have shown that the so-called heterocyclic nitrogen may be derived originally from α-NH_2-nitrogen of amino acids. It can be demonstrated that the chemical composition and the amount of functional groups, as well as the relationship of humic to fulvic acids, depend upon the conditions of soil organic matter formation under different environmental conditions, such as climate, types of plants, and inorganic soil constituents. The main reactions are caused by the action of microorganisms present in the soil.

The acidic function of the fractions of the fulvic and humic acids are explained by the content of carboxyl and to minor extent of phenolic groups. The higher content of carboxyl groups in fulvic acids may be due to their lower particle weight, because decarboxylation did not occur before polymerization or condensation to higher molecular humic acids, or on the other hand due to oxydative degradation of humic acids.

Potentiometric and conductometric titration point to polyelectrolyte behavior of humic acids, whereby different acidic groups participate. The complex formation with ions of heavy metals depends also upon these groups. The polyelectrolytic character of humic acids is confirmed by measurements of viscosity in the presence of electrolytes and investigations of the influence of hydrogen ion concentration on diffusion and sedimentation.

According to all investigations the humic acids are a mixture of similar substances. Therefore, various attempts are made to separate fractions by partial precipitation by neutral salts (brown and gray humic acids), chromatographic methods, gel filtration, or electrophoresis. Comparative investigations with gel filtration and ultracentrifuge demonstrated that the separation by chromatographic methods does not lead to a uniform molecular weight but to a distribution of the polymolecular humic acids. The investigations with ultracentrifuge, light scattering, small-angle X-ray scattering, and electron microscopy reveal humic acids as polydisperse, spherical colloids with particle diameters of 80 to 100 Å and an average particle weight of about 5×10^4.

Furthermore, electron and X-ray diffraction of carefully purified humic acids show only broad diffraction bands (halos); therefore, humic acids must be assumed to have little crystallographic structure.

Other methods for characterization of humic substances, such as electronic spectra, show no specific absorption bands, but only more or less monotonously rising curves in the direction of shorter wavelengths. The slope depends upon the origin of humic acids and increase from chernozem humic acids to these of podzol and finally to fulvic acids. To a certain extent they are useful for characterization of isolated fractions but not for elucidation of chemical constitution.

The few results of investigations of thermal degradation promise new knowledge of the composition of humic acids.

To our present knowledge humic acids as a fraction of soil organic matter separated by conventional methods may be defined as polydispersed mixtures of higher molecular, structureless polyanions, which are mainly but not exclusively formed by reactions of phenolic and nitrogenous compounds.

REFERENCES

Abel, O., 1965. Untersuchungen zum Vergleich zum Entgasungs-, Schrumpf-, und Brikettierverhaltens von Braunkohlen. Habilitationsschrift zur Erlangung der venia legendi der Fakultät für Bergbau und Hüttenwesen der Bergakademie Clausthal-Technische Hochschule.

——, and H. Luther, 1962. Gravimetrische Untersuchungen zum Zersetzungsverhalten jüngerer Brennstoffe. *Erdöl Kohle* 15:90.

Achard, F., 1786. Chemische Untersuchung des Torfs. *Grell's Chem. Ann.* 2:391.

Adams, G. A., and A. E. Castagne, 1948. Holocelluloses from straw. *Can. J. Res., sect. B* 26:325.

Adler, E., 1957. Newer views of lignin formation. *Tappi* 40:294–301.

——, 1961. Über den Stand der Ligninforschung. *Z. Papier* 15:604–609.

——, and K. Lundquist, 1961. Estimation of uncondensed phenolic units in spruce lignin. *Acta Chem. Scand.* 15:223–224.

Dell'Agnola, G., and A. Maggioni, 1965a. Gel filtrazione dell'humus. Nota V. Effetto della soluzione eluante sul fraxionamento della sostanza organica del terreno con Sephadex. *Bollettino* 23:321–332.

——, 1965b. Gel filtrazione dell'humus. Nota VI. Polimerizzazione delle sostanze umiche a basso peso melecolare. *Bollettino* 23:333–345.

——, G. Ferrari, and A. Maggioni, 1964. Gelfiltration of humus. 2. Fractionation of soil organic matter by filtration across gels of Sephadex type. *Ric. Sci., Part* 2, B4:347–352.

——, and G. Ferrari, 1965a. Gelfiltration of humus. 3. Distribution of nitrogen forms in fractions obtained by means of various types of Sephadex. *Agrochimica* 9:169–177.

——, and G. Ferrari, 1965b. Gelfiltration of humus. 4. Evaluation of different techniques of humus extraction through fractionation by means of Sephadex. *Agrochimica* 9:224–231.

Aleksandrova, L. N., 1954. The nature and properties of the products of reaction of humic acids and humates with sesquioxides. *Pochvovedenie* 1:14–29.

——, 1960. The use of sodium pyrophosphate for isolating free humic substances and their organic mineral compounds from the soil. *Soviet Soil Sci.* 2:190–197.

——, 1967. Organomineral humic acid derivatives and methods of studying them. *Soviet Soil Sci.* 7:903–913.

——, and M. Nad, 1958. The nature of organo-mineral colloids and methods for their study. *Pochvovedenie* 10:21–27.

Anderson, G., 1958. Identification of derivatives of deoxyribonucleic acid in humic acid. *Soil Sci.* 86:169–174.

Anderson, M. S., and H. G. Byers, 1936. U.S. Dept. Agr. Tech. Bull. No. 542.

Andrzejewski, M., 1964. Wirkung von Stickstoff auf die Spektren von synthetischen Huminsäuren. *Jahresber.* Landwirtschaftswissenschaften 88-A-2, 165–172.

——. 1965. Der Einfluß des Stickstoffs auf die Veränderung der Absorptionsspektren und auf die Anzahl der während der Gärung des Strohs entstehenden Humusverbindungen. *Roczniki Nauk Rolniczych, Ser. A.* 89:387–410.

Anslow, W. K., and H. Raistrick, 1938a. Studies in the biochemistry of microorganisms. 59. Spinulosin (3.6-dihydroxy-4-methoxy-2.5-toluquinone), a metabolic product of a strain of aspergillus fumigatus fresenius. *Biochem. J.* 32:228.

——, and H. Raistrick, 1938b. Studies in the biochemistry. 57. Fumigatin (3-hydroxy-4-methoxy-2.5-toluquinone) and spinulosin (3.6-dihydroxy-4-methoxy-2.5-toluquinone), metabolic products respectively of aspergillus fumigatus fresenius and penicillium spinulosum Thom. *Biochem. J.* 32:687.

——, and H. Raistrick, 1938c. 108. Studies in the biochemistry of microorganisms. 58. Synthesis of spinulosin (3.6-dihydroxy-4-methoxy-2.5-toluquinone), a metabolic product of penicillium spinulosum Thom. *Biochem. J.* 32:803.

——, J. Breen, and H. Raistrick, 1940. Studies in the biochemistry of microorganisms. 64. Emodic acid (4.5.7-Trihydroxyanthraquinone-2-carboxylic acid) and hydroxy-emodin (4.5.7-trihydroxy-

2-hydroxymethyl-anthraquinone), metabolic products of a strain of penicillium cyclopium Westling. *Biochem. J.* 34:159–168.

Bailey, N. T., G. G. Briggs, G. J. Lawson, J. M. Scruton, and S. G. Ward, 1965. Observation on the structure of humic acid. 6. Intern. Kohlenwiss. Tagung, Münster, W., 1.–3.6.1965. Beitrag Nr. 3, 1.

Balint, J. and J. Treiber, 1959. Zur Frage der Differentialthermoanalyse der organischen Boden-bestandteile. *Agrochem. Bodenkunde* 8:59–64.

Baker, W., and D. Miles, 1955. Red color given by coal-tar-phenols and aqueous alkalies-isolation of a quinone derived from a tetrahydroxytetramethylbiphenyl. *J. Chem. Soc. (London)* 1955:2089–2096.

Baltzly, R., and E. Lorz, 1948. Addition of dimethylamine to benzoquinone. *J. Am. Chem. Soc.* 70:861–862.

Bartlett, J. B., 1939. The effect of decomposition of the lignin of plant materials. *Iowa State Coll. J. Sci.* 14:11–13.

————, and A. G. Norman, 1938. Changes in the lignin of some plant material as a result of decomposi-tion. *Proc. Soil Sci. Soc. Am.* 3:210–216.

————, F. B. Smith and P. E. Brown, 1937. Lignin decomposition in soils. *Proc. Iowa Acad. Sci.* 44:97–101.

El-Basyouni, Said Z., D. Chen, R. K. Ibrahim, A. C. Neish, and G. H. N. Towers, 1964. The bio-chemistry of hydroxybenzoic acids in higher plants. *Phytochemistry* 3:485–492.

Baumann, A., and E. Gully, 1910. Untersuchungen über die Humussäuren. 2. Die freien Huminsäuren des Hochmoores. *Mitt. K. Bayr. Moorkult.* 4:31–170.

Beckwith, R. S., 1955. Metal complexes in soils. *Australian J. Agr. Res.* 6:685–698.

————, 1959. Titration curves of soil organic matter. *Nature* 184:745–746.

Beecken, H., E. M. Gottschalk, U. van Gizycki, H. Krämer, D. Maasen, H. G. Matthies, H. Musso, C. Rathjen, and U. I. Zahorsky, 1961. Orcein und Lackmus. *Angew. Chem.* 73:665–673.

Belkevic, P. J., and K. A. Gayduk, 1964. Über die stufenweise Zersetzung von Huminsäuren. *Dokl. Akad. Nauk SSSR* 8:650–652.

Bennett, E., 1949. Fixation of ammonia by lignin. *Soil Sci.* 68:399–400.

Beres, T., and I. Kiraly, 1957. Iron compounds of fulvic acids. *Agrochem. Talajt.* 6:167–176.

Berthelot, M., and G. Andrè, 1892. *Ann. Chim. Phys.* 25:Ser. 6, 362.

Berzelius, J., 1833. Undersökning of vattnet i Porla källa. *Ann. Phys. Chem.* 105:3, 238.

Beutelspacher, H., 1950. unveröffentlicht. loc. cit. Kononova, M. M.: Die Humusstoffe des Bodens. VEB Deutsch. Verl. Wiss., Berlin, S. 72 (1958).

————, 1952. Physikalisch-chemische Beiträge zur Humusforschung. *Z. Pflanzenernähr. Düng. Bodenk.* 57:57–65.

————, 1955. Wechselwirkung zwischen anorganischen und organischen Kolloiden des Bodens. *Z. Pflanzenernähr. Düng. Bodenk.* 69:108–115.

Bhatia, B. S., N. S. Kapur, D. S. Bhatia, and L. Girdhari, 1957. Studies on the nonenzymatic browning of foods—organic acid-amino model systems. *Food Res.* 22:266.

Bisle, H. E., 1959. Zur Kennzeichnung der Huminsäuren. Universität München, Germany.

Björkman, A., 1954. Isolation of lignin from finely divided wood with neutral solvents. *Nature* 174:1057

————, 1956. Studies on finely divided wood. 1. Extraction of lignin with neutral solvents. *Svensk papperstidning* 59:477–485.

Blois, S., 1955. Observation of the semiquinone of benzoquinone. *J. Chem. Phys.* 23:1351.

Blois, M. S., 1965. Random polymers as a matrix for chemical evolution. In *The origins of prebiological systems*, Sidney W. Fox, ed. New York–London: Academic Press.

Börner, H., 1955. Untersuchungen über phenolische Verbindungen aus Getreidestroh und Getrei-derückständen. *Naturwissenschaften* 42:583.

————, 1956. Der papierchromatographische Nachweis von Ferulasäure in wässrigen Extrakten von Getreidestroh und Getreiderückständen. *Naturwissenschaften* 43:129.

————, 1957. Die Abgabe organischer Verbindungen aus den Karyopsen, Wurzeln, und Ernterück-ständen von Roggen (Secale cereale L.), Weizen (Triticum aestivum L.) und Gerste (Hordeum vulgare L.) und ihre Bedeutung bei der gegenseitigen Beeinflussung der höheren Pflanzen. *Beitr. Biol. Pfl.* 33–83.

Boldyrev, H. I., and S. N. Aleshin, 1963. Extinktionskoeffizienten und die spektrophotometrische Bestimmung von Huminverbindungen im Boden (russ.). *Dokl. TSCHA.* 89.

Bondy, A., and H. Meyer, 1948. Lignin in young plants. *Biochem. J.* 43:248.

Brauns, F. E., 1939. Native lignin. 1. Its isolation and methylation. *J. Amer. Chem. Soc.* 61:2120–2127.

———, 1952. *The chemistry of lignin.* New York: Academic Press.

———, and D. A. Brauns, 1960. *The chemistry of lignin,* supplement volume covering the literature for the years 1949–1958. New York and London: Academic Press.

Bremner, J. M., 1949. Studies on soil organic matter. Part 3. The extraction of organic carbon and nitrogen from soil. *J. Agric. Sci.* 39:280–282.

———, 1952. The nature of soil-nitrogen complexes. *J. Sci. Food Agr.* 3:497–500.

———, 1954. A review of recent work on soil organic matter. Part 2. *J. Soil Sci.* 5:214–232.

———, 1955a. Recent work on soil organic matter at Rothamsted. *Z. Pflanzenernähr. Düng. Bodenk.* 69:32–38.

———, 1955b. Studies in soil humic acids. 1. The chemical nature of humic nitrogen. *J. Agric. Sci.* 46:247–256.

———, 1955c. Nitrogen distribution and amino-acid composition of fractions of a humic acid from a chernozem soil (Hildesheimer Schwarzerde). *Z. Pflanzenernähr. Düng. Bodenk.* 71:63–66.

———, and T. Harada, 1959. Release of ammonium and organic matter from soil by hydrofluoric acid and effect of hydrofluoric acid treatment on extraction of soil organic matter by neutral and alkaline reagents. *J. Agric. Sci.* 52:2.

———, and H. Lees, 1949. Studies on soil organic matter. Part 2. The extraction of organic matter from soil by neutral reagents. *J. Agric. Sci.* 39:274–279.

———, and K. Shaw, 1954. Studies on the estimation and decomposition of amino sugars in soil. *J. Agric. Sci.* 44:152–159.

———, W. Flaig, and E. Küster, 1955. Zur Kenntnis der Huminsäuren. 9. Mitteilung. Der Gehalt an Aminosäuren in Streptomyceten-Huminsäuren. *Z. Pflanzenernähr. Düng. Bodenk.* 71:58–63.

Breyhan, T., 1956. Eine Mikromethode zur Stickstoffbestimmung. *Z. Anal. Chem.* 152:412–417.

Broadbent, F. E., 1953. The soil organic fraction. *Advan. Agron.* 5:153–183.

———, 1954. Modification in chemical properties of straw during decomposition. *Proc. Soil Sci. Soc. Amer.* 18:165–169.

———, 1957. Soil organic matter-metal complexes. 2. Cation exchange chromatography of copper and calcium complexes. *Soil Sci.* 84:127–131.

———, and G. R. Bradford, 1952. Cation-exchange groupings in the soil organic fractions. *Soil Sci.* 74:447–457.

———, and J. B. Ott, 1957. Soil organic matter-metal complexes. 1. Factors affecting retention of various cations. *Soil Sci.* 83:6,419–427.

———, W. D. Burge, and T. Nakashima, 1960. Factors influencing the reaction between ammonia and soil organic matter. Seventh International Congress of Soil Science, Madison, Wisconsin, Vol. 3, 509–516.

Brockmann, H., and E. Meyer, 1953. Äquivalent- und Molekulargewichtsbestimmungen durch potentiometrische Mikrotitration in nichtwässrigen Lösungsmitteln. *Chem. Ber.* 86:1514–1523.

Bromfield, A. R., C. B. Coulson, and R. I. Davies, 1959. Humic acid investigations: Fraction studies. *Chem. Ind.:* 1959: 601–602.

Brooks, J. D., R. A. Durie, and S. Sternhell, 1958. Chemistry of brown coals. 3. Pyrolytic reactions. *Australian J. Agric. Res.* 9:303.

Brown, J. K., 1955a. The infrared spectra of coals. *J. Chem. Soc. (London)* 1955:744–752.

———, 1955. Infrared studies of carbonized coals. *J. Chem. Soc. (London),* 752–757.

Brown, S. A., and A. C. Neish, 1955. Shikimic acid as a precursor in lignin biosynthesis. *Nature* 175: 688–689.

Bruckert, S., F. Jacquin, and M. Metche, 1967. Contribution a l'étude des acides phénols présents dans les sols. *Bulletin de l'Ecole Nationale Supérieure Agronomique de Nancy* 9:73–92.

Burges, A., 1960. Physico-chemical investigations of humic acid. *Seventh International Congress of Soil Science,* Madison, Wisconsin. 128–133.

Burges, Alan, and P. Latter, 1960. Decomposition of humic acid by fungi. *Nature* 186:404–405.

————, H. M. Hurst, and S. B. Walkden, 1963. Nature of humic acids. *Nature* 199:696.

————, H. M. Hurst, and B. Walkden, 1964. The phenolic constituents of humic acid and their relation to the lignin of the plant cover. *Geochim. Cosmochim. Acta* 28:1547–1554.

Burkat, S. E., 1959. Solubility of peat humic acids. *J. Appl. Chem. USSR* 32:227–230.

Burton, H., and H. B. Hopkins, 1952. The oxidation of 4-methylcatechol in aqueous solution by ferric ion. *J. Chem. Soc.* 1952: 2445–2447.

Butenandt, A., H. Bieckert, and W. Schäfer, 1960. Über Ommochrome. 19. Mitt.: Versuche zur Konstitution der Ommochrome: Anillinochinone als Zwischenstufen der Phenoxazonsynthese. *Liebigs Ann. Chem.* 632:143–157.

Cain, R. B., 1961. The metabolism of protocatechuic acid by a vibrio. *Biochem. J.* 79:298–312.

Carles, J., and J. Decau, 1960a. Variations in the amino acids of soil hydrolysates. *Sci. Proc. Roy. Dublin Soc., Ser. A1,* 177–182.

——————, 1960b. Some conditions likely to modify the proportion of amino acids in soil. *Ann. Agronom. Paris.* 11:557–575.

Cartwright, N. J., and R. D. Haworth, 1944. The constituents of natural phenolic resins. 19. The oxidation of ferulic acid. *J. Chem. Soc. (London)* 1944:535–537.

Cassidy, H. G., 1949. Electron exchange polymers. *J. Am. Chem. Soc.* 71:402.

Ceh, M., and D. Hadzi, 1956. Infrarotspektren von Huminsäuren und deren Derivaten. *Fuel* 35:77–83.

Chakrabartry, S. K., B. K. Mazumdar, S. N. Roy, and A. Lahiri, 1960. Structural parameters of coal based on oxidation experiments. *Brennstoff-Chem.* 41:138–142.

Chatterjee, B., and S. Bose, 1952. The electrochemical properties of humic acids. *J. Colloid Sci.* 7:414–427.

Chermin, H. A. C., and D. W. van Krevelen, 1957. Chemical structure and properties of coal. 17. A mathematical model of coal pyrolysis. *Fuel* 36:85.

Choudri, M. B., and F. J. Stevenson, 1957. Chemical and physicochemical properties of soil humic colloids. 3. Extraction of organic matter from soils. *Proc. Soil. Sci. Soc. Amer.* 21:508–513.

Chudakov, M. I., S. I. Sukanovskii, and M. P. Akimova, 1959. Benzenoid structure of hydrolytic lignin. *Zhur. Priklad. Khim.* 32:608–613.

Coffin, D. E., and W. A. DeLong, 1960. Extraction and characterization of organic matter of a podzol B-horizon. Seventh International Congress of Soil Science, Madison, Wisconsin, 2:91–97.

Coulson, C. B., R. I. Davies, and E. J. A. Khan, 1959. Humic acid investigations. 2. Studies in the fractionation of humic acids. *J. Soil Sci.* 10:271–283.

————, R. I. Davies, and D. A. Lewis, 1960a. Polyphenols in plant, humus, and soil. 1. Polyphenols of leaves, litter and superficial humus from mull and morsites. *J. Soil Sci.* 11:20–29.

————, 1960b. Polyphenols in plant, humus, and soil. 2. Reduction and transport by polyphenols of iron in model columns. *J. Soil Sci.* 11:30–44.

Cromartie, R. I. T., and J. Harley-Mason, 1953. The structure of tyrosine melanin. *Chem. Ind.* 1953: 972–973.

Dagley, S., W. C. Evans, and D. W. Ribbons, 1960. Metabolism of aromatic compounds by microorganisms. *Nature* 188:560.

Danchy, J. P., and W. W. Pigman, 1951. Reactions between sugars and nitrogenous compounds and their relationship to certain food problems. *Advan. Food Res.* 3:241.

Davies, R. I., and C. B. Coulson, 1959. Humic acid. *Soils Fert.* 22:159.

Deb, B. C., 1949–50a. The movement and precipitation of iron oxides in podzol soils. *Soil Sci.* 1:112–121.

————, 1949–50b. The estimation of free iron oxides in soils and clays and their removal. *Soil Sci.* 1:212–220.

Debye, P., 1944. Light scattering in solutions. *J. Appl. Phys.* 15:338–344.

Demumbrum, L. E., and M. L. Jackson, 1956. Copper and zinc exchange from dilute neutral solutions by soil colloidal electrolytes. *Soil Sci.* 81:353–357.

Deuel, H., and P. Dubach, 1958a. Decarboxylierung der organischen Substanz des Bodens. 2. Nachweis von Uronsäuren. *Z. Pflanzenernähr. Düng. Bodenk.* 82:97–106.

————, and P. Dubach, 1958b. Decarboxylierung der organischen Substanz des Bodens. 3. Extraktion und Fraktionierung decarboxylierbarer Humusstoffe. *Helv. Chim. Acta.* 41:1310–1321.

————, and K. Hutschneker, 1955. Über den Aufbau und die Wirkungsweise von Ionenaustauschern. *Chimia* 9:49–65.

————, P. Dubach, and R. Bach, 1958. Decarboxylierung der organischen Substanz des Bodens. 1. Decarboxylierung der gesamten Humusstoffe. *Z. Pflanzenernähr. Düng. Bodenk.* 81:189.

Diebler, H., M. Eigen, and P. Matthies, 1961. Kinetische Untersuchungen über die Bildung von p-Benzochinon aus Chinon und Hydrochinon in alkalischer Lösung. *Z. Naturfosch.* 16b: 629–637.

Diels, O., and R. Kassebart, 1937. Zur Kenntnis der durch Pyridin bewirkten Polymerisationsvorgänge. 1. Polymerisation des p-Chinons. *Liebigs Ann. Chem.* 530:51–67.

Dijk, van H., 1959. Zur Kenntnis der Basenbindung von Huminsäuren. *Z. Pflanzenernähr. Düng. Bodenk.* 84:150–155.

————, 1960. Electrometric titrations of humic acids. *Sci. Proc. Roy. Dubl. Soc.*, Ser. A1:163.

————, 1962. Analyse von Säurefunktionen bei Huminsäuren mittels elektronischer Titration. *Chem. Weekblad* 34, 58:426–427.

————, 1967. Torf Kolloquium, Peat-Colloquium in Rostock, Deutsche Demokratische Republik, Poland 1/3.

Dirks, B., 1940. Das Redoxsystem des Bodens, ein neuer Wachstumsfaktor von ausschlaggebender Bedeutung und zwei weitere Wachstumsfaktoren. *Bodenkunde Pflanzenernähr.* 21/22: 684–697.

Dragunov, C. C., H. H. Zhelokhovtseva, and E. I. Strelkova, 1948. A comparative study of soil and peat humic acids. *Pochvovedenie* 7:409–420.

Drozdova, T. V., 1955. Application of the method of chromatography for studying fulvic acids. *Pochvovedenie* 1:83–87.

————, 1959. Chitin und dessen Umwandlung unter natürlichen Bedingungen. Bildung der Melan-oidine. *Fortschr. gegenwärt. Biol.* 47:277.

————, 1963. Role of humic acids in soil geochemistry. *Soviet Soil Sci.* 8:753–758.

————, and M. P. Yemel'yanova, 1960. Copper complexes with humic acids. *Dokl. Akad. Nauk SSSR* 131:3.

————, R. P. Kravtsova, and K. I. Tobelko, 1962. Studies of complex compounds of phenols with germanium by paper electrophoresis and X-ray analysis. *Izv. Akad. Nauk SSSR, Otd. khim. Nauk* 1:38–44.

Dubach, P., and N. C. Mehta, 1963. The chemistry of soil humic substances. *Soils Fertilizers* 26:293–300.

————, N. C. Mehta, and H. Deuel, 1961. Extraktion von Huminstoffen aus dem B-Horizont eines Podsols mit ÄDTE. *Z. Pflanzenernähr. Dung. Bodenk.* 95:119–123.

————, N. C. Mehta, and H. Deuel, 1963. Schonende Extraktion von Huminstoffen und Isolierung der Fulvosäure-Fraktion aus verschiedenen Bodentypen. *Z. Pflanzenernähr. Düng. Bodenk.* 102: 1–7.

————, N. C. Mehta, T. Jakab, F. Martin, and N. Roulet, 1964. Chemical investigations on soil humic substances. *Geochim. Cosmochim. Acta* 28:1567–1578.

Duchaufour, Ph., and F. Jacquin, 1963. Recherche d'une méthode d'extraction et de fractionnément des composés humiques contrôlée par l'électrophorése. *Ann. Agron.* 14:885–918.

Ebertova, H., 1959. Eine Methode zur Ermittlung der Oxidations-Reduktions-Potentiale in Humat-lösungen, im Kompost und im Boden. *Sbornik Ceskoslov. akad. zemedel skych ved. rostlinna vyroba* 5:479–494.

Eigen, M. and P. Matthies, 1961. Über Kinetik und Mechanismus der Primärreaktionen der Zerset-zung von Chinon in alkalischer Lösung. *Chem. Ber.* 94:3309–3317.

Einstein, A., 1906. Eine neue Bestimmung der Moleküldimensionen. *Ann. Phys.* 19:289–306.

————, 1911. Berichtigung zu: Eine neue Bestimmung der Moleküldimensionen. *Ann. Phys.* 34:591–592,

Eistert, B., 1948. Synthesis with diazomethane. In *Newer Methods of Preparative Chemistry*, New York, Interscience Publishers, Inc. p. 513.

Eller, W., 1921. Künstliche und natürliche Huminsäuren. *Brennstoffchem.* 2:129–133.

————, 1923. Studien über Huminsäuren. 4. Darstellung und Eigenschaften künstlicher und natürlicher Huminsäuren. *Liebigs Ann. Chem.* 431:133–161.

————, 1925a. Studien über Huminsäuren. Einige Eigenschaften und Reaktionen der Huminsäuren und Humine. *Liebigs Ann. Chem.* 442:160–180.

————, 1925b. Zur Frage der Konstitution der natürlichen Huminsäuren. *Brennstoffchem.* 6:54.

————, and K. Koch, 1920. Synthetische Darstellung von Huminsäuren. *Ber. Dtsch. Chem. Ges.* 53:1469–1476.

————, E. Herdickerhoff, and H. Saenger, 1923. Einwirkung von Chlor auf Huminsäuren. *Liebigs Ann. Chem.* 431:177–186.

————, H. Meyer, and H. Saenger, 1923. Einwirkung von Salpetersäure auf Huminsäuren. *Liebigs Ann. Chem.* 431:162–175.

Ellis, G. P., 1959. The maillaird reaction. *Advan. Carbohydr. Chem.* 14:63.

Enders, C., 1942a. Über den Chemismus der Huminsäurebildung unter physiologischen Bedingungen. 1. *Mitt. Biochem. Z.* 312:339–348.

————, 1942b. Über den Chemismus der Huminsäurebildung unter physiologischen Bedingungen. 3. Mitt.: Die Abhängigkeit von Bildungsgeschwindigkeiten, Eigenschaften und Zusammensetzung der aus Methylglyoxal und Glykokoll gebildeten Huminsäuren von verschiedenen Faktoren. *Biochem. Z.* 313:352–371.

————, 1943. Wie entsteht der Humus in der Natur? *Die Chemie* 56:281–292.

————, and S. Sigurdsson, 1942. Über den Chemismus der Huminsäurebildung unter physiologischen Bedingungen. 2. Mitt. Über das Vorkommen von Methylglyoxyl in Erde. *Biochem. Z.* 313:174–181.

————, and S. Sigurdsson, 1948. Über den Chemismus der Huminsäurebildung unter physiologischen Bedingungen. 7. *Biochem. Z.* 318:44–46.

————, and K. Theiss, 1938a. Die Melanoidine und ihre Beziehung zu den Huminsäuren. *Brennstoff-Chem.* 19:360.

————, 1938b. Die Melanoidine und ihre Beziehung zu den Huminsäuren. *Brennstoff-Chem.* 19:402.

————, 1938c. Die Melanoidine und ihre Beziehung zu den Huminsäuren. *Brennstoff-Chem.* 19:439.

————, M. Tschapek, and R. Glawe, 1948. Vergleichende Untersuchungen einiger kolloider Eigenschaften von natürlichen Huminsäuren und synthetischen Melanoidinen. *Kolloid Z.* 110:3.

Ensminger, L. E., and J. E. Gieseking, 1939. The absorption of proteins by montmorillonitic clays. *Soil Sci.* 48:467–474.

Erdtman, H. G. H., 1933a. Studies on the formation of complex oxidation and condensation products of phenols. A contribution to the investigation of the origin and nature of humic acid. 1. Studies of the reactivity of simple monocyclic quinones. *Proc. Roy. Soc.* (London) A143:177–191.

————, 1933b. Studies on the formation of complex oxidation and condensation products of phenols. 2. Coupling of simple phenols and quinones to diphenyl derivatives. *Proc. Roy. Soc.* (London) A 143:191–222.

————, 1933c. Studies on the formation of complex oxidation and condensation products of phenols. 3. Re-arrangements of oxidation-reduction type in the diquinone group. *Proc. Roy. Soc.* (London) A 143:223–228.

————, 1933d. Studies on the formation of complex oxidation and condensation products of phenols. 4. Termolecular polymerisation products of p-benzoquinone, and α-naphthoquinone. *Proc. Roy. Soc.* (London) A 143:228–241.

————, 1933e. Dehydrierungen in der Coniferylreihe. 1. Dehydrodieugenol und Dehydrodiisoeugenol. *Biochem. Z.* 252:172–180.

————, 1933f. Dehydrierungen in der Coniferylreihe, 2. Dehydrodiisoeugenol. *Liebigs Ann. Chem.* 503:283–294.

————, 1935. Phenoldehydrierungen. 6. Dehydrierende Kupplung einiger Gaujacolderivate. *Svensk kem. Tidskr.* 47:223–230.

————, 1955. Die Reaktion von Benzochinon mit Alkali. *Z. Pflanzenernähr. Düng. Bodenk.* 69:38–43.

————, 1962. The chemistry of forests, humic acids, lichens, lignans, lignins, and conifers. *Tappi* 45:4.

Erdtman, H. G. H., and M. Granath, 1954. Studies on humic acids. 5. The reaction of p-benzoquinone with alkali. *Acta Chem. Scand.* 8:811–816.

————, M. Granath, and G. Schultz, 1954. Studies on humic acids. 6. Triquinone and tretraquinone. *Acta Chem. Scand.* 8:1442–1450.

————, and N. E. Stjernström, 1959. Studies on humic acids. 7. The structure of the dihydroxy-dioxidoterphenyl obtained by the action of sulphuric acid on p-benzoquinone. *Acta Chem. Scand.* 13:653–658.

————, and C. A. Wachtmeister, 1957. Phenoldehydrogenation as a biosynthetic reaction. Festschrift Prof. Dr. W. Stoll, 144–165.

Esh, C. C., and S. S. Guha-Siecar, 1940. An investigation on the soil and peat humic acids. 2. Oxidation with hydrogen peroxide, hot alkali and chlorine dioxide solution. *J. Indian Chem. Soc.* 17:6, 405–411.

Evans, L. T., 1959. The use of chelating reagents and alkaline solutions in soil organic matter extractions. *J. Soil Sci.* 10:110–118.

Evans, W. C., 1947. Oxidation of phenol and benzoic acid by some soil bacteria. *Bio chem. J.* 41:373–382.

————, B. S. W. Smith, R. P. Linstead, and J. A. Elvidge., 1951. Chemistry of the oxidative metabolism of certain aromatic compounds by micro-organisms. *Nature* 168:772–775.

Even-Haim, A., 1966. Etude sur l'absorption de la matière organique par les plantes supérieurs au moyen de leurs racines, essais a l'aide d'acide vanillique contenant carbone radioactif. In "The use of isotopes in soil organic matter studies", Report of the FAO/IAEA Technical Meeting. Oxford:Pergamon Press, Ltd., 49–59.

Farmer, V. C., and R. I. Morrison, 1960. Chemical and infrared studies on phragmites peat and its humic acid. *Sci. Proc. Roy. Dublin Soc.* Ser. A1:85–104.

————, and R. I. Morrison, 1964. Lignin in Sphagnum and phragmites and in peats derived from these plants. *Geochim. Cosmochim. Acta* 28:1537–1546.

————, M. E. K. Henderson, and J. D. Russell, 1959. Reduction of certain aromatic acids to aldehydes and alcohols by Polystictus versicolor. *Biochim. Biophys. Acta* 35:202–211.

Felbeck, G. T., Jr., 1965. Studies on the high pressure hydrogenolysis of the organic matter from a Muck soil. *Proc. Soil Sci. Soc. Am.*:29:48–55.

Ferrari, G., and G. Dell'Agnola, 1963. Fractionations of the organic matter of soil by gel filtration through Sephadex. *Soil Sci.* 96:418–421.

Feustel, I. C., 1936. Transactions of the Third International Congress of Soil Science. Comm. 6B: 345.

————, and H. G. Byers, 1936. The behaviour of lignin and humic acid preparations toward a bromination treatment. *Soil Sci.* 42:11–21.

Finkle, B. J., 1965. Soil humic acid as a hydroxypolystyrene: A biochemical hypothesis. *Nature* 207: 604.

Fischer, G., 1953. Untersuchungen über den biologischen Abbau des Lignins durch Mikroorganismen. *Arch. Mikrobiol.* 18:397–424.

Fischer, F., and H. Schrader, 1932. Neue Beiträge zur Entstehung und chemischen Struktur der Kohle. *Brennstoff-Chem.* 3:65–72.

Fischer, W., G. Schlungbaum, and L. Belau, 1964. Zur Kenntnis der chemischen Zusammensetzung von Torfen aus Vorkommen in der Deutschen Demokratischen Republik. 6. Orientierende Untersuchungen von Humusfraktionen. 3. Poln. Torfkolloquium, Warschau, 11.–14.6.1964. Naczelna Organizscja, Techniczna, Stowarzyszenie Inzynierow Technikow Wodoych i Meliorac, 34–51.

————, G. Schlungbaum, and R. Kadner, 1964. Chemische Untersuchungen von Braunkohlen aus Vorkommen in der DDR. 3. Zur Untersuchung der Carbonyl- und Methoxylgruppen in Braunkohlen und Torfen. *Z. Chem.* 4:394–395.

————, L. Belau, G. Schlungbaum, and E. Preu, 1965. Zur Kenntnis der chemischen Zusammensetzung von Torfen aus Vorkommen der Deutschen Demokratischen Republik. 7. CH_2- und Carbonyl-banden in IR-Spektren eines stark zersetzten Hochmoortorfes und der daraus isolierten Fraktionen. *Bergbautechnik* 15:429–430.

190   W. Flaig, H. Beutelspacher, and E. Rietz

Flaig, W., 1950. Zur Kenntnis der Huminsäuren. 1. Zur chemischen Konstitution der Huminsäuren. *Z. Pflanzenernähr. Düng. Bodenk.* 51:193–212.

———, 1955. Zur Bildungsmöglichkeit von Huminsäuren aus Lignin. *Holzforschung* 9:1–4.

———, 1956. Zur Chemie der Huminsäuren und deren Modellsubstanzen. 6e. Congrès Internat. de la Science du Sol. Paris, 2:471–478.

———, 1958. Die Chemie organischer Stoffe im Boden und deren physiologische Wirkung. Verhandl. d. II. und IV. Komm. d. Internat. Bodenkundl. Ges., Hamburg, Vol. II.

———, 1959. Sobre la formacion del humus y su influencia sobre el crecimiento de las plantas. *Anales de Edafologia y Fisiologia Vegetal* 18:505–539.

———, 1960a. Chemie der Humusstoffe. *Suomen Kemistilehti* A33:229–251.

———, 1960b. Comparative chemical investigations on natural humic compounds and their model substances. *Sci. Proc. Roy. Dublin Soc.* ser. A1, 4:149.

———, 1962. Zur Umwandlung von Lignin in Humusstoffe. Freiberger Forschungshefte A254:39–56.

———, 1964a. Chemische Untersuchungen an Humusstoffen. *Z. Chem.* 4:253–265.

———, 1964b. Gedanken zur Nomenklatur der im Boden vorhandenen organischen Stoffe. Eighth International Congress of Soil Science. Transaction, Vol. 3. Bucharest–Romania, 389–399. Discussions regarding the terminology in the field of humus, 405–413.

———, 1966. The chemistry of humic substances. In "The use of isotopes in soil organic matter studies." Report of the FAO/IAEA Technical Meeting. Oxford: Pergamon Press, Ltd. 103–127.

Flaig, W., and H. Beutelspacher, 1950. Elektronenmikroskopische Untersuchungen in der Bodenkunde. *Optik* 7:237–240.

———, and H. Beutelspacher, 1951. Zur Kenntnis der Huminsäuren. 2. Elektronenmikroskopische Untersuchungen an natürlichen und synthetischen Huminsäuren. *Z. Pflanzenernähr. Düng. Bodenk.* 52:1–21.

———, and H. Beutelspacher, 1953. Elektronenmikroskopische Untersuchungen über die Gestalt der Tonminerale in Böden. *Grundlagen der Landtechnik* 5:87–95.

———, and H. Beutelspacher, 1954. Physikalische Chemie der Huminsäuren. *Landbouwk. Tijdschr.* 66:306–336.

———, and H. Beutelspacher, 1968. Investigations of humic acids with the analytical ultracentrifuge. In: Isotopes and radiation in soil organic matter studies. Proceedings of the Symposium on the Use of Isotopes and Radiation in Soil Organic Matter Studies. International Atomic Energy Agency, Vienna 23–30.

———, H. Beutelspacher, and H. Söchtig, 1954. Humus als Kationenaustauscher. *Kalium-Symposium* 81–107.

———, and T. Breyhan, 1956. Über das Vorkommen von Indolverbindungen in Schwarzerde-Huminsäuren. *Z. Pflanzenernähr. Düng. Bodenk.* 75:132–135.

———, and K. Haider, 1961a. Reaktionen mit oxydierenden Enzymen aus Mikroorganismen. *Planta Med.* 9:123–139.

———, and K. Haider, 1961b. Die Verwertung phenolischer Verbindungen durch Weißfäulepilze. *Arch. Mikrobiol.* 40:212–223.

———, and K. Reinemund, 1955. Zur Kenntnis der Huminsäuren. Studien zur Beschleunigung von Dehydrierungsreaktionen durch Modellsubstanzen von Huminsäurevorstufen bzw. -abbauprodukten. *Z. Pflanzenernähr. Düng. Bodenk.* 68:255–269.

———, and J.Chr. Salfeld, 1958. UV-Spektren und Konstitution von p-Benzochinonen. *Liebigs Ann. Chem.* 618:117–139.

———, and J.Chr. Salfeld, 1959. Infrarotspektren von p-Benzochinonen. *Liebigs Ann. Chem.* 626:215–224.

———, and J.Chr. Salfeld, 1960a. Zwischenstufen bei der Bildung von Huminsäuren aus Phenolen. Transactions Vol. 2, Seventh International Congress of Soil Science, Madison, Wisconsin, 648–656.

———, and J.Chr. Salfeld, 1960b. Nachweis der Bildung von Hydroxy-p-benzochinon als Zwischenprodukt bei der Autoxydation von Hydrochinon in schwach alkalischer Lösung. *Naturwissenschaften* 47:516.

Flaig, W., and H. L. Schmidt, 1957. Über die Einwirkung von Huminsäuren auf das Wachstum einiger Penicilliumarten. *Arch. Mikrobiol.* 27:1–32.

——, and H. Schulze, 1952. Über den Bildungsmechanismus der Synthesehuminsäuren. *Z. Pflanzenernähr. Düng. Bodenk.* 58:59–67.

——, Th. Ploetz, and A. Küllmer, 1955. Über Ultraviolettspektren einiger Benzochinone. *Z. Naturforsch.* 10b:668–676.

——, J. Chr. Salfeld, and K. Haider, 1963. Zwischenstufen bei der Bildung von natürlichen Huminsäuren und synthetischen Vergleichssubstanzen. *Landw. Forsch.* 16:85–96.

——, J. Chr. Salfeld, and G. Rabien, 1963. Die Bildung von 1.2-Di-(benzochinon-1.5)-äthanen aus Additionsprodukten von Diazomethan an p-Benzochinon. *Makromol. Chem.* 59:206–212.

——, F. Scheffer, and B. Klamroth, 1955. Zur Kenntnis der Huminsäuren. 8. Zur Charakterisierung der Huminsäuren des Bodens. *Z. Pflanzenernähr. Düng. Bodenk.* 71:33–37.

——, U. Schobinger, and H. Deuel, 1959. Umwandlung von Lignin in Huminsäuren bei der Verrottung von Weizenstroh. *Chem. Ber.* 92:1973–1982.

——, H. Söchtig, and H. Beutelspacher, 1963. Einfluß der Humusstoffe auf die Umtauschkapazität der Böden. *Landbauforsch. Völkenrode* 13:13–20.

Forsyth, W. G. C., 1947a. Studies on the more soluble complexes of soil organic matter. 1. A method of fractionation. *Biochem. J.* 41:176–181.

——, 1947b. The characterization of the humic complexes of soil organic matter. *J. Agric. Sci.* 37:132–138.

——, 1950. Studies on the more soluble complexes of soil organic matter. 2. The composition of the soluble polysaccharide fraction. *Biochem. J.* 46:141.

——, and V. C. Quesnel, 1957. Intermediates in the enzymic oxidation of catechol. *Biochem. Biophys. Acta* 25:155–160.

——, V. C. Quesnel, and J. B. Roberts, 1960. Diphenylenedioxide-2. 3-quinone: An intermediate in the enzymic oxidation of catechol. *Biochem. Biophys. Acta* 37:322–326.

Frankel, M., and A. Katschalsky, 1941. The interaction of aldose with α-amino acids or peptides. 4. The percentage combination. *Biochem. J.* 35:1028.

Freudenberg, K., 1956. Lignin im Rahmen der polymeren Naturstoffe. *Angew. Chem.* 68:84.

——, 1959. Über die Biosynthese und Konstitution des Lignins. *Chem. Ber.* 92:89.

——, 1962. Forschung am Lignin. *Forschr. Chem. Org. Naturstoffe* 20:41.

——, 1964. Entwurf eines Konstitutionsschemas für das Lignin der Fichte. *Holzforschung* 18:3–9.

——, and J. M. Harkin, 1960. Modelle für die Bindung des Lignins an die Kohlenhydrate. *Chem. Ber.* 93:2814–2819.

——, and J. M. Harkin, 1964. Ergänzung des Konstitutionsschemas für das Lignin der Fichte. *Holzforschung* 18:166.

Freytag, H. E., 1955. Absorptionsphotometrische Änderungen der Huminstoffe im Verlaufe des Wurzelbaues. 1. Über stofflich begründbare Zusammenhänge von Farbtyp- und Konzentrationsänderungen während der Huminstoffbildung. *Z. Pflanzenernähr. Düng. Bodenk.* 71:67–76.

——, 1961a. Ein Beitrag zur Kenntnis der Huminsäuresynthese. 1. Über die Zerlegung von Huminstoffextrakten mit Hilfe der Diffusion und Zusammenhänge zwischen Kopplungsgrad und Farbtypsteilheit. *Albrecht Thaer Archiv* 5:583–603.

——, 1961b. Ein Beitrag zur Kenntnis der Huminsäuresynthese. 2. Fraktionsübergänge synthetischer Huminstoffe. *Albrecht Thaer Archiv* 5:643–654.

——, 1961c. Ein Beitrag zur Kenntnis der Huminsäure-synthese. 3. Fraktionsübergänge von natürlichen Huminstoffen. *Albrecht Thaer Archiv* 5:730–743.

——, 1962. Über einige Analogien im Verlauf pflanzlicher und synthetischer Humifizierungsprozesse. Studies about Humus, Symposium on Humus and Plants, Prague, Czechoslovakia. 75–87.

Frömel, W., 1938a. Über Absorptionsspektren von Huminsäuren in Lösungen. *Bodenk. Pflanzenernähr.* 6:93–119.

——, 1938b. Über UV-Absorptionsspektren von Huminsäuren in Lösungen. *Bodenk. Pflanzenernähr.* 11:129–144.

——, 1941. Über Fulvosäuren. *Bodenk. Pflanzenernähr.* 25:345–358.

Fuchs, W., 1928. Über die sogenannte "Nitro"-Huminsäure. *Brennstoff-Chem.* 9:178–182.

——, and O. Horn, 1930. Über die Salze und die Ester der Huminsäuren. *Brennstoff-Chem.* 11:372–375.

——, and W. Stengel, 1930. Über den Abbau der Huminsäuren zu Benzolcarbonsäuren und Nitrophenolen. *Liebigs Ann. Chem.* 478:267–283.

Führ, F., 1962. Untersuchungen zur Aufnahme von Kohlendioxid und Strohabbauprodukten durch die Pflanzenwurzel. Doctoral dissertation, University of Bonn, Germany.

——, and D. Sauerbeck, 1966. The uptake of straw decomposition products by plant roots. In "The use of isotopes in soil organic matter studies." Report of the FAO/IAEA Technical Meeting, Oxford: Pergamon Press, Ltd. 73–83.

Fukuzumi, R., K. Minami, and S. Shibamoto, 1959. Metabolic products from aromatic compounds by the wood-rotting fungus "Polystictus sanguineus." 2. Acetoacetic acid and α-ketoglutaric acid from the medium containing benzoic acids. *J. Japan. Wood. Res. Soc.* 5:96.

Gätke, C. R., 1961. Investigations on the extraction of humus typical of the substrate. *Albrecht Thaer Archiv* 5:256–265.

Gale, E. F., 1946. The bacterial amino acid decarboylases. *Advan. Enzymol.* 6:1–32.

Ganz, E., 1944. Contribution a l'étude de la structure des acides humique naturels. *Ann. Chim.* 19:202.

Gillam, W. S., 1940. A study on the chemical nature of humic acids. *Soil Sci.* 49:433–453.

Giza, Y. Ch., K. A. Kun, and H. G. Cassidy, 1962. Electron exchange polymers. 16. The oxidation behaviour of dimethyl ethers of hydroquinones and methyl substituted hydroquinones. *J. Org. Chem.* 27:679–682.

Gortner, R. A., 1916. The organic matter of the soil. 1. Some data on humus, humus carbon, and humus nitrogen. *Soil Sci.* 2:395–441.

Gottlieb, S., and S. B. Hendricks, 1945. Soil organic matter as related to newer concepts of lignin chemistry. *Proc. Soil Sci. Soc. Am.* 10:117–125.

——, W. C. Day, and M. J. Pelczar, 1950. The biological degradation of lignin. 2. The adaption of white rot fungi to growth on lignin media. *Phytopathology* 40:926–935.

Gottschalk, A., and S. M. Partridge, 1950. Interaction between simple sugars and amino acids. *Nature* 165:684.

Goulden, J. D. S., and D. S. Jenkinson, 1959. Studies on the organic material extracted from soils and composts. 2. The infrared spectra of ligno-proteins isolated from compost. *J. Soil Sci.* 10:264–270.

Greene, G., and G. Steelink, 1962. Structure of soil humic acid. 2. Some copper oxide oxidation products. *J. Org. Chem.* 27:170–174.

Grisebach, H., 1957. Zur Biogenese des Cyanidins. 1. Versuche mit Acetat-(1-C14) und Acetat (2-C14). *Z. Naturforsch.* 12b:227–231.

Gross, S. R., R. D. Gafford, and E. L. Tatum, 1956. The metabolism of protocatechuic acid by Neurospora. *J. Biol. Chem.* 219:781–796.

Gündel, W., and R. Pummerer, 1937. Über die Wirkung aromatischer Nitrosoverbindungen auf Chinone. *Liebigs Ann. Chem.* 529:11–32.

Guinier, A., 1939. Beugung von Röntgenstrahlen bei sehr kleinen Winkeln. Anwendung auf die Untersuchung ultramikroskopischer Phänomene. *Ann. Phys.* 12:161–237.

Hagihara, B., 1954. Crystalline bacterial amylase and proteinase. *Ann. Rept. Sci. Works Fac. Sci. Osaka Univ.* 2:35–80.

Haider, K. 1965. Untersuchungen über den mikrobiellen Abbau von Lignin. *Zbl. Bakteriol., Parasitenkunde Infektionskrankh., Hyg.,* I. Abt., Orig. 198:308–316.

——, 1967. Ligninabbau durch mesophile Organismen in Ackerböden. Landwirtsch. Forsch. 21. Sonderheft.

——, and K. Grabbe, 1967. Die Rolle der Phenoloxydasen beim Ligninabbau durch Weißfäulepilze. *Zbl. Bakteriol., Parasitenkunde, Infektionskrankh., Hyg.,* I. Abt., Orig. 205:91–96.

——, and J. P. Martin, 1967. Synthesis and transformation of phenolic compounds by Epicoccum nigrum in relation to humic acid formation. *Proc. Soil Sci. Soc. Am.* 31:766–772.

——, L. R. Frederick, and W. Flaig, 1965. Reactions between amino acid compounds and phenols during oxidation. *Plant and Soil* 22:49–64.

Haider, K., S. Lim, and W. Flaig, 1962. Untersuchungen über die Einwirkung von Mikroorganismen auf C¹⁴-markierte phenolische Verbindungen. *Landwirtsch. Forsch.* 15:1–9.

———, S. Lim and W. Flaig, 1964. Experimente und Theorie über den Ligninabbau bei der Weißfäule des Holzes und bei der Verrottung pflanzlicher Substanz im Boden. *Holzforschung* 18:81–88.

Halla, F., and W. Ruston, 1955. Leitfähigkeitsanomalien bei Huminsäuresolen. *Z. Elektrochem.* 59:525–530.

Hargital, L., 1962. Untersuchungen über den Zusammenhang zwischen optischen Eigenschaften und Stickstoffgehalt der Huminsäuren (ung.). *A Keszthelyi Mezögazdasági Akadémia Kiadványai.* 3–20.

Hathway, D. E., 1957. Autoxidation of polyphenols. 1. Autoxidation of methyl gallate and its o-methyl ethers in aqueous ammonia. *J. Chem. Soc.* 1957, 519–523.

———, 1958. An approach to the study of vegetable tannins by the oxidation of plant phenolics. *J. Soc. Leath. Tr. Chem.* 42:109–121.

———, and J. W. Sekins, 1957. Enzymic oxidation of catechin to a polymer structurally related to some phlobatannins. *Biochem. J.* 67:239–245.

Haugard, G., L. Tumerman, and H. Silvestri, 1951. A study on the reaction of aldoses and amino acids. *J. Amer. Chem. Soc.* 73:4594.

Hayashi, T., and T. Nagai, 1956a. On the nitrogen constituents of A-type humic acids. *J. Fac. Agric. Tottori Univ.* 2:105.

———, 1956b. On the components of soil humic acids (part 4) on the absorption spectra of the components of B-type humic acids. *Soil and Plant Food* 2:171–175.

———, 1956c. Untersuchung über die Bestandteile der Bodenhuminsäuren. 3. Mitt. Über die N-haltigen Bestandteile der Huminsäuren vom Typ. A. *J. Sci. Soil Manure, Japan* 26:371–375.

———, 1960. On the components of soil humic acids. 7. Quantitative determination and distribution of components. *Soil and Plant Food* 5:153–160.

———, 1961. On the components of soil humic acids (part 8). The oxidative decomposition by alkaline potassium permaganate and nitric acid. *Soil and Plant Food* 6:170–175.

Hayaishi, O., M. Katagiri, and S. Rothberg, 1955. Mechanism of the pyrocatecase reaction. *J. Am. Chem. Soc.* 77:5450–5451.

Henderson, M. E. K., 1955. Release of aromatic compounds from birch and spruce sawdusts during decomposition by white-rot fungi. *Nature* (London) 175:634–635.

———, 1957. Metabolism of methoxylated aromatic compounds by soil fungi. *J. Gen. Microbiol.* 16:686–695.

Henrich, F., W. Schmidt, and F. Rossteutscher, 1915. Über ein Oxydationsprodukt des Orcins. *Ber. dtsch. chem. Ges.* 48:483–489.

Herrmann, K., 1954. Über die Bildung der Purpurogallincarbonsäure durch fermentative Oxydation der Gallussäure. 2. Über gerbstoffartige Substanzen. *Arch. Pharmazie* 287:497–503.

Higushi, T., 1953. Biochemical study of wood-rotting fungi. 1. The enzymes responsible for the Bavendamm reaction. *J. Japan. Forest Soc.* 35:77–85.

———, I. Kawamura, and I. Hayashi, 1956. Biochemical study of wood-rotting fungi. *J. Japan. Wood. Res. Soc.* 2:31–35.

———, I. Kawamura, and H. Kawamura, 1955. Properties of the lignin in decayed wood. *J. Japan. Forest Soc.* 37:298–302.

Himes, F. L., and S. A. Barber, 1957. Chelating ability of soil organic matter. *Proc. Soil Sci. Soc. Am.* 21:368–373.

Hobson, R. P., and H. J. Page, 1932. Studies on the carbon and nitrogen cycles in the soil. 7. The nature of the organic nitrogen compounds of the soil: "Humic" nitrogen. *J. Agric. Sci.,* 22:497–515.

Hock, A., 1937. Farbtiefen- und Farbtonwerte als charakteristische Kennzeichen für Humusform und Humustyp in Böden nach neuen Verfahren. *Bodenk. Pflanzenernähr.* 2:304–315.

———, 1938a. Beziehungen zwischen Konzentration und Farbwerten von Huminsäurelösungen. *Bodenk. Pflanzenernähr.* 7:99–117.

———, 1938b. Grundsätzliches bei Farbmessungen in Huminlösungen. *Bodenk. Pflanzenernähr.* 7:279–302.

Hock, A., 1942. Beitrag zum Humusproblem in der Bodenkunde. *Jb. Reichsst. Bodenforsch.* 6:817–822.

Hodge, J. E., 1953. Dehydrated foods, chemistry of browing reactions in model systems. *J. Agric. Food Chem.* 1:928.

————, and C. H. Rist, 1953. The amadori rearrangement under new conditions and its significance for non-enzymatic browning reactions. *J. Am. Chem. Soc.* 75:316.

Hoffmann, F., 1964. Farbmessung an Humin- und Fulvosäuren. *Arch. Forstwesen* 13:1159–1180.

Holmberg, B., 1934. Thioglykolsäure als Ligninreagenz. *Ing. Ventenskaps Akad.* 131:5–15.

Hoppe-Seyler, F., 1889. Über Huminsubstanzen, ihre Entstehung und ihre Eigenschaften. *Hoppe-Seyler's Z. Physiol. Chem.* 13:66–121.

Hori, S., and A. Okuda, 1961. Purification of humic acid by the use of ion exchange resin. *Soil Sci. Plant Nutrition* 7:4.

Horner, L., and W. Dürckheimer, 1959. Zur Kenntnis der o-Chinone. 12. o-Chinone aus Brenz-catechin-Derivaten. *Z. Naturforsch.* 14B:741.

————. and K. Sturm, 1957. Zur Kenntnis der o-Chinone. 10. Modellreaktionen zur Melaninbildung. *Liebigs Ann. Chem.* 608:128–139.

Hough, L., J. R. N. Jones, and E. L. Richards, 1952. The reaction of amino-compounds with sugars 1. The action of ammonia on d-glucose. *J. Chem. Soc.* (London): 3854–3857.

Huisman, H. O., 1950. Investigations on quinone and quinone derivatives. Preparation and antibiotic properties of some substituted p-benzo- and p-toluquinones, hydroquinones and hydroquinone diesters. *Rec. Trav. Chim.* 69:1133–1156.

Irving, M., and R. I. P. Williams, 1948. Order of stability of metal complexes. *Nature* 162:746–747.

Ishikawa, H., W. J. Schubert, and F. F. Nord, 1963a. Investigations on lignins and lignification. 27. The enzymic degradation of softwood lignin by white-rot fungi. *Arch. Biochem. Biophys.* 100:131–139.

————, W. J. Schubert, and F. F. Nord, 1963b. Investigations on lignins and lignification. 30. Enzymic degradation of guajacolglycerol and related compounds by white-rot fungi. *Biochem. Z.* 338:153–163.

Jacquin, F., 1960. Chromatographic study of various types of humic acids. *C. R. Hebd. Séances Acad Sci.* 250:1892–1893.

Jacquin, R., 1963. Contribution à l'étude des processus de formation et d'évolution des divers compóses humiques. Thèse Docteur ès Sciences. *Bull. ENSAN* V:1–156.

Jakab, T., P. Dubach, N. C. Mehta, and H. Deuel, 1962. Abbau von Huminstoffen. 1. Hydrolyse mit Wasser und Mineralsäuren. *Z. Pflanzenernähr. Düng. Bodenk.* 96:213–217.

————, 1963. Abbau von Huminstoffen. 2. Abbau mit Alkali. *Z. Pflanzenernähr. Düng. Bodenk.* 102:8–17.

James, T. H., and A. Weissberger, 1938. Oxidation processes. 11. The autoxidation of durohydroquinone. *J. Am. Chem. Soc.* 60:98–104.

James, T. H., J. M. Snell, and A. Weissberger, 1938. Oxidation processes. 12. The autoxidation of hydroquinone and of the mono-, di- and trimethylhydroquinone. *J. Am. Chem. Soc.* 60:2084–2093.

Jenkinson, D. S., and J. Tinsley, 1959. Studies on the organic material extracted from soils and composts. Part I. The isolation and characterization of ligno-proteins from compost. *J. Soil Sci.* 10:245–263.

————, and J. Tinsley, 1960. A comparison of the ligno-protein isolated from a mineral soil and from a straw compost. *Sci. Proc. Roy Dublin Soc.* A1:141–147.

Jodl, R., 1941a. Über Chemie und Technologie natürlicher und künstlicher Huminstoffe. *Brennstoff-Chem.* 22:78–85.

————, 1941b. Über Kohlehydrathuminsäuren und ihre Beziehungen zu natülichen und Phenol-Huminsäuren. *Brennstoff-Chem.* 22:217–220.

————, 1941c. Über Grösse und Form von Huminsäurekristalliten. *Brennstoff-Chem.* 22:256–257.

————, 1941d. Röntgenographische Untersuchungen über natürliche und künstliche Huminstoffe. *Brennstoff-Chem.* 22:157–161.

Jones, E. J. Jr., 1949. The ultraviolet absorption spectra of complex hydroxyaromatic compounds and derivatives, with particular reference to lignin. *Tappi* 32:311–315.

Jung, E., 1938. Physikalisch-chemische Untersuchungen an Braunhuminsäuren. *Bodenk. Pflanzenernähr.*, 9/10 (54/55) 248–273.

Juste, C., 1966. Modification in electrophoretic mobility and infrared spectrum of an iron or aluminium enriched humic acid. *C. R. Acad. Sci. Paris*, Ser. D, 262:2692–2695.

Kaila, A., 1952. Humification of straw at various temperatures. *Acta agral. Fennica* 78:3–32.

Kasatochkin, V. I., 1951. The structure of carbonized substances. *Izv. Akad. Nauk SSSR, Otd. tekh. Nauk* 9.

———, and O. I. Zilberbrand, 1956. X-ray and infrared spectroscopy applied to the study of the structure of humic substances. *Pochvovedenie* 5:80–85.

———, M. M. Kononova, and G. I. Zilberbrand, 1958. Infrarotspektren von Huminsäuren des Bodens (russ.) *Dokl. Akad. Nauk* 119:785–788.

———, M. M. Kononova, N. K. Larina, and O. I. Egorova, 1964. Theses on "Spectral and X-ray investigation of chemical structure of humic substances of soil". Transaction, C. R., Ber. -III. Eighth International Congress Soil Science, Bucharest-Romania 1964, 81–86. Also pages 195–205.

Katchalsky, A., and O. Spitnik, 1947. Potentiometric titration of polymethacrylic acid. *J. Polymer. Sci.* 2:432–446.

Kaurichev, I. S., and YE. M. Nozdrunova, 1960. Calculation of migration of certain compounds in the soil by means of lysimetric chromatographic columns. *Pochvovedenie* 12.

———, E. A. Fedorov, and I. A. Shnabel, 1960. Separation of humic acids by the method of continuous electrophoresis on paper. *Pochvovedenie* 10:31–36.

———, E. V. Kulakov, and E. M. Nozdrunova, 1960. The nature of complex ferro-organic compounds in soil. *Dokl. sovet. Pochvovedenie* 7:137–143.

Kawaguchi, K., and K. Kyuma, 1959. On the complex formation between soil humus and polyvalent cations. *Soil and Plant Food* 5:54–63.

Khanna, S. S., and F. J. Stevenson, 1962. Metallo-organic complexes in soil. 1. Potentiometric titration of some soil organic matter isolates in the presence of transition metals. *Soil Sci.* 93:298–305.

Kilby, B. A., 1948. The bacterial oxidation of phenol to β-keto-adipic acid. *Biochem. J.* 43:5.

———, 1951. The formation of β-ketoadipic acid by bacterial fission of aromatic rings. *Biochem. J.* 49:671–674.

Klamroth, B., 1954. Beitrag zur Charakterisierung der Huminsäuren des Bodens. Dissertation Universität Göttingen.

Kleinhempel, D., and W. Hieke, 1965. Zur Trennung von Huminsäuren durch Geldiffusion. *Albrecht Thaer Archiv* 9:165–172.

Kleist, H., 1963. Untersuchungen über den Ladungszustand einiger Metallionen in Gegenwart von Huminsäuren. *Acta Biol. Med. Ger.* 11:156–163.

———, 1965. Untersuchungen über Metall-Huminsäure-Verbindungen. *Wiss. Z. Univ. Rostock* 14:415.

———, 1967. Stabile freie Radikale in Huminsäuren und ihren Metallverbindungen. 4. Torf-Kolloquium DDR-VR Polen, I/4 (1967) (Rostock).

———, and D. Mücke, 1966a. Stabile freie Radikale in Huminsäuren. *Experienta* 22:136–137.

———, 1966b. Optische Untersuchungen an Huminsäuren. *Albrecht Thaer Archiv* 10:471–484.

———, 1966c. ESR-Untersuchungen von Huminsäure-Metall-Verbindungen. *Z. Chem.* 6:380–381.

Kobo, K. and R. Tatsukawa, 1959. On the coloured material of fulvic acids. *Z. Pflanzenernähr. Düng. Bodenk.* 84:137–147.

Kodama, H., and M. Schnitzer, 1967. X-ray studies of fulvic acids, a soil humic compound. *Fuel, J. Fuel Sci.* 46:87–94.

Kolenbrander, G. J., 1955. Die Verluste an organischer Substanz im Stalldünger. *Z. Pflanzenernähr. Düng. Bodenk.* 69:125–134.

Kolthoff, I. M., and U. U. Lingame. Polarography, London 1952.

Konetzka, W. A., E. T. Woodings, and J. Stove, 1957. Microbial dissimilation of methoxylated aromatic compounds. *Bact. Proc.* 85.

Kononova, M. M., 1943. Organic matter of soils of the arid steppes of the trans-Volga and the processes of its transformation under irrigated conditions. D.Sc. Thesis, Tashkent: Sredne-Aziat. gos. Univ.

Kononova, M. M., 1947. Grundzüge der Forschungen auf dem Gebiet der Humusstoffe der Böden in der SSR der letzten 30 Jahre. *Pochvovedenie* 10:590–599.

―――, 1956. Humus der Hauptbodentypen der UdSSR, seine Natur und Bildungsweisen. Rapports 6. Congr. Internat. de la Science de Sol, Chemie du Sol, S. 5, Moskau.

―――, 1958. Die Humusstoffe des Bodens. VEB. Deutscher Verlag der Wissenschaften, Berlin (1958), übersetzt und bearbeitet von H. Beutelspacher.

―――, 1961. Soil organic matter. Its nature, its role in soil formation and in soil fertility. Pergamon Press: Oxford, London, New York, Paris.

―――, 1964. Gedanken zur Nomenklatur der im Boden vorhandenen organischen Stoffe. Eighth International Congress of Soil Science, Transaction, Vol. 3. Bucharest–Romania, 401–404 (1964). Discussions regarding the terminology in the field of humus, 405–413.

―――, 1966. Soil organic matter, its nature, its role in soil formation and in soil fertility. 2nd ed. Pergamon Press, Oxford, London, Edinburgh, New York, Toronto, Sydney, Paris, Braunschweig.

―――, and I. V. Alexandrova, 1956a. The use of paper partition chromatography for a study of the forms of nitrogen in humic substances. *Pochvovedenie* 5:86.

―――, and I. V. Alexandrova, 1956b. La biochimie des processus de formation des matières humiques. 6e. Congrès International de la Science du Sol. Paris–1956. Vol. C, 133–137.

―――, and N. P. Belchikova, 1950. Versuch einer Charakteristik der Art von Huminsäuren im Boden mit Hilfe der Spektrophotometrie (russ.) *Dokl. AN UdSSR*, 72:1.

―――, and N. P. Belchikova, 1956. Humus der Böden der UdSSR, seine Natur und Rolle in Bodenbildungsprozessen. 6th International Congr. Sci. Sol. Paris, Vol. B, 557–565.

―――, and N. P. Belchikova, 1960. Investigation of the nature of soil humus substances by fractionation. *Soviet Soil Sci.* 11:1149–1155.

―――, and N. P. Belchikova, 1961. Rapid methods of determining the humus composition of mineral soils. *Pochvovedenie* 10:75.

―――, and N. A. Titova, 1961. Die Anwendung der Papierelektrophorese zur Fraktionierung der Bodenhumustoffe und das Studium ihrer komplexen Verbindungen mit Eisen. *Pochvovedenie* 11:81.

―――, N. P. Belchikova, and V. K. Nikiforov, 1958. Use of the chromatographic method in studying humic substances of the soil. *Soviet Soil Sci.* 3:285–292.

Kosaka, J., Ch. Honda, and A. Izeki, 1961. Fractionation of humic acid by organic solvents. *Soil and Plant Food* 7:48.

Kratky, O., 1938. Die Berechnung der Micelldimensionen von Faserstoffen aus den unter kleinsten Winkeln abgebeugten Interferenzen. *Naturwissenschaften* 26:94.

―――, 1960. Diffuse Röntgenkleinwinkelstreuung. Bestimmung von Größe und Gestalt von Kolloidteilchen und Makromolekülen. *Angew. Chem.* 72:467–482.

Kratzl, K., and P. Claus, 1962. Zur quantitativen Bestimmung der monomeren Äthanolysenprodukte aus dem Lignin monocotyler und dicotyler Angiospermen. *Monatsh. Chem.* 93:219–229.

―――, and E. Risnyovszky, 1964. Zum Mechanismus der alkalischen Hydrolyse der Ligninsulfosäure. Chimie et Biochimie de la Lignine de la Cellulose et des Hèmicelluloses. *Actes dus Symposium International de Grenoble Juillet.*

―――, E. Risnyovszky, P. Claus, and E. Wittmann, 1966. Über den Mechanismus der alkalischen Hydrolyse von Guajacylglyzerin-β-guajacyläther-α-sulfosäure. 1. Mitteilung. Modellversuche zur Fragmentierung der Ligninsulfosäure in alkalischem Medium. *Holzforschung* 20:21–27.

Kretovich, V. L., and R. R. Tokareva, 1948. Wechselwirkung zwischen Aminosäuren und Zuckern bei erhöhter Temperatur. *Biochimia* 13:508.

―――, and R. R. Tokareva, 1949. Volatile aroma compounds in bread and Kwass. *Dokl. Akad. Nauk, SSSR*, 69:231.

Krevelen, van D. W., 1950. Graphical statistical method for study of structure and reaction processes of coal. *Fuel* 29:269–284.

―――, C. van Heerden, and F. J. Huntjes, 1951. Physicochemical aspects of the pyrolysis of coal and related organic compounds. *Fuel* 30:253–259.

Kröger, C., 1966. Die chemische Konstitution der Braunkohlen-Humussubstanzen. *Brennstoff-Chem.* 47:249–252.

————, and R. Brücker, 1961. Über die physikalische und chemischen Eigenschaften der Steinkohlengefügebestandteile. 14. Die Hochvakuumzersetzung. *Brennstoff-Chem.* 42:245–254.

————, G. Darsow, and K. Führ, 1965. Physikalisch-chemische Eigenschaften von Braunkohlen und Braunkohlenkomponenten. 5. Die Carbonylgruppen- und Äthersauerstoffbestimmung. *Erdöl Kohle, Erdgas, Petrochemie* 18:701–705.

Küster, E., 1952. Umwandlung von Mikroorganismen-Farbstoffen in Huminstoffe. *Z. Pflanzenernähr. Düng. Bodenk.* 57:51–57.

————, 1953. Mikrobielle Polyphenoloxydasen und Humusbildung. *Zbl. Bakteriol., Parasitenkunde, Infektionskrankh. Hyg. I. Abt. Orig.* 160:207–213.

————, 1956. Beiträge zur Physiologie der Streptomyceten. 6. Congrès International de la Sci. du Sol. Paris 3:67–72.

————, 1958. Der Einfluss der C- und N-Quelle auf die Peptonbraunfärbung durch Streptomyceten. *Zbl. Bakteriol. Parasitenkunde, Infektionskrankh. Hyg. Abt. II,* 111:227–234.

Kuhn, R., and H. Beimert, 1944. Über die Umsetzung von Cystein mit Chinon. *Ber. dtsch. chem. Ges.* 77:606.

Kukharenko, T. A., and T. E. Kredenskaja, 1956. Erschöpfende Spaltung der Huminsäuren mit metallischem Natrium und flüssigem Ammoniak. *Chem. Techn. Brennstoffs* 6:25.

————, and A. S. Savelev, 1951. Hydrierung von Huminsäuren am Nickelkatalysator. *Dokl. Akad Nauk. USSR* 76:77.

————, A. S. Savelev, 1952. Neutrale Stoffe der Hydrogenesis der Huminsäuren verschiedener Herkunft. *Dokl. Akad. Nauk. SSSR* 86:729–732.

Kumada, K., 1955. Absorption spectra of humic acids. *Soil and Plant Food* 1:29–30.

————, 1965. Studies on the colour of humic acids. Part 1. On the concepts of humic substances and humification. *Soil Sci. Plant Nutr.* 11:11–16.

————, and A. Aizawa, 1958. The infra-red spectra of humic acids. *Soil and Plant Food* 3:152–159.

————, and K. Aizawa, 1959. The infrared absorption spectra of soil components. *Soil and Plant Food* 4:101–188.

————, and Y. Kawamura, 1965. Intrinsic viscosity and the functional groups of humic acids. *Nippon Dojo-Hiryogaku Zasshi* 36:367–372.

————, and Y. Kawamura, 1968. Viscosimetric characteristics of humic acids. *Soil Sci. Plant Nutr.* 14:5, 190–197.

————, and O. Sato, 1962. Chromatographic separation of green humic acid from podzol humus. *Soil Sci. Plant Nutr.* 8:31–33.

————, A. Suzuki, and K. Aizawa, 1961. Isolation of anthraquinone from humus. *Nature* (London) 191:415–416.

Kuo, M. H., and W. V. Bartholomew, 1966. On the genesis of organic nitrogen in decomposed plant residue. In "The use of isotopes in soil organic matter studies." Report of the FAO/IAEA Technical Meeting. Oxford: Pergamon Press, Ltd., 329–335.

Kyuma, K., 1964. A fractional precipitation technique applied to soil humic substances. *Soil Sci. Plant Nutr.* 10:33–35.

Laatsch, W., 1957. *Dynamik der deutschen Acker- und Waldböden.* 4. Dresden: Auflage.

————, 1958/59. Die Stickstoffernährung des Waldes. Jahresbericht Bayer. Forstverein 3–17.

————, L. Hoops, and O. Bieneck, 1952. Über Huminsäuren des Pilzes Spicaria elegans. *Z. Pflanzenernähr. Düng. Bodenk.* 58:258–268.

Ladd, J. N., and J. H. A. Butler, 1966. Comparison of properties of synthetic and natural humic acids. In: "The use of isotopes in soil organic matter studies," Report of the FAO/IAEA Technical Meeting. Oxford: Pergamon Press, Ltd., 49–59.

Langel, L. P., and M. Dekker, 1954. A browning reaction between thiamine and glucose. *Nature* 173:4041.

Langenbeck, W., R. Schaller, and K. Armberg, 1942. Über organische Katalysatoren. 25. Mitt. Katalytische Wirkungen von o-Chinonen. *Ber. dtsch. chem. Ges.* 75:1483–1488.

Langenbeck, W., O. Tarnan, G. Kertel, and B. Hirsch, 1948. Über organische Katalysatoren. 27. Mitt. Katalytische Wirkungen von o-Chinonen. 2. Mitteilung. *Chem. Ber.* 81:356–361.

Lantz, R., and H. Michel, 1961. Action de l'ammoniaque ou des amines primaires sur le benzène substitué en 1, 2, et 4 par des groupes amino ou hydroxyles, non substitués ou substitués. *Bull. Soc. Chim. France* 12:2402–2408.

Lehner, A., and U. Springer, 1950. Über den Stoffabbau und Humusaufbau bei der Zersetzung von Stroh unter dem Einfluß verschiedener N-Formen. *Pflanzenbau Pflanzenschutz* 1:99–118.

Leopold, B., 1950. Aromatic keto- and hydroxy-polyethers as lignin models. 3. *Acta Chem. Scand.* 4:1523–1537.

———, and I. L. Malmström, 1951. Constitution of resin phenols and their biogenetic relations. 15. Nitrobenzene oxidation of compounds of the lignan type. *Acta Chem. Scand.* 5:936–940.

Levashkevich, G. A., 1966. Interaction of humic acids with iron and aluminium hydroxides. *Soviet Soil Sci.* 4:422–427.

Levesque, M., and M. Schnitzer, 1967a. Extraction of soil organic matter by base and chelating agent. *Canadian J. Soil Sci.* 47:76–78.

———————, 1967b. Organo-metallic interactions in soils. 6. Preparation and properties of fulvic acid-metal phosphates. *Soil Sci.* 103:183–190.

Lewis, T. E., and F. E. Broadbent, 1961a. Soil organic matter-metal complexes. 3. Exchange reactions of model compounds. *Soil Sci.* 91:341–348.

———————, 1961b. Soil organic matter-metal complexes. 4. Nature and properties of exchange sites. *Soil Sci.* 9:393–399.

Lim, S., 1965. Beiträge zur Aufklärung der Zusammenhänge zwischen dem mikrobiellen Abbau des Lignins und der Bildung von Humusstoffen. Dissertation, University of Bonn, Germany.

Lindbeck, M. R., and J. L. Young, 1966. Polarography and coulometry, in dimethylsulfoxide, of nitric acid oxidation products from soil humic acid. *Soil Sci.* 101:366–372.

Lindgren, B. O., 1960a. Dehydrierung von Phenolen. 1. Mitt. Dehydrierung von 2, 6-Dimethylphenol mit Silberoxid. *Acta Chem. Scand.* 14:1203–1210.

———, 1960b. Dehydrierung von Phenolen. 2. Mitt. Polymere durch Dehydrierung von Guajacol. *Acta Chem. Scand.* 14:2089–2096.

Lucena-Conde, F., and A. Gonzalez-Crespo, 1960. The polarographic behaviour of the three fractions of the organic matter of the soils. Seventh International Congress of Soil Science, Vol. II. Madison, Wisconsin, 59–65.

Lüers, H., 1953. Melanoidine. *Schweiz. Brauerei-Rundschau* 64:141.

Luther, H., 1960. Über das Erweichungs- und Entgasungsverhalten von Kohlen. Methoden und Ergebnisse. *Erdöl, Kohle, Erdgas, Petrochemie* 13:947.

———, W. Eisenhut, and O. Abel, 1966. Zur DTA von Steinkohle I. *Brennstoff-Chem.* 47:258–264.

———, O. Abel, U. Kett, and K. Chr. Traenckner, 1960. Gravimetrische und volumetrische Messungen zum Entgasungsverhalten von Kohlen. *Brennstoff-Chem.* 41:257–263.

Lynch, D. L., L. M. Wright, and H. O. Olney, 1957. Qualitative and quantitative chromatographic analyses of the carbohydrate constitutents of the acid-insoluble fraction of soil organic matter. *Soil Sci.* 84:405–411.

Maeder, H., 1960. Chemische und pflanzenphysiologische Untersuchungen mit Rottestroh. Dissertation, University of Giessen, Germany.

Maillard, L. C., 1912. Formation d'humus et de combustibles minèraux sans intervention de l'oxygène atmosphérique, des microorganisms, ou de haute temperatures ou de forte pressions. *C. R. Acad. Sci. Paris* 154:66 and 155:1554.

———, 1931. Formation de matières humiques per action de polypeptides sur les sucres. *C. R. Acad. Sci. Paris* 156:1159.

Majima, F., and G. Takayama, 1920. Über den Hauptbestandteil des Japanlacks. 7. Der Urushiol-monomethyläther und der Mechanismus des Urushiols. *Ber. dtsch. chem. Ges.* 53:1907–1916.

Manskaja, S. M., and T. W. Drosdova, 1964. Geochemie der organischen Stoffe. Verlag Nauka Moskau.

Manskaja, S. M., T. W. Drosdova, and M. P. Emel'Yanova, 1958. Copper fixation by various forms of natural-organic compounds. *Soviet Soil Sci.* 6: 623–629.

Markuze, E., 1960. Browning and reducing power of sugar solutions heated with amino acids. *Acta Chim. Acad. Sci. Hung.* 23:247.

Martell, A. E., and M. Calvin, 1952. Chemistry of the metal chelate compounds. Prentice-Hall, Inc., New York.

Martin, A. E., 1960. Chemical studies of podzolic illuvial horizons. 5. Flocculation of humus by ferric and ferrous iron and by nickel. *J. Soil Sci.* 11:382–393.

———, and R. Reeve, 1955. The extraction of organic matter from podzolic B-horizons with organic reagents. *Chem. and Ind.* 1955, 356.

———, 1957a. Chemical studies on podzolic illuvial horizons. 1. The extraction of organic matter by organic chelating agents. *J. Soil Sci.* 8:268–278.

———, 1957b. Chemical studies on podzolic illuvial horizons. 2. The use of acetylacetone as extractant of translocated organic matter. *J. Soil Sci.* 8:279–286.

———, 1958. Chemical studies on podzolic illuvial horizons. 3. Titration curves of organic matter suspensions. *J. Soil Sci.* 9:89–100.

———, 1960. Chemical studies of podzolic illuvial horizons. 4. The flocculation of humus by aluminium. *J. Soil Sci.* 11:369–381.

Martin, F. E., and R. R. Jay, 1962. Manometric determination of active hydrogen by reaction with diborane. *Anal. Chem.* 34:1007–1009.

———, P. Dubach, N. C. Mehta, and H. Deuel, 1963. Bestimmung der funktionellen Gruppen von Huminstoffen. *Z. Pflanzenernähr., Düng. Bodenk.* 103:29–39.

Martin, J. P., S. J. Richards, and K. Haider, 1967. Properties and decomposition and binding action in soil of "humic acids" synthesized by Epicoccum nigrum. *Soil Sci. Soc. Amer., Proc.* 31:657–662.

Mason, H. S., 1948. Chemistry of melanin. 3. Mechanism of oxidation of 3, 4-dihydroxyphenylalanine by tyrosinase. *J. Biol. Chem.* 172:83–99.

———, 1955. Reactions between quinones and proteins. *Nature* 175:771–772.

———, 1957. Mechanisms of oxygen metabolism. *Advan. Enzymol.* 19:79–231.

———, W. L. Fowlks, and E. W. Peterson, 1955. Oxygen transfer and electron transport by the phenolase complex. *J. Amer. Chem. Soc.* 77:2914–2915.

Mayaudon, J., and P. Simonart, 1958. Étude de la décomposition de la matière organique dans le sol au moyen de carbone radioactif. 2. Décomposition du glucose radioactif dans le sol. A. Repartition de la radioactivité dans fractions humique du sol. *Plant and Soil* 9:376–380.

———, 1959a. Étude de la décomposition de la matière organique dans le sol au moyen de carbone radioactif. 3. Décomposition des substances solubles dialysables des protéines et des hémicelluloses. *Plant and Soil* 11:170.

———, 1959b. Étude de la décomposition de la matière organique dans le sol au moyen de carbone radioactif. 5. Décomposition de cellulose et de lignine. *Plant and Soil* 11:181–192.

Mazumdar, B. K., S. K. Chakrabartty, and A. Lahiri, 1957. Aromaticity and oxidation of coal. *Sci. Ind. Res.* (India) 16:B, 275.

Mehta, N. C., P. Dubach, and H. Deuel, 1963a. Abbau von Huminstoffen. 2. Oxydation mit Chlordioxyd, Wasserstoffperoxyd und Perjodat. *Z. Pflanzenernähr. Düng. Bodenk.* 101:147–152.

———, 1963b. Untersuchungen über die Molekulargewichtsverteilung von Huminstoffen durch Gelfiltration an Sephadex. *Z. Pflanzenernähr., Düng. Bodenk.* 102:128–137.

Meinschein, W. G., and G. S. Kenny, 1957. Analyses of a chromatographic fraction of organic extracts of soils. *Analyt. Chem.* 29:1153–1161.

Menger, A., 1953. Die Maillardsche Reaktion und ihre Bedeutung beim Backprozess. *Brot. und Gebäck* 1:85.

Meyer, L., Die Entwicklung der pflanzlichen Agrikulturchemie seit Julius Kühn und die Leistungsfähigkeit unserer Böden als Forschungsaufgabe der Gegenwart. Kühn-Archiv 50 (Festschrift 1933) 229–249.

Meyer, W., 1962. Die Bestimmung funktioneller Gruppen an Huminsubstanzen aus Böden. Dissertation E.T.H. Zurich, Switzerland.

200   W. Flaig, H. Beutelspacher, and E. Rietz

Michaelis, L., 1933. Oxydations-Reduktionspotentiale, Berlin: Springer-Verlag.

———, 1951. Theory of oxidation-reduction. "The Enzymes," II, Part 1, New York.

Micheel, F., and R. Büning, 1957. N-Glukoside des Polyvinylamins und deren Amadori-Umlagerung. *Chem. Ber.* 90:1606.

———, and B. Schleppinghoff, 1956. Zur Kenntnis der Amadori-Umlagerung. *Chem. Ber.* 82:1702.

Misterski, W., and W. Loginov, 1959. A study of some physicochemical properties of humic acids. *Soviet Soil Sci.* 2:170–181.

Mohtadi, S., 1962. Beitrag zur Isolierung von pflanzenphysiologisch wirksamen Stoffen während der Rotte von Stroh. Dissertation University of Giessen, Germany.

Morita, H., 1962. Composition of peat humus and its derivatives. Oxidative conversion to lignin model compounds. *J. Org. Chem.* 27:1079–1080.

Morrison, R. I., 1958. The alkaline nitrobenzene oxidation of soil organic matter. *Soil Sci.* 9:130–140.

———, 1963. Products of the alkaline nitrobenzene oxidation of soil organic matter. *J. Soil Sci.* 14:201–216.

Mortensen, J. L., 1960. Physico-chemical properties of a soil polysaccharide. Seventh International Congress of Soil Science, Madison, Wisconsin, 2:98–104.

———, 1963. Complexing of metals by soil organic matter. *Soil Sci. Soc. Amer., Proc.* 27:179.

Mücke, D., and R. Obenaus, 1961. Eine Methode zur Isolierung wasserlöslicher Huminsäuren. *Naturwissenschaften* 48:478.

Müller, E., K. Ley, and G. Schlechte, 1957. Zur Dehydrierung von Phenolen. *Angew. Chem.* 69:204.

———, R. Mayer, H. Spanagel, and K. Scheffler, 1961. Über Sauerstoffradikale. 19. Dehydrierung Hydroxyphenyl-substituiertes Äthylen. *Liebigs Ann. Chem.* 645:53–65.

Mukherjee, J. N., 1920. The origin of the charge of a colloidal particle and its neutralisation by electrolytes. *Trans. Faraday Soc.* 16:103–115.

———, 1922. The adsorption of ions. *Phil. Mag.* 44:321–346.

———, and A. Lahiri, 1958. Polyelectrolytisches Verhalten von Huminsäuren. *Fuel* 37:220–226.

Mulder, G. J. "Chemie der Ackerkrume." J. Muller, Berlin, 1860–1861.

Murphy, D., and A. W. Moore, 1960. A possible structural basis of natural humic acid. *Sci. Proc. Roy. Dubl. Soc.,* Ser. A1:191.

Musso, H., 1958. Über Orceinfarbstoffe. 7. Synthese, Konstitution und Lichtabsorption des Henrichschen Chinons. *Chem. Ber.* 91:349–363.

———, 1963. Über Phenol-Oxydationen. *Angew. Chem.* 75:965–977.

———, U.V. Gizycki, U. I. Zahorsky, and D. Bormann, 1964. Die Bildung von Hydroxyl-Chinonen durch Addition von Phenolen an Chinone. *Liebigs Ann. Chem.* 676:10–20.

———, U.V. Gizycki, H. Krämer, and H. Döpp, 1965. Über Orceinfarbstoffe. 24. Über den Autoxydationsmechanismus bei Resorcinderivaten. *Chem. Ber.* 98:3952–3963.

Natkina, A. I., 1940. Investigations on the composition and properties of humic acids from chernozem and podzolic soil. *Trudy pochv. Inst. Dokutchaeva Akad. Nauk SSSR* 23:

Naucke, W., 1967. Möglichkeiten zur Analyse von Torfinhaltstoffen mit physikalisch-chemischen Methoden. 4. Torf-Kolloquium DDR-VR Polen, 1/6. (Rostock).

———, 1968. Die Untersuchung des Naturstoffes Torf und seiner Inhaltsstoffe. *Chemiker-Zeitung/Chemische Apparatur* 92:261–280.

Nehring, K., 1955. Untersuchungen an aus verschiedenen Bodentypen isolierten Huminsäuren. *Z. Pflanzenernähr. Düng. Bodenk.* 69:71–86.

———, and R. Schiemann, 1952a. Untersuchungen zum Humusproblem. 1. Mitt. Beiträge zur Kenntnis der Vorgänge bei der Rotte von Stallmist und Komposten sowie zur Kenntnis der Huminsäuren *Z. Pflanzenernähr. Düng. Bodenk.* 57:97–113.

———, 1952b. Untersuchungen zum Humusproblem. 1. Mitt. Beiträge zur Kenntnis der Vorgänge bei der Rotte von Stallmist und Komposten sowie zur Kenntnis der Huminsäuren. 2. Teil. Beiträge zur Kenntnis der Huminsäuren. *Z. Pflanzenernähr. Düng. Bodenk.* 57:193–215.

Netzsch-Lehner, A., and J. Klee, 1962. Anwendung der Differentialthermoanalyse zur Untersuchung von Huminsäuren und Vergleichssubstanzen. *Bay. Landwirtsch. Jahrbuch* 39:113–121.

Nicolaus, R. A., 1962. Biogenese der Melanine. Conf. VII Corso Chim. Acad. naz. Lincei, Milano.

Norman, A. G., and S. H. Jenkins, 1934a. The determination of lignin. 1. Errors introduced by the presence of certain carbohydrates. *Biochem. J.* 28:2147–2159.

————, 1934b. The determination of lignin. 2. Errors introduced by the presence of proteins. *Biochem. J.* 23:2160–2168.

Obenaus, R., and H. J. Neumann, 1965. Entsalzung von Huminsäuren an Sephadex G 25. *Naturwissenschaften* 52:131.

Oden, S., 1912. Zur Kenntnis der Huminsäuren des Sphagnum-Torfes. *Ber. dtsch. chem. Ges.* 35:651.

————, 1919. Die Huminsäuren, chemische, physikalische und bodenkundliche Forschung. *Kolloidchem. Beihefte* 11:75–260.

————, 1922. Die Huminsäuren, Leipzig, Verlag Steinkopff.

Odintsov, P., and Z. Kreitsberg, 1953. Oxidation of isolated lignins and of wood lignin. *Voprosy Leoskhim. i Khim. Drevesiny, Trudy Inst. Lesokhoz. Problem, Akad. Nauk Latv. SSSR* 6:63–76.

Öholm, 1904. Die Hydrodiffusion der Elektrolyte. *Z. physik. Chem.* 50:309–349.

Ogg, C. L., W. L. Porter, and C. O. Willits, 1945. Determining the hydroxyl groups of certain organic compounds. *Ind. Eng. Chem. Anal. Ed.* 17:394.

Okuda, A., and S. Hori, 1954. Chromatographic investigation of amino acids in humic acid and alkaline alcohol lignin. Trans. 5th Internat. Congr. Soil Sciences. 2:255.

————————, 1955. Identification of amino acids in humic acid. *Soil and Plant Food* 1: 39–40.

Orlov, D. S., 1959. Absorption of light by soil humus substances in the visible region of the spectrum. *Nauch. Dokl. vyssh. Shkol. Biol. Nauki* 4:192–197.

————, 1960. Zur Erforschung der optischen Eigenschaften der Humusstoffe. *Wiss. Hochschulber. biol. Wiss.* 1:204–207.

————, 1966. Spektrophotometrische Analyse von Humussubstanzen. *Pochvovedenie* 11:84–95.

————, and N. L. Erosiceva, 1967. Zur Frage der Wechselwirkung von Huminsäuren mit einigen Metall-Kationen. Vestnik-Moskovskogo Universiteta. Biologia, *Pochvovedenie* 1:98–105.

————, and E. I. Gorskova, 1965. Große und Gestalt der Huminsäurepartikel aus Schwarzerde und Rasenpodsolböden. *Nauchenye doklady ryssej skoly Biologielskie nauki Moskva* 1:207–212.

————, and N. M. Grindel, 1967. Spectrophotometric determination of the humus content in soil. *Pochvovedenie* 1:112–122.

————, and N. V. Nesterenko, 1960. The formation of cobalt nickel, copper and zinc humates. *Nauch. Dokl. vyssh. Shkol. Biol., Nauki* 3:195–198.

————, O. H. Rozanova, and S. G. Matyukhina, 1962. Infrarotabsorptionsspektren der Huminsäuren. *Pochvovedenie* 1:17–25.

————, and V. T. Zub, 1964. Optical properties and elementary composition of fulvic acids of various origin. *Soviet Soil Sci.* 13:1439–1444.

Pallmann, H., 1938/39. Über starre und elastische Umtauschkörper. *Soil Res.* 6:21–48.

————, 1942. Dispersoidchemische Probleme in der Humusforschung. *Kolloid-Z.* 101:73–81.

Parsons, J. W., and J. Tinsley, 1960. Extraction of soil organic matter with anhydrous formic acid. *Soil Sci. Soc. Amer., Proc.* 24:198–201.

Paulik, F., and M. Weltner, 1958. Über die derivative thermogravimetrische Analyse von Torfen und Torfbestandteilen. *Acta Chim. Acad. Sci. Hung. Tomus* 16, Fasc. 2, 159.

Pavel, L., 1959. On the knowledge of humus substances. 4. Watersoluble metal-chelates formed by the dark-coloured fulvic acids fraction. *Sborn. csl. Akad. Zemd. Ved. Rostl. Vyroba* 32:639–650.

————, 1960. Some observations on the high-frequency titration of highly dispersed acidoids in soil. *Sborn. csl. Akad. Zemed. Ved. Rostl. Vyroba* 6:91.

————, and M. Zazvorka, 1965. Spectrophotometric determination of the concentration of solution of soil humic acids of various origin. *Soviet Soil Sci.* 10:1237–1241.

————, J. Kolousek, and V. Smatlak, 1954. Soil humus problem. 2. Amino-acid composition of the hydrolysates of humic acids in different soil types. *Sborn. Ceskoslov. Akad. Zemed. Ved.* 27A:207–212.

Pelczar, M. J., S. Gottlieb, and W. C. Day, 1950. Growth of Polyporus versicolor in a medium with lignin as the sole carbon source. *Arch. Biochem.* 25:449–451.

Pepper, J. M., M. Manolopoulo, and R. Burton, 1962. Gas-liquid chromatographic analysis of lignin oxidation products. *Can. J. Chem.* 40:1976–1980.

Phillips, M., 1934. The chemistry of lignin. *Chem. Rev.* 14:103–170.

Philpot, J. St. L., 1940. Direkte Photographie von Ultrazentrifugensedimentationskurven. *Nature* 141:283–284. vgl. auch Svensson.

Pinck, L. A., 1962. Absorption of proteins, enzymes and antibiotics by montmorillonite. *Clays and Clay Minerals* 9:520–529.

———, F. W. Holton, and F. E. Allison, 1961a. Antibiotics in soils. 1. Physico-chemical studies of antibiotic clay complexes. *Soil Sci.* 91:22–28.

———, D. A. Soulides, and E. F. Allison, 1961b. Antibiotics in soils. 2. Extent and mechanism of release. *Soil Sci.* 91:94–99.

Piret, E. L., G. White, H. C. Walther, Jr., and A. J. Madden, Jr., 1960. Some physico-chemical properties of peat humic acids. *Sci. Proc. Roy. Dubl. Soc. A1*:4, 69–79.

Ploetz, Th., 1940. Beiträge zur Ligninbestimmung mit starker Schwefelsäure. *Cellulosechemie* 18:49–57.

———, 1955. Polymere Chinone als Huminsäuremodelle. *Z. Pflanzenernähr. Düng. Bodenk.* 69:50–58.

Plotho, O. van, 1950. Die Humusbildung der Mikroorganismen. *Z. Pflanzenernähr. Düng. Bodenk.* 51:212.

———, 1951. Weitere Untersuchungen zur Humusbildung der Mikroorganismen. *Z. Pflanzenernähr. Düng. Bodenk.* 55:151–169.

Pommer, A. M., and I. A. Breger, 1960a. Potentiometric titration and equivalent weight of humic acid. *Geochim. Cosmochim. Acta* 20:30–44.

———, 1960b. Equivalent weight of humic acid from peat. *Geochim. Cosmochim. Acta* 20:45–50.

Ponomareva, V. V., 1947. Methods for the isolation of fulvic acids and their chemical nature. *Pochvovedenie* 12:714–723.

Posner, A. M., 1963. Importance of electrolyte in the determination of molecular weights by "Sephadex" gel filtration with especial reference to humic acid. *Nature* 198:1161–1163.

———, 1964. Titration curves of humic acids. Eighth International Congress of Soil Science, Bucharest, Romania, 161–173.

———, 1966. The humic acid extracted by various reagents from a soil. Part 1. Yield, inorganic components, and titration curves. *J. Soil Sci.* 17:65–78.

———, B. K. G. Theng, and J. R. H. Wake, 1968. The extraction of soil organic matter in relation to humification. Ninth International Congress of Soil Science, Adelaide, Australia, Transaction III, 153–162.

Pospisil, F., 1962a. Electrophoresis of humus substances and their complexes with copper. *Soviet Soil Sci.* 12:1357–1362.

———, 1962b. Einige physikalisch-chemische Eigenschaften der Humusstoffe. Studies about Humus, Symp. Humus and Plant, Praha, 209–222.

Prokh, I., 1961. The application of ion exchange resins in the study of organic matter. *Pochvovedenie* 1:95–98.

Quillico, A., 1951. Récent dévelopment de la chimie des produits naturals: La chimie desmoisissures et des bactéries. *Chim. Ind.* 66:205.

Rajalakshmi, N., S. R. Sivarajan, and R. D. Vold, 1959. Die Viskosität der Lösungen von Natrium-huminate aus Kohle. *J. Colloid Sci.* 14:419–429.

Randhawa, N. S., and F. E. Broadbent, 1965a, Soil organic matter-metal complexes 5. Reactions of zinc with model compounds and humic acids. *Soil Sci.* 99:295–300.

———, 1965b. Soil organic matter-metal complexes. 6. Stability constants of zinc humic acid complexes at different pH-values. *Soil Sci.* 99:362–366.

Raper, H. S., 1927. Die Einwirkung von Tyrosinase auf Tyrosin. *Fermentforschung* 9:206.

———, 1928. The aerobic oxidases. *Physiol. Rev.* 8:245–282.

Read, D. E., and C. B. Purves, 1952. Isolation of penta- and 1.2.4.5-benzenetetracarboxylic acids from wood lignins oxidized with alkaline permanganate. *J. Amer. Chem. Soc.* 74:120.

Reinders, W. and P. Dingemans, 1934. Die Oxydationsgeschwindigkeit von Hydrochinon mit Luftsauerstoff. 1. Recueil des Travaux chimiques des Pays-Bas 53, 209–230.

Reinemund, K., and W. Flaig, 1955. Zur Kenntnis der Huminsäuren. 7. Chinone als Zwischenkatalysatoren bei Redoxsubstanzen von biologischer Bedeutung und bei Fermentsystemen. *Z. Pflanzenernähr. Düng. Bodenk.* 68:269–275.

Rex, R. W., 1960. Electron paramagnetic resonance (EPR) studies of stable free radicals in lignins and humic acids. *Nature* 188:1185–1186.

Richtzenhain, H., 1949. Enzymatische Versuche zur Entstehung des Lignins. 4. Dehydrierungen in der Guajacolreihe. *Chem. Ber.* 82:447–453.

Ritter, G. J., R. M. Seborg, and R. L. Mitchell, 1932. Factors affecting quantitative determination of lignin by 72 per cent sulfuric acid method. *Ind. Eng. Chem. Analyt. Edit.* 4:202.

Roadhouse, F. E., and D. Macdougall, 1956. A study of the nature of plant lignin by means of alkaline nitrobenzene oxidation. *Biochem. J.* 63:33–39.

Robinson, C., 1911. Organic nitrogenous compounds in peat soil. *Mich. Agric. Exp. Sta. Techn. Bull.* 7.

Robinson, D. O., W. P. Martin, and J. B. Page, 1951. *Ohio Experimental Station Bulletin* 705:18 (69th Ann. Rep.)

Roulet, N., N. C. Mehta, P. Dubach, and H. Deuel, 1963. Abtrennung von Kohlehydraten und Stickstoffverbindungen aus Huminstoffen durch Gelfiltration and Ionenaustausch-Chromatographie. *Z. Pflanzenernähr., Düng., Bodenk.* 103:1–9.

Roy, M. M., Elektrometrische Titration von aus Kohle gewonnener Huminsäure. *Kolloid-Z.* 153:174–178.

Ruchti, J., 1962. Fraktionierung von organischen Substanzen aus dem B-Horizont eines Podsols. Dissertation E.T.H. Zürich.

Rydalevskaya, M., and V. Tishchenko, 1944. The cationic exchange of humic acids. *Pochvovedenie* 10:491–499.

Saalbach, E., 1958. Beitrag der Grundlagenforschung zum Problem der Strohdüngung. *Landwirtsch. Forsch.* 8:83–84.

Saeki, H., 1939. Studies on humus-clay complexes. *Mem. Faculty Sci. Agr., Taihoku Imp. Univ.* 25, No. 1:186.

Salfeld, J. Chr., 1957. Zum Reaktionsmechanismus der Purpurogallinbildung. *Angew. Chem.* 69:723–724.

———, 1960. Über die Oxydation von Pyrogallol und Pyrogallolderivaten. 2. Die Konstitution dimerer 3-Hydroxy-o-benzochinone. *Chem. Ber.* 93:737–745.

———, 1961. Zur Chemie der Huminsäurebildung. *Landbauforsch. Völkenrode* 11:22–24.

———, 1964. Fraktionierung eines Huminstoffpräparates mit wasserhaltigen Lösungsmitteln. *Landbauforsch. Völkenrode* 14:131–136.

———, 1965. Die Charakterisierung von Huminstoffen durch Differenzen-Spektrogramme. *Mitt. Dtsch. Bodenkundl. Ges.* 4:289–290.

———, 1968. Untersuchungen zur Klassifizierung der organischen Bodensubstanz. *Mitt. Dtsch. Bodenkundl. Ges.* 8:133–135.

———, and E. Baume, 1960. Über die Oxydation von Pyrogallolderivaten. 3. Die Umlagerung von dimerem 3-Hydroxy-4, 6-di-tert. butyl-o-benzochinon durch Alkali. *Chem. Ber.* 93:745–751.

———, and E. Baume, 1964. Über die Oxydation von Pyrogallol und Pyrogallolderivaten. 4. Die Konstitution der Purpurogallincarbonsäure-(9). *Chem. Ber.* 97:307–311.

Sanger, F., and E. O. P. Thompson, 1953. The amino acid sequence in the glycyl chain of insulin. *Biochem. J.* 53:353–366.

Sauerlandt, W., and O. Graff, 1959. Die Strohdecke auf dem Ackerboden. *Landbauforsch. Völkenrode* 9:57–60.

Savage, S., and F. J. Stevenson, 1961. Behavior of soil humic acids towards oxidation with hydrogen peroxide. *Proc. Soil Sci. Soc. Amer.* 25:35–39.

Schachtschabel, P., 1940. Untersuchungen über die Sorption der tonmineralischen und organischen Bodenkolloide und die Bestimmung des Anteils dieser Kolloide an der Sorption im Boden. *Kolloidbeihefte* 51:199–276.

Schànel, L., 1962. Laccase activity and decomposition of lignin by white-rot fungi. DSc. Thesis, Fac. Nat. Sci. UJEP Brno.

Scharpenseel, H. W., 1960. Papierchromatographische Untersuchungen an Humussubstanzen sowie Huminsäureaufschlüssen und Modellsubstanzen. Z. *Pflanzenernähr. Düng. Bodenk.* 88:97–115.

———, and W. Albersmeyer, 1960. Infrarotspektroskopische Untersuchungen an Huminsäuren, Huminsäureaufschlüssen und phenolisch-chinoiden Vergleichsubstanzen. Z. *Pflanzenernähr. Düng. Bodenk.* 88:3, 203–211.

———, and R. Krausse, 1962. Aminosäureuntersuchungen an verschiedenen organischen Sedimenten, besonders Grau- und Braunhuminsäurefraktionen verschiedener Bodentypen (einschließlich C^{14}-markierter Huminsäuren). Z. *Pflanzenernähr. Düng. Bodenk.* 96:11–34.

———, and R. Krausse, 1963. Radiochromatographische Untersuchungen zum Umsatz von Sulfat sowie der S-Aminosäuren Cystein und Methionin in Boden und Huminsäuren. Z. *Pflanzenernähr. Düng. Bodenk.* 101:11–23.

———, E. König, and E. Menthe, 1964. Infrarot- und Differentialthermo-Analyse an Huminsäure-proben aus verschiedenen Bodentypen, aus Wurmkot und Streptomyceten. Z. *Pflanzenernähr. Düng. Bodenk.* 106:134–150.

Schatz, A., V. Schatz, E. B. Schalscha, and J. J. Martin, 1964. Soil organic matter as a natural chelating material. Part 1: The chemistry of chelation. *Compost Sci.* 25–28.

Scheele, W., and L. Steinke, 1936. Über Humussäuren. 4. Mitteilung. *Kolloid Z.* 77:312–320.

Scheffer, F., and R. Kickuth, 1961a. Chemische Abbauversuche an einer natürlichen Huminsäure. 1. Z. *Pflanzenernähr., Düng., Bodenk.* 94:180–188.

———, 1961b. Chemische Abbauversuche an einer natürlichen Huminsäure. 2. Z. *Pflanzenernähr., Düng., Bodenk.* 94:189–198.

———, and B. Ulrich, 1960. Lehrbuch der Agrikulturchemie und Bodenkunde. 3. Teil. Humus und Humusdüngung 1: Enke Stuttgart, 266.

———, and W. Ziechmann, 1958. Über Synthesehuminsäuren aus Polyoxybenzolen. Z. *Pflanzenernähr. Düng., Bodenk.* 80:126–127.

———, B. Meyer, and A. Niederbudde, 1959. Huminstoffbildung unter katalytischer Einwirkung natürlich vorkommender Eisenverbindungen im Modellversuch. Z. *Pflanzenernähr., Düng., Bodenk.* 87:26–44.

———, O. van Plotho, and E. Welte, 1950. Untersuchungen über die Bildung von Humusstoffen durch Actinomyceten. *Landwirtsch. Forsch.* 1:81–92.

———, E. Welte, and W. Ziechmann, 1955. Über Untersuchungen an künstlichen Huminsäuren. Z. *Pflanzenernähr., Düng., Bodenk.* 69:58–65.

———, W. Ziechmann, and H. Schlüter, 1955. Die Papierelektrophorese als Möglichkeit der Auftrennung von Huminsäuren. Z. *Pflanzenernähr., Düng., Bodenk.* 70:260.

———, W. Ziechmann, and H. Schlüter, 1958. Über die Sedimentation von Huminsäuren in der Ultrazentrifuge. *Kolloid Z.* 160:1–7.

———, W. Ziechmann, and H. Schlüter, 1959. Über Huminsäuren aus Polyoxybenzolen. 4. Mitt. Über die Anwendung der Papierelektrophorese zur Trennung von Huminsäuren und die Charakterisierung der gewonnenen Fraktionen. Z. *Pflanzenernähr., Düng., Bodenk.* 85:32–43.

———, W. Ziechmann, and J. Schmidt, 1958. Die einleitende Phase der Oxydation von Hydrochinon zu Huminsäuren im homegenon System mit alkalischer Reaktion. Z. *Pflanzenernähr., Düng., Bodenk.* 80:127–133.

———, W. Ziechmann, and H. Scholz, 1958. Über Synthesehuminsäuren aus Polybenzolen. Z. *Pflanzenernähr., Düng., Bodenk.* 80:134–149.

———, B. Ulrich, B. Meyer, and J. Oehmichen, 1956. Natronlauge-Extraktion von Modellsubstanzen und Waldhumusformen. Ein Beitrag zur Frage der Huminsäureextraktion. Z. *Pflanzenernähr., Düng., Bodenk.* 74:157–161.

Schiff, H., 1882. Über Proctocatechugerbsäure und über Anhydride aromatischer Oxycarbonsäuren. *Ber. dtsch. chem. Ges.* 15:2588–2592.

Schlichting, E., 1953a. Zur Kenntnis des Heidehumus. 1. Fraktionierung und Untersuchung des ganzen Humuskörpers. Z. *Pflanzenernähr., Düng., Bodenk.* 61:1–12.

Schlichting, E., 1953b. Zur Kenntnis des Heidehumus. 2. Die Fulvosäuren. *Z. Pflanzenernähr., Düng., Bodenk.* 61:97–107.

Schlüter, H., 1959. Über Sedimentationsmessungen an Huminsäuren. *Z. Pflanzenernähr., Düng., Bodenk.* 84:169–174.

Schmidt, E., and M. Atterer, 1927. Zur Kenntnis der Huminsubstanzen. *Ber. dtsch. chem. Ges.* 60: 1671.

Schmidt, O. Th., R. Lademann, 1951. Corilagin, ein weiterer kristallisierter Gerbstoff aus Dividivi. 10. Mitteilung über natürliche Gerbstoffe. *Liebigs Ann. Chem.* 571:232–237.

————, and D. M. Schmidt, 1952. Über das Vorkommen von Corilagin in Myrobalanen. 15. Mitteilung über natürliche Gerbstoffe. *Liebigs Ann. Chem.* 578:31–33.

Schnitzer, M., 1965. The application of infrared spectroscopy to investigations on soil humic compounds. *Canad. Spectroscopy* 10, No. 5, 121–127.

————, and J. G. Desjardins, 1962. Molecular and equivalent weights of the organic matter of a podzol. *Soil Sci. Soc. Amer., Proc.* 26:362–365.

————————, 1964. Further investigations on the alkaline permanganate oxidation of organic matter extracted from a podzol Bh horizon. *Can. J. Soil Sci.* 44:272–279.

————————, 1965. Carboxyl and phenolic hydroxyl groups in some organic soils and their relation to the degree of humification. *Can. J. Soil. Sci.* 45:257–264.

Schnitzer, M., and U. C. Gupta., 1964. Some chemical characteristics of the organic matter extracted from the O and B2 horizons of a gray wooded soil. *Soil Sci. Soc. Amer., Proc.* 28:374–377.

————————, 1965. Determination of acidity in soil organic matter. *Soil Sci. Soc. Amer., Proc.* 29:274–277.

Schnitzer, M., and I. Hoffmann, 1961. Thermogravimetry of the organic matter of a podzol. *Soil Chem. Ind.* 1397–1398.

————————, 1965. Thermogravimetry of soil humic compounds. *Geochim. Cosmochim. Acta* 29:359–370.

————————, 1966. A thermogravimetric approach to the classification of organic soils. *Soil Sci. Soc. Amer., Proc.* 30:63–66.

————————, 1967. Thermogravimetric analysis of the salts and metal complexes of soil fulvic acids. *Geochim. Cosmochim. Acta* 31:7–16.

Schnitzer, M., and S. U. Khan, 1972. Humic substances in the environment, New York: Marcel Dekker, Inc.

Schnitzer, M., and S. I. M. Skinner, 1963a. Organo-metallic interactions in soils: 1. Reactions between a number of metal ions and the organic matter of a podzol B_h-horizon. *Soil Sci.* 96:86–93.

————————, 1963b. Organo-metallic interactions in soils: 2. Reactions between different forms of iron and aluminium and the organic matter of a podzol. B_h-horizon. *Soil Sci.* 96:181–186.

————————, 1964. Organo-metallic interactions in soils: 3. Properties of iron- and aluminium-organic matter complexes, prepared in the laboratory and extracted from a soil. *Soil Sci.* 98:197–203.

————————, 1965a. Organo-metallic interactions in soils: 4. Carboxyl and hydroxyl groups in organic matter and metal retention. *Soil Sci.* 99:273–284.

————————, 1965b. The carboxyl group in a soil organic matter preparation. *Soil Sci. Soc. Amer., Proc.* 29:400–405.

————————, 1966a. Organo-metallic interactions in soils: 5. Stability constants of Cu^{2+}-, Fe^{2+}-, and Zn^{2+}-fulvic acids complexes. *Soil Sci.* 102:361–365.

————————, 1966b. A polarographic method for the determination of carbonyl in soil humic compounds. *Soil Sci.* 101:120–124.

————————, 1967. Organo-metallic interactions in soils. 7. Stability constants of Pb^{2+}-, Ni^{2+}-, Mn^{2+}-, Co^{2+}-, Ca^{2+}-, and Mg^{2+}-fulvic acid complexes. *Soil Sci.* 103:247–252.

————————, 1968. Gel filtration of fulvic acid, a soil humic compound. Isotopes and radiation in soil organic matter studies. International Atomic Energy Agency, Vienna 41–55.

Schnitzer, M., and J. R. Wright, 1956. Note on the extraction of organic matter from the B horizon or a podzol soil. *Can. J. Agric. Sci.* 36:511–512.

Schnitzer, M. and J. R. Wright, 1957. Extraction of organic matter from podzolic soils by means of dilute inorganic acids. *Can. J. Soil. Sci.* 37:89–95.

——————, 1960. Studies on the oxidation of the organic matter of the Ao and Bh horizons of a podzol. Seventh International Congress of Soil Science, Madison, Wisconsin, 2:112–119.

Schnitzer, M., D. A. Shearer, and J. R. Wright, 1959. A study in the infrared of high molecular weight organic matter extracted by various reagents from a podzolic horizon. *Soil Sci* 87:252–257

——————, R. C. Turner, and I. Hoffman, 1964. A thermogravimetric study of organic matter of representative Canadian podzol soils. *Can. J. Soil Sci.* 44:7–13.

——————, J. R. Wright, and J. G. Desjardins, 1958. A comparison of the effectiveness of various extractants for organic matter from two horizons of a podzol profile. *Can. J. Soil Sci.* 38:49–53.

Schobinger, U., 1958. Chemische Untersuchungen über die Umwandlung von Weizenstrohlignin im Laufe der Verrottung. Dissertation, ETH Zürich.

Schreiner, O., and E. Shorey, 1910. Chemical nature of soil organic matter. *United States Department of Agriculture Bureau of Soils Bulletin* 74:5–48.

Schuffelen, A. C., and G. Bolt, 1950. Some notes on the synthesis of humus compounds. *Landbouwk. tijdschr.* 62:333.

Sedletzky, J., and B. Brunowsky, 1935. Der Aufbau der Huminsäure und ihre Beziehung zum Lignin und zu Kohlen. *Kolloid Z.* 73:90–91.

Seifert, K., 1962. Die chemische Veränderung der Holzzellwand-Komponenten unter dem Einfluß pflanzlicher und tierischer Schädlinge. 2. Abbau von Pinus sylvestris L. durch Coniophora cerebella Pers. *Holzforschung* 16:102–113.

Shmuk, A. A., 1924. The chemistry of soil organic matter. *Trudy kuban. s.-kh. Inst.* 1-2.

——————, 1930. The chemical nature of soil organic matter. *Byull. Pochvoveda* 5–7.

Shorey, E., 1913. Some organic soil constituents. *U.S. Department of Agriculture Bureau of Soils Bull.* 88:5–41.

Simon, K., 1929. Über die Herstellung von Humusextrakten mit neutralen Mitteln. *Z. Pflanzenernähr. Düng. Bodenk.* 14:252–257.

——————, and H. Speichermann, 1938. Beiträge zur Humusuntersuchungsmethodik. *Bodenk. Pflanzenernähr.* 8:129–152.

Simonart, P., and J. Mayaudon, 1958a. Étude de la décomposition de la matière organique dans le sol au moyen de carbons radioactif. 1. Cinetique de l'oxydation en CO_2 de divers substrate radioactifs. *Plant Soil* 9:367–375.

——————, 1958b. Étude de la décomposition de la matière organique dans le sol au moyen de carbone radioactif. 2. Décomposition du glucose radioactif dans le sol. B. Répartition de la radioactivité dans l'α-humus. *Plant Soil* 9:381–384.

——————, and L. Batistic, 1959. Étude de la décomposition de la matière organique dans le sol au moyen de carbone radioactif. 4. Décomposition des pigments foliaires. *Plant Soil* 11:176–180.

Slyke, D. D. van, D. A. MacFadyen, and P. Hamilton, 1941. Determination of free amino acids by titration of the carbon dioxide formed in the reaction with ninhydrin. *J. Biol. Chem.* 141:671–680.

Smith, D. G., and J. W. Lorimer, 1964. An examination of the humic acids of sphagnum peat. *Can. J. Soil Sci.* 44:76–87.

Smith, F. B., W. H. Stevenson, and P. E. Brown, 1930. The production of artificial manures. *Agr. Exp. Sta. Res. Bull.* 126.

Sneath, P. H. A., 1957. The application of computers to taxonomy. *J. Gen. Microbiol* 17:201–226.

Söchtig, H. 1961a. Zur Erfassung der bei der Strohrotte gebildeten Ligninabbauprodukte: Kennzeichnung durch UV-Spektren. *Landbauforsch. Völkenrode* 11:13–15.

——————, 1961b. Inwieweit können Strohrotteprodukte das Pflanzenwachstum beeinflussen? Über Veränderungen von Ligninabbauprodukten. *Landbauforsch. Völkenrode* 11:16–18.

——————, 1965. Auftrennung von Huminstoffen durch Gelfiltration. *Mitt. Dtsch. Bodenkundl. Ges.* 4:297–300.

——————, 1966. Zur Fraktionierung von Humusstoffen durch Gelfiltration. 1. Mitt. Das Verhalten von anorganischen Ionen aus der Asche und den Lösungen von Humusstoffen auf Sephadex-Gelen. *Landbauforsch. Völkenrode* 16:25–30.

Sørensen, H., 1963. Studies on the decomposition of ^{14}C-labelled barley straw in soil. *Soil Sci.* 95:45–51.

———, 1966. Formation of soil organic matter during decomposition of plant components. In "The use of isotopes in soil organic matter studies." Report of the FAO/IAEA Technical Meeting, Oxford: Pergamon Press, Ltd., 495–498.

Sowden, F. J., 1955. Estimation of amino acids in soil hydrolysates by the Moore and Stein method. *Soil Sci.* 80:181–188.

———, 1957. Note on the occurrence of amino groups in soil organic matter. *Can. J. Soil Sci.* 37:143–144.

———, and H. Deuel, 1961. Fractionation of fulvic acids from the B horizon of podzol. *Soil Sci.* 91:44–47.

———, and D. I. Parker, 1953. Amino nitrogen of soils and of certain fractions isolated from them. *Soil Sci.* 76:201–208.

Sprengel, C., 1826. Über Pflanzenhumus, Humussäure und humussäure Salze. *Kastners Arch. Ges. Naturlehre* 8:145.

———, 1828. Von den Substanzen der Ackerkrume und des Untergrundes insbesondere, wie solche durch die chemische Analyse entdeckt und voneinander geschieden werden können. *Erdmanns J. techn. ökonom. Chem.* 2:423–474 and 3:42–99, 313–351, 397–421.

Springer, U., 1934. 1. Farbtiefe und Farbcharakter von Humusextrakten in ihrer Abhängigkeit von der Alkalikonzentration, zugleich ein Beitrag zur Kenntnis der Humustypen. *Z. Pflanzenernähr. Düng. Bodenk.* 34:1–14.

———, 1938. Der heutige Stand der Humusuntersuchungsmethodik mit besonderer Berücksichtigung der Trennung, Bestimmung und Charakterisierung der Huminsäuretypen und ihre Anwendung auf charakteristische Humusformen. *Bodenk. Pflanzenernähr.* 6:312–373.

———, 1944/45. Stoffabbau und Humusaufbau untersucht an einem Strohmist und Strohligninmist (Laboratoriumsversuch). *Prakt. Blätter Pflanzenbau Pflanzenschutz* 21/22:1–57.

———, 1955. Über Komposthuminsäuren aus verschiedenen pflanzlichen Ausgangsstoffen. *Z. Pflanzenernähr. Düng. Bodenk.* 69:66–71.

———, and Klee, 1958. Die Charakterisierung und Unterscheidung von Waldhumusformen mittels der Natronlauge- und Natriumphosphat-Extraktion und der Stickstoff-Fraktionierung. *Z. Pflanzenernähr. Düng. Bodenk.* 80:109–126.

———, J. Klee, and A. Wagner, 1962. Über die quantitativen Veränderungen der organischen Substanz sowie der organischen und anorganischen Stickstofformen bei der Verrottung eines Gemisches von Stroh und Grünmasse in Gegenwart verschiedener Stickstoffdünger. *Bayr. Landw. Jahrbuch* 39:528–551.

———, and A. Lehner, 1941/42. Einfluß einiger Stickstoffsalze und Handelsdünger auf die Verrottung von Stroh. Prakt. *Blätter Pflanzenbau Pflanzenschutz* 19:20–41.

———, and A. Lehner, 1952a. Stoffabbau und Humusaufbau bei der aeroben und anaeroben Zersetzung landwirtschaftlich und forstwirtschaftlich wichtiger organischer Stoffe. 1. *Z. Pflanzenernähr. Düng. Bodenk.* 58:193–231.

———, and A. Lehner, 1952b. Stoffabbau und Humusaufbau bei der aeroben und anaeroben Zersetzung landwirtschaftlich und forstwirtschaftlich wichtiger organischer Stoffe. 2. *Z. Pflanzenernähr. Düng. Bodenk.* 59:1–27.

———, and F. Selschab, 1961. Zur Kenntnis der bei der Kompostbereitung auftretenden stofflichen Veränderungen. Mit Ergebnissen eines dreijährigen Gefässversuches. *Bayr. Landw. Jahrbuch* 38:250–300.

Stadnikov, G., 1937. Über Huminsäuren. *Kolloid Z.* 80:60–68.

———, 1963. Die Analyse von Phenolgemischen vermittels ihrer Sauerstoffabsorption und die Verwendung dieses Verfahrens zur Bestimmung von Phenol-Hydroxylen in Ligninen und Huminsäuren und zum Nachweis von Ligninderivaten in fossiler Kohle. *Brennstoff-Chem.* 44:242–245.

———, 1965. Analyse von Phenolgemischen vermittels ihrer Sauerstoffabsorption und die Verwendung dieses Verfahrens zur Bestimmung von Phenolhydroxylgruppen in Ligninen und Huminsäuren und zum Nachweis von Ligninderivaten. *Brennstoff-Chem.* 46:209–215.

———, K. Ssyskoff, and A. Uschakowa, 1936. Über die thermische Zersetzung der Huminsäuren. *Brennstoff-Chem.* 17:383.

Stanier, R. Y., 1947. Simultaneous adaption: A new technique for the study of metabolic pathways. *J. Bact.* 54:339–348.

————, and J. L. Ingraham, 1954. Protocatechuic oxidase. *J. Biol. Chem.* 210:799–808.

Staudinger, H., 1930. Über hochpolymere Verbindungen Viskösitätsuntersuchungen an Molekül-kolloiden Kolloid Z. 51:71–89.

Staudinger, H. and E. Husemann, 1935. Über hochpolymere Verbindungen. *Ber. dtsch. chem. Ges.* B68:1618–1634.

Steelink, C., 1959. Colour reactions of certain phenols with Ehrlich's Reagent. *Nature* 184:720.

————, 1964. Free radical studies of lignin, lignin degradation products and soil humic acids. *Geochim. Cosmochim. Acta* 28:1615–1622.

————, and G. Tollin, 1962. Stable free radicals in soil humic acid. *Biochim. Biophys. Acta* 59:25–34.

————, J. W. Berry, A. Ho, and H. E. Nordby, 1960. Alkaline degradation products of soil humic acid. *Sci. Proc. Roy. Dublin Soc. Ser.* A1, 59–67.

Stein, H. N., and H. J. C. Tendeloo, 1955. On the kinetics of the oxidation of phloroglucinol. *Recueil Trav. Chim. Pays. Bas.* 74:905–918.

————, and H. J. C. Tendeloo, 1956a. Some observations on the oxidation products of phloroglucinol. *Recueil Trav. Chim. Pays. Bas.* 75:1233–1239.

————, and H. J. C. Tendeloo, 1956b. Evidence for free radicals as intermediates in the oxidation of phloroglucinol. *Recueil Trav. Chim. Pays. Bas.* 75:1240–1242.

————, and H. J. C. Tendeloo, 1958. The kinetics of the oxidation of 2.4-Dimethylphloroglucinol. *Recueil Trav. Chim. Pays. Bas.* 77:487–490.

————, and H. J. C. Tendeloo, 1959. The oxidation of phloroglucinol as a model for humification processes. *Plant Soil* 11:131–138.

Steinmetz, A., 1956. Modellversuche zur Beteiligung von Polyphenoloxydasen bei der Bildung natürlicher Huminsauren. Dissertation TH Braunschweig, Germany.

Stern, R., J. English, Jr., and H. G. Cassidy, 1957. Electron exchange polymers. 10. A general method for the preparation of phenolic polystyrene. *J. Am. Chem. Soc.* 79:5792–5797.

Stevenson, F. J., 1960. Chemical nature of the nitrogen in the fulvic fraction of soil organic matter. *Soil Sci. Soc. Am., Proc.* 24:472–477.

————, Q. van Winkle, and W. P. Martin., 1953. Physicochemical investigations of clay adsorbed organic colloids. 2. *Soil Sci. Soc. Am., Proc.* 17:31–34.

————, J. D. Marks, J. W. Varner, and W. P. Martin, 1952. Electrophoretic and chromatographic investigations of clay adsorbed organic colloids. 1. Preliminary investigations. *Soil Sci. Soc. Am., Proc.* 16:69–73.

Stöckli, A., 1952. Über den Abbau von Lignin, Cellulose und Hemicellulose durch Pilze. Versuche mit Streumaterialien, Holz und Ligninsulfonsäure. Dissertation ETH Zürich, Switzerland.

Stuart, H. A., 1953. *Die Physik der Hochpolymeren. Bd. II: Das Makromolekül in Lösungen.* Springer-Verlag: Berlin, Göttingen, Heidelberg.

Sundmann, V., and K. Haro, 1966. On the mechanism by which cycloligninolytic agrobacteria might cause humification. *Finska Kemists Medd.* 75:111–118.

Suzuki, S., 1906–1908a. On the formation of humus. *Bull. Coll. Agric., Japan* 7:95–99.

————, 1906–1908b. Studies on humus formation. 2. *Bull. Coll. Agric. Japan.* 7:419–423.

————, 1906–1908c. Studies on humus formation. 3. *Bull. Coll. Agric., Japan* 7:513–529.

Svedberg, T., and K. O. Pedersen, 1940. Die Ultrazentrifuge. Verlag Steinkopf: Dresden and Leipzig.

————, and K. O. Pedersen, 1940, 1960. *The ultracentrifuge.* Oxford: Clarendon Press.

Svensson, H., 1940. Theorie der Beobachtungsmethoden der gekreuzten Spalte. *Kolloid Z.* 90:141–156.

Swaby, R., 1956–58. Soil organic matter. 8th, 9th and 10th C.S.I.R.O. Ann. Reports. Government Printers, Sydney.

————, and J. M. Ladd, 1962. Chemical nature, microbial resistance and origin of soil humus. *Trans. Internat. Soc. Soil Sci. Comm. IV, New Zealand.*

Täufel, K., and H. Iwainsky, 1952. Analytische und chromatographische Studien zur Maillard-Reaktion. *Biochem. Z.* 325:299.

Tanaka, T., Ch. Yamashita, and T. Yokoyama, 1957. Synthesis and polymerization of o-hydroxystyrene. *Kôgyô Kagaku Zasshi* 60:1595–1600.

Tanford, C., 1962. The interpretation of hydrogen ion titration curves of proteins. *Adv. Protein Chem.* 17:69–165.

Teuber, H. J., and C. Jellinek, 1952. Reaktionen mit Nitrosodisulfonat. 1. Einwirkung auf Phenole und aromatische Stickstoffverbindungen. *Chem. Ber.* 85:95–103.

———, and W. Rau, 1953. Reaktionen mit Nitrosodisulfonat. 2. Über die Bildung von Chinonen aus einwertigen Phenolen. *Chem. Ber.* 86:1036–1047.

Theng, B. K. G., and A. M. Posner, 1967. Nature of the carbonyl groups in soil humic acids. *Soil Sci.* 101:191–201.

———, J. R. H. Wake, and A. M. Posner, 1966. Infrared spectrum of humic acids. *Soil Sci.* 102:70–72.

Thiele, H., and H. Kettner, 1953. Über Huminsäuren. *Kolloid Z.* 130:131–160.

Thompson, G. O. P., 1958. Further observations on the N-terminal amino acids of bovine serum albumin. *Biochim. Biophys. Acta* 29:643–644.

Tinsley, J., and A. Salam, 1961. Extraction of soil organic matter with aqueous solvents. *Soils Fert.* 24:81–84.

———, and M. K. Zin, 1954. The isolation of ligno-protein from soil. Trans. Fifth International Congress of Soil Science, Albertsville 2:324–347.

Tishenko, V., and M. Rydalevskaya, 1936. An attempt at the chemical study of humic acids of different soil types. *Dokl. Akad. Nauk SSSR* 4.

Titova, N. A., 1962. Iron humus complexes of certain soils. *Soviet Soil Sci.* 12:1351–1356.

Todd, A. R., 1962. Die Farbstoffe der Blattläuse (Aphididae). *Experientia* 18:433–438.

Tokudome, S., and I. Kanno, 1965. Nature of the humus of humic allophane soils in Japan. 2. Some physicochemical properties of humic and fulvic acids. *Soil Sci. Plant Nutr.* 11:9–15.

———, and I. Kanno, 1968. Nature of the humus of some Japanese soils. Ninth International Congress of Soil Science, Adelaide, Australia, Transactions III, 163–173.

Tokuoka, M., and I. Ruizicka, 1934. Polarographische Untersuchungen über die Humussäure, Hymatomelansäure und Torf mittels der tropfenden Quecksilberkathode. *Z. Pflanzenernähr. Düng. Bodenk.* 35:79.

Trautner, E. M., and E. A. H. Roberts, 1950. The chemical mechanism of the oxidative deamination of amino acids by catechol and polyphenoloxidase. *Australian J. Sci. Res. Ser. B3*:356–380.

Traynard, Ph., and A. Eymery, 1956. Delignification par les solutions hydrotropiques. 2. Étude des lignines hydrotripiques. *Holzforschung* 10:6–11.

Trippett, S., S. Dagley, and D. A. Stopher, 1960. Bacterial oxidation of protocatechuic acid. *Biochem. J.* 76:9.

Trojanowski, J., A. Leonowicz, and B. Hampel, 1966. Exoenzymes in fungi degrading lignin. 2. Demethoxylation of lignin and vanillic acid. *Acta Microbiol. Polon.* 15:17–22.

———, A. Leonowicz, and M. Wojtas, 1967. Exoenzymes in fungi degrading lignin. 3. The effect of peroxidase on veratric acid. *Acta Microbiol. Polon.* 15:215–222.

Turner, R. C., and M. Schnitzer, 1962. Thermogravimetry of the organic matter of a podzol. *Soil Sci.* 93:225–232.

Tyurin, I. V., 1937. *Die organischen Stoffe der Böden.* Selchosgis Moskau.

———, 1940. The nature of the fulvic acid of soil organic matter. *Trans. Dokuchaev Soil Inst. Moscow, USSR* 23:23–40.

Ubaldini, I., 1937. Beitrag zur Kenntnis der Huminsäuren. *Brennstoff-Chem.* 18:273.

Updegraff, I. H., and H. Cassidy, 1949. Electron exchange-polymers. 2. Vinylhydroquinone monomer and polymer. *J. Am. Chem. Soc.* 71:407.

Urbanski, T., W. Kuczynski, W. Hofman, H. Urbanik, and M. Witanowski, 1959. Infrared absorption spectra of products of carbonization of lignin. *Bull. Acad. Polon. Sci., Ser. sci. chim. geol. geogr.* 7:207.

Valek, B., 1962. Das Humusmaximum der Differentialthermokurven aus Bodenproben von verschiedenen Pflanzenbeständen. *Studies about Humus, Symp. Humus and Plant, Praha* 341–345.

Venkataraman, B., and G. K. Fraenkel, 1955. Proton hyperfine interactions in paramagnetic-resonance of semiquinones. *J. Am. Chem. Soc.* 77:2707–2713.

Visser, S. A., 1964. Oxydation-reduction potentials and capillary activities of humic acids. *Nature (London)* 204:581, 7–11.

Vosburgh, W. C., and G. R. Cooper, 1941. Complex ions. 1. The identification of complex ions in solution by spectrophotometric measurements. *J. Am. Chem. Soc.* 63:437–442.

Wagner, G. H., and F. J. Stevenson, 1965. Structural arrangement of functional groups in soil humic acids as revealed by infrared analysis. *Soil Sci. Soc. Am., Proc.* 29:43–48.

Waksman, S. A., 1936. Humus, origin, chemical composition, and importance in nature. London: Ballière, Tindall and Cox.

———, and I. J. Hutchings, 1936. Decomposition of lignin by microorganisms. *Soil Sci.* 42:119–130.

———, and H. W. Smith, 1934. Transformation of methoxyl group in lignin in the process of decomposition of organic residues by microorganisms. *J. Am. Chem. Soc.* 56:1225.

———, and K. R. Stevens, 1930. A system of proximate chemical analysis of plant materials. *Ind. Eng. Chem.* 2:167–173.

———, and F. G. Tenney, 1927a. The composition of natural organic materials and their decomposition in the soil. 1. Methods of quantitative analysis of plant materials. *Soil Sci.* 24:275–283.

———, and F. G. Tenney, 1927b. The composition of natural organic materials and their decomposition in the soil. 2. Influence of age of plant upon the rapidity and nature of its decomposition—rye plants. *Soil Sci.* 24:317–333.

———, F. C. Tenney, and R. A. Diehm, 1929. The chemical and microbiological principles underlaying the transformation of organic matter in the preparation of artificial manures. *J.A.S.A.* 21:533–545.

Waldron, A. C., and J. L. Mortensen, 1961. Soil nitrogen complexes: II. Electrophoretic separation of organic components. *Soil Sci. Soc. Am. Proc.* 25:29–32.

Walker, N., and W. C. Evans, 1952. Pathways in the metabolism of the monohydroxybenzoic acid by soil bacteria. *Biochem. J.* 52:23.

Walter, H. H., 1960. Beitrag zur Kenntnis der Oxydation des Guajacols unter physiologischen Bedingungen. Dissertation TH Braunschweig, Germany.

Walters, E. H., 1917. The isolation of p-hydroxy-benzoic acid from soils. *J. Am. Chem. Soc.* 39.

Wehrmann, J. 1955. Zur Kenntnis der biologischen und chemischen Umformung von Roggenstroh-Lignin zu Huminsäuren. *Z. Pflanzenernähr. Düng. Bodenk.* 70:34–44.

Welte, E., 1952. Über die Entstehung von Huminsäuren und Wege ihrer Reindarstellung. *Z. Pflanzenernähr. Düng. Bodenk.* 56:105–139.

———, 1953. Über Humuskolorimetrie. Vortrag auf der Dtsch. Bodenkundl. Ges., Bonn.

———, 1956. Zur Konzentrationsmessung von Huminsäuren. *Z. Pflanzenernähr. Düng. Bodenk.* 74:219–227.

———, M. Schatz, and W. Ziechmann, 1954. Über Synthesehuminsäuren (Oxydationsmessungen). *Naturwissenschaften* 41:213–214.

Weltner, M. 1961. Die derivatographische Analyse der thermischen Zersetzung von Kohlen. *Brennstoff-Chem.* 42:40–46.

Wershaw, R. L., P. J. Burcar, C. L. Sutula, and B. J. Wiginton, 1967. Sodium humate solution studied with small angle X-ray scattering. *Science.* 157:1429–1431.

Whitehead, D. C., 1964. Identification of p-hydroxybenzoic, vanillic, p-coumaric and ferulic acids in soils. *Nature* 202:417–418.

———, and J. Tinsley, 1964. Extraction of soil organic matter with dimethylformamide. *Soil Sci.* 97:34–42.

Wiesemüller, W., 1965. Untersuchung über die Fraktionierung der organischen Bodensubstanz. *Albrecht Thaer Archiv* 9:419–436.

Wold, F., 1961. Some properties of cross-linked bovine serum albumin. *Biochim. Biophys. Acta* 54:604–606.

Wolfrom, M. L., R. D. Schuetz, and L. F. Cavalieri, 1948. Chemical interaction of amino compounds and sugars. 3. The conversion of D-glucose in 5-(hydroxymethyl)-2-furaldehyde. *J. Amer. Chem. Soc.* 70:514.

————, R. D. Schuetz, and L. F. Cavalieri, 1949. Chemical interaction of amino compounds and sugars. 4. The significance of furan derivatives in color formation. *J. Am. Chem. Soc.* 71:3518.

Wood, J. C., S. E. Moschopedis, and W. den Hertog, 1961. Untersuchungen der Huminsäuren-Chemie. 2. Mitt. Huminsäurenhydride. *Fuel* 40:491–502.

Wright, J. R., and M. Schnitzer, 1959. Oxygen containing functional groups in the organic matter of a podzol soil. *Nature* 184:1462–1463.

————, and M. Schnitzer, 1960. Oxygen containing functional groups in the organic matter of the Ao and Bh horizon of a podzol. 7th Internat. Congr. Soil Sci. Madison, Wisconsin 2:120–127.

————, and M. Schnitzer, 1961. An estimate of the aromaticity of the organic matter of a podzol soil. *Nature* 190:4777, 703–704.

————, and M. Schnitzer, 1963. Metallo-organic interactions associated with podzolization. *Soil Sci. Soc. Am. Proc.* 27:171–176.

————, M. Schnitzer, and R. Levick, 1958. Some characteristics of the organic matter extracted by dilute acids from a podzolic B horizon. *Can. J. Soil Sci.* 38:14.

Young, J., and J. L. Mortensen, 1958. Soil nitrogen complexes. 1. Chromatography of amino compounds in soil hydrolysates. *Ohio Agric. Exp. Sta. Res. Circ.* 61:18.

Zadmard, H., 1939. Zur Kenntnis der kolloidchemischen Eigenschaften des Humus. *Kolloid-Beihefte* 49: 315–364.

Zetzsche, F., and H. Reinhart, 1937. Über die Huminsäuren des Torfes. *Brennstoff-Chem.* 18:393–395.

Ziechmann, W., 1954. Untersuchungen über das Teilchengewicht, die Einheitlichkeit und den Bildungsmechanismus synthetischer Huminsäuren. Dissertation, University of Göttingen, Germany.

————, 1958. Infrared spectra of humic acids. *Brennstoff-Chem.* 39:353.

————, 1959. Die Darstellung von Huminsäuren im heterogen System mit neutraler Reaktion. *Z. Pflanzenernähr. Düng. Bodenk.* 84:155–159.

————, 1960a. Über Modellreaktionen zur Bildung synthetischer Huminsäuren. 1. Die Synthese von Huminsäuren im alkalischen Milieu. *Brennstoff-Chem.* 41:289–320.

————, 1960b. Über Modellreaktionen zur Bildung synthetischer Huminsäuren. 2. Die Synthese von Huminsäuren im neutralen Milieu. *Brennstoff-Chem.* 41:334–340.

————, 1961. Die Bildung von 2.5-Dioxy-p-benzochinon bei der Autoxydation des Hydrochinons im dreiphasigen System. *Naturwissenschaften* 48:456–457.

————, 1964. Spectroscopic investigations of lignin, humic substances and peat. *Geochim. Cosmochim. Acta* 28:1555–1566.

————, 1967. Huminstoffe als Bodenkolloide. *Kali-Briefe, Fachgeb. 1, Bodenk. 6. Folge*, Februar.

————, and R. Kickuth, 1961. Die Struktur der Huminsäuren (Zur Theorie der Huminstoffe). *Kolloid Z.* 174:38–46.

————, and H. Scholz, 1960. Spektroskopische Untersuchungen an Huminsäuren. *Naturwissenschaften* 47:193–196.

Chapter 2

Saccharides

D. J. Greenland and J. M. Oades

Contents

A. Saccharides Added to the Soil

Almost 200 polysaccharides are now known to occur naturally. It is probable that the majority are at some time added to the soil. They do not, however, remain in the soil indefinitely, but are attacked by the soil fauna and flora, which in turn will contribute other saccharides to the soil. The nature of polysaccharides in soils must, therefore, be deduced in part from the known properties of the polysaccharides added to the soil and produced in it and, in part, from a knowledge of the saccharides that can be isolated from the soil itself. As yet, the techniques for the isolation of saccharides and polysaccharides from soils are very inadequate. This isolation is difficult largely because of the strong interaction between the inorganic materials and much of the saccharide fraction of soil organic matter. To date, little more than half of the carbohydrates known to occur in the soil have been separated from the inorganic materials. In this review the major groups of saccharides added to the soil are described first (Table 1), and the factors involved in their decomposition in the soil are discussed. Then the amounts of polysaccharides found in soils, their isolation from soils, and the properties of the isolates are considered. Finally, a brief account of the relationship between saccharides in the soil and other soil properties is presented.

I. Saccharides of the Higher Plants

1. *Cellulose*

Most of the carbohydrate added to the soil is in the form of cellulose. This comprises approximately 50 to 70% or more of the dry weight of plants, mostly occurring as cell-wall material. It is usually associated with smaller amounts of hemicelluloses, pectic substances, and lignin. Its chemistry and occurrence have been described in detail in several books (NORMAN [1937], HERMANS [1949], OTT, SPURLIN, and GRAFFLIN [1954], HONEYMAN [1959]). Cellulose is composed of β-D-glucose molecules joined through the 1 and 4 positions. The chains of glucose molecules so formed lie parallel to each other, in a more or less regular array, to form crystallites. These are further grouped into fibrils. The size and degree of order in fibrils and crystallites vary considerably with different celluloses isolated from different plant materials. X-ray diffraction methods have shown that the crystallinity may vary considerably along the length of a single fibril. Considerable effort has been expended in studying the structures of cellulose crystals, and much is now known of the organization of the individual cellulose molecules within the crystal. The degree of crystallinity of cellulose is important in relation to its accessibility to decomposition by either chemical reagents or by enzymes. Attack on the molecule occurs considerably more readily in the amorphous regions than in the crystalline regions. Cellulose is highly insoluble in most solvents. Water may enter the crystal structure and cause extensive swelling, but the cellulose does not dissolve. Chemical hydrolysis occurs only in strong acid. Thus, it seems that decomposition in soil must be brought about entirely by enzymic reactions.

2. *Hemicelluloses*

"Hemicellulose" has been widely used to refer to those cell-wall polysaccharides normally found in association with cellulose but soluble in dilute alkali. Although the term has come into common use because of its convenience, it does not refer to any specific group of polysaccharides. Polysaccharides can be broadly divided into those containing uronic acids and those devoid of specific residues. The non-uronide hemicelluloses are usually closely bound to cell-wall cellulose. Xylans are the predominant constituent, but mannans, glucomannoglycans, galactans, arabinogalactoglycans, arabinoxyloglycans, and other polysaccharides may also be included (SCHUERCH and THOMPSON [1963]). The polyuronide hemicelluloses are composed principally of glucuronoxyloglycans, but galacturonoarabinoglycans have also been isolated. The proportion of uronic acid residues to neutral sugar residues in these polysaccharides differs according to the source. Most of the figures reported lie between 7 and 18 neutral sugars to each uronic acid residue. In addition to the principal sugars, it is probable that rhamnose, fucose, and various O-methylated neutral sugars are present (TIMELL [1964]). The structures of several hemicelluloses have now been determined in some detail (ASPINALL [1959], TIMELL [1964]). Most are linear chains with a degree of polymerization of the order of 150 to

Table 1. Principal polysaccharides added to soils.

Origin	Polysaccharide	Function	Component sugars	General
	Cellulose	Cell-wall material	Glucose	Forms 30 to 70% of tissue of higher plants. Linear polymer of β-(1→4)-D-glucopyranose units.
	Hemicelluloses, which include glucuronoxylo-glycans	Associated cell-wall material	Glucuronic acid, xylose	Forms 2 to 5% of higher plants. Linear xylose polymer with attached uronic acid units.
	galacturono-arabino-xyloglycans		Galacturonic acid, arabinose, xylose	Linear xylose polymer with other units attached.
	xylans		Xylose	Linear polymers of β-(1→4)-D-anhydroxylopyranose units.
	glucomannans		Glucose, mannose, occasionally some galactose	Mannose usually predominates in β-(1→4)-D-pyranose chain.
Higher plants	arabino-galactans		Arabinose, galactose	Highly branched structures.
	Pectic substances, predominantly galacturonans associated with some arabans and galactans	In cell walls, some inter-cellular regions, and possibly in mucilages	Galacturonic acid	Linear polymer of α-(1→4)-D-pyranose units partially or fully methylated.
	Gums	Exudates from damaged tissue	Large number recorded	Generally complex, and frequently branched.
	Starches composed of amylose and	In storage organs	Glucose	Linear α-(1→4)-D-glucopyranose polymer.
	amylopectin		Glucose	Branched glucopyranose structure.
	Exocellular, include	Form protective capsules		
	dextrans		Glucose	α-(1→6)-D-polyglucosans
	levans		Fructose	β-(2→6)-D-polyfructosans.
	microbial gums		Large number recorded	
	Structural,			
	chitin	Cell-wall material in fungi	Glucosamine	Linear polymer of 2-N-acetamido-2-deoxy-β-D-glucopyranose units, linked (1→4).
	mucopeptides	Cell-wall material in bacteria	Amino sugars with amino acids	
Micro-organisms	mucopoly-saccharides		Various sugars and amino sugars	

Table 1. (*continued*)

Origin	Polysaccharide	Function	Component sugars	General
	teichoic acids		Glucose, glucosamine, ribitol, and glycerol phosphates and alanine	
	Somatic starches levans glycogens	Energy storage materials	Glucose Fructose Glucose	
Animals	chitin	Structural materials in lower animals	Glucosamine	Linear polymer of 2-N-acetamido-2-deoxy-β-D-glucopyranose units linked (1→4).
	hyaluronic acid	Intercellular cement	Glucosamine, glucuronic acid	
	glycogen	Energy storage material	Glucose	Branched chain.

200. One highly branched galactan, however, has been isolated from beech tension wood (MEIER [1962]). Such branched structures appear to be highly exceptional.

3. *Pectic Substances, Gums, and Mucilages*

Pectic substances occur in all plant tissues (KERTESZ [1951], DEUEL and STUTZ [1958]). The amounts found in plants are, however, much less than those of cellulose and hemicelluloses. Usually they comprise between 0.1 and 2 % of plant material. They are often present in higher proportion in material from young plants than that from older plants.

The pectins are primarily cell-wall materials forming a rigid matrix around young cellulose fibrils. Their function is taken over by lignin and cellulose itself as the plants age. The principal constituent of all pectins is D-galacturonic acid. The galacturonic acid residues are linked through the 1 and 4 positions to form polymer chains that are largely linear. The carboxyl groups are frequently methylated. Pectic substances have been specified as protopectin, which is a water-insoluble constituent of cell walls; pectin, in which the carboxyl groups are very largely esterified; and pectic acids, which are the water-soluble polygalacturonic acids largely devoid of methylester groups. Some galactose and arabinose units are almost invariably found when pectic substances are hydrolyzed. It is not certain whether these are included within the pectic molecules or whether their occurrence is due to the fact that some galactans and arabans occur together with the polygalacturonides and are extremely difficult to separate from them. The insolubility of protopectin has been attributed to its association with cellulose. The number of methylester groupings in the pectins and their molecular weights are very variable, depending on the source of the pectin. Degrees of polymerization of 1,000 and upward have commonly been obtained.

Root hairs of many plants are surrounded by a mucilaginous gel (mucigel), which is thought to be composed of pectic substances (SCOTT *et al.* [1958], DAWES and BOWLER [1959], CORMACK [1962], SCOTT, BYSTROM, and BOWLER [1963], JENNY and GROSSENBACHER [1963]), although its chemical nature has not yet been definitely established. It shows a weak reaction to staining with ruthenium red (SCOTT *et al.* [1958]), which supports the suggestion that it is a pectic substance. However, SCOTT *et al.* [1958] state that this mucilaginous layer is visibly heterogeneous. Thus, it may contain other materials such as mucopolysaccharides, which LIPPMAN [1965] has suggested are present at the external surfaces of all cells. In the soil this mucilaginous layer will normally be invaded by microorganisms and will be in contact with soil particles. The invasion by microorganisms makes it probable that in addition to mucilaginous material derived from the root, other polysaccharides of microbial origin are also present. The materials of plant origin are apparently excreted through the root (SCOTT *et al.* [1958]). They may be particularly important in the soil, not only as a source of polysaccharides but also in relation to plant nutrition and the composition of the soil.

Although little work has been done on root mucilages, those found in the parts of plants above the ground have been quite extensively studied (MANTELL [1947], HOWES [1949], SMITH and MONTGOMERY [1959]) because of their importance as gums. The gums are characteristically composed of a number of saccharides, usually including uronic groups, some of which are esterified. Gum arabic, for instance, contains D-galactose, L-arabinose, L-rhamnose, and D-glucuronic acid; and L-fucose has been found in other *Acacia* gums (SMITH and MONTGOMERY [1959]). Proof of the separation of pure, individual polysaccharides from gums is difficult, and hence some uncertainties exist regarding their chemical structure. Much of the evidence available is presented by SMITH and MONTGOMERY [1959].

4. *Starches and Inulin*

Starch is the second most common carbohydrate, providing the major form of energy reserve material. The amount added to soils is, however, considerably less than that of cellulose. It is again a polymer of glucose, but linked primarily by α-D-$(1\rightarrow4)$ linkages. Starches are composed of two major components: amyloses, which contain linear molecules, and amylopectins, which have highly branched molecules with α-D-$(1\rightarrow6)$ linkages at the point of branching. In contrast to cellulose, starches are soluble in hot water and are readily hydrolyzed. Amylose will form an acetate which, on stretching, gives fiber-type X-ray patterns, which confirms its linear structure. On the other hand, amylopectin and its derivatives cannot be obtained in crystalline forms. Physical studies on solutions of the polymer indicate that the molecules or molecular aggregates exist in a more or less spherical form.

Inulin is the major reserve carbohydrate of compositae. The molecule consists of a chain of β-D-$(2\rightarrow1)$ linked fructofuranose residues. It is usually associated with a small amount of D-glucose, which is thought to occur as the end member of the chain, being attached to the fructose units by a sucrose-type linkage. A number of other fructans are known to occur in plant materials.

II. Saccharides of Microorganisms

The polysaccharides of microorganisms have been discussed in detail by STACEY and BARKER [1960] and BARKER [1963]. They have been classified as cell-wall materials, exocellular and capsular materials, and intracellular or somatic polysaccharides, which are apparently similar in function to storage materials in plants. In most instances they are much more complex than the glucans, which are the predominant constituents of higher plants. They have been less extensively studied, and consequently their chemistry is less well known. It is important to note that most existing knowledge of microbial polysaccharides is derived from studies on pure cultures. Changes in culture conditions, and notably changes of carbon substrate, are known to lead to the production of polysaccharides of different compositions. Thus, some of the polysaccharides produced within the natural soil environment may differ considerably from those obtained from pure cultures.

A number of the polysaccharides found in higher plants also occur in microorganisms. Cellulose is synthesized by a number of species of *Acetobacter* (STACEY and BARKER [1960]) and polysaccharides very closely similar to starch by *Corynebacterium diphtheriae* and *Clostridium butyricum* (WHISTLER and SMART [1953]). In addition, chitin (poly-2-N-acetamido-2-deoxyglucose, or poly-N-acetylglucosamine), a major component of insect cuticle, is found in the cell walls of most fungi and in some algae (SALTON [1960]).

As well as these homopolysaccharide dextrans (α-(1→6)-linked polyglucosans) and levans (β-(2→6)-linked polyglucosans), polyfructosans, polymannans, and polygalactans are synthesized by various microorganisms (BARKER [1963]). However, the complex exocellular and capsular polysaccharides synthesized by many microorganisms are perhaps of most interest in relation to saccharides in the soil, since they are synthesized by several of the more common soil organisms (MARTIN [1945b], FORSYTH and WEBLEY [1949], FORSYTH [1954], WEBLEY *et al.* [1965]) and have been shown to be very effective in binding soil particles into stable aggregates (MARTIN [1945a, 1946], McCALLA [1945], HAWORTH, PINKARD, and STACEY [1946], GEOGHEGAN and BRIAN [1946, 1948], SWABY [1949], CLAPP, DAVIS, and WAUGAMAN [1962], MARTIN and RICHARDS [1963], MARTIN, ERVIN, and SHEPHERD [1965]). Although the component sugars in many of the "slimes" produced by bacteria isolated from the soil have been identified, and those of gram-negative bacteria in general have been listed rather comprehensively, very little information exists about the more detailed structures (DAVIES [1960], FORSYTH [1954], FORSYTH and WEBLEY [1949], MARTIN [1945b], STACEY and BARKER [1960], WEBLEY *et al.* [1965], WILKINSON [1958]). They mostly contain between two and five different sugar residues, and frequently include a uronic acid moiety. More than 20 monosaccharides have been reported in these polymers. Isolation of single polysaccharides from the microorganisms is difficult. The chemical structure of the polysaccharide synthesized by *Azotobacter chroococcum* has, however, been determined (LAWSON and STACEY [1954]). It contains glucose and galactose, mostly linked through the 1 and 3 positions, together with glucuronic acid and a trace of mannose. Some glucose residues were linked 1:4, and some were linked 1:4:6 and formed branch points. The glucuronic acid residues were linked 1:4. Most of the physico-chemical properties such as size, shape, and flexibility, which are important in relation to effects on soil aggregation, remain to be studied. The determination of these properties for polysaccharides from other sources has been reviewed by GREENWOOD [1952, 1956].

In addition to the exocellular polysaccharides synthesized by microorganisms, it is likely that cell-wall components may remain in the soil (WHITEHEAD and TINSLEY [1963]). In fungi chitin is the main constituent. Its occurrence and properties have been reviewed by FOSTER and WEBBER [1960]. In bacteria other mucopolysaccharides replace chitin in the cell wall. The rather complex structure of bacterial cell walls is discussed by SALTON [1960]. For the gram-positive bacteria these consist essentially of mucopeptides (polymers composed of amino acids and amino sugars), mucopolysaccharides (polymers composed of sugars and amino sugars), and teichoic acids (polymers of ribitol and glycerol phosphates together with alanine, glucose, and glucosamine residues). In the cell walls of gram-negative bacteria, polysaccharides are a lesser component, but nevertheless mucopolysaccharides again seem to be present, as well as proteins and lipids (SALTON [1960]). The amino sugars in the mucopolysaccharides include muramic acid (3-0-carboxyethyl-D-glucosamine), as well as glucosamine and galactosamine, the latter two being present mostly as the acetyl derivatives.

III. Animal Polysaccharides

Many animal polysaccharides may be added to the soil, but it is unlikely that the amounts added will be comparable to those added to the soil by plants or synthesized in the soil by microorganisms. However, it is possible that any having a high resistance to breakdown in the soil may tend to accumulate. The high proportion of chitin in insect cuticle may in part account for the relatively large amounts of glucosamine found in soils (see Section C.II 6). Its occurrence in fungi has been noted above.

Starches are the food reserves of plants. Glycogens serve the same function in the animal kingdom. They are chemically very similar to amylopectin but are rather more highly branched. The amounts of glycogen added to soils will be small, originating when various components of the soil fauna die in the soil. Its rapid decomposition makes its persistence in the soil unlikely.

B. Metabolism of Saccharides in the Soil

I. Decomposition of Organic Materials in the Soil

1. Factors Affecting the Rate of Decomposition

Many studies have been made of the decomposition of organic materials in the soil, and it is well known that most organic materials when added to the soil will be rapidly decomposed under aerobic conditions, with the conversion of a large part of the added carbon to carbon dioxide. The remainder of the added carbon (1) is incorporated into the tissue of the soil organisms and (2) becomes a component of the "humus" materials of the soil (SCHEFFER and ULRICH [1960], RUSSELL [1961], ALEXANDER [1961], BEAR [1964], KONONOVA [1966]). The rate at which decomposition occurs, and the distribution of the added carbon between carbon dioxide, tissue, and humus, are very dependent on the physical state and the chemical composition of the organic material and the precise soil conditions. In general, decomposition proceeds most rapidly in soils close to their field capacity (*i.e.*, where pores smaller than about 30 microns in diameter are filled with water and those larger than this are filled with air), whose temperatures are between 25 and 35°C, which have a pH between 5 and 8, and which are adequately supplied with nutrient ions. Under any one set of conditions the rates become dependent on the characteristics of the organic material added.

One set of such data is illustrated in Figure 1. The decomposition of glucose proceeded more rapidly than that of starch, which decomposed more rapidly than cellulose. Lignin was broken down only very slowly. These results are typical of many such experiments. Hemicelluloses and pectins fall somewhere between starch and cellulose in their rate of decomposition in soils. The rates at which microbially synthesized polysaccharides are attacked vary considerably. Thus, those synthesized by *Azotobacter indicus* and *Chromobacterium violaceum* have been found to be very resistant to decomposition, whereas many others are subject to extremely rapid breakdown (MARTIN and RICHARDS [1963], MARTIN, ERVIN, and SHEPHERD [1965]).

From these data it would seem unlikely that many plant polysaccharides would remain for long in the soil, except perhaps for well-crystallized celluloses and any mucilages which, like certain microbial gums, are peculiarly resistant to breakdown. Cellulose might be expected to be found in the soil, not only because it is moderately resistant to breakdown, but also because it is added to the soil in much larger amounts than other plant polysaccharides. The great differences in the rate of breakdown of microbial polysaccharides, and the uncertainties regarding the composition and relative activities of the soil population at any one time make speculations regarding the probable predominance of individual microbial polysaccharides in soil of limited value.

The situation is further complicated by the possibility that the resistance of some poly-

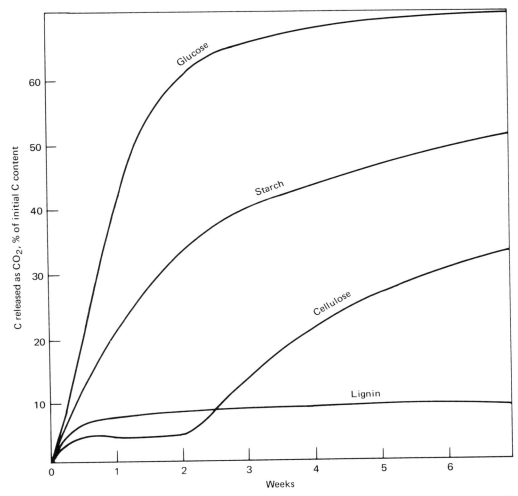

Figure 1. Decomposition of saccharides and lignin during incubation (from data of Schmalfuss reproduced by SCHEFFER and ULRICH [1960]).

saccharides to breakdown may be enhanced by their adsorption on the surfaces of clay particles (LYNCH and COTNOIR [1956], LYNCH, WRIGHT, and COTNOIR [1956], GREENLAND [1965]). It has been shown that a mixture of polysaccharides isolated from the soil is strongly adsorbed by montmorillonite (GREENLAND [1956]), so that it is quite probable that some of the soil polysaccharides are stabilized in this way. It is also possible that some polysaccharides are present in parts of the soil inaccessible to soil organisms (ROVIRA and GREACEN [1957]).

2. Decomposition of Individual Polysaccharides

(a) Cellulose

The decomposition of cellulose by microorganisms has been extensively studied (see, for example, SIU [1951], GASCOIGNE and GASCOIGNE [1960], and KING [1961]). The probable course of decomposition is an initial promotion of disaggregation of the fibril and micelle into molecular chains by an extracellular cellulase enzyme. This enzyme, or another associated with it, then brings about a hydrolytic cleavage of the β-glycosidic linkages to produce molecular fragments which can be ingested and further decomposed by an intracellular enzyme.

Numerous microorganisms are able to decompose cellulose in this way. Those commonly occurring in soils are shown in Table 2. Although HORTON and WOLFROM [1963] state categoric-

Table 2. Principal genera of microorganisms involved in the decomposition of cellulose in the soil. (Modified from ALEXANDER [1961]. Much more extensive lists of organisms producing cellulases are given by SIU [1951, pp. 129–158] and GASCOIGNE and GASCOIGNE [1960, pp. 53–68]).

Fungi	Bacteria	Actinomycetes

Probably species of most fungal genera are able to decompose cellulose, although they do so only under certain conditions. Particularly active are species of
    Aspergillus
    Fusarium
    Myrothecium
    Chaetomium
    Penicillium
    Pseudocoprinus
    Sordaria
    Trichoderma
	Achromobacter
	Bacillus
	Cellulomonas
	Cellvibrio
	Clostridium
	Corynebacterium
	Cytophaga
	Micrococcus
	Polyangium
	Pseudomonas
	Sporocytophaga
	Vibrio
		Micromonospora
		Nocardia
		Streptomyces
		Streptosporangium

ally that "*Actinomyces* species are mainly responsible for the aerobic breakdown of cellulose in the soil," in most soil conditions fungi and true bacteria are much more rapid in their attack on cellulose, and are more likely to be the main agents responsible. However, considerable differences may occur in soils of different faunal and microbial population. Where an active termite population exists, much of the decomposition may occur in the digestive tracts of these insects, which have been shown to contain protozoa producing large amounts of cellulase enzymes (SIU [1951]). Snails are also well known for their ability to digest carbohydrates.

(b) Hemicelluloses

In many investigations of the decomposition of plant material in soil the rate of disappearance of the hemicelluloses has been followed by determination of the change in dilute acid-hydrolyzable polysaccharides (FORSYTH [1948]). In most instances a rapid loss was observed, followed by a more gradual decrease. Thus, the rate of disappearance was initially greater, and subsequently slower, than that of cellulose (WAKSMAN and HUTCHINGS [1935], SHEWAN [1938], ASHWORTH [1942]). However, in these investigations no differentiation was made between the original plant hemicelluloses and related materials synthesized by the microorganisms causing the decomposition. It seems likely that the apparently more-resistant hemicelluloses were newly synthesized materials produced by the microorganisms decomposing the plant material.

In accordance with the general rapidity with which hemicelluloses are attacked when plant material is added to soil, hemicelluloses have been found to be widely distributed in microorganisms (ALEXANDER [1961]). Very few studies of the activity of specific hemicellulases in soils have been reported. Much of the data available has been summarized by GASCOIGNE and GASCOIGNE [1960]. SORENSON [1955] has discussed xylanase activity in one soil, and has shown that it relates to the history of the soil in terms of previous xylan additions. The difficulties involved in isolating individual polysaccharides from hemicelluloses have largely precluded systematic study of their relative rates of decomposition. Nevertheless, it seems probable that differences occur, since TENNEY and WAKSMAN [1929, 1930] found very marked differences in the rates at which hemicelluloses of different species were lost from soil-inoculated cultures.

(c) Other Polysaccharides

A very wide range of soil organisms has the ability to decompose starches, inulin, and pectic substances (ALEXANDER [1961]). In general, the rate of attack is faster than with cellulose, probably because the materials are less well organized (less crystalline), so that the enzyme causing the decomposition has a greater number of points at which to attack.

The considerable amount of information now available relating to the enzymic breakdown of starch and pectic substances is summarized elsewhere (WILLIAMS [1953], LAMBINA [1956], DEUEL and STUTZ [1958]). Much less information is available regarding the breakdown of gums and mucilages. However, from the data presented by MARTIN and RICHARDS [1963] and MARTIN, ERVIN, and SHEPHERD [1965], it can be seen that many soil organisms must be able to produce enzymes capable of attacking a wide range of soil polysaccharides, since when incubated with various soils, about 75% of the carbon in polysaccharides synthesized by *Bacillus subtilis* and *Azotobacter chroococcum*, as well as various gums from higher plants, was converted to carbon dioxide in four weeks. In contrast about 10% of the carbon of the polysaccharides from *Azotobacter indicus* (*Beijerinckia*) was lost in the same time. The rate of decomposition of other microbial gums was closer to that of the readily decomposed materials than to that of *Azotobacter indicus*.

Chitin and other cell-wall mucopolysaccharides are generally very resistant to hydrolysis by acids and alkali, and so might be expected to accumulate in soils. However, as with cellulose, there is a large number of microorganisms able to produce an exocellular enzyme that will hydrolyze chitin into readily digested fragments (N-N-diacetyl chitobiose, N-acetyl glucosamine, etc.) (ALEXANDER [1961]). When chitin is added to soils, most of the breakdown is due to actinomycetes, although numerous species of fungi and bacteria are also able to digest chitin. It is interesting to note that the rate of production of mineral nitrogen from chitin is not much less than that from glucosamine, when both are added to the soil in equivalent amounts (BREMNER and SHAW [1954]).

There is no information available at present on the decomposition of other mucopolysaccharides in soils, although it seems probable that they are utilized by various organisms in a manner similar to chitin.

II. Transformations of Saccharides in the Soil

1. General Information

The transformation of polysaccharides in the soil involves an initial hydrolytic cleavage into oligosaccharides and/or simple sugars. These units may be utilized as such in the subsequent synthesis of tissue materials or exocellular polysaccharides, or they may be further metabolized to aliphatic acids and alcohols, carbon dioxide, and water. It is also possible that the products of initial decomposition may be involved in nonbiological chemical reactions to form "humus" compounds. HODGE [1953] has discussed the reaction of sugars with amino acids, which he believes leads to the formation of melanoidins. This reaction could contribute to humus formation. SCHEFFER and KICKUTH [1961] have reported that a humic acid from an "intermediate bog" soil contained 15% of glucoside bound into the structure, in addition to 25% of cellulose in a slightly modified state. The presence of such a large amount of carbohydrate in a humic acid is contrary to what has been found for other soils (see Section C), but if this is generally true for such wet soils, it indicates that there is a specific mechanism by which sugars can be incorporated into humic acids.

The synthesis of polysaccharides by microorganisms provided with simple sugars as substrates is now a relatively well-known phenomenon (STACEY and BARKER [1960]). In some instances the enzyme involved in the coupling reaction has been isolated. Addition of substrates such as sucrose to soils frequently leads to an increase in the stability of soil aggregates, and is presumably due to the production of gums by organisms using the sucrose as substrate (McCALLA, HASKINS, and FROLIK [1957]). Such syntheses in soils must occur intracellularly and will be dependent on the ingestion of simple sugars released from other polysaccharides. The synthesis of polysaccharides by microorganisms provided with various plant residues, forest litter, or other microbial products and residues as substrates has been recorded in a number of instances (MARTIN [1946], BERNIER [1958b]).

2. Studies with ^{14}C-labeled Saccharides

The transformations of a saccharide added to the soil can be followed if radioactive carbon is incorporated in the saccharide prior to its addition to the soil. A number of such studies have been made using labeled sugars, cellulose, and hemicelluloses (SIMONART and MAYAUDON [1958], MAYAUDON and SIMONART [1958, 1959a, 1959b], SORENSON [1963], KEEFER and MORTENSEN [1963], MACURA, SZOLNOKI, and VANCURA [1963]). In each of these studies the labeled carbon is found to be distributed very rapidly between humic acids, fulvic acids, and humin (Table 3). It therefore appears that part at least of each of these components of soil organic matter is in dynamic equilibrium with newly added carbohydrate substrates. KEEFER and MORTENSEN [1963] examined the distribution of radioactivity between different sugars in a polysaccharide extracted from the soil after incubation with labeled glucose or alfalfa tissue. All the sugars isolated were labeled, although there were marked differences in the proportions of radioactive carbon in the different sugars, glucose, galactose, and mannose being extensively labeled. Changes in the activity of the different sugars with time indicated that the soil polysaccharides undergo continual degradation and resynthesis. These results establish clearly that soil organisms are able to synthesize from glucose or plant tissue polysaccharides containing each of the labeled sugars found (glucose, arabinose, galactose, fucose, mannose, rhamnose, xylose, and uronic acids).

C. Total Carbohydrate Contents of Soil

I. Methods of Estimation

1. Hydrolysis Procedures

A large number of methods have been used for the determination of total carbohydrate contents of soils. They are all dependent on an initial hydrolysis of the soil to release monosaccharides from the polysaccharides or other polymeric molecules in which they may be incorporated. The free saccharides may then be estimated by any one of several methods discussed below. However, choosing hydrolysis conditions is difficult. One of the earliest estimates of carbohydrate contents of soil was by WAKSMAN and HUTCHINGS [1935]. They postulated that the reducing substances brought into solution by dilute acid were hemicelluloses and that those by concentrated acid were cellulose, by analogy to the proximate analysis of plant materials. Cellulose or hemicellulose was calculated as reducing sugar $\times 0.9$. This idea is not necessarily applicable to soils where the polysaccharides are not all in the form of plant material.

Table 3.   Distribution of ^{14}C between soil organic matter fractions after decomposition of labeled cellulose, hemicellulose, and glucose in the soil, for four weeks.

Authors	Substrate added	Percent ^{14}C added evolved as CO_2 in four weeks	Percent ^{14}C remaining with soil		
			Fulvic acid	Humic acid	Humin
MAYAUDON and SIMONART [1959b]	Glucose	89	20	35	45
,, ,,	Cellulose	80	25	30	45
,, ,,	Hemicellulose	80	22	42	36
SORENSON [1963]	Cellulose	64	15	13	63
,, ,,	Hemicellulose	72	26	15	44

Recently, exhaustive studies of hydrolysis conditions have been made by IVARSON and SOWDEN [1962], GUPTA and SOWDEN [1965], and CHESHIRE and MUNDIE [1965]. In addition to a reflux in normal sulfuric acid for 16 hours, Ivarson and Sowden showed that an initial treatment with 72% sulfuric acid could increase the amount of glucose released by a factor of 3. This was thought to be due to the hydrolysis of cellulose or a conditioning of the cellulose, which enabled complete hydrolysis to take place during the subsequent reflux. More recently, GUPTA and SOWDEN [1965] reported an increased extraction of almost all sugars, particularly arabinose and the hexoses, due to the 72% sulfuric acid pretreatment. Losses of pentoses during hydrolysis were occasionally shown by lower pentose contents in hydrolysates that had received the pretreatment. Larger quantities of uronic acids were also extracted using the 72% sulfuric acid pretreatment. The larger amounts of sugars released by this pretreatment may be derived from complex carbohydrate-containing polymers or from carbohydrate adsorbed on minerals which are dissolved by the more concentrated acid. Cellulose, however, cannot be ruled out as a possible source (GUPTA and SOWDEN [1964]). CHESHIRE and MUNDIE [1966] examined the above procedure in more detail and showed that Ivarson and Sowden's procedure does not necessarily extract maximum monosaccharide contents. In a "sandy granitic loam" the amounts of pentoses, 6-deoxyhexoses, and hexoses estimated by Ivarson and Sowden's method were, respectively, 84, 87, and 97% of the maximum contents obtained by altering the duration of the 72% sulfuric acid pretreatment and the reflux in 1 N acid. Pretreatment for 16 hours with 72% sulfuric acid before hydrolysis with sulfuric acid for 5 hours was found to be the best compromise for estimating reducing sugars in their soil. The Ivarson and Sowden method extracted more than 95% of the anthrone-reactive materials in an Australian red-brown earth containing about 2.5% of total carbon (OADES [1967]). Conditions for maximum extraction of monosaccharides probably need to be worked out for each individual soil, as is the case with amino sugars (STEVENSON [1957a]) and will vary somewhat for each class of monosaccharides. Pentoses present as furanosides are particularly susceptible to hydrolysis, and FORSYTH [1950] showed that a large proportion of the pentoses could be removed from a soil polysaccharide by autohydrolysis in distilled water at 100°C.

The effect of a mild hydrolysis on soils is well illustrated by the results of LYNCH, OLNEY, and WRIGHT [1958] shown in Table 4. The relatively high proportions of pentoses and 6-deoxyhexoses are probably due to incomplete cleavage of polymers containing the three major hexoses: glucose, galactose, and mannose. Continuous monitoring of monomers released during hydrolysis could yield useful information on the constitution of some of the polysaccharides in soil.

Information currently available suggests that the optimum hydrolysis conditions for

Table 4. Monosaccharide composition of some soils, composts, and decomposing litter-soil mixtures.

Sample	Total organic matter	Galactose	Glucose	Mannose	Arabinose	Xylose	Ribose	Rhamnose	Fucose	Other sugars	Total neutral sugars	Uronic anhydride	Hexosamines	Authors
Sassafras loam	1.4	14.7	12.3	6.4	25.7	15.5	—	9.7	18.8	—	5.6	—	—	LYNCH, OLNEY, and WRIGHT [1958]
Sassafras loam + 5% lucerne meal	1.9	11.1	9.3	4.1	30.5	11.8	—	10.1	22.9	—	4.8	—	—	
Cultivated Pocomoke	4.8	19.3	21.5	5.8	17.6	13.1	—	9.7	15.7	—	2.4	—	—	
Virgin Pocomoke	4.4	16.9	23.8	4.7	21.6	10.3	—	8.7	14.1	—	2.3	—	—	
Compost (11 months)	58.9	16	54	12	8	10	—	—	—	—	5.0	5.4†	1.9	PARSONS and TINSLEY [1961]
Compost (48 months)	—	17	48.6	8.6	8.6	8.6	—	—	—	—	3.5	6.1†	1.6	
Sandy loam	3.5	21.8	29.9	19.5	13.8	11.5	—	3.4	—	—	3.7	16.6†	3.5	
Meadow soil	17.6	21.0	31.0	18.0	13.0	13.0	—	4.0	—	—	10.0	13.3†	4.0	
Peat	31.9	24.4	28.7	16.4	12.4	12.4	—	5.7	—	—	3.5	11.9†	1.6	
Podzol A_0	39.7	22.3	46.1	16.3	7.6	4.0	—	3.7	—	—	6.7	11.1†	1.2	
Podzol B	2.3	21.1	30.1	22.6	11.6	13.1	—	1.5	—	—	6.6	14.3†	1.1	
Brown forest soil	15.6	15.1	40.7	12.4	9.7	8.6	1.2	7.7	—	4.6	10.9	1.7	5.2	SOWDEN and IVARSON [1962a]
Podzol	16.4	13.6	44.1	10.2	11.6	9.7	1.1	5.7	—	4.2	12.6	2.2	3.8	
Dark-brown soil	9.6	17.2	36.0	15.3	10.7	11.1	0.9	5.3	—	3.5	14.6	1.7	5.2	
Black soil	10.6	20.1	36.5	12.5	10.1	10.6	1.1	5.3	—	3.7	11.9	1.3	4.5	
Coniferous litter podzol mixture 0 days	67.8	8.0	57.4	14.9	9.6	7.4	0.1	1.6	—	1.1	27.7	2.5	—	SOWDEN and IVARSON [1962b]
25 "	79.1	6.8	58.6	12.3	10.9	8.2	0.3	1.8	—	1.4	27.8	2.9	—	
53 "	70.0	10.6	53.6	15.6	7.8	8.4	0.3	2.2	—	1.1	25.6	2.9	—	
81 "	78.0	8.1	55.9	15.6	8.5	8.5	0.3	1.8	—	0.9	27.1	2.2	—	
137 "	75.1	8.1	53.4	15.7	9.9	9.0	0.3	2.2	—	1.3	29.7	2.4	—	
165 "	70.9	6.7	53.9	16.1	9.8	9.8	0.5	1.6	—	1.6	27.2	2.8	—	
1200 "	62.4	14.0	47.3	12.9	11.8	9.9	0.3	3.3	—	1.1	14.9	1.6	—	

Soil														Reference
Deciduous litter-brown forest soil mixture														
0 days	66.5	7.1	60.1	3.6	6.5	17.9	0.5	3.6	—	0.6	25.3	4.8	—	GUPTA, SOWDEN, and STOBBE [1963]
25 ″	57.0	6.6	59.6	4.8	6.0	19.3	0.3	2.4	—	0.6	23.3	5.8	—	
53 ″	59.0	7.5	54.9	5.3	6.0	21.1	0.5	3.0	—	1.5	21.7	4.9	—	
81 ″	56.8	7.8	53.9	4.7	5.5	22.7	0.2	3.9	—	1.6	22.5	3.9	—	
137 ″	55.1	9.0	46.1	7.9	10.1	19.1	0.8	4.4	—	2.2	16.2	2.9	—	
165 ″	50.8	11.0	44.0	6.6	11.0	18.7	0.6	5.5	—	2.2	17.9	3.1	—	
1200 ″	35.9	12.0	44.2	9.3	12.0	12.0	1.2	6.9	—	2.3	12.0	1.4	—	
Orthic podzol O	73.2	15	54	15	5	4	3*	4	—	+	16.1	—	—	
B₂	13.9	16	35	16	9	9	6*	9	—	—	8.0	—	—	
Podzol with permafrost O₁	80.2	15	42	15	10	8	3*	7	—	+	11.1	—	—	
O₂	59.7	13	50	14	10	7	3*	4	—	+	10.3	—	—	
B₂	1.8	9	30	17	13	12	9*	10	—	—	9.0	—	—	
C₁	2.1	16	22	20	8	11	10*	13	—	+	10.7	—	—	
Orthic grey wooded O	52.4	15	44	12	12	9	4*	5	—	—	16.4	—	—	
A₁	4.1	14	48	13	10	8	2*	5	—	—	12.6	—	—	
A₂	1.1	15	37	14	13	10	4*	7	—	—	12.3	—	—	
B₂	2.5	19	33	14	13	8	4*	9	—	—	10.6	—	—	
C₂	1.3	17	28	15	17	11	3*	13	—	+	7.1	—	—	
Solodized solonetz A₁	7.2	16	28	18	18	9	4*	8	—	+	13.5	—	—	
B₂	1.9	14	24	18	18	10	8*	8	—	—	12.3	—	—	
Orthic black A₁	6.2	14	36	16	15	8	3*	8	—	—	9.7	—	—	
B	2.0	15	34	14	12	8	5*	12	—	—	5.9	—	—	
C	0.4	20	26	20	12	10	4*	8	—	—	5.0	—	—	
Orthic dark grey gleysolic Ap	2.4	15	31	15	15	11	5*	8	—	+	15.6	—	—	
As above, heavily manured	4.0	14	28	20	15	10	5*	8	—	—	15.6	—	—	
Sandy granitic loam	—	13.8	40.8	12.1	10.8	13.3	1.4	5.6	2.3	—	—	—	—	CHESHIRE and MUNDIE [1965]
Orthic podzol L–H	79.4	14.9	54.0	14.9	5.0	3.7	3.1	4.3	—	—	16.1	2.6	—	GUPTA and SOWDEN [1965]
Northern podzol F	—	14.8	41.7	14.8	10.4	7.8	2.7	5.4	—	—	11.5	3.0	—	
Orthic grey wooded L–H	51.2	15.2	43.9	11.6	12.2	8.5	3.7	4.9	—	—	16.4	—	—	
Brown forest Aₕ	—	14.7	31.9	12.9	14.9	10.4	4.8	10.4	—	—	4.0	—	—	

* = ribose plus fucose.
† = by decarboxylation.
+ = present but not determined.

release of hexuronic acids from soils are probably the same as those described above for the neutral sugars.

The hexosamines, on the other hand, require more stringent hydrolysis conditions, as it has been shown (STEVENSON [1957a]) that maximum yields are obtained by use of 6 N hydrochloric acid. The optimum duration of hydrolysis varied from one soil to another, but in general, soils with high organic matter contents required longer hydrolysis times.

2. Total Reducing Sugars

WAKSMAN and HUTCHINGS [1935] estimated total reducing sugars in soil hydrolysates by the Hagedorn–Jensen procedure, which is based on the reduction of ferricyanide. SINGH and BHANDARI [1963] have used the FOLIN–WU [1920] method, which employs reduction of alkaline copper solutions. However, many organic compounds and in particular soil organic compounds other than sugars possess reducing properties, and results may always be erroneously high (HEWITT [1938], PHILIPSON [1943], DEUEL and DUBACH [1958a, b], BERES and KIRALY [1959]). According to MEHTA, DUBACH, and DEUEL [1961], soil organic materials do not interfere with the Somogyi method for the determination of reducing sugars, and PARSONS and TINSLEY [1961] have used a modification of this procedure to estimate reducing sugars in soil extracts and hydrolysates. The Somogyi procedure is based on the measurement of reducing substances before and after enzymic breakdown of sugars by yeast. The results are much higher than those obtained for total sugars by paper chromatography (PARSONS and TINSLEY [1961]).

Recently, CHESHIRE and MUNDIE [1966] have automated the alkaline ferricyanide procedure as described by HOFFMAN [1937] for use on soil hydrolysates. This procedure does not appear to give the high results obtained by the original Hagedorn–Jensen procedure, and figures obtained on soil hydrolysates are comparable with those given by the anthrone colorimetric procedure.

Total soil carbohydrate estimations have also been attempted using the phenol-concentrated sulfuric acid method of DUBOIS et al. [1956] on soil hydrolysates (ACTON, PAUL, and RENNIE [1963]). The phenol-concentrated sulfuric acid reagent reacts with all the reducing sugars and their derivatives, with the exception of 2-amino-2-deoxy sugars. α- and β-keto acids and aliphatic aldehydes and ketones also give a yellow color with this reagent, but amino acids and proteins do not interfere (MONTGOMERY [1961]). When used on soil hydrolysates the phenol-concentrated sulfuric acid method indicates that the amount of carbohydrate carbon present is 150 to 200% of that estimated using the anthrone method (ACTON, PAUL, and RENNIE [1963], OADES [1967]). It is possible that pentoses could account for some of this increase along with the other nonsaccharide carbohydrates mentioned. However, at the wavelength used for optical density measurements in the phenol-concentrated sulfuric acid method (490 nm), nonspecific absorption makes a large contribution to the optical density. This absorption is intensified by addition of concentrated sulfuric acid. Reaction of noncarbohydrates with phenol to give yellow products can also occur.

Another approach to the measurement of the polysaccharide content of soils is the gravimetric estimation of "microbial gum" precipitated from fulvic acid fractions by means of acetone (RENNIE, TRUOG, and ALLEN [1954], CHESTERS, ATTOE, and ALLEN [1957], SALOMON, [1962], ACTON, PAUL, and RENNIE [1963]). However, this method cannot be recommended as a measure of total carbohydrate in the soil, as the proportion of carbon extracted by 0.5 N sodium hydroxide varies widely (1/3 to 2/3) from soil to soil and the microbial gum has initially about 60% ash, which can be reduced to about 20% by purification. In addition, the chemical composition varies considerably from soil to soil and it in fact contains < 10% of the soil carbohydrate (ACTON, PAUL, and RENNIE [1963]).

3. *Hexoses*

One of the methods most commonly used to give a guide to total carbohydrate contents of soils is the anthrone procedure. TOENNIES and KOLB [1964] have presented an appraisal of this reagent for sugar estimations. When carbohydrates are heated in sulfuric acid with anthrone, a series of reactions occurs. Polysaccharides are hydrolyzed to monosaccharides, which then react to form furfural derivatives. These condense with anthrone to give a green complex that slowly decomposes with further heating. In general, noncarbohydrate materials do not interfere, and colors produced by different saccharides have similar absorption spectra, but the rate of appearance of color, its maximum intensity, and its rate of fading vary widely. Colors with pentoses develop rapidly and fade rapidly and equimolar intensities are small compared with hexoses (BAILEY [1958], JOHANSON [1953]). The same is true of methylhexoses, except that the intensity is initially much higher than for hexoses. The color with hexoses develops slowly but also fades slowly. In general, the anthrone procedure based on glucose standards will give a good measure of hexoses present in a mixture, but will underestimate pentoses, hexuronic acids, and hexosamines. When the anthrone procedure was applied directly to a soil extract, it was found to be of little use because of interference (MACAULAY INSTITUTE 1953/54). However, MACLEAN and DELONG [1956] apparently used the reagent successfully to estimate hexoses in leaf extracts and the surface horizons of soils, and since then a number of procedures based on anthrone have been used on soil hydrolysates. BRINK, DUBACH, and LYNCH [1960] suggested the use of 0.2% anthrone after hydrolysis with 3 N sulfuric acid and neutralization with calcium carbonate. No serious interference from the noncarbohydrate materials occurred except with a prairie soil, when the difference in the results obtained using anthrone before and after neutralization of the hydrolysate with calcium carbonate was large. These authors suggested that the use of 0.2% anthrone gives a measure of hexoses, methylhexoses, and 6-deoxyhexoses only, and thus provides a reliable relative measure of carbohydrate content for the purpose of comparison of soils. Although equimolar quantities of the various carbohydrate constituents of soils do not give equal color intensities with anthrone, paper chromatography has shown the spectrum of sugars in different soils and different soil fractions to be similar, so that their suggestion would seem to be well founded. ACTON, PAUL, and RENNIE [1963] and GRAVELAND and LYNCH [1961] have also used this procedure.

During an examination of the use of anthrone and benzidine for the estimation of sugars in sulfuric acid hydrolysates of Australian red-brown earths it has been found that neutralization with barium hydroxide removed the yellow-brown materials by precipitation. However, at the same time sugars were precipitated with the yellow interfering materials (OADES [1967]). This loss was sometimes quite large, as was shown by both benzidine and anthrone colorimetric procedures. Similar losses presumably occurred in the studies on the prairie soil described by BRINK, DUBACH, and LYNCH [1960]. This phenomenon does not occur when the unknown yellow materials are absent. Quantitative recoveries of sugars from normal sulfuric acid can be obtained by the use of the barium hydroxide neutralization method, which suggests that coprecipitation or perhaps formation of acetal linkages between the sugars and interfering substances normally occurs in the soil hydrolysates.

This problem was overcome by removing the interfering materials by sorption on a charcoal pad. The hydrolysates were leached through a 1:1 mixture of activated charcoal and celite, and the sugars completely eluted with 8.5 N acetic acid and estimated colorimetrically using anthrone. Provided that standards were prepared in equivalent concentrations of acetic acid, no interference was encountered except with hydrolysates from podzol B horizons which were not completely clarified. However, the amount of interference above 600 nm was negligible.

Hexoses have also been estimated by the chromotropic acid procedure, as described by DERIAZ [1961] and IVARSON and SOWDEN [1962]. Glucose oxidase has been used as a specific procedure for glucose (CHESHIRE and MUNDIE [1965]), and these authors (CHESHIRE and MUNDIE [1966]) have also used the orcinol procedure of MILLER [1960].

4. Pentoses

Estimation of pentoses in soils is not as well documented as estimation of other groups of sugars. In the early 1900's pentosans were determined quantitatively in organic matter by hydrolysis and transformation of the monomer to furfuraldehyde by boiling in 12% hydrochloric acid (WAKSMAN [1936]). The furfuraldehyde was then precipitated using phloroglucinol. However, this reagent is not specific for furfuraldehyde and gives precipitates with other aldehydes. More recently, the orcinol-ferric chloride procedure of Bial has been used to estimate pentoses in soils (MEHTA and DEUEL [1960], THOMAS and LYNCH [1961], CHESHIRE and MUNDIE [1965, 1966]) and also the aniline acetate procedure of TRACEY [1950] WEN and CHEN [1962], IVARSON and SOWDEN [1962]. According to MEHTA and DEUEL [1960], other aldehydes present in the hydrochloric acid distillate of soils do not interfere in the furfural determination by the orcinol procedure. However, the orcinol procedure is not specific. 6-Deoxyhexoses, hexuronic acids, heptoses, trioses, and in high concentrations D-mannose and D-galactose produce a green color, with absorption maxima at about 660 nm. Also, carbohydrates other than pentoses may be transformed to furfural, and a significant source of error present in such measurements is the furfural derived from uronic acids. This can be overcome by removing the uronic acids, prior to the colorimetric estimation, by means of an ion exchange resin; for example, THOMAS and LYNCH [1961] used Amberlite IRA-400. The aniline acetate procedure is based on the formation of furfural but is not completely specific, since hydroxymethyl-furfural heated with acid produces small amounts of furfural.

An estimate of pentoses can also be obtained by estimation of furfural by the method of ADAMS and CASTAGNE [1948]. This method involves distilling the furfural from 12% hydrochloric acid containing sodium chloride. The pentoses are known to yield about 90% of the theoretical yield of furfural (MEHTA and DEUEL [1960]). This is similar to the recovery of pentoses after acidic hydrolysis procedures (SAEMAN et al. [1954]). IVARSON and SOWDEN [1962] have compared the furfural determined by the use of this procedure with a calculated furfural figure derived from separate determinations of pentoses and uronic acids. Pentoses were estimated using aniline acetate after removal of uronic acids by passage through Dowex 1. The uronic acids were then estimated using carbazole after elution with an ammonium chloride–ammonium hydroxide buffer. In general, there was good agreement between the determined and the calculated furfural figures. DUBACH and LYNCH [1957] have also determined furfural contents of fulvic acid fractions of soils and found that, except for podzol B horizons, the determined furfural figure was always higher than that calculated from a uronic acid content as estimated using carbazole. IVARSON and SOWDEN [1962] have shown that over half the furfural determined in soil hydrolysates is derived from pentoses. The remainder arises from uronic acids (vide infra).

CHESHIRE and MUNDIE [1965, 1966] have used the cysteine-sulfuric acid method (DISCHE [1962]) for estimation of 6-deoxyhexoses.

5. Uronic Acids

The decarboxylation method of LEFEVRE and TOLLENS [1907] was applied to the determination of uronic acids in soils by SHOREY and MARTIN [1930] and since then has been widely used in organic matter studies (NORMAN and BARTHOLOMEW [1943], FULLER [1947], KOSAKA and HONDA [1957], PARSONS and TINSLEY [1961]). Results obtained using this method indicated that in some podzol horizons almost half the carbon present was apparently present in uronic acid moieties. Values obtained for other soils indicated that 10 to 30% carbon was present in this form. BREMNER [1950c] criticized the results obtained on the basis of the great discrepancy with the small amounts of uronic acid containing polysaccharides which could be isolated from soils. Subsequently, DEUEL, DUBACH, and BACH [1958], DEUEL and DUBACH [1958a, b], and DEUEL, DUBACH, and MEHTA [1960] have shown that the decarboxylation procedure is not suited for the determination of uronic acids in soil organic materials, partially decomposed plant materials, and even some fresh plant materials. Decarboxylation was found to be a general property of colored humic materials and preparations of humic materials known to be free of uronic acids and sugars evolved carbon dioxide at 70°C, even in a neutral medium. Uronic acids will not readily decarboxylate under these conditions. In addition, DUBACH [1958] found that furfural formation from such carbohydrate-free preparations was far less than would be expected if the uronic acid content as determined by decarboxylation was correct.

The determination of uronic acids in soil hydrolysates by means of carbazole was first done by LYNCH, HEARNS, and COTNOIR [1957]. The uronic acids were estimated in a classical fulvic extract of soil by a modification of the method of STARK [1950]. To avoid interference by ferric iron it was removed by reduction with stannous chloride. Paper chromatography showed that only about 10% of extracted uronides was precipitated with the humic acid. However, the efficiency of the extraction of uronides using 0.5 N sodium hydroxide is not known. DUBACH and LYNCH [1959] compared the Lefevre and Tollens decarboxylation procedure with the carbazole procedure as described above. Furfural was also determined. The decarboxylation procedure gave much higher apparent uronic acid figures than carbazole (by a factor of 10 for the fulvic extract of a podzol B horizon). However, the determined furfural contents and relative carbohydrate contents as estimated using anthrone agreed reasonably well with uronic acids as estimated by carbazole. GRAVELAND and LYNCH [1961] modified the carbazole procedure by addition of a pretreatment with normal hydrofluoric–hydrochloric acid. With one exception an increased yield of uronic acids resulted. The decarboxylation procedure in general showed large increases in the proportions of apparent uronic acids present in organic matter in the lower soil horizons. The carbazole method indicated only small increases with depth.

In spite of the improved extraction of uronic acids using normal hydrofluoric–hydrochloric acid extraction prior to 0.5 N sodium hydroxide, there is no means of testing the efficiency of such procedures unless the uronic acids in the soil residues can be estimated. IVARSON and SOWDEN [1962], DORMAAR and LYNCH [1962], and GUPTA and SOWDEN [1965] have used acid hydrolysis prior to uronic acid estimation by means of carbazole and the phenol-concentrated sulfuric acid method. Uronic acids may be isolated from soil and plant hydrolysates by means of Dowex-1 resin in the chloride form (IVARSON and SOWDEN [1962], GUPTA and SOWDEN [1965]). After the neutral sugars had been washed through with distilled water, the uronic acids were eluted with 0.25 N ammonium chloride adjusted to pH 10 with concentrated ammonia. The separation of uronic acids from neutral sugars before application of the carbazole colorimetric method is well worthwhile, since neutral sugars in sufficiently large amounts also produce a color with carbazole (*e.g.*, 100 γ glucose will give the equivalent extinction of 8 γ glucuronic acid). The hydrolysis procedure appears to be the most satisfactory method available at present, even though less than half of a uronic acid added to hydrolysates can be recovered. This is true of the estimation of uronic acids in any biological materials and not soil alone (see PERRY and HULYALKAR [1965]).

6. *Hexosamines*

In contrast to other groups of sugars, the hexosamines were not identified in soils until Bremner showed the presence of amino sugars in soil hydrolysates by colorimetric methods and by paper chromatography (BREMNER [1949, 1950a, b]). Subsequently, a number of workers have identified glucosamine and galactosamine in soil hydrolysates by paper chromatography (STEVENSON [1954, 1956a, b], BREMNER [1958], YOUNG and MORTENSEN [1958], SOWDEN [1959]). BREMNER [1958] has isolated these compounds as their hydrochlorides. The presence of the N-acetyl derivatives of glucosamine and galactosamine has been reported (STEVENSON [1954, 1956a], YOUNG and MORTENSEN [1958]). However, their detection has not been confirmed in other work (BREMNER [1958], STEVENSON [1960]). No other amino sugars have been detected in soil hydrolysates, and quantitative estimation of glucosamine and galactosamine after paper or column chromatography by the method of MOORE and STEIN [1951] has shown that these two sugars account for almost all amino sugar as estimated colorimetrically (BREMNER [1958], SOWDEN [1959]).

Several methods have been used for estimations of hexosamines in soils. In general, there is good agreement on the optimum conditions for hexosamine release from soils. Maximum yields have been given by hydrolysis in 6 N hydrochloric acid at 100°C for 6 to 9 hours. Soils high in organic matter may require a longer period. Losses of both sugars occur during the hydrolysis and values are usually multiplied by a correction factor of about 1.25 (BREMNER and SHAW [1954], STEVENSON [1957a]), although a value of 2.0 has also been used (SOWDEN and IVARSON [1959]). For critical studies the release with time can be measured so that the optimum duration of treatment can be defined and the loss during hydrolysis can be corrected by extrapolation back to zero hydrolysis time. A constant rate of destruction must be reached before such an extrapolation can be made (SOWDEN [1959]). The correction factor given by this procedure was 1.25 and 1.26 for glucosamine and galactosamine, respectively.

BREMNER [1949, 1955], BREMNER and SHAW [1954], and SOWDEN [1959] have measured ammonia released by alkaline deamination as a measure of amino sugars. While this method appears reliable, the Elson–Morgan colorimetric method has been more popular (BREMNER and SHAW [1954], STEVENSON [1954, 1956a, 1957a, b], BREMNER [1958], SOWDEN [1959], SINGH and SINGH [1960], SALOMON [1963]). SOWDEN [1959] and SOWDEN and IVARSON [1959] have used the method of EASTOE [1954] after the Moore and Stein procedure to separate hexosamines from other amino compounds. Results were similar to those given by the Elson–Morgan reagent. The Elson–Morgan method is based on the condensation of amino sugars with acetylacetone by heating in alkaline conditions to give several chromogens that will react with N, N-dimethyl-p-aminobenzaldehyde in acid conditions to form red complexes. Interference from large amounts of hexoses and amino acids, particularly L-lysine, can occur; otherwise the procedure is specific for hexosamines.

The proportion of the soil nitrogen present as hexosamines appears constant in surface soils and ranges from 5 to 10% (BREMNER and SHAW [1954], STEVENSON [1957a, b, c], SOWDEN [1959], SINGH and SINGH [1960], CHENG and KURTZ [1963], SALOMON [1963], KEENEY and BREMNER [1964]). In lower horizons the proportion increases and may be as high as 25% of the total nitrogen in clay rich B horizons of acidic soils. No accumulation in subsurface layers was noted in Indian soils with neutral to high pH (SINGH and SINGH [1960]). If the hexosamine contents are referred to a carbon base, it becomes apparent that they make a large contribution to the total saccharide content of a soil.

BREMNER [1958], SOWDEN [1959], and SOWDEN and IVARSON [1959] have shown that the ratio of glucosamine to galactosamine varies from 1.2 to 4.6. The highest ratios are associated with acidic forest litter and podzolic soils, where fungi play a more significant role than in alkaline environments. BREMNER [1958] and SOWDEN [1959] found that glucosamine and galactosamine accounted for almost all the hexosamine in soil hydrolysates as determined colorimetrically. Although other amino sugars were not detected, the possibility that several others were present cannot be excluded.

7. Total Saccharide Content of Soils

Although the colorimetric procedures described for estimation of hexoses, pentoses, uronic acids, and hexosamines in soil hydrolysates may be considered well tried and reliable, particularly for survey purposes, they cannot be regarded as accurate for two main reasons. (1) None of the colorimetric procedures described gives the same color intensity with equimolar concentrations of different sugars. This is particularly true of the anthrone and carbazole methods. (2) None of the hydrolysis conditions for the separate groups of sugars is entirely satisfactory, and any hydrolysis is a compromise between a sufficiently high acid concentration, and temperature and time of hydrolysis to give good glycosidic cleavage and sufficiently mild conditions to prevent breakdown of the monomers during hydrolysis. The pentoses and uronic acids are particularly difficult in this respect.

Estimates of total carbohydrate contents have been obtained by summation of the different groups of sugars as estimated colorimetrically (IVARSON and SOWDEN [1962]). ROULET, MEHTA, DUBACH, and DEUEL [1963] used uronic acids × 2.5 plus pentoses × 5 as a measure of total carbohydrate. A more satisfactory determination of total carbohydrate content of soils is given by summation of individual sugars after either paper or column chromatography. PARSONS and TINSLEY [1961] measured monosaccharides as reducing sugars by a modified NELSON [1944] procedure after separation by paper chromatography. SOWDEN and IVARSON [1962a, b], LYNCH, OLNEY, and WRIGHT [1958], and LYNCH, WRIGHT, and OLNEY [1957] used the Dubois method after paper and cellulose column chromatography. GUPTA and SOWDEN [1965] used the Dubois method after paper chromatography, and CHESHIRE and MUNDIE [1966] used alkaline ferricyanide for all sugars, anthrone for hexoses, orcinol-FeCl₃ for pentoses, cysteine-sulfuric acid for 6-deoxyhexoses, and glucose oxidase for glucose after separation of monomers by paper chromatography. The results of this work are collected in Table 4, and show clearly the overall changes in monosaccharide composition from plant material through various stages of decomposition to the more stable situation in a soil.

II. Amounts of Saccharides in Soils

1. Free Saccharides

Until paper chromatographic techniques for the separation and identification of sugars were applied to soils, little was known of the nature of saccharides in the soil solution. ALVSAKER [1948] determined reducing sugars in cold-water extracts of pine forest soils in Norway and estimated that the sugars constituted less than 1 percent of the total soil organic matter. This figure is probably an overestimate, as it is well known that a variety of organic materials in soils possess reducing properties. FORSYTH [1948] identified glucose, xylose, rhamnose, galactose, and possibly hexosamines (positive Elson–Morgan test) in hot-water extracts of Scottish soils. Extraction of the uppermost layer of pine forest soil (the F layer) with ether and then water below 20°C yielded a variety of sugars, as shown by chromatography after ethanol precipitation of polysaccharides (ALVSAKER and MICHELSON [1957]). Arabinose, fructose, and ribose were detected in addition to the sugars found by Forsyth. The fructose was identified by X-ray analysis of its (2, 4-dinitrophenyl) hydrazone. Sugars and uronic acids have also been detected in solutions pressed out of peat (ALEKSANDROVA [1960], KRYM [1962]).

Since 1960 investigations have been extended to a range of soils by GROV [1963], NAGAR

[1962], and GUPTA and SOWDEN [1963]. Nagar found free glucose only in four widely different soils. Grov extended the work of Alvsaker and Michelson. Combined data for the pine forest soils are shown in Table 5. There is a decrease of free monosaccharides from the F layer

Table 5. Water-soluble monosaccharides in a pine forest soil expressed as percent of organic matter (105°C) (ALVSAKER and MICHELSON [1957], GROV [1963]).

Horizon	Galact-uronic acid	Galact-ose	Glucose	Mann-ose	Fruct-ose	Arabin-ose	Xylose	Ribose	Rhamn-ose	Total
F	—	0.02	0.22	—	0.035	0.04	0.03	0.001	—	0.346
H	+	Trace	0.189	—	0.010	0.027	0.004	0.008	Trace	0.238
A	—	Trace	0.182	Trace	—	0.037	Trace	Trace	—	0.219

downward, but the reverse is true of water-soluble polysaccharides. The fact that the method of handling of the soil samples is important for free sugar estimations was shown by GUPTA and SOWDEN [1963], who extracted a variety of soils with ethanol. A northern Canadian podzol yielded a range of sugars totalling 0.0245% of the organic matter, but other soils including a peat contained 0.0030% or less and glucose was the predominant sugar. Air-dried samples of peat reduced the free glucose content by half compared with storage at 2°C. Drying also decreased the free sugar content in other soils.

The greatest amounts of free sugars are associated with cool climatic conditions where surface litter is abundant. Free monosaccharides are difficult to detect in many soils, which is not surprising, considering the rapidity with which they are likely to be utilized by micro-organisms.

Another source of free carbohydrates with a transient existence in soils is exudates from plant roots. The presence of free carbohydrates in these exudates has been demonstrated by the examination of plants grown under sterile conditions. The subject has been reviewed by ROVIRA [1965]. A range of carbohydrates is released, including glucose, fructose, sucrose, xylose, maltose, rhamnose, arabinose, raffinose, and oligosaccharides. VANCURA [1964] has also detected galactose, ribose, deoxyribose and some other deoxy sugar, glucuronic and galacturonic acids, glucosamine, galactosamine, and two other amino sugars. Under non-sterile conditions, as in soil, these sugars would be less likely to be detected, as they form a very readily available substrate for microorganisms and are involved in the formation and maintenance of the rhizosphere, and probably also in the attraction of plant pathogens toward roots and seeds.

2. Total Reducing Sugars

The methods described above yielded figures which indicate that from 5 to 45% of the organic matter in soils is present as reducing sugars (Table 6). Generally the higher the carbon content of the sample, the higher the proportion of the reducing materials. However, the figures must be regarded as overestimates. MEHTA, DUBACH, and DEUEL [1961] claim that humic substances do not interfere in the Somogyi method for the determination of reducing sugars, but figures obtained by this procedure do not appear to bear any close relationship to estimates obtained by quantitative paper chromatography of monosaccharides in the same hydrolysates (PARSONS and TINSLEY [1961]).

Contents of microbial gum as precipitated from fulvic fractions using acetone cannot be regarded as useful indications of carbohydrate content because of difficulties inherent in the method, the high ash content, and incomplete extraction of carbohydrate.

Table 6. Examples of carbohydrate contents of soils determined as reducing sugars.

Soil	Organic matter content percent	Carbohydrate as reducing sugar percent organic matter	Method	Authors
Raw humus podzol				
Litter	96.17	33.95	Hagedorn–Jensen	WAKSMAN and HUTCHINGS [1935]
F layer	78.19	20.18		
H layer	74.20	14.42		
A_1 horizon	18.46	8.00		
A_2 horizon	4.33	10.30		
B_1 horizon	8.36	8.70		
B_2–C horizon	5.94	6.22		
5 Chernozems	1.1–2.29	4.28–15.81	Hagedorn–Jensen	WAKSMAN and HUTCHINGS [1935]
28 Pine forest podzols	< 10	5.2–21.6	Hagedorn–Jensen	ALVSAKER [1948]
38 Pine forest podzols	> 10	11.8–45.2		
7 Rajasthan soils	< 5	9.8–19.3	Folin–Wu	SINGH and BHANDARI [1963]

3. *Hexoses*

Hexoses represent the major portion of the soil carbohydrates (Table 4) and are also the principal group of materials in soils which react with anthrone. Thus, figures obtained using the anthrone procedure provide an estimate of hexose contents of soils as well as the most generally useful indication of comparative total carbohydrate contents. From 5 to 20 % of soil organic matter is present as such compounds (Table 7). Studies on soil profiles suggest that there is an increase in the proportion of organic matter present as carbohydrate with increase in depth in the profile. However, such trends must be viewed with caution, as many of the organic matter figures are based on carbon estimations by wet oxidation with dichromate, which becomes a less efficient measure of organic carbon in subsoils—and the figures do not appear to have been appropriately corrected.

In any one soil type the proportion of the organic matter present as carbohydrate remains reasonably constant. SALOMON [1962], using the anthrone procedure of BRINK, DUBACH, and LYNCH [1960], estimated that the carbohydrate contents in the Bridgehampton silt loam accounted for 8.8 to 11.8 % of the soil organic matter irrespective of rotation or aggregate size. Similar results were obtained by WEBBER [1965], who showed that in the Haldimand clay under continuous corn or a rotation including a pasture, the organic matter content varied from 3.8 to 5.5 % but the proportion of organic matter present as carbohydrate remained relatively constant between 9.0 and 10.9 %. More recently, OADES [1967] has found that the proportion of carbohydrate is higher under permanent pasture than under continuous arable cropping. The Urrbrae fine sandy loam, a red-brown earth, contained 4.3 % organic matter under permanent pasture and 1.9 % organic matter under continuous wheat-after-fallow. The corresponding proportions of carbohydrate were 9.4 and 7.7 %, respectively. However, if partially decomposed plant materials were removed from the soils (GREENLAND and FORD [1964]) prior to carbohydrate estimation, it was found that irrespective of rotation a relatively constant proportion of the organic matter remaining was present as carbohydrate (6.6 to 7.6%).

4. Pentoses

The amounts of pentoses in soils as determined by colorimetric procedures represent less than 5% of the total organic matter, but tests for pentoses are always positive. Estimates obtained by boiling with 12% HCl are much higher (up to 28%) than those presented in Table 8 (SHOREY and LATHROP [1910], ALVSAKER [1948]), but as discussed above are unreliable.

5. Uronic Acids

Most figures for uronic acids obtained by the carbazole procedure fall in the range of 1 to 5% of the organic matter (Table 9). These estimations are far from satisfactory and must be regarded as minimal amounts.

The Lefevre–Tollens decarboxylation procedure led to apparent uronic acid contents that represented 6 to 40% of the organic matter (SHOREY and MARTIN [1930], NORMAN and BARTHOLOMEW [1943], FULLER [1947], KOSAKA and HONDA [1957], PARSONS and TINSLEY [1961]). In general this proportion increased with depth in the profile, but such figures mean little in terms of actual uronic acid contents.

Table 7. Examples of carbohydrate contents of soils determined by colorimetric procedures.

Soil	Organic matter content percent	Saccharides as percent organic matter	Method	Authors
Silt loam	3.2	9.1	Anthrone	BRINK, DUBACH,
Sandy loam	4.8	5.8	,,	and LYNCH [1960]
Prairie	5.2	11.0	,,	
Podzol A_1	7.9	8.9	,,	
Podzol A_2	1.4	11.6	,,	
Podzol B	10.0	5.1	,,	
Solodized solonetz				
A_1 horizon	9.8	9.6	Anthrone	GRAVELAND and
A_2 horizon	2.1	12.4	,,	LYNCH [1961]
B_2 horizon	1.4	19.0	,,	
B_{Sa} horizon	0.5	20.0	,,	
Solod				
A_1 horizon	6.9	10.6	,,	
A_2 horizon	1.7	15.3	,,	
B_2 horizon	1.6	14.7	,,	
B_3 horizon	1.0	17.5	,,	
Orthic				
A_1 horizon	10.5	9.5	,,	
B_2 horizon	1.7	17.0	,,	
B_3 horizon	1.4	16.0	,,	
Brown forest	14.9	8.3	Anthrone	IVARSON and
Podzol	18.7	8.0	,,	SOWDEN [1962]
Dark brown	8.6	12.7	,,	
Black	10.6	7.2	,,	
Red-brown earth	4.3	13.1	Phenol-H_2SO_4	OADES [1967]
,, ,, ,,	4.3	8.7	Anthrone	
,, ,, ,,	4.3	5.8	Benzidine	

6. *Hexosamines*

Hexosamines represent 5 to 10% of the total nitrogen contents of soils, and 2 to 5% of the total carbon content.

7. *Total Sugars*

The data in Table 4 show that the amount of glucose decreases during decomposition of plant material in the soil and that a concomitant increase in the proportions of ribose, rhamnose, and other sugars, which are mainly high Rf sugars, occurs. However, there is no absolute increase in the amounts of such sugars and it is not possible from these data to decide whether this increase in proportion is due to microbial synthesis, to greater stability to breakdown, or both.

D.   Isolation and Properties of Soil Polysaccharides

I. Extraction of Polysaccharides from Soils

The estimation of soil carbohydrate by hydrolytic procedures as described above and identification of individual monosaccharides yields little information about the manner in

Table 8.   Pentose contents of soils.

Soil	Organic matter content percent	Pentose as percent organic matter	Method	Authors
Solod				
A$_1$ horizon	6.9	5.3	Orcinol-FeCl$_3$	THOMAS and
A$_2$ horizon	1.7	6.0	,,	LYNCH [1961]
B$_2$ horizon	1.6	4.8	,,	
Orthic				
A$_1$ horizon	10.5	4.2	,,	
B$_2$ horizon	1.7	4.0	,,	
Solodized solonetz				
A$_1$ horizon	9.8	4.8	,,	
Brown earth	5.4	3.6 4.2*	Orcinol-FeCl$_3$	MEHTA and
Podzol	2.1	3.1 4.6*	,,	DEUEL [1960]
Podzol	10.2	3.1 3.9*	,,	
Podzol	3.2	1.6 3.1*	,,	
Brown forest	14.9	2.2 1.9†	Aniline acetate	IVARSON and
Podzol	18.7	2.2 1.8†	,,	SOWDEN [1962]
Dark brown	8.6	4.4 3.2†	,,	
Black	10.6	2.7 1.8†	,,	
Chestnut earth	3.8	3.6	Aniline acetate	WEN and CHEN
Black earth	3.5	2.2	,,	[1962]
Black earth under				
grass	5.2	3.6	,,	
Paddy soil	1.7	7.8	,,	
Red loam paddy soil	1.2	11.9	,,	
Red loam	1.5	7.8	,,	
Red loam (baked?)	6.0	5.2	,,	

* By the phloroglucinol precipitation procedure.
† Furfural by method of ADAMS and CASTAGNE [1948].

which the sugars are linked. From the large numbers of saccharides identified in soil hydrolysates it is obvious that a complex of polysaccharides is present in soils, as it is well known that most naturally occurring polysaccharides do not contain more than two or three different monosaccharides.

To obtain information about the soil polysaccharides they must be extracted from soil.

Table 9.   Uronic acid contents of soils.

Soil	Percent organic matter	Uronic acids as percent organic matter	Method	Authors
Podzol				
A_0 horizon	4.9	2.4	Hydrolysis of	DUBACH and
A_1 horizon	1.9	0.9	NaOH extract	LYNCH [1959]
A_2 horizon	1.1	0.8	and carbazole	
B	4.7	0.4		
Silt loam	1.1	1.2		
Sandy loam	1.2	0.6		
Prairie	1.3	1.2		
Solodized solonetz				
A_1 horizon	9.8	1.8	Hydrolysis	GRAVELAND and
A_2 horizon	2.1	3.3	after HF/HCl	LYNCH [1961]
B_2 horizon	1.4	3.0	and NaOH	
B_{sa} horizon	0.5	2.3	extraction	
Solod			and carbazole	
A_1 horizon	6.9	2.0		
A_2 horizon	1.7	5.1		
B_2 horizon	1.6	5.1		
B_3 horizon	1.0	4.7		
Orthic				
A_1 horizon	10.5	1.3		
B_2 horizon	1.7	3.4		
B_3 horizon	1.4	1.8		
Brown forest	14.9	1.8	Hydrolysis of	IVARSON and
Podzol	18.7	1.9	whole soil and	SOWDEN [1962]
Dark brown	8.6	1.9	carbazole	
Black	10.6	1.3		
Black solodic	13.6	3.9	Hydrolysis of	DORMAAR and
Degraded black	8.0	3.3	whole soil and	LYNCH [1962]
Grey wooded	8.2	1.2	carbazole	
Black solodized				
solonetz	2.0	3.9		
Northern podzol				
F layer	46.1	4.4	Hydrolysis plus	GUPTA and
H layer	34.3	5.8	chromatography	SOWDEN [1965]
Grey wooded			and	
L–H layer	30.1	5.0	phenol-H_2SO_4	
Ah or A_1 horizon	2.4	4.6		
Solodized solonetz				
Ah or A_1 horizon	4.1	5.6		
Orthic podzol				
Bh or B_2 horizon	8.0	5.4		

The ideal extractant should (1) be equally effective for all soils, (2) extract selectively carbohydrate materials, (3) be nondegradative, and (4) give a sufficiently complete extraction for the material extracted to be representative of the total. It is doubtful if any of the many extractants used fulfills one of these requirements, let alone all of them. The main aim of the majority of investigators has been to obtain relatively pure soil polysaccharides for chemical and physicochemical characterization. Extractants used can be listed as water, buffer solutions, complexing reagents, acids, alkalis, and organic solvents. Each of these groups of extractants has advantages and disadvantages, and at this stage none can be stated to be vastly superior to the others.

Water has been a popular extractant for soil polysaccharides because of the simplicity of extraction, the relatively low simultaneous extraction of humic materials, and the ease of subsequent purification. Most of the procedures involve extraction with hot water by shaking, or by use of Soxhlet's apparatus. What quantitative information there is shows that the procedure extracts up to about 2% of the soil organic matter, which could represent up to one-quarter of the total soil carbohydrates. The quantitative figures presented in Table 10 are little more than a guide, as the organic materials extracted are not all carbohydrate material and many of the extracts include a high proportion of inorganic materials due to widely different purification procedures. It is possible that during extraction with hot water, hydrolysis of soil polysaccharides takes place, as FORSYTH [1950] found that during autohydrolysis of a soil polysaccharide in distilled water on a boiling water bath for 24 hours about 90% of the arabinose and considerable amounts of ribose were released.

Extraction of soil polysaccharides with inorganic acids has two advantages: (1) very little humic material is extracted, and (2) subsequent purification procedures are simplified to some extent. In addition, BARKER et al. [1965] suggested that treatment with acid hydrolyzes bonds between polysaccharides and humic materials, so that separation of polysaccharide from these materials is simplified. Barker and his colleagues took precautions against acid hydrolysis of polysaccharides by carrying out the extraction in a cold room.

The use of formic acid has been investigated by PARSONS and TINSLEY [1961], and the proportions of organic matter (3.5 to 17.5%) extracted as polysaccharide are some of the highest recorded. The extraction procedure was a double reflux with 0.2 N lithium bromide in anhydrous formic acid. The hydrolytic effects of this extractant were not investigated and could be considerable.

Caustic soda, the classical extractant for soil organic matter, has proved useful for extraction of polysaccharides. It is considerably more efficient than most other extractants, and the risk of hydrolysis is small although degradation of some sugars is possible. One problem is the amount of humic materials simultaneously extracted. Up to 50% of soil carbohydrates have been removed from a red-brown earth by serial extractions with 0.5 N sodium hydroxide at room temperature (SWINCER, OADES, and GREENLAND [1968]). However, it is not as efficient in all soils.

Buffer solutions and complexing reagents have been used in attempts to obtain polysaccharides with a minimum of alteration from their natural state. In general, yields are low and it is impossible to know whether the material is representative of the total.

The use of organic solvents has recently been applied to the extraction of soil polysaccharides, and these methods may be of much value. Dimethylformamide was used to advantage by WHITEHEAD and TINSLEY [1964]. Yields can be quite high, but comparison of the molecular weight distribution of polysaccharides extracted by this reagent with those extracted by 0.5 N sodium hydroxide suggests that degradation may be occurring during the extraction.

Acetylation mixtures such as acetic anhydride with sulfuric acid, pyridine, or sodium acetate are also useful extractants of soil polysaccharides. The acid mixture is extremely efficient

Table 10. Yields of crude polysaccharides obtained from soils by various procedures.

Soil	Percent organic matter	Extractant	Method	Yield* Percent of soil	Yield* Percent of organic matter	Authors
Scottish	n.d.	Water	2 × 4 hr at 85°C	0.3	—	Duff [1952a, b]
Sphagnum peats	98.3–99.3	Water	3 × 3 hr at 60–65°C after benzene and methanol	1.1–2.0	1.1–2.0	Theander [1954]
Scottish peats	96.3–99.1	Water	1 × 4 hr at 100°C	n.d.	—	Black, Cornhill, and Woodward [1955], and
American	3.38	Water	1 × 4 hr + 1 × 6 hr at 85°C	0.05	1.45 (after purification with charcoal, 0.2)	Whistler and Kirby [1956]
American muck soil	n.d.	Water	1 × 24 hr in Soxhlet	—	1.8	Clapp [1957]
Fen peat	80	Water	n.d.	0.008	0.009	Barker et al. [1965]
Red-brown earth	4.3	Water	1 × 16 hr at 70°C	0.06	1.5	Swincer, Oades, and Greenland [1968]
Five forest soils	n.d.	pH 7 buffers with 5% KCl	10 hr at 20°C	0.01–0.039	—	Bernier [1958a]
Fen peat	80	7% disodium EDTA	n.d.	0.45	0.56	Barker et al. [1965]
Scottish peats	96.3–99.1	0.09 N HCl	1 hr at 60–70°C	0.15	0.15	Black, Cornhill, and Woodward [1955]
Five British soils	2.3–58.9	Anhydrous HCOOH containing 0.2 N LiBr	"2 hot extractions"	0.4–5.6	4.0–11.5	Parsons and Tinsley [1961]
Fen peat	80	0.6 N H₂SO₄	24 hr at 3°C	0.03	0.04	Barker et al. [1965]
Fen peat	80	Amberlite IRC H⁺-form	n.d.	0.13	0.17	Barker et al [1965]
Red-brown earth	4.3	1 N HCl	16 hr at 20°C	0.04	1.0	Swincer, Oades, and Greenland [1968]
Four Scottish soils	n.d.	0.5 N NaOH	48 hr at 15°C	—	1.32–1.94	Forsyth [1947a, b; 1950]
Wisconsin soils	0.8–2.6	0.5 N NaOH	3 hr at 20°C	0.02–0.25	—	Rennie, Truog, and Allen [1954]
Swiss soils	3.0–32.5	0.5 N NaOH	"Extracted at room temperature"	0.02–0.13	0.1–1.0	Dubach et al. [1955]
Saskatchewan soils	4.4–9.8	0.5 N NaOH	3 hr at 20°C	0.14–0.78	3.2–14.0	Acton, Paul, and Rennie [1963]
Red-brown earth	4.3	0.5 N NaOH	16 hr at 20°C	0.13	3.2	Swincer, Oades, and Greenland [1968]
Canadian soils	n.d.	Schweitzer's reagent	2 × 6 hr at 20°C	—	0.3–1.9	Gupta and Sowden [1964]
Fen peat	80	N-methyl-2-pyrrolidone	n.d.	0.03	0.04	Barker et al. [1965]
Fen peat	80	8 M urea	n.d.	0.46	0.57	Barker et al. [1965]
Fen peat	80	Dimethylformamide	n.d.	3.0	3.8	Barker et al. [1965]

* These data are for impure polysaccharides, some containing up to 80% of ash and other nonsaccharide materials.

n.d. = no data given.

but acetolysis takes place. Even so, this procedure may well yield information about the carbohydrates which are not extractable by other procedures, as the acetolysis is by no means complete and the polymer fragments obtained are larger than oligosaccharides. The other acetylation procedures are not as efficient but the risk of acetolysis is small.

II. Purification of Soil Polysaccharides

DUFF [1952a, b] and WHISTLER and KIRBY [1956] removed soil residues from hot-water extracts of soil by centrifugation. The supernatant was concentrated and then dialyzed against water. The crude polysaccharide preparation obtained by adding this solution to three volumes of acetone was further purified by reprecipitation from water with a mixture of alcohol and acetic acid (3/1). An essentially similar procedure was used by BERNIER [1958a] with phosphate buffer extracts. Solution and reprecipitation using ethanol were carried out until the supernatant was colorless.

MORTENSEN [1960], KEEFER and MORTENSEN [1963], and THOMAS [1963] precipitated humic materials from water extracts by acidification to pH 2.0 with hydrochloric acid. The polysaccharides were again precipitated using acetone. Removal of inorganic materials was attempted by desalting on anion and cation resins, before further purification as above. A more stringent attempt to remove inorganic materials from an extracted soil polysaccharide is described by MORTENSEN and SCHWENDINGER [1963], who treated a dialyzed pyrophosphate extract with a mixture 0.3 M with respect to both hydrofluoric and hydrochloric acids for 10 minutes at 4°C. This treatment removed clay materials which had yielded X-ray diffraction patterns in the original product. The treatment samples appeared amorphous to X-rays, and the infrared spectra were practically unchanged by the treatment. The effects of such a treatment on polysaccharides are unknown, but hydrofluoric acid may effect reorientation of hydroxyl groups in saccharides.

The first step in the treatment of acid extracts involves neutralization. BLACK, CORNHILL, and WOODWARD [1955] used 40% sodium hydroxide prior to dialysis and precipitation of the crude polysaccharide with ethanol. Reprecipitation from water with ethanol was used for further purification. BARKER et al. [1965] neutralized the extract with sodium bicarbonate and examined the polysaccharides in both the supernatant and the precipitate. The latter was brought into solution by 0.3 N hydrochloric acid.

PARSONS and TINSLEY [1961] used di-isopropyl ether to precipitate organic materials from anhydrous formic acid extracts of soil.

Where caustic soda has been used for polysaccharide extraction, the first step after removal of soil residues is acidification to pH 2–3 to precipitate humic acids. This treatment removes a large proportion of the extracted organic materials of high molecular weight which cannot be separated subsequently from high molecular weight polysaccharides by gel filtration. However, the manner in which the pH is lowered is important, particularly from the point of view of the amount of polysaccharide that is coprecipitated or even bound into the humic acid molecule. Normally, acid is added rapidly to give an almost immediate dark brown flocculent precipitate of humic acid, which may contain up to 12% of the total soil carbohydrate (ACTON, PAUL, and RENNIE [1963]). The humic acid described by SCHEFFER and KICKUTH [1961] does not have the same composition as humic acids extracted from better aerated soils. The large amount of carbohydrate present is unusual. Humic acid from a red-brown earth contained only about 2% of the total soil carbohydrate. This humic acid was obtained by passing a 0.5 N sodium hydroxide extract upward through a column of H^+ Dowex-50 (20–30 mesh). This procedure precipitates humic acids which are retained in the column and simultaneously removes cations from the preparation. The clear pale yellow fulvic extract obtained has a pH between 2 and 4. The amount of high molecular weight brown materials in the fulvic fraction is less than after rapid precipitation with acid. This aids in the purification of the high molecular weight polysaccharides by gel filtration (SWINCER, OADES, and GREENLAND [1968]).

High molecular weight polysaccharides in acid (ROULET *et al.* [1963], BARKER *et al.* [1965]) and caustic extracts (SWINCER, OADES, and GREENLAND [1968]) have been purified by removal of lower molecular weight humic materials by means of Sephadex gel filtration. ROULET *et al.* [1963] used Sephadex G-75 and found that acetic acid promoted sorption of humic materials and enabled a good separation to be obtained between polysaccharides and lower molecular weight humic materials. BARKER and his coworkers used Sephadex G-100 to remove humic materials from high molecular weight polysaccharides. Gel filtration has also been used to obtain molecular weight distributions of polysaccharides in various extracts in order to compare the degradation brought about by different chemicals during the extraction (SWINCER, OADES, and GREENLAND [1968]). Coarse grade G-25 enabled rapid separation of the major portion of the humic materials from the polysaccharides. In general, higher grades of Sephadex permitted better separation of polysaccharides from brown materials but pure white preparations have not yet been obtained. The possibility of a covalent link between the polysaccharides and humic materials must be considered, as 8 M urea has no effect on these complexes.

Filtration through charcoal has also been widely used as a purification procedure. FORSYTH [1947b] introduced the use of acid-washed animal charcoal for sorption of organic materials from classical fulvic acid extracts. The charcoal removed most of the organic matter and all the color from such extracts. The adsorbed materials were then eluted to give four fractions. Fraction A contained a colorless filtrate and 0.1 N HCl washings from the column. This fraction contained nearly all the simple organic substances, such as amino acids, purine bases, and sugars. This fraction gave positive tests using α-naphthol, Fehling's solution, phloroglucinol, and orcinol. Fraction B consisted of colored materials eluted with acetone containing 10% water. It contained little polysaccharide material. The major proportion was purified by ether precipitation from ethanol, dried *in vacuo*, and ground to a red powder. These powders turned black when left to stand in air. Fresh powders gave positive Molisch, phloroglucinol, and orcinol tests and also dark red to purple colors with ferric chloride solution, indicating the presence of phenolic compounds. Forsyth suggested that these substances were phenolic glycosides, although SCHLICHTING [1953] could not detect phenolic glycosides in fraction B obtained from a podzol B horizon. Fraction C was removed with distilled water and was shown by various sugar tests to be polysaccharide in nature. It was purified by reprecipitation from water. Fraction D contained all the substances removed by 0.5 N sodium hydroxide that were not eluted by other procedures. They were mainly colored humic substances.

The losses of polysaccharide material on such charcoal columns are considerable. For example, WHISTLER and KIRBY [1956] recovered a little over 14% of the polysaccharide from a hot-water extract after charcoal filtration, and SWINCER, OADES, and GREENLAND [1968] recovered only 12% from caustic soda extracts. The recovery of carbohydrate from the charcoal can be increased by the use of 8.5 N acetic acid as eluant but about two-thirds of the carbohydrate cannot be removed from the charcoal except by sodium hydroxide extraction, which also removes colored humic materials. Deactivated charcoal, treated with stearic acid, as described by DALGLEISH [1955], has also been used to remove humic materials from fulvic acid fractions (SWINCER, OADES, and GREENLAND [1968]), and recoveries of polysaccharides can be increased up to about 65% by 8.5 N acetic acid elution. This method appears promising but is difficult to carry out with reproducible results. However, BERNIER [1958a] found that polysaccharides could be removed almost quantitatively from activated charcoal columns by elution with water followed by 1mM sodium tetraborate solution.

Polysaccharides eluted from charcoal are free from colored humic materials, and in this respect are satisfactory for further fractionation and characterization.

The removal of nitrogen from polysaccharide preparations can to some extent be regarded as a separate step. No preparations have been obtained which are free from nitrogen. Amino acids can be found and sometimes hexosamines in purified fractions. However, much protein material has been removed from preparations by the emulsification procedure of Sevag (BERNIER [1958a], MORTENSEN [1960]) and by cadmium sulfate precipitation and sorption on Fuller's earth in N-acetic acid (BERNIER [1958a]). A more extensive attempt to remove protein materials from polysaccharides was carried out by ROULET, MEHTA, DUBACH, and DEUEL [1963], who used Sephadex G-75, SE-Sephadex A-50, DEAE-Sephadex A-50, SM-cellulose, and DEAE-cellulose. Differences in the uronic acid and nitrogen contents in fractions from Sephadex G-75 were obtained, which suggested that the proteins and carbohydrates may not have been associated. The fractions obtained from the charged supports were all of similar sugar and amino acid composition. Further purification of polysaccharides has also been carried out by precipitation of polysaccharides as copper complexes (FORSYTH [1950]).

The more common procedures for initial purification can thus be listed as: (1) Acidification of extracts to precipitate high molecular weight humic materials with possible loss of carbohydrates by coprecipitation or as an integral part of the humic acid molecules. This loss is generally fairly small. (2) Dialysis to remove salts, solvents, and low molecular weight materials. This procedure causes loss of carbohydrate materials in two ways. Up to 40% of the carbohydrate in the fulvic acid fraction is lost during dialysis in Visking tubing against distilled water. Half of this material can be recovered in the dialysis water, presumably as small molecular weight fragments, but not as monosaccharides. The other half is sorbed on the tubing and cannot be recovered from inside or outside the dialysis tubing (CLAPP [1957], MÜLLER et al. [1960], SWINCER, OADES and GREENLAND [1968]). (3) Precipitation from aqueous solution by acetone, alcohol, or ether. These precipitation procedures are not specific for polysaccharides but leave polysaccharides in the supernatant and also precipitate noncarbohydrate materials. Large losses can be expected during purification involving these materials. (4) Charcoal filtration. This enables relatively pure carbohydrates to be obtained but quantitative recovery is difficult. (5) Deproteinization by Sevag's procedure, which removes considerable amounts of protein, but complete separations are never obtained. (6) Desalting by ion exchange, which can combine the removal of acids or alkalis and humic acids.

Some properties of purified polysaccharide preparations are shown in Table 11.

III. Fractionation of Purified Soil Polysaccharides

1. *General Information*

Fractionation of soil polysaccharides has proved an extremely difficult problem. Fractions obtained from soil organic matter always contain the same spectrum of monosaccharides in the same proportions, as shown by paper chromatography (LYNCH, OLNEY, and WRIGHT [1958], LYNCH, WRIGHT, and OLNEY [1957], ACTON, PAUL, and RENNIE [1963], SWINCER, OADES, and GREENLAND [1968]). Several workers (e.g., WHISTLER and KIRBY [1956], BERNIER [1958a], PARSONS and TINSLEY [1961]) have found that fractional precipitation techniques using ethanol or quaternary ammonium compounds yield fractions that are also essentially similar in terms of sugar components and their relative amounts. Similarly, electrophoresis has indicated extreme heterogeneity of soil polysaccharide preparations without giving unequivocal separations (BERNIER [1958a], WHISTLER and KIRBY [1956], CLAPP [1957], MORTENSEN [1960], THOMAS [1963], FINCH [1965]). The most promising results so far obtained have involved

Table 11. Properties of soil polysaccharides obtained by various procedures.

Authors	Forsyth [1947b, 1950]	Duff [1952]	Black, Cornhill, and Woodward [1955]	Whistler and Kirby [1956]	Mortensen [1960]
Soil { Description	Scottish agricultural soil	Scottish soil	Scottish peat	Indiana soil	Paulding clay
Organic matter content, %	8.5	9.5	About 99	3.38	—
Extractant	0·5 N NaOH	Water	0·9 N HCl	Water	Hot water
Purification procedure	Charcoal filtration ethanol and copper precipitation	Acetone precipitation	Ethanol precipitation	Acetone precipitation	Acidification and ethanol/glacial acetic acid precipitation
Ash, %	*	5	3.48	8.3	5.64
Carbohydrate, %	80 (reducing sugars)	About 50	89·8	About 50	—
Nitrogen, %	0.34	1.6	—	0.34	2.59
Uronic anhydride, %	15.8†	20.1†	5.6	9.1	7′
Hexosamine, %	0	0	—	+	10′
Mean equivalent weight	1,185	1,000	3,150	1,945	3,050
Glucose	20.8	39.3	66.0	21.2	8′
Galactose	20.0		8.2	16.6	5′
Mannose	21.9	33.0	15.0	18.5	1′
Arabinose	11.7		—	10.4	4′
Xylose	23.6	11.2	9.8	12.6	9′
Ribose	1.5	4.9	—	+	6′
Rhamnose	0	12.0	+	14.2	2′
Fucose	0	—	—	0	3′
Other sugars	0	+	—	—	—

Individual monosaccharides are expressed as % total neutral sugars in the preparation.
+ = present but not estimated.
* = not determined.
† = by decarboxylation.
1′ ——— 10′ sugars in order of decreasing amounts.

gel filtration on Sephadex (Roulet, Mehta, Dubach, and Deuel [1963], Swincer, Oades, and Greenland [1968], Barker et al. [1965]) and chromatography on charged celluloses and Sephadex (Müller, Mehta, and Deuel [1960], Roulet, Mehta, Dubach, and Deuel [1963], Thomas [1963], Finch [1965]).

2. Fractional Precipitation

Fractional precipitation of soil polysaccharide preparations from aqueous solutions has been attempted by Whistler and Kirby [1956] and Bernier [1958a], but there were no differences in the optical rotations and sugar components of the separates obtained. Fractional precipitation from formamide using ethanol was also unsuccessful (Bernier [1958b]). Similarly, little success has been obtained using quaternary ammonium compounds (Bernier [1958a]). Parsons and Tinsley [1961] dissolved polysaccharide preparations in water adjusted to pH 7.0 with lithium carbonate. The acidic materials present were precipitated with hexadecyltrimethyl-ammonium bromide (cetavalon). The cetavalon precipitate was extracted with 0.5 M lithium chloride to bring acidic polysaccharides into solution. However, the supernatant from the cetavalon precipitation, the portion of the precipitate brought into solution by lithium chloride, and the residue were all similar in saccharide composition. The only noted

Table 11 (*Continued*)

Sandy loam	Meadow soil	Peat	Podzol A_0	Podzol B	Compost 11 months	Compost 48 months	Fen peat			
			PARSONS and TINSLEY [1961]				BARKER *et al.* [1965]			
3.5	17.6	31.9	39.7	2.3	58.9	—	80.0			
Anhydrous formic acid containing LiBr							0.6 N H_2SO_4 Neutralization, dialysis, and fractionation on Sephadex G–100 and DEAE-Sephadex-A-50 to give four fractions			
Cetavalon precipitation										
11.7	6.0	4.4	5.7	3.9	8.3	9.2	—	—	—	—
26.3	37.8	22.9	24.7	23.1	17.5	19.4	—	—	—	—
3.16	2.65	3.67	3.02	3.49	4.47	5.08	—	—	—	—
15.8†	17.9†	16.0†	16.3†	13.5†	8.4†	6.7†	0.11	0.32	1.67	1.13
5.2	5.0	3.3	3.1	4.3	5.0	4.3	Not detected			
—	—	—	—	—	—	—	17,100	1,860	1,220	600
33.6	37.7	26.8	33.6	31.8	42.3	47.2	27.0	14.8	0	4.6
19.1	20.1	20.8	23.1	18.2	15.4	16.7	13.0	0	20.9	1.4
18.0	18.2	15.8	17.9	20.5	15.4	13.9	46.8	59.0	49.3	59.0
7.9	7.6	7.9	6.7	6.8	11.5	8.3	} 2.8	9.3	14.6	15.1
7.9	8.2	8.9	7.5	9.1	15.4	13.9				
13.5	8.2	18.8	11.2	13.6	—	—	} 10.4	16.4	15.2	19.9
—	—	—	—	—	—	—				
—	—	—	—	—	—	—				

difference was that the methoxyl content of the insoluble cetavalon precipitate was higher than that of the other two fractions.

Limited success was obtained by fractionation of acetylated polysaccharides from chloroform using light petroleum (boiling point, 60 to 80°C). The same sugars were found in each fraction, but there were indications of changes in the relative proportions of sugars, particularly glucose and arabinose. Precipitation of a polysaccharide preparation from approximately N sodium hydroxide using Benedict's solution yielded fractions that showed different mobilities during free boundary electrophoresis (BERNIER [1958a]).

The poor results obtained by such procedures with soil polysaccharides is probably a reflection of the complexity of the preparations, as these methods have been used successfully on polysaccharides obtained from other sources (*e.g.*, ERSKINE and JONES [1956], WHISTLER and LAUTERBACH [1958], PALMSTIERNA, SCOTT, and GARDELL [1957]).

3. *Chromatography*

The ion exchange resins so widely used in chemistry for separation, purification, concentration, etc. of inorganic and small organic materials cannot be used successfully for macromolecules, as the charges on such materials are not accessible to large molecules. The same is true to some extent of the charged Sephadex products. For example, ROULET *et al.* [1963] found SE Sephadex-A-50 and DEAE-Sephadex-A-50 to exclude much polysaccharide material, thus preventing separations of the larger polysaccharides on a charge basis. However, BARKER *et al.* [1965] were able to fractionate a soil polysaccharide preparation obtained as high molecular weight material on Sephadex-100 using DEAE-Sephadex-A-50. Stepwise elution of soil polysaccharides from a column of DEAE-Sephadex-A-50 with phosphate and chloride buffer solutions yielded four fractions. In general, the pentose content increased from fractions 1 to 4, whereas the glucose content decreased. A more logical choice of chromatographic support would seem to be charged cellulose, where more of the charges are present on accessible surfaces. ROULET *et al.* [1963] used DEAE- and SE-cellulose and were able to obtain a fractionation by elution from a DEAE-cellulose column with phosphate buffer and an increasing sodium hydroxide gradient. Five fractions were obtained with increasing uronic acid content and decreasing neutral sugar content. The largest fraction was lowest in uronic acids, whereas the final much smaller fraction was high in uronic acids. The same sugars were present in all fractions, but relative amounts differed. THOMAS [1963] has also successfully fractionated soil polysaccharides by chromatography on a DEAE-cellulose column. Fractions were again eluted with phosphate buffers and sodium hydroxide. The recovery was 62.7%; thus a third of the polysaccharides was not eluted from the column. MÜLLER *et al.* [1960] and BARKER *et al.* [1965] reported 100% recoveries.

MÜLLER *et al.* [1960] found that the uronic acids were concentrated in a minor fraction of their soil polysaccharides; THOMAS [1963] and BARKER *et al.* [1965] also obtained a concentration of their uronic components, although not to such a marked extent.

This approach, which is proving successful for fractionation of polysaccharides from different origins, is promising, as pointed out by MEHTA *et al.* [1961], and should help considerably in attempts to isolate homogeneous polysaccharides, a very necessary step before the chemical and physicochemical properties of soil polysaccharides and their origin can be usefully studied.

IV. Properties of Soil Polysaccharides

1. *General Information*

It is doubtful if anyone has obtained a homogeneous soil polysaccharide, except perhaps the cellulose isolated by GUPTA and SOWDEN [1964]. Therefore, most of the determined properties are those of a complex of polysaccharides. Table 11 shows data for various polysaccharide preparations. Although most have been subjected to the purification procedures described, very few are free from noncarbohydrates and ash.

2. *Composition*

Most workers have hydrolyzed their polysaccharide preparations for investigation of the component monomers. The same sugars have been identified as in total soil hydrolysates— namely, glucose, galactose, mannose, arabinose, xylose, ribose, rhamnose, fucose, uronic acids, amino sugars, and several unknown sugars with high Rf values.

Apart from the sugars listed in Table 11 a number of minor components of extracted polysaccharides have been identified. DUFF [1952a, b] separated three high Rf sugars from a soil polysaccharide hydrolysate by chromatography on cellulose and showed that they contained methoxyl groups. Periodate oxidation and reduction followed by oxidation (DUFF [1954]) indicated that two of the fractions were 2-0-methylaldoses. Subsequently, by elution from a highly active vegetable charcoal (Ultrasorb SC120/240) seven separate high Rf fractions were isolated. DUFF [1961] identified 2-0-methyl-L-rhamnose and 4-0-methyl-D-galactose. THOMAS [1963] identified glucosamine, galactosamine, and N-acetyl glucosamine. Although several authors have identified hexosamines in extracted polysaccharides (*e.g.*, WHISTLER and KIRBY [1956], PARSONS and TINSLEY [1961], SWINCER, OADES and GREENLAND [1968]) and other workers were not able to detect them (FORSYTH [1950], DUFF [1952], BARKER *et al.* [1965]). Amino acids have also been found in hydrolysates of soil polysaccharides (ROULET *et al.* [1963], BERNIER [1958b], WHISTLER and KIRBY [1956], MÜLLER *et al.* [1960], THOMAS [1963]). It is likely that many other monosaccharides will be identified in soil hydrolysates as more refined techniques are applied.

No structural work has been carried out on soil polysaccharides except for some enzymological studies by THOMAS [1963], who obtained evidence that his polysaccharide preparation did not contain the β–$(1\rightarrow4)$–glycosidic linkage of cellulose, the methylester linkage of pectin, or the $(1\rightarrow4)$–glycosidic linkage of polygalacturonic acid.

Infrared absorption studies (MORTENSEN [1960], MORTENSEN and SCHWENDINGER [1963], THOMAS [1963], SWINCER, OADES and GREENLAND [1968]) of extracted soil polysaccharides have yielded spectra typical of carboxylated polymers and absorption characteristic of the bonds expected in a polysaccharide molecule. However, although infrared spectroscopy has yielded useful general knowledge about the polysaccharides, little specific information has as yet been obtained by this procedure.

3. *Physicochemical Properties*

Free-boundary and continuous-paper electrophoresis have been used to characterize soil polysaccharides. Free-boundary electrophoresis in borate and phosphate buffers has usually shown a single broad peak with several superimposed ridges (WHISTLER and KIRBY [1956], CLAPP [1957], BERNIER [1958b], THOMAS [1963], FINCH [1965]). Boundaries became diffuse in most cases, indicating a heterogeneous mixture of polysaccharides. THOMAS [1963] used free-boundary electrophoresis to examine polysaccharide fractions eluted from a DEAE-cellulose column. Fractions with high electrophoretic mobilities showed high uronic acid contents. Thus, in phosphate buffer at pH 7.65 the uronic acid content had a direct influence on the electrophoretic mobility.

Continuous-flow paper electrophoresis in phosphate and borate buffers was carried out by MORTENSEN [1960] and MORTENSEN and SCHWENDINGER [1963]. Good separation was not obtained, but peaks corresponding to uronic-acid-containing components were obtained. Complex formation in borate buffer increased electrophoretic mobilities of polysaccharides and obscured separation of uronic-acid-containing components. Purification of the polysaccharide preparation improved the resolution of a slow-moving component, but other peaks were still superimposed on this major peak. MORTENSEN and SCHWENDINGER [1963] suggested that a highly mobile polysaccharide fraction might be associated with brown humic materials.

Studies with the ultracentrifuge (OGSTON [1958], SWINCER, OADES, and GREENLAND [1968]) have shown polysaccharide preparations to be extremely polydisperse, making molecular weights difficult to calculate. Mean molecular weights of 130×10^3 and 124×10^3 were obtained by OGSTON [1958] and BERNIER [1958a], respectively, by sedimentation and viscometry procedures. Ogston suggested that the average particle approximates to an elongated stiff and little solvated rod with an axial ratio of 80. Studies on molecular weight distributions of polysaccharide preparations carried out by SWINCER, OADES, and GREENLAND [1968] by Sephadex gel filtration showed a bimodal distribution, with roughly 40% with molecular weights greater than 100,000 and more than 50% having molecular weights less than 40,000. The large difference between the molecular weights of these two fractions was confirmed by ultracentrifuge studies, but molecular weights were not calculated because the samples were too polydisperse. MORTENSEN [1960] obtained a molecular weight of 450,000 from viscometry studies.

According to MORTENSEN [1960], temperature and salt effects on the shape of titration curves indicated that a soil polysaccharide was probably a hydrogen bonded polyuronide containing contaminants. Minima in viscosity curves suggested an isoelectric point near pH 4.9.

The known characteristics of the polysaccharide preparations so far obtained yield little information about individual polysaccharides in terms of their constitution. However, GUPTA, and SOWDEN [1964] have isolated cellulose from soils using Schweitzer's reagent. The cellulose accounted for 20 to 40% of glucose estimated in soil hydrolysates, and that isolated from a podzolic soil had an infrared spectrum essentially similar to Whatman no. 1 filter paper. X-ray diffraction showed the material to be amorphous.

The properties of extracted soil polysaccharides in terms of aggregate stabilizing ability have been examined by several authors, and generally an increase in stability has been reported. Apart from this, little information concerning the polysaccharides is obtained from such studies.

E. Effects of Saccharides on Soil Properties

I. Chemical Properties

Very little is known of the influence that saccharides have on the chemical properties of the soil. Uronic acids obviously contribute to the cation exchange capacity of the soil. In quantitative terms, however, the contribution will normally be less than that due to carboxyl groups present on the humic acids. In more acid soils the amino groups of mucopolysaccharides may contribute to anion retention. Formation of specific complexes with inorganic ions very probably occurs, but no information relating to the occurrence of such complexes in the soil seems to be available at present. DUFF and WEBLEY [1959] have shown that certain bacteria isolated from soils produce 2-ketogluconic acid. This powerful chelating agent may well be important in the weathering of minerals and in the release of phosphate from insoluble forms in the soil (DUFF, WEBLEY, and SCOTT [1963]). Other workers (BRADLEY and SIELING [1953]) have shown that carbohydrates can hinder the fixation of phosphate by iron and aluminum. They may also be involved in the translocation of iron and aluminum in the soil profile (SCHNITZER and DELONG [1955]) and sorption of borate (CLAPP [1957]).

II. Physical Properties

The relationship between the polysaccharides in the soil and the physical properties of soils has been studied by a number of investigators. Much of the work derived from the observation that microbially produced gums will bind soil particles into stable aggregates (WINOGRADSKY [1929], McCALLA [1943, 1945], MARTIN [1945a, 1946], GEOGHEGAN and BRIAN [1946, 1948], SWABY [1949], CLAPP, DAVIS, and WAUGAMAN [1962], HARRIS, ALLEN, CHESTERS, and ATTOE [1963], MARTIN and RICHARDS [1963], MARTIN, ERVIN, and SHEPHERD [1965]). The presence of organisms in the soil which produce these polysaccharides when cultured in the laboratory (MARTIN [1945a, b], FORSYTH and WEBLEY [1949], FORSYTH [1954], BERNIER [1958b], WEBLEY et al. [1965]) suggests that these materials will also occur in soils and so would be expected to increase the stability of natural soil aggregates. As was discussed in Section D, mixtures of polysaccharides have been isolated from a wide range of soils which had properties suggesting that microbial gums were one of the components of the mixtures. In some instances it has been shown that the extracted polysaccharides are able to stabilize soil aggregates (RENNIE, TRUOG, and ALLEN [1954], DUBACH et al. [1955], WHISTLER and KIRBY [1956], MEHTA *et al.* [1960]). Statistical correlations have also been demonstrated to exist between estimates of polysaccharide content and the degree of aggregation of the soil (RENNIE, TRUOG, and ALLEN [1954], CHESTERS, ATTOE, and ALLEN [1957]). The correlations obtained were not particularly close, but this is not unexpected, since the estimates of polysaccharide content were somewhat crude. It is also probable that only certain of the polysaccharides present are involved in aggregate stabilization, since the bacterially produced gums differ very considerably in their effectiveness (MARTIN and RICHARDS [1963], MARTIN, ERVIN, and SHEPHERD [1965]) and some plant products such as cellulose exert no effect on aggregate stability. Furthermore, the distribution of the polysaccharides within and around aggregates is important (WILLIAMS [1965]) so that only a portion of an effective polymer may actually be controlling the stability.

MEHTA *et al.* [1960] showed that artificial aggregates stabilized by adding dextrans or soil polysaccharides to dispersed soils lost their stability when treated with dilute (0.01 M) sodium periodate followed by sodium borate. This treatment cleaves the polymer molecules at several points, so that they no longer act as bridges between the particles forming the aggregate. Natural aggregates treated in the same way did not lose their stability and MEHTA *et al.* [1960] concluded that other agents were stabilizing these natural aggregates, possibly in addition to the polysaccharides. Subsequently, GREENLAND, LINDSTROM, and QUIRK [1961, 1962] and CLAPP and EMERSON [1965] have shown that in a number of soils the stability of aggregates is very largely dependent on materials cleaved by the dilute periodate-borate treatment, such materials presumably being polysaccharides. HARRIS, ALLEN, CHESTERS, and ATTOE [1963] and WATSON and STOJANOVIC [1965] have shown that the increase in the stability of aggregates in soils incubated with added organic materials is largely associated with the development of periodate sensitive materials.

In most of the studies discussed above the relationship between polysaccharides and aggregate stability as determined by wet sieving has been discussed. Aggregate stability is important in most soils except very sandy ones, since it is only by virtue of aggregate formation that a satisfactory continuity of pores in the soil is maintained, whereby adequate air and water movement can occur for optimum plant growth. In some instances more direct measurements have been made of the changes which occur in the physical properties of soils when polysaccharides are added or removed. GREENLAND, LINDSTROM, and QUIRK [1962] showed that the permeability of beds of aggregates could be reduced by periodate treatment, and CLAPP and EMERSON [1965] that clays slaked and dispersed more readily after this treatment. ALLISON [1947], however, found that production of microbial gums in soils could lead to an undesirable reduction in permeability because the microbial products were blocking some of the more important coarse pores. RUSSELL [1961] has suggested that the permeability of soils heavily dressed with readily digested organic materials might be reduced because the products of microbial metabolism can act as waterproofing agents, particularly after drying of the soil.

This evidence and the other data discussed above leave little room for doubt that polysaccharides in soils exert an important influence on the stability of their physical structure. It is also clear, however, that other organic and inorganic materials can stabilize soil aggregates, and where such materials are present, the polysaccharides may be of little additional benefit. The polysaccharides are probably of greatest importance in cultivated soils of relatively low total organic matter content.

III. Biological Properties

Saccharides are probably the most important materials readily available as an energy substrate for soil organisms. The amounts in soils, therefore, exert an overall influence on the level of biological activity in the soil. It is also probable that more specific influences are exerted on specific components of the population, but little is known of this subject.

For further information on the structural relationships of monosaccharides and polysaccharides the following texts could be consulted.

Pigman, W. *The Carbohydrates.* New York: Academic Press, 1957.

Whistler, R. L., and Smart, C. L. *Polysaccharide Chemistry.* New York: Academic Press, 1953.

REFERENCES

Acton, C. J., E. A. Paul, and D. A. Rennie. 1963. Measurement of the polysaccharide content of soils. *Can. J. Soil Sci.* 43:141–150.

Acton, C. J., D. A. Rennie, and E. A. Paul. 1963. The relationship of polysaccharides to soil aggregation. *Can. J. Soil Sci.* 43:201–209.

Adams, G. A., and A. E. Castagne. 1948. Some factors affecting the determination of furfural. *Can. J. Res.* B26:312–324.

Aleksandrova, I. V. 1960. A method of studying the qualitative composition of organic matter in soil solutions. *Pochvovedenie* 11:85–87.

Alexander, M. 1961. *Introduction to Soil Microbiology.* New York: Wiley.

Allison, L. E. 1947. Effect of microorganisms on permeability of soil under prolonged submergence. *Soil Sci.* 63:439–450.

Alvsaker, E. 1948. A modified Waksman procedure and its application to soil samples from western Norway. *Univ. Bergen Skr.* 23.

———, and K. Michelsen. 1957. Carbohydrates in a cold water extract of a pine forest soil. *Acta Chem. Scand.* 11:1794–1795.

Ashworth, M. R. F. 1942. The fractionation of the organic matter, including nitrogen, of certain soils and its relation to their quality. *J. Agric. Sci. Camb.* 32:349–359.

Aspinall, G. O. 1959. Structural chemistry of the hemicelluloses. *Adv. Carbohyd. Chem.* 14:429–468.

Baily, R. W. 1958. The reaction of pentoses with anthrone. *Biochem. J.* 68:669–672.

Barker, S. A. 1963. Polysaccharides of bacteria, moulds, yeasts and protozoa. In *Comprehensive Biochemistry*, Vol. 5, M. Florkin and E. H. Stotz, eds. New York: Elsevier.

———, P. Finch, M. H. B. Hayes, R. G. Simmonds, and M. Stacey. 1965. Isolation and preliminary characterisation of soil polysaccharides. *Nature, Lond.* 205:68–69.

Bear, F. E. 1964. *Chemistry of the Soil*, 2d ed. New York: Reinhold.

Beres, T., and I. Kiraly. 1959. Untersuchungen über die Reduktionwirkung der Torf-fulvosaüre auf dreiwertiges Eisen und Fulvosaureeisen verbindungen. *Z. Pfl. Ernähr. Düng. Bodenk.* 87:16–26.

Bernier, B. 1958a. Characterisation of polysaccharides isolated from forest soils. *Biochem. J.* 70:590–598.

———. 1958b. The production of polysaccharides by fungi active in the decomposition of wood and forest litter. *Can. J. Microbiol.* 4:195–204.

Black, W. A. P., W. J. Cornhill, and F. N. Woodward. 1955. A preliminary investigation on the chemical composition of sphagnum moss and peat. *J. Appl. Chem., Lond.* 5:484–492.

Bradley, D. B., and D. H. Sieling. 1953. Effect of organic anions and sugars on phosphate precipitation by iron and aluminium as influenced by pH. *Soil Sci.* 76:175–179.

Bremner, J. M. 1949. Studies on soil organic matter. III. The extraction of organic carbon and nitrogen from soil. *J. Agric. Sci. Camb.* 39:280–282.

———. 1950a. Amino-acids in soil. *Nature, Lond.* 165:367.

———. 1950b. The amino acid composition of the protein material in soil. *Biochem. J.* 47:538–542.

———. 1950c. A review of recent work on soil organic matter. I. *J. Soil Sci.* 2:67–82.

———. 1955. Nitrogen transformations during the biological decomposition of straw composted with inorganic nitrogen. *J. Agric. Sci. Camb.* 45:469–475.

———. 1958. Amino sugars in soil. *J. Sci. Fd. Agric.* 9:528–532.

Bremner, J. M., and K. Shaw. 1954. Studies on the estimation and decomposition of amino sugars in soil. *J. Agric. Sci. Camb.* 44:152–159.

Brink, R. H., P. Dubach, and D. L. Lynch. 1960. Measurement of carbohydrates in soil hydrolysates with anthrone. *Soil Sci.* 89:157–166.

Cheng, H. H., and L. T. Kurtz. 1963. Chemical distribution of added nitrogen in soils. *Proc. Soil Sci. Soc. Am.* 27:312–316.

Cheshire, M. V., and C. M. Mundie. 1965. The extraction of carbohydrate from soils. *Biochem. J.* 96:38P.

————. 1966. The hydrolytic extraction of carbohydrates from soil by sulphuric acid. *J. Soil Sci.* 17:372–381.

Chesters, G., O. J. Attoe, and O. N. Allen. 1957. Soil aggregation in relation to various soil constituents. *Proc. Soil. Sci. Soc. Am.* 21:272–277.

Clapp, C. E. 1957. High molecular weight water-soluble muck. Isolation and determination of constituent sugars of a borate complex forming polysaccharide, employing electrophoretic techniques. Ph.D. thesis, Cornell University, New York.

————, R. J. Davis, and S. H. Waugaman. 1962. The effect of rhizobial polysaccharides on aggregate stability. *Proc. Soil Sci. Soc. Am.* 26:466–469.

————, and W. W. Emerson. 1965. The effect of periodate oxidation on the strength of soil crumbs. I. Qualitative studies. II. Quantitative studies. *Proc. Soil. Sci. Soc. Am.* 29:127–134.

Cormack, R. G. H. 1962. Developments of root hairs in angiosperms. II. *Bot. Rev.* 28:446–464.

Dalgleish, C. E. 1955. Isolation and examination of urinary metabolites containing an aromatic system. *J. Clin. Path.* 8:73–78.

Davies, D. A. L. 1960. Polysaccharides of gram-negative bacteria. *Adv. Carbohyd. Chem.* 15:271–340.

Dawes, C. J., and E. Bowler. 1959. Light and electron microscope studies on the cell wall structure of the root hairs of *Raphanus sativus. Am. J. Bot.* 46:561–565.

Deriaz, R. E. 1961. Routine analysis of carbohydrates and lignin in herbage. *J. Sci. Fd. Agric.* 12:152–160.

Deuel, H., and P. Dubach. 1958a. Dekarboxylierung der organischen substanz des Bodens. III. Extraktion und Fraktionierung decarboxylierbarer Humusstoffe. *Helv. Chim. Acta* 41:1310–1321.

————. 1958b. Decarboxylation of the organic matter in soils. II. Identification of uronic acids. *Z. Pfl. Ernähr. Düng Bodenk.* 82:97–106.

————, P. Dubach, and R. Bach. 1958. Dekarboxylierung der organischen Substanz des Bodens. I. Dekarboxylierung der gesamten Humusstoffe. *Z. Pfl. Ernähr. Düng. Bodenk.* 81:189–201.

————, P. Dubach, N. C. Mehta. 1960. Decarboxylation of uronic acids and soil humic substances. *Sci. Proc. R. Dubl. Soc.* Ser Als: 115–121.

————, and E. Stutz. 1958. Pectic substances and pectic enzymes. *Adv. Enzymol.* 20:341–382.

Dische, Z. 1962. Color reactions of 2-deoxysugars. In "Methods in carbohydrate chemistry. I. Analysis and Preparation of Sugars." p. 503. New York and London: Academic Press.

Dormaar, J. F., and D. L. Lynch. 1962. Amendments to the determination of "uronic acids in soils with carbazole." *Proc. Soil Sci. Soc. Am.* 26:251–253.

Dubach, P. 1958. Dekarboxylierung der organischen Substanz des Bodens. Thesis Eidg. Techn. Hochschule, Zürich.

————, and D. L. Lynch. 1959. Comparison of the determination of "uronic acids" in soil extracts with carbazole and by decarboxylation. *Soil Sci.* 87: 273–275.

————, G. Zweifel, R. Bach, and H. Deuel. 1955. Examination of the fulvic acid fraction of some Swiss soils. *Z. Pfl. Ernähr. Düng. Bodenk.* 69:97–108.

Dubois, M., K. A. Gilles, J. K. Hamilton, P. A. Rebers, and F. Smith. 1956. Colorimetric method for determination of sugars and related substances. *Analyt. Chem.* 28:350–356.

Duff, R. B. 1952a. The occurrence of L-fucose in soil, peat and in a polysaccharide synthesised by soil bacteria. *Chemy. Ind.* 1104.

————. 1952b. The occurrence of methylated carbohydrates and rhamnose as components of soil polysaccharides. *J. Sci. Fd. Agric.* 3:140–144.

————. 1954. The partial identification of 0-methyl sugars occurring in soil, peat and compost. *Chemy. Ind.* 1513.

Duff, R. B. 1961. Occurrence of 2-0-methylrhamnose and 4-0-methylgalactose in soil and peat. *J. Sci. Fd. Agric.* 12:826–831.

———, and D. M. Webley. 1959. 2-ketogluconic acid as a natural chelator produced by soil bacteria. *Chemy. Ind.* 1376–1377.

———, D. M. Webley, and R. O. Scott. 1963. Solubilisation of minerals and related materials by 2-ketogluconic acid producing bacteria. *Soil Sci.* 95:105–114.

Eastoe, J. E. 1954. Separation and estimation of chitosamine and chondrosamine in complex hydrolysates. *Nature, Lond.* 173:540–541.

Erskine, A. J., and J. K. N. Jones. 1956. Fractionation of polysaccharides. *Can. J. Chem.* 34:821–826.

Finch, P. 1965. Physico-chemical studies on soil polysaccharides. M.Sc. thesis, University of Birmingham.

Folin, O., and H. Wu. 1920. A simplified and improved method for determination of sugar. *J. Biol. Chem.* 41:367–374.

Forsyth, W. G. C. 1947a. Characterisation of the humic complexes of soil organic matter. *J. Agric. Sci. Camb.* 37:132–138.

———. 1947b. Studies on the more soluble complexes of soil organic matter. I. A method of fractionation. *Biochem. J.* 41:176–181.

———. 1948. Carbohydrate metabolism in the soil. *Chemy. Ind.* 515–518.

———. 1950. Studies on the more soluble complexes of soil organic matter. 2. The composition of the soluble polysaccharide fraction. *Biochem. J.* 46:141–146.

———. 1954. Synthesis of polysaccharides by bacteria isolated from soil. *Trans. 5th Intern. Congr. Soil Sci.* 3:119–122.

———, and D. M. Webley. 1949. The synthesis of polysaccharides by bacteria isolated from the soil. *J. Gen. Microbiol.* 3:395–399.

Foster, A. B., and J. M. Webber. 1960. Chitin. *Adv. Carbohyd. Chem.* 15:371–393.

Fuller, W. H. 1947. Investigations on the separation of uronides from soils. *Soil Sci.* 64:403–411.

Gascoigne, J. A., and M. M. Gascoigne. 1960. *Biological Degradation of Cellulose.* London: Butterworths.

Geoghegan, M. J., and R. C. Brian. 1946. Influence of bacterial polysaccharides on aggregate formation in soils. *Nature, Lond.* 158:837.

———. 1948. Aggregate formation in soil. 2. Influence of various carbohydrates and proteins in aggregation of soil particles. *Biochem. J.* 43:14.

Graveland, D. N., and D. L. Lynch, 1961. Distribution of uronides and polysaccharides in the profiles of a soil catena. *Soil Sci.* 91:162–165.

Greenland, D. J. 1956. The adsorption of sugars by montmorillonite. II. Chemical studies. *J. Soil Sci.* 7:329–334.

———. 1965. Interaction between clays and organic compounds in soils. II. *Soils Fertil., Harpenden* 28:521–532.

———, and G. W. Ford. 1964. Separation of partially humified organic materials from soils by ultrasonic dispersion. *Trans. 8th Intern. Congr. Soil Sci.,* Bucharest 3:137–148.

———, G. R. Lindstrom, and J. P. Quirk. 1961. Role of polysaccharides in stabilization of natural soil aggregates. *Nature, Lond.* 191:1283–1284.

———. 1962. Organic materials which stabilize natural soil aggregates. *Proc. Soil Sci. Soc. Am.* 26:366–371.

Greenwood, C. T. 1952. The size and shape of some polysaccharide molecules. *Adv. Carbohyd. Chem.* 7:289–332.

———. 1956. Aspects of the physical chemistry of starch (addendum). *Adv. Carbohyd. Chem.* 11:385–393.

Grov, A. 1963. Carbohydrates in cold water extracts of a pine forest soil. *Acta Chem. Scand.* 17:2301–2306.

Gupta, U. C., and F. J. Sowden. 1963. Occurrence of free sugars in soil organic matter. *Soil Sci.* 96:217–218.

Gupta, U. C. and F. J. Sowden. 1964. Isolation and characterisation of cellulose from soil organic matter. *Soil Sci.* 97:328–333.

—————.1965. Studies on methods for the determination of sugars and uronic acids in soils. *Can. J. Soil Sci.* 45:237–240.

—————, and P. C. Stobbe. 1963. The characterisation of carbohydrate constituents from different soil profiles. *Proc. Soil. Sci. Soc. Am.* 27:380–382.

Harris, R. F., O. N. Allen, G. Chesters, and O. J. Attoe. 1963. Evaluation of microbial activity in soil aggregate stabilisation and degradation by the use of artificial aggregates. *Proc. Soil Sci. Soc. Am.* 27:542–545.

—————, G. Chesters, O. N. Allen, and O. J. Attoe. 1964. Mechanisms involved in soil aggregate stabilisation by fungi and bacteria. *Proc. Soil Sci. Soc. Am.* 28:529–532.

Haworth, W. N., R. W. Pinkard, and M. Stacey. 1946. Function of bacterial polysaccharides in soil. *Nature, Lond.* 158:836–837.

Hermans, P. H. 1949. *Physics and Chemistry of Cellulose Fibers.* New York: Elsevier.

Hewitt, L. F. 1938. The polysaccharide content and reducing power of proteins and of their digest products. *Biochem. J.* 32:1554–1560.

Hodge, J. 1953. Chemistry of browning reactions in model systems. *J. Agric. Fd. Chem.* 1:928–943.

Hoffman, W. S. 1937. A rapid photo-electric determination of glucose in blood and urine. *J. Biol. Chem.* 120:51–55.

Honeyman, J. 1959. *Recent Advances in the Chemistry of Cellulose and Starch.* London: Heywood and Co.

Horton, D., and M. L. Wolfrom. 1963. Polysaccharides. Chapter 7 of *Comprehensive Biochemistry*, Vol. 5. M. Florkin and E. H. Stotz, eds. New York: Elsevier.

Howes, F. N. 1949. *Vegetable Gums and Resins.* Massachusetts: Chronica Botanica Co.

Ivarson, K. C., and F. J. Sowden. 1962. Methods for the analysis of carbohydrate material in soil. *Soil Sci.* 94:245–250.

Jenny, H., and K. Grossenbacher. 1963. Root–soil boundary zones as seen by the electron microscope. *Proc. Soil Sci. Soc. Am.* 27:273–277.

Johanson, R. 1953. Interference of pentose in the estimation of hexose sugars with anthrone. *Nature. Lond.* 171:176–178.

Keefer, R. F., and J. L. Mortensen. 1963. Biosynthesis of soil polysaccharides. 1. Glucose and alfalfa tissue substrates. *Proc. Soil Sci. Soc. Am.* 27:156–160.

Keeney, D. R., and J. M. Bremner. 1964. Effect of cultivation on the nitrogen distribution in soils. *Proc. Soil Sci. Soc. Am.* 28:653–656.

Kertesz, Z. I. 1951. *The Pectic Substances.* New York: Interscience.

King, K. W. 1961. Microbial degradation of cellulose. *Va. Polytech. Inst. Tech. Bull.* 154. p. 55.

Kononova, M. M. 1966. *Soil Organic Matter*, 2d ed. Oxford: Pergamon Press.

Kosaka, J., and C. Honda. 1957. Uronic acid in humus. *Soil Pl. Fd. Tokyo* 2:142–147.

Krym, I. Ya. 1962. Use of the method of partition chromatography for determining carbohydrates in soils. *Pochvovedenie* 6:101–103.

Lambina, V. A. 1956. Bacteria decomposing plant protopectin. *Mikrobiologiya* 25:629–638.

Lawson, G. J., and M. Stacey. 1954. Immunopolysaccharides. I. Preliminary studies of a polysaccharide from *Azotobacter chroococcum*, containing a uronic acid. *J. Chem. Soc.*: 1925–1931.

Lefévre, K. V., and B. Tollens. 1907. Investigation of glucuronic acid; its quantitative determination and its colour reactions. *Ber. dt. chem. Ges.* 40:4513–4523.

Lippman, M. 1965. Proposed role for mucopolysaccharides in the initiation and control of cell division. *Trans. N.Y. Acad. Sci.* 27:342–360.

Lynch, D. L., and L. J. Cotnoir, Jr. 1956. The influence of clay minerals on the breakdown of certain organic substances. *Proc. Soil Sci. Soc. Am.* 20:367–370.

—————, E. E. Hearns, and L. J. Cotnoir. 1957. Determination of polyuronides in soils with carbazole. *Proc. Soil Sci. Soc. Am.* 21:160–162.

—————, H. O. Olney, and L. M. Wright. 1958. Some sugars and related carbohydrates found in Delaware soils. *J. Sci. Fd. Agric.* 9:56–60.

Lynch, D. L., L. M. Wright, and L. J. Cotnoir. 1956. The adsorption of carbohydrates and related compounds on clay minerals. *Proc. Soil Sci. Soc. Am.* 20:6–9.

————, L. M. Wright, and H. O. Olney. 1957. Qualitative and quantitative chromatographic analyses of the carbohydrate constituents of the acid-insoluble fraction of soil organic matter. *Soil Sci.* 84:405–411.

Macaulay Institute for Soil Research Annual Report. 1953/54. p. 26.

McCalla, T. M. 1943. Influence of biological products on soil structure, and infiltration. *Proc. Soil Sci. Soc. Am.* 7:209–214.

————. 1945. Influence of microorganisms and some organic substances on soil structure. *Soil Sci.* 59:287–297.

————, F. A. Haskins, and E. F. Frolik. 1957. Influence of various factors on aggregation of Peorian loess by microorganisms. *Soil Sci.* 84:155–161.

Maclean, K. D., and W. A. DeLong. 1956. On the carbohydrate component in leaf extracts and in leachates obtained under forest canopy. *Can. J. Agric. Sci.* 36:267–275.

Macura, J., J. Szolnoki, and V. Vancura. 1963. Decomposition of glucose in soil. In *Soil Organisms.* J. Doeksen and J. Van der Drift, eds. pp. 231–237.

Mantell, C. L. 1947. *The Water-Soluble Gums*, New York: Reinhold.

Martin, J. P. 1945a. Microorganisms and soil aggregation. I. Origin and nature of some of the aggregating substances. *Soil Sci.* 59:163–174.

————. 1945b. Some observations on the synthesis of polysaccharides by soil bacteria. *J. Bact.* 50:349–360.

————. 1946. Microorganisms and soil aggregation. II. Influence of bacterial polysaccharides on soil structure. *Soil Sci.* 61:157–166.

————, J. O. Ervin, and R. A. Shepherd. 1965. Decomposition and binding action of polysaccharides from *Azotobacter indicus* (*Beijerinckia*) and other bacteria in soil. *Proc. Soil Sci. Soc. Am.* 29:397–400.

————, and S. J. Richards. 1963. Decomposition and binding action of a polysaccharide from *Chromobacterium violaceum* in soil. *J. Bact.* 85:1288–1294.

Mayaudon, J., and P. Simonart. 1958. A study of the decomposition of organic matter in the soil using radioactive carbon. II. Decomposition of radioactive glucose in the soil. A. Distribution of radioactivity in the humic fractions of the soil. *Pl. Soil* 9:376–380.

————————. 1959a. A study of the decomposition of organic matter in the soil using radioactive carbon. III. Decomposition of dialysable substances and of proteins and hemicelluloses. *Pl. Soil* 11:170–175.

————————. 1959b. A study of the decomposition of organic matter in the soil using radioactive carbon. V. Decomposition of cellulose and lignin. *Pl. Soil* 11:181–192.

Mehta, N. C., and H. Deuel. 1960. Zur pentosanbestimmung in Böden. *Z. Pfl. Ernähr. Düng. Bodenk.* 90:209–218.

Mehta, N. C., P. Dubach, and H. Deuel. 1961. Carbohydrates in the soil. *Adv. Carbohyd. Chem.* 16:335–355.

Mehta, N. C., H. Streuli, M. Müller, and H. Deuel. 1960. Role of polysaccharides in soil aggregation. *J. Sci. Fd. Agric.* 11:40–47.

Meier, H. 1962. Galactan from tension wood of beech (*Fagus silvatica L.*). *Acta Chem. Scand.* 16:2275–2283.

Miller, G. L. 1960. Microcolumn chromatographic method for analysis of oligosaccharides. *Analyt. Biochem.* 1:133–140.

Montgomery, R. 1961. Further studies of the phenol-sulfuric acid reagent for carbohydrates. *Biochim. Biophys. Acta* 48:591–593.

Moore, S., and W. H. Stein. 1951. Chromatography of amino acids on sulfonated polystyrene resins. *J. Biol. Chem.* 192:663–681.

Mortensen J. L. 1960. Physico-chemical properties of a soil polysaccharide. *Trans. 7th Intern. Congr. Soil Sci.* 2:98–104.

————, and R. B. Schwendinger. 1963. Electrophoretic and spectroscopic characterization of high molecular components of soil organic matter. *Geochim. Cosmochim. Acta* 27:201–208.

Müller, M., N. C. Mehta, and H. Deuel. 1960. Chromatographic fractionation of soil polysaccharides by means of cellulose ion-exchangers. 14. Report on ion-exchangers. *Z. Pfl. Ernähr. Düng. Bodenk.* 90:139–145.

Nagar, B. R. 1962. Free monosaccharides in soil organic matter. *Nature, Lond.* 194:896–897.

Nelson, N. J. 1944. A photometric adaption of the Somogyi method for the determination of glucose. *J. Biol. Chem.* 153:375–380.

Norman, A. G. 1937. *The Biochemistry of Cellulose, the Polyuronides, Lignin, etc.* London: Oxford University Press.

——, and W. V. Bartholomew. 1943. The chemistry of soil organic matter. I. Distribution of uronic carbon in some soil profiles. *Soil Sci.* 56:143–150.

Oades, J. M. 1967. Carbohydrates in some Australian soils. *Aust. J. Soil Res.* 5:103–115.

Ogston, A. G. 1958. Sedimentation in the ultracentrifuge (addendum to Bernier). *Biochem. J.* 70:598–599.

Ott, E., H. M. Spurlin, and M. W. Grafflin. 1954. *Cellulose and Cellulose Derivatives.* 2d ed. New York: Interscience.

Palmstierna, H., J. E. Scott, and S. Gardell. 1957. The precipitation of neutral polysaccharides by cationic detergents. *Acta Chem. Scand.* 11:1792–1793.

Parsons, J. W., and J. Tinsley. 1961. Chemical studies of polysaccharide material in soils and composts based on extraction with anhydrous formic acid. *Soil Sci.* 92:46–53.

Perry, M. B., and R. K. Hulyalkar. 1965. The analysis of hexuronic acids in biological materials by gas-liquid partition chromatography. *Can. J. Biochem.* 43:573–584.

Philipson, T. 1943. Über die quantitative Bestimmung reduzierender Zucker. *Ark. Kemi Miner. Geol.* 16A, No. 22.

Rennie, D. A., E. Truog, and O. N. Allen. 1954. Soil aggregation as influenced by microbial gums, level of fertility and kind of crop. *Proc. Soil Sci. Soc. Am.* 18:399–403.

Roulet, N., P. Dubach, N. C. Mehta, M. Muller-Vonmoos, and H. Deuel. 1963. Distribution of organic matter and of carbohydrates in the separation of "root free" soil material by flotation sieving. *Z. Pfl. Ernähr. Düng. Bodenk.* 101:210–214.

——, N. C. Mehta, P. Dubach, and H. Deuel. 1963. Separation of carbohydrates and nitrogen compounds from humic substances by gel filtration and ion exchange chromatography. *Z. Pfl. Ernähr. Düng. Bodenk.* 103:1–9.

Rovira, A. D. 1965. Plant root exudates and their influence upon soil microorganisms. In *Ecology of Soil Borne Plant Pathogens; Prelude to Biological Control.* K. F. Baker and W. C. Snyder, eds. International Symposium, Berkeley, April 7–13. 1963.

——, and E. L. Greacen. 1957. The effect of aggregate disruption on the activity of microorganisms in the soil. *Aust. J. Agric. Res.* 8:659–673.

Russell, E. W. 1961. *Soil Conditions and Plant Growth*, 9th ed. London: Longmans.

Saeman, J. F., W. E. Moore, R. L. Mitchell, and M. A. Millett. 1954. Techniques for the determination of pulp constituents by quantitative paper chromatography. *Tech. Pap. Addr. Tech. Pulp. Pap. Ind.* 37:336–343.

Salomon, M. 1962. Soil aggregation organic matter relationships in red-top potato rotations. *Proc. Soil Sci. Soc. Am.* 26:51–54.

——. 1963. Hexosamines in soil aggregates. *Nature, Lond.* 200:500.

Salton, M. R. J. 1960. *Microbial Cell Walls.* New York: John Wiley.

Scheffer, F., and R. Kickuth. 1961. Chemical decomposition experiments on a natural humic acid. *Z. Pfl. Ernähr. Düng. Bodenk* 94:180–198.

——, and B. Ulrich. 1960. *Humus und Humusdungung.* Stuttgart: Enke.

Schlichting, E. 1953. Zur kenntnis der Heidehumus. 11. Die Fulvosaure Fraktion. *Z. Pfl. Ernähr. Düng. Bodenk.* 61:97–107.

Schnitzer, M., and W. A. DeLong. 1955. Investigations on the mobilisation and transport of iron in forested soils. II. The nature of the reaction of leaf extracts and leachates with iron. *Proc. Soil Sci. Soc. Am.* 19:363–368.

Schuerch, C., and N. S. Thompson. 1963. The hemicelluloses. In *The Chemistry of Wood*, pp. 191–247. B. L. Browning, ed. New York: Interscience.

Scott, F. M., B. G. Bystrom, and E. Bowler. 1963. Root hairs, cuticle and pits. *Science, N.Y.* 140: 63–64.

———, K. C. Hamner, E. Baker, and E. Bowler. 1958. Electronic microscope studies of the epidermis of *Allium cepa. Am. J. Bot.* 45:449–461.

Shewan, J. M. 1938. The proximate analysis of the organic constituents in north-east Scottish soils. *J. Agric. Sci. Camb.* 28:324–340.

Shorey, E. C., and E. C. Lathrop. 1910. Pentosans in soils. *J. Am. Chem. Soc.* 32:1680–1683.

———, and J. B. Martin. 1930. Presence of uronic acids in soils. *J. Am. Chem. Soc.* 52:4907–4915.

Simonart, P., and J. Mayaudon. 1958. A study of the decomposition of organic matter in the soil using radioactive carbon. II. Decomposition of radioactive glucose in the soil. B. Distribution of the radioactivity in α–humus. *Pl. Soil* 9:381–384.

Singh, S., and G. S. Bhandari. 1963. Investigations on acid stable reducing organic fractions in some soils of Rajasthan. *J. Indian Soc. Soil Sci.* 11:293–298.

———, and P. H. Singh. 1960. Distribution of hexosamines in some soils of Uttar Pradesh. *J. Indian Soc. Soil Sci.* 8:125–128.

Siu, R. G. H. 1951. *Microbial Decomposition of Cellulose.* New York: Reinhold.

Smith, F., and R. Montgomery. 1959. *The Chemistry of Plant Gums and Mucilages.* New York: Reinhold.

Sorenson, H. 1955. Xylanase in the soil and the rumen. *Nature, Lond.* 176:74.

———, 1963. Studies on the decomposition of ^{14}C-labelled barley straw in soil. *Soil Sci.* 95:45–51.

Sowden, F. J. 1959. Investigations on the amounts of hexosamines found in various soils and methods for their determination. *Soil Sci.* 88:138–143.

———, and K. C. Ivarson. 1959. Decomposition of forest litters. II. Changes in the nitrogenous constituents. *Pl. Soil* 11:249–261.

—————. 1962a. Methods for the analysis of carbohydrate material in soil. 2. Use of cellulose column and paper chromatography for determination of the constituent sugars. *Soil Sci.* 94: 340–344.

—————. 1962b. Decomposition of forest litters. III. Changes in the carbohydrate constituents. *Pl. Soil* 16:389–400.

Stacey, M., and S. A. Barker. 1960. *Polysaccharides of Microorganisms.* London: Oxford University Press.

Stark, S. M. 1950. Determination of pectic substances in cotton — Colorimetric reaction with carbazole. *Analyt. Chem.* 22:1158–1160.

Stevenson, F. J. 1954. Ion exchange chromatography of the amino acids in soil hydrolysates. *Proc. Soil Sci. Soc. Am.* 18:373–377.

———. 1956a. Isolation and identification of some amino compounds in soils. *Proc. Soil Sci. Soc. Am.* 20:201–204.

———. 1956b. Effect of some long-term rotations on the amino acid composition of the soil. *Proc. Soil Sci. Soc. Am.* 20:204–208.

———. 1957a. Investigations of aminopolysaccharides in soils. 1. Colorimetric determination of hexosamines in soil hydrolysates. *Soil Sci.* 83:113–122.

———. 1957b. Investigations of amino polysaccharides in soils. 2. Distribution of hexosamines in some soil profiles. *Soil Sci.* 84:99–106.

———. 1957c. Distribution of the forms of nitrogen in some soil profiles. *Proc. Soil Sci. Soc. Am.* 21:283–287.

———. 1960. Chemical nature of the nitrogen in the fulvic fraction of soil organic matter. *Proc. Soil Sci. Soc. Am.* 24:472–477.

Swaby, R. J. 1949. The relationship between microorganisms and soil aggregation. *J. Gen. Microbiol.* 3:236–254.

Swincer, G. D., J. M. Oades, and D. J. Greenland. 1968. Studies on soil polysaccharides. I. The isolation of polysaccharides from soil. II. The composition and properties of polysaccharides in soils under pasture and under fallow-wheat rotation. *Australian J. Soil Res.* 6:211–239.

Tenney, F. G., and S. A. Waksman. 1929. Composition of natural organic materials and their decomposition in the soil. IV. Nature and rapidity of decomposition of various organic complexes in different plant materials under aerobic conditions. *Soil Sci.* 28:55–84.

————————. 1930. Composition of natural organic materials and their decomposition in the soil. V. Decomposition of various chemical constituents in plant materials under anaerobic conditions. *Soil Sci.* 30:143–160.

Theander, O. 1954. Studies on sphagnum peat. III. A quantitative study on the carbohydrate constituents of sphagnum mosses and sphagnum peat. *Acta Chem. Scand.* 8:989–1000.

Thomas, R. L. 1963. Fractionation and characterization of the polysaccharides extracted from the Brookston soil. Ph.D. thesis, Ohio State University.

————, and D. L. Lynch. 1961. A method for the quantitative estimation of pentoses in soil. *Soil Sci.* 91:312–316.

Timell, T. E. 1964. Wood hemicelluloses. Part 1. *Adv. Carbohyd. Chem.* 19:247–302.

Toennies, G., and J. J. Kolb. 1964. Carbohydrate analysis of bacterial substances by a new anthrone procedure. *Analyt. Biochem.* 8:54–69.

Tracey, M. V. 1950. A colorimetric method for the determination of pentoses in the presence of hexoses and uronic acids. *Biochem. J.* 47:433–436.

Vancura, V. 1964. Root exudates of plants. 1. Analysis of root exudates of barley and wheat in their initial phases of growth. *Pl. Soil* 21:231–248.

Waksman, S. A., 1936. *Humus.* London: Tindall and Cox.

————, and I. J. Hutchings. 1935. Chemical nature of organic matter in different soil types. *Soil Sci.* 40:347–363.

Watson, J. H., and B. J. Stojanovic. 1965. Synthesis and bonding of soil aggregates as affected by microflora and its metabolic products. *Soil Sci.* 100:57–62.

Webber, L. R. 1965. Soil polysaccharides and aggregation in crop sequences. *Proc. Soil Sci. Soc. Am.* 29:39–42.

Webley, D. M., R. B. Duff, J. S. D. Bacon, and V. C. Farmer. 1965. A study of polysaccharide-producing organisms occurring in the root region of certain pasture grasses. *J. Soil Sci.* 16:149–157.

Wen, C. H., and L. L. Chen. 1962. The determination of pentoses in soils by means of aniline. *Acta Pedol. Sin.* 10:220–226.

Whistler, R. L., and K. W. Kirby. 1956. Composition and behaviour of soil polysaccharides. *J. Am. Chem. Soc.* 78:1755 1759.

————, and G. E. Lauterbach. 1958. Isolation of two further polysaccharides from corn cobs. *Arch. Biochem. Biophys.* 77:62–67.

————, and C. L. Smart. 1953. *Polysaccharide Chemistry.* New York: Academic Press.

Whitehead, D. C., and J. Tinsley. 1963. The biochemistry of humus formation. *J. Sci. Fd. Agric.* 14:849–857.

————————. 1964. Extraction of soil organic matter with dimethylformamide. *Soil Sci.* 97:34–42.

Wilkinson, J. P. 1958. The extracellular polysaccharides of bacteria. *Bact. Rev.* 22:46–73.

Williams, B. G. 1965. Influence of PVA on physical properties of soil aggregates. Ph.D. thesis. University of Adelaide, Australia.

Williams, R. T. 1953. *Biological Transformations of Starch and Cellulose.* Cambridge: Cambridge University Press.

Winogradsky, S. 1929. Soil microbiology. IV. Degradation of cellulose in the soil. *Annls. Inst. Pasteur, Paris* 43:549–633.

Young, J. L., and J. L. Mortensen, 1958. Soil nitrogen complexes. 1. Chromatography of amino compounds in soil hydrolysates. *Ohio Agric. Exptl. Sta. Res. Circ.* 61.

Appendix: Haworth formulas for monosaccharides commonly found in soils

Hexoses, Hexosamines, and Hexuronic Acids

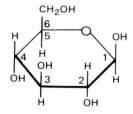

β-D-glucopyranose.

β-D-mannopyranose is the 2-epimer of β-D-glucopyranose.

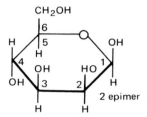

β-D-galactopyranose is the 4-epimer of β-D-glucopyranose.

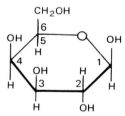

4 epimer

Epimerization involves transposition of the —H and —OH groups on carbon atoms 2, 3, and 4 in the aldoses and carbon atoms 3 and 4 in the ketoses. For example, the structure of β-D-mannopyranose is represented by the structure above, except that the —OH group at carbon atom 2 appears above the ring and the —H group below.

Anomerization involves transposition of the —H and —OH groups on carbon atom 1 in the aldoses or transposition of —CH_2OH and —OH groups about carbon atom 2 in the ketoses, for example, fructofuranose. The —OH group is above the ring for β-D-sugars, but below the ring for β-L-sugars and vice versa for the α-anomers.

2-amino-2-deoxy-β-D-glucopyranose, or glucosamine, has the structure above with replacement of the —OH group on carbon atom 2 by —NH_2.

2-amino-2-deoxy-β-D-galactopyranose, or galactosamine, is the 4-epimer of 2-amino-2-deoxy-β-D-glucopyranose.

Galacturonic acid (β-D-galacturonopyranose) has the same structure as β-D-galactopyranose, with replacement of —H_2OH by —OOH on carbon atom 6.

Glucuronic acid (β-D-glucuronopyranose) has the same structure as β-D-glucopyranose, with replacement of —H_2OH by —OOH on carbon atom 6.

Pentoses

HOH$_2$C   OH

β-D-xylofuranose

β-D-ribofuranose is the 3-epimer

    If the —CH_2OH group is drawn below the ring in the structure above, the sugar represented is α-L-arabinofuranose.

6-Deoxy-hexoses

6-deoxy-α-L-glucopyranose or α-L-quinovose
(not identified in soils)

6-deoxy-α-L-mannopyranose, or α-L-rhamnose, is the 2-epimer of 6-deoxy-α-L-glucopyranose.

6-deoxy-α-L-galactopyranose, or α-L-fucose, is the 4-epimer of 6-deoxy-α-L-glucopyranose.

Structures of Some Common Polysaccharides

Cellulose—a linear polymer of β-$(1\rightarrow4)$-D-glucopyranose units.

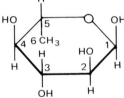

Starch (amylose)—a linear polymer of α-(1→4)-D-glucopyranose units.

Amylopectin has the same structure except that an α-(1→6) link occurs about every 25 glucose units to produce a branch.

Xylan—a linear polymer of β-(1→4) linked xylopyranose units.

Galacturonan as in pectic substances—a linear polymer of α-(1→4) linked D-galacturono-pyranose units in which the carboxyl groups are partially or fully methylated.

Dextran—α-(1→6) linked D-glucopyranose units.

Levan—β-(2→6) linked D-fructofuranose units.

Chitin—a linear polymer of 2-N-acetamido-2-deoxy-β-D-glucopyranose units linked (1→4).

Chapter 3

Nitrogenous Substances

J. W. Parsons and J. Tinsley

Contents

1. Introduction: Origin and Distribution of Combined Nitrogen in Soils

(a) The Nitrogenous Compounds of Plant, Animal, and Microbial Tissues

The organic nitrogen compounds account for well over 90% of the total nitrogen in most soils and are mainly the products of microbial decomposition of plant and animal remains. Hence, it is convenient to begin with a review of the nitrogen-containing compounds found in the tissues of plants, animals, and microorganisms.

In terms of bulk the largest contributor to the humification process is plant material, the nitrogen content of which commonly varies from 1 to 4% of the dry matter, although there is some variation with age and species, and between different parts of the plant. In the seed, well over 90% of the nitrogen is in the form of storage protein, whereas in the leaves and stems only about 60% of the nitrogen is present as enzyme or membrane protein, most of the remainder being accounted for as free amino acid nitrogen. The nitrogen content of microbial cells is somewhat higher, amounting to 5 to 10% of the dry matter, and much of this is in the form of structural protein. That of faunal tissues, composed mainly of protein, may be still higher. Most proteins are composed of the same 18 amino acids and 2 amides, and VEGOTSKY and FOX [1962] emphasize the overall similarity in their compositition, from the simplest to the most complex organisms.

The cell walls of higher plants, green algae, and fungi are predominantly polysaccharide in nature, cellulose being one of the major components, although it is often replaced by, or found along with, chitin in algae and fungi. Protein is not a component of the cell walls of higher plants, but it may be present as contaminating cellular membranes. In many microbial species protein or peptides are present in the wall as a structural component and usually in close association with polysaccharide. CROOK and JOHNSTON [1962] identified amino acids, amino sugars, and sugars as the monomer components of acid hydrolysates of the cell wall material of 16 species of fungi and suggested the possibility of the existence of a carbohydrate-protein complex. Earlier, KESSLER and NICKERSON [1959] identified glucan- and mannan-protein complexes in yeast cell walls, and these were confirmed by KORN and NORTHCOTE [1960]. The normal pattern of protein amino acids was found in all three cases, along with glucosamine (2-deoxy-2-amino D-glucose) and galactosamine (2-deoxy-2-amino D-galactose), but the linkage between the two polymers is not known. KESSLER and NICKERSON [1959] suggest that an ester linkage may be formed between a carboxyl group on the protein and a carbohydrate hydroxyl group, whereas DYKE [1964] suspects that the hydroxy amino acids threonine and serine, which are present in very high proportions, may be involved. CROOK and JOHNSTON [1962] and KORN and NORTHCOTE [1960] suggest that amino sugars may be responsible for linking the polymers. Glycoproteins are relatively widespread in animal tissues, and GOTTSCHALK, MURPHY, and GRAHAM [1962] consider that any of the functional groups may be involved in the linkage. Recently, a glycoprotein containing bound glucosamine has been reported in the seeds of some higher plants by PUSZTAI [1964].

Although both protein and polysaccharides are found in certain bacterial cell walls, the main structural component of both gram-positive and gram-negative bacteria is a hetero-polymer, usually referred to as a mucopeptide. The structure of the bacterial cell wall has been comprehensively discussed by SALTON [1964], who prefers the term "glycosamino-peptides" for these components. Their chemical structure and biosynthesis have also been reviewed by WORK [1961]. The mucopeptides appear to be highly polymerized, which accounts for their rigidity and insolubility, and they are resistant to proteolytic enzymes, although many are attacked by lysozyme (lysis). The monomeric components of these polymers are two acetylated amino sugars: N-acetylglucosamine and N-acetylmuramic acid; and three amino acids: D-glutamic acid, D-alanine, and either L-lysine or meso-α,ε-diaminopimelic acid. Muramic acid, 3-0-carboxyethyl glucosamine, is a 3-0-lactyl ether of glucosamine and has been identified only in bacteria, while α,ε-diaminopimelic acid has been found only in bacteria and blue-green algae. The commonly occurring protein amino acids are the L-isomers, but both glutamic acid and alanine have been found in the D-configuration, both in the bacterial mucopeptides and in some antibiotics. Recently, CUMMINS [1965] and PERKINS and CUMMINS [1964] have reported the presence of two other diamino acids: ornithine and 2,4 diaminobutyric acid in the muco-peptides of some gram-positive organisms.

The exact structure of the mucopeptides has not been elucidated, but it is now generally agreed that they possess a "backbone" of N-acetylglucosamine glycosidically linked to N-acetylmuramic acid, with peptides attached through the amino group of alanine and the carboxyl group of muramic acid. The dibasic acids may provide an opportunity for cross-linking between adjacent chains.

Other wall components are thought to be attached to this mucopeptide backbone. In the gram-positive bacteria they are polysaccharide and teichoic acids, but in the walls of gram-negative organisms, which are rather more complex, protein and lipopolysaccharides are also present.

Teichoic acids, which may account for as much as 50% of some bacterial cell walls, are polymers of either glycerol or ribitol phosphate units joined through phosphodiester linkages. Either glucose or N-acetylglucosamine may be attached glycosidically to the polyol, while alanine is attached through a hydroxyl group, either of the sugar or — where that is absent — directly to the polyol. It is interesting to note that N-acetylgalactosamine, which was reported by CUMMINS and HARRIS [1956] to be a common constituent of bacterial polysaccharides, has been identified in the teichoic acid of *Staphylococcus albus* but has never been found in bacterial mucopeptides.

Chitin, whose chemistry and structure have been reviewed by FOSTER and WEBBER [1960] is an insoluble unbranched polymer of β (1 – 4) linked N-acetyl-D-glucosamine found not only in the cell walls of fungi and green algae but also in the exoskeleton of insects and crus-taceans. It accounts for only 3 to 5% of fungal cell walls and is generally found in close associa-tion with other polysaccharides and protein. On the other hand, 25 to 50% of the cuticle of arthropods may consist of chitin, again in close association with protein. The outer layer of the cuticle is a thin layer of protein, hardened either by deposition of calcium carbonate (as in crustaceans) or by a cross-linking tanning reaction with polyphenolic compounds (as in insects). The inner layer is composed of chitin and protein probably weakly bound.

The other nitrogen-containing compounds common to all types of organisms are the nucleotides (coenzymes) and the nucleic acids DNA and RNA. In the cells nucleic acids are probably linked to basic proteins of small molecular weight, histones and protamines, but it is extremely difficult to isolate the complex without disrupting the linkage. In plants and micro-organisms the nucleic acids account for less than one-tenth of the total nitrogen.

Of the wide range of monomeric nitrogen compounds found in various tissues, the most important are the free amino acids that accumulate in higher plants and may account for

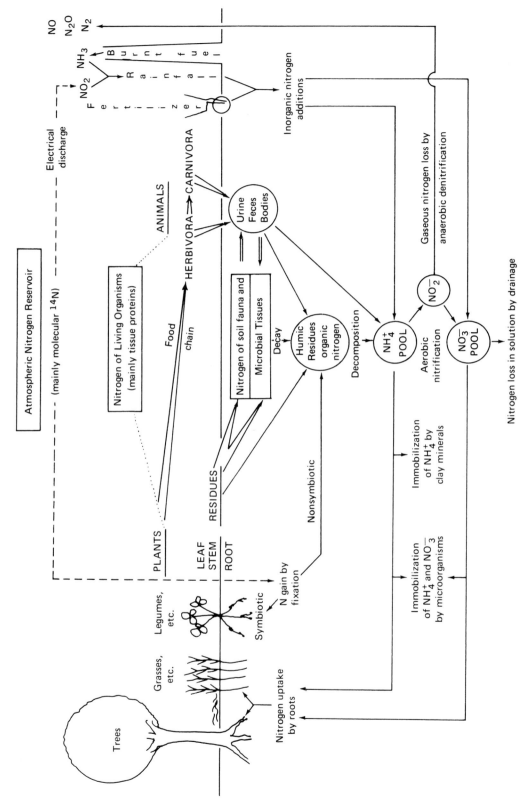

Figure 1. The nitrogen cycle.

one-third of the total nitrogen. In addition to the 18 common protein α-amino acids, a very large number of amino and imino acids including some non-α-amino acids, have been identified (FOWDEN [1958] and SYNGE [1955]). They appear to form a metabolic pool of compounds that accumulates during the synthesis or degradation of the protein amino acids and other nitrogen compounds.

Apart from cell constituents a wide range of extracellular compounds are produced. Higher plants and some microorganisms secrete amino acids into the soil. The capsules that envelop certain bacteria are mainly polysaccharide in nature in which amino sugars have occasionally been identified, but peptides have also been found. Many of the antibiotics produced in trace amounts by fungi and actinomycetes contain amino acids as specific peptides; amino sugars in various macrolides, glycosides, and other compounds; and a variety of other nitrogen compounds whose chemistry has been reviewed by UMIZAWA [1964].

This brief review indicates the types of nitrogen compounds that occur in the starting material of the humification process. New compounds are continually being identified by modern sensitive techniques, but with minor constituents it may be difficult to determine whether they are of wide occurrence or limited to a few species. At one time the only known naturally occurring amino sugars were glucosamine and galactosamine, but recently, according to SHARON [1965], D-fucosamine, mannosamine, and 2-aminomannuronic acid have been identified in bacterial cell walls.

(b) Decomposition of Tissue Nitrogen Compounds in Soils

The turnover of nitrogen compounds between higher plants and soil depends almost entirely on the multifarious activities of the many different populations of organisms that live on and in the soil.

Most obvious are the activities of grazing animals and other vertebrates, whether the wild species native to virgin lands or the domestic species of the farmer. Although these play an important part in the nitrogen cycle, as indicated in Figure 1, particularly in relation to mankind's food supply, it should be appreciated that the meso– and microfauna and the microflora of the soil play the dominant role. Digestion of plant tissues by herbivores and of animal tissues by carnivores ultimately results in some portion of the nitrogen being returned to the soil in their excreta and carcasses. The nitrogen in the partially humified feces is mainly insoluble compounds relatively resistant to digestive and microbial enzyme action. The nitrogen in the urine is mainly dissolved urea and other relatively simple compounds of nitrogen that readily become available to plants and microorganisms. In general, the uneven deposition of fecal droppings and urine by large animals tends to a heterogeneous distribution of nitrogen in the upper horizons of soils, as may readily be observed from the growth of grass in pastures grazed by cattle. Workers in New Zealand have studied the effects of the urine and feces from cattle and sheep on pasture growth and composition. DOAK [1952] found that ammonia formed rapidly in surface soils after wetting with urine, sometimes to such an extent that, if the pH rose above 9 for a time, nitrite accumulated temporarily until the normal nitrification processes were restored. No doubt elephants feeding on tropical grass or forest land have similar effects even though they are more scattered.

But the food chain does not end at this stage. Fecal droppings attract peculiar populations of meso- and microfauna which, together with the rich microflora, rapidly bring about disintegration and decomposition with further release of mineral forms of nitrogen.

It is difficult to assess quantitatively the activities of the invertebrate fauna under field conditions. Since the time of Darwin, soil biologists have continued to study the action of earthworms, but closer attention is now being given to the many other species of mesofauna that are particularly abundant in the surface organic horizons of grassland and forest soils under moist aerated conditions. These belong mainly to the phyla Mollusca and Arthropoda, and collectively their main action is to disintegrate dead plant tissues, whether they be leaves, twigs, or roots—though some feed on fresh plant material, including roots.

According to KÜHNELT [1961], the mesofauna may be broadly grouped into the primary and secondary decomposers. The primary ones, which include Isopoda, Diplopods, and some larger Diptera larvae, consume plant tissues, with their associated microorganisms, at a rapid rate and produce large quantities of dark-colored excrement. The contents of the ruptured plant cells including cytoplasmic proteins and nucleic acids are largely digested, but the cellulosic and ligneous cell walls are little affected, apart from being fragmented and softened to render them more easily attacked by microorganisms. The secondary decomposers, which are generally smaller organisms (e.g., mites and Collembola, small Diptera larvae) consume the macerated debris and in turn produce microscopic pellets of humified coprogenic material.

Some species prefer to feed almost exclusively at certain times on the mycelia or spores of fungi when these are abundant in the soil.

The interrelationships of the primary and secondary decomposers find expression in the L, F, and H layers at the surface of forest soils, especially on podzolic sites. The litter (L) layer provides a foraging ground for the larger fauna of primary decomposers and, beneath it, in the fermentation (F) layer, the secondary decomposers and microorganisms are active in producing the humus of the H layer. According to BURGES [1965], the H layer is particularly rich in chitin from the exoskeletons of animals and the remains of fungal hyphae.

Chemically, the overall effect of these faunal decomposers is to metabolize carbon compounds, releasing CO_2 and excreting various nitrogen compounds along with other mineral elements. According to GILMOUR [1961], most terrestrial insects excrete about 80% of their nitrogen as uric acid by urinary function along with smaller, variable amounts of ammonia, urea, or allantoin. Many soil arthropods have a similar pattern of nitrogen excretion. On the other hand, most nematodes and some aquatic larvae excrete nitrogen principally as ammonia, though some species produce free amino acids or even peptides. The net result is that the C:N ratio of the residues from faunal digestion is considerably narrower than that of the original plant material. Thus, it is clear that the nitrogen of fresh plant tissues follows a pathway through the digestive tracts of grazing animals which is paralleled by that passing through the mesofauna beneath their feet. In both cases the simple urinary nitrogen compounds are rapidly metabolized or transformed to ammonia by soil microorganisms, but whereas the grazing animals produce scattered patches of soil with high ammonia contents, the mesofauna produce a more even distribution and mixing of their excreta within the soil.

The development of a mull-type organic matter in moist, near-neutral soils depends greatly on the silent, persistent work of the earthworms in masticating the organic debris in intimate mixture with the finer mineral grains. The combined action of the digestive enzymes of the worms and the microorganisms within their guts achieves a considerable breakdown of cellulose and chitin, with some release of nitrogen from the latter. LAVERACK [1963] has given a detailed account of nitrogen excretion by earthworms, and it appears that, from the common species, normally about half is in the form of mucus proteins and the rest mainly as ammonia, with smaller amounts of urea, uric acid, and allantoin. DUBACH and GANTI [1964] have shown that small amounts of free amino acids are also excreted, but it should be emphasized that the total amount of these nitrogen compounds is quite small in proportion to the volume of soil involved. BARLEY and JENNINGS [1959] found that only about 6% of the "nonavailable" nitrogen fed to Allolophora caliginosa (Sav.) was rendered "available" by passage through the gut.

In tropical savannah grasslands communities of termites perform much the same functions as earthworms in a more localized manner, where they collect soil into mounds.

However, it should be appreciated that all the soil fauna, whatever their size, are only intermediaries in the nitrogen cycle. Whereas in near-neutral aerated soils any surplus of ammonia released by microorganisms is rapidly oxidized to nitrate, which then may be absorbed by plants or lost by leaching, in strongly acid soils the nitrifying organisms are inhibited and the plant cycle operates largely through ammonia as the available form. The conditions governing the release of mineral forms of nitrogen from the reservoir of humus compounds have been discussed in detail by HARMSEN and VAN SCHREVEN [1955], JANSSON [1958], and HARMSEN and KOLENBRANDER [1965].

From the foregoing it is evident that in a fresh sample of soil taken at a particular time, some of the nitrogen may be present in living plant tissues, particularly roots, some in the living animal tissues of the soil fauna, some in the living tissues of the soil microorganisms, some perhaps in resting spores or cysts, the bulk in the accumulated humus compounds, and some in simple organic or inorganic compounds representing the end products of nitrogen metabolism from the whole population of organisms. One of the experimental difficulties in studying the nitrogen compounds of the soil organic matter is to separate or distinguish between the various organic compounds of undecomposed tissues. As BURGES [1965] and his coworkers have demonstrated, decomposition of plant tissues, particularly senescent leaves, begins with their invasion by fungi before they fall to the ground. When each animal species has taken its fill of plant debris, fungal tissue, animal tissue, or excrement, ultimately it is the microflora that complete the cycle of changes in the nitrogen compounds, particularly the bacteria and actinomycetes.

The activities of the microflora, like those of the fauna, are controlled by the available food supply and by other environmental factors, namely the temperature, the water supply, and the air supply. Apart from extreme cold or heat or drought, the two factors that have a very widespread influence on the ecology of the living organisms in soils are excess water and excess acidity. Many terrestrial fauna cannot survive under waterlogged conditions, and furthermore, the activities of microorganisms are largely confined to those that operate by anaerobic metabolism, so that in peat, the pattern of humification encourages the preservation of the more resistant structural plant components.

At pH values around 4 or less earthworms cannot survive, so in podzolic soils organic matter tends to accumulate as a mor-type litter overlying the mineral horizons and the population of microorganisms is different from that in mull soils. STÖCKLI [1946] has recorded the following weights for live fauna and microflora (collectively termed the "biomass") in a Swiss meadow soil to a depth of 15 cm. However, these weights do not include any estimate of living root tissue.

Bacteria, actinomycetes, mycelia, algae	2022 g/m^2
Earthworms	400
Arthropods (larger species)	10
Mollusca	80
Protozoa	38
Nematodes	5
Enchytraeids	2
Mites, Collembola, Protura, Diplura	2
Total	2559

Assuming that these tissues contain an average of 1% nitrogen, the total amount of nitrogen in the biomass was about 25 g/m^2, compared with a total amount of about 700 g/m^2 in the whole soil. Clearly the nitrogen in the undecayed tissue normally represents only a small proportion of the total soil nitrogen, and most chemical investigations have ignored it.

By far the major portion of the biomass is in the cells of microflora. Next in order of contribution are the earthworms in those soils where they flourish. It is comparatively easy to separate earthworms and other larger mesofauna and also larger roots when preparing soil samples for laboratory examination, but it is virtually impossible to separate the undecomposed cells of the microflora. So, for practical reasons, it is sensible to consider the latter as part of the humus compounds of the soil.

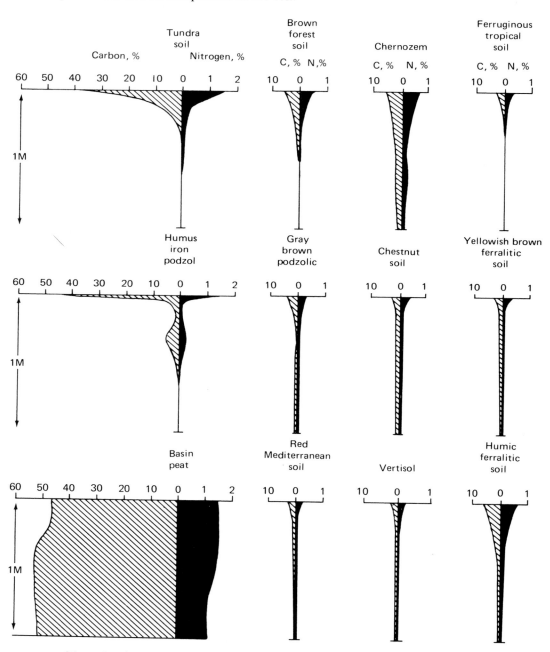

Figure 2. Distribution of carbon and nitrogen in representative soil profiles.

(c) Nitrogen Accumulation in Soils

When studying the amount and distribution of combined nitrogen in soils, it is usual to determine the total nitrogen content in the first instance. This is commonly done by the Kjeldahl method, with suitable precautions where necessary to prevent loss of nitrate.

Mineral forms of nitrogen, namely ammonium, nitrate, and occasionally nitrite are determined separately or collectively as required, with precautions where necessary to take account of fixed ammonium retained by certain clay minerals. The amount of mineral nitrogen can then be subtracted from the total to obtain the content of true organic nitrogen. However, this correction is frequently not made when the content of mineral nitrogen is expected to be small in relation to the total.

Details of suitable analytical procedures for total and for mineral forms of nitrogen are given by BREMNER [1965a, b].

α. Distribution of Organic Carbon and Nitrogen

Besides total nitrogen, it is common practice to determine total organic carbon and to calculate the C:N ratios. Details of suitable analytical procedures for organic carbon are given by ALLISON [1965] and TINSLEY [1950].

By sampling field soils in depth, horizon by horizon, not only can the amounts of total nitrogen and organic carbon per unit area or volume of soil be assessed, but also their distribution within the profile can be determined. Then it is apparent, as the data in Figure 2 show, that considerable differences exist between soil types.

The most noteworthy feature is that organic matter tends to accumulate in tundra soils under cold conditions, in the surface litter of podzol soils under acid conditions, and often to depths of several meters under anaerobic conditions. These accumulations represent large reservoirs of combined nitrogen.

In free-draining nonacid conditions the abundance and distribution of organic matter in the profile are conditioned by the prevailing climate and the type of vegetation. Thus, the chestnut soil contains less organic matter to a shallower depth than the deep chernozem, and the vertisol still less, although what is present is distributed to a great depth and imparts a dark color to the soil, giving an impression of abundance.

Tropical soils other than peats are generally low in organic carbon and nitrogen and even the humic ferralitic soil, No. 12 in Table 1, has less than 0.4% nitrogen in the top 20 cm.

β. Carbon:Nitrogen Ratios

Another prominent feature is the variation in the C:N ratios reported for different soils and for succeeding horizons within single soils. The data in Table 1 demonstrate that where organic matter accumulates in quantity, the C:N ratio is generally wide, being about 30 in the surface litter of the podzol and throughout the top meter of the peat, indicating an arrestment of the decay processes. In the other soils where the carbon content of the surface horizon is about 5% or less and the drainage is free, the C:N ratios vary mostly between 13 or less in the surface and about 8 in the deeper horizons. However, in the red Mediterranean soil (No. 6 in Table 1) the ratio at a depth of 80 cm is only 4, and values as low as 1 were reported by RODRIGUES [1954] for the lower horizons of certain tropical soils. His results for four soils are presented in Table 2 and show that, after correction for the amounts of exchangeable and fixed ammonium nitrogen, the ratios of organic carbon/organic nitrogen show a decline with depth in three cases, but in no case was the value less than 5. The same decline of the organic carbon/organic nitrogen ratio was reported by STEVENSON [1959].

Table 1. Carbon:Nitrogen Ratios of Representative Soils.

Depth, cm	Types of Soil											
	1	2	3	4	5	6	7	8	9	10	11	12
0												
	25			12	13		11					
	20							11		16	11	
10			31	11	10	13						11
	31	29										
20									13		10	
	16				10	8	13	9				
30				11								9
		12										
			31		10						7	
40									8			
	12											
				11			12	9				
50		23				5						
					12					12	8	
60									10			
												8
70			32									
		6					12					
80						4						
					9			10			8	
90									9	12		
100												

1. Tundra soil, Alaska, RIEGER [1966]
2. Humus iron podzol, England, DAVIES and OWEN [1934]
3. Peat, Scotland, GLENTWORTH and MUIR [1963]
4. Brown forest soil (cultivated), England, KAY [1936]
5. Gray brown podzolic, U.S.A., BROWN and THORPE [1942]
6. Red Mediterranean, Spain, GUERRA et al. [1966]
7. Chernozem (cultivated), U.S.S.R., SHUVALOV [1964]
8. Chestnut soil (cultivated), Rumania, CERNESCU [1964]
9. Vertisol (cultivated), Spain, GUERRA et al. [1966]
10. Ferruginous tropical soil, Angola, D'HOORE [1964]
11. Yellowish brown ferralitic soil, Congo, D'HOORE [1964]
12. Humic ferralitic soil, Congo, D'HOORE [1964]

(d) Inorganic nitrogen

As has already been indicated in Section (b), inorganic nitrogen occurs principally as ammonium and nitrate, and only under exceptional circumstances does nitrite accumulate temporarily in soils.

α. *Free and Fixed Ammonium*

The chain of microbiological transformations that normally results in release of ammonia is very much influenced by environmental conditions within the soil, particularly the moisture status, temperature, reaction, and proportion of decomposable carbon compounds. Consequently, the rate at which ammonium becomes available varies greatly from soil to soil, and in any one soil, with seasonal climatic or cultivation changes.

For a given rate of release, the fate of the ammonia depends on whether conditions are favorable to its absorption directly by plants or to its oxidation to nitrate, and thus the amount and distribution of ammonia nitrogen in soils fluctuate widely.

The course of nitrification is certainly checked by acidity, as in podzolic forest soils, where virtually all the mineral nitrogen is found in ammonium form. Nitrification is also inhibited by anaerobic conditions, and therefore little nitrate is found in peat soils unless they are drained and are not too acid. Temporary accumulation of ammonia to high levels can follow partial sterilization of a soil or compost, due to the spores of ammonifying bacterial species being generally more resistant to heat or chemical treatment than those of the nitrifying species, with the result that the former proliferate more rapidly and build up a concentration of ammonia until the balance of the nitrifying population is restored.

WALKER and THOMPSON [1949] reported increases in glasshouse soils from about 10 ppm ammonia nitrogen to 150 ppm or more a few weeks after steaming. High levels can also result from heavy applications of ammonium salts, urea, or free ammonia as fertilizers, and it

Table 2. Organic Carbon/Organic Nitrogen Ratios of Tropical Soils (after correction for exchangeable and fixed NH_4) RODRIGUES [1954].

Soil	Depth, cm	C:N Ratio	Organic C/Organic N
British Honduras Yellow podzolic on alluvium	10–25 65–105 135–150	9 3 1	12 7 7
British Honduras Yellow Podzolic on alluvium	0–15 60–75 135–150	9 8 8	10 12 16
Trinidad Rendzina on calcareous colluvium	0–15 15–38 38–53	9 9 3	14 16 10
British Guiana Low humic gley on alluvium (Much $(NH_4)_2SO_4$ applied for sugar cane)	0–45 58–70 100–125 133–165	10 4 3 2	15 14 9 5

is now apparent that ammonium nitrogen can be sequestered in soils not only by fixation in "nonexchangeable" form within the lattice structures of certain clay minerals, but also by a chemical interaction between ammonia and soil organic matter components.

Both forms of ammonia fixation have been reviewed by NOMMIK [1965], and it is clear that, while the mechanism of ammonia retention by the mineral fractions of soils is well understood and the amounts of exchangeable and nonexchangeable ammonium can be determined with precision, the mechanism of its reaction with the organic fractions is still obscure. According to JANNSON [1960], ammonia condenses with phenolic constituents involving autoxidation under alkaline conditions, resulting in the formation of quinone-imines. Other possible reactions have been discussed in detail by MORTLAND and WOLCOTT [1965].

Certainly in detailed studies of the organic nitrogen compounds of intensively manured soils, the possibility of such compounds being present should be borne in mind.

β. Nitrite and Nitrate

FRAPS and STERGES [1947] showed that nitrite-nitrogen can accumulate to levels of 200 to 300 ppm in calcareous soils following the application of ammonium sulfate. Since then, the studies of STOJANOVIC and ALEXANDER [1958] and other workers have confirmed that at pH values above neutral, free ammonia in the soil inhibits normal microbiological oxidation of nitrite to nitrate, so that the former may sometimes accumulate to levels toxic to plants.

Nitrite is possibly also produced as an intermediate in the anaerobic reduction of nitrate to nitrogen oxides or nitrogen during denitrification, but the concentration does not build up as it does in the nitrification process.

BROADBENT and CLARK [1965] have discussed the mechanism of denitrification; and the possibility of chemical reactions between nitrite, nitrous acid, nitrosonium free radicals — NO^+, or nitrogen oxides with organic components, especially aromatic constituents, of the soil under natural conditions have been considered by BREMNER [1965c] and by MORTLAND and WOLCOTT [1965]. It appears unlikely that such products accumulate to a significant extent, but this possibility is worthy of more critical examination.

The amounts of nitrate formed naturally in soils depend very much on the environmental conditions that affect the nitrifying organisms, and the amount present at any particular time depends on the balance between rate of formation, rate of uptake by plants, and rate of loss through other causes, such as leaching. Over and above these factors is the direct addition of nitrate as fertilizers or in organic manures.

Being freely soluble in the soil-water and not absorbed by electronegative soil colloids, nitrate moves in any flow of water within the soil profile. Hence, in humid climates nitrate formed in or applied to the surface horizons tends to be leached downward by continuing rains, but in structured clay soils the nitrate may be unevenly distributed in consequence. Thus, WEBSTER and GASSER [1959] reported laboratory evidence in support of the opinion expressed much earlier by LAWES, GILBERT, and WARINGTON [1882] that following dry weather the first-drainage from the Rothamsted soil occurred through the cracks and macro-pores and contained a lower concentration of nitrate than subsequent runnings.

Under arid climatic conditions nitrate may be concentrated to high levels in the upper layers as a result of upward water flow in response to surface evaporation. WETSELAAR [1961] found maximum accumulation of more than 50 ppm nitrate-nitrogen at a depth of about 1 in. in a red clay loam of northern Australia, but in saline spots the concentration may be much higher. In tropical regions with a monsoon climate high concentrations of nitrate may accumulate after a long period of drought with the onset of the first rains and vigorous nitrification.

GRIFFITH [1951] reported that sometimes the top 15 cm of fallow soil in Uganda temporarily contains more than 100 ppm nitrate-nitrogen.

If soil samples are dried slowly in a warm atmosphere microbiological mineralization and nitrification may lead to considerable production of nitrate, so that the whole pattern of the simpler organic nitrogen compounds is changed. This point should be appreciated when studies of the more mobile organic nitrogen compounds are contemplated. Immediate analysis of the fresh samples or storage in deep freeze may be necessary. The extent of variations in the distribution of organic and mineral forms of nitrogen with season and within the profile should also be considered.

2. Simple Monomeric Organic Nitrogen Compounds

In a soil with an active flora and fauna one would expect to find a wide range of monomeric compounds arising partly from the breakdown of organic polymers and partly as normal byproducts of metabolic processes. Of the simple nitrogen-containing compounds, the presence of free amino acids in soils has now definitely been established, although their existence is transient and consequently their concentration is low but variable. They originate in a number of ways: as excretory products of invertebrates, secreted by microorganisms and the roots of higher plants, as intermediates in the decomposition of protein, and by plant cells sloughing off.

(a) Methods of Study

SCHREINER and SHOREY [1910a], at the beginning of this century, using classical organic chemical techniques, were able to identify arginine and histidine in a 0.2% sodium hydroxide extract of soil and the nitrogen bases cytosine, xanthine, and hypoxanthine in the same extract. POTTER and SNYDER [1915] were the first to make a quantitative estimate of unbound amino acids in soils. Using a copper precipitation method, the quantity was lower than the ammonia-nitrogen content, ranging between 5 and 15 ppm amino acid-nitrogen of the air-dry soil; it decreased on incubation. Neither case provides uncontroversial evidence for the presence of free amino acids, because an alkaline extraction was used in both and this can cause hydrolysis of protein. In fact, BREMNER [1950a] was unable to locate any free amino acids in concentrated cold-water extracts of several air-dried soils, and this he attributed to rapid microbial breakdown of the amino acids. Their short life in soils has been demonstrated by QUASTEL and SCHOLEFIELD [1949] and GREENWOOD and LEES [1956], and consequently success in isolation and identification must depend to a large extent on the soil treatment and the handling of the extract.

The usual procedure for isolating free amino acids from biological material is to extract with water or 80% ethanol, at room temperature. But when this procedure is applied to soils such extracts are very dilute and need to be concentrated by evaporation of the solvent. To avoid this initial dilution and also to obviate the effects of microbiological decomposition or chemical change, DADD, FOWDEN and PEARSALL [1953] expressed the natural soil solution from a range of nine fresh soils and plant residues, and then concentrated the clear extracts *in vacuo* without heating. They identified 13 common amino acids by paper chromatography, in amounts up to 4 μg N per ml for individual acids and totaling from 2 to 35 μg N per ml of soil solution. ALEXANDROVA [1960] reported similar results for Russian soils, but more recent studies by KOTER, MAZAR and CHODÁN [1964] have indicated that some amino acids are absorbed by the clay fraction of soils.

PUTNAM and SCHMIDT [1959], using 80% ethanol, determined the yields of nine amino acids and found that these ranged from 5 to 31 μg per kg soil. Later, work by PAUL and SCHMIDT [1960] demonstrated that both 0.1 N barium hydroxide and 0.5 N ammonium acetate solution at pH 6.8 gave higher recoveries of amino acids added to soils, especially basic amino acids, than either water or ethanol. Recently, PAUL and TU [1965] reported that increasing the concentration of ammonium acetate to 1 N gave less variable and more quantitative results.

Concentration of the extracts can cause the loss of amino acids by reactions of the Maillard type with sugars and polyphenols present in the extract. In a comparison of concentration methods, PAYNE, ROUATT, and KATZNELSON [1956] noted a darkening in color during evaporation *in vacuo* at 40°C and obtained a higher recovery by freeze-drying the extract. BISWAS and DAS [1957] and NYKVIST [1963] removed sugars before concentration by absorbing the amino acids on a cation exchange resin and then eluting them with ammonia.

(b) Free Amino Acids and Amino Sugars

The concentrated extract is desalted, where necessary, and can then be applied either to a paper or thin-layer chromatogram and developed in one or two directions. Two-dimensional chromatography provides better resolution of the amino acids, but the spots are too diffuse for reliable quantitative estimation. GROV and ALVSAKER [1963] separated amino acids in a concentrated aqueous extract from a pine forest soil on a circular paper chromatogram using multiple development technique. After treatment with ninhydrin the spots were cut out and eluted with ethanol, and the color intensity of the amino acid–ninhydrin-copper complex was measured. Some of the results are reproduced in Table 3.

The first accurate quantitative estimate of free amino acids was made by PUTNAM and SCHMIDT [1959] on an 80% ethanol extract of soil using the MOORE and STEIN [1951] ion ex-

Table 3. Distribution of Free Amino Acids in Soils (expressed as μg per g of dry soil material).

	Pine forest soil GROV [1963a]			Waukegon silt loam PAUL and SCHMIDT [1961]	
	F layer	H layer	A layer	Non-rhizosphere	Rhizosphere
Aspartic acid	74.1	66.7	68.0	0.2	0.59
Glutamic acid	77.6	73.3	67.2	0.31	0.32
Serine	28.0	38.1	41.4	0.16	0.09
Threonine	23.8	49.5	82.0	0.20	0.19
Glycine	53.1	51.4	69.0	0.21	0.32
α-Alanine	62.2	63.8	103.4	0.18	0.45
Valine	51.1	44.3	25.8	0.20	*
Leucine	—	84.3	28.2	0.13	*
Isoleucine	45.5	20.5	7.1	0.12	0.32
β-Alanine	Trace	Trace	—	0.09	0.07
Tyrosine	—	—	—	0.27	0.46
Phenylalanine	16.1	24.8	Trace	0.12	0.18
γ-Aminobutyric acid	—	Trace	—	0.08	*
Lysine	Trace	—	—	0.52	0.99
Histidine	Trace	—	—	0.33	0.41
Hydroxyproline	16.1	—	—		
Methionine	42.6	65.7	48.3		
Ethanolamine	22.7	43.8	Trace		

change technique, and the same procedure was later applied to a 0.5 N ammonium acetate extract by PAUL and SCHMIDT [1961]. These latter results are presented in Table 3, and it is apparent that the distribution conforms with the semiquantitative data, obtained with the aid of paper chromatographic methods, by DADD et al. [1953], BISWAS and DAS [1957], SINGH and BHANDARI [1963], SIMONART and PEETERS [1954], ALEKSANDROVA [1960], NYKVIST [1963], and ROBERT [1964]. With some variations, the most prominent amino acids are aspartic, glutamic, serine, threonine, glycine, α-alanine, valine, and phenylalanine. SULOCHANA [1962] has commented that this pattern agrees with the general distribution of amino acids in microbial and plant tissues and with those secreted by the roots of higher plants.

The concentration of free amino acids extracted from the acid pine forest soil (pH < 4.0) is very high compared to that of the Waukegan silt loam under a soybean crop (Table 3). It is interesting to note that JONES et al. [1962] reported the level of biotin in the surface layer of a coniferous forest soil to be over 10 times the level in a New Zealand surface pasture soil. The quantitative and qualitative distribution of amino acids in the F and H layers corresponds closely to that in an aqueous extract of fresh pine needles obtained by MUIR et al. [1964]. GROV [1963a] considers that some hydrolysis of protein may occur at the low pH of this soil with the liberation of amino acids, and in fact, DADD et al. [1953] found an inverse relationship between the concentration of free amino acids and soil pH. The relatively high values attributed to the A layer are probably due to leaching down the profile.

PAUL and SCHMIDT [1961] found that the levels of amino acids in the rhizosphere and nonrhizosphere samples of the cultivated soil were approximately the same as each other, but that they were increased considerably by incubation in the presence of glucose and potassium nitrate. Maximum levels were obtained at three days, and this appears to represent the equilibrium state between microbial synthesis and breakdown. The high levels reached in the amended soil, approaching those in the forest soil, are explained by the presence of available carbohydrate-stimulating microbial activity. Although the concentration of amino acids may be very small and their presence short-lived, LOCHHEAD and THEXTON [1947] have suggested that one of the characteristic rhizosphere effects is the preferential encouragement of organisms for which amino acids are essential or stimulative. In studies of the amino acids in the rhizosphere great care must be taken in the treatment and handling of samples if the results are to represent truly the environment in the natural soil. PAYNE et al. [1956] reported an increase in the level of amino acids due to air drying, while NYKVIST [1963] suggested that the pattern of amino acids changes if the sample is stored moist. PAUL and TU (1965) found that air drying for one week produced a sharp increase in the quantities of certain amino acids but that on extended drying the levels fell. Both steaming and oven drying raised the level tenfold; the least detrimental treatment appeared to be freeze-drying of the sample.

Up to the present, most of the work has been directed toward the isolation and identification of free amino acids, although as might be expected, the presence of small peptides in soils has been reported by GROV and ALVSAKER [1963] and PAUL and TU [1965]. Neither galactosamine nor glucosamine, both of which are present in soil in the bound form, have yet been identified in the monomeric form.

Breakdown of amino acids in soils is rapid, and only trace amounts of added samples can be detected after two or three days incubation. There is an initial deamination step followed by rapid oxidation of the carbon chain, although very little keto acid has been detected. QUASTEL and SCHOLEFIELD [1949] found the rate of nitrification of most amino acids to be the same as for ammonium chloride, the exception being DL-methionine, which was not oxidized and in fact inhibited nitrification of ammonium chloride.

According to GREENWOOD and LEES [1956], some of the carbon and nitrogen of amino acids added to soil is retained in a form whose C:N ratio approaches that of protein, presumably of microbial origin. They reported that methionine and threonine were resistant to decomposition under both aerobic and anaerobic conditions, but SCHMIDT, PUTNAM, and PAUL [1960] did not find that these two were more stable than other amino acids. The latter authors noted the appearance of β-alanine after 24 hours incubation, which may arise from the decarboxylation of aspartic acid or by resynthesis by soil microorganisms.

3. Nitrogen Compounds Derived by Hydrolysis and Other Degradative Methods

Monomeric nitrogen compounds account for only a small proportion of the soil nitrogen, but the chemical composition of the remainder is still largely unresolved. This is not due to lack of interest on the part of the soil chemist, but rather to the nature of the material with which he works. Considerable progress has been achieved in recent years with the aid of modern analytical techniques such as partition and ion-exchange chromatography. Further elucidation of the chemistry of this fraction should help not only to explain the role of the organic nitrogen compounds in the nitrogen cycle but also to throw some light on the origin and structure of soil organic matter as a whole.

(a) Methods of Study

Soil organic nitrogen was long considered to be present largely as protein, since it is mainly derived from plant and microbial tissues. The early work on this subject has been adequately reviewed by MORROW and GORTNER [1917], and subsequent studies have confirmed that 40 to 50% can be accounted for as α-amino acid and amide nitrogen, following treatment of soils, according to the standard biochemical procedure for protein hydrolysis with hot 6 N hydrochloric acid. Nevertheless, neither pure protein nor "protein"-free material have yet been isolated from soils by normal biochemical techniques.

There are two possible approaches to this subject. Either the soil as a whole can be subjected to hydrolysis or other degradative treatments and the component units separated and identified, or the organic material can be extracted and fractionated before treatment; but separation of the organic from the inorganic soil constituents with which they are so intimately associated is a major problem in itself.

Acid hydrolysis of the whole soil brings into solution metal ions which are difficult to remove and which may catalyze the degradation of organic compounds or interfere with colorimetric determinations. Hydrolysis of pure protein in 6 N hydrochloric acid causes little decomposition of amino acids, but in the presence of carbohydrate material deamination may occur, and certainly the amount of acid-insoluble material is increased. Useful information has been obtained in this way, but the yields are thought to be on the low side.

α. Extraction

The search for a "universal" solvent for soil organic matter has been going on for over a century. The requirement is one that will extract a representative sample, if not all, of the organic material from soils without producing artefacts. The likelihood of finding a universal solvent is remote because of the wide spectrum of molecular size and composition of the polymers involved. No specific solvent for the extraction of nitrogen-containing polymers has proved applicable to soils. The use of aqueous solvents has been reviewed by TINSLEY and SALAM [1961]. Of these, dilute alkali is the more widely used, and although it has been criticized, in terms of yield of material extracted, it is generally the most efficient. The system of SPRENGEL [1826], with few modifications, is still widely applied. Commonly a dilute acid pretreatment is given to remove metals before extraction with dilute alkali, and after removing the residual soil, the extract is acidified to precipitate the humic acid fraction, leaving the fulvic acid fraction in solution. The organic material remaining undissolved in the soil residue is generally termed "humin."

The objections to alkaline extraction are that both humified and nonhumified materials dissolve and that hydrolysis, oxidation, and condensation can occur in such a medium. The division between humified and nonhumified material is in practice completely arbitrary; nevertheless, there is an obvious danger that nonhumified plant or microbial protein dissolved along with the humified material may appear to be an integral part of the humified material.

Hydrolysis in an alkaline medium is a real danger. The continual dissolution of soil nitrogen on extended shaking with dilute sodium hydroxide was interpreted by BREMNER [1949a] as being due to depolymerization, while MEHTA, DUBACH, and DEUEL [1963] observed that the molecular weight of humic acid estimated by the Sephadex gel filtration technique decreased on standing for one month in dilute alkali. Condensation reactions of the browning-type between ammonia, amines, or amino groups and other reactive groups may also occur at elevated pH values. BREMNER [1950b] found that alkaline solutions of soil organic matter take up oxygen from the atmosphere to form carbon dioxide by autoxidation. For this reason some workers prefer to carry out alkaline extractions under an inert atmosphere of nitrogen, but there is no clear evidence that this is an advantage.

Further disadvantages of an alkaline extractant are that it dissolves silica, while fine clay particles peptized in alkali are flocculated and coprecipitated with humic acid in acid medium. Consequently, it becomes difficult to remove these contaminating inorganic components.

Neutral complexing extractants that complex metal cations, such as sodium fluoride and pyrophosphate, are preferable on chemical grounds to sodium hydroxide, but in general they extract less organic nitrogen, as the following results obtained by BREMNER and LEES [1949] show:

Percent of nitrogen extracted from four soils by sodium hydroxide, sodium pyrophosphate, sodium carbonate, and sodium fluoride.

Extract/Soil	Garden loam	Broadbalk clay loam	Hoosfield clay loam	Burwell fen peat
0.1 M $Na_4P_2O_7$ (pH 7)	9.3	11.0	11.2	22.0
0.5 M NaF	7.3	7.6	10.0	21.9
0.5 M NaOH	26.0	26.0	20.0	24.9
0.5 M Na_2CO_3	9.0	11.2	11.8	25.5

These figures indicate that the neutral and mildly alkaline extractants dissolve considerably less nitrogen than sodium hydroxide, except in the organic fen peat.

Other chelating agents such as the disodium salt of EDTA at pH 7 have proved to be effective extractants, though the disodium salt of EDTA is a nitrogen-containing reagent. DUBACH, MEHTA, and DEUEL [1961] found that it dissolved almost all the organic matter from the B horizon of a podzol, but according to SCHNITZER, WRIGHT, and DESJARDINS [1958], very little is dissolved at pH 4 from the A_0 horizon.

A more recent development along these lines has been the use of a chelating resin, Dowex A–1 (Na^+ form), by BREMNER and HO [1961] to take up metal cations and disperse the organic polymers in aqueous solution. This is a mild extractant with an efficiency equivalent to sodium hydroxide but with the added advantage that the extracting reagent is very easily and completely separated from the solution by centrifuging. This method holds great promise for future studies where labile bonds are present which might be easily ruptured by less mild solvents.

Highly organic soils present peculiar problems, as they contain a large proportion of nonhumified material, but little mineral matter. In such cases it may be more convenient to remove the inorganic component with dilute hydrochloric and hydrofluoric acids, leaving the organic fraction intact, according to the procedure described by BROADBENT and BRADFORD [1952].

CHOUDRI and STEVENSON [1957] found that washing a range of soils with dilute hydrochloric acid or with a mixture of dilute hydrochloric and hydrofluoric acids prior to extraction with sodium hydroxide or sodium pyrophosphate increased the quantity of nitrogen dissolved. The acid pretreatment itself brought some nitrogen into solution which may be accounted for as fixed ammonia or may be due to acid hydrolysis.

The use of organic solvents for the extraction of soil organic matter has been neglected. Alcohol, benzene, and ether were employed by WAKSMAN and STEVENS [1930] for the removal of fats, waxes, and resins, while German workers, notably SPRINGER [1943], used anhydrous acetic acid to dissolve "humified" material. MARTIN and REEVE [1957] found aqueous-acetone solutions of cupferron, oxine, and acetylacetone to be very effective solvents for the podzol B horizon, but again little was dissolved from other soils.

PARSONS and TINSLEY [1960] examined the use of anhydrous formic acid as a possible extractant because of its good solvent properties for plant components and other large polymers. The yield of nitrogen, which followed closely that of carbon, dissolved by two 30-minute boiling periods, varied between 27 and 43% from a wide range of soils. Dissolved material was recovered by precipitation with di-isopropyl ether, and purification was attained by a number of precipitations in the presence of 1% acetyl chloride. Addition of acetyl acetone (10% v/v) to formic acid has greatly increased the yield of organic material dissolved. Dimethyl formamide is a neutral polar organic solvent containing nitrogen, which by itself dissolves comparatively little organic matter from mineral soils, WHITEHEAD and TINSLEY [1964], however, found that addition of various metal-complexing compounds brought a higher proportion of organic matter into solution than did formic acid without acetyl acetone.

β. Fractionation

Separation into humic acid, fulvic acid, and insoluble humin fractions has become established as the classical fractionation technique with aqueous solvents. Unfortunately, on top of all the criticisms directed at alkaline extractments there is some variation in the strength of sodium hydroxide used, the pH of fractionation, and the type of purification procedure employed, which makes comparison between results of various workers extremely difficult. Apart from these variations the division itself is rather arbitrary. For instance, HOBSON and PAGE [1932a] found that total nitrogen extracted with sodium hydroxide increased with the time of shaking, but the ratio of humic to fulvic acids remained approximately constant. This, they suggested, was due to an association between fulvic and humic acids in soil which allows the former to come into solution along with humic acid as it is dissolved. Heavy metal ions are thought to aggregate soil organic polymers, and it is noticeable in the laboratory that as the inorganic contamination is decreased, the solubility in water increases. WRIGHT and SCHNITZER [1960] noted that an interaction of mineral and low molecular weight organic constituents prevented the passage of the latter through a dialysis membrane.

If soil nitrogen or part of it is present as protein which is mixed and not attached to humified material, it should be possible to separate the two polymers. Although there is a distribution of nitrogen between the humic and fulvic fractions, in both cases approximately one-third can be identified as α-amino acid nitrogen following acid hydrolysis. Attempts to remove nitrogen from humic acid by dilute acid and enzymatic hydrolysis, notably by HOBSON and PAGE [1932b] and FORSYTH [1947a], met with only partial success.

BREMNER [1955a] examined the nitrogen distribution between grey and brown humic acids of a chernozem, the latter being soluble in 5% sodium carbonate, the former remaining flocculated. The brown humic acid contained more nitrogen, a higher proportion of acid-soluble nitrogen, and more α-amino acid nitrogen than the grey humic acid. However, the amino acid composition of the two hydrolysates was very similar. These findings were later confirmed for a wide range of soils by SCHARPENSEEL and KRAUSSE [1962].

DUBACH and MEHTA [1963] argue on the basis of recent evidence provided by DUBACH, MEHTA, and DEUEL [1961] and ROULET, MEHTA, DUBACH, and DEUEL [1963] that in the B_h horizon of a podzol, carbohydrate and protein are not covalently bound to humic substances but exist as a series of separate polymers. They do, however, suggest that there may be extremely labile bonds linking the polymers which are broken by dilute acid or alkaline hydrolysis. The podzol B horizon is purposely chosen by many investigators because the organic matter present is completely humified and is readily extracted by most solvents, but it is a rather special case since this horizon has developed under acid leaching conditions possibly involving chromatographic separation down the profile. WRIGHT and SCHNITZER [1960] were able to dissolve 30 and 95% of the organic material with 0.5 N sodium hydroxide from the A_0 and B_h horizons, respectively, while the humic acid fraction accounted for 70 and 15% of that extracted from the A_0 and B_h horizons, respectively.

A great many attempts have been made, using a wide variety of biochemical techniques, to isolate or even concentrate protein material from soil extracts, but they have met with little success. JENKINSON and TINSLEY [1960] used the Sevag technique to remove protein from sodium sulfite and sodium pyrophosphate extracts of a soil and a compost. Protein in neutral aqueous solution is absorbed at the interface of droplets formed when the solution is emulsified with carbon tetrachloride. The stable gel can be centrifuged off; the amounts of this fraction, called ligno-protein by TINSLEY and ZIN [1954], were as follows:

	Ligno-protein-nitrogen as percent of total nitrogen in extracts	Percent nitrogen content of ligno-protein-nitrogen	α-amino acid-nitrogen as per cent of ligno-protein-nitrogen
Soil	9.1	4.79	64.0
Compost	18.3	4.23	58.0

This fraction thus represents only a small portion of the soil nitrogen, but some concentration of protein was attained. Amide I and II bands, characteristic of peptides and proteins, were observable in the infrared spectra but disappeared after acid hydrolysis.

Long-chain quaternary ammonium compounds will precipitate anions from aqueous solution leaving protein in the supernatant liquid. DAVIES, COULSON, and LUNA [1957] applied this principle as a means of fractionating an alkaline extract of peat. They also found that only a minor portion of the protein fragments was held loosely, the major fraction being strongly bound to the humic material. COULSON, DAVIES, and KHAN [1959a] used partition and ion-exchange chromatography, high- and low-voltage electrophoresis, and gelatine gel diffusion plates in an attempt to fractionate humic acid extracted with pyrophosphate from peat. Some methods failed completely and none produced material free of acid-hydrolyzable amino acid polymers. In addition to the above techniques, SWABY and LADD [1962] applied the Sevag technique and fractional precipitation to humic material obtained from a range of soils by alkaline extraction. Again no protein was isolated from any of the humic acids. STEVENSON *et al.* [1952] separated pyrophosphate extract from a clay soil into a major and minor component by moving boundary electrophoresis. In the acid hydrolysate of the dark-colored main component they identified a number of sugars and amino acids. They suggest that the polysaccharide had been adsorbed on the humic material and the protein had been protected from decomposition by association with the clay. MORTENSEN and SCHWENDINGER [1963] were able to separate three bands from a pyrophosphate extract of the same soil type by continuous flow paper electrophoresis. Infrared spectra indicated more hydrogen-bonded NH groups and $CO\overline{O}$ in the dark-colored material with high electrophoretic mobility and fewer OH and C-H than in the fractions of lower mobility. More recently, SIMONART, BATISTIC, and MAYAUDON [1967] have successfully isolated a small amount of protein with a nitrogen content of 14.8% by treatment of humic acid with aqueous phenol solution, which they suggest breaks hydrogen bonds between humic acid and protein.

The fulvic acid fraction has frequently been discarded because of its high concentrations of iron, aluminium, and salt, depending on the extractant and precipitant used. Nevertheless, FORSYTH [1947b] adsorbed the organic components of this fraction on a charcoal column and separated four fractions by successive elution. The first, the effluent fraction A, accounted for 20 to 30% of the nitrogen and contained simple nitrogen compounds. The second fraction, B, contained phenyl glycosides, and the third fraction, C, a polyuronide free of hexosamine. The final fraction, D, eluted with dilute sodium hydroxide, contained most of the nitrogen, but gave no positive test for protein. In view of the high phosphorus and pentose contents it was thought to contain nucleic acid. STEVENSON [1960] later modified this procedure by dialyzing the fractions. Acid hydrolysis of the dialyzable and nondialyzable, or "colloidal," fractions contained both amino acids and amino sugars, although there was some concentration of the latter in fraction B.

A fraction of the fulvic acid from a humus carbonate soil was obtained by ROULET, MEHTA, DUBACH, and DEUEL [1963] using Sephadex G-75. The carbohydrate content was reduced from 25 to 5%, and the nitrogen from 2.1 to 1.7%; the latter was reduced further to 1.3% on passing the solution through a cellulose anion exchange resin, although no change in the amino acid pattern of the various fractions was observed. The molecular weight of a considerable part of the carbohydrate and nitrogen-containing fraction was over 40,000.

γ. *Hydrolysis*

The hydrolytic procedure commonly adopted for most studies on soil amino acid containing polymers is boiling with 6 N hydrochloric acid for a specific period of time. Modifications have been introduced by adjusting the main variables: volume and concentration of acid, temperature, and time. BREMNER [1949b], in one of the few studies on conditions required for optimum hydrolysis, found that the quantity of α-amino acid – nitrogen liberated on refluxing with 6 N hydrochloric acid reached a maximum at 12 hours and then slowly decreased. A preliminary investigation had shown that both 6 N hydrochloric acid and 8 N sulfuric acid used in the proportion of 3 ml acid per gram of soil gave maximum extraction of nitrogen. This ratio of acid to soil is rather narrow and was increased to 25:1 by SOWDEN [1955] to reduce degradation of amino acids. FELBECK [1965], while attempting to remove protein from a humic acid fraction, found that 3 N sulfuric at 90°C dissolved approximately twice as much nitrogen (50.2%) from soil as 72% sulfuric acid at room temperature (26.9%). The temperature of hydrolysis is usually maintained at 100°C or at reflux temperature, slightly higher. STEVENSON [1957a] suggested that optimum conditions must be established for each type of soil when investigating the amino polysaccharides in soils. He found that the time of maximum liberation of hexosamines with 6 N hydrochloric acid varied with the soil type, but was generally shorter when the soil was autoclaved with 6 N hydrochloric acid at 120°C and 15 lb/sq in. Length of hydrolysis varies from 6 to 48 hours and obviously depends on establishing a compromise between hydrolysis and degradation. SOWDEN [1956] hydrolyzed three soils for 72 and 96 hours with no increase in soluble nitrogen.

Partial hydrolysis has been used by SOWDEN and PARKER [1953] in the detection of free amino groups with the dinitrofluorobenzene reaction of SANGER [1945]. Boiling with 1% hydrochloric acid, 5% sodium bicarbonate at 50°C, or a mixture of concentrated acetic acid and 10 N hydrochloric acid at 37°C for 60 hours each, liberated small amounts of amino acids and a complex mixture of peptides. The acetic-hydrochloric acid treatment appeared to produce more DNP derivatives of hydroxyamino acids.

The advantage of hydrochloric acid over other mineral acids is the ease of removal by evaporation *in vacuo* over potassium hydroxide. However, BREMNER [1949b] reported that up to 50% of the soil as a whole is dissolved by acid hydrolysis. The presence of so much inorganic material makes estimation of and identification of amino acids difficult. LYNCH, HUGHES, and RHODES [1959] hydrolyzed soil with 6 N sulfuric acid and, after filtering, neutralized the hydrolysate with calcium carbonate. This not only removed most of the sulfate ion but precipitated iron and aluminum. Iron in particular interferes with the Elson and Morgan colorimetric estimation of amino sugars, and it was removed by STEVENSON [1957a] along with chloride anions by passing the hydrolysate through an anion exchange resin in the carbonate form and eluting the hexosamines in dilute bicarbonate solution. For amino acid separation STEVENSON [1954] used a cation exchange resin to remove metals and eluted the amino compounds with dilute ammonia.

Similar hydrolysis procedures have been applied to material isolated from soils by alkaline or neutral extractants. Usually a wider ratio of acid to solid is employed, and as the sample is smaller, it is convenient to complete the hydrolysis in a sealed tube. Extracted material can be purified by reprecipitation; consequently, there will be less interference from metals but other organic components may still cause some degradation of amino compounds.

Humic acid by definition is insoluble in acid, and most of it remains insoluble during acid hydrolysis. This must reduce the efficiency of hydrolysis, and it would therefore be preferable to use a reagent in which the material dissolved. Some investigators have used a mixture of formic acid and mineral acid for the hydrolysis of protein because of its solubility in polar-organic acid solvent. Formylation has also been used for the protection of amino and hydroxy groups during the hydrolysis of small peptides. PARSONS and TINSLEY [1961] found that optimum liberation of hexosamines from material isolated from soil by formic acid extracts occurred with a 1:1 mixture of 90% formic acid and 6 N hydrochloric acid held at 100°C for 2 to 4 hours. This gave better recovery than hydrolysis with either aqueous 4 N or 6 N hydrochloric acids over periods of 6 to 16 hours.

Sodium hydroxide has also been used for protein hydrolysis because it does not cause destruction of tryptophan and thus does not produce humin–nitrogen. BREMNER [1949b] found little difference in the amount of humin–nitrogen formed by either acid or alkaline hydrolysis under both reducing and nonreducing conditions, but the color of the alkali hydrolysates was appreciably darker than that of the acid or alkali-stannite hydrolysates. BREMNER [1949b] and KOJIMA [1947a] concluded from the absence of tryptophan in soil alkali hydrolysates that it plays little part in the formation of humin–nitrogen. They consider that this nitrogen is nonprotein, but is possibly in the form of complex heterocyclic compounds.

Using a combination of continuous flow paper electrophoresis and paper chromatography WALDRON and MORTENSEN [1962] obtained 50 ninhydrin positive spots from an acid hydrolysate (6 N hydrochloric acid at 100°C for 18 hours) from a silty clay loam and 53 from a 0.5 N sodium hydroxide hydrolysate (at 105°C for 31 hours). They were able to identify 25 spots in the acid and 24 spots in the alkaline hydrolysate, but there was little qualitative difference between the two patterns. Many of the unidentified spots must be due to small peptides, indicating incomplete hydrolysis.

Perchloric acid has been used for the hydrolysis of nucleic acids, but drastic conditions are required for the release of pyrimidine bases of DNA. The release of nitrogen bases from humic acid preparations is usually accomplished by hydrolysis with either 72% (12 N) or 7.5 N perchloric acid for 1 hour at 100°C.

Having chosen the most suitable hydrolytic agent for the work at hand and having performed the hydrolysis, the component amino acids can be separated either by two-dimensional paper chromatography, by thin-layer chromatography, by ion-exchange chromatography or by electrophoresis. An initial step for the removal of acid and metal cations must be performed before applying the hydrolysate to a chromatogram; otherwise the spots will be poorly defined. For ion-exchange chromatography either the MOORE and STEIN [1954] method of stepwise elution with buffers, or the gradient elution technique with hydrochloric acid of LYNCH, HUGHES, and RHODES [1959] can be used. For the definite confirmation of individual amino acids, paper or thin-layer chromatography must be used in conjunction with the ion-exchange procedure.

(b) Hydrolysis Products

α. Amino Acids

The first reliable quantitative estimates of bound amino acids in soils were reported by KOJIMA [1947a] and BREMNER [1949b]. Kojima, using both the Van Slyke decarboxylation and nitrous acid methods for amino acid determination, found that acid hydrolysis of a muck soil brought into solution 73% of the nitrogen, half of which was in the amino form; *i.e.*, 37% of the total. Adding the value obtained for amide nitrogen, she estimated that half the soil nitrogen could be in the form of amino acid polymers with an upper maximum of 66 to 75%. In a later paper KOJIMA [1947b] described the isolation and characterization of five individual acids by forming crystalline derivatives.

At the same time BREMNER [1949b], working at Rothamsted and using similar analytical methods, found that 68 to 87% of the nitrogen was dissolved from a range of soils by acid hydrolysis and that the amino acid nitrogen content ranged from 24 to 37% of the total. These results agreed with and confirmed Kojima's, but Bremner considered them minimal because of the destruction that occurs during hydrolysis, as indicated by the high ammonia contents.

In a study of the distribution of nitrogen compounds in a number of soil profiles, STEVENSON [1957b] reported amino acid–nitrogen contents mostly between 30 and 40% of the total nitrogen, with extremes of 8.9 and 58.5%. In all the soils examined, except a podzol and a groundwater podzol, the amino acid content decreased down the profile. GROV [1963b] found that the amino acid content expressed as a percentage of the organic matter increased down the profile of a pine forest soil.

According to BREMNER [1955b] the distribution of nitrogen in the acid hydrolysates of humic acids extracted with sodium hydroxide is similar to that of whole soils. In general, alkali-humic acid had a higher nitrogen content, a higher proportion of acid-soluble nitrogen and of α-amino acid–nitrogen than pyrophosphate humic acid. This is in line with the observation that alkali extracts material with a lower C:N ratio than pyrophosphate. CHENG and VAN HOVE [1964] found that for acid soils the ratio of humic to fulvic acid was the same in both sodium hydroxide extracts and for extracts obtained with a chelating resin, but for alkaline soils the results were rather more variable. The proportions of amino acid–nitrogen in the acid hydrolysates of the total soil, the extract, and the soil residue were similar, ranging from 31 to 58 % of the total nitrogen. This suggests that the resin does not extract any distinct fraction of the organic matter but merely a portion similar in composition to that remaining insoluble.

Introduction of paper chromatography in 1944 provided for the first time a reliable and relatively rapid semiquantitative method of identifying individual amino acids. With this technique BREMNER [1950a] compared the amino acid distribution in acid hydrolysates of a number of soils, and identified 20 amino acids and glucosamine. The strongest spots were due to valine, α-alanine, glycine, threonine, lysine, and aspartic and glumatic acids. This pattern agrees very well with that of free amino acids in soils and with the normal distribution in plant, animal, and microbial protein. A similar pattern was obtained from hydrolysates of humic acid preparations by BREMNER [1955b], confirming his earlier results. The most interesting acids identified are the non-α-amino acids β-alanine and γ-aminobutyric acids, the dibasic acid, α,ε-diaminopimelic acid, and α-aminobutyric acid. These are not generally found in the proteins of higher plants, but except for α,ε-diaminopimelic acid, they are fairly widely distributed either in the free form or in peptides. The presence of α,ε-diaminopimelic acid and glucosamine suggests that much of the nitrogen fraction is of microbial origin.

Using a combination of ion-exchange and paper chromatography, STEVENSON [1954] was able to detect 32 amino compounds, identifying and determining 20 of them. In a later

Table 4. Amounts of Amino Acids and Amino Sugars in Soils (mg/100 g of soil).

	Black soil from Lacombe	Flanagan silt loam	Wooster silt loam
	SOWDEN [1955]	STEVENSON [1954]	YOUNG and MORTENSEN [1958]
N-Acetylglucosamine		—	
Glucosamine		25–35	
Galactosamine		6–15	
α-Alanine	46	50.6	17
Glycine	81	35.9	10
Lysine		31.5	13
Threonine	79	39.0	11
Valine	58	26–28	11
Glutamic acid	117	61.9	22
Aspartic acid	146	48.6	29
Serine	48	34.2	
β-Alanine		8.3	1
γ-Amino-n-butyric acid		6.0	0.3
Proline	40	17.0	9
Hydroxyproline		—	4
Isoleucine	34	19–23	7
Leucine	39	20–25	7
Tyrosine	26	10–12	1.0
Histidine		5–8	0.3
Arginine		10–12	1.0
Phenylalanine	36	5–6	4
Ornithine		Trace	7

Other acids identified: cysteine, cysteic acid, methionine, methionine sulfone, α,ε-diaminopimelic acid, and α-amino butyric acid.

publication STEVENSON [1956a] was able to identify 9 more of the remainder. These results are presented in Table 4 and show a qualitative distribution which agrees with that of Bremner.

The presence of nonprotein amino acids, cysteic acid, methionine sulfoxide, and methionines ulfone—probably artefacts from hydrolysis of cystine and methionine—was confirmed. For the first time, ornithine, galactosamine, and N-acetyl glucosamine were identified in soil hydrolysates, and these were later confirmed by WALDRON and MORTENSEN [1962]. The latter authors identified cystine and cysteine, which are generally unstable to acid hydrolysis.

To date, from the many studies on the amino acid distribution in a large number of soils reported by BREMNER [1950a], SOWDEN [1955], OKUDA and HORI [1955], KONONOVA and ALEXANDROVA [1956], YOUNG and MORTENSEN [1958], GROV [1963b], SCHARPENSEEL and KRAUSSE [1962], and WANG, YANG, and CHENG [1964], the interesting conclusion to be drawn is the overall similarity in the pattern of the main amino acids. Obviously the sensitivity of the method limits the number of the less common amino acids identified.

Despite the similarity in the general amino acid pattern of soils, there are certain minor differences that appear to be due to cultural techniques and climatic conditions. RENDIG [1951] found the non-basic amino acid–nitrogen of a virgin forest soil was 23% of the total but only 15% on an adjacent area of the same soil under cultivation. STEVENSON [1956b] reported the following amino acid distribution in samples from the Morrow plots at the University of Illinois, which have been subjected to the same cropping systems since 1904:

Rotation	Percent of amino acid nitrogen of total nitrogen	Percent of basic amino acid nitrogen of total amino acid nitrogen
Untreated corn	30.1	22.0
Untreated corn–oats	30.5	20.6
Treated corn–oats–clover	36.2	15.2
Grass border	39.9	12.2

For the purpose of calculating these results, the basic amino acids were taken as arginine, histidine, lysine, ornithine, and α,ϵ-diaminopimelic acid. The inference to be drawn from these results is that, as the organic reserves are depleted, the amino acid content falls but the basic amino acids are more persistent than the others. Because of their basicity, they may become attached to other soil constituents. Accumulation of basic amino acids has been confirmed by YAMASHITA and AKIYA [1963] in Japanese soils.

Climatic conditions, especially temperature, appear to influence the proportion of basic amino acids in soils. CARLES, SOUBIÉS, and GADET [1958] found lysine and leucine to be the dominant acids in organic soils of the humid tropics, but in cooler regions the lysine content was less. On electrophoretograms of soil hydrolysates CARLES and DECAU [1960a] obtained the largest basic amino acid peak for a West African soil and the smallest for a sample from Versailles. They suggest [1960b] that conditions of high fertility and high temperature favor microbial activity and thus encourage the decarboxylation of α,ϵ-diaminopimelic acid to lysine. However, the lysine content of plant, animal, and microbial tissues from which the humic material is derived will be very much higher than that of α,ϵ-diaminopimelic acid and consequently decarboxylation would have little effect on the proportions of the two acids. The content of both will, of course, be very much higher where there is an active microflora, but it is more likely that they accumulate because of their relative resistance to microbial decomposition. However, this does not mean that microbial decarboxylation does not occur. It is highly likely that the nonprotein amino acids are largely derived from this process because, although some may result as artefacts from acid hydrolysis, especially in the presence of metals, BREMNER [1955c] has shown that both alanine and α,ϵ-diaminopimelic acid are produced during the rotting of straw.

The commonly occurring form of amino acids is the L-isomer. Occurrence of the D-form is extremely rare in plants and animals, but more frequent in bacterial cell wall material. BREMNER [1950a] was able to detect 1.4 to 1.7% of the total amino acid–nitrogen in the D-form by measuring oxygen uptake in the presence of D-amino acid oxidase. Some racemization does occur during acid hydrolysis and this could account for these small amounts, but on the other hand, D-amino acid oxidase which shows absolute specificity to the D-form sometimes shows absolute or restricted substrate specificity. So far it is impossible to separate the D- and L-isomers of most amino acids by paper chromatography, although LL-, α,ϵ-diaminopimelic acid can be separated from the DD- and *meso*-forms.

Free amino groups of proteins, either as end groups or on dibasic amino acids, can be tagged by Sanger's technique. A number of dibasic amino acids have been identified in soil hydrolysates, but SOWDEN and PARKER [1953] found no free amino groups in three different soils or in the humic acid fractions isolated from them. Partial hydrolysis with 5% sodium bicarbonate or 1% hydrochloric acid released very little, but a mixture of acetic and 10 N hydrochloric acids at 37°C released approximately 10% of the nitrogen as free amino groups, mainly in complex peptides. DNP derivatives of hydroxy acids were identified in the latter hydrolysates, indicating that peptide bonds in which these are located may be broken more easily. OKUDA and HORI [1954] and BREMNER [1957] also failed to detect free amino groups by this method. It must be assumed that the dibasic amino acids are involved in cross-linkages between polymers or that steric hindrance prevents tagging. JENKINSON and TINSLEY [1959] were able to detect DNP-lysine following acid hydrolysis of a ligno-protein fraction after DNFB treatment.

KOJIMA [1947a] and BREMNER [1957] obtained anomalous results with the Van Slyke nitrous acid manometric method for the determination of free amino groups. Bremner found that both lignin and humic acid produced a gas other than nitrogen with nitrous acid, lost methoxyl groups, and fixed nitrogen which could not be fully recovered by acid hydrolysis. STEVENSON and SWABY [1963, 1964] identified carbon dioxide and nitrous oxide in the gases evolved from humic acid, while from lignin STEVENSON and KIRKMAN [1964] obtained methyl nitrite.

The presence in soils of amino acids that are generally associated with bacterial cellular material suggests that they are of microbial rather than plant origin. BREMNER [1955c] found the amino acid patterns of straw and composted straw to be similar except for the presence of β-alanine and α,ϵ-diaminopimelic acid in the latter; further supporting evidence has been provided by SØRENSEN [1963]. Incubation of ^{14}C-labeled carbohydrate from barley straw with soil produced an even distribution of the label between humic acid, fulvic acid, and humin with the highest specific activity in the α-amino acids liberated from the humic fraction by acid hydrolysis.

β. Amino Sugars

The amino sugars found in acid hydrolysates of soil are also interesting because they are rarely found in the tissues of higher plants. N-acetylglucosamine is the structural component of insect and fungal chitin, but insect integuments will be of less importance in arable soils. Glucosamine, galactosamine, and other amino sugars and their derivatives commonly occur in the cell wall material or microorganisms, along with amino acids, sugars, and uronic acids. As time goes on, it is becoming more and more obvious that many of these constituent polymers are covalently linked.

The presence of amino sugars in soils was first demonstrated by BREMNER [1949b. 1950a]. and now it is apparent that they account for a relatively high proportion of the total nitrogen in surface soils. STEVENSON [1957a] has reported from 7.6 to 12.9%; BREMNER [1958], from 4.9 to 10.2%; SOWDEN [1959], from 5 to 10%; SINGH and SINGH [1960], from 2.6 to 6%; SALOMAN [1963], from 5 to 7%; and CHENG and VAN HOVE [1964], from 5 to 11%. These values are possibly on the low side, although a correction factor of 1.25 is generally applied, based on the extent of deamination of glucosamine and chitin during acid hydrolysis.

The variation of amino sugar–nitrogen content between surface soils is not large, but there is a greater variation down the profile. STEVENSON [1957c] found a relative accumulation in the lower horizons of a wide range of soils, except for podzols and a solonetz, which is difficult to explain other than by an association with clays to increase their stability. The highest value obtained by Stevenson was 24.3% of the total nitrogen for the B_3 horizon of a planosol where clay had accumulated. SOWDEN [1959], on the other hand, found the highest value of 24.1% in the B_{21} horizon of a podzol, while SINGH and SINGH [1960] found that the maximum amount of amino sugar occurred in the surface horizons of tropical soils with no accumulation in the lower horizons.

According to STEVENSON [1957c], the proportion of amino sugars in the lower horizons appeared to increase with maturity of the soil, corresponding to the relative stability of basic amino acids.

SOWDEN [1959] found the content of amino sugars in whole soils to be higher than that in the humic acids extracted from them. This is in agreement with the results obtained by CHENG and VAN HOVE [1964]. Material extracted by PARSONS and TINSLEY [1961] with formic acid and precipitated with di-isopropyl ether contained 4.8 to 12.4% of the total nitrogen in the form of amino sugars, compared with 1.3 to 4.4% in the material insoluble in formic acid. This suggests that formic acid dissolves more of the amino sugar containing polymers than alkali.

Amino sugars are also present in the acid-soluble fulvic acid fraction. Using Forsyth's fractionation technique, STEVENSON [1960] obtained a value as high as 25.4% for the B fraction, whereas HAYASHI and NAGAI [1961] found most glucosamine in the polysaccharide C fraction.

Although glucosamine appears to be the dominant amino sugar, both galactosamine and N-acetylglucosamine have been identified in soil hydrolysates by STEVENSON [1956a] and WALDRON and MORTENSON [1962]. The presence of N-acetylglucosamine is unexpected, as the N-acyl group is generally removed by 6 N hydrochloric acid hydrolysis. Using weaker 0.4 N sulfuric acid, LYNCH, OLNEY, and WRIGHT [1958] detected glucosamine and N-acetylglucosamine in four Delaware soils. The only other amino sugar identified in soils is talosamine [2-deoxy 2-amino D-talose] in a peat, reported from Germany by WANG and CHENG [1964]. SOWDEN [1959] separated glucosamine and glactosamine and found the sum of these two approximately equal to the total amino sugar content, suggesting that if any other amino sugars remain undetected they must be there in very small amounts.

The form in which amino sugars are present in soil can only be surmised, but their variety suggests they are of microbial origin. It is unlikley that much chitin exists in soils, as both VELDKAMP [1952] and BREMNER and SHAW [1957] have shown that it is fairly rapidly decomposed. Although BREMNER and SHAW [1954] found that glucosamine added to soils is readily deaminated, in polymer form it appears to be stabilized or to be continuously synthesized, as BREMNER [1955c] reported that its concentration remained approximately constant over a period of 90 days in straw composted with ammonium carbonate. Chitin would not be extracted from soil by alkali, but it is soluble in anhydrous formic acid. JACQUIN [1960] observed that a decrease in amino acids was accompanied by an increase in the amino sugar content of incubated soils. especially where the microbial population was rich in bacteria. According to

SOWDEN [1959], the ratio of glucosamine to galactosamine attains a maximum in soils where fungi predominate.

γ. Amides and Ammonia

Ammonia is produced during acid hydrolysis of whole soil, the amount liberated increasing with time. The rate of release is rapid for the first two or three hours and then declines. Some is formed from the deamination of amino acids, especially the hydroxy acids and amino sugars; some from amides; and some is due to the liberation of ammonia by clays; but the origin of the remainder is unknown. Volatile amines would be determined along with ammonia, but no trace of these was detected by BREMNER [1949b] and [1955b].

Most proteins contain the amides asparagine and glutamine, which are deaminated during acid hydrolysis. As aspartic and glutamic acids are two of the dominant acids in hydrolysates of soil, there is the possibility that they were originally present as their amides and thus could account for some of the ammonia produced during hydrolysis. However, it is extremely difficult to determine this. BREMNER [1955b] found that the ammonia liberated during acid hydrolysis of humic acid reached a maximum at six hours, and took this to be the amide nitrogen value. He estimated that 7.3 to 12.6 % of the total nitrogen of the preparation was in the form of amide nitrogen, but this value will include some nonamide nitrogen.

Methylation followed by reduction with lithium borohydride converts free carboxyl groups from dicarboxylic amino acids or end groups to the corresponding alcohol. Where an amide is present in the protein, the carboxyl group is protected, and consequently on acid hydrolysis it appears as the free acid. Thus, the amount of glutamic and aspartic acids after reduction and hydrolysis gives the amide content. SOWDEN [1958] used this technique to determine the amide content of four horizons of a podzol. The results for the largely undecomposed organic layer were doubtful, as the recovery of most amino acids following reduction was low, but those for the other three layers suggest that both acids are present as their amides. The sum of aspartic nitrogen and glutamic nitrogen varied between 7.1 and 9.5 % of the total nitrogen and decreased down the profile. The value for amide nitrogen did not account for all the ammonia nitrogen released by acid hydrolysis, the discrepancy increasing with depth in the profile and length of time of hydrolysis. Other methods that rely on the liberation of amide ammonia in alkaline solution gave higher values, the discrepancy being greatest for the lower horizons.

Narrowing of the C:N ratio down the profile is a common feature, but the reasons are not always clear. STEVENSON [1957b] and SOWDEN [1956] have shown that amino acid–nitrogen and amide nitrogen as a percentage of the total nitrogen decrease down the profile, thus disproving the suggestion that protein material accumulates in the lower horizons. As discussed in Section C, β, some of the increase in the nitrogen content is due to ammonia fixed by clays, part of which is released by sodium hydroxide extraction and appears in the fulvic acid fraction (STEVENSON [1960]). Much of the acid-labile nitrogen appears to be in the fulvic acid fraction, the proportion of which increases in the lower horizons and is associated with the pigmented fractions obtained by Forsyth's procedure. (STEVENSON [1960]).

δ. Nitrogen Bases

Both protein and nucleic acids are fairly rapidly decomposed in soils but, just as amino acid polymers have been found, it would not be surprising if traces of nucleic acids or their derivatives are also present. Nitrogen bases have been detected in hydrolysates of soils and of fractions isolated from soils by SCHREINER and SHOREY [1910b], BOTTOMLEY [1919], and WRENSHALL and DYER [1941]. These studies were confined to the fulvic acid fraction, which is likely to include only derivatives of RNA but not acid-insoluble DNA. ADAMS, BARTHOLO-

MEW, and CLARK [1959] resolved fulvic acid into a number of fractions which they examined by ultraviolet spectrophotometric analysis. But none of the extracts showed a distinct peak in the 255 to 375 nm region, although one small peak was observed in the position of cytidylic acid. Even assuming that all the absorption was due to the nitrogen base, the soil extracts contained very little nucleic acid. ANDERSON [1958] examined the hydrolysates of humic acid fractions from three Scottish soils for derivatives of DNA. After passing the hydrolysates through a column of cation exchange resin, the fractions were spotted on paper chromatograms. The purine bases guanine and adenine, and the pyrimidines cytosine and thymine were identified, guanine and cytosine being predominant. This distribution and the absence of 5-methylcytosine in humic acids suggests that the bases were probably in the form of polynucleotides derived from DNA of bacterial origin. According to ANDERSON [1961] the nucleic acid–nitrogen accounted for only 0.06 to 0.88% of the total nitrogen of a number of soils.

4. Chemical Composition and Biochemical Reaction Mechanisms of Nitrogen Compounds in Soils

Amino acids liberated by acid hydrolysis of soils were for a long time thought to come from protein of either plant or microbial origin, but this view was inconsistent with the known stability of the nitrogen in humus compounds to microbial decomposition. Consequently, a number of theories have been proposed to account for the stability of protein in soils or to explain the chemical structure of the bound amino acid polymers. When pure proteins are incubated with soils, they are rapidly decomposed, so the persistence of soil "protein" has been explained by postulating an association with either the inorganic or other organic components.

(a) Stabilization by Adsorption on Soil Minerals

The existence of clay-organic matter complexes in soils is now well established from X-ray and electron miscroscope studies, but as yet there is no direct evidence to indicate that protein is involved. However, under mild conditions in the laboratory, clay-protein complexes are easily produced in which the protein appears to be so orientated that the active groups are inaccessible to enzymes. McLAREN [1954] and ARMSTRONG and CHESTERS [1964] have shown that adsorption of proteins by both kaolinite and montmorillonite is rapid, mostly within a few minutes, with the maximum occurring at a pH just below the isoelectric point of the protein. This suggests that positively charged amino groups are involved in the linkage. ENSMINGER and GIESEKING [1939] showed that destruction of these groups with nitrous acid reduced the amount of protein adsorbed. In one of the few studies on whole soil BRYDON and SOWDEN [1959] were able to identify free amino groups after removal of free iron oxides by dithionite or EDTA treatments. That such complexes can form in soils is highly probable, but there is some disagreement over their stability to enzymatic and microbial hydrolysis.

ENSMINGER and GIESEKING [1942] found that clays with a high base exchange capacity, such as Wyoming bentonite, do protect protein from hydrolysis by pepsin in acidic and pancreatin under alkaline conditions. PINCK, DYAL, and ALLISON [1954] demonstrated the resistance of a montmorillonite-gelatin complex to decomposition by a soil infusion when absorbed as a monolayer in the interlayer spacing. A double layer was hydrolyzed, with some of the products being held by the clay.

On the other hand, ESTERMANN, PETERSON, and McLAREN [1959] reported that monolayers of denatured lysozyme on both kaolinite and montmorillonite are hydrolyzable by enzymes, although the rate of decomposition of the montmorillonite complex was retarded.

According to McLaren and Estermann [1956], hydrolysis results from the formation of an enzyme-substrate-clay complex rather than enzyme activity in solution. The presence of clay did not affect the rate of proteolysis of lysozyme by either a mixed soil infusion or pure bacterial cultures. Under certain conditions Estermann and McLaren [1959] found that the presence of kaolinite stimulated the release of ammonia from protein by microorganisms, the clay probably acting as a concentrating surface.

The availability of active groups in the protein to enzyme hydrolysis is likely to be influenced by the physical nature of the complex, resulting from the treatment of the two components before and during complex formation. Those complexes which had first been thoroughly dried over magnesium perchlorate or phosphorus pentoxide exhibited great stability, whereas Estermann, Peterson, and McLaren [1959] used undried suspensions. They found that alternate drying and wetting of a montmorillonite-lysozyme complex increased its resistance, while Pinck, Dyal, and Allison [1959] reported that a mixture of clay and protein suspensions was much less stable than the dried complex. It is interesting to note that Estermann, Peterson, and McLaren [1959] found protein–silica gel complexes to be very much more stable than those involving clays.

If such complexes do occur in soils, the protein moiety should be extracted with a solvent having a pH above its isoelectric point. Assuming that no major alteration occurs during extraction with sodium hydroxide, protein would be precipitated in the humic acid fraction. However, true protein has never been identified in this fraction, although approximately half the nitrogen can be accounted for as bound amino acid–nitrogen. The generally accepted explanation is that the nitrogen compounds are stabilized by reaction with other organic components during the normal course of humification in soils, and such stable complexes are then further protected by adsorption on clay minerals or on hydrated oxides of aluminum and iron.

(b) Stabilization by Reaction with Organic Components

A number of biosynthetic and chemical reaction mechanisms have been postulated for the formation of such complexes which probably constitute a part at least of the resistant core of the soil organic matter. Some authorities, notably Dubach and Deuel, Schnitzer, and Burges, claim that nitrogen is not an integral part of humic acid. They consider the portion that cannot be accounted for as amino acid–nitrogen to be an artefact of the hydrolysis procedure, but their work has been concerned predominantly with podzol B material. Undoubtedly in the upper layers of other soils, a fraction of the organic nitrogen is virtually biologically inert.

α. *Ligno-Protein Complex*

Waksman and Iyer [1932, 1933] suggested that soil protein is stabilized by the formation of a complex with lignin of the type previously prepared by Jensen [1931]. They estimated that the sum of lignin plus protein in six soils accounted for an average of 78% of the total organic material. They went on to show that the complex formed on acidification of an alkaline mixture of lignin and protein possessed properties very similar to those of humic acids. The brown polymer was soluble in alkali with a nitrogen content of 2 to 3%, of which less than one-third was liberated as amino acid–nitrogen by acid hydrolysis, although the conditions employed were milder than in the usual procedure. The complex was resistant to microorganisms, a characteristic that was verified by Estermann, Peterson, and McLaren [1959], who showed it to be more stable than clay-protein complexes.

Waksman's theory gained wide support, and in fact it had much to commend it since it accounted for a major portion of humus and at the same time explained the stability of proteins in soils. Most of the evidence for it was obtained from laboratory studies on synthetic complexes but in the ensuing years attempts to isolate ligno-protein from soils have not been very successful. MORRISON [1958, 1963] and SWABY and LADD [1962] found that alkaline nitrobenzene oxidation of soils yielded only traces of the products obtainable from true lignin and the infrared spectra show certain points of difference. TINSLEY and ZIN [1954] and JENKINSON and TINSLEY [1959] isolated less than 10% of the soil nitrogen in the form of a "protein" complex using the Sevag emulsion technique, although a higher proportion was obtained from a straw compost.

Clearly the major part of the soil organic matter is not a simple complex of residual plant lignin and microbial protein.

β. Tannin Complexes

To explain a mor-type of organic matter HANDLEY [1954] postulated that leaf protein is stabilized by a process analogous to the tanning of hide with vegetable tannins, the chemistry of which has been reviewed by GUSTAVSON [1949].

According to this hypothesis, the tanning reaction occurs in the mesophyll cells and the resulting amorphous material forms a protective coating to the residual cell walls. The polyphenolic compounds in plants associated with mull formation produce tanned proteins, which are more easily decomposed by microorganisms. However, the greater microbial activity in mull soils contributes a considerable amount of organic matter by way of microbial synthesis, and Handley ascribes the stability of nitrogen compounds in mull soils to the interaction of microbial protein and hydroxy aromatic compounds also of microbial origin.

γ. Polyphenol Complexes

The concept of a chemical or enzymatic oxidation of phenolic compounds and the polymerization with amino compounds is common to a number of theories on the stabilization of organic nitrogen compounds in soils. Reactions of this type are known to occur in both plant and animal cells—on the one hand, the so-called browning reactions in the tissues of certain fruits and vegetables when exposed to the atmosphere, and on the other, the production of melanins and the hardening or sclerotization of arthropod cuticles. The biochemistry of the enzymes responsible for hydroxylation of phenols and dehydrogenation of diphenols, the phenolase complex, was reviewed by MASON [1955]. Quinones and especially ortho-quinones produced by phenolase enzymes are extremely reactive, although the reacting species is probably a semiquinone free radical.

Quinones polymerize in alkaline solution to form brown insoluble polymers of relatively large molecular size. FLAIG [1950], taking ortho- and p-dihydroxybenzene as starting materials, reviewed the possible polymerization reactions and compared the polymers produced to humic acids. Including ammonia in the reaction mixture, he was able to produce polymers with nitrogen incorporated in the structure.

The principal reactions of diphenols under oxidizing conditions and in the presence of amino acids are typified for catechol in Figure 3.

A molecule of amino acid reacts with an ortho-quinone by nuclear substitution in the

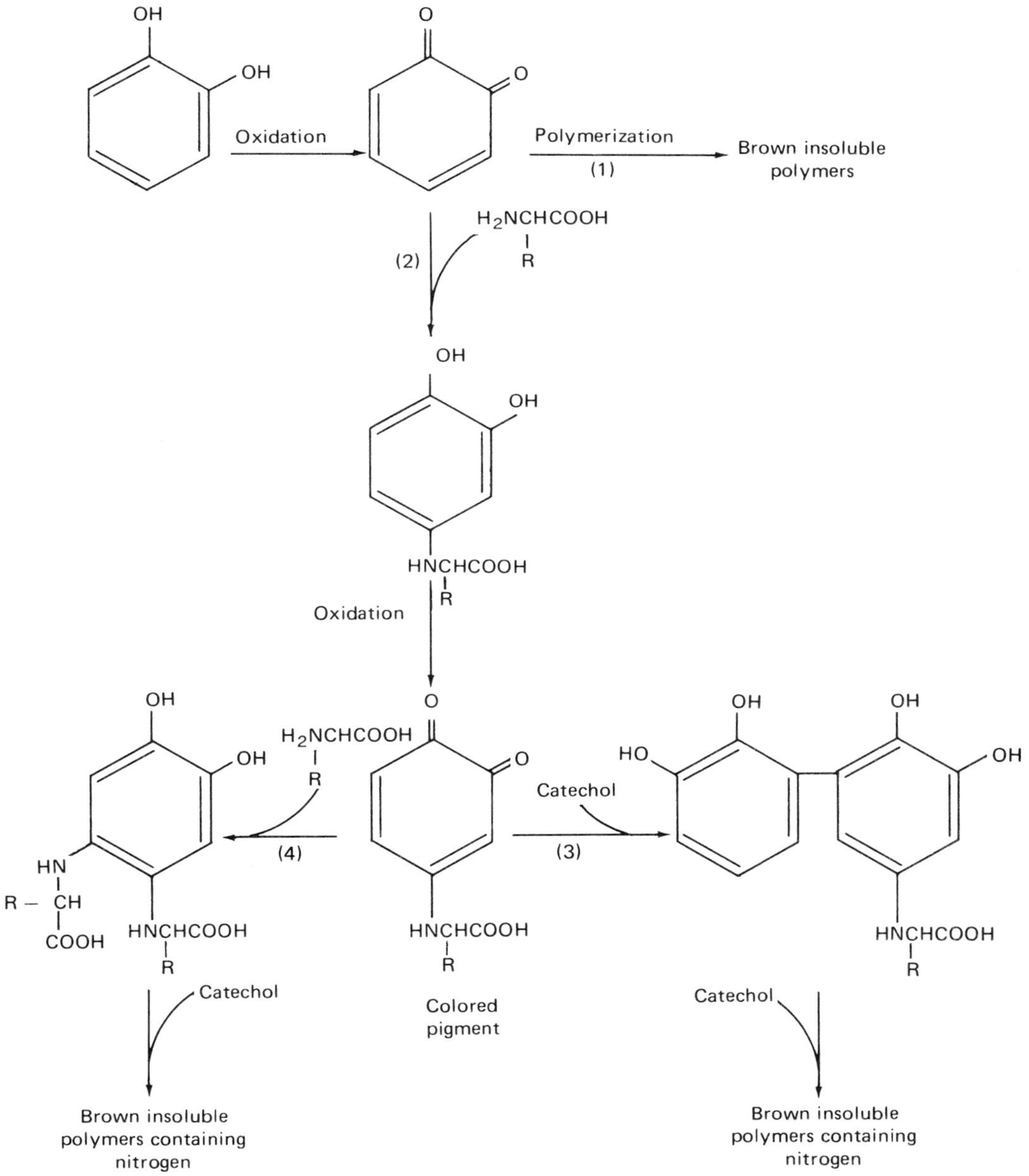

Figure 3. Oxidative polymerization of catechol with amino acids.

4 position, as in reaction (2). The product can be further oxidized to a colored quinone pigment, and this in turn can polymerize with catechol, as in reaction (3), or it can take up a second molecule of amino acid in the 5 position to form a di-imino compound, which can itself polymerize with catechol.

The situation is far more complicated than it appears in Figure 3. The course of the reactions will be determined by the level of oxidation and the pH of the system, and the substituents on the quinone ring, but there will be a tendency for reaction (2) to proceed faster than reaction (1). Amino acids can react with both ortho- and paraquinones, and also with more complicated polymers from reaction (1), provided the quinone grouping has not been lost. Nor is the reaction limited to the α-amino group. Other nucleophilic groups such as amine, imino, and sulfydryl groups will react in the same way, as will the side chain amino groups of which the ε-lysylamino group is an example. Dibasic or sulfydryl amino acids can form cross-linkages between quinone rings. Terminal amino and side chain amino groups of both peptides and proteins will also condense with ortho- or paraquinones; in fact MASON and PETERSON [1955] found the rate of reaction of ortho-quinones and amino groups to increase, within limits, with an increase in the length of the peptide chain.

The early Russian thinking on the chemical nature of humic acids and the possibility of this type of reaction being involved has been well reviewed by KONONOVA [1961]. She is of the opinion that the starting materials are plant and microbial phenols, quinones, amino acids, and peptides, which condense to form complex polymers of varying molecular size.

A wide range of reactants of plant and microbial origin for this type of polymerization is likely to occur in soils. FLAIG [1960] considers lignin to be an important if not an essential parent material for the formation of humic acids. HAIDER, FREDERICK, and FLAIG [1965] examined the conditions of phenol oxidase catalyzed reactions between amino acids, peptides, and protein and various phenols formed during lignin decomposition and compared the properties of the polymers formed with those of humic acids. 6 N hydrochloric acid hydrolysis of the peptide complexes released amino acids other than the nitrogen-terminal amino acid, indicating that peptide bonds are split, but the C–N bond between the ring and the terminal acid remained. The incomplete release of nitrogen is comparable to the incomplete release of soil nitrogen using the same methods. SWABY and LADD [1962] have made the point that polymerization must proceed through free radicals formed by enzymes and in relatively high concentrations and that the only possible environment that could possibly provide these conditions is the plant or microbial cell just before death. During normal metabolism the formation of quinones is repressed by the low oxidation–reduction potential within the cell, but upon the death of the cell, phenols, tannins, and flavanoids will be oxidized and may polymerize. At the same time autolytic proteolysis of cell proteins liberates amino acids, thus providing all the components for the formation of dark-colored heterogeneous copolymers.

SWABY and LADD [1962] first suggested that only quinones and amino acids were involved, because no peptide bonds were detected, but later LADD and BUTLER [1963] accepted the possibility of peptides and protein entering the polymerization process.

When synthetic polymers produced from amino acids, peptides, or protein condensed chemically with catechol or p-benzoquinone were subjected to acid hydrolysis, the proportion of α-amino acid to total nitrogen in the hydrolysate increased with increasing length of the peptide chain, from 10% or less in the case of the glycine polymer to 70% for a casein polymer. This amount was considerably lower than the 30% reported by COULSON, DAVIES, and KHAN [1959].

FELBECK [1965] suggested that the central resistant core to which polyphenols and amino acids are linked is a microbially produced polymer of 4-pyrone units linked together by methylene bridges at the 2,6 positions. The evidence for this was provided by hydrogenolysis of the residue from acid hydrolysis of a muck soil which produced a pure product in the form of an n-C_{25} or n-C_{26} hydrocarbon.

All the structures postulated so far are polymers containing amino acid nitrogen, but are devoid of any regular recurring unit and will, therefore, be resistant to enzyme attack. The pattern of soil amino acids suggests that they are of microbial origin, so it is conceivable that mucopeptides found in cell walls may be incorporated. Glucosamine also condenses with catechol and p-benzoquinone, and may therefore act as the point of attachment of carbohydrates to the polymers.

Certainly the hydrolytic studies suggest a similarity between this type of polymer and humic acids, especially where peptides are included. The data obtained on protein-quinone complexes is limited, but HAIDER, FREDERICK, and FLAIG [1965] suggested that they can be hydrolyzed by enzymes. However, the stability of tanned protein is influenced by the molecular size and complexity of the tanning agent and the degree of cross-linkage attained.

The amino acids split off by acid appear to come from peripheral groups, possibly peptides, because the bulk of the model polymer does not pass into solution during acid hydrolysis. However, neither dibasic- nor sulfur-containing amino acids were present to provide cross-linkages.

Proof that peptide bonds are abundant in soils is still lacking, although JENKINSON and TINSLEY [1959] report that up to 15% of the nitrogen of a ligno-protein complex isolated from a straw compost was hydrolyzed by papain, while SCHARPENSEEL and KRAUSSE [1962] identified one-third of the nitrogen of ^{14}C- and ^{3}H-labeled humic acid in peptide and amide linkages following papain hydrolysis. LADD and BRISBANE [1967] found pronase, a proteolytic enzyme from *Streptomyces griseus* a more effective enzyme than papain for the hydrolysis of humic acid. Pronase released between 9.9 and 16.1% of humic acid nitrogen from four soils as amino acid nitrogen, but no dibasic amino acids were identified. During the slow decomposition of organic matter, any peptide bonds present would be the first to be attacked by microorganisms leaving the nitrogen-terminal acids attached to the ring structure. This view is supported by the findings of KUMADA [1956] that the proportion of acid hydrolyzable-nitrogen decreases as humification increases.

A number of dibasic amino acids have been identified in hydrolysates of soils including lysine, arginine, histidine, α,ε-diaminopimelic acid, yet free amino groups have not so far been identified in soil organic matter. It is plausible to assume they have condensed with quinones, but this would mean that C-N bonds are split by acid hydrolysis, whereas such bonds are relatively stable. There is no evidence that these bonds are susceptible to enzyme attack. LADD [1964] studied the decomposition of N-O-carboxyphenylglycine by *Achromobacter* spp. and found that the amino group disappeared but without the liberation of ammonia. If these bonds are resistant to microbial decomposition, this would explain the apparent accumulation of dibasic acids in highly humified soils. It would be interesting to prepare model polymers from dibasic- and sulfur-containing amino acids and then study the products derived by mineral acid and microbial hydrolysis.

The chemical nature of the 40 to 50% of soil nitrogen not accounted for as amino acids is still undetermined, but certainly a portion of it may be associated with ring structures. BREMNER [1949b] and KOJIMA [1947a] expressed the view that some of it may be in the form of heterocyclic rings. More recently, BREMNER [1965c] has suggested a possible mechanism for their formation.

There is a possibility of ring formation during polymerization of quinones and amino compounds. RINDERKNECHT and JURD [1958] report a nonenzymic browning reaction at pH 7 between phloroglucinol and glycine with the formation of 3-hydroxyindole, possibly an intermediate in melanin formation.

Indole ring formation has been suggested by HACKMAN and TODD [1953] as a result of a rearrangement following a condensation reaction between ortho-quinone and the terminal amino group of a protein. This may be followed by oxidative polymerization.

Indole has been identified by FLAIG and BREYHAN [1956] in an alkali fusion mixture of the residual material from 6 N hydrochloric acid hydrolysis. ORLOV and DENISOVA [1962] oxidized humic acids with alkaline potassium permanganate and found a high nitrogen content in the fraction containing aromatic acids, which they consider to be caused by the presence of purine or pyrimidine derivatives. SAVAGE and STEVENSON [1961] obtained a fraction of low molecular weight acids with a high nitrogen content from the oxidation of humic acid with hydrogen peroxide, but were unable to identify the products by ultraviolet absorption, although they could be ring-type compounds.

δ. *Acid-Labile Nitrogen*

A comparatively large proportion of the soil nitrogen is liberated as ammonia during acid hydrolysis, even when allowance is made for fixed ammonia. Some undoubtedly arises from the deamination of amino acids and amino sugars, possibly catalyzed by metal ions dissolved from the soil. Some may arise from deamination of the type of compounds mentioned by MATTSON and KOULTER-ANDERSSON [1943] and SWABY and LADD [1963]. FÜHR and BREMNER [1964] have recently shown that in acid and neutral soils, soil organic matter reacts with nitrite on air drying and much of this fixed nitrogen is liberated as ammonia by acid. Because much of the labile nitrogen occurs in the fulvic acid fraction which also contains carbohydrates, STEVENSON [1960] suggests that it may be present in polymers produced as a result of the Maillard reaction between amino groups and sugars.

Although considerable progess has been made in recent years toward an understanding of the chemical nature of soil organic matter, it is still not possible to state unequivocably the form in which bound amino acids occur in soil, nor is much known about the nature of the remaining nitrogen compounds. There is adequate circumstantial evidence to suggest that true protein is not present in soils as a major component, although there will be a small quantity in the form of extracellular enzymes, possibly bound to clay mineral surfaces. Many of the reaction mechanisms for nitrogen compounds discussed in Section 4 are possible under the conditions obtaining in soils and many of the chemical properties of the products formed are similar to those of soil humic acids. Thus, by analogy the assumption has been made that these reactions are responsible for the formation of those soil organic nitrogen polymers that are relatively stable to microbial decomposition. However, as yet, there is no direct evidence to support this view.

The heterogeneous nature of the copolymers described would render them virtually resistant to enzymatic hydrolysis, and in soil their stability may be enhanced by adsorption on clays and hydrated oxides, particularly under conditions of alternate wetting and drying.

Further chemical, microbial, or enzymatic degradation studies are required to produce simpler units in which the principal chemical bonds can be characterized. While this step is in the process of being achieved, it must be borne in mind that, because soil organic matter consists of a mixture of complex polymers varying both in composition and in molecular size and because these polymers probably have "life spans" varying from a few hours to thousands of years in a microbiological environment, no purely static chemical model of the nitrogen compounds will suffice to represent the changing patterns of composition that occur under field conditions.

REFERENCES

Adams, A. P., W. V. Bartholomew, and F. E. Clark, 1959. Measurement of nucleic acid components in soil. *Proc. Soil Sci. Soc. Amer.* 18:40–46.

Aleksandrova, I. V., 1960. A method for studying the qualitative composition of organic matter in soil solutions. *Pochvovedeniye* 11:85–87.

Allison, L. E., 1965. Organic carbon in "Methods of soil analysis." Section 90 (Part 2), Amer. Soc. Agron., Wisconsin.

Anderson, G. 1958. Indentification of derivatives of dexoyribonucleic acid in humic acid. *Soil Sci.* 86:169–174.

———. 1961. Estimation of purines and pyrimidines in soil humic acid. *Soil Sci.* 91:156–161.

Armstrong, D. E., and G. Chesters, 1964. Properties of protein–bentonite complexes as influenced by equilibrium conditions. *Soil Sci.* 98:39–52.

Barley, K. P., and A. C. Jennings, 1959. Earthworms and soil fertility III. The influence of earthworms on the availability of nitrogen. *Aust. J. Agric. Res.* 10:364–370.

Biswas, T. D., and N. B. Das, 1957. Amino acids in soils growing berseem. *J. Indian Soc. Soil Sci.* 5:31–37.

Bottomley, W. B., 1919. The isolation from peat of certain nucleic acid derivatives. *Proc. R. Soc. B,* 90:39–44.

Bremner, J. M., 1949a. Studies on soil organic matter. Part 3. The extraction of organic carbon and nitrogen from soil. *J. Agric. Sci.* 39:280–282.

———. 1949b. Studies on soil organic matter. Part 1. The chemical nature of soil organic nitrogen. *J. Agric. Sci.* 39:183–193.

———. 1950a. The amino acid composition of the protein materials in soil. *Biochem. J.* 47:538–542.

———. 1950b. Some observations on the oxidation of soil organic matter in the presence of alkali. *J. Soil Sci.* 1:198–205.

———. 1955a. Nitrogen distribution and amino acid composition of fractions of a humic acid from a chernozem soil [Hildesheimer Schwarzede]. *Z. Pfl. Ernähr. Düng.* 71:63–66.

———. 1955b. Studies on soil humic acids, I, The chemical nature of humic nitrogen. *J. Agric. Sci.* 46:247–256.

———. 1955c. Nitrogen transformations during the biological decomposition of straw composted with inorganic nitrogen. *J. Agric. Sci.* 45:469–475.

———. 1957. Studies on humic acids, II, Observations on the estimation of free amino groups. Reactions of humic acid and lignin preparations with nitrous acid. *J. Agric. Sci.* 48:352-360.

———. 1958. Amino sugars in soil. *J. Sci. Fd. Agric.* 9:528–532.

———. 1965a. Total nitrogen in "Methods of soil analysis," Section 83 (Part 2). Amer. Soc. Agron., Wisconsin.

———. 1965b. Inorganic forms of nitrogen in "Methods of soil analysis," Section 84 (Part 2) Amer. Soc. Agron., Wisconsin.

Bremner, J. M., 1965c. Organic nitrogen in soils in "Soil nitrogen." Chapter 3, Amer. Soc. Agron., Wisconsin.

———— and C. L. Ho, 1961. Use of ion-exchange resins for extraction of soil organic matter. *Agron. Abstr.* Pp 15, Amer. Soc. Agron., Wisconsin.

———— and H. Lees, 1949. Studies on soil organic matter. 2. The extraction of organic matter from soil by neutral reagents. *J. Agric. Sci.* 39:274–279.

———— and K. Shaw, 1954. Studies on the estimation and decomposition of amino sugars in soil. *J. Agric. Sci.* 44:152–159.

————————. 1957. The Mineralisation of some nitrogenous materials in soil. *J. Sci. Fd. Agric.* 8:341–347.

Broadbent, F. E., and G. R. Bradford, 1952. Cation exchange grouping in the soil organic fraction. *Soil Sci.* 74:447–457.

———— and F. Clark, 1965. Denitrification in "Soil nitrogen." Chapter 9. Amer. Soc. Agron., Wisconsin.

Brown, I. C., and J. Thorpe, 1942. Morphology and composition of some soils of the Miami family and the Miami catena. *U.S.D.A. Tech. Bull. 834.*

Brydon, J. E., and F. J. Sowden, 1959. A study of the clay-humus complexes of a chernozemic and a podzol soil. *Canad. J. Soil Sci.* 39:136–143.

Burges, A. 1965. The soil microflora — its nature and biology. Ecology of soil-borne plant pathogens. Int. Symp. Univ. Calif. 1963, 21–32.

Carles, J., and J. Decau, 1960a. De quelques conditions susceptibles de modifier les proportions des acides aminés du sol. *Ann. Agron. Paris* 11:557–575

————————. 1960b. Variations in the amino acids of soil hydrolysates. *Sci. Proc. R. Dublin Soc. Ser. A.* 1:177–182.

————, L. Soubries, and R. Gadet, 1958. Les acides aminés du sol et leurs variations. *C. R. Acad. Sci.* 247:1229–1232.

Cernescu, N., 1964. Guidebook of excursions 1:168–179. Eighth International Congress of Soil Science, Bucharest.

Cheng, H. M., and J. Van Hove, 1964. Characterisation of organic matter in European soils by nitrogen fractionation. *Pédologie* 14:8–23.

Choudri, M. B., and F. J. Stevenson, 1957. Chemical and physical properties of soil humic colloids: III. Extraction of organic matter from soils. *Proc. Soil Sci. Soc. Amer.* 21:508–513.

Coulson, C. B., R. I. Davies, and E. J. A. Khan, 1959a. Humic acid investigations. 2. Studies in the fractionation of humic acids. *J. Soil Sci.* 10:271–283.

————————. 1959b. Humic acid investigations. 3. Studies on the chemical properties of certain humic acid preparations. *Soil Sci.* 88:191–195.

Crook, E. M., and I. R. Johnston, 1962. The qualitative analysis of the cell walls of selected species of fungi. *Biochem. J.* 83: 325–331.

Cummins, C. S., 1965. Ornithine in mucopeptide of gram-positive cell walls. *Nature* 206:1272.

————, and H. Harris, 1956. The chemical composition of the cell wall in some gram-positive bacteria and its possible value as a taxonomic character. *J. Gen. Microbiol.* 14: 583–600.

Dadd, C. C., L. Fowden, and W. H. Pearsall, 1953. An investigation of the free amino acids in organic soil types using paper partition chromatography. *J. Soil Sci.* 4:69–71.

Davies, R. I., C. B. Coulson, and C. Luna, 1957. Humic acid investigations. *Chemy. Ind.* 1544–1545.

Davies, W. M., and G. Owen, 1934. Soil survey of North Shropshire. II. Classification of series and types. *Emp. J. Exp. Agric.* 2:359–379.

Doak, B. W., 1952. Some chemical changes in the nitrogeneous constituents of urine when voided on pasture. *J. Agric. Sci.* 42:162–171.

Dubach, P., and N. C. Mehta, 1963. The chemistry of soil humic substances. *Soils Fert.* 25:83–84.

————, N. C. Mehta, and H. Deuel, 1961. Extraktion von Huminstoffen aus dem B-Horizont eines Podsols mit Äthylendiamintetraessigsäuse. *Z. Pfl. Ernähr. Düng.* 95:119–123.

Dubach, P. J., and S. S. Ganti, 1964. Earthworms and amino acids in soil. *Curr. Sci.* 33:219–220.

Dyke, K. G. H., 1964. The chemical composition of the cell wall of yeast *Nadsonia elongata. Biochem. Biophys. Acta* 82:374–384.

Ensminger, L. E., and J. E. Gieseking, 1939. The adsorption of proteins by montmorillonitic clays. *Soil Sci.* 48:467–474.

————. 1942. Resistance of clay-adsorbed proteins to proteolytic hydrolysis. *Soil Sci.* 53:205–209.

Estermann, E. F., and A. D. McLaren, 1959. Stimulation of bacterial proteolysis by adsorbents. *J. Soil Sci.* 10:64–78.

————, G. H. Peterson, and A. D. McLaren, 1959. Digestion of clay-protein, lignin-protein, and silica-protein complexes by enzymes and bacteria. *Proc. Soil Sci. Soc. Amer.* 23:31–36.

Felbeck, T. G., 1965. Studies on the high pressure hydrogenolysis of organic matter from a muck soil. *Proc. Soil Sci. Soc. Amer.* 29:48–55.

Flaig, W., 1950. Zur Kenntnis der Huminsäuren. 1. Zur Chemischen Konstitution der Huminsäuren. *Z. Pfl. Ernähr. Düng* 51:193–212.

————. 1960. Comparative chemical investigations on natural humic compounds and their model substances. *Sci. Proc. R. Dublin Soc. Ser. A.* 1:149–162.

————. 1963. The chemistry of humic substances. The use of isotopes in soil organic matter studies, Special supplement to the *J. Appl. Radiat. Isotopes.* Oxford: Pergamon Press.

————, and Th. Breyhan, 1956. Über das Vorkammen von Indolverbindungen in Schwarzerde-huminsäuren. *Z. Pfl. Ernähr. Düng.* 75:132–135.

Forsyth, W. C. G. 1947a. The characterisation of the humic complexes of soil organic matter. *J. Agric. Sci.* 37-132–138.

————. 1947b. Studies on the more soluble complexes of soil organic matter. I. A method of fractionation. *Biochem. J.* 41:176–181.

Foster, A. B., and J. M. Webber, 1960. Chitin. *Carbohyd. Chem.* 15:371–393.

Fowden, L., 1958. New amino acids of plants. *Biol. Rev.* 33:393–441.

Fraps, G. S., and A. J. Sterges, 1947. Nitrification capacities of Texas soil types and factors which affect nitrification. *Texas Ag. Exp. Sta. Bull. 693.*

Führ, and J. M. Bremner, 1964. Beeinflussende Faktoren in der Fixierung des Nitrit-Stickstoffs durch die Organische Masse des Bodens. *Atompraxis* 10:109–113.

Gilmour, D., 1961. *The Biochemistry of Insects.* New York: Academic Press.

Glentworth, R., and J. W., Muir, 1963. The soils of the country round Aberdeen, Inverurie and Fraserburgh. Edinburgh: H.M.S.O.

Gottschalk, A., W. H. Murphy, and E. R. B. Graham, 1962. Carbohydrate-peptide linkages in glycoproteins and methods for their elucidation. *Nature* 194:1051–1053.

Greenwood, D. J., and H. Lees, 1956. Studies on the decomposition of amino acids in soils. I. A preliminary survey of techniques. *Plant Soil* 7:253–268.

Griffiths, G. ap: 1951. Factors affecting nitrate accumulation in Uganda soil. *Emp. J. Exp. Agric.* 19:1–12.

Grov, A., 1963a. Amino acids in soil. II. Distribution of water-soluble amino acids in a pine forest soil profile. *Acta Chem. Scand.* 17:2316–2318.

————. 1963b. Amino acids in soil. III. Acids in hydrolysates of water extracted soil and their distribution in a pine forest soil profile. *Acta Chem. Scand.* 17:2319–2324.

————, and E. Alvsaker, 1963. Amino acids in soil. I. Water-soluble acids. *Acta Chem. Scand.* 17:2307–2315.

Guerra, A., 1966. Guide to the excursion in Spain. Meeting of Comm. V. Int. Soc. Soil Sci., Madrid.

Gustavson, K. H., 1949. Some protein-chemical aspects of tanning processes. *Adv. Prot. Chem.* 5:353–421.

Hackmann, R. H., and A. R. Todd, 1953. Some observations on the reaction of catechol derivatives with amines and amino acids in presence of oxidising agents. *Biochem. J.* 55:631–637.

Haider, K., L. R. Frederick, and W. Flaig, 1965. Reactions between amino acid compounds and phenols during oxidation. *Plant Soil* 22:49–64.

Handley, W. R. C., 1954. Mull and mor formation in relation to forest soils. *For. Comm. Bull. No. 23.* H.M.S.O. London.

Harmsen, G. W., and G. J. Kolenbrander, 1965. Soil Inorganic Nitrogen in "Soil Nitrogen," Chapter 2., Amer. Soc. Agron., Wisconsin.

Harmsen, G. W., and D. A. Van Schreven, 1955. Mineralization of organic nitrogen in soil. *Adv. Agron.* 7:299–398.

Hayashi, T., and T. Nagai, 1961. The components of soil humic acids. 9. Distribution of carbohydrates in humic acids. *Soil Pl. Fd. Tokyo* 7:129.

Hobson, R. P., and P. J. Page, 1932a. Studies on the carbon and nitrogen cycles in the soil. VI. The extraction of the organic nitrogen of the soil with alkali. *J. Agric. Sci.* 22:297–299.

————. 1932b. Studies on the carbon and nitrogen cycles in the soil. VII. The nature of the organic nitrogen of the soil "humic" nitrogen. *J. Agric. Sci.* 22:497–515.

d'Hoore, J. L., 1964. Soil map of Africa — scale 1 to 5 million, explanatory monograph, publication 93, *Comm. Tech. Co-operation Africa*. Lagos.

Jacquin, F. 1960. Evolution des acides aminés lors de la decomposition de la matière organique du sol. *C. R. Acad. Sci. Paris* 251:1810–1811.

Jansson, S. L., 1958. Tracer studies on nitrogen transformation in soil with special attention to mineralization–immobilization relationships. *Kungl LantbraksHögsk Ann.* 24:101–361.

————. 1960. On the humus properties of organic manure I and II. *Kungl LantbraksHögsk Ann.* 26:51–75, 135–172.

Jenkinson, D. S., and J. Tinsley, 1959. Studies on the organic material extracted from soils and composts. I. The isolation and characterisation of ligno-proteins from compost. *J. Soil Sci.* 10:245–263.

————. 1960. A comparison of the ligno-protein isolated from a mineral soil and from a straw compost. *Sci. Proc. R. Dublin. Soc. Ser. A* 1:141–147.

Jenson, H. L., 1931. The microbiology of farm yard manure decomposition in soil. I. Changes in the microflora, and their relation to nitrification. *J. Agric. Sci.* 21:38–80.

Jones, P. D., V. Graham, L. Segal, W. J., Baillie, and M. H. Briggs, 1962. Forms of soil biotin. *Life Sci.* 1:645–648.

Kay, F. F., 1936. A survey of the university farm. *Bull. 49*, University of Reading.

Kessler, G., and W. J. Nickerson, 1959. Glucomannan–protein complexes from cell walls of yeasts. *J. Biol. Chem.* 234:2281–2285.

Kojima, R. T., 1947a. Soil organic nitrogen. I. Nature of the organic nitrogen in a muck soil from Geneva, New York. *Soil Sci.* 64:157–165.

————. 1947b. Soil organic nitrogen. II. Some studies on the amino acids of protein material in a muck soil from Geneva, New York. *Soil Sci.* 64:245–252.

Kononova, M. M., 1961. *Soil Organic Matter*. Oxford: Pergamon Press.

————, and I. V. Aleksandrova, 1956. The use of paper partition chromatography for a study of the forms of nitrogen in humic substances. *Pochvovedeniye* 5:88–92.

Korn, E. D., and D. H. Northcoate, 1960. Physical and chemical properties of polysaccharides and glycoproteins of the yeast-cell wall. *Biochem. J.* 75:12–17.

Koter, M., T. Mazur, and J. Chodan, 1964. Amino acid sorption in soil. *Roczn. Nauk Rol.* 88A, 763–771. *Soils Fert.* 28:787 (1965).

Kühnelt, W., 1961. *Soil biology with special reference to the animal kingdom*. Translated by N. Walker. London: Faber & Faber.

Kumada, D. 1956. Several properties of humic acid. *Soil Pl. Fd., Tokyo* 2:44–48.

Ladd, J. N., 1964. Studies on the metabolism of model compounds related to humic acids. I. The decomposition of N-0-carboxyphenylglycine. *Aust. J. Biol. Sci.* 17:153–169.

————, and P. G. Brisbane, 1967. Release of amino acids from soil humic acids by proteolytic enzymes. *Aust. J. Soil Res.* 5:161–171.

————, and J. H. A. Butler, 1963. Comparison of properties synthetic and natural humic acids. The use of isotopes in soil organic matter studies. Supplement to the *J. Appl. Radiat. Isotopes*. Oxford: Pergamon Press.

Laverick, M. S., 1963. *The physiology of earthworms*. Oxford: Pergamon Press.

Lawes, J. B., J. H. Gilbert, and R. Warrington, 1882. On the amount and composition of drainage waters collected at Rothamsted. Part III. *J. Roy. Agri. Soc. Ser. II*, 18:1.

Lochhead, R. H., and A. G. Thexton, 1947. Qualitative studies of soil microorganisms. VII. Rhizosphere effect in relation to the amino acid nutrition of bacteria. *Canad. J. Res. C.* 25:20–26.

Lynch, D. L., D. H. Hughes, and Y. E. Rhodes, 1959. Pressure and gradient elution in ion exchange chromatography of the amino acids in soils. *Soil Sci.* 87:339–344.

————, H. O. Olney, and L. M. Wright, 1958. Some sugars and related carbohydrates found in Delaware soils. *J. Sci. Fd. Agric.* 9:56–60.

Martin, A. E., and R. Reeve, 1957. Chemical studies on podzolic illuvial horizons. I. The extraction of organic matter by organic chelating agents. *Soil Sci.* 8:268–278.

Mason, H. S., 1955. Comparative biochemistry of the phenolase complex. *Adv. Enzymol.* 16:105–184.

————, and E. W. Peterson, 1955. Interactions of quinones with proteins. Congr. Intern. Biochem. Résumés communs., 3rd Cong., Brussels.

Mattson, S., and E. Koulter-Anderson, 1943. The acid-base conditions in vegetation, litter, and humus. VI. Ammonia fixation and humus nitrogen. *Lantbr. Högsk Ann.* 11:107–134.

McLaren, A. D., 1954. The adsorption and reactions of enzymes and proteins on kaolinite. I. *J. Phys. Chem.* 58:129–137.

McLaren, A. D., and E. F. Estermann, 1956. The adsorption and reaction of enzymes and proteins on kaolinite. II. The isolation of enzyme–substrate complexes. *Arch. Biochem. Biophys.* 61:158–173.

Mehta, N. C., P. Dubach, and H. Deuel, 1963. Untersuchungen über die Molekulargewichtsverteilung von Huminstoffen durch Gelfiltration on Sephadex. *Z. Pfl. Ernähr. Düng.* 102:128–137.

Moore, S., and W. H. Stein, 1951. Chromatography of amino acids on sulphonated polystyrene resins. *J. Biol. Chem.* 192:663–681.

Moore, S. and W. H. Stein, 1954. Procedures for the chromatographic determination of amino acids on four per cent cross-linked sulfonated polystyrene resins. *J. Biol. Chem.* 211:893–906.

Morrison, R. I., 1958. The alkaline nitrobenzene oxidation of soil organic matter. *J. Soil Sci.* 9:130–140.

————. 1963. Products of the alkaline nitrobenzene oxidation of soil organic matter. *J. Soil Sci.* 14:201–216.

Morrow, C. A., and R. A. Gortner, 1917. The organic matter of the soil. V. A study of the nitrogen distribution in different soil types. *Soil Sci.* 3:297–331.

Mortenson, J. L., and R. B. Schwendinger, 1963. Electrophoretic and spectroscopic characterisation of high molecular weight compounds of soil organic matter. *Geochim. Cosmoch. Acta* 27:201–208.

Mortland, M. M., and A. R. Wolcott, 1965. Sorption of inorganic nitrogen compounds by soil materials, in "Soil nitrogen," Chapter 4, Amer. Soc. Agron., Wisconsin.

Muir, J. W., R. I. Morrison, C. J. Bown, and J. Logan, 1964. The mobilisation of iron by aqueous extracts of plants. I. Composition of the amino acid and organic acid fractions of an aqueous extract of pine needles. *J. Soil Sci.* 15:220–225.

Nommik, H. 1965. Ammonium fixation and other reactions involving a nonenzymatic immobilization of mineral nitrogen in soil, in "Soil nitrogen," Chapter 5, Amer. Soc. Agron., Wisconsin.

Nykvist, N. 1963. Leaching and decomposition of water soluble organic substances from different types of leaf and needle litter. *Stud. For. Suec.* 3:11–31.

Okuda, A., and S. Hori, 1954. Chromatographic investigation of amino acids in humic acid and alkaline alcohol lignin. Trans. Int. Congr. Soil Sci. 5th Congr. Léopoldville, 2:255–258.

————. 1955. Identification of amino acids in humic acid. *Soil Pl. Fd. Tokyo* 1:39–40.

Orlov, D. S., and M. F. Denisova, 1962. Aromatic nature of humic acid nucleus from chernozem and soddy podzols. *Nauch. Dokl. vyssh. Shkoly, Bio. Nauki* 3:184–188 (1962). *Soils Fert.* 26:2829 (1963).

Parsons, J. W., and J. Tinsley, 1960. Extraction of soil organic matter with anhydrous formic acid. *Proc. Soil Sci. Soc. Amer.* 24:198–201.

————. 1961. Chemical studies of polysaccharide material in soils and composts based on extraction with anhydrous formic acid. *Soil Sci.* 92:46–53.

Paul, E. A., and E. L. Schmidt, 1960. Extraction of free amino acids from soil. *Proc. Soil Sci. Soc. Amer.* 24:195–198.

Paul, E. A., and E. L. Schmidt, 1961. Formation of free amino acids in rhizosphere and nonrhizosphere soil. *Proc. Soil Sci. Soc. Amer.* 25:359–362.

————, and C. M. Tu, 1965. Alteration of microbial activities, mineral nitrogen and free amino acid constituents of soils by physical treatment. *Plant Soil* 22:207–219.

Payne, J. M. B., J. W. Rouatt, and H. Katznelson, 1956. Detection of free amino acids in soil. *Soil Sci.* 82:521–524.

Perkins, H. R., and C. S. Cummins, 1964. Chemical structure of bacterial cell walls. Ornithine and 2, 4-diamino butyric acid as components of the cell walls of plant pathogenic *Corynebacteria*. *Nature* 201:1105–1107.

Pinck, L. A., R. S. Dyal, and F. E. Allison, 1954. Protein-montmorillonite complexes, their preparation and the effects of soil microorganisms on their decomposition. *Soil Sci.* 78:109–118.

Potter, R. S., and R. S. Snyder, 1915. The amino acid nitrogen of soil. *J. Ind. Eng. Chem.* 7:1049–1053.

Pusztai, A. 1964. Hexosamines in the seeds of higher plants (spermatophytes). *Nature* 201:1328–1329.

Putnam, H. D., and E. L. Schmidt, 1959. Studies on the free amino acid fractions of soils. *Soil Sci.* 87:22–27.

Quastel, J. H., and P. G. Scholefield, 1949. Influence of organic nitrogen compounds on nitrification in soil. *Nature* 164:1068–1072.

Rendig, V. V., 1951. Fractionation of soil nitrogen and factors affecting distribution. *Soil Sci.* 71:253–267.

Rieger, S., 1966. Dark, well-drained soils of tundra regions in western Alaska. *J. Soil Sci.* 17:264–273.

Rinderknecht, H., and L. Jurd, 1958. A novel non-enzymatic browning reaction. *Nature* 181:1268–1269.

Robert, M., 1964. Étude biologique des sols au cours de l'epreuve d'incubation. 1. Acides aminés libres. *Ann L'Inst. Pasteur* 106:319–325.

Rodrigues, G., 1954. Fixed ammonia in tropical soils. *J. Soil Sci.* 5:264–274.

Roulet, N., N. S. Mehta, P. Dubach, and H. Deuel, 1963. Abtrennung von Kohlehydraten und Stickstoffverbindungen aus Huminstoffen durch Gelfiltration und Ionenaustausch — Chromatographie. *Z. Pfl. Ernähr. Düng.* 103:1–9.

Saloman, M. 1963. Hexosamines in soil aggregates. *Nature* 200:500.

Salton, M. R. J., 1964. The bacterial cell wall. Amsterdam: Elsevier.

Sanger, F., 1945. Free amino groups of insulin. *Biochem. J.* 39:507–515.

Savage, S. M., and F. J. Stevenson, 1961. Behavior of soil humic acids towards oxidation with hydrogen peroxide. *Proc. Soil Sci. Soc. Amer.* 25:35–39.

Scharpenseel, H. W., and R. Krausse, 1962. Aminosäureuntersuchungen an Verschiedenen Organischen Sedimenten, besonders Grau-und Braunhuminsäurefraktionen Verschliedener Bodentypen (einschliesslich C¹⁴ Markierter Huminsäuren). *Z. Pfl. Ernähr. Düng.* 96:11–34.

Schmidt, E. L., H. D. Putnam, and E. A. Paul, 1960. Behavior of free amino acids in soil. *Proc. Soil Sci. Soc. Amer.* 24:107–109.

Schnitzer, M., J. R. Wright, and J. G. Desjardins, 1958. A comparison of the effectiveness of various extractants for organic matter from two horizons of a podzol profile. *Canad. J. Soil Sci.* 38:49–53.

Schriener, O., and E. C. Shorey, 1910a. The presence of arginine and histidine in soils. *J. Biol. Chem.* 8:381–384.

————. 1910b. Pyrimidine derivatives and purine bases in soils. *J. Biol. Chem.* 8:385–393.

Sharon, N., 1965. The amino sugars. Vol. 2A. E. A. Balazs and R. W. Jeanloz, eds. New York: Academic Press.

Simonart, P., L. Batistic, and J. Mayaudon, 1967. Isolation of protein from humic acid extracted from soil. *Plant Soil* 27:153–161.

————, and F. Peeters, 1954. Acides aminés libres dans l'humus. *Trans. Int. Congr. Soil Sci. 5th Congr. Leopoldville* 3:132–135.

Singh, S., and G. S. Bhandari, 1963. A study on the amino compounds in acid hydrolysates of aqueous leachates of some soils of Rajasthan. *J. Indian Soc. Soil Sci.* 11:1–7.

Lingh, S., and P. K. Singh, 1960. Distribution of hexosamines in some soils of Uttar Pradesh. *J. Indian Soc. Soil Sci.* 8:125–128.

Sørenson, H., 1963. Studies on the decomposition of C^{14}-labelled barley straw in soil. *Soil Sci.* 95:45–51.

Sowden, F. J., 1955. Estimation of amino acids in soil hydrolysates by the Moore and Stein method. *Soil Sci.* 80:181–183.

———. 1956. Distribution of amino acids in selected horizons of soil profiles. *Soil Sci.* 82:491–496.

———. 1958. The forms of nitrogen in the organic matter of different horizons of soil profiles. *Canad. J. Soil Sci.* 38:147–154.

———. 1959. Investigations on the amounts of hexosamine found in various soils and methods for their determination. *Soil Sci.* 88:135–143.

———, and D. I. Parker, 1953. Amino nitrogen of soils and of certain fractions isolated from them. *Soil Sci.* 76:201–208.

Sprengel, C., 1826. Uber Pflanzenhumus, Humussaure und Humussaure Salze. *Kastners. Arch. Ges. Naturlehre* 8:145–220.

Springer, U. 1943. Beitrag zur Fraktionierung der echten Humusstoffe. *Bodenk Pfl. Ernähr* 32:129–146.

Stevenson, F. J., 1954. Ion exchange chromatography of the amino acids in soil hydrolysates. *Proc. Soil Sci. Soc. Amer.* 18:373–377.

———. 1956a. Isolation and identification of some amino compounds in soils. *Proc. Soil Sci. Soc. Amer.* 20:201–204.

———. 1956b. Effect of long-time rotations on the amino acid composition of the soil. *Proc. Soil Sci. Soc. Amer.* 20:204–208.

———. 1957a. Investigations of aminopolysaccharides in soils. I. Colorimetric determination of hexosamines in soil hydrolysates. *Soil Sci.* 83:113–122.

———. 1957b. Distribution of the forms of nitrogen in some soil profiles. *Proc. Soil Sci. Soc. Amer.* 31:283–287.

———. 1957c. Investigations of aminopolysaccharides in soils. 2. Distribution of hexosamines in some soil profiles. *Soil Sci.* 84:99–106.

———. 1959. Carbon-nitrogen relationships in soil. *Soil Sci.* 88:201–208.

———. 1960. Chemical nature of the nitrogen in the fulvic acid fraction of soil organic matter. *Proc. Soil Sci. Soc. Amer.* 24:472–477.

———, and M. A. Kirkman, 1964. Identification of methyl nitrite in the reaction product of nitrous acid and lignin. *Nature* 201:107.

———, J. D. Marks, J. E. Varner, and W. P. Martin, 1952. Electrophoretic and chromatographic investigations of clay-adsorbed colloids. I. Preliminary investigations. *Proc. Soil Sci. Soc. Amer.* 16:69–73.

———, and R. J. Swaby, 1963. Occurrence of a previously unobserved nitrogen gas in the reaction product of nitrous acid and lignin. *Nature* 199:97–98.

———. 1964. Nitrosation of soil organic matter. I. Nature of gases evolved during nitrous acid treatment of lignins and humic substances. *Proc. Soil Sci. Soc. Amer.* 28:773–778.

Stöckli, A., 1946. Der Boden als Heimstätte des Lebens. *Schweiz Z. Forstw.* 97:356–378.

Stojanovic, B. J., and M. Alexander, 1958. Effect of inorganic nitrogen on nitrification. *Soil Sci.* 86:208–215.

Sulochana, C. B., 1962. Amino acids in root exudates of cotton. Plant and Soil 16:312–326.

Swaby, R. J., and J. N. Ladd, 1962. Chemical nature, microbial resistance and origin of soil humus. *Trans. Int. Soc. Soil Sci., N.Z., Comm. IV*: 197–202.

———. 1963. Stability and origin of soil humus. The use of isotopes in soil organic matter studies. Supplement to *J. Appl. Radiat Isotopes.* Oxford: Pergamon.

Synge, R. L. M., 1955. Modern methods of plant analysis. Vol. 4. K. Paech and M. V. Tracey, eds. Berlin: Springer-Verlag.

Tinsley, J. 1950. The determination of organic carbon in soils by dichromate mixtures. *Trans. Int. Congr. Soil Sci. 4th Congr. Amsterdam* 1:161–164.

Tinsley, J., and A. Salam, 1961. Extraction of soil organic matter with aqueous solvents. *Soils Fert.* 24:81–84.

————, and M. K. Zin, 1954. The isolation of ligno-proteins from soil. *Trans. Int. Congr. Soil Sci. 5th Congr. Léopoldville* 2:342–347.

Umezawa, H., 1964. *Recent advances in chemistry and biochemistry of antibiotics.* Tokyo: Microbial Chemistry Research Foundation.

Vegotsky, A., and S. W. Fox, 1962. *Comparative biochemistry.* M. Florkin and H. S. Mason, eds. 4: 185–244, Academic Press, New York.

Veldkamp, H., 1952. Aerobic decomposition of chitin by micro-organisms. *Nature* 169:500.

Waksman, S. A., and K. R. N. Iyer, 1932. Contribution to our knowledge of the chemical nature and origin of humus. *Soil Sci.* 34:43–69.

————. 1933. Contribution to our knowledge of the chemical nature and origin of humus. IV. Fixation of proteins by lignin and formation of complexes resistant to microbial decomposition. *Soil Sci.* 36:69–82.

————, and K. R. Stevens, 1930. A critical study of the methods for determining the nature and abundance of soil organic matter. *Soil Sci.* 30:97–116.

Waldron, A. C., and J. L. Mortensen, 1961. Soil nitrogen complexes. II. Electrophoretic separation of organic components. *Proc. Soil Sci. Soc. Amer.* 25:29–32.

————. 1962. Soil nitrogen complexes. III. Distribution and identification of nitrogenous constituents in electrophoretic separates. *Soil Sci.* 93:286–293.

Walker, T. W., and R. Thompson, 1949. Some observations on the chemical changes effected by the steam sterilization of glasshouse soils. *J. Hortic. Sci.* 25:19–35.

Wang, T. S. C., and S. C. Cheng, 1964. Amino sugars in soil hydrolysates. *Rep. Taiwan Sug. Exp. Sta.* 34:77–86.

————, T. K. Yang, and S. C. Cheng, 1964. Investigation of amino acids in soil hydrolysates by means of ion exchange resin. *Rep. Taiwan Sug. Exp. Sta.* 34:57–75.

Webster, R., and J. K. R. Gasser, 1959. Soil nitrogen. V. Leaching of nitrate from soils in laboratory experiments. *J. Sci. Fd. Agric.* 10:584–588.

Wetselaar, R., 1961. Nitrate Distribution in tropical soils. I. Possible causes of nitrate accumulation near the surface after a long dry period. *Plant and Soil* 15:110–120.

Whitehead, D. C., and J. Tinsley, 1964. Extraction of soil organic matter with dimethylformamide. *Soil Sci.* 97:34–42.

Work, E. 1961. The mucopeptides of bacterial cell walls. *A review. J. Gen. Microbiol.* 25:167–189.

Wrenshall, C. I., and W. J. Dyer, 1941. Organic phosphorus in soils. II. The nature of the organic phosphorus compounds. A. Nucleic acid derivatives. B. Phytin. *Soil Sci.* 51:235–248.

Wright, J. R., and M. Schnitzer, 1960. Oxygen-containing functional groups in the organic matter of the A_0 and B_h horizons of a podzol. *Trans. Int. Congr. Soil Sci. 7th Congr., Wisconsin* 2:120–127.

Yamashita, T., and T. Akiya, 1963. Amino acid composition of soil hydrolysates. *J. Soil Sci. Tokyo* 34:255–258.

Young, J. L., and J. L. Mortensen, 1958. Soil nitrogen complexes. I. Chromatography of amino compounds in soil hydrolysates. *Ohio Agri. Exp. Sta. Res. Circ. 61.*

Chapter 4

Other Organic Phosphorus Compounds

G. Anderson

Contents

A. Introduction

The chemistry and significance of soil organic phosphate, which often accounts for a high proportion of the total phosphate, has been the subject of extensive investigation for well over half a century, and detailed reviews have at intervals summarized the advances achieved (*e.g.*, BLACK and GORING [1953], ULRICH and BENZLER [1955]).

The soil chemist in devising methods for isolating and estimating phosphate esters has been faced with systems of great complexity, and about half of the organic phosphate has not yet been identified. Some of the techniques evolved are of necessity lengthy and not well suited to routine application, so that data on the distribution of different esters are very limited. On the other hand, great progress has been made in assessing the distribution of organic phosphate as a whole, and many of the factors affecting its transformation in soils are more fully appreciated.

B. The Nature of Soil Organic Phosphate

I. Types of Phosphate Esters Identified in Soil

Many of the biochemical reactions that occur in living cells involve the participation of phosphate esters at some stage, and a very large number of these compounds have been isolated from plant, animal, and microbial sources. It is possible to divide the phosphate esters of biological importance into several groups, and compounds belonging to three of these, the inositol phosphates, the nucleic acids (Chapter 3), and the phospholipids, have so far been identified in soils.* There is some evidence that phosphoprotein and small amounts of sugar phosphates may also be present.

II. Inositol Phosphates

1. *Occurrence and Main Properties*

Salts of *myo*-inositol hexaphosphate (phytic acid), (formula I), are common constituents

---

* An alternative classification of phosphate esters is based on the type of substitution on the acid molecule, for example, "monoesters of orthophosphoric acid," "diesters of orthophosphoric acid," etc., (KHORANA, [1961]).

of plant material, and occur principally in the seeds, which are the usual source of commercial phytin, the calcium-magnesium salt. Phytic acid is a very stable substance, most of whose salts (phytates), with the exception of those of the alkali metals, are only sparingly soluble in water. The insolubility of the ferric salt in hydrochloric acid has been used to distinguish inositol hexaphosphate (IHP) from other phosphate esters and orthophosphate, and as a basis for estimation in plant extracts. The ester is very stable in strong alkali, but is gradually hydrolyzed in acid solution to give a range of lower phosphate esters of inositol. Some of the lower esters have also been isolated from plant and animal tissue, and are released by hydrolysis of a class of phospholipids known as phosphoinositides (FOLCH and LeBARON [1956]). The properties of the inositol phosphates have been reviewed by COURTOIS [1951].

Phytic acid

2. Isolation from Soil

(a) Extraction and Precipitation

Although SHOREY [1913] suspected the presence of inosite (inositol) in soils, it was not until 1940 that its occurrence was proved by its isolation from hydrolyzates of material containing organic phosphate, and it seemed likely that it was in the form of a phosphate ester (YOSHIDA [1940]).

Soil IHP is usually very insoluble in dilute acids but is extracted with sodium hydroxide after a preliminary acid leaching. In the technique of DYER, WRENSHALL, and SMITH [1940] and

WRENSHALL and DYER [1941], much of the alkali-soluble organic matter is destroyed by hypobromite oxidation, to which IHP is very stable. Excess bromine and oxalic acid produced as an artefact are removed by acidification and ether extraction, and IHP is precipitated as a ferric salt. Phytate preparations obtained in this way have a composition and rate of hydrolysis similar to plant products (WRENSHALL and DYER [1941], BOWER [1945]) and have been converted to sodium salts of sufficient purity to be crystallized from water and matched with authentic specimens (PEDERSEN, [1953]).

Lower phosphate esters of inositol cannot be quantitatively precipitated as ferric salts. but give insoluble calcium salts in an alkaline medium (BOWER [1945]).

With the advent of chromatographic and electrophoretic techniques, it has been found that some of the products obtained from soil extracts by precipitation are impure, and the use of these techniques has led to a more precise characterization of the soil inositol phosphates. In particular, they have shown the presence of a number of isomers not previously isolated from any biological source.

(b) Separation of Inositol Phosphates by Chromatography or Electrophoresis

Phosphate esters are strong acids with pK values of about 1 and 6 for the first and second dissociation, and are readily adsorbed by anion-exchange resins. SMITH and CLARK [1952] were the first to describe a method for the separation of IHP from orthophosphate and lower phosphate esters of inositol by ion-exchange chromatography. After partial hydrolysis of IHP, phosphates were adsorbed on a De-Acidite weak-base anion-exchange resin and resolved by development with increasing concentrations of hydrochloric acid. A number of esters were separated, and although not all were identified, they included IHP, inositol penta-phosphate, and inositol tetraphosphate. The method of Smith and Clark was adapted and shortened by CALDWELL and BLACK [1958]. Later, SCHORMÜLLER and BRESSAU [1960] and COSGROVE [1963] described techniques using gradient development with hydrochloric acid from Dowex-2 or Dowex-1 strong-base anion-exchange resins (Figure 1). ANDERSON [1964] also used Dowex-1 anion-exchange resin, but developed with ammonium formate at pH 7 (Figure 2).

A very good separation of inositol phosphates can be achieved by paper partition chromatography using n-propanol, concentrated ammonium hydroxide, water (5:4:1) as solvent (DESJOBERT and PETEK [1956]). The hexa- and pentaphosphates are not distinguished. Other solvents have been used for the separation of certain inositol phosphates (e.g., ANDERSON [1956], BIELESKI and YOUNG [1963]), but propanol-ammonium hydroxide systems are the most valuable for general use. Phosphate zones on chromatograms can be detected by the spray reagents of HANES and ISHERWOOD [1949] or WADE and MORGAN [1955].

ARNOLD [1956] described a separation of inositol phosphates by paper electrophoresis, with a potential gradient of about 3 v/cm in an acetate buffer of pH 3.65.

(c) Chromatography of Soil Products

Prior to chromatographic analysis inositol phosphates are extracted from soils with acid and alkali and are separated from the mixed extracts (CALDWELL and BLACK [1958a], THOMAS and LYNCH [1960]) or the alkaline extracts alone (SMITH and CLARK [1951], ANDERSON [1956], COSGROVE [1963]). They are usually precipitated as ferric salts from acid medium or barium salts from alkaline medium, in the presence of ethanol. The first procedure is more selective, and coprecipitates less impurity, but if the concentration of iron is high and inositol phosphate is low, quantitative yields may not be obtained, due to the formation of soluble complexes (ANDERSON [1963]). The lower phosphate esters of inositol are not readily recovered as ferric salts. Precipitation of barium salts, on the other hand, is likely to give good recoveries, but in

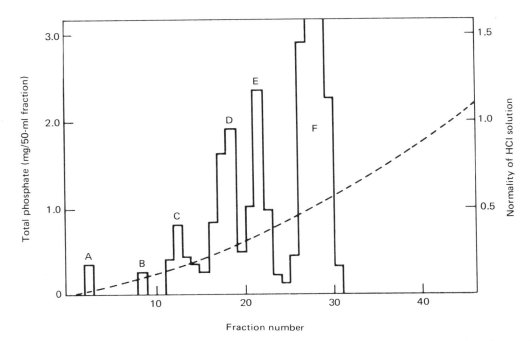

Figure 1. Elution of *myo*-inositol phosphates and orthophosphate with HCl from a column of Dowex-1-chloride (× 8), 200–400 mesh 28 × 1.2 cm. (COSGROVE [1963]). A = orthophosphate; C = inositol tetraphosphate; D and E = inositol pentaphosphates; F = inositol hexaphosphate.

some soils so much material is coprecipitated that when the precipitate is redissolved and fractionated by ion exchange, columns are quickly contaminated and resolution of the phosphates is impaired (ANDERSON [1964]). If extracts are subjected to the hypobromite oxidation of Wrenshall and Dyer (SMITH and CLARK [1951], COSGROVE [1963]) prior to precipitation of the iron salts, then better resolution is sometimes obtained, but some loss may occur.

An association between IHP and other soil organic constituents which alters the chromatographic characteristics of the ester can be broken by prolonged hydrolysis with hot alkali (ANDERSON and HANCE [1963], ANDERSON [1964]).

For chromatographic analysis the ferric or barium salts are converted to the corresponding sodium salts or free acids.

By ion-exchange chromatography SMITH and CLARK [1951, 1952] were able to show that only a minor proportion of the phosphate in soil ferric "phytate" preparations was in fact IHP. Other esters were not fully characterized, but analysis of the relative amounts of inositol and phosphorus showed that pentaphosphates were probably present. Authentic mixtures of the mono-, di-, and triphosphates did not survive the extraction and precipitation procedures. Another well-defined peak which had no analogue among the inositol phosphates from wheat bran was obtained in all cases. This was later shown by CALDWELL and BLACK [1958b] to give positive reactions to the chemical tests of Scherer and Salkowski (see FLEURY and BALATRE [1947]), indicating that it contained inositol, but the compound released on hydrolysis did not promote growth of *Saccharomyces carlsbergensis* or an inositol-requiring mutant of *Neurospora crassa* and was, therefore, not *myo*-inositol (formula II), the only form hitherto found combined with phosphate in nature. Caldwell and Black assumed that the unknown ester was an isomer of *myo*-IHP, and in 49 soils from the U.S.A. and Canada they found a quantitative

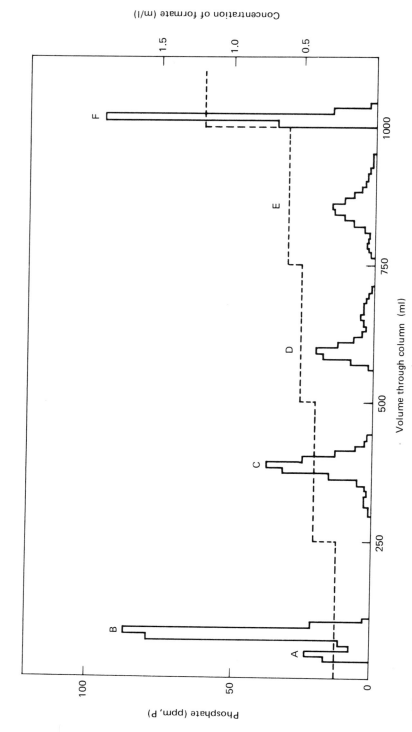

Figure 2. Elution of *myo*-inositol phosphates and orthophosphate from a column of Dowex-1-formate (×8), 100–200 mesh, 16 × 1.2 cm. (ANDERSON [1964]). A = inositol monophosphate; B = orthophosphate; C = inositol diphosphate; D = inositol triphosphate; E = inositol tetraphosphate; F = inositol hexaphosphate. More than one isomer of some of the esters may have been present. Pentaphosphates were not tested, but it is now known that one pentaphosphate, at least, would be eluted concurrently with the hexaphosphate.

relationship between the two isomers, the unknown form amounting, on average, to 46% of the *myo*-IHP. The total IHP accounted for 4 to 98 ppm of phosphorus in the soil, and up to 52% of the total organic phosphate.

THOMAS and LYNCH [1960], using an identical technique for the analysis of Canadian soils, also found both isomers, but in most cases the amount of the unknown exceeded that of the *myo*-IHP.

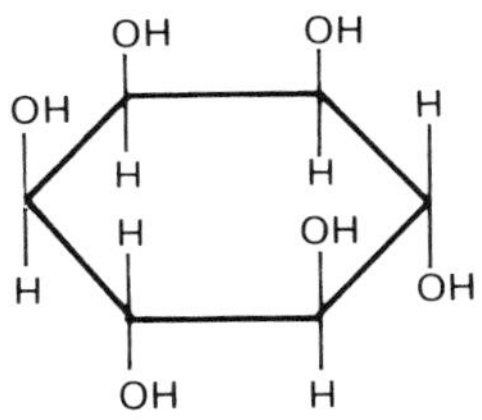

II

myo-Inositol

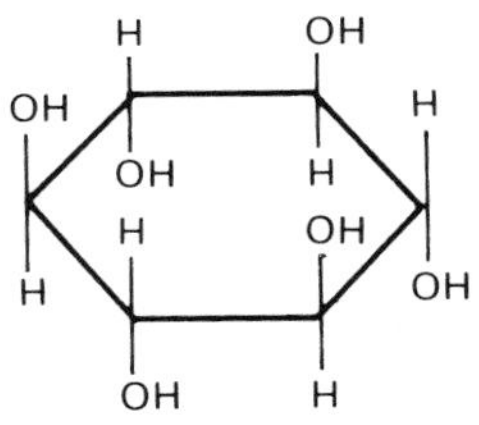

III

scyllo-Inositol

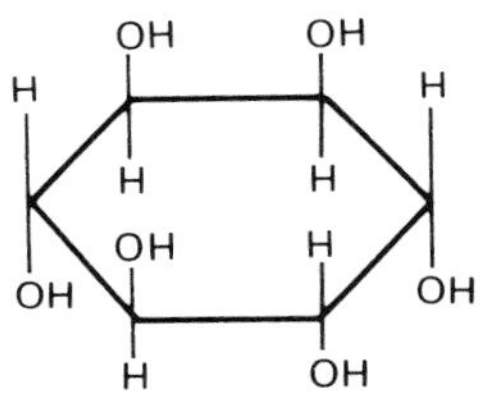

IV

D-chiro-Inositol

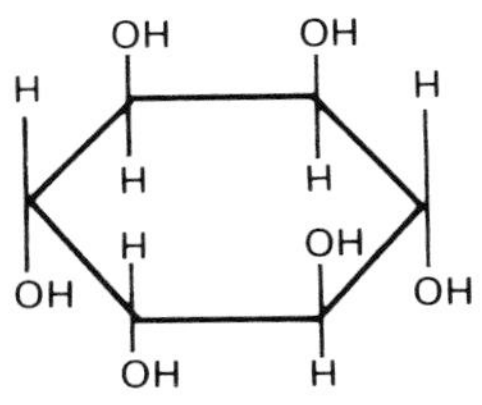

V

L-chiro-Inositol

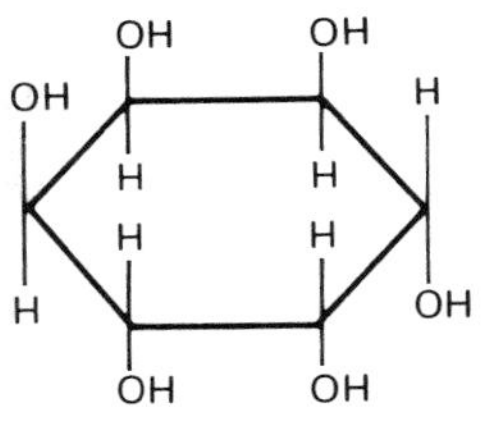

VI

neo-Inositol

This isomer was shown by COSGROVE [1962, 1963] to be the hexaphosphate of *scyllo*-inositol (formula III). Cosgrove also confirmed the presence in soils of the hexaphosphate of *chiro*-inositol (formulae IV and V). and with Tate (COSGROVE and TATE [1963]) that of *neo*-inositol (formula VI). The identity of the esters was confirmed by paper chromatography before and after hydrolysis and by the preparation of inositol derivatives such as hexa-benzoates or hexa-acetates, which were compared with authentic samples. The pentaphosphates of *myo*-inositol, *chiro*-inositol, and *scyllo*-inositol were also detected. The fact that the racemate, DL-inositol, was isolated, and not one of the enantiomorphs led Cosgrove at first to believe that its presence might be due to inversion of *myo*-inositol. However, neither the extraction techniques nor the conditions used for subsequent hydrolysis of the phosphate preparations caused the inversion of authentic *myo*-inositol. About one-fifth of the IHP fraction was present as *scyllo*-IHP in three soils.

By paper partition chromatography ANDERSON [1956] showed the presence in British soils of inositol tri- or tetraphosphates, but the bulk of the inositol phosphate fraction consisted of the hexa- and pentaphosphates. By ion-exchange and paper chromatographic techniques it was later found that the last two esters accounted for an average of 40% of the organic phosphate in 17 soils. Detailed analysis of three of them confirmed the presence of esters of *myo*-, *chiro*- and *scyllo*-inositol, of which *myo*-IHP was by far the most abundant. The esters were first thought to be hexaphosphates, on the basis of their properties during chromatography on paper or by anion exchange in a formate system (ANDERSON [1964]), but resolution by ion exchange in a chloride system has since shown the presence of minor amounts of penta-phosphates (Anderson, unpublished observations). Up to 460 ppm of phosphorus in the soil occurred as inositol hexa- and pentaphosphates. Similar values were found for these esters by two contrasting extraction techniques, one involving isolation of the ferric salts and the other of the barium salts.

Recently, MARTIN [1964a, b] has separated inositol phosphates from alkaline soil extracts by direct adsorption on columns of strong-base anion-exchange resins. The total amount of organic phosphate adsorbed was critically affected by the degree of cross-linkage in otherwise identical resins, and a much better recovery of added IHP was obtained from De-Acidite FF (3% cross-linked) than from Dowex-1 (8% cross-linked). Adsorption of IHP by the Dowex resin was quantitative in the absence of other soil components, and the effect with soil extracts was attributed to complex formation. Elution of the columns with increasing concentrations of potassium hydroxide yielded first inorganic phosphate and then an organic fraction. The latter in one case was rechromatographed by the method of COSGROVE [1963], showing that one-third of the phosphate in the fraction was probably IHP, corresponding to 20% of the organic phosphate in the soil.

3. *Sources of Inositol Phosphates in Soil*

There is no evidence to show to what extent soil IHP is synthesized *in situ* by soil micro-organisms, or alternatively reaches the soil from decomposing plant material and animal excreta. While *myo*-, *scyllo*-, and *chiro*-inositols have been isolated from one or more of these sources (ANGYAL and ANDERSON [1959]), the phosphate esters of only *myo*-inositol have been detected up to the present time.

Although *myo*-IHP is of wide occurrence in plants, most of the available data relate to the seeds, and little is known of its distribution in the roots, stems, and leaves. Attempts by COSGROVE [1964] to identify *scyllo*-IHP in acorns, which contain more *scyllo*- than *myo*-inositol, were not successful, and he concluded that the enzyme systems incorporating inositol into IHP in this material were specific for *myo*-inositol.

Analysis of fresh farmyard manures from various sources has shown that up to 40% of the total phosphate is in the form of organic material soluble in hydrochloric acid and much of this may be IHP (KAILA [1948]). PEPERZAK *et al.* [1959] examined the nature of the acid-soluble phosphate by ion-exchange chromatography, confirming the presence of IHP in fresh manures from many sources. Hen manure yielded the highest amount (12% of the total phosphate), the average from all sources being about 5%.

Although inositol hexaphosphates have not yet been isolated from pure cultures of microorganisms, a product yielding both inositol and phosphate on hydrolysis has been obtained from fungal mycelium (KITA and PETERSON [1953]) and a *meso*-inositol phosphate has been reported among the products formed by yeast in the presence of glucose (AUBERT and MILHAUD [1955]). SMITH [1951] had evidence that he had found phytate phosphorus in the yeast *S. carlsbergensis*, but the methods he used are now known to be unspecific, and COSGROVE [1964] has been unable to confirm its presence. There is, however, no doubt that IHP can be synthesized in the soil. It was shown by SMITH and CLARK [1951] that if soils were incubated with inorganic phosphate labeled with P^{32}, some of the activity was incorporated into fractions corresponding to *myo*-IHP and what is now known to be *scyllo*-IHP. To obtain more direct evidence of microbial synthesis, CALDWELL and BLACK [1958b] inoculated sands, clay-sand mixtures, and the C horizons of three soils with a water extract of surface soil, adding mineral and organic nutrients at intervals. Only one of the samples contained any organic phosphate initially, but after incubation at 28 or 30°C for 3 to 9 months, 9 to 96 ppm of organic phosphorus had accumulated. Fractionation by ion-exchange chromatography showed the presence of *myo*-IHP and its isomer (*scyllo*-IHP), the latter being present in considerably greater quantity. The sum of the isomers accounted for 7% of the total organic phosphate. A similar investigation by COSGROVE [1964] yielded detectable IHP in only one out of three cases, and only the *myo*-isomer was identified.

III. Phospholipids

The term "lipid" does not imply a particular chemical structure, and in view of the many types of substances that have been classified in this way a precise definition is difficult. In describing lipids, BLOOR [1925] classified them as substances that are insoluble in water but soluble in "fat solvents" such as ether, chloroform, or benzene, are actual or potential esters of fatty acids and are utilizable by living organisms. This definition has probably been the most widely accepted, but it is very imprecise in some respects and excludes certain compounds now considered to be lipids.

The lipids containing phosphorus, the phospholipids or phosphatides, can be divided into distinct groups, one of the most important being the glycerophosphatides (general formula VII), which are present in plants, animals, and microorganisms. The corresponding lipids containing no phosphate, glycerides of fatty acids, have been extracted from soils (SCHREINER and SHOREY [1910]).

Choline (formula VIII), a degradation product of many lipids, has been identified in soil extracts by ASO [1905] and SHOREY [1913], who characterized the compound by examination of the crystalline precipitates obtained with gold and platinum chlorides. Shorey considered that the choline, present in only small amounts, might have been added to the soil as such in plant residues, or formed in the soil by decomposition of more complex substances, but it might alternatively have been an artefact, released by alkaline hydrolysis of lipids during extraction.

Extraction of the soil with alcohol or ether has been used as a basis for measurement of the phospholipids present, and values ranging from about 1 to 3 ppm of phosphorus have been obtained in this way, although as much as 34 ppm has been reported (BLACK and GORING [1953]). WRENSHALL and MCKIBBIN [1937], who used ether followed by alcohol for extraction, accounted for 0.2 to 0.3% of the total organic phosphate, and SOKOLOV [1948] by a similar method found up to 1.5%. VINCENT [1937] extracted about 6 ppm of phosphorus from soils by treating successively with acetone, chloroform, and ethanol.

Many of the lipids occurring in plant or animal tissue are associated with other substances such as proteins or carbohydrates, and complexes of this kind are generally insoluble in fat solvents. The lipids can usually be liberated by treatment with ethanol, but it has been found

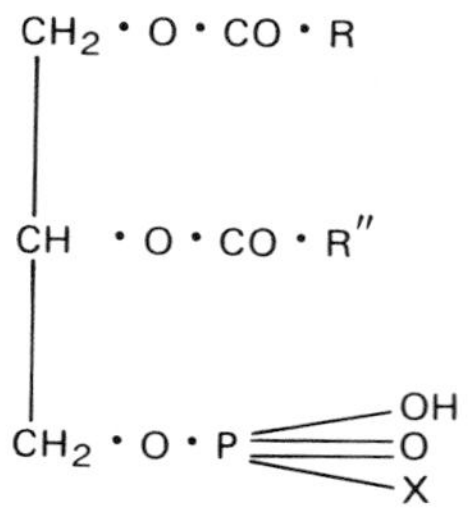

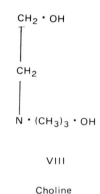

CH$_2 \cdot$ O $\cdot$ CO $\cdot$ R

CH $\cdot$ O $\cdot$ CO $\cdot$ R$''$

CH$_2 \cdot$ O $\cdot$ P

OH
O
X

VII

General structure of a glycerophosphatide.
R$\cdot$CO $-$, R$'\cdot$CO $- =$ long-chain fatty acyl groups
x = choline, ethanolamine, etc.

CH$_2 \cdot$ OH

CH$_2$

N $\cdot$ (CH$_3$)$_3 \cdot$ OH

VIII

Choline

that subsequent extraction with a number of different solvents is then necessary to dissolve them completely. In the soil the extraction of lipids is likely to be further complicated by the presence of reactive inorganic constituents, as demonstrated by GORING and BARTHOLOMEW [1949], who found that the amount of lipid phosphorus extracted from microbial tissue with a mixture of ethanol and ether was progressively reduced in the presence of increasing amounts of clay minerals, and with 10% of kaolinite present, only 34% was recovered.

It was found by HANCE and ANDERSON [1963a] that more phospholipids could be extracted from a fresh than from a dried soil, but the difference could be eliminated by pretreatment of the latter with a mixture of dilute hydrofluoric and hydrochloric acids. Extraction could also be improved by using a range of solvents, and a procedure was adopted in which the soil was pretreated with acid, washed with water, and extracted successively with acetone, light petroleum, ethanol-benzene, and methanol-chloroform. Since ethanol and acetone can to some extent dissolve inorganic or nonlipid organic substances containing phosphorus, the extracts were evaporated and the residues re-extracted with ether-light petroleum and chloroform. Lipid values obtained in this way, though considerably higher than those obtained with alcohol and ether alone, accounted for only 3 to 7 ppm of phosphorus, less than 1% of the soil organic phosphate.

Attempts to characterize soil lipid material have shown the presence of glycerophosphate, released by alkaline hydrolysis, and choline and ethanolamine, released by acid hydrolysis (HANCE and ANDERSON [1963b]. The amounts detected indicate that phosphatidyl choline may be one of the predominant phospholipids in the soil, which is in keeping with the fact that it is found in many plants and microorganisms.

While variations in extraction techniques could probably raise the soil phospholipid values to some extent, these substances would appear to constitute a very small proportion of the total soil organic phosphate.

IV. Other Esters

A large part of the organic phosphate in soils has not yet been characterized. About 5 to 50 % of the total is insoluble in acid at low pH values, occurring in the humic acid fraction of the organic matter (YOSHIDA [1940], DYER and WRENSHALL [1941a], BOWER [1949], ADAMS et al. [1954], ANDERSON [1961], MARTIN [1964a]). A small proportion of this is in the form of nucleic acid derivatives (Chapter 3), but the remainder has not been identified. Some of the phosphate in this fraction is mineralized by incubation in dilute alkali, a characteristic property of phosphoproteins (ANDERSON [1961]).

There is some evidence that glucose-1-phosphate may be a constituent of the organic matter (ROBERTSON [1958]), but its amount must be very small in most soils since it is readily mineralized in acid, unlike the bulk of the organic phosphate.

During the chromatographic analysis of soil ferric phytate preparations, COSGROVE [1963] obtained two unidentified phosphate ester fractions that contained no inositol and did not respond to tests for nucleotides, amino acids, or carbohydrates.

C. The Estimation of Soil Organic Phosphate

I. Basis of Estimation

Two contrasting types of methods are used for determining soil organic phosphate. In the first, organic phosphate is extracted from the soil and measured by the difference between the total and inorganic phosphate in the extract. In the second, the organic phosphate is mineralized by ignition of the soil and measured by the difference in inorganic phosphate extracted from comparable ignited and unignited samples. In other methods, similar in principle to ignition, phosphate ester linkages are destroyed by hydrolysis or oxidation, usually with hydrogen peroxide. The latter, however, is not a particularly effective reagent for destroying soil organic matter (PURI and SARUP [1937]), and phosphate esters have been detected in soil after peroxide treatment (BLACK and GORING [1953]).

Values obtained by differences in this way are subject to relatively large errors, particularly if inorganic phosphate constitutes a high proportion of the total, but no direct method has yet been described in which the organic phosphate can be determined by a single measurement. Early workers thought that organic and inorganic phosphates could be selectively extracted from soil, but such a separation seems unlikely to be achieved. More promising are the observations of GORING [1955] and ANDERSON [1963] that a high degree of separation of organic and inorganic phosphates can be made in alkaline extracts by acidification and selective adsorption on charcoal. Quantitative separations have been obtained with ammonium hydroxide extracts, but not with sodium hydroxide, probably due to greater degradation of the organic fractions.

II. Extraction Methods

The most effective extractants of soil organic phosphate so far used are alkaline solvents,

Table 1. An outline of some extraction methods for measuring total soil organic phosphate.

Authors	Dean [1938]	Pearson [1940]	Ghani [1943a]	Williams [1950]	Mehta et al. [1954]	Saunders and Williams [1955]	Kaila and Virtanen [1955]
Pretreatment	CH$_3$COONa	0·1 N HCl	0·5 N CH$_3$COOH	2·5% CH$_3$COOH + 1% oxine	12 N HCl	0·1 N HCl	4 N H$_2$SO$_4$
	R.T.* Leach	R.T. 5 min → Leach	R.T. 2 hr	20 °C 16 hr	R.T.→70°C 10 min → 12 N HCl R.T. 1 hr	c. 90°C 1 hr → Leach	R.T. 18 hr
Alkaline extraction	0·25 N NaOH 95°C 16 hr	0·5 N NH$_4$OH 90°C 18 hr	0·25 N NaOH R.T. 2 hr → Repeated	0·1 N NaOH 20°C 16 hr	0·5 N NaOH R.T. 1 hr → 0·5 N NaOH 90°C 8 hr	0·1 N NaOH R.T. 16 hr → Repeated once	0·5 N NaOH R.T. 18 hr → 0·5 N NaOH 90°C 4 hr

* R.T. = room temperature.

but pretreatment of the soil is first necessary to remove cations, such as calcium, which render phosphates insoluble in alkali. Extraction methods will give accurate values if the organic phosphate is quantitatively removed from the soil without hydrolysis, and if the inorganic and total phosphate in the extract can be precisely measured. In early methods inorganic phosphate was determined by precipitation with magnesia mixture and ammonium molybdate, a procedure that does not give quantitative recoveries at low phosphate concentrations and is not specific in the presence of some phosphate esters. This technique has been superseded by colorimetric methods, which are specific for orthophosphate and can be used in the presence of phosphate esters, provided no hydrolysis of these takes place during the estimation. The most common alkaline extractants have been ammonium or sodium hydroxide, but potassium and lithium hydroxides and sodium carbonate and bicarbonate have also been used (SCHOLLEN-BERGER [1918]). Although sodium hydroxide can cause more hydrolysis of organic phosphate than ammonium hydroxide under comparable conditions, it is also a more effective extractant and has largely replaced the latter.

The procedures of Dean, Ghani, and Williams (Table 1) formed part of a broad scheme embracing also a fractionation of inorganic phosphate. Ghani modified Dean's method, pretreating with acetic acid rather than sodium acetate, primarily so that calcium phosphates as well as exchangeable calcium would be dissolved. To estimate "available" phosphate in soils he later proposed extraction with acetic acid containing 8-hydroxyquinoline (oxine) to block readsorption of dissolved phosphate and prevent precipitation of iron and aluminum phosphates (GHANI [1943b]). This was incorporated as a pretreatment into the extraction method of WILLIAMS [1950a], who found that the presence of oxine increased not only the total phosphate extracted by acetic acid, but also the organic phosphate extracted subsequently with alkali. However, the organic phosphate which might have been present in the initial acid extracts could not be measured. Ghani found a single hot extraction with sodium hydroxide superior to a single cold extraction, but it caused some hydrolysis. If the cold extractions were repeated until no more organic phosphate dissolved, up to five successive treatments were required. The hydrolytic action of hot sodium hydroxide was also noted by MEHTA et al. [1954], but they found no such effect if successive cold and hot extractions were used, and they concluded that the most easily hydrolyzed phosphate esters in soil were also the most soluble. Higher values for organic phosphate were obtained if alkaline extraction followed pretreatment with hot concentrated as opposed to cold dilute hydrochloric acid. It was suggested that the organic matter was protected by coatings of mineral substances such as ferric oxides or clays.

While the method of Mehta and coworkers is one of the most effective described, it will undoubtedly cause some hydrolysis due to the severity of the acid pretreatment. Monoesters of orthophosphoric acid differ markedly in their stability in acid media. Many are hydrolyzed quite slowly, the rate depending upon the pH of the medium as typified by inositol hexaphosphate (DESJOBERT and FLEURENT [1954]). There is a rate maximum about pH 4 where the reactant species is the mono-anion and a decrease in rate at higher and lower pH values due, respectively, to the presence of the less reactive di-anion and the undissociated neutral species (VERNON [1957]). There is an increase in rate at high acid concentrations giving a minimum at pH values about 0 to 1. Some sugar phosphates, on the other hand, are very susceptible to acid hydrolysis, the rate increasing rapidly as the pH of the medium decreases.

Since the precise nature of a large part of the soil organic phosphate is still obscure, recovery tests with specific compounds are of limited value, but they help to give a general picture of the possible extent of hydrolysis. In a number of cases soils have been treated with representatives of the main groups of esters whose presence is known or suspected, and their recoveries by various extraction methods measured (ANDERSON [1960], HARRAP [1963],

DORMAAR and WEBSTER [1963]). Some, such as inositol hexaphosphate, are recovered in high yields, even after treatment with hot concentrated acid, but others are hydrolyzed — in some cases completely. In consequence, several pretreatments have been proposed which are designed to reduce acid hydrolysis and are effective in recovering acid-labile esters from the soil (Table 2). Comparisons of these with other methods have not shown any marked advantages, although increases have been obtained in some cases (HANCE and ANDERSON [1962], DORMAAR [1964]), and it would appear that, in general, esters which are readily hydrolyzed in acid do not long survive in the soil.

Table 2. Methods of recovering acid-labile esters during extraction of soil organic phosphate.

Authors	ANDERSON [1960]	HARRAP [1963]	DORMAAR and WEBSTER [1963]
Pretreatment	0.3 N NaOH R.T.* 16 hr	5% Na$_2$EDTA R.T. 2 hr	1:1 acetone water R.T. 4 hr ↓ 0.3 N NaOH R.T. 4 hr
Subsequent extraction (see Table 1)	Method of MEHTA et al. (complete) [1954]	Alkaline extraction of MEHTA et al. [1954]	Method of KAILA and VIRTANEN (complete) [1955]

* R.T. = room temperature.

Although many phosphate esters present in biological tissue are more stable in alkaline than comparable acid media, there are notable exceptions, including phosphoproteins and phosphorylated amino acids, which are rapidly mineralized in dilute sodium hydroxide (SCHMIDT and THANNHAUSER [1945], BAMANN et al. [1955]). If substances of this type are present in the soil they will suffer considerable hydrolysis during extraction, even with dilute alkali. While sodium hydroxide is the most effective extractant yet used for soil organic phosphate, a considerable proportion can be removed under much milder conditions. For example, BOSWALL and DELONG [1959] achieved 75 and 86% extraction from two soils by the use of a chelating agent at pH 9.2. The soils were first extracted with hydrochloric acid and then with dilute ammonium hydroxide and a solution of oxine in benzene; the metal chelates passed largely into the benzene phase, while the organic phosphate dissolved in the aqueous phase. A mild extractant of this type could have value in recovering esters that are hydrolyzed in strong alkali.

A number of comparisons of the methods meanwhile in common use have shown that no single one can be considered superior to a number of others, and the success of a particular method appears to depend on soil-type and environment (KAILA and VIRTANEN [1955], VAN DIEST and BLACK [1958], HANCE and ANDERSON [1962], KAILA [1962] DORMAAR [1964]). Several give very similar results and are adequate for the general characterization of soils. Where in individual cases values as precise as possible are desired, it is advisable to use more than one method, and any striking difference can then be assessed by further investigation.

III. Ignition Methods

1. *Requirements of an Ignition Technique*

Early soil chemists noticed that after ignition of the soil, a greater amount of phosphorus could be extracted with dilute acids, and the increase was attributed to the destruction of organic matter containing phosphorus. Ignition techniques were soon used for measuring soil organic phosphate and a variety of methods have been described. If valid results are to be obtained, ignition must quantitatively mineralize organic phosphate yet not alter the acid solubility of the native inorganic phosphate; the mineralized organic phosphate must be completely extractable; and no organic phosphate must be lost by hydrolysis during extraction of the ignited soil or by volatilization during ignition. In practice it is certain that some of these demands are not completely satisfied and that both positive and negative errors will arise, but during the development of ignition techniques various precautions have been taken to minimize them.

2. *Sources of Error*

(a) Incomplete Mineralization of Organic Phosphate and Loss by Volatilization

Both of these sources of error are dependent on the temperature and duration of ignition. Short ignition periods of one to two hours are commonly used with ignition temperature of around 500 to 600°C, and under these conditions mineralization should be complete in most soils. In mineral soils where sufficient bases are present there will be negligible losses by volatilization until temperatures of about 800°C are reached (SAUNDERS and WILLIAMS [1955]), but DORMAAR and WEBSTER [1964] consider that with some peats volatilization occurs at about 400°C. MEHLICH [1963], on the other hand, found that maximum extraction of phosphate from a peat and a manure was achieved after ignition at 550 and 750°C, respectively.

(b) Change in Solubility of Native Inorganic Phosphate

The acid solubility of the phosphate in minerals such as wavellite and dufrenite, and in some other naturally occurring aluminum or iron phosphates, increases after ignition (FRAPS [1911]), while with synthetic aluminum or iron phosphates increases or decreases have been reported, depending on the experimental conditions (DOUGHTY [1935], GHANI [1944]).

Recently, MEHLICH [1963] compared the solubility in 1 N sulfuric acid of the phosphate in a number of minerals and soils after various heat treatments. With montmorillonite the solubility did not vary much after heating but with goethite, kaolinite, limonite, and allophane it increased substantially, the extractable phosphate varying considerably at different ignition temperatures. Similar changes were found with soils, and clearly some of these were related to the behavior of inorganic rather than organic constituents. In several cases there was a pronounced increase in solubility above an ignition temperature of 550°C, reaching a maximum at 750°C, and it was suggested that this might be due to the presence of kaolinite. The majority of the soils examined were subsurface samples, low in organic matter.

With many surface soils from temperate regions, on the other hand, virtually all the phosphate is extractable with 1 N or 2 N sulfuric acid after ignition to 550°C (WALKER and ADAMS [1958], HANCE and ANDERSON [1962]), and similar values have been reported for organic phosphate at ignition temperatures over the range 400 to 650°C (SAUNDERS and WILLIAMS [1955]), 300 to 600°C (LEGG and BLACK [1955]), and 350 to 600°C (GHANI [1944]).

The effect of ignition on the solubility of phosphate in minerals is likely to be closely associated with their state of hydration, their physical characteristics, and their association with other substances, and it is difficult to predict to what extent a change occurring in a pure mineral will take place in a soil containing it. Changes will almost certainly occur, but in many cases their magnitude is likely to be small.

(c)   Incomplete Extraction of Mineralized Organic Phosphate

Many soils, particularly acid soils rich in reactive compounds of iron and aluminum, will adsorb phosphate in weakly acid media. While maximum sorption is likely to take place at pH values about 3 to 4, instances have been cited where sorbed phosphate resists extraction with 1 N sulfuric acid (THOMSON, BLACK, and CLARK [1948]). If more dilute acid extractants are used, tests are usually carried out to ensure that increasing the concentration of extractant does not increase the organic phosphate values (SAUNDERS and WILLIAMS [1955], KAILA and VIRTANEN [1955], WALKER and ADAMS [1958]).

(d)   Hydrolysis of Organic Phosphate during Extraction

Phosphate esters may be hydrolyzed in acid solution, the extent depending upon the ester, the nature of the extractant, and the temperature and duration of extraction (see under "extraction methods," Section C II). Hydrolysis is likely to be small where cold dilute mineral acids are used, but esters as unstable as glucose-1-phosphate will not be recovered (ANDERSON [1960]).

3.   *Methods of Analysis*

The most frequently used ignition techniques are typified by those outlined in Table 3.

Although there are several possible sources of error it has been found in practice that, with many soils, methods similar to that of Odynsky give values for organic phosphate which are in reasonable agreement with extraction values (SAUNDERS and WILLIAMS [1955], WILLIAMS, WILLIAMS, and SCOTT [1960], HANCE and ANDERSON [1962], KAILA [1962]).

Table 3.   An outline of typical ignition methods for measuring soil organic phosphate.

Authors	ODYNSKY [1936]	SAUNDERS and WILLIAMS [1955]	LEGG and BLACK [1955]
Ignition temperature and duration	600°C 1 hr	550°C 1 hr	240°C 1 hr
Extractant	2 N H$_2$SO$_4$ R.T.* 1 hr repeated	0.2 N H$_2$SO$_4$ R.T. 16.5 hr	12 N HCl R.T.→70°C 10 min ↓ 12 N HCl R.T. 1 hr

* R.T. = room temperature.

Where differences occur, the ignition values are usually higher and, on the assumption that extraction values are the more accurate, LEGG and BLACK [1955] recommended an ignition procedure designed to give similar values to the extraction method of MEHTA et al. By lowering the ignition temperature to 240°C they introduced a negative error, due to incomplete combustion of organic phosphate, to offset the supposed positive errors caused by ignition. Extraction methods are likely to cause hydrolysis, however, and the real organic phosphate values probably lie between the ignition and extraction values in many cases.

D. Accumulation and Breakdown of Organic Phosphate

I. In Composts and Microbial Tissue

When biological materials decompose, the phosphate transformations that take place depend on many factors, including the nature and age of the materials and their phosphate content (LOCKETT [1938], CHANG [1939, 1940], KAILA [1948, 1949, 1954]; BIRCH [1961]). Phosphate is an essential nutrient for the microorganisms which develop, and if its supply is limited, breakdown of the material will be restricted. Even if it is in plentiful supply, the assimilation of phosphate by microorganisms can be influenced by the availability or ease of breakdown of the bulk of the organic constituents.

Studies of the transformations of inorganic and organic phosphate which occur in cultures of microorganisms and in composted plant residues and animal manures have led to the conclusion that when the decomposing materials contain less than about 0.2 to 0.3% organic phosphorus, a net synthesis will initially take place, but if the content is higher there will be a net mineralization (KAILA [1948, 1949]). In very resistant residues mineralization will occur at a lower phosphate level (KAILA [1954]). With most green manures and straws, and some animal manures, there will be an initial increase in the amount of organic phosphate during decomposition.

Ultimately, as the energy source is used up, the microbial tissue will itself break down due to autolysis and attack by other microorganisms, resulting in mineralization of organic phosphate (CHANG [1940], BARTHOLOMEW and GORING [1948], BIRCH [1961]).

II. In Soils

When an organic substance containing little or no phosphate decomposes in the soil, the microorganisms attacking it derive the necessary phosphate from readily soluble soil fractions. Thus, CHOUCHAK [1929] found that the addition of dried blood to soil low in available phosphate decreased millet yields unless phosphate was also added. In laboratory experiments TAYLOR [1946], FULLER and McGEORGE [1950], and HANNAPEL et al. [1964] have obtained evidence that decomposition of simple compounds such as urea, polypeptides, and carbohydrates causes some conversion of inorganic phosphate to organic forms.

The transformations that occur when materials rich in phosphate esters decompose in intimate contact with the soil may be influenced by reaction of the individual esters with the clay-organic fractions. Decomposition of added phytin can be very slow in soils, particularly if they are acid (DYER and WRENSHALL [1941b], PEARSON et al. [1941]), but rapid hydrolysis occurs in inoculated sand cultures (AUTEN [1923]). The iron and aluminum salts of IHP are much more resistant to attack by the enzyme phytase than are more soluble salts (WRENSHALL and DYER [1941], CASIDA [1959]), and the stability of the ester in acid soils has been attributed to combination with iron and aluminum, possibly in the form of the active sesquioxide constituents of soil colloids. The retarding effect of soluble aluminum salts on the decomposition of the phosphate in wheat bran has been demonstrated by KAILA [1948]. The insolubility of IHP in the presence of ferric or aluminum ions has been confirmed over a wide range of acid concentrations (JACKMAN and BLACK [1951], ANDERSON [1963]), and it is strongly adsorbed by clay minerals and finely divided sesquioxides (GORING and BARTHOLOMEW [1950], ANDERSON and ARLIDGE [1962]). Reactions of this kind could account not only for the stability of the ester in acid soils, but also for its insolubility in strong acid extractants. In calcareous soils the ester is probably stabilized by the formation of insoluble calcium salts (JACKMAN and BLACK [1951]).

Clay minerals can also inhibit the enzymatic hydrolysis of phosphate esters by adsorbing the enzymes (MORTLAND and GIESEKING [1952]).

Phosphate esters are capable of forming complexes with other organic substances such as proteins and with metal ions (COURTOIS [1951], MORTENSEN [1963], ANDERSON [1963]). The tendency is particularly strong in the case of IHP and it is sometimes difficult to remove the ester during the purification of vegetable proteins. The complexes with proteins are very insoluble in the pH range 2 to 6. There is evidence that soil IHP is strongly associated with organic constituents containing amino acids and carbohydrates (ANDERSON and HANCE [1963]).

Nucleic acids have a very limited solubility in acid solution, particularly in the presence of certain clay minerals (GORING and BARTHOLOMEW [1952]). Under field conditions, breakdown of the acids can be slow (McCONAGHY [1960]), and though in laboratory tests initial breakdown is rapid, it tends to reach a limit (DYER and WRENSHALL [1941b], PEARSON et al. [1941]) probably due to the formation of "limit polynucleotides," highly resistant esters produced by enzyme attack on nucleic acids (CHARGAFF and DAVIDSON [1955]).

The amount of organic phosphate in soils is usually far in excess of that likely to be present in fresh plant or microbial residues, and the esters must have accumulated for a large number of years. The forms reaching the soil in most abundance are probably the nucleic acids and the simple sugar or nucleoside phosphates involved in cellular metabolism, but these — with the possible exception of the nucleic acids — will be quickly mineralized or otherwise utilized by soil microorganisms. Numerous strains are able to break down the phosphate esters in plant, animal, or microbial residues (KAILA [1948]), and many soil organisms have been isolated which mineralize the phosphate in phytin, nucleic acid, or phospholipids (CASIDA [1959], SZEMBER [1960], GREAVES et al. [1963], GREAVES and WEBLEY [1965]). The soil as a whole exhibits phytase or other phosphatase activity (JACKMAN and BLACK [1952], DROBNIKOVA [1961]), which is predominantly of microbial origin, though some persists for a time after sterilization of the soil with an electron beam (McLAREN et al. [1962]). Nevertheless, esters such as IHP, although added in much smaller amounts than nucleic acids, are so stabilized that they gradually accumulate until they form a large part of the organic phosphate in the soil.

III. Other Factors Affecting the Amount of Organic Phosphate in Soils

1. *Organic Matter Level, Total Phosphate, and Other Soil Components*

Although the organic phosphate as a whole is very stable, some of its constituents are involved in a continuous biological turnover (JOHNSON and BROADBENT [1952]) and the equilibrium level will depend on the nature of the soil and its environment.

The behavior of the organic phosphate is to some extent analogous to that of the soil organic matter considered as a whole, and its amount is correlated with the contents of carbon and/or nitrogen (SCHOLLENBERGER [1920], PEARSON and SIMONSON [1939], KAILA [1948, 1963], THOMSON et al. [1954], JACKMAN [1955b], WILLIAMS and DONALD [1957], WILLIAMS et al. [1960]). However, it is a much more variable constituent than, for example, is organic sulfur, and wide variations have been reported in the ratios of carbon to organic phosphorus. The possible factors influencing the proportion of phosphorus in soil organic matter have been discussed in a detailed review by BARROW [1961] and include depth of soil, pH, climate, drainage, cultivation, and phosphorus supply. In a survey of Finnish soils KAILA [1963] found that the ratio of carbon to organic phosphorus ranged from 61 to 276 in cultivated mineral soils, 141 to 526 in cultivated humus soils, and 67 to 311 in virgin soils. Relatively high organic phosphate values have been reported from some parts of Britain and New Zealand, with mean C:P ratios of about 60 (WILLIAMS et al. [1960], WALKER and ADAMS [1958]).

Relationships have been noted between organic phosphate and inorganic phosphate "fixing capacity" (JACKMAN [1955b], WILLIAMS [1959], KAILA [1963]) and also oxalate-soluble aluminum (KAILA [1963]). The phosphate sorption capacity of Scottish soils is related to oxalate-soluble aluminum and iron, and organic carbon (WILLIAMS *et al.* [1958]), and it has been suggested that the active sesquioxides are present mainly as humate complexes comprising a fairly constant proportion of the total organic matter. The relationships observed by Jackman, Williams, and Kaila may be a reflection of the association of organic phosphate with other organic constituents or of the similarity between the sorption characteristics of inorganic and organic phosphates. From the results of extensive studies on phosphate transformations KAILA [1948], and KAILA and MISSILA [1956] consider that microbial attack on the organic matter of peats may result in biological absorption of readily soluble inorganic phosphate, and considerable conversion of fertilizer phosphate to organic forms can take place.

The organic phosphate content of soils is also closely associated with the total phosphate (SCHOLLENBERGER [1920], JACKMAN [1955b], WALKER and ADAMS [1958], KAILA [1963]).

2. *Soil Temperature*

Net transformations of organic phosphate in soil are slow, at least under temperate conditions, and seasonal variations are very small (SAUERLANDT *et al.* [1957]). Laboratory incubation studies have confirmed that in the temperature range -14 to $30°C$ the rate of breakdown is small and independent of temperature, but at higher temperatures it progressively increases (THOMSON and BLACK [1947, 1949]).

The organic matter in peat soils develops in a very restricted microbial environment, and in the vastly different conditions met with during laboratory incubation mineralization of organic phosphate can be rapid (McCALL, DAVIS, and LAWTON [1956]). In a soil examined by BIRCH and FRIEND [1961] containing 8.5% carbon, the organic phosphate was hydrolyzed more rapidly than carbon and nitrogen during repeated cycles of moist incubation followed by drying at $100°C$.

3. *Cultivation, Liming, and pH*

With continuous cultivation the amount of organic phosphate gradually decreases (DeTURK [1938]) and the level in virgin soils is usually considerably higher than in comparable cultivated areas (SCHOLLENBERGER [1920], GARMAN [1948], THOMSON and BLACK [1949]). If grass is sown the amount rises again (DeTURK [1938]) and in a survey of New Zealand pastures JACKMAN [1955a] found that top-dressed phosphate was partly converted into organic forms, mainly in the top inch of soil. Considerable increases in organic phosphate, and organic matter generally, have occurred in a group of Australian podzolic soils under a pasture of subterranean clover top-dressed with superphosphate (WILLIAMS and DONALD [1957]). A considerable part of the increase under pasture may be gained from the excreta of grazing animals (BROMFIELD [1961]).

Liming of acid soils causes a decrease of organic phosphate (DAMSGAARD-SORENSEN [1946], KAILA [1948]), probably by increasing its solubility and by stimulating microbial activity. More hydrolysis of phosphate esters usually occurs in acid soils on incubation if lime is added (PEARSON *et al.* [1941], GHANI and ALEEM [1942], KAILA [1948], HALSTEAD *et al.* [1963]) though exceptions have been noted where pH levels are already reasonably high (KAILA [1961]). THOMSON *et al.* [1954] found that the rate of hydrolysis increased with pH in a range of soils that had received no lime immediately prior to incubation. A significant negative correlation of organic phosphate content with pH has been found in Finnish virgin soils (KAILA [1963]) but with cultivated soils of various types in Finland (KAILA [1963]) and loess soils in Germany (SCHACHTSCHABEL and HEINEMANN [1964]) only small and insignificant relationships with pH have been noted.

In a group of British calcareous soils the amount of organic phosphate relative to total phosphate has been correlated with the calcium carbonate content (SEN GUPTA and CORNFIELD [1962]).

4. *Soil Type, Texture, and Drainage*

Phosphate esters in soil are found mainly in the clay and silt fractions (WILLIAMS and SAUNDERS [1956], McDONNELL and WALSH [1957]) and clays and loams tend to have higher levels than comparable lighter soils (SAXENA and KASINATHAN [1956], KAILA [1963]).

In New Zealand, JACKMAN [1955b] attributed higher levels in yellow brown loams, compared with podzolic soils derived from sedimentary rocks, to the presence of allophane. The influences of various pedological factors have been noted by SAUERLANDT et al. [1957].

Drainage differences can have a pronounced effect on the organic phosphate level. Thus, WILLIAMS and SAUNDERS [1956] and WILLIAMS [1959] observed that poorer drainage in a number of Scottish argicultural soils was reflected in a much lower level of organic phosphate, and an abrupt fall in organic phosphate with depth. Similar effects were noted in Ireland by McDONNELL and WALSH [1957].

E. Utilization by Plants

It has been shown by numerous investigators that esters such as inositol hexaphosphate and nucleic acid supply phosphate readily to plants grown in water or in sand (*e.g.*, WHITING and HECK [1926], WEISSFLOG and MENGDEHL [1933], ROGERS et al. [1940]). Plant roots exhibit considerable phosphatase activity, part of which is associated with microorganisms on or near the root surface and part with enzymes from the root itself (ROGERS et al. [1940], FLAIG and KAUL [1955], RATNER and SAMOILOVA [1958], SZEMBER [1960], ESTERMANN and McLAREN [1961], GREAVES and WEBLEY [1965]), and it is probable that mineralization precedes uptake of phosphate in most cases. While there is evidence that some esters, including inositol hexaphosphate, can be absorbed by plants without prior mineralization (ROGERS et al. [1940], TOKARSKAYA-MERENOVA [1956]), tests by FLAIG et al. [1960] on the utilization of P^{32}-labeled phytate by barley in nonsterile culture led them to conclude that the rate of decomposition of phytate was probably the limiting factor in the uptake of phosphate from this compound.

In some soils the native organic phosphate can have an important effect in supplying phosphate to growing crops. The total organic phosphate in a variety of soil types in East Africa has been found to give a better correlation with responses of wheat than a number of inorganic values (FRIEND and BIRCH [1960]). When the amount of organic phosphate was considered together with the phosphate retention capacity of the soil, giving a measure of "available mineralized phosphate," the correlation was improved. In soils of temperate regions the contribution of the organic fraction is likely to be very small relative to that of inorganic phosphate (*e.g.*, WILLIAMS [1950b]). After relating crop yields from a large number of Norwegian field experiments on rotational leys with laboratory measurements of different categories of soil phosphate, SEMB and UHLEN (1954) concluded that below pH 5.5 the phosphorus in organic compounds did not seem to be of any importance, but above pH 5.5 it had a significant influence on crop yield. In Iowa, VAN DIEST and BLACK [1959a, b] have worked out equations relating the amounts of phosphorus available to plants grown in soil-sand and measurements of labile inorganic phosphate and various categories of organic phosphate in the soil. In alkaline soils they found a significant regression of phosphate yield on the organic phosphate which was mineralized by alkaline hypobromite, or by an incubation technique, but not on total organic phosphate. In acid soils a significant regression was obtained only when the phosphate mineralized on incubation was multiplied by the fractional recovery of added orthophosphate, giving a measure of the labile quantity of mineralized organic phosphate. Their results showed that the phosphate present in organic form at the beginning of the season may contribute substantially to the phosphorus nutrition of plants grown during the season.

The use of bacterial fertilizers which may promote the hydrolysis of soil organic phosphate has become a common agricultural practice in the Soviet Union (COOPER [1959]). In particular, inoculation of seed with preparations of *Bacillus megatherium* var. *phosphaticum*, otherwise called "phosphobacterin," has had widespread application, and yield increases have been reported in some areas. Until a greater number of critically controlled field experiments have been carried out under a variety of conditions, however, the value of this approach cannot be assessed.

REFERENCES

Adams, A. P., W. V. Bartholomew, and F. E. Clark, 1954. Measurement of nucleic acid components in soil. *Soil Sci. Soc. Am. Proc.* 18:40.

Anderson, C. A., 1963. Quantitative chromatographic separation of total organic phosphorus from soil extracts. *Dissertation Abstract.* 23:3053.

Anderson, G., 1956. The identification and estimation of soil inositol phosphates. *J. Sci. Food Agr.* 7:437.

———. 1960. Factors affecting the estimation of phosphate esters in soil. *J. Sci. Food Agr.* 11:497.

———. 1961. A partial fractionation of alkali-soluble soil organic phosphate. *J. Soil Sci.* 12:276.

———. 1963. Effect of iron/phosphorus ratio and acid concentration on the precipitation of ferric inositol hexaphosphate. *J. Sci. Food Agr.* 14:352.

———. 1964. Investigations on the analysis of inositol hexaphosphate in soils. *Trans. Intern. Congr. Soil Sci.*, 8th, Bucharest, 4:563.

———. 1970. The isolation of nucleoside diphosphates from alkaline extracts of soil. *J. Soil Sci.* 21:96.

———, and E. Z. Arlidge, 1962. The adsorption of inositol phosphates and glycerophosphate by soil clays, clay minerals and hydrated sesquioxides in acid media. *J. Soil Sci.* 13:216.

———, and R. J. Hance, 1963. Investigation of an organic phosphorus component of fulvic acid. *Plant Soil* 19:296.

Angyal, S. J., and L. Anderson, 1959. The cyclitols. In *Advances in Carbohydrate Chemistry*, Vol. 14, New York: Academic Press.

Arnold, P. W., 1956. Paper ionophoresis of inositol phosphates, with a note on the acid hydrolysates of phytic acid. *Biochim. Biophys. Acta* 19:552.

Aso, K., 1905. On organic compounds of phosphoric acid in the soil. *Tokyo Imp. Univ. Coll. Agr., Bull.* 6, 277.

Aubert, J. P., and G. Milhaud, 1955. Étude du métabolisme de la levure de boulangerie à l'aide de glucose radioactif. *Compt. Rend. Acad. Sci.* (Paris) 240:1943.

Auten, J. T., 1923. Organic phosphorus of soils. *Soil Sci.* 16:281.

Bamann, E., H. Trapmann, and A. Schuegraf, 1955. Metallkatalytische Spaltung von Phosphoproteiden. *Chem. Ber.* 88:1726.

Barrow, N. J., 1961. Phosphorus in soil organic matter. *Soils Fert.* 24:169.

Bartholomew, W. V., and C. A. I. Goring, 1948. Microbial products and soil organic matter. I. Some characteristics of the organic phosphorus of microorganisms. *Soil Sci. Soc. Am. Proc.* 13:238.

Bieleski, R. L., and R. E. Young, 1963. Extraction and separation of phosphate esters from plant tissues. *Anal. Biochem.* 6:54.

Birch, H. F., 1961. Phosphorus transformations during plant decomposition. *Plant Soil* 15:347.

———, and M. T. Friend, 1961. Resistance of humus to decomposition. *Nature* 191:731.

Black, C. A., and C. A. I. Goring, 1953. Organic phosphorus in soils. *Agronomy* 4:123.

Bloor, W. R., 1925. Biochemistry of the fats. *Chem. Rev.* 2:243.

Boswall, G. W., and W. A. DeLong, 1959. The use of 8-hydroxyquinoline (oxine) in the extraction of soil organic phosphorus. *Can. J. Soil Sci.* 39:20.

Bower, C. A., 1945. Separation and identification of phytin and its derivatives from soils. *Soil Sci.* 59:277.

————. 1949. Studies on the forms and availability of soil organic phosphorus. *Iowa Agr. Exp. Sta. Res., Bull. 362.*

Bowman, B. T., R. L. Thomas, and D. E. Elrick, 1967. The movement of phytic acid in soil cores. *Soil Sci. Soc. Am. Proc.* 31:477.

Bromfield, S. M., 1961. Sheep faeces in relation to the phosphorus cycle under pastures. *Austral. J. Agr. Res.* 12:111.

————, and O. L. Jones, 1970. The effect of sheep on the recycling of phosphorus in hayed-off pastures. *Austral. J. Agr. Res.* 21:699.

Caldwell, A. G., and C. A. Black, 1958a. Inositol hexaphosphate. 1. Quantitative determination in extracts of soils and manures. *Soil Sci. Soc. Am. Proc.* 22:290.

————. 1958b. Inositol hexaphosphate. 2. Synthesis by soil microorganisms. *Soil Sci. Soc. Am. Proc.* 22:293.

————. 1958c. Inositol hexaphosphate. 3. Content in soils. *Soil Sci. Soc. Am. Proc.* 22:296.

Casida, L. E., 1959. Phosphatase activity of some common soil fungi. *Soil Sci.* 87:305.

Chang, S. C., 1939. The transformation of phosphorus during the decomposition of plant materials. *Soil Sci.* 48:85.

————. 1940. Assimilation of phosphorus by a mixed soil population and by pure cultures of soil fungi. *Soil Sci.* 49:197.

Chargaff, E., and J. N. Davidson, 1955. *The nucleic acids.* New York: Academic Press.

Chouchak, D. 1929. La lutte entre les plantes cultivées et les microorganismes du sol pour leur nutrition minérale; action du sang desséché sur l'engrais phosphaté. *Compt. Rend. Acad. Sci.* (Paris) 189:262.

Cooper, R., 1959. Bacterial fertilizers in the Soviet Union. *Soils Fert.* 22:327.

Cosgrove, D. J., 1962. Forms of inositol hexaphosphate in soils. *Nature* 194:1265.

————. 1963. The chemical nature of soil organic phosphorus. 1. Inositol phosphates. *Austral. J. Soil Res.* 1:203.

————. 1964. An examination of some possible sources of soil inositol phosphates. *Plant Soil* 21:137.

————. 1966. Detection of isomers of phytic acid in some Scottish and Californian soils. *Soil Sci.* 102:42.

————. 1969. The chemical nature of soil organic phosphorus. 2. Characterization of the supposed DL-*chiro*-inositol hexaphosphate component of soil phytate as D-*chiro*-inositol hexaphosphate. *Soil Biol. Biochem.* 1:325.

————, G. C. J. Irving, and S. M. Bromfield, 1970. Inositol phosphate phosphatases of microbiological origin. The isolation of soil bacteria having inositol phosphate phosphatase activity. *Austral. J. Biol. Sci.* 23:339.

————, and M. E. Tate, 1963. Occurrence of *neo*-inositol hexaphosphate in soil. *Nature* 200:568.

Courtois, J., 1951. Les esters phosphoriques de l'inositol. *Bull. Soc. Chim. Biol.* 33:1075.

Damsgaard-Sorensen, P., 1947. Studier over Jordens Fosforsyreindhold. 4. Det Organisk Bundne Fosfor. *Tidsskr. Planteavl.* 50, 653 (1946). From *Soils Fertil.* 10, 131 (1947).

Dean, L. A., 1938. An attempted fractionation of the soil phosphorus. *J. Agr. Sci.* 28:234.

Desjobert, A., and P. Fleurent, 1954. Influence de la réaction du milieu sur l'hydrolyse chimique de l'inositolhexaphosphate. Considérations sur la constitution de ce dérivé. *Bull. Soc. Chim. Biol.* 36:475.

————, and F. Petek, 1956. Chromatographie sur papier des esters phosphoriques de l'inositol. Application a l'etude de la dégradation hydrolytique de l' inositolhexaphosphate. *Bull. Soc. Chim. Biol.* 38:871.

DeTurk, E. E., 1938. Changes in the soil of the morrow plots which have accompanied long-continued cropping. *Soil Sci. Soc. Am. Proc.* 3. 83.

Dormaar, J. F., 1964. Evaluation of methods for determination of total organic phosphorus in chernozemic soils of Southern Alberta. *Can. J. Soil Sci.* 44:265.

————. 1967. Distribution of inositol phosphates in some chernozemic soils of Southern Alberta. *Soil Sci.* 104:17.

————. 1970. Phospholipids in chernozemic soils of Southern Alberta. *Soil Sci.* 110:136.

————, and G. R. Webster, 1963. Determination of total organic phosphorus in soils by extraction methods. *Can. J. Soil. Sci.* 43:35.

————. 1964. Losses inherent in ignition procedures for determining total organic phosphorus. *Can. J. Soil Sci.* 44:1.

Doughty, J. L., 1935. Phosphate fixation in soils, particularly as influenced by organic matter. *Soil Sci.* 40:191.

Drobníková, V., 1961. Factors influencing the determination of phosphatases in soil. *Folia Microbiol.* (Prague) 6:260.

Dyer, W. J., and C. L. Wrenshall, 1941a. Organic phosphorus in soils. 1. The extraction and separation of organic phosphorus compounds from soil. *Soil Sci.* 51:159.

————. 1941b. Organic phosphorus in soils. 3. The decomposition of some organic phosphorus compounds in soil cultures. *Soil Sci.* 51:323.

————, and G. R. Smith, 1940. The isolation of phytin from soil. *Science* 91:319.

Estermann, E. F., and A. D. McLaren, 1961. Contribution of rhizoplane organisms to the total capacity of plants to utilize organic nutrients. *Plant Soil* 15:243.

Flaig, W., and R. Kaul, 1955. Zur Aufnahme der Phosphorsäure aus Nucleinsäure durch die Pflanze. *Phosphorsäure* 15:208.

————, G. Schmid, E. Wagner, and H. Keppell, 1960. Die Aufnahme von Phosphor aus Inosithexaphosphat. *Landw. Forsch. Sonderh.* 14:43. From *Soils Fert.* 23:345 (1960).

Fleury, P., and P. Balatre, 1947. *Les Inositols.* Paris: Masson et Cie.

Folch, J., and F. N. LeBaron, 1956. The chemistry of the phosphoinositides. *Can. J. Biochem. Physiol.* 34:305.

Fraps, G. S., 1911. Effect of ignition on solubility of soil phosphates. *Ind. Eng. Chem.* 3:335.

Friend, M. T., and H. F. Birch, 1960. Phosphate responses in relation to soil tests and organic phosphorus. *J. Agr. Sci.* 54:341.

Fuller, W. H., and W. T. McGeorge, 1950. Phosphates in calcareous Arizona soils. I. Solubilities of native phosphates and fixation of added phosphates. *Soil Sci.* 70:441.

Garman, W. L., 1948. Organic phosphorus in Oklahoma soils. *Oklahoma Acad. Sci. Proc.* 28:89. From *Soils Fert.* 12:244 (1949).

Ghani, M. O., 1943a. Fractionation of soil phosphorus. 1. Method of extraction. *Indian J. Agr. Sci.* 13:29.

————. 1943b. The use of 8-hydroxyquinoline as a means of blocking active iron and aluminium in the determination of available phosphoric acid of soils by dilute acid extraction. *Indian J. Agr. Sci.* 13:562.

————. 1944. Fractionation of soil phosphorus. 3. The organic phosphorus fraction. *Indian J. Agr. Sci.* 14:261.

————, and S. A. Aleem, 1942. Effect of liming on the transformation of phosphorus in acid soils. *Indian J. Agr. Sci.* 12:873.

Goring, C. A. I., 1955. Biological transformations of phosphorus in soil. 1. Theory and methods. *Plant Soil* 6:17.

————, and W. V. Bartholomew, 1949. Microbial products and soil organic matter. 2. The effect of clay on the decomposition and separation of the phosphorus compounds in microorganisms. *Soil Sci. Soc. Am. Proc.* 14:152.

————. 1950. Microbial products and soil organic matter. 3. Adsorption of carbohydrate phosphates by clays. *Soil Sci. Soc. Am. Proc.* 15:189.

————. 1952. Adsorption of mononucleotides, nucleic acids, and nucleoproteins by clays. *Soil Sci.* 74:149.

Greaves, M. P., G. Anderson, and D. M. Webley, 1963. A rapid method of determining phytase activity of soil microorganisms. *Nature* 200:1231.

————, G. Anderson, and D. M. Webley, 1967. The hydrolysis of inositol phosphates by *Aerobacter Aerogenes. Biochim. Biophys. Acta* 132:412.

————, and D. M. Webley, 1965. A study of the breakdown of organic phosphates by microorganisms from the root region of certain pasture grasses. *J. Appl. Bacteriol.* 28:454.

——————, 1969. The hydrolysis of *myo*-inositol hexaphosphate by soil microorganisms. *Soil Biol. Biochem.* 1:37.

————, and M. J. Wilson, 1970. The degradation of nucleic acids and montmorillonite-nucleic-acid complexes by soil microorganisms. *Soil Biol. Biochem.* 2:257.

Halstead, R. L., and G. Anderson, 1970. Chromatographic fractionation of organic phosphates from alkali, acid, and aqueous acetylacetone extracts of soils. *Can. J. Soil Sci.* 50:111.

————, J. M. Lapensee, and K. C. Ivarson, 1963. Mineralization of soil organic phosphorus with particular reference to the effect of lime. *Can. J. Soil Sci.* 43:97.

Hance, R. J., and G. Anderson, 1962. A comparative study of methods of estimating soil organic phosphate. *J. Soil Sci.* 13:225.

————, and G. Anderson, 1963a. Extraction and estimation of soil phospholipids. *Soil Sci.* 96:94.

————, and G. Anderson, 1963b. Identification of hydrolysis products of soil phospholipids. *Soil Sci.* 96:157.

Hanes, C. S., and F. A. Isherwood, 1949. Separation of the phosphoric esters on the filter paper chromatogram. *Nature* 164:1107.

Hannapel, R. J., W. H. Fuller, S. Bosma, and J. S. Bullock, 1964. Phosphorus movement in a calcareous soil. 1. Predominance of organic forms of phosphorus in phosphorus movement. *Soil Sci.* 97:350.

Harrap, F. E. G., 1963. Use of sodium EDTA in the determination of soil organic phosphorus. *J. Soil Sci.* 14:82.

Jackman, R. H., 1955a. Organic phosphorus in New Zealand soils under pasture. 1. Conversion of applied phosphorus into organic forms. *Soil Sci.* 79:207.

————. 1955b. Organic phosphorus in New Zealand soils under pasture. 2. Relation between organic phosphorus content and some soil characteristics. *Soil Sci.* 79:293.

————, and C. A. Black, 1951a. Solubility of iron, aluminium, calcium and magnesium inositol phosphates at different pH values. *Soil Sci.* 72:179.

————, and C. A. Black, 1952. Phytase activity in soils. *Soil Sci.* 73:117.

Johnson, D. D., and F. E. Broadbent, 1952. Microbial turnover of phosphorus in soil. *Soil Sci. Soc. Am. Proc.* 16:56.

Kaila, A. 1948. Viljelysmaan orgaanisesta fosforista. Valtion Maatalouskoetoiminnan Julkaisuja No. 129.

————. 1949. Biological absorption of phosphorus. *Soil Sci.* 68:279.

————. 1954. Über mikrobiologische Festlegung und Mineralisierung des Phosphors bei der Zersetzung organischer Stoffe. *Z. Pfl. Ernähr. Düng.* 64:27.

————. 1961. Effect of incubation and liming on the phosphorus fractions in soil. *Maataloust. Aikak.* 33:185.

————. 1962. Determination of total organic phosphorus in samples of mineral soils. *Maataloust. Aikak.* 34:187.

————. 1963. Organic phosphorus in Finnish soils. *Soil Sci.* 95:38.

————, and H. Missilä, 1956. Accumulation of fertilizer phosphorus in peat soils. *Maataloust. Aikak.* 28:168.

————, and O. Virtanen, 1955. Determination of organic phosphorus in samples of peat soils. *Maataloust. Aikak.* 27:104.

Khorana, H. G., 1961. *Some recent developments in the chemistry of phosphate esters of biological interest.* New York: Wiley.

Kita, D. A., and W. H. Petersen, 1953. Forms of phosphorus in the penicillin-producing mold *Penicillium chrysogenum* Q-176. *J. Biol. Chem.* 203:861.

Kowalenko, C. G., and R. B. McKercher, 1970. An examination of methods for extraction of soil phospholipids. *Soil Biol. Biochem.* 2:257.

Legg, J. O., and C. A. Black, 1955. Determination of organic phosphorus in soils. 2. Ignition method. *Soil Sci. Soc. Am. Proc.* 19:139.

Lockett, J. L., 1938. Nitrogen and phosphorus changes in the decomposition of rye and clover at different stages of growth. *Soil Sci.* 45:13.

McCall, W. W., J. F. Davis, and K. Lawton, 1956. A study of the effect of mineral phosphates upon the organic phosphorus content of organic soil. *Soil Sci. Soc. Am. Proc.* 20:81.

McConaghy, S., 1960. Soil phosphates, with special reference to organic forms and their availability to plants. *Agr. Progr.* 35:82.

McDonnell, P. M., and T. Walsh. 1957. The phosphate status of Irish soils with particular reference to farming systems. *J. Soil Sci.* 8:97.

McKercher, R. B., 1968. Studies on soil organic phosphorus. *Trans. Intern. Congr. Soil Sci.*, 9th, Adelaide, Australia.

————, and G. Anderson, 1968. Characterization of the inositol penta- and hexaphosphate fractions of a number of Canadian and Scottish soils. *J. Soil Sci.* 19:302.

McKercher, R. B., and G. Anderson, 1968. Observations on the accuracy of an ignition and an extraction method for measuring organic phosphate in some Canadian soils. *Soil Sci.* 105:198.

McLaren, A. D., R. A. Luse, and J. J. Skujins, 1962. Sterilization of soil by irradiation and some further observations on soil enzyme activity. *Soil Sci. Soc. Am. Proc.* 26:371.

Martin, J. K., 1964a. Soil organic phosphorus. 1. Methods for the extraction and partial fractionation of soil organic phosphorus. *New Zealand J. Agr. Res.* 7:723.

————. 1964b. Soil organic phosphorus. 2. The nature of soil organic phosphorus. *New Zealand J. Agr. Res.* 7:736.

————, and A. J. Wicken, 1966. Soil organic phosphorus. 4. Fractionation of organic phosphorus in alkaline soil extracts and the identification of inositol phosphates. *New Zealand J. Agr. Res.* 9:529.

Mehlich, A., 1963. Die Wirkung der Temperatur auf den durch Säure extrahierbaren Phosphor unter dem Einfluss von Tonmineralien und in Beziehung zum organisch gebundenen Phosphor im Boden. *Phosphorsäure* 23:330.

Mehta, N. C., J. O. Legg, C. A. I. Goring, and C. A. Black, 1954. Determination of organic phosphorus in soils. 1. Extraction method. *Soil Sci. Soc. Am. Proc.* 18:443.

Mortensen, J. L., 1963. Complexing of metals by soil organic matter. *Soil Sci. Soc. Am. Proc.* 27:179.

Mortland, M. M., and J. E. Gieseking, 1952. The influence of clay minerals on the enzymatic hydrolysis of organic phosphorus compounds. *Soil Sci. Soc. Am. Proc.* 16:10.

Moyer, J. R., and R. L. Thomas, 1970. Organic phosphorus and inositol phosphates in molecular size fractions of a soil organic matter extract. *Soil Sci. Soc. Am. Proc.* 34:80.

Odynsky, W., 1936. Solubility and distribution of phosphorus in Alberta soils. *Sci. Agr.* 16:652.

Omotoso, T. I., and A. Wild, 1970. Content of inositol phosphates in some English and Nigerian soils. *J. Soil Sci.* 21:216.

————, and A. Wild, 1970. Occurrence of inositol phosphates and other organic phosphate components in an organic complex. *J. Soil Sci.* 21:224.

Pearson, R. W., 1940. Determination of organic phosphorus in soils. *Ind. Eng. Chem. (Anal.)* 12:198.

————, A. G. Norman, and C. Ho, 1941. The mineralization of the organic phosphorus of various compounds in soil. *Soil Sci. Soc. Am. Proc.* 6:168.

————, and R. W. Simonson, 1939. Organic phosphorus in seven Iowa soil profiles: Distribution and amounts as compared to organic carbon and nitrogen. *Soil Sci. Soc. Am. Proc.* 4:162.

Pedersen, E. J. N., 1953. On phytin phosphorus in the soil. *Plant Soil* 4:252.

Peperzak, P., A. G. Caldwell, R. R. Hunziker, and C. A. Black, 1959. Phosphorus fractions in·manures. *Soil Sci.* 87:293.

Puri, A. N., and A. Sarup, 1937. The destruction of organic matter in the preliminary treatment of soils for mechanical analysis. *Soil Sci.* 44:87.

Ratner, E. I., and S. A. Samoilova, (1958). Assimilation of nucleic acids by plants and extracellular phosphatase activity of roots. *Fiziologiia Rastenii* 5:209. Translation, *Plant Physiology* (U.S.S.R.) 5:207.

Robertson, G., 1958. Determination of phosphate in citric acid extracts. *J. Sci. Food Agr.* 9:288.

Rogers, H. T., R. W. Pearson, and W. H. Pierre, 1940. Absorption of organic phosphorus by corn and tomato plants and the mineralizing action of exo-enzyme systems of growing roots. *Soil Sci. Soc. Am. Proc.* 5:285.

Sauerlandt, W., F. Scheffer, and H. J. Banse, 1957. Untersuchungen über organisch gebundenen Phosphor in verschiedenen Böden. *Z. Pfl. Ernähr. Düng.* 76:244.

Saunders, W. M. H., and E. G. Williams, 1955. Observations on the determination of total organic phosphorus in soils. *J. Soil Sci.* 6:254.

Saxena, S. N., and S. Kasinathan, 1956. Organic phosphorus content as an index of fertility of some Indian soils. *Proc. Nat. Acad. Sci. India* 25A:452.

Schachtschabel, P., and G. Heinemann, 1964. Beziehungen zwischen P-Bindungsart und pH Wert bei Lössböden. *Z. Pfl. Ernähr. Düng.* 105:1.

Schmidt, G., and S. J. Thannhauser, 1945. A method for the determination of deoxyribonucleic acid, ribonucleic acid, and phosphoproteins in animal tissues. *J. Biol. Chem.* 161:83.

Schollenberger, C. J., 1918. Organic phosphorus of soil: Experimental work on methods for extraction and determination. *Soil Sci.* 6:365.

———. 1920. Organic phosphorus content of Ohio soils. *Soil Sci.* 10:127.

Schormüller, J., and G. Bressau, 1960. Phosphate und organische Phosphorverbindungen in Lebensmitteln. 7. Trennung und Charakterisierung von Inositphosphorsäureestern. *Z. Lebensm. Untersuch. Forsch.* 113:387.

Schreiner, O., and E. C. Shorey, 1910. Chemical nature of soil organic matter. *U.S. Dept. Agr. Bur. Soils Bull. 74.*

Semb, G., and G. Uhlen, 1954. A comparison of different analytical methods for the determination of potassium and phosphorus in soil based on field experiments. *Acta Agr. Scand.* 5:44.

Sen Gupta, M. B., and A. H. Cornfield, 1962. Phosphorus in calcareous soils. 2. Determination of the organic phosphorus content of calcareous soils and its relation to soil calcium carbonate content. *J. Sci. Food Agr.* 13:655.

Shorey, E. C., 1913. Some organic soil constituents. *U.S. Dept. Agr. Bur. Soils Bull. 88.*

Smith, D. H., and F. E. Clark, 1951. Anion-exchange chromatography of inositol phosphates from soil. *Soil Sci.* 72:353.

———. 1952. Chromatographic separations of inositol phosphorus compounds. *Soil Sci. Soc. Am. Proc.* 16:170.

Smith, R. H., 1951. A study of the role of inositol in the nutrition of *Nematospora gossypii* and *Saccharomyces carlsbergensis*. *J. Gen. Microbiol.* 5:772.

Sokolov, D. F., 1948. The presence of several organic phosphorus compounds in the soil. *Pochvovedenie* 502 (1948). *Chem. Abs.* 43:4409g (1949).

Szember, A., 1960. Influence on plant growth of the breakdown of organic phosphorus compounds by microorganisms. *Plant Soil* 13:147.

Taylor, C. B., 1946. Loss of available phosphate in soil due to microorganisms. *Nature* 158:447.

Thomas, R. L., and D. L. Lynch, 1960. Quantitative fractionation of organic phosphorus compounds in some Alberta soils. *Can. J. Soil Sci.* 40:113.

Thomson, L. M., and C. A. Black, 1947. The effect of temperature on the mineralization of soil organic phosphorus. *Soil Sci. Soc. Am. Proc.* 12:323.

———, 1949. The mineralization of organic phosphorus, nitrogen and carbon in Clarion and Webster soils. *Soil Sci. Soc. Am. Proc.* 14:147.

———, C. A. Black, and F. E. Clark, 1948. Accumulation and mineralization or microbial organic phosphorus in soil materials. *Soil Sci. Soc. Am. Proc.* 13:242.

———, C. A. Black, and J. A. Zoellner, 1954. Occurrence and mineralization of organic phosphorus in soils, with particular reference to associations with nitrogen, carbon and pH. *Soil Sci.* 77:185.

Tokarskaya-Merenova, V. I., 1956. The direct utilization by higher plants of organic phosphorus compounds. *Pochvovedenie* 12:17 *Soils Fert.* 20:160 (1957).

Ulrich, B., and J. H. Benzler, 1955. Der organisch gebundene Phosphor im Boden. *Z. Pfl. Ernähr. Düng.* 70:220.

Van Diest, A., and C. A. Black, 1958. Determination of organic phosphorus in soils. 3. Comparison of methods. *Soil Sci. Soc. Am. Proc.* 22:286.

——, and C. A. Black, 1959a. Soil organic phosphorus and plant growth. 1. Organic phosphorus hydrolyzed by alkali and hypobromite treatments. *Soil Sci.* 87:100.

——, and C. A. Black, 1959b. Soil organic phosphorus and plant growth. 2. Organic phosphorus mineralized during incubation. *Soil Sci.* 87:145.

Vernon, C. A., 1957. The mechanisms of hydrolysis of organic phosphates. In "Phosphoric Esters and Related Compounds," *Chem Soc., London, Spec. Publ. No.* 8:17.

Vincent, V., 1937. Combinasions organiques du phosphore des sols acides: assimilation par les vegetaux. *Compt. Rend. 17th Congr. Chim. Ind., Paris* 2:861.

Wade, H. E., and D. M. Morgan, 1955. Fractionation of phosphates by paper ionophoresis and chromatography. *Biochem. J.* 60:264.

Walker, T. W., and A. F. R. Adams, 1958. Studies on soil organic matter. 1. Influence of phosphorus content of parent materials on accumulations of carbon, nitrogen, sulfur and organic phosphorus in grassland soils. *Soil Sci.* 85:307.

Weissflog, J., and H. Mengdehl, 1933. Studien zum Phosphorstoffwechsel in der höheren Pflanze. 3. Aufnahme und Verwertbarkeit organischer Phosphorsäureverbindungen durch die Pflanze. *Planta.* 19:182.

Whiting, A. L., and A. F. Heck, 1926. The assimilation of phosphorus from phytin by oats. *Soil Sci.* 22:477.

Wild, A., and O. L. Oke, 1966. Organic phosphate compounds in calcium chloride extracts of soils: Identification and availability to plants. *J. Soil Sci.* 17:356.

Williams, C. H., 1950a. Studies on soil phosphorus. 1. A method for the partial fractionation of soil phosphorus. *J. Agr. Sci.* 40:233.

——. 1950b. Studies on soil phosphorus. 3. Phosphorus fractionation as a fertility index in South Australian soils. *J. Agr. Sci.* 40:257.

——, and G. Anderson, 1968. Inositol phosphates in some Australian soils. *Australian J. Agr. Res.* 6:121.

——, and C. M. Donald, 1957. Changes in organic matter and pH in a podzolic soil as influenced by subterranean clover and superphosphate. *Australian J. Agr. Res.* 8:179.

——, E. G. Williams, and N. M. Scott, 1960. Carbon, nitrogen, sulphur and phosphorus in some Scottish soils. *J. Soil Sci.* 11:334.

Williams, E. G., 1959. Influences of parent material and drainage conditions on soil phosphorus relationships. *Agrochimica* 3:279.

——, N. M. Scott, and M. J. McDonald, 1958. Soil properties and phosphate sorption. *J. Sci. Food Agr.* 9:551.

——, and W. M. H. Saunders, 1956. Distribution of phosphorus in profiles and particle-size fractions of some Scottish soils. *J. Soil Sci.* 7:90.

Wrenshall, C. L., and W. J. Dyer, 1941. Organic phosphorus in soils. 2. The nature of the organic phosphorus compounds. A. Nucleic acid derivatives. B. Phytin. *Soil Sci.* 51:235.

——, and R. R. McKibbin, 1937. Pasture studies. 12. The nature of the organic phosphorus in soils. *Can. J. Res.* 15B:475.

Yoshida, R. K., 1940. Studies on organic phosphorus compounds in soil; isolation of inositol. *Soil Sci.* 50:81.

Chapter 5

Sulfur in Soil Organic Substances

G. Anderson

Contents

A. The Nature of Soil Organic Sulfur

There is a growing interest in the nature and properties of the sulfur-containing components of soil organic matter which constitute the bulk of the total sulfur in many soils, particularly in humid regions. Their characterization is of particular importance in the numerous areas where there is a deficiency of available inorganic sulfate and the organic sulfur is the only reserve source of supply.

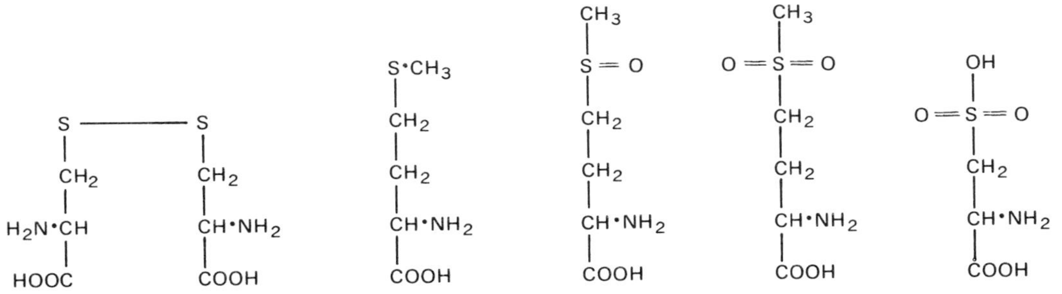

I	II	III	IV	V
Cystine	Methionine	Methionine sulfoxide	Methionine sulfone	Cysteic acid

333

Among the organic sulfur compounds that occur in greatest amount in plants and micro-organisms are the proteins, and since the N:S ratio in many soils is similar to that found in proteins, some investigators for a time thought that most of the soil sulfur was in this form (*e.g.*, MADANOV [1946]). The amino acids cystine (I) and methionine (II) and derivatives such as methionine sulfoxide (III), methionine sulfone (IV), and cysteic acid (V) have been obtained from soil hydrolyzates (BREMNER [1950], SOWDEN [1955, 1958], STEVENSON [1956]) and are probably in the form of proteins or polypeptides, but amino nitrogen usually accounts for less than half the total soil nitrogen and amino acid sulfur is also likely to constitute a minor proportion of the total sulfur. The amounts of sulfur-containing amino acids in some soils from Canada and the U.S.A. are shown in Table 1. Traces of free cystine have also been detected in soils (PUTNAM and SCHMIDT [1959]). By hydrolysis of a soil which had been incubated with ^{35}S-labeled sulfate, cystine, or methionine, SCHARPENSEEL [1962] obtained a number of radioactive fractions which were separated by ion-exchange chromatography. From the various flow diagrams obtained he concluded that sulfate, cysteic acid, cysteine sulfinic acid (VI), taurine (VII), methionine sulfoxide, methionine sulfone, cystine, and methionine are the main sulfur-containing constituents in soil.

$$O = S - OH$$
$$|$$
$$CH_2$$
$$|$$
$$CH \cdot NH_2$$
$$|$$
$$COOH$$

VI

Cysteine sulfinic acid

$$OH$$
$$|$$
$$O = S = O$$
$$|$$
$$CH_2$$
$$|$$
$$CH_2 \cdot NH_2$$

VII

Taurine

The organic sulfur compound isolated in greatest quantity from any soil is probably trithiobenzaldehyde, but it cannot be considered a typical soil component. SHOREY [1913] extracted it from a salt marsh soil from Georgia, U.S.A., and in greater quantity from a California sandy loam, where it constituted about 30% of the total soil sulfur, but it was not detected in the many other soils examined in the same way. Trithiobenzaldehyde has not been found in living organisms, and Shorey believed that it was formed in the soil by reaction of benzaldehyde and hydrogen sulfide, both derived from decomposing plant residues.

The sulfur-containing vitamins biotin (VIII) and to a lesser extent thiamine (IX) are among the compounds that have been detected in the exudates from plant roots by microbiological assay techniques (WEST [1939], ROVIRA and HARRIS [1961]). Evidence for their occurrence in soil has been obtained in the same way, but very small amounts are present (ROULET and SCHOPFER [1950]). Analysis of a group of New Zealand soils has shown that most of the biotin is in a water-insoluble form but is released by acid hydrolysis — the highest amount (15 ppm) being found in a surface soil under coniferous forest (JONES *et al.* [1962]).

While the bulk of the sulfur in soils remains uncharacterized, some progress has been made in subdividing it into broad categories that should serve as a useful preliminary to

Table 1. Sulfur-containing amino acids in the hydrolyzates of some North American soils.

| Soil | Vegetation | Horizon | p.p.m. | | | | | | Reference |
			Cystine*	Cysteic acid	Methionine	Methionine sulfoxide	Methionine sulfone	Amino acid S.	
Brunizem (Illinois)	Grass	"Plow layer"	T†	159	T	227	T	74	STEVENSON (1956)
	Cultivated (corn-oats)	"Plow layer"	T	64	T	104	T	32	
Podzol (Nova Scotia)	Heath	A_0. 10–0 in.	120	—	110	—	—	56	SOWDEN [1956]
		B_{21}. 4–6 in.	60	—	50	—	—	27	
Podzol (Nova Scotia)	Maple and Spruce	A_{01}. 6–4 in.	180	—	360	—	—	125	
		B_{21}. 4–6 in.	90	—	60	—	—	37	

* Identity of cystine not confirmed by Sowden.
† Trace.

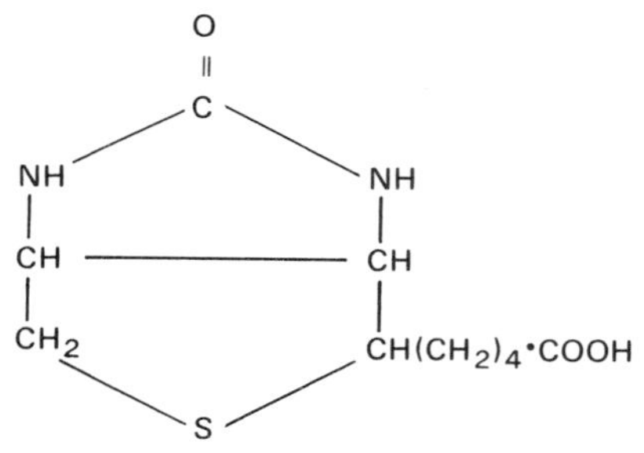

$$
\begin{array}{c}
O \\
\parallel \\
C \\
\diagup \quad \diagdown \\
NH \qquad NH \\
| \qquad\qquad | \\
CH \rule{2cm}{0.4pt} CH \\
| \qquad\qquad | \\
CH_2 \qquad CH(CH_2)_4 \cdot COOH \\
\diagdown \quad \diagup \\
S
\end{array}
$$

VIII

Biotin

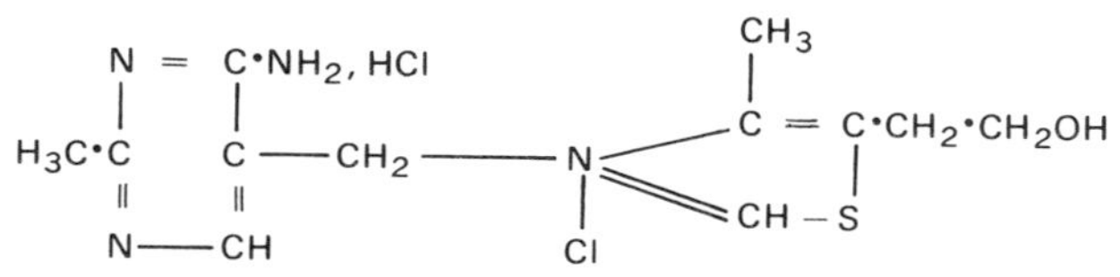

$$
\begin{array}{c}
\qquad\qquad\qquad\qquad\qquad CH_3 \\
\qquad\qquad\qquad\qquad\qquad | \\
N = C\cdot NH_2, HCl \qquad\qquad C = C\cdot CH_2\cdot CH_2OH \\
| \qquad\quad | \qquad\qquad\qquad | \\
H_3C\cdot C \qquad C\rule{0.6cm}{0.4pt}CH_2\rule{0.8cm}{0.4pt}N \qquad\qquad | \\
\parallel \qquad\quad \parallel \qquad\qquad\qquad CH - S \\
N\rule{0.6cm}{0.4pt}CH \qquad\qquad Cl
\end{array}
$$

IX

Thiamine

more precise identification. A considerable proportion can be reduced to hydrogen sulfide with a mixture of hydriodic acid, formic acid, and hypophosphorous acid (Freney [1961], Johnson and Nishita [1952]), a reagent that does not liberate sulfur directly bonded to carbon. In 24 contrasting surface soils from Australia, from 24 to 77% (average 52%) of the sulfur was found to be reducible in this way. Of this, 8% was in inorganic forms, but the remainder was apparently organic. Reducible organic sulfur is largely extractable from these soils with sodium hydroxide, but is partially mineralized during extraction; neutral sodium oxalate extracts much less, but without conversion to inorganic sulfate. Part can be mineralized by hydrolysis with 6 N hydrochloric acid, and there is apparently a very labile fraction which is mineralized by heating or grinding (Williams and Steinbergs [1959], Spencer and Freney [1960], Barrow [1961], Freney [1961]). Much of the reducible sulfur is likely to be in the form of sulfate esters (Freney [1961]), possibly associated in part with soil polysaccharides (Freney et al. [1962]). A sulfated polysaccharide-protein complex has been reported among the extracellular components produced by some soil bacteria (Taylor and Novelli [1961]), and substances of this type may be found in the soil in some quantity. Other sulfate esters probably reaching the soil include choline sulfate, also produced by microorganisms (Freney [1961]), and the ethereal sulfates, found in the urine of animals (Starkey [1950], Walker [1957]).

Attempts have been made to assess the amount of carbon-bonded sulfur by desulfurization with Raney nickel in alkaline medium (DeLong and Lowe [1962], Lowe and DeLong [1963], Lowe [1964]). The C-S bond in many compounds is ruptured in this way with the formation of nickel sulfide, which on acidification releases hydrogen sulfide but there are exceptions, including saturated aliphatic sulfonic acids and sulfones, which do not react (FEIGL [1961]). In the few Canadian soils so far examined the proportion of the sulfur in carbon-bonded form is higher where the organic matter is high. With the organic soils nearly all the material containing the C-S bonds can be extracted with acid followed by alkali apparently without major change of the sulfur moiety, but with some mineral soils Raney nickel desulfurization gives appreciably lower values after extraction. A small residue remains in the soil, but it is possible that alkaline treatment causes some conversion of aliphatic sulfoxides or sulfides to unreactive sulfones (LOWE and DeLONG [1963]). In four out of five soils examined by LOWE [1964] the hydriodic acid-reducible and carbon-bonded sulfur together accounted for over 95% of the total soil sulfur.

It has been suggested (WHITEHEAD [1964]) that sulfur can be incorporated into soil organic matter by condensation reactions between thiols such as cysteine and methyl mercaptan, and quinones, aldehydes, or sugars. The products of such reactions would no doubt be much more resistant to microbial or chemical attack, and this might account for the high amount and stability of the soil organic sulfur.

B. Estimation of Organic Sulfur

Both direct and indirect methods have been used for measuring total organic sulfur. In the direct methods of VINOKUROV [1937] and EVANS and ROST [1945] soils were pre-extracted with sodium chloride and alcohol, or water and hydrochloric acid, respectively, and the soil residues oxidized with hydrogen peroxide. The sulfate released was precipitated as the barium salt and taken as a measure of organic sulfate. Values obtained in this way are likely to be low, principally due to incomplete oxidation of organic matter by the peroxide treatment (PURI and SARUP [1937]), but also in some cases to a slight loss of organic sulfur in the initial extracts (WILLIAMS and STEINBERGS [1959]).

More commonly the organic sulfur is measured by the difference between the total soil sulfur and the inorganic sulfate, although the latter may be difficult to estimate if insoluble forms are present. FRENEY [1961] took into account the adsorbed and readily soluble sulfate which could be extracted with potassium phosphate solution, and also inorganic categories in a lower state of oxidation than sulfate. The total sulfur less these inorganic forms gave values that correlated well with soil carbon and nitrogen values, but if samples that contained much calcium sulfate were included, the correlations were weakened.

In many cases where the bulk of the sulfur is known to be in organic combination total sulfur values have been used in calculating the composition of the organic matter. Significant correlations have been found between the carbon and sulfur or nitrogen and sulfur contents of soils or the organic matter (DONALD and WILLIAMS [1954], WALKER and ADAMS [1958], WILLIAMS and STEINBERGS [1958], WILLIAMS, WILLIAMS, and SCOTT [1960], FRENEY [1961]) and several workers have quoted mean values around 7 or 8 for the relative proportions of nitrogen to sulfur.

C. Sulfur Transformation in Soil

In pure culture a given organic sulfur compound may be attacked in a variety of ways by different organisms, sometimes yielding different end products. Excellent reviews on the transformations of sulfur by microorganisms have been written by STARKEY [1950, 1956]. In soil the microbial population is so heterogeneous that it is not practicable to consider all the reactions that might take place and it is sometimes more appropriate to treat the whole system as a single organ or tissue capable of certain overall transformations (LEES and QUASTEL [1946], McLAREN [1963]).

The initial forms of sulfur in developing soils are inorganic (mainly sulfates and sulfides) derived from the parent rock. In well-aerated soils elemental sulfur, sulfides, and most other inorganic sources of sulfur are eventually oxidized to sulfate, which is the usual inorganic end product of sulfur metabolism in soils under aerobic conditions (GLEEN and QUASTEL [1953]). Purely chemical oxidations can to some extent occur, but most of the transformations are likely to be effected by microorganisms. In humid environments as the sulfates dissolve they will be rapidly lost by leaching unless they are used by living organisms, and thus involved in a biological cycle with the production of organic sulfur compounds.

The breakdown of some organic sulfur compounds in soils is very rapid. The aerobic decomposition of cysteine or cystine usually ends in the formation of sulfate. The conversion of cysteine to cystine and its "disulfoxide" in a perfusion apparatus has been demonstrated by FRENEY [1958], who confirmed their presence by paper chromatography. When a soil was perfused with a solution of cysteine, there was at first a lag phase in the production of sulfate until the soil was enriched with microorganisms capable of oxidizing cysteine to sulfate (see GLEEN and QUASTEL [1953]). If sulfate was then removed and the enriched soil perfused with cystine or cystine disulfoxide, there was no lag phase, showing that the organisms decomposing cysteine were also able to decompose these compounds and supporting the contention that both are intermediates in the oxidation of cysteine. Taurine, an intermediate in the metabolism of cysteine in animals, was not immediately oxidized to sulfate under these conditions, and could not be detected during the perfusion with cysteine. In further tests with eight soils from Australia and the U.S.A. FRENEY [1960] showed that cysteine sulfinic acid and cysteic acid were also probable intermediates and suggested the following pathway:

$$\text{cysteine} \rightarrow \text{cystine} \rightarrow \text{cystine disulfoxide} \rightarrow$$
$$\text{cysteine sulfinic acid} \rightarrow \text{cysteic acid} \rightarrow \text{sulfate}$$

The same intermediates were detected during the breakdown of cysteine in the field.

Comparison of the amounts of sulfate produced during decomposition of different sulfur compounds in soil led FREDERICK, STARKEY, and SEGAL [1957] to the conclusion that the rate of decomposition could not be related to the type of sulfur linkage. Cystine was fairly quickly broken down in a slightly acid sandy loam incubated at 28°C, with conversion of 57% of the sulfur to sulfate in four weeks and 85% in 21 weeks. Thiamine hydrochloride released sulfa e much more slowly with 3% conversion in four weeks and 65% in 21 weeks. Decomposition of methionine in perfused soil yielded the volatile compounds methyl mercaptan and dimethyl disulfide, and no sulfate could be detected. While BARJAC [1952] noted the formation of mercaptans during breakdown of methionine in soil, the sulfate content of the soil increased during later stages of decomposition, and HESSE [1957] found complete conversion to sulfate after 30 days incubation in an East African forest soil. The rate of decomposition of methionine in perfused soil is much slower than that of most other amino acids, and has been found to vary markedly even in different samples of soil from the same area (GREENWOOD and LEES [1956]).

Under anaerobic conditions decomposition of organic sulfur compounds results in the formation of hydrogen sulfide, methyl mercaptan, and dimethyl sulfide (McLAREN [1963]), and both hydrogen sulfide and methyl mercaptan have been detected in paddy soils (TAKAI and ASAMI [1962], ASAMI and TAKAI [1963]).

The decomposition of more complex materials such as plant residues can also be rapid in soil. Incubation of soils treated with a range of materials from sown and "native" Australian pastures has demonstrated that, while the amount of sulfur mineralized depends on the sulfur content and C:S ratio of the added material, for any given C:S ratio it can vary within quite wide limits (BARROW [1960]). HESSE [1957] found that decomposition of ground litter in an East African forest soil led to an increase in soluble sulfate within 30 days, while decomposition of fresh leaves resulted in a small loss. Both materials had a C:S ratio in the region of 300. The addition to soil of easily decomposable material containing little or no sulfur can result in reduction in crop yields unless sulfate is also applied (CONRAD [1950]).

The sulfur in the soil organic matter is generally much more stable than that present in fresh organic residues, but as in the latter, its rate of mineralization is likely to be affected by its amount relative to other organic constituents. While the contents of carbon and sulfur in soil are usually well correlated, the C:S ratio can nevertheless vary between wide limits. In a group of New Zealand soils, where clovers respond to sulfur, values of 80 to 200 have been reported (WALKER, ADAMS, and ORCHISTON [1955], WALKER [1957]). Any changes in the organic matter as a whole are likely to be reflected in the sulfur fractions, and incubation tests of New Zealand pasture soils have indicated that mineralization of soil sulfur may follow a similar pattern to that of nitrogen (WHITE [1959]). The rate relative to the total amount present tends to be somewhat lower for sulfur than for nitrogen, however, (WHITE [1959]) and generally the amount of sulfate released on incubation is very small (HESSE [1957], FRENEY and SPENCER [1960], BARROW [1961]). In a number of Australian soils deficient in available sulfur, mineralization of organic sulfur is greater in the presence of growing plants, even when mineral sulfate is added (FRENEY and SPENCER [1960]). In the absence of plants considerable net immobilization of added inorganic sulfate takes place.

Liming of soils can stimulate the mineralization of organic sulfur. In lysimeter studies ELLETT and HILL [1929] found that the loss of sulfate in drainage water was greater when lime was added, and during incubation some acid soils release more sulfate in the presence of lime (WHITE [1959]). There is some evidence that lime may cause a relatively greater mineralization of sulfur than of nitrogen (GIOVANNINI, [1964]), resulting in the formation of organic matter with a lower sulfur content.

REFERENCES

Asami, T., and Y. Takai, 1963. Formation of methyl mercaptan in paddy soils. 2. *Soil Sci. Plant Nutr. (Tokyo)* 9:65.

Barjac, H. De, 1952. Contribution a l'étude du métabolisme des acides aminés soufré et spécialement de la méthionine dans le sol. *Annal. Inst. Pasteur* 82:623.

Barrow, N. J., 1960. A comparison of the mineralization of nitrogen and of sulphur from decomposing organic materials. *Austral. J. Agr. Res.* 11:960.

———. 1961. Studies on mineralization of sulphur from soil organic matter. *Austral. J. Agr. Res.* 12:306.

Bremner, J. M., 1950. The amino-acid composition of the protein material in soil. *Biochem. J.* 47:538.

Conrad, J. P., 1950. Sulfur fertilization in California and some related factors. *Soil Sci.* 70:43.

DeLong, W. A., and L. E. Lowe, 1962. Note on carbon-bonded sulphur in soil. *Can. J. Soil Sci.* 42:223.

Donald, C. M., and C. H. Williams, 1954. Fertility and productivity of a podzolic soil as influenced by subterranean clover (*Trifolium subterranean* L.) and superphosphate. *Australian J. Agr. Res.* 5:664.

Ellet, W. B., and H. H. Hill, 1929. Effect of lime materials on the outgo of sulphur from Hagerstown silt loam soil. *J. Agr. Res.* 38:697.

Evans, C. A., and C. O. Rost, 1945. Total organic sulfur and humus sulfur of Minnesota soils. *Soil Sci.* 59:125.

Feigl, F., 1961. Spot tests based on redox reactions with Devarda's alloy and Raney alloy. *Anal. Chem.* 33:1118.

Frederick, L. R., R. L. Starkey, and W. Segal, 1957. Decomposability of some organic sulfur compounds in soil. *Soil Sci. Soc. Am. Proc.* 21:287.

Freney, J. R., 1958. Aerobic transformation of cysteine to sulphate in soil. *Nature (London)* 182:1318.

——, 1960. The oxidation of cysteine to sulphate in soil. *Australian J. Biol. Sci.* 13:387.

——, 1961. Some observations on the nature of organic sulphur compounds in soil. *Australian J. Agr. Res.* 12:424.

——, 1967. Sulfur-containing organics. In '*Soil Biochemistry*,' Vol. 1. New York: Marcel Dekker.

——, N. J. Barrow, and K. Spencer, 1962. A review of certain aspects of sulphur as a soil constituent and plant nutrient. *Plant Soil* 17:295.

——, G. E. Melville, and C. H. Williams, 1969. Extraction, chemical nature, and properties of soil organic sulphur. *J. Sci. Food Agr.* 20:440.

——————. 1970. The determination of carbon bonded sulfur in soil. *Soil Sci.* 109:310.

——, and K. Spencer, 1960. Soil sulphate changes in the presence and absence of growing plants. *Australian J. Agr. Res.* 11:339.

——, and F. J. Stevenson, 1966. Organic sulfur transformations in soils. *Soil Sci.* 101:307.

Giovannini, E., 1964. Le Varie Forme dello Zolfo e loro Evoluzione in Alcuni Terreni Tipici della Piana di Catania. *Agrochimica* 8:275.

Gleen, H., and J. H. Quastel, 1953. Sulphur metabolism in soil. *Appl. Microbiol.* 1:70.

Greenwood, D. J., and H. Lees, 1956. Studies on the decomposition of amino acids in soils. 1. A preliminary survey of techniques. *Plant Soil* 7:253.

Hesse, P. R., 1957. Sulphur and nitrogen changes in forest soils of East Africa. *Plant Soil* 9:86.

Johnson, C. M., and H. Nishita, 1952. Microestimation of sulfur in plant materials, soils and irrigation waters. *Anal. Chem.* 24:736.

Jones, P. D., V. Graham, L. Segal, W. J. Baillie, and M. H. Briggs, 1962. Forms of soil biotin. *Life Sci.* 1:645.

Lees, H., and J. H. Quastel, 1946. Biochemistry of nitrification in soil. 1. Kinetics of, and the effects of poisons on, soil nitrification, as studied by a soil perfusion technique. *Biochem. J.* 40:803.

Lowe, L. E., 1964. An approach to the study of the sulphur status of soils and its application to selected Quebec soils. *Can. J. Soil Sci.* 44:176.

Lowe, L. E., 1969. Sulfur fractions of selected Alberta profiles of the gleysolic order. *Can. J. Soil Sci.* 49:375.

——, and W. A. DeLong, 1963. Carbon bonded sulphur in selected Quebec soils. *Can. J. Soil Sci.* 43:151.

McLaren, A. D., 1963. Biochemistry and soil science. *Science* 141:1141.

Madanov, P. 1946. Nitrogen-sulphur ratio in the humus of steppe soils. *Pedology* 9:517.

Melville, G. E., J. R. Freney, and C. H. Williams, 1969. Investigation on the use of a chelating resin for the extraction of sulphur from soil. *J. Sci. Food Agr.* 20:203.

Puri, A. N., and A. Sarup, 1937. The destruction of organic matter in the preliminary treatment of soils for mechanical analysis. *Soil Sci.* 44:87.

Putnam, H. D., and E. L. Schmidt, 1959. Studies on the free amino acid fraction of soils. *Soil Sci.* 87:22.

Roulet, M. A., and W. H. Schopfer, 1950. Les vitamines du sol et leur signification. *Trans. Intern. Congr. Soil Sci. 4th, Amsterdam* 1:202.

Rovira, A. D., and J. R. Harris, 1961. Plant root excretions in relation to the rhizosphere effect. 5. The exudation of B-group vitamins. *Plant Soil* 14:199.

Scharpenseel, H. W., 1962. Studies with ^{14}C- and ^{3}H- labeled humic acids as to the mode of linkage of their amino-acid fractions and with $^{35}SO_4{}^{2-}$, ^{35}S – methionine and ^{35}S – cystine regarding the sulphur turnover in humic acid and soil. From "Radioisotopes in Soil-Plant Nutrition Studies," *International Atomic Energy Agency*, p. 115.

Shorey, E. C., 1913. Some organic soil constituents. *U.S. Dept. Agri. Bur. Soils Bull.* 88.

Sowden, F. J., 1955. Estimation of amino acids in soil hydrolyzates by the Moore and Stein method. *Soil Sci.* 80:181.

——, 1956. Distribution of amino acids in selected horizons of soil profiles. *Soil Sci.* 82:491.

Sowden, F. J., 1958. The forms of nitrogen in the organic matter of different horizons of soil profiles. *Can. J. Soil Sci.* 38:147.

Spencer, K., and J. R. Freney, 1960. A comparison of several procedures for estimating the sulphur status of soils. *Australian. J. Agr. Res.* 11: 948.

Starkey, R. L., 1950. Relations of microorganisms to transformations of sulphur in soils. *Soil Sci.* 70:55.

————, 1956. Transformations of sulfur by microorganisms. *Ind. Eng. Chem.* 48:1429.

Stevenson, F. J., 1956. Effect of some long-time rotations on the amino-acid composition of the soil. *Soil Sci. Soc. Am. Proc.* 20:204.

Tabatabai, M. A., and J. M. Bremner, 1970. Arylsulfatase activity of soils. *Soil Sci. Soc. Am. Proc.* 34:225.

Takai, Y., and T. Asami, 1962. Formation of methyl mercaptan in paddy soils. 1. *Soil Sci. Plant Nutr. (Tokyo)* 8:132.

Taylor, M. B., and G. D. Novelli, 1961. Analysis of a bacterial sulfated polysaccharide. *Bacteriol. Proc.*: 190.

Vinokurov, M. A., 1937. The content of sulfur in the organic part of soil and a method for its extraction. *Pedology* 32:493.

Walker, T. W., 1957. The sulphur cycle in grassland soils. *J. Brit. Grassland Soc.* 12:10.

————, and A. F. R. Adams, 1958. Studies on soil organic matter. 1. Influence of phosphorus content of parent materials on accumulations of carbon, nitrogen, sulfur and organic phosphorus in grassland soils. *Soil Sci.* 85:307.

————, A. F. R. Adams, and H. D. Orchiston, 1955. The effects and interactions of sulphur, phosphorus and molybdenum on the growth and composition of clovers. *New Zealand J. Sci. Technol.* A.36:470.

West, P. M., 1939. Excretion of thiamin and biotin by the roots of higher plants. *Nature (London)* 144:1050.

White, J. G., 1959. Mineralization of nitrogen and sulphur in sulphur-deficient soils. *New Zealand J. Agr. Res.* 2:255.

Whitehead, D. C., 1964. Soil and plant-nutrition aspects of the sulphur cycle. *Soils Fert.* 27:1.

Williams, C. H., and A. Steinbergs, 1958. Sulphur and phosphorus in some eastern Australian soils. *Australian J. Agr. Res.* 9:483.

————, and A. Steinbergs, 1959. Soil sulphur fractions as chemical indices of available sulphur in some Australian soils. *Australian J. Agr. Res.* 10:340.

————, E. G. Williams, and N. M. Scott, 1960. Carbon, nitrogen, sulphur and phosphorus in some Scottish soils. *J. Soil Sci.* 11:334.

Chapter 6

Fats, Waxes, and Resins in Soil

O. C. Braids and R. H. Miller

Contents

A. Introduction

A survey of the literature pertaining to soil fats, waxes, and resins (often called bitumens or simply lipids) indicates that these substances are probably the least studied of soil organic matter components. Soil organic matter chemists have largely ignored these materials in preference to studies on the true humic materials. This neglect is probably based on the fact that fats, waxes, and resins comprise but a small percentage of the total organic matter of mineral soils (1 to 5%). However, 10 to 20% of the total organic matter of organic soils may come under the lipid classification. Three review articles deal with the soil lipid fraction: HOWARD and HAMER [1960], STEVENSON [1966], and MORRISON [1969]. WOLLRAB and STREIBL [1969] also review the literature on peat and lignite waxes.

Knowledge about soil lipids has been derived from a number of specific unrelated interests. The possible commercial use of peat wax as a substitute for Montan wax (a commercially valuable wax obtained from lignite) has been the impetus for much of the research. Additional research has been done by coal and petroleum chemists interested in peat as a precursor of coal and petroleum. Except for the classical work of Schreiner, Shorey, and associates, conducted on organic substances in soils that might affect soil fertility, agronomic interest in lipids as soil components has not developed extensively. Since their work, which included numerous reports on the isolation and identification of fats, waxes, and resinous substances from soils, relatively few studies have been conducted on the effect of these materials on the nature and properties of soils. Occasional reports in the literature since that time, however, implicate these hydrophobic materials as influencing soil behavior in a number of interesting ways. These are discussed in section E.

Fats, waxes, and resins form a convenient analytical group rather than a structural group having only the common property of being variably soluble in numerous organic solvents (*e.g.*, benzene, methanol, ethanol, ether, petroleum ether, acetone, chloroform, gasoline, etc.) or mixtures of these solvents. The resins are more polar than fats and waxes and thus show greater solubility in methanol or ethanol, a property often used to separate the groups (see Tables 1 and 2). Structurally these lipid materials are very diverse, ranging from the relatively simple materials such as fatty acids and glycerol to complex compounds such as chlorophyll, complex terpenes, sterols, and polynuclear hydrocarbons.

Table 1.   Quantity of ether-soluble (fats and waxes) and alcohol-soluble (resins) materials in various mineral soils.

Mineral soils (WAKSMAN and STEVENS [1930])	Ether-soluble	Alcohol-soluble
	Percent*	Percent
Summit soil (Missouri), A horizon	3.56	0.58
Chernozem soil (Kansas), A horizon	4.71	1.53
Chernozem soil (Alberta, Canada), 1–25 cm	0.80	0.82
Chernozem soil (Manitoba, Canada), 1–20 cm	0.46	0.84
Chernozem soil (Manitoba, Canada), 25–50 cm	0.52	0.63
Brown soil (Saskatchewan, Canada), 1–20 cm	1.02	0.88
Prairie soil, dark colored, A horizon	0.62	0.61
Alpine humus, Pikes Peak (13,800 ft elevation)	0.94	3.10

Mineral soils (WAKSMAN [1936])	Ether- and alcohol-soluble
	Percent
Podzol (Michigan), A_1 horizon	3.29
Podzol (Michigan), A_2 horizon	4.58
Chestnut soil, 0–15 cm	3.14
Serozem (Arizona), 0–15 cm	7.63

* Percent on the basis of total soil organic matter (C × 1.72).

Widespread usage of pesticides and the extensive urbanization of large areas of the world within the past decade have caused extensive casual contamination of soils with hydrocarbons and other chemical compounds extractable with organic solvents. Future chemical investigators of soil lipids must carefully consider the possible unnatural origin of chemical compounds which would appear as components of the soil lipid fraction.

B. Extraction of Soil Lipids

Although by definition lipids are materials soluble in "fat solvents," their extraction from soil may be complicated by a number of factors. These factors as they apply to lipid extraction from mineral soils have been considered by HANCE and ANDERSON [1963a] and from peat by HOWARD and HAMER [1960]. A few of the more important considerations are discussed in the following paragraphs.

Table 2. Quantity of ether-soluble and alcohol-soluble material in various peat profiles.

Organic soils (FEUSTEL and BYERS [1930])		Ether-soluble	Alcohol-soluble
		Percent*	Percent
Sphagnum peat (Orono, Maine)	2–4 in.	2.39	4.78
	8–12 in.	2.83	4.95
	12–16 in.	3.27	5.16
Sphagnum peat (Cherryfield, Maine)	2–4 in.	1.98	4.25
	5–8 in.	2.55	5.24
	8–12 in.	2.39	5.25
	12–16 in.	3.25	6.31
Heath peat (Beaufort, North Carolina)	0–3 in.	1.29	6.11
	3–9 in.	2.11	7.02
Heath peat with sedimentary admixture	12–18 in.	5.16	8.30
Sedimentary peat	24–30 in.	9.36	7.08
White cedar forest peat	36–42 in.	2.29	6.80
Saw grass peat (Belle Glade, Florida)	0–4 in.	0.44	1.79
	4–6 in.	0.45	2.04
	32 in.	0.42	1.92
Saw grass peat, sedimentary admixture	49 in.	0.49	1.93
Saw grass peat	94–96 in.	0.75	1.98
Sedimentary peat (Miami Canal Lock, Florida)	15–30 in.	0.77	2.70
	42–50 in.	0.47	1.71
	62–68 in.	0.52	2.07
Saw grass (Clewiston, Florida)	0–4 in.	0.60	1.89
	10–16 in.	0.55	2.01
	20–26 in.	0.70	2.27
	30–36 in.	1.47	3.25
Woody sedge peat (Monroe, Washington)	0–6 in.	1.82	7.43
Herbaceous sedge peat	10–16 in.	2.01	5.67
Tule sedge peat	18–24 in.	1.75	3.85

* Percent on the basis of 100 parts air dry peat.

The choice of solvent influences both the yield and the type of extract. Consideration must be given to the insolubility of various lipid components in some of the common lipid solvents. For example, lipids linked in combination with protein or carbohydrate, as well as heavy metal salts of high molecular weight fatty acids, are largely insoluble in nonpolar organic solvents such as ether. GALLOPINI and RIFFLADI [1969] reported fatty acids and unsaponifiable constituents as the major components of an ether extraction of forest soil. Methanol or other polar organic solvents in an extraction mixture will commonly break lipoprotein or glycoprotein complexes, but their presence has the disadvantage of limiting our knowledge of these complexes *in situ*. Likewise, methanol has been found an effective solvent for extracting heavy metal salts of fatty acids from soil (WANDER [1949a,b]). Inclusion of methanol or ethanol in the extraction mixture, or extraction with these solvents individually, results in higher yields but also removes a large percentage of resinous materials from soil that would not be extracted with the nonpolar solvents (REILLY et al. [1937], REILLY et al. [1939], KWIATKOWSKI [1963], SUNDGREN [1949]). One other advantage of polar solvents such as methanol, ethanol, or acetone is their ability to extract some inorganic and nonlipid organic compounds. There are clearly many possible lipid solvents, and their selection will depend on the character and nature of lipids. Since the nature of soil lipids remains somewhat obscure, the most effective solvent for their extraction is difficult to predict and the use of a combination of solvents would seem essential.

Changing the physical conditions of extraction can also aid recovery. BONE and TEI [1934] and SUNDGREN [1949] reported that extraction of peat at elevated temperature and pressure gave significant increases in yield.

Another factor influencing the yield of soil lipids is pretreatment before extraction. HANCE and ANDERSON [1963a] found that the solubility of soil lipids was greater in field-moist soils than in dried soils, although this difference was eliminated by pretreatment with a HCl:HF mixture. They attributed the decreased solubility of lipids on drying of soil to adsorption by clay minerals. HOWARD and HAMER [1960] also refer to numerous reports on decreased yields of peat wax after oven drying prior to extraction by an apparent polymerization process. This information would certainly indicate that extraction of soil lipids would be more efficient when performed on field-moist samples rather than on oven-dried or even air-dried samples. Pretreatment of the soil with hot mineral acid (FEUSTEL and BYERS [1930]) or dilute acid (STADNIKOFF and WAHNER [1931]) before extraction with organic solvents had been shown to increase the yield of lipid materials. The effectiveness of the hot acid pretreatment was attributed to release of "combined" fats and fatty acids, while the dilute acid treatment removed metal cations responsible for formation of insoluble salts with fatty acids.

STEVENSON [1966] hypothesized that ultrasonification of soil-extraction solution mixtures might increase yields of lipids by disrupting chemical bonds that hold them in complexed form.

C. Index of Compounds Identified from the Lipid Fraction of Organic and Mineral Soils

As the lipid fraction of soil has been investigated, a number of compounds have been isolated and identified. The n-alkane series and long-chain fatty acids have recently been separated by means of gas-liquid chromatography. Because identification of compounds of these two homologous series is based on chromatographic behavior or mass spectra, many of their physical constants are omitted in the index presented below. This index of compounds and their physical constants has been compiled to aid the research worker in identifying compounds and to bring together in an up-to-date list those components which have been identified in the soil lipid fraction. Only those compounds that occur naturally have been included. Compounds that are produced from rigorous chemical or heat treatments of soils have been omitted.

Hydrocarbons

Phenanthrene $C_{14}H_{10}$ m.p. 100°C, b.p. 340°C, UV λ_{max} 242, 250, 260, 281, 286, 289, 293, 301 nm (alcohol). Picrate ($C_{20}H_{13}O_7N_3$) yellow needles, m.p. 145°C (corr.). Extracted from soil with benzene-methanol (1:1) by BLUMER [1961]. Physical data from JOSEPHY [1946].

Fluoranthene $C_{16}H_{10}$ m.p. 111°C, UV λ_{max} 235, 260, 277, 287, 397, 323, 340, 357 nm. 1,3,5-trinitrobenzene adduct $C_{22}H_{13}O_6N_3$ m.p. 205°C. Extracted from soil with benzene-methanol (1:1) by BLUMER [1961]. Physical data from JOSEPHY [1946].

10,11-Benzofluoranthene $C_{20}H_{12}$ m.p. 165–167°C, picrate red needles m.p. 195°C. Extracted from beech and spruce forest soil by BORNEFF and KANTE [1963]. Physical data from JOSEPHY [1946].

3,4 Benzofluoranthene $C_{20}H_{12}$ m.p. 167°C. Extracted from beech and spruce forest soil by BORNEFF and KANTE [1963]. Physical data from JOSEPHY [1946].

Pyrene $C_{16}H_{10}$ m.p. 150.9–151.1°C., colorless when pure, very sharp melting point. UV λ_{max} 231.5, 240.4, 251.1, 261.7, 268.7, 272.7, 292.8, 305.5, 313.4, 318.7, 329.8, 334.3, 343.9, 351.1, 357.0, 361.4, 364.2, 371.8, 377.4 nm (alcohol). Picrate $C_{22}H_{13}O_7N_3$ m.p. 220–222°C. Extracted from soil with benzene-methanol (1:1) by BLUMER [1961]. Physical data from JOSEPHY [1946].

Chrysene $C_{18}H_{12}$ m.p. 255.8–256.3°C, UV λ_{max} 258.6, 268.3, 283.4, 294.4, 306.3, 320.0, 336.8, 344.1, 350.2, 353.3, 361.0 nm. Extracted from soil with benzene by KERN [1947]. Physical data from JOSEPHY [1946].

1,2-Benzanthracene $C_{18}H_{12}$ m.p. 162.0–162.5°C, b.p. 435°C/760, UV λ_{max} 282, 291, 335, 358, 386 nm (in chloroform). Extracted from beech and spruce forest soil by BORNEFF and KANTE [1963]. Physical data from JOSEPHY [1946].

3,4-Benzpyrene (Benzo[α]pyrene) $C_{20}H_{12}$ m.p. 178.8–179.3°C, pale yellow monoclinic needles, carcinogen of high potency. UV λ_{max} 224, 255, 264, 296, 347, 364, 384, 403 nm. Extracted from soil by BLUMER [1961], BORNEFF and KANTE [1963], and SMIRNOV [1970]. Physical data from JOSEPHY [1946].

Perylene $C_{20}H_{12}$ m.p. 273–274°C, yellow crystals, UV λ_{max} 225, 254, 268, 365, 385 nm. 1,3,5-trinitrobenzene adduct $C_{26}H_{15}O_6N_3$ m.p. 245–252°C, monopicrate $C_{26}H_{15}O_7N_3$ m.p. 223–224.5°C. Extracted from peat with benzene-ethanol (7:3) by BERGMANN et al. [1964]; also reported from peat by GILLILAND et al. [1960]. Physical data from JOSEPHY [1946].

Dehydroabietene $C_{19}H_{28}$. An oil at room temperature. n^{25} 1.536, $[\alpha]_D^{25} + 84.5°$. Dinitrate m.p. 178°C uncorr. Extracted from a British Columbia forest soil by SWAN [1965].

Hentriacontane $C_{31}H_{64}$ m.p. 67.2–67.3°C, b.p. 266.2°C/3 mm Hg, d^{70} 0.7827. A member of the n-alkane series. Extracted from soil with ethanol by SCHREINER and SHOREY [1911a]. Physical data from ASINGER [1956].

$CH_3(CH_2)_{29}CH_3$

n-Alkane series C_{18}–C_{34}. Odd carbon numbers predominated, but even carbon numbers were present. Extracted and fractionated by MORRISON and BICK [1967].

Alcohols

Alcohol $C_{20}H_{40}O_4$ m.p. 124–130°C. Isolated after saponification from an ethanol vapor extract of peat by ZALOZIEKI and HAUSMAN [1907].

Alcohol $C_{21}H_{35}O_7$ m.p. above 160°C. Isolated after saponification from an ethanol vapor extract of peat by ZALOZIEKI and HAUSMAN [1907].

Ceryl alcohol $C_{26}H_{54}O$ m.p. 79.5–79.8°C. Isolated from the unsaponifiable fraction of peat wax extracted with hot petrol under pressure by REILLY and WILSON [1940].

Alcohol $C_{27}H_{56}O$ m.p. 86.5°C. Isolated after saponification from a crude benzine, benzene extract of peat by TITOV [1932]. Melting point determined by WARTH [1956].

Glycerol $C_3H_8O_3$ m.p. 17.9°C., b.p. 290°C. Isolated after saponification of a hot 95% ethanol extract of soil humic acid by SCHREINER and SHOREY [1911b].

$$CH_2OHCHOHCH_2OH$$

n-Alkanol series C_{16}–C_{30}. Even carbon numbers predominated, but odd numbers were present. Extracted and fractionated by MORRISON and BICK [1967].

Acids

Acid $C_{16}H_{25}O_5$ m.p. 184°C. Isolated after saponification from an ethanol vapor extract of peat by ZALOZIEKI and HAUSMAN [1907].

α-Mono-hydroxystearic acid (11-hydroxy-octadecanoic acid) $C_{18}H_{36}O_3$ m.p. 84–85°C. Isolated from a boiling 95% ethanol extract of soil humic acid by SCHREINER and SHOREY [1910].

$$CH_3(CH_2)_6CHOH(CH_2)_9CO_2H$$

Dihydroxystearic acid (9,10-dihydroxy-octadecanoic acid) $C_{18}H_{36}O_4$ m.p. 98–99°C. Extracted from soil with 2% sodium hydroxide and recovered by ether extraction of the sodium hydroxide solution. Reported by SCHREINER and SHOREY [1908, 1909b].

$$CH_3(CH_2)_7(CHOH)_2(CH_2)_7CO_2H$$

Humoceric acid $C_{19}H_{24}O_2$ m.p. 72–73°C. Extracted from peat with benzol by ASCHAN [1921]. An unsaturated acid that reacts readily with $KMnO_4$ solution. Forms colorless crystals from petroleum ether and methanol. Reacts with Balzer's reagent.

Triacosanoic acid $C_{23}H_{46}O_2$ m.p. 79.1°C. Extracted and identified by SUNDGREN and RAUHALA [1949]. Also identified in gas-liquid chromatograms by IVANOVA et al. [1968] and KIAN et al. [1968]. The 22- and 24-methyl derivatives identified with gas-liquid chromatography by WANG et al. [1969].

$$CH_3(CH_2)_{21}CO_2H$$

Lignoceric acid (tetracosanoic acid) $C_{24}H_{48}O_2$ m.p. 80°C. Extracted from a peat soil with boiling 95% ethanol by SCHREINER and SHOREY [1910]. Identified in gas-liquid chromatograms by IVANOVA et al. [1968] and KIAN et al. [1968].

$$CH_3(CH_2)_{22}CO_2H$$

Paraffinic acid $C_{24}H_{48}O_2$ m.p. 45–48°C. Isolated from a boiling 95% ethanol extract of soil humic acid by SCHREINER and SHOREY [1910].

Pentacosanoic acid $C_{25}H_{50}O_2$ m.p. 83°C. Separated from the resin fraction of a benzene-ethanol (1:1) extract of peat by RAKOWSKI and EDELSTEIN [1932] and identified in a gas-liquid chromatogram by IVANOVA et al. [1968]. Physical data from RALSTON [1948].

$$CH_3(CH_2)_{23}CO_2H$$

Cerotic acid (hexacosanoic acid $C_{26}H_{52}O_2$ m.p. 87.7°C, d_4^{100} 0.8198, n_8^{100} 1.4301. Ethyl ester m.p. 59.5°C, anhydride m.p. 89.3°C. Extracted from peat with benzene-ethanol (1:1) by RAKOWSKI and EDELSTEIN [1932]. REILLY and WILSON [1940] also isolated this acid after saponification of Mona wax obtained by extracting peat with hot petrol under pressure. Identified by KIAN et al. [1968], IVANOVA et al. [1968], and GALLOPINI and RIFFLADI [1969] in gas-liquid chromatograms of extracted lipids. Physical data from RODD [1951].

$$CH_3(CH_2)_{24}CO_2H$$

Carboceric acid (heptacosanoic acid) $C_{27}H_{54}O_2$ m.p. 87.5–87.7°C. Isolated after saponification from a benzine-benzene extract of peat by TITOV (1932). REILLY and

$$CH_3(CH_2)_{25}CO_2H$$

WILSON [1940] isolated this acid after saponification of Mona wax obtained by extracting peat with hot petrol under pressure. Identified as a component of peat bitumen wax by ROGINSKAYA [1936].

Montanic acid (octacosanoic acid) $C_{28}H_{56}O_2$ m.p. 90.3–90.5°C, d_4^{100} 0.8191, n^{100} 1.4313. Isolated from peat by RYAN and DILLON [1909], ROGINSKAYA [1936], SUNDGREN and RAUHALA [1949, 1965], BRAIDS [1966], and GALLOPINI and RIFFLADI [1969]. Identified in a gas-liquid chromatogram by IVANOVA *et al.* [1968]. Physical data from RALSTON [1948].

$$CH_3(CH_2)_{26}CO_2H$$

Sterols

β-Sitosterol (Δ^5-stigmasten-3β-ol) $C_{29}H_{50}O$ m.p. 137–137.5°C. Acetate $C_{31}H_{52}O_2$ softens 116°C, m.p. 127–128°C (JOSEPHY [1946]), 122–123°C (BEL'KEVICH *et al.* [1968]). Exists as hydrate or hemihydrate, converted to anhydrous in a vacuum or by heating to 100°C. Extracted from peat with benzene-ethanol by IVES and O'NEILL [1958], also reported in peat wax by BEL'KEVICH *et al.* [1963], GILLILAND and HOWARD [1963], BLACK *et al.* [1955], McLEAN *et al.* [1958], and KSENOFONTOVA *et al.* [1969]. Physical data from JOSEPHY [1946].

β-Sitostanol (dihydro - "β" sitosterol) $C_{29}H_{52}O$ m.p. 137°C (hemihydrate). Acetate $C_{31}H_{54}O_2$ m.p. 129–129.5°C. Extracted from peat with benzene-ethanol by IVES and O'NEILL [1958] and reported by IKAN and KASHMAN [1963]. Physical data from JOSEPHY [1946].

Agrosterol $C_{26}H_{44}O$ m.p. 237°C. Extracted from soil with 95 % ethanol by SCHREINER and SHOREY [1909a]. Gives a positive Libermann's reaction test. Other sterol tests are negative. The formula and melting point are not typical for known sterols.

Terpenes

Friedelin $C_{30}H_{50}O$ m.p. 265–266°C, $[\alpha]_D^{25}$ −20° (c = 0.85). Friedelane $C_{30}H_{52}$ m.p. 233–234°C, $[\alpha]D +41.8°$. Extracted from peat by GILLILAND and HOWARD [1968] and IKAN and KASHMAN [1963]. Physical data from SIMONSEN [1957].

Friedelin-3β-ol $C_{30}H_{52}O$ m.p. 301.4°C, $[\alpha]D + 21.9°$. Acetate m.p. 315–316°C, benzoate m.p. 250–251°C. Extracted from peat with light petroleum by McLEAN et al. [1958]. Reported by GILLILAND and HOWARD [1963] and IKAN and KASHMAN [1963]. Physical data from SIMONSEN [1957].

Taraxerol (skimmiol) $C_{30}H_{50}O$ m.p. 282–283°C, $[\alpha]_D \pm 0°$. Benzoate m.p. 292–293°C, $[\alpha]_D + 37°$. Extracted from peat with benzene-ethanol by IVES and O'NEILL [1958]. Physical data from SIMONSEN [1957].

Taraxerone (skimmione) $C_{30}H_{49}O$ m.p. 240°C, $[\alpha]_D +12°$. Extracted from peat with benzene-ethanol by IVES and O'NEILL [1958]. Physical data from SIMONSEN [1957].

α-Amyrin $C_{30}H_{50}O$ m.p. 186.5–187°C, b.p. 243°/0.7 mm Hg, $[\alpha]_D +91.6°$ (in benzene). Acetate m.p. 227°C, $[\alpha]D + 83.3°$. Extracted from peat with benzene-ethanol by IVES and O'NEILL [1958]. Physical data from SIMONSEN [1957].

Pigments

Carotene (β) $C_{40}H_{56}$ m.p. 183°C (evacuated tube), λ_{max} 497, 466 in chloroform. Extracted from soil with ether and alcohol by JOHNSON and THIESSEN [1934] and with ether by BAUDISCH and VON EULER [1935].

Chlorophyll (a) $C_{55}H_{92}MgN_4O_5$ m.p. 117–120°C, $[\alpha]_D^{20} - 262°$ (acetone). λ_{max} 660, 613, 577, 531, 498, 429, 409 nm in ether. Forms a green solution in ether. Chlorophyll (b) $C_{55}H_{70}MgN_4O_6$ sinters at 86–92°C, becomes a viscous liquid at 120–130°C, $[\alpha]_D^{20} - 267°$ (acetone-methanol, λ_{max} 642, 593, 565, 545, 453, 429 nm in ether. Also forms a green solution in ether. Extracted from soil with ether and alcohol by JOHNSON and THIESSEN [1934] and with ether by BAUDISCH and VON EULER [1935]. Physical data from STECHER [1960].

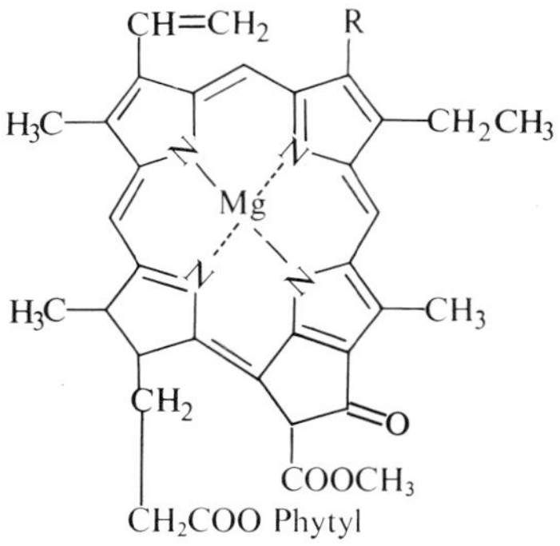

(a) R=CH$_3$

(b) R=CHO

Red pigment $C_{20}H_4O_5Cl_6$. Extracted from lateritic and podzolic soils. The pigment is considered to be a hexachloro-polynuclear-quinone of the dihydroxyperylene-quinone or dihydroxydinaphthyl-quinone type. Reported by BUTLER et al. [1964].

Xanthophyll $C_{40}H_{56}O_2$ m.p. 190°C (corr.). A higher melting point indicates impure material $[\alpha]_{Cd}^{18} + 165°$ (c = 0.7 in benzene). λ_{max} 477.5 nm in petroleum ether. Yellow prisms from ether and methanol. Extracted from soil with ether and alcohol by JOHNSON and THIESSEN [1934]. Physical data from STECHER [1960].

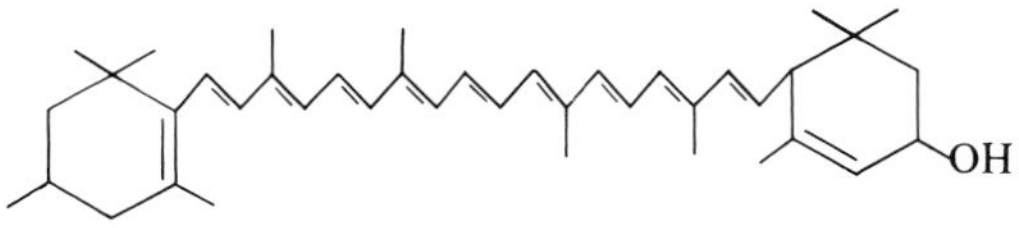

Heterocyclics

Carbazole $C_{10}H_9N$ m.p. 245°C, sublimes, λ_{max} 237, 257, 292.5, 323, 336 nm in ether. Natural occurrence in the anthracene fraction of coal tar. Extracted with benzene-methanol (10:1) from peat by BRAIDS et al. [1967].

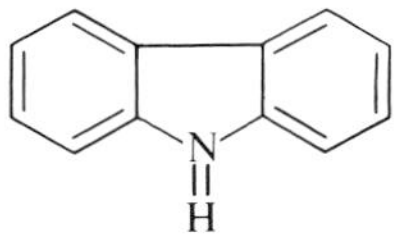

Phospholipids

HANCE and ANDERSON [1963b] reported the isolation of glycerophosphate, choline, and ethanolamine from an organic solvent extract of soil. These materials provide evidence that phosphorus found in the extract is in the form of phospholipids. It was concluded that phosphatidyl choline was one of the predominant lipid components. KOWALENKO and MCKERCHER [1971] report phosphatidyl ethanolamine and phosphatidyl choline as comprising 40 and 30%, respectively, of the total phospholipids in soil. BHANDARI et al. [1969] report that 5 to 10% of the lipids in the humic fraction of soil occur as phospholipids.

Miscellaneous

SIMONART and BATISTIC [1966] report that several aromatic hydrocarbons were extracted from meadow soil by ether. By means of gas-liquid chromatography, benzene, toluene, ethylbenzene, p/m-xylene, and naphthalene were identified.

HIMES and BLOOMFIELD [1967] reported the isolation and identification of triacontylstearate from an orchard soil. Minor components in the extract were octacosanyl and hexacosanyl esters of palmitic and arachidic acids. Another ester was identified in the extracts of tropical soils by WANG et al. [1971]. They identified dioctylphthalate and hypothesized it to be of microbial origin. BEL'KEVICH et al. [1965] reported on the presence of stearine in peat wax.

MORRISON and BICK [1966] identified long-chain methyl ketones in garden soil and peat extracts. The ketone fractions ranged from C-17 to C-37, with odd numbers predominating. The ketone yield from soil was 0.021 % of the total organic matter; from peat it was 0.008 %.

Waxes

Wax with the following properties was extracted from White Russian pine-cotton grass peat with gasoline in an 8:1 ratio (BEL'KEVICH [1960]). The yield was 6 %.

Wax type	Wax, %	Paraffin, %	Resin, %	Acid number	Ester number	Saponification number	Iodine number
Crude	52.5	20.4	24.6	50.1	75.2	125.2	28.3
Deresined	88.5	5.6	5.5	51.4	113.1	61.7	18.8
Purified	92.8	6.4	0.8	116.3	168.4	52.2	11.2

Waxes from various sources were compared according to the following properties by CAWLEY and KING [1945].

Wax type	Wax, %	Resin, %	Acid number	Ester number	Saponification number	Specific gravity 20°
German montan	74–82	14	33	43	76	1.05
Devon lignite	73–83	40	30	45	75	1.04
English peat wax	63–68	—	50	69	119	1.03
Scottish peat wax	64–73	23	48	65	113	—

Bitumens m.p. 70–100°C extracted from several peats which were 40 to 55 % decomposed show the following chemical composition (KATKOUSKI and KARASIK [1954]).

Carbon, %	Hydrogen, %	Wax, %	Acid number	Saponification number	Iodine number
70–72	10	44–48	35–80	60–200	15–30

REILLY and EMLYN [1940] compared Irish peat waxes according to the following properties.

Wax type	Specific gravity 20°	Melting point	Acid number	Saponification number	Ester number	Iodine number	Yield
Petroleum extract	1.01	69°C	35	93	58	34	8–10 %
Petroleum/ alcohol extract	1.03	69–75°C	42	126	84	37	11–14 %

RYAN and DILLON [1909] compared montan and montanin waxes from peat according to the following properties.

Wax type	Melting point	Acid number	Saponification number	Ester number	Iodine number
Montan	76°C	73.3	73.9	—	16.0
Montanin*	95–97°C	56.8	57.9	1.0	—

* Montanin wax is composed of 41.33 % montanic acid and 23.8 % of its sodium salt.

Waxes from 29 peat deposits were extracted with benzene and were fractionated into resins, asphalts, etc., by CAWLEY et al. [1948]. Characteristics of the fractions are given below.

	Crude	De-asphalted	Deresined	De-asphalted-resined
Melting point, °C	65–70	61–66	65–72	63–71
Specific gravity, D_{20}	1.033	1.024	1.024	1.024
Resin, %	22	19	3	3
Asphalt, %	12	3	14	3
Acid number	45	45	40	30
Saponification number	120	120	135	125
Ester number	75	75	95	95
Br number	7	9	11	10

D. Origin and Decomposition of Soil Lipids

I. Origin

The bulk of lipid material in soils undoubtedly is a product of partially or undecomposed plant and animal residues, although lipid materials of microbial origin certainly contribute to the total.

Plant residues contain a variety of lipids, both saponifiable and unsaponifiable. Among the saponifiable lipid materials are fatty acids, the triglycerides and other fatty acid esters, phospholipids, and glycolipids. The unsaponifiable lipid materials are long-chained hydrocarbons and alcohols, phloroglucinol derivatives, polynuclear quinones, terpenoids, and sterols (ROBINSON [1963]). Many of the terpenoid alcohols and sterols do not exist as free alcohols but rather as glycosides or esters with fatty acids. Triterpenoid acids, the so-called resin acids, are also common plant constituents and are frequently associated with polysaccharide gums in gum resins.

Consideration of lipid materials which have been extracted from organic and mineral soils with organic solvents (summarized in section C) leaves little doubt that all could be of plant origin, or of plant origin slightly modified by microbial activity. However, more definitive studies evaluating the long-term contribution of plant material to the soil lipid fraction are certainly needed.

A paucity of information also exists in the area of microbial contribution to soil lipids. The observation of SCHREINER and LATHROP [1911] that the appearance of dihydroxystearic acid in soils is at least in part associated with the growth of fungi is one of only two references that relate the appearance of a particular soil lipid component with microbial activity. The other reference (MALLET and TISSIER [1969]) reports synthesis of 3,4-benzopyrene in soils inoculated with *Escherichia coli* or *Clostridium putride*. Lipid materials are present as normal constituents of microbial cells and are necessary for the normal integrity (cell wall and membrane) and function of the cell. Thus, within the cell we would expect to find fats, waxes, and complex lipids (glycolipids, lipoproteins, etc.) which are not unlike those in plants (HAUSER and KARNOVSKY [1954]). These lipids would be released into the soil environment upon death and lysis of the microbial cells. A reflection of the biological origin of lipids is the fact that they exhibit optical activity (NAGY [1966]).

Bacterial cells on the average contain 5 to 10% lipid, with fungi containing slightly more (about 10 to 25%). Certain bacteria, notably *Mycobacterium* and *Azotobacter*, contain as much as 20 to 40% on the dry-weight basis. An examination of the normal lipid components of fungi and bacteria reveals some important differences. Evidence for the existence of true steroids within bacterial cells has not been found; fungi, however, universally contain steroids, both free and bound (FOSTER [1949]). A large number of bacterial species, on the other hand, contain a unique lipid, poly-β-hydroxybutyrate, which functions as an energy storage compound for the bacterial cells (DAWES and RIBBONS [1964]). The quantity of poly-β-hydroxybutyrate within bacterial cells varies widely, with contents of up to 50% of the dry weight having been recorded. No evidence for the isolation of poly-β-hydroxybutyrate from soils has been found in the literature, although its common occurrence within bacterial cells would suggest its presence in soils. Extracellular synthesis of lipid by at least one species of bacteria has also been documented. JARVIS and JOHNSON [1949] isolated a glycolipid from a medium on which *Pseudomonas aeruginosa* had been grown. Chemical analysis showed the glycolipid to consist of two moles of L-rhamnose and two moles of 1-β-hydroxydecanoic acid. HAUSER and KARNOVSKY [1954] worked with the same organism and obtained yields of crystalline glycolipid as high as 3.16 g/l. of medium.

Among fungi, yeasts including various soil-inhabiting yeasts of the genera *Lipomyces* (STARKEY [1946], KONONENKO [1958]), *Cryptoccoccus* (PEDERSEN [1958]), and *Torulopsis* (KLEINZELLER [1944]) are particularly efficient at converting carbohydrates into lipids. Lipids of these genera are contained in large fat globules within the cell which commonly make up 30 to 40% of the dry weight of the cell, but may, as in the case of *Lipomyces* (STARKEY [1946]), reach 50 to 65%. Of special interest in the area of lipid synthesis by yeasts is the work of DI MENNA [1958], who noticed that a strain of *Rhodotorula graminis* isolated from the phyllosphere (plant foliage) of pasture grasses synthesized copious amounts of extracellular lipid in culture. Since that time, a number of other yeasts from the phyllosphere, including strains of *Pullularia*, *Candida*, and *Cryptococcus*, have been studied (RUINEN and DEINEMA [1964], DEINEMA *et al.* [1964], DEINEMA and LANDHEER [1960]) and have been found to have similar synthetic capabilities. Examination of these extracellular lipids (DEINEMA [1961], and RUINEN and DEINEMA [1964]) has shown their composition to differ greatly from the intracellular lipids of the same strain. To date, investigations on the phenomenon of extracellular lipid production by soil yeasts have not been performed, but it seems feasible that because these genera occur in soil (ALEXANDER [1961], extracellular lipid production may be a capability of soil-inhabiting yeast. DEINEMA [1961] has prepared an excellent review on the factors affecting extracellular and intracellular lipid production by yeasts; it includes a chemical analysis of these lipids.

II. Decomposition

The abundance of lipid in any soil represents the net of all additive processes, such as residue additions or microbial synthesis, and degradative processes. The extent of decomposition depends upon the nature of the lipid material and upon the environmental conditions that affect microbial activity, since the microbial population is primarily responsible for losses that occur. That compounds of the lipid fraction are resistant to decomposition and can, under proper environments, persevere for long periods of time is shown by ORO et al. [1965]. They showed the presence of paraffins of biological origin in Precambrian chert 1.9 billion years old. MAIR [1964] cites evidence for terpenoids and related plant steroids as being source materials for petroleum hydrocarbons.

Plant residues differ greatly in lipid content (WAKSMAN et al. [1928]) as well as in their susceptibility to decomposition. These differences are not surprising in light of the diverse nature of plant lipids and the complex associations within the plant tissue, both of which depend on the species and age of the plant. For example, fatty acids and triglycerides are considered to be readily but somewhat slowly decomposed, whereas plant waxes are much more resistant to decomposition. TENNEY and WAKSMAN [1929] found that the ether-soluble fraction of composted alfalfa or corn stalks decomposed readily, whereas the ether-soluble component of straw proved much more resistant to decomposition. SPRINGER and LEHNER [1952a,b] found the rate of decomposition of a number of agricultural and forest residues to be green fodder > straw > leaves > needles > peat mull > high moor peat. The same sequence was observed for the decomposition of the constituent fats, oils, resins, waxes, chlorophyll, tannins, and polysaccharides of these residues. There is also evidence that the accumulation of lipids in peats is influenced in part by the plant species from which the peat deposit developed (TSYBUL'KIN and BEL'KEVICH [1964]). Moreover, RAKOVSKII and LUKOSHKO [1965] showed that peat-forming plants changed chemical composition during the growing season. Generally, the easily hydrolyzable substances decreased with plant maturity. Using ^{14}C-labeled rye tissue, CHAHAL et al. [1966] investigated plant decomposition in peat soil. Lipid materials comprised the smallest fraction of carbon extracted (1.9%), but they contained the largest fraction of ^{14}C (40.3%). TENNEY and WAKSMAN [1930] noted that low moor peats formed from herbacious (Cladium, Carex, Phragmites) and woody plants are low in ether-soluble components, while high moor peats formed largely from sphagnum mosses are very high in ether-soluble constituents. Even more striking was the fact that almost one-third of the organic matter of pollen peats formed from pondweeds, pollen, plankton, etc., was ether- and alcohol-soluble.

Environmental factors within the soil exert an influence on the accumulation of soil lipids. Residue decomposition studies by WAKSMAN et al. [1928], WAKSMAN and STEVENS [1929], SPRINGER and LEHNER [1952a,b], and LUKOSHKO [1965] were carried out under aerobic and anaerobic conditions. It was evident from these studies that the majority of the lipid materials were decomposed after one year aerobically, while under anaerobic conditions decomposition occurred at a much lower rate. Certainly waterlogging with concomitant anaerobic conditions is a primary reason that soil fats and waxes accumulate noticeably in organic soils but not appreciably in well-drained mineral soils.

Soil pH also influences the decomposition of soil fats and waxes. WAKSMAN and HUTCHINGS [1935] presented data on the ether- and alcohol-soluble fractions of soils limed for 25 years versus untreated soils. Liming decreased the ether-soluble fraction to 0.7%—in contrast to 5.7% in the untreated soil. Interestingly, the alcohol-soluble fraction (resinous material) remained unchanged. TURFITT [1943], after investigating the factors influencing the accumulation of sterols in different soil types, concluded that extreme acidity, poor aeration, and excess water content were factors that individually or collectively inhibited the decomposition of sterols. These environmental conditions were considered responsible for high sterol contents in bog and fen peats. Furthermore, TURFITT [1944a] found that a single genus of actinomycetes, Proactinomyces (Nocardia), was responsible for decomposing added cholesterol in all the soils studied. A survey of the distribution of this microorganism in soils indicated that it was found infrequently under conditions of waterlogging, extreme acidity, and lack of aeration, but abundantly in fertile areas. A graphical summary of the relationship of fats–waxes–resins content and soil pH is given by STEVENSON [1966].

These broad generalizations about the fate of lipid materials in soil can be supplemented by a few studies on the transformations of certain groups of lipid compounds by micro-organisms, some of which are soil isolates. Most of these studies were conducted in solution culture under laboratory conditions, so it becomes difficult to evaluate the performance of these microorganisms in a soil environment where adsorption reactions on clay minerals, salt formation, or other phenomena certainly would be influential. Little information on the decomposition of specific groups of lipid materials in soil is available, but this information will be included in the following paragraphs where applicable.

Fatty acids are readily metabolized by numerous bacteria (RANDLES [1950], SILLIKER and RITTENBERG [1951, 1952], and IVLER et al. [1955]), including some isolated from soil (SILLIKER and RITTENBERG [1951], IVLER et al. [1955]). The soil isolates of SILLIKER and RITTENBERG [1951] were obtained by soil enrichment with stearic acid, and without exception proved to be gram-negative rods, apparently of the genus Pseudomonas. Tests with these soil isolates, as well as with other strains of Pseudomonas, Escherichia coli, Bacillus, Neisseria, and Serratia, indicated that the ability to oxidize fatty acids and use them as a sole source of carbon is a common property of aerobic bacteria. The fatty acid oxidizing enzymes of all the bacteria except Serratia marcescens appeared to be constitutive, requiring no adaptation prior to oxidation. A strain of Serratia marcescens studied by BISHOP and STILL [1961], however, did possess a constitutive enzyme for fatty acid oxidation. Research on the pathway of fatty acid degradation yielded the conclusion that β-oxidation was the common pathway (RANDLES [1950], IVLER et al. [1955]).

Studies on the ability of actinomycetes to oxidize fatty acids are limited. Laboratory experiments by WEBLEY [1954] and WEBLEY et al. [1955] confirmed the ability of one species, Nocardia opaca, to grow effectively on a large number of fatty acids. Evidence showed β-oxidation to be the mechanism of degradation for this microorganism. ADELSON et al. [1957] studied the ability of a soil actinomycete, apparently Streptomyces griseus, to utilize a large variety of lipids and lipid derivatives, including stearic and palmitic acids.

Utilization of individual fatty acids by fungi is likewise not extensively documented. Most of the information on this subject pertains to studies on the nature of rancidity of plant fats such as coconut and palm oils (FOSTER [1949]). Rancidity is due chiefly to the oxidative break-down by fungi of fatty acids liberated by prior lipolytic action yielding a series of methyl ketones containing one carbon less than the original fatty acid. Strains of Penicillium glaucum, Aspergillus niger, and Aspergillus fumigatus were capable of forming methyl ketones when grown in pure culture on individual fatty acids as a source of carbon. STERN et al. [1954] indicated that a strain of Mucor mucedo could utilize stearic and oleic acids and methyl linoleate as carbon sources, while palmitic acid and methyl linolenate were not utilized. KRAUSE and LANG [1965] found vigorous visible growth of Fusarium solani, Fusarium diversisporum, and Fusarium equiseti in soils after the addition of the straight-chained stearic and behenic acids. Branched fatty acids produced no visible growth.

The occurrence and microbiological degradation of steroids in soils were studied extensively by TURFITT [1944a,b, 1947]. The microorganisms considered solely responsible for steroid degradation in all soils studied were actinomycetes of the genus Proactinomyces (Nocardia). Turfitt noted, however, that if the C_{17} side chain of steroids was modified or lacking, certain gram-negative bacteria were capable of using the molecule. SCHATZ et al. [1949] questioned Turfitt's work, which limited steroid decomposition primarily to a single genus of micro-organism, since 10 different soils, manures, and composts studied were found to harbor cholesterol oxidizing bacteria. Altogether 19 microorganisms capable of using cholesterol as cholesterol acetate and cholesterol palmitate were isolated. Ergosterol was toxic to all but one of the isolates, whereas stilbestrol was toxic to all. Moreover, limited work with soil fungi indicated that under suitable conditions these microorganisms possess the capability of oxidizing ergosterol. Recently, GREGORS-HANSEN [1964) studied the decomposition in soil of diethyl-stilbestrol-1-^{14}C, a synthetic steroid growth hormone. From 12 to 28% of the added ^{14}C was recovered as CO_2 after six months, indicating that the soil microorganisms possess the capability for degradation, but at a slow rate. Turfitt's study (2) with *Proactinomyces erythropolis* affords some information on part of the degradative mechanism of the steroid, cholesterol. In pure culture in buffered mineral-salts media, the initial stage of attack on cholesterol was an oxidation to 4-cholestenone. This involved an oxidation of the hydroxyl group at carbon 3 and a rearrangement of the double bond to the 4:5 position. He assumed that since bacterial growth continued actively after this conversion and because disappearance of 0.45 g cholesterol in 100 ml of medium was almost complete in four months, a secondary reaction in which ring cleavage occurred followed the initial oxidation. Turfitt did not propose a mechanism for the ring cleavage.

Soil microorganisms also possess the enzymatic capability for degrading other lipid compounds that would normally be incorporated into soils. One compound of special interest is 3,4-benzopyrene, which has carcinogenic properties and which, in addition to its natural occurrence, occurs as an air pollutant in combustion products of motor vehicles and some industrial flue gases. SHCHERBAK [1969] reported that soil samples that were collected close to an industrial plant known to be emitting 3,4-benzopyrene contained as much as 8.35 mg/kg of that compound. He concluded that removal of 3,4-benzopyrene from soil was accomplished by transport to lower depths or by absorption by plants. However, KHESINA *et al.* [1969] showed that microbial decomposition of 3,4-benzopyrene in "heavily polluted" soil accounted for as much as 50% reduction within three months. "Slightly polluted" soil effected no decomposition. ROGOFF and WENDER [1957] report that another aromatic polynuclear hydrocarbon, phenanthrene, was decomposed in culture medium by a species of *Pseudomonas*.

Among other lipid compounds utilized by one or more soil microorganisms are aliphatic hydrocarbons and long-chained aliphatic alcohols (WEBLEY [1954], KRAUSE and LANG [1965]); phospholipids (lecithin and caphalin), inositol lipids, and complex proteolipids (ADELSON *et al.* [1957]); and a variety of different fatty acid esters (KRAUSE and LANG [1965]).

E. Influence of Lipids on Soil Physical Properties and Plant Growth

The lipid fraction of soil organic matter is seldom considered to have an important influence on soil properties or plant growth. The literature, however, contains a number of studies which imply that under certain conditions these materials are or have the potential to be of considerable importance.

I. Influence on the Physical Properties of Soils

Soil lipids are distinctly hydrophobic and thus could be expected to alter such soil physical properties as soil aggregate stability and degree of wetting.

SCHREINER and SHOREY [1910] made reference to a California soil that could not be readily wetted by rain, irrigation, or movement of water from the subsoil, thus making it unfit for agriculture. The problem was attributed to wax bodies present in the soil organic matter. JAMISON [1942, 1945] and WANDER [1949a,b] reported on sandy soils in Central Florida that also resisted wetting. The water repellent condition existed primarily in the darker surface layer high in organic matter. The earlier studies by JAMISON [1942, 1945] had seemed to eliminate soil fats and waxes as the causative agents, because extraction of the problem soils with numerous water-free organic solvents did not appreciably improve wettability. WANDER [1949a,b] presented a logical explanation for the problem. Data obtained indicated that the repellent nature of these soils was associated with the application of ground limestone or magnesium-containing fertilizer. The added calcium or magnesium formed insoluble soaps with soil fatty acids, which were responsible for the water repellent character of the soils. These soaps, once formed, were only slightly soluble in ether and other nonpolar solvents, which explained why Jamison could not improve soil wettability by ether extraction. Wander was able to duplicate the water repellent conditions in soils in the laboratory with either fatty acid mixtures from soil or with stearic acid, provided dilute aqueous solutions of calcium or magnesium hydroxide were also added. Without the addition of calcium or magnesium hydroxide, neither the soil fatty acid mixture nor the stearic acid had any effect on soil wettability.

Recently BOND and HARRIS [1964] and BOND [1964] described some sandy soils in southern Australia with wetting characteristics similar to those in southern Florida. These Australian soils had a thin waterproof film of organic material on the surface and consistently contained a zone of extensive fungal proliferation, although not necessarily on the surface. The bulk of the mycelium was believed to be from basidiomycetes, but attempts to culture the mycelium were not successful. No information was given on the nature of these waterproofing materials, and it remained uncertain whether the basidiomycetes were the sole or even the main source of these substances. The extreme water repellent condition of these soils suggested that lipids were involved.

Considerable research has been conducted in the past two decades on the nature of organic constituents responsible for stabilizing soil aggregates. Since considerable improvement of soil aggregation occurs upon addition of readily decomposable organic materials to soils, the increase in aggregate stability is attributed to microbial activity. Among the microbial products most commonly considered effective are the various extracellular polysaccharides synthesized by a large variety of soil microorganisms.

The possibility that materials other than polysaccharides can, under certain conditions, be responsible for stable aggregation was suggested in studies by GREENLAND et al. [1961, 1962]. These authors worked with fine-textured soils and measured aggregate stability before and after treatment with periodate—a procedure that should destroy the integrity of polysaccharides. This treatment collapsed aggregates from young pastures and cultivated soils, but aggregates from old pasture soils high in organic matter remained stable. Presumably some organic material other than polysaccharides stabilized the aggregates of these old pasture soils. Soil lipids could serve such a purpose.

Studies have been conducted to determine the effectiveness of fats and waxes in stabilizing soil aggregates. McCALLA [1945, 1946] conducted experiments with various vegetable oils and fats, two waxes, resin, and paraffin. All the materials were effective in stabilizing lumps of Peorian loess subsoil at concentrations as low as 0.25 percent in benzene-alcohol solvent. Particularly effective were the vegetable oils and fats. McCALLA [1945] also ascertained the influence of compounds extracted from wheat straw with alcohol-benzene on the stability of Peorian loess. This material had a negligible effect on aggregate stability at concentrations less than 1 percent but became extremely effective at concentrations of 2 and 4 percent. An attempt to show the importance of naturally occurring fats and waxes on soil aggregate stability was carried out by treating soil lumps of Peorian loess subsoil and Marshall topsoil with an excess of various organic solvents. None of the solvents altered the stability of the loess, but all decreased slightly the stability of the topsoil lumps.

As was discussed in Section D on the origin of fats and waxes in soil, many microorganisms, especially certain species of fungi, synthesize lipid materials. GEOGHEGAN and ARMITAGE [1949] and GEOGHEGAN [1950] compared the aggregating ability of alcohol-benzene extracts of mycelium of *Penicillium notatum* and *Aspergillus niger* with substances extracted from wheat straw with the same solvents and an ether extract of soil. The authors concluded that these materials produced poor soil aggregation when compared with microbial levans and dextrans and became reasonably effective only when sprayed on the surface of soil aggregates. Aggregates formed with these products were waterproofed sufficiently to float on the surface of water. It was concluded that waterproofing in this manner could be detrimental to crop production, but partial waterproofing might in certain cases delay the breakdown of crumb structure.

MARTIN *et al.* [1959] incubated two sandy loam soils with ether extracts of mycelium from 24 different species of fungi. The ether-soluble material from all except one fungus species significantly increased soil aggregation after one day's incubation. After 20 days a very high degree of aggregation was attained from all fungal extracts and aggregation remained at this high level, even after 100 days' incubation. Ether extraction of the soils after 20 days' incubation recovered about 50 percent of the added material and reduced aggregation by 15 to 20 percent. This indicated that the original material contributed to the marked binding action but that substances produced by the activity of the soil microbial population were also of considerable importance.

Conversion by microorganisms of lipid compounds to compounds imparting stability to soil was demonstrated in experiments of FEHL and LANGE [1965]. Long-chain n-alkanes and their l-substituted derivatives were added as a substrate for 3 *Fusarium* species. The resulting microbial growth led to greatly increased soil strength which outlasted the carbon nutrient source.

Fats and waxes from biological forms other than soil inhabiting microorganisms may also have an influence on soil structure. AGARWAL *et al.* [1958] noted that heavy infestations of some earthworms (*Allobophora*) adversely affected soil productivity by excreting a waxy fluid that cemented the worm castings together and ultimately resulted in a cloddy structureless soil.

II. Influence on Plant Growth

(1) *Nutrient Availability*

The orientation of soil lipids on the surface of soil particles may have a pronounced influence on the decomposition of soil organic matter and subsequent mineralization of plant nutrients. This possibility was considered early in the twentieth century by GREIG-SMITH [1910], who suggested that the soil waxy materials waterproof soil particles, thereby limiting the free solution of soil nutrients for plant growth and microbial activity. He felt that the addition of fat solvents to soil would segregate this waterproofing material and make nutrients more available to plants and microorganisms. A similar conclusion was reached by WAKSMAN and STEVENS [1929] to explain why peat treated with toluene or ether, even without extraction, decomposed more rapidly than untreated peat. These authors did not feel that the effect of the ether or toluene was one of altering the microbial population by partial sterilization or stimulation.

A report by ROMASHKEVICH in the U.S.S.R. [1964] would seem to support in part Greig-Smith's earlier idea. He found that plants grown in soil with peat amendments from which bitumens had been removed by alcohol:benzene extraction produced larger yields and took up more nitrogen than plants grown in the unextracted peat-soil mixtures. A laboratory experiment in which peat with and without bitumens was composted with soil indicated that bitumens had a detrimental effect on nitrification. Reference was also made to an unpublished report by Logvinova where the removal of bitumens from peat increased mineralization of inorganic nitrogen in peat soil composts. According to EDIGAROVA [1963], bituminous material of petroleum origin also affected soil nitrification. After introduction to soil of petroleum substances (spent gumbrin, bituminous rock) nitrate-nitrogen decreased and sharply increased a month later. The solubility of difficulty-soluble phosphoric acid was also increased.

(2) *Direct Influence on Plant Growth*

In recent years a considerable interest has been shown in the possible stimulatory or phytotoxic effects of organic compounds in soils (WINTER [1961], McCALLA *et al.* [1963], McCALLA [1964]). Consideration of the influence of "organic nutrition" is not new, however, having been given sporadic attention throughout this century. Among the early studies were those by SCHREINER *et al.* [1908–1911] on soil organic compounds deleterious to plant growth. Among the organic compounds isolated from mineral and organic soils and tested for phytotoxic activity were a large number soluble in organic solvents, all of which were listed previously in Section C. Of this group, dihydroxystearic acid was particularly interesting because it was toxic to wheat seedlings growing in solution at a concentration of 10 to 50 ppm, reducing plant transpiration and presumably interfering with the normal oxidative metabolism of plant roots. Further studies on the occurrence of dihydroxystearic acid in soils (SCHREINER and LATHROP [1911]) demonstrated its presence to be largely associated with poor unproductive soils and seldom in fertile soils. The investigators considered dihydroxystearic acid to be one agent responsible for infertile or exhausted soils. This function was questioned by FRAPS [1915], · who found that dihydroxystearic acid at concentrations as high as 500 ppm had but a slight injurious effect on corn and sorghum grown in pots containing soil. Corn and sorghum seedlings grown in water culture with as little as 10 ppm dihydroxystearic acid were noticeably injured.

In addition to the work of Schreiner and Shorey just discussed, other studies have considered the possible influences of lipid materials on plant growth. McCALLA *et al.* [1963] found that materials extracted from stubble-mulched soils with methanol and acetone inhibited the growth of the roots and shoots of wheat seedlings. GUENZI and McCALLA [1962] also showed that ethanol-soluble substances extracted from wheat straw inhibited germination and growth of wheat. Studies of MISHUSTIN and EROFEEV [1966] further elucidate the mechanism of inhibition. They found that, especially under anaerobic conditions, volatile fatty acids produced in the decomposition of wheat straw inhibited the growth of rye grass. Free phenolic acids such as ferulic and coumaric which were also produced were slightly less toxic. In a series of studies by PRILL *et al.* [1949a,b] and BARTON and SOLT [1948], evidence was presented that phospholipids from an extract of pole bean seeds (*Phaseolus* sp.) could inhibit wheat root growth. A later investigation by PRILL *et al.* [1949a] testing the inhibitory action of various organic acids on the growth of wheat roots in solution included a number of the shorter chain fatty acids, butyric, caproic, capric, and crotonic. All of the fatty acids were found inhibitory at the 0.2 mM level. It was suggested that the inhibitory effect of these acids might be related to their ability to lower the surface tension of water. Continued work by PRILL *et al.* [1949a], using a number of commercially available strongly surface active agents, supported this idea. Among the compounds used were two plant lipid materials Astec. H. S. (a water-soluble product from soybean phosphatides) and Saponin (a glycoside of a steroid or triterpene common to the plant kingdom). Saponin and Astec H. S. were effective in markedly reducing wheat root growth at concentrations as low as 5 ppm. Consideration of these results suggests surface tension effects as an explanation for the toxicity of dihydroxystearic acid when plants were grown in solution.

Even though lipid materials have been shown to be phytotoxic at relatively low concentrations in solution, it is presently difficult to evaluate their effectiveness in the natural soil environment. Only the study of FRAPS [1915] with dihydroxystearic acid indicates the attenuating effect which soil may exert on a phytotoxic substance. Until further research is completed, we must consider the possibility that under special circumstances the concentration of soil lipids might reach levels detrimental to plant growth.

REFERENCES

Adelson, L. M., A. Schatz, and G. S. Trelawny, 1957. Metabolism of lipids and lipid derivatives by a soil actinomycete. *J. Bacteriol.* 73:148.

Agarwal, G. S., K. S. K. Rao, and L. S. Negi, 1958. Influence of certain species of earthworms on the structure of some hill soils. *Curr. Sci.* 27:213.

Alexander, M., 1961. *Introduction to soil microbiology.* New York: John Wiley.

Aschan, O., 1921. Humoceric acid. *Finska Kem. sanfundets Medal.* 30:37. *Chem. Abstr.* 16:1567 (1921).

Asinger, F., 1956. *Chemie und Technologie der Paraffin—Kohlenwasserstoffe.* Berlin: Akademie.

Barton, L. V., and M. L. Solt, 1948. Growth inhibitors in seeds. *Contrib. Boyce Thompson Inst.* 15:259.

Baudisch, O., and H. Von Euler, 1935. Über den Gehalt einiger Moor-Erdarten an Carotinoiden. *Arkiv Kemi. Miner. Geol.* 11:21, A, 10.

Bel'kevich, P. I., 1960. Production of wax from peat. *Tr. Inst. Torfa, Akad. Nauk Belarussk.S.S.R.* 9:19. *Chem. Abstr.* 58:6620P (1963).

———, G. P. Verkholitova, F. L. Kaganovich, and L. V. Targov, 1963. β-Sitosterol from peat wax. *Izv. Akad. Nauk. S.S.S.R., Otd. Khim Nauk.* 112. *Chem. Abstr.* 58:10011 (1963).

———, F. L. Kaganovich, E. V. Trubilko, A. V. Bystraya, and E. A. Yurkevich., 1965. The presence of stearins in peat waxes. *Kompleksn. Ispol'z Torfa, Uses. Nauchn.-Issled. Inst. Torfa* 73. *Chem. Abstr.* 64:7929 (1966).

Bergmann, E. D., R. Ikan, and J. Kashmann, 1964. The occurrence of perylene in Huleh peat. *Israel J. Chem.* 2:171.

Bhandari, G. S., M. S. Maskina, and N. S. Randhawa, 1969. Characterization of lipids in some soils and their humic acids formed under different agro-climatic conditions. *Sci. Cult.* 35:68. *Chem. Abstr.* 71:111908 (1969).

Bishop, D. G., and J. L. Still, 1961. Fatty acid metabolism in *Serratia marcescens.* 1. Oxidation of saturated fatty acid by whole cells. *J. Bacteriol.* 82:370.

Black, W. A. P., W. J. Cornhill, and F. N. W. Woodward, 1955. A preliminary investigation on the chemical composition of sphagnum moss and peat. *J. Appl. Chem.* 5:484.

Blumer, M., 1961. Benzypyrenes in soil. *Science* 134:474.

Bond, R. D., 1964. The influence of the microflora on the physical properties of soils. 2. Field studies on water repellent sands. *Australian J. Soil Res.* 2:123.

———, and J. R. Harris, 1964. The influence of the microflora on physical properties of soils. 1. Effects associated with filamentous algae and fungi. *Australian J. Soil Res.* 2:111.

Bone, W. A., and L. J. Tei, 1934. Researches on the chemistry of coal. Part 7. An investigation of German brown coals and Irish peat. *Proc. Royal Soc. (London) A* 147:58.

Borneff, I., and H. Kante, 1963. Kanzerogene Substanzen in Wasser und Boden. Weitere Untersuchungen uber polyzyklische, aromatische Kohlenwasserstoffe in Erdproben. *Arch. Hyg. Bakt.* 147:401.

Braids, O. C., 1966. A study of the components of the lipid fraction of Rifle peat. Ph.D. Dissertation, The Ohio State University.

———, F. L. Himes, and G. W. Volk, 1967. The occurrence of carbazole in a peat soil. *Soil Sci. Soc. Amer. Proc.* 31:435.

Butler, J. H. A., D. T. Downing, and R. S. Swaby, 1964. Isolation of chlorinated pigment from green soil. *Aust. J. Chem.* 17:717.

Cawley, C. M., and J. G. King, 1945. Ester waxes from British lignite and peat. *J. Soc. Chem. Ind. (London)* 64:237.

———, J. H. G. Carlile, and C. C. Naaks, 1948. Ester waxes from British peat. *Petroleum (London)* 11:77. *Chem. Abstr.* 42:520r (1948).

Chahal, K. S., J. L. Mortensen, and F. L. Himes, 1966. Decomposition products of carbon-14 labelled rye tissue in a peat profile. *Soil Sci. Soc. Amer. Proc.* 30:217.

Dawes, E. A., and D. W. Ribbons, 1964. Some aspects of the endogenous metabolism of bacteria. *Bact. Rev.* 28:126.

Deinema, M. H., 1961. Intra- and extracellular lipid production by yeast. *Meded. Landbouwhogeschool, Wageningen* 61:1.

————, and C. A. Landheer, 1960. Extracellular lipid production by a strain of *Rhodotorula graminis Biochem. Biophys. Acta* 37:178.

————, M. Van Ammers, G. A. Landheer, and M. H. M. Van Rooyen, 1964. Note on the isolation of β-hydroxypalmitic acid from the extracellular lipids of *Rhodotorula glutinis. Rec. Trav. Chim. Pays-Bas* 83:708.

Di Menna, M. E., 1958. Two new species of yeasts from New Zealand. *J. Gen. Microbiol.* 18:269.

Edigarova, N. N., 1963. Behavior of organic substances of petroleum origin in the soil. *Neft. Udobreniya i Stimulyatory (Baku: Akad. Nauk Azerb. S.S.R.) Sb.* 190. *Chem. Abstr.* 60:13, 822a (1964).

Fehl, A. J., and W. Lange, 1965. Soil stabilization induced by growth of microorganisms on high calorie mold nutrients. *Soil Sci.* 100:368.

Feustel, I. C., and H. G. Byers, 1930. The physical chemical characteristics of certain American peat profiles. *U.S. Dept. Agr. Tech. Bull. 214.*

Fraps, G. S., 1915. The effect of organic compounds in pot experiments. *Texas Agr. Exp. Sta. Bull. 174.*

Foster, J. W., 1949. *Chemical activities of fungi.* New York: Academic Press.

Gallopini, C., and R. Riffladi, 1969. Composition of ether extracts of soil. *Agrochimica* 13: 207.

Geoghegan, M. J., 1950. Aggregate formation in soil. Influence of some microbial metabolic products and other substances on aggregation of soil particles. *Trans. 4th Int. Congr. Soil Sci.* 1:198.

————, and E. R. Armitage, 1949. Influence of some lipoidal substances on aggregate formation in soils. *Nature, Lond.* 163:29.

Gilliland, M. R., and A. J. Howard, 1968. Some constituents of peat wax separated by column chromatography. *Transactions of the 2nd International Peat Congress*, Leningrad, 1963. Vol. II, p. 877. Edinburgh: H.M.S.O.

————, A. J. Howard, and D. Hamer, 1960. Polycyclic hydrocarbons in crude peat wax. *Chem. Ind.* 1357.

Greenland, D. J., G. R. Lindstrom, and J. P. Quirk, 1961. Role of polysaccharides in stabilization of natural soil aggregates. *Nature, Lond.* 191: 1283.

————, G. R. Lindstrom, and J. P. Quirk, 1962. Organic materials which stabilize natural soil aggregates. *Soil Sci. Soc. Amer. Proc.* 26:366.

Gregors-Hansen, B., 1964. Decomposition of diethylstilboestrol in soil. *Plant Soil* 20:50.

Greig-Smith, R., 1910. Contributions to our knowledge of soil fertility. 1. The action of wax solvents and the presence of thermolabile bacteriotoxins in soil. *Proc. Linn. Soc. N.S. Wales* 35:808.

Guenzi, W. D., and T. M. McCalla, 1962. Inhibition of germination and seedling development by crop residues. *Soil Sci. Soc. Amer. Proc.* 26:456.

Hance, R. J., and Anderson, G., 1963a. Extraction and estimation of soil phospholipids. *Soil Sci.* 96:94.

————, and G. Anderson, 1963b. Identification of hydrolysis products of soil phospholipids. *Soil Sci.* 96:157.

Hauser, G., and M. Karnovsky, 1954. Studies on the production of glycolipids by *Pseudomonas aeruginosa. J. Bacteriol.* 68:645.

Himes, F. L., and C. Bloomfield, 1967. Extraction of triacontyl stearate from a soil. *Plant Soil* 26:383.

Howard, A. J., and D. Hamer, 1960. The extraction and constitution of peat wax. Review of peat wax chemistry. *J. Amer. Oil Chem. Soc.* 37:478.

Ikan, R., and J. Kashman, 1963. Steroids and triterpenoids of Hula peat as compared to other humoliths. *Israel J. Chem.* 1:502.

Ivanova, L. A., P. I. Bel'kevich, F. L. Kaganovich, and P. D. Shepetovskii, 1968. Acids of peat wax. 2. Separation and identification of methylates from insoluble sodium salts by gas-liquid chromatography. *Vestsi. Akad. Navuk Belarus. SSR, Ser. Khim. Navuk* 121. *Chem. Abstr.* 70:79824 (1969).

Ives, A. J., and A. N. O'Neill, 1958. The chemistry of peat. Part 1. The sterols of peat moss (Sphagnum). *Can. J. Chem.* 36:434.

Ivler, D., J. B. Wolfe, and S. O. Rittenberg, 1955. Studies on the aerobic oxidation of fatty acids by bacteria. 5. Caprate oxidation by cell-free extracts of *Pseudomonas fluorescens. J. Bacteriol.* 70:99.

Jamison, V. C., 1942. The slow reversible drying of soil beneath citrus trees in Central Florida. *Soil Sci. Soc. Amer. Proc.* 7:36.

————, 1945. Penetration of irrigation and rain water into sandy soils of central Florida. *Soil Sci. Soc. Amer. Proc.* 10:25.

Jarvis, F. G., and M. J. Johnson, 1949. A glyco-lipid produced by *Pseudomonas aeruginosa. J. Am. Chem. Soc.* 71:4124.

Johnson, R. C., and R. Thiessen, 1934. Studies on peat alcohol and ether-soluble matter of certain soils. *Fuel* 8:44. *Chem. Abstr.* 28:4205 (1934).

Josephy, E., and F. Radt, editors. 1946. *Elsevier's Encyclopedia of Organic Chemistry.* New York: Elsevier.

Katkouski, A. P., and N. Ts. Karosik., 1954. Preparation of bitumens from peat. *Vestsi Akad. Navuk, Belarus, S.S.R. No. 2,* 78. *Chem. Abstr.* 49:15211 (1955).

Kern, W., 1947. The occurrence of chrysene in soil. *Helv. Chim. Acta* 30:1595.

Khesina, A. Ya., N. P. Shcherback, L. M. Shabad, and I. S. Vostrov, 1969. Destruction of benzo [α] pyrene by soil microflora. *Byull. Eksp. Biol. Med.* 68:70. *Chem. Abstr.* 72:42240 (1969).

Kian, R., G. Stahl, and E. D. Bergmann, 1968. Constituents of Huleh peat. 3. Acids. *Israel J. Chem.* 6:485.

Kleinzeller, A., 1944. Fat formation in *Torulopsis lipofera. Biochem. J.* 38:480.

Kononenko, E. V., 1958. Soil yeasts of the genus *Lipomyces. Mikrobiologiya* 27:605.

Kowalenko, C. G., and R. B. McKercher, 1971. Phospholipid components extracted from Saskatchewan soils. *Can. J. Soil Sci.* 51:19.

Krause, F. P., and W. Lang, 1965. Vigorous mold growth in soils after addition of water insoluble fatty substances. *Appl. Microbiol.* 13:160.

Ksenofontova, E. V., M. V. Mukhina, A. M. Khaletskii, F. L. Kaganovich, and P. I. Bel'kevich, 1969. Quantitative determination of β-sito-sterol in peat wax resin. *Dokl. Akad. Nauk Beloruss. SSR* 13:143. *Chem. Abstr.* 70:99569 (1969).

Kwiatkowski, A., 1963. Effect of the extracting solvent on the composition of peat bitumen. *Zeszyty Nauk. Politech. Gdansk. Chem.* 33:53. *Chem. Abstr.* 60:7839 (1964).

Lukoshko, E S., 1965. Changes in the chemical composition of peat-forming plants during decomposition of peat-forming layers under aerobic and unaerobic conditions. *Vestis Akad. Navuk Belarusk. SSR. Ser: Khim. Navuk* 90. *Chem. Abstr.* 64:4212d (1966).

Mair, B. J., 1964. Terpenoids, fatty acids and alcohols as source materials for petroleum hydrocarbons. *Geochim. Cosmochim. Acta* 28:1303.

Mallet, L., and M. Tissier, 1969. Biosynthesis of polycyclic hydrocarbons of the benzoα-pyrene type in forest soil. *C. R. Soc. Biol.* 163:63.

Martin, J. P., J. D. Ervin, and R. A. Shepard, 1959. Decomposition and aggregating effect of fungus cell material in soil. *Soil Sci. Soc. Amer. Proc.* 23:217.

McCalla, T. M., 1945. Influence of microorganisms and some organic substances on soil structure. *Soil Sci.* 59:287.

————. 1946. The biology of soil structure. *J. Soil Water Cons.* 1:71.

————. 1964. Phytotoxic substances from soil microorganisms and crop residue. *Bact. Rev.* 28:181.

————, W. D. Guenzi, and F. A. Norstadt, 1963. Microbial studies of phytotoxic substances in the stubble-mulch system. *Z. Allgem, Microbiol.* 3:202.

McLean, J., G. H. Rettie, and F. S. Spring, 1958. Triterpenoids from peat. *Chem. Ind.* 1515.

Mishustin, E. N., and N. S. Erofeev, 1966. Nature of toxic substances accumulating during straw decomposition in soil. *Mikrobiol.* 35:150.

Morrison, R. I., 1969. Soil lipids. In *Organic geochemistry,* Chap. 23, p. 558. New York: Springer-Verlag.

————, and W. Bick, 1966. Long-chain methyl ketones in soils. *Chem. Ind. (London)* 596.

Morrison, R. I., and W. Bick, 1967. The wax fraction of soils; separation and determination of some components. *J. Sci. Food Agr.* 18:351.

Nagy, B., 1966. The optical rotation of lipids extracted from soils, sediments, and the Orgueil carbonaceous meteorite. *Proc. Natl. Acad. Sci. U.S.* 56:389.

Oro, J., D. W. Nooner, A. Zlatkis, S. A. Wikstrom, and E. S. Barghoorn, 1965. Hydrocarbons of biological origin in sediments about 2 billion years old. *Science* 148:77.

Pedersen, T. A., 1958. *Cryptococcus terricolus* Nov. spec. A new yeast isolated from Norwegian soils. *C. R. Trav. Lab. Carlsberg* 31:93.

Prill, E. A., L. V. Barton, and M. L. Solt, 1949a. Effects of some surface-active agents on the growth of wheat roots in solutions. *Contrib. Boyce Thompson Inst.* 15:311.

———, L. V. Barton, and M. L. Solt, 1949b. Effect of some organic acids on the growth of wheat roots in solutions. *Contrib. Boyce Thompson Inst.* 15:429.

Rakovskii, V. E., and E. S. Lukoshko, 1965. Changes in the chemical composition of peat-forming plants during the growth period. *Kompleksn. Ispol'z. Torfa, Vses. Nauchn.-Issled. Inst. Torfa.* 24. *Chem. Abstr.* 64:7042h (1966).

Rakowski, E. W., and N. G. Edelstein, 1932. Peat bitumens. 1. Fatty acids. *Brennstoff—Chem.* 13:46. *Chem. Abstr.* 26:1751 (1932).

Ralston, A. W., 1948. *Fatty acids and their derivatives.* New York: John Wiley.

Randles, G. I., 1950. The oxidation of fatty acids by *Neisseria catarrholis. J. Bacteriol.* 60:627.

Reilly, J., and J. A. Emlyn, 1940. Studies in peat. 8. Preliminary note on Irish peat wax (mona wax). *Sci. Proc. Royal Dublin Soc.* 22:263.

———, and J. P. Wilson, 1940. Studies in peat. 9. The cerotic and carboceric acid fractions of mona wax. *Sci. Proc. Royal Dublin Soc.* 22:321.

———, D. F. Kelly, and J. Duffy, 1939. Extraction of peat with azeotrope-like petroleum mixed solvents. *Sci. Proc. Royal Dublin* 22:149.

———, D. F. Kelly, and D. J. Ryan, 1937. Mixtures of constant boiling point for solvent-extraction purposes. Extraction of waxes from peat. *J. Soc. Chem. Ind.* 56:231.

Robinson, T., 1963. *The organic constituents of higher plants.* Minneapolis: Burgess Publishing Co.

Rodd, E. H., ed. 1951. *Chemistry of the carbon compounds, Vol. Ia.* Amsterdam: Elsevier.

Roginskaya, E. V., 1936. The composition of the acids of high molecular weight from the bitumen wax of peat. *J. Applied Chem.* (*USSR*) 9:108. *Chem. Abstr.* 30:5441 (1936).

Rogoff, M. H., and I. Wander, 1957. The microbiology of coal. 1. Bacterial oxidation of phenanthrene. *J. Bacteriol.* 73:764.

Romashkevich, I. F., 1964. Role of bitumens in delaying mobilization of nitrogen compounds in peats and uptake of nitrogen by plants. *Soviet Soil Sci.* 1:81. Translation of *Pochvovednie* 1:102 (1964).

Ruinen, J., and M. H. Deinema, 1964. Composition and properties of the extra-cellular lipids of yeast species from the phyllosphere. *Antonie van Leeuwenhoek* 30:377.

Ryan, H., and T. Dillon, 1909. Montanin and montana (montan) waxes. *Proc. Dublin Soc.* 12:202.

Schatz, A., K. Savard, and I. J. Pintner, 1949. The ability of soil microorganisms to decompose steroids. *J. Bacteriol.* 58:117.

Schreiner, O., and E. C. Lathrop, 1911. Examination of soils for organic constituents, especially dihydroxystearic acid. *U.S. Dept. Agr. Bur. Soils Bull.* 80.

———, and E. C. Shorey, 1908. The isolation of dihydroxystearic acid from soils. *J. Amer. Chem. Soc.* 30:1599.

———, and E. C. Shorey, 1909a. The presence of a cholesterol substance in soils. Agrosterol. *J. Amer. Chem. Soc.* 31:116.

———, and E. C. Shorey, 1909b. The isolation of harmful organic substances from soils. *U.S. Dept. Agr. Bur. Soils, Bull.* 53.

———, and E. C. Shorey, 1910. Some acid constituents of soil humus. *J. Amer. Chem. Soc.* 32:1674.

———, and E. C. Shorey, 1911a. Cholesterol bodies in soils: Phytosterol. *J. Biol. Chem.* 9:9.

———, and E. C. Shorey, 1911b. Glycerides of fatty acids in soils. *J. Amer. Chem. Soc.* 33:78.

———, and E. C. Shorey, 1911c. Paraffin hydrocarbons in soils. *J. Amer. Chem. Soc.* 33:81.

Shcherbak, N. P., 1969. Fate of benzo[α]pyrene in soil. *Vop. Onkol.* 15:75. *Chem. Abstr.* 71:48877 (1969).

Silliker, J. H., and S. C. Rittenberg, 1951. Studies on the aerobic oxidation of fatty acids by bacteria. 1. The nature of the enzymes constitutive or adaptive. *J. Bacteriol.* 61:653.

————, and S. C. Rittenberg, 1952. Studies on the aerobic oxidation of fatty acids by bacteria. 3. The effect of 2,4-dinitrophenol on the oxidation of fatty acids by *Serratia marcescens*. *J. Bacteriol.* 64:197.

Simonart, P., and L. Batistic, 1966. Aromatic hydrocarbons in soils. *Nature* 212:1461.

Simonsen, J., and W. C. J. Ross, 1957. *The terpenes.* Vol. 4. The triterpenes and their derivatives. Cambridge: Cambridge University Press.

Smirnov, G. A., 1970. Benzo[α] pyrene content in soil and vegetation near an airport. *Vop. Onkol.* 16:83. *Chem. Abstr.* 73:59045 (1970).

Springer, V., and A. Lehner, 1952a. Decomposition and the synthesis of humus by aerobic and anaerobic decomposition of organic substances important in agriculture and forestry. 1. *Z. Pflernahr. Dung.* 58:193.

————, and A. Lehner, 1952b. Decomposition and the synthesis of humus by aerobic and anaerobic decomposition of organic substances important in agriculture and forestry. 2. *Z. Pflernahr. Dung.* 59:1.

Stadnikoff, G., and R. Wahner, 1931. Über die Natur der Kohlenbitumina. *Brennstoff-Chem.* 12:23.

Starkey, R. L., 1946. Lipid production by a soil yeast. *J. Bacteriol.* 51:33.

Stecher, P. G., Ed. 1960. *The Merck index.* Rahway, N. J.: Merck and Company.

Stern, A. M., Z. J. Ordal, and H. O. Halverson, 1954. Utilization of fatty acids by and lipolytic activities of *Mucor mucedo. J. Bacteriol.* 68:25.

Stevenson, F. J., 1966. Lipids in soil. *J. Amer. Oil Chem. Soc.* 43:203.

Sundgren, A., 1949. Investigations on extraction of peat and manufacture of wax and resinous substances from the peat bitumen obtained. *Tek. Foren. Finland. Fork.* 69:29. *Chem. Abstr.* 43:5570 (1949).

————, and V. T. Rauhala, 1949. Preliminary note on fatty acids in peat. *Suomen Kemistilekti* 228:24 *Chem. Abstr.* 44:3270 (1950).

————, and V. T. Rauhala, 1965. Free acids of peat wax. *Valtion Tek. Tutkimuslaitos, Julkaisu No. 92,* 27 pp. *Chem. Abstr.* 65:3610 (1966).

Swan, E. P., 1965. Identity of a hydrocarbon found in a forest soil. *Forest Prod. J.* 15:272.

Tenney, F. G., and S. A. Waksman, 1929. Composition of natural organic materials and their decomposition in the soil. 4. The nature and rapidity of decomposition of the various organic complexes in different plant materials, under aerobic conditions. *Soil Sci.* 28:55.

————, and S. A. Waksman, 1930. Composition of natural organic materials and their decomposition in the soil. 5. Decomposition of various chemical constituents in plant materials, under anaerobic conditions. *Soil Sci.* 30:143.

Titov, N., 1932. Bitumens of sphagnum peat. *Brennstoff-Chem.* 13:266. *Chem. Abstr.* 26:5195 (1932).

Tsybul'kin, V. M., and P. I. Bel'kevich, 1964. Comparative study of bitumen-forming substances of some species of plant and peat bitumens. *Vestis Akad. Tavik. Belarusk. SSSR, Ser. Fiz-Telshan Navik,* 101. *Chem. Abstr.* 61:1671 (1964).

Turfitt, G. E., 1943. The microbiological degradation of steroids. 1. The sterol content of soils. *Biochem. J.* 37:115.

————. 1944a. The microbiological agencies in the degradation of steroids. 1. The cholesterol-decomposing organisms of soils. *J. Bacteriol.* 47:487.

————. 1944b. The microbiological degradation of steroids. 2. Oxidation of cholesterol by *Proactinomyces* spp. *Biochem. J.* 38:492.

————. 1947. Microbiological agencies in the degradation of steroids. 2. Steroid utilization by the microflora of soils. *J. Bacteriol.* 54:557.

Waksman, S. A., 1936. *Humus.* Baltimore: Williams and Wilkins.

————, and I. J. Hutchings, 1935. Chemical nature of organic matter in different soil types. *Soil Sci.* 40:347.

Waksman, S. A., and K. R. Stevens, 1929. Contribution to the chemical composition of peat. 5. The role of microorganisms in peat formation and decomposition. *Soil Sci.* 28:315.

——, and K. R. Stevens, 1930. A critical study of the methods for determining the nature and abundance of soil organic matter. *Soil Sci.* 30:97.

——, F. G. Tenny, and K. R. Stevens, 1928. The role of microorganisms in the transformations of organic matter in forest soils. *Ecology* 9:126.

Wander, I. W., 1949a. An interpretation of the cause of water-repellent sandy soils found in citrus groves in central Florida. *Science* 110:299.

——. 1949b. An interpretation of the cause of resistance to wetting in Florida soils. *Proc. Fla. Hort. Soc. Nov.*, 92.

Wang, T. S. C., Pau-Tsung Hwang, and Chung-Yi Chen, 1971. Soil lipids under various crops. *Soil Sci. Soc. Amer. Proc.* 35:584.

——, Yu-Cheng Liang, and Wei-Chiang Shen, 1969. Method of extraction and analysis of higher fatty acids and triglycerides in soils. *Soil Sci.* 107:181.

Warth, A. H., 1956. *The chemistry and technology of waxes.* New York: Reinhold.

Webley, D. M. 1954. The morphology of *Nocardia opaca* Waksman and Henrici (*Proactinomyces opacus* Jensen) when grown on hydrocarbons, vegetable oils, fatty acids and related substances. *J. Gen. Microbiol.* 12:420.

——, R. B. Duff, and V. C. Farmer, 1955. Beta-oxidation of fatty acids by *Nocardia opaca. J. Gen. Microbiol.* 13, 361.

Winter, A. G., 1961. New physiological and biological aspects in the interrelationships between higher plants. *Symposia Soc. Exptl. Biol.* 15:229.

Wollrab, V., and M. Streibl, 1969. Earth waxes, peat, montan wax and other organic brown coal constituents. In *Organic geochemistry*, Chap. 24, p. 576. New York: Springer-Verlag.

Zalozieki, R., and J. Hausman, 1907. Zer Kenntnis des Torfwachses. *Z. Angew. Chem.* 20:1141.

Chapter 7

Micromorphology of Soil Organic Matter*

U. Babel

DEDICATION

I dedicate this work in gratitude to the memory of Prof. Dr. h. c. Dr. W. Kubiena.

I wish to express my thanks to some colleagues for their critical reading of the manuscript. Among them I am particularly indebted to Dr. G. Zachariae, then Freiburg i. Br., who has given me many valuable hints, especially concerning the sections which touch on zoological problems. My thanks are also due to Dr. N. Walker, Rothamsted, for the translation of the manuscript into English.

Contents

I. Introduction*

A. Fundamentals of Humus Micromorphology

Natural humus formations look very different morphologically to the naked eye. These differences provide the basis for a division into groups, which are called humus forms. The morphology of humus forms is determined by environmental factors.

A great variety of morphological forms exist also on the microscopic level. Even humus

---

* The ground thin sections were produced more or less according to the method of ALTEMÜLLER [1956b] (Vestapol embedding) in the Bundesforschungsanstalt für Forst- und Holzwirtschaft, Reinbek bei Hamburg, Germany (Department of Soil Science, of which W. Kubiena was then the Director). The sliced thin sections, unless otherwise indicated, are paraffin sections covered with Caedax. All photographs were produced by the author. The term "natural size" indicates natural height of the illustration. For horizon symbols compare Section II C (BABEL [1971a]).

formations which appear homogeneous to the naked eye are composed of numerous microscopic constituents that clearly differ in size, color, and form (including internal structure). The constituents may be arranged in a great number of characteristic fabrics. These facts, first observed by MÜLLER [1887], make it possible for microscopic methods to be used in investigations of humus.

The morphological characteristics of the microscopic humus constituents are determined by the physical, chemical, and biotic factors of the microenvironment in which they have developed. The processes brought about by these factors take place on the surface and in the interior of the microscopic constituents. The morphological characteristics of the constituents are thereby altered. Thus the constituents reveal the processes of humus formation.

For example, small aggregates with regular egg-shaped contours indicate the presence and activity of soil animals; they are droppings (German "Losungen"). (This term, borrowed from hunting language, is used in micromorphology for well-preserved excrements of soil animals.)

From the form, size, and spatial arrangement of the droppings one can often draw conclusions about the animal group, and the composition of the droppings may reveal the material that the animals have consumed. By comparing the material of the droppings with the material not yet consumed by the animals, changes caused by the animals can be ascertained.

The organic constituents influence the physical, chemical, and biological interactions in soil. That is to say, they are not only effects but also causes of processes of soil dynamics. Thus, the presence of fungal mycelium suggests the action of fungal ectoenzymes (*e.g.*, Figure 28). A loose arrangement of the solid constituents, which results in a well-developed porous system, facilitates air access and promotes oxidation in terrestrial humus formations.

From the morphology of the constituents, general conclusions about their extramorphological properties can be drawn. Thus, differences in color point to chemical differences. In cork tissue residues, cells with brownish-red contents, as well as those with black contents, often occur. This is the result of chemical—especially oxidative—changes in the cork (Figure 16).

Therefore, in humus research, as in other fields of science, morphology does not only serve the description of an instantaneous state. In the sense used by GOETHE [1807], who in his scientific writings was one of the first to employ the term "morphology," its primary objective is the research on the development of the natural bodies and the principles effective therein. According to KUBIENA [1972], the final object of soil micromorphology is micromorphogenesis.

Since the organic neoformations of the soil develop largely from residues of plants, humus morphological investigations are bound to make use of *botanical morphology*. Analogous to microscopic step by step investigations of mineral weathering (*e.g.*, MEYER and KALK [1964]), the decomposition of the plant residues can be followed step by step from the plant tissues.

This procedure leads to the classification of the organic microconstituents of soils. A botanical identification of the tissue and cell parts can be made, from which the constituents are derived and the description of the type and extent of their morphological transformation is made. But correlations with the categories of humus chemistry, which are deduced from extraction and fractionation experiments, cannot be rigidly made. The results of these two branches of humus research often cannot be compared with each other without special, additional study, because of the diverse methods used in each. Such lignin is detected under the microscope usually by the phloroglucinol reaction. However, in humus chemistry it is generally separated as an acid-insoluble residue.

Because the transformations of plant tissue occupy such an important position in microscopic investigations, restricting the term "humus" to a limited sense would be inconvenient in humus micromorphology. The term here will include not only the more or less stable substances newly produced in the soil, but all the dead organic matter of the soil (SCHEFFER and ULRICH [1960]). Examples of "humus" are a leaf that has fallen to the ground or a newly cast-off piece of root bark. The microscope shows still another reason why the term "humus" must be used in a broad sense. In certain cases new formations in the soil are scarcely distinguishable morphologically from formations in or on the living plant; they can also be chemically very similar.

Thus, phlobaphenes, which are produced in the living plant, are similar morphologically and chemically to the leaf-browning substances, a great part of which are formed only on the ground and which are very stable humus substances.

It is important in the investigation of the breakdown of plant residues that the development of intermediate substances and end products is linked *topochemically* with the starting structures. This pertains particularly to formations of holorganic humus layers. There the bulk of the organic substances retains most of the basic morphological features (size, form, color, arrangement). With mull, on the other hand, solution processes predominate, and dissolved organic compounds from the plant structures migrate to the mineral fabric components. These organic substances are only recognizable from their color (as components of a color combination). They form pigments of the mineral substance.

B. Applicability of Humus Micromorphology

Micromorphological investigations of humus require solid starting substances and solid topochemically formed end products for the most effective results. Displacements in the solid form, perhaps by soil animals, may also be investigated satisfactorily. The possibilities of humus micromorphological studies are limited when displacement processes in solution or in gaseous form play a role. Little possibility is left for investigation of processes in the molecular region, when these are not manifested in the microscopic region. However, even in such cases indirect relationships may exist.

Thus, the condition and processes of ion sorption on the organic substances are not visible microscopically. There are, however, general empirical relationships between the micromorphologically characterized humus form and its sorption behavior. Correlations between sorption and microscopic features of decomposing plant residues are given by PAULI [1961] and PAULI and GROBLER [1961].

Consequently, humus micromorphological investigations can be applied mainly to organic layers. With mull, on the contrary, the possibilities of their application are more limited, as holorganic constitutents are of less importance in mull. However, fabric investigations of mull may lead to important results, particularly in the field of soil biology.

Humus micromorphological studies can be *preliminary* to other investigations. They can contribute to the knowledge of a material to be treated in some other way and provide clues as to the suitable preparation of this material. The following is an example: In studies on carbohydrates in the organic matter of soil, thin sections of fractions of different grain sizes showed that a considerable part of the extracted pentosans must have originated from fine roots. On the basis of these observations, washed fractions were then prepared and examined separately (ROULET *et al.* [1963]).

Micromorphological investigations are useful in *designing* further investigations in humus biology and humus chemistry.

For the micromorphologically described aggregate formation in mull, soil zoological field studies provided important information. The large rounded, loose-lying aggregates are by no means always derived from excrements (*inter alia*, of earthworms); to a large extent they are produced by the movements of a great variety of small animals, among them some which do not consume their food in the soil (ZACHARIAE [1962]).

Micromorphological work can be employed to *check* other investigations or be performed to *support* them. VON BUCH [1962] has examined the mechanism of extraction methods by using ground thin sections of fresh and extracted material. KULLMANN [1958] studied the changes of decaying plant roots micromorphologically along with analysis of substance groups.

Micromorphological investigations can be used *independent* of others for studies on *humus formation*. Chemical problems of the transformation of plant substances, for example, can be studied by microscopic methods, especially when histochemical and histophysical techniques are added (*e.g.*, BABEL [1965a]). Model and decay experiments often need not be employed for this, as the multiplicity of factors, which are active in natural humus formations, can be extensively analyzed with the help of the microscope and their effects determined. Therefore micromorphology can test results of chemical or microbiological studies, obtained from model experiments or extracts, with respect to their actual significance in natural humus formation and determine their localization.

Paleopedology and archeology are other fields in which humus morphological studies have been employed. SMOLIKOVA and LOZEK [1969] detected traces of earthworm activity in middle Pleistocene paleosols. ZACHARIAE [1967], by zooecological interpretation of micromorphological findings contributed to archeological discoveries.

Further possibilities of applying humus microscopy are in the field of *forest and agricultural ecology*. It was in studies for classification of humus forms as indicators of site conditions that microscopic methods have been applied for the first time in humus research (MÜLLER [1887]).

BARRATT [1967, 1968a, 1968b] gives micromorphological descriptions of humus forms in relation to agricultural management of grasslands as well as in relation to fertilizing. BABEL and STRAUB [1969] found correlations between stereomicroscopic features and forest management. BENECKE and BABEL [1969], by micromorphological investigations, found differences in the activity of soil organisms in relation to vegetation (spruce forest and deciduous forest). They were interpreted as the result of differences in water and nutrient cycle. Also decomposition processes in composts have been studied by micromorphological methods (ALTEMÜLLER and TIETJEN [1967]).

In close relation to this field is the application of humus microscopy in studies on the *ecology of soil organisms*. It is microscopic investigations of humus profiles that reveal the habitat of the decomposer organisms and the scene of the decomposing processes in ecosystems. Soil zoologists apply thin section techniques for studies on the positions and feeding behavior of soil animals (*e.g.*, ANDERSON and HEALEY [1970]). Soil microbiologists perform direct microscopic observations of soil particles for studies on localization of bacteria and fungi in their natural habitats (*e.g.*, GADGIL *et al.* [1967]).

The following account emphasizes the results of direct investigations of soil organic matter by light microscopy. In some cases results of electron microscopy are included. Other investigations, such as chemical, zoological, and microbiological, are only cited for discussion of morphological findings.

II. Survey of the Microscopic Structure of Humus Formations

Generally, ground thin sections give the best picture of the microscopic structure of a humus formation. In a typical case the following groups of constituents are found. At low or medium magnification fairly large particles are seen, which are often strikingly colored and easily recognizable as plant residues. Between them lies a material that is distinguished from the plant residues by its much more finely divided, irregular, and often denser structure; it usually has darker colors. This material is often formed from regular egg-shaped or cylindrical aggregates, sometimes with racemose knobbly contours. These are the droppings of soil animals. At a somewhat greater magnification thin brown fungal hyphae can be seen. In addition, mineral grains are found; they are usually pale or colorless. Often they are accompanied by a yellow-to-brown fine substance. Crossed polarizers show that it consists of mineral particles bound together by colored isotropic materials. Within and between these different groups of solid substances, air-filled pores of the most varied shapes and sizes are developed, (*e.g.*, Figures 1, 2). These parts are not distributed at random but exhibit a more or less obvious pattern.

Investigations with numerous illustrations of microscopic humus preparations are: Bal [1973] (69 microphotographs, 16 in color); Babel [1972b] (21 microphotographs of moder horizons); Barratt [1968/69] (12 semidiagrammatic drawings); Geyger [1967] (19 small microphotographs); Frercks and Puffe [1964] (36 small microphotographs of secondarily decomposed peats); Grosse-Brauckmann and Puffe [1964] (21 small microphotographs of peats); Hartmann [1952] (about 45 microphotographs); Hartmann [1965] (79 micro- and semimicrophotographs, 56 of which are colored); Kubiena [1944] (32 illustrations); Kubiena [1948] (about 25, some of them colored illustrations); Kubiena [1953] (16 semidiagrammatic drawings); Kubiena [1970] (about 25 colored microphotographs; Puffe [1965] (36 small microphotographs of peats); Puffe and Grosse-Brauckmann [1963] (50 small microphotographs of peats); Van der Drift [1964] (16 small microphotographs); and Zachariae [1965] (18 large microphotographs).

A. The Constituents

Constituents are defined here as individual constituents (structural units of the lowest level). The three groups already mentioned are described under the terms "plant residues," "organic fine substance," or, where no confusion can arise, simply as "fine substance" and "mineral constituents."

1. *Plant Residues*

The term "plant residue" means coherent tissue parts, which are recognizable, as such, microscopically. They show distinct cell structures or distinct contours of plant organs or both. The plant residues could be designated purely morphologically as organic skeletons. This is mechanically correct. However, from the viewpoint of soil dynamics, plant residues are not inert, but undergo much more vigorous and more rapid conversions than does, as a rule, a mineral soil skeleton.

The plant residues reveal quite clearly their origins from the different plant organs and tissues. Leaf and needle residues are naturally frequent. Residues from fine roots also often occupy a large share of the microscopic picture. All other plant residues that reach the soil can also be recognized in the first stages of disintegration from histology. Most frequently among these are pieces of wood and bark residues.

The structures of the tissues in plant residues are deformed to varying extents and in various ways. Yellow, brown, or red parts appear. These "rotting products" were often termed "humolignins" in the older literature (Maiwald [1939]). Their identification and interpretation can be given on an histological basis (see Part III).

2. *Organic Fine Substance*

The term "organic fine substance" is used here to mean organic material without definite

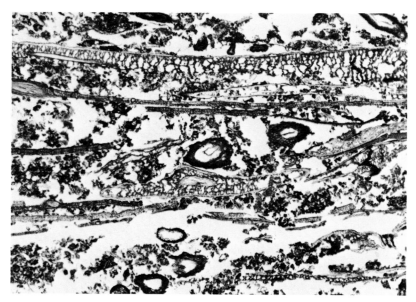

Figure 1. Decay layer (F_m) of a moder. Horizontally lying leaf residues (narrow) and needle residues (broad, thus top, one occurring longitudinally; somewhat below the center, an obliquely occurring needle). Root residues (*e.g.*, in the center, two examples transversely; mainly cortical parts are intact). A part of the fine substance shows indistinct (granular), more rarely, egg-shaped droppings. Colors of the fine substance, from bright reddish brown to black. In the fine substance fungal hyphae (indistinctly recognizable at this magnification). The residues of needles and leaves still form a laminated fabric in places (bridges of fine substance); in places greater areas of fine substance occur, forming loose to dense clusters. Wet moder under *Abies*, *Pinus* silv., *Fagus* forest (Black Forest, Germany). Ground thin section (25 μm), bright field, natural size 4 mm.

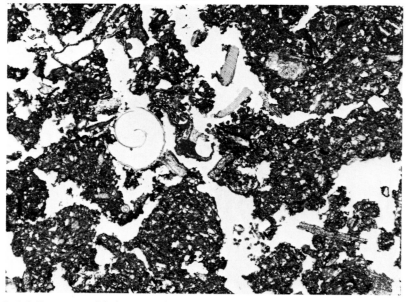

Figure 2. Mull, spongy fabric. Irregularly formed, combined aggregates (more or less altered droppings ("castings") of earthworms) with many plant residues, which are mixed into the mineral soil largely by earthworms (pieces of wood, leaf fragments (below, right-hand side), root residues, etc.). The plant residues have been disintegrated partly by mites (droppings, *e.g.*, in the center.) The mineral substance contains quartz and calcite grains (the quartz grains here appear like holes in the fine substance). The mineral fine substance is colored by organic matter. Rendzina-mull under beech, A_{hh} (4 cm) (Swiss Jura). Ground thin section (25 μm), bright field, natural size 5.5 mm.

microscopically recognizable structures. Cell wall fragments, single cells, fungal hyphae and very small pieces of tissue which comprise only a few cells are also included in the definition.

The organic fine substance is produced from plant residues by disintegration and shows little or none of the original structures, or it may be formed from precipitates of dissolved organic substances. The parts have irregular fine-grained and fine-foliated forms and usually an irregular flocculent arrangement. Their colors are sometimes pale, like those of the initial tissue, less frequently bright (yellowish red, brownish red), and mostly dull, dark, and often black. As to the sizes of the individual particles, the size of the particles of the fine substances is usually in sharp contrast with the plant residues. Nevertheless, a definitive threshold size cannot be given, since the size is dependent on the disintegrating agents. In general, the threshold size lies at about 100 μm.

3. *Mineral Constituents*

As a rule, mineral constituents are of essential importance to natural humus formations. This is true for the mineral skeleton, which influences especially the physical conditions of humus formation and on which chemical effects of humus can also become manifest. Mineral fine substance may react with dissolved organic substances to give a stable complex, the so-called clay-humus complex. In this the humus cannot be recognized in microscopic constituents of its own.

In addition to these solid constituents there are also pores filled with air or water.

B. The Fabric

The term "fabric," which was first introduced by KUBIENA [1935, 1938], is defined by ALTEMÜLLER [1962b] as the arrangement of the constituents with respect to form, function, and genesis.

The natural humus fabrics are built of varying proportions of the constituents which may be arranged in many different ways. Fabric analysis describes the mutual arrangement of the constituents (elementary fabric) as well as the size, shape, and arrangement of aggregates and pores. Plant residues often are interconnected with each other by fine substance or fungal hyphae. The fine substance is often aggregated in droppings. Droppings can be interconnected with each other. Coherent masses of fine substance without aggregation are found also. Mineral constituents can be arranged without connection with the organic fine substance (moder, Part V D 1); in other cases mineral and organic constituents are loosely aggregated with each other (mull-like moder, Part V D 4); finally clay can be pigmented by organic substance, without intrinsic organic particles being visible microscopically (mull, Part V D 5).

Natural humus horizons show a number of fabric types, which are found repeatedly. Horizons built up by plant residues prevailingly often show a laminated fabric. Horizons of organic fine substance sometimes reveal a dropping fabric. This is found in humus mineral soil horizons also. Under different conditions there, a spongy fabric is formed.

Fabrics of humus soil horizons are formed by action of biological and nonbiological forces. By fabric analysis, therefore, physical and biological soil conditions can be investigated.

C. The Profile

Most humus formations have a profile differentiated into horizons that can be detected macroscopically. This is particularly distinct with terrestrial superficial humus forms (raw humus and different forms of moder). The horizons of such a humus profile illustrate the successive transformation stages of the litter. Increasing depth in the profile corresponds to increasing age of the organic components. Since the breakdown of the plant residues can be detected microscopically, no model decay experiments need be constructed.

Naturally there are always minor discrepancies in this correlation between age and depth in the profile. They are usually caused by irregularities in the subsoil or by rather gross mechanical effects which are brought about, among other things, by large animals. Besides, there are some spatial fluctuating differences which are attributable to the uneven distribution of soil animals and microorganisms. With humus forms other than the terrestrial superficial humus forms, for example, mull and fen-mull, the correlation between age and depth is not so clear, because the processes are intrinsically more rapid and because of the mixing action of the soil animals.

Fabric analysis shows variations in proportions of constituents within the profile. In a beech raw humus, for example, the amount of recognizable leaf residues decreases and the amount of fine substance and fine roots increase with increasing depth. In the deeper parts, mineral grains begin to appear (BABEL [1965a]). Correspondingly, the fabric type changes altogether. The laminated fabric of the upper parts becomes transformed into a fine substance fabric and finally into an agglomeratic fabric.

The constituent analysis reveals other changes with increasing depth. In the plant residues, progressively larger amounts of brown substances are generally formed, the tissue structures are gradually lost, and some tissue parts disappear altogether. The fine substance also changes its color with depth (see Part III).

Macromorphologically these differences lead to the differentiation of the humus profile into three layers, as first described by HESSELMAN [1926]:

L layer (or L horizon, litter layer, förna): The more or less loose deposit of the plant cover litter, which, after its accumulation on the soil, is not yet or is only slightly changed morphologically.

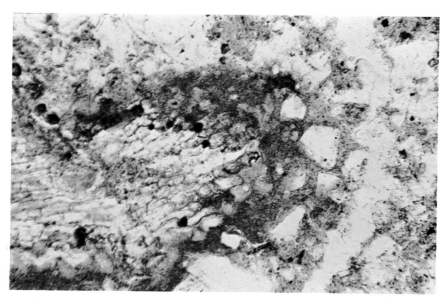

Figure 3. Iron oxide encrusted root residue in pseudo-gley, A_h. The iron oxide penetrates into the tissue. (The root is sectioned through the cortical tissue, oblique.) Colorless regions in the iron crusts are quartz grains. Ground thin section (30 μm), bright field, natural size 500 μ.

F layer (fermentation layer, decay layer): The layer in which more or less disintegrated plant residues as well as increasing amounts of organic fine substance occur.

H layer (humus substance layer): The layer consisting almost entirely of fine substance. Except root residues, plant residues are not recognizable macroscopically and microscopically or only in small amounts.

Below there is a mineral humus horizon (A_h) in which organic as well as mineral constituents, or even organic substance as a pigment of the mineral, are present.

According to a new proposal of BABEL [1971a, 1972b], these horizons can be subdivided further: L_n (n: new), L_v (v: German "verändert" analogous C_v), F_r (r: plant residue), F_m (m: medium ratio plant residue: fine substance), H_r (r: plant residue), H_f (f: fine substance), A_{hh} (hh: rich in humus), A_{hu} (part of A_h under A_{hh}). These horizons were defined by their plant residue/fine substance ratio (except for L_n with L_v and the A_h horizons) (see Part V D 1a).

The horizons of the humus profile, which are defined purely morphologically, have characteristic dynamics (*e.g.*, LEHMANN, MEYER, and SCHONLAU [1969]), detailed work on which is one of the chief objectives of humus micromorphology (*e.g.* BABEL [1972b]).

The sequence of L, F, H, and A_h horizon is not developed in all humus profiles. Any of these four horizons may be missing.

In a special case, namely with secondary humus formations from peats, a fermentation layer can follow below the humus substance layer (see Part VB2).

III. The Organic Constituents of Humus Soil Horizons

A. Animal Residues

While the activity of animals in the breakdown of plant residues has been extensively studied, little is known about the occurrence of residues of soil animals as constituents of natural humus formations or about the phenomena of their decomposition. Therefore, only few data can be presented here. On the whole, animal residues are seldom found in microscopic preparations of humus formations.

MÜLLER [1887] found pieces of chitin from soil animals in debris preparations of the humus substance layer of moder. Shells of thekamoebae can often be observed in ground thin sections of moder and raw humus formations. Mollusk shells may be found fairly often, especially in submerged and semiterrestrial humus formations (see Figure 2). According to KÜHNELT [1961], brown substances can be observed occasionally in their interior as decomposition products of the soft parts. Chitin residues of insects in the soil are often attacked by actinomycetes and by molds, according to KUBIENA [1938, p. 226 ff]. See also microbiological studies on this subject by OKAFOR [1966], where bacteria and protozoa have been found growing on insect wings in the soil. Chitin residues, being influenced by microorganisms, become colored dark brown to black in this process. The decomposition products of animal plankton also give very dark colorations. The details of the colonization of animal residues by microorganisms and the type of their decomposition, as in the case of plant residues, are dependent on the humus form (KUBIENA [1938]). In the hydrobiological literature, there are studies on the breakdown of animal chitin in submerged soils (RUTTNER [1962]). In thin sections sporadic particles occur that may probably be assumed to be animal residues, especially chitin fragments. They are usually colored yellow, with regular outlines, and often show primary fluorescence. (Colorless particles sometimes become recognizable only by their fluorescence, BABEL [1972a].) Many animal residues show (with reference to the longitudinal axis) partly positive and partly negative birefringence. Round particles, which might be the cases of eggs or cysts, may also be found. Cysts were identified by SEKERA [1956] in debris preparations. In such preparations, which are better suited to investigations of this kind than ground thin sections, other animal residues (*e.g.*, tarsal joints of arthropods) may also be recognized.

Animals that were still living when the sample was taken are found very seldom in normal ground thin sections of soils. BABEL [1972a], by fluorescence microscopy, found the body of an enchytraeid. Soil zoologists have developed agar or gelatine-embedding techniques for studying animals in their natural positions (*e.g.*, ANDERSON and HEALEY [1970]).

The number of animals and animal residues that may be found in ground thin sections of humus formations is much smaller than would be expected, considering the short life span of most animals and the large animal biomass, which is about 15 to 500 g/m², according to KÜHNELT [1961].

This cannot be explained simply by the fact that many of the large soil animals escape during sampling and the smallest forms can scarcely be recognized in thin sections. Arthropods, most of which should be found in thin sections, are present in amounts of the order of 10 g/m² (MACFADYEN [1963]); but they are also found more seldom in thin sections than would be expected from calculations.

This circumstance indicates the ease of decomposition of animal residues. Many animals shrink beyond recognition during drying which precedes the embedding of the sample in synthetic resin (MINDERMAN [1956]). In the soil, animal corpses are consumed often very soon by carrion feeders (KÜHNELT [1961]). The decomposition of animal chitin by the microflora of a humus substance layer has been demonstrated in laboratory experiments (GRAY and BELL [1963]). It may be assumed that breakdown products of animals and new formations from those in the organic fine substance or in the organic pigment of mineral soil are quantitatively significant as stable humus substances. However, this cannot be proved micromorphologically, at least in investigations of natural humus formations. In model experiments on the decay of combined plant and animal residues, SEKERA [1956] was able to show the importance of animal residues in the formation of stable humus substances.

B.   Plant Organ and Tissue Residues

Plant organ and tissue residues are here briefly termed "plant residues" (see Part II). Their descriptions are related to the *histology* and *organography* of the fresh parent material. Information is thereby obtained, as discussed in the introductory sections, not only on the morphological changes, but also on the material transformations that occur in the course of humus formation (see Section I A).

The general structure of the fresh tissues and organs is quite well known from botanical studies. The unique structure of the original materials in a given instance, however, often needs to be determined, because botanists' investigations are usually limited to either teaching examples or technologically important material.

METCALFE and CHALK [1950] and METCALFE [1960], who deal with the anatomy of the taxonomic groups of plants in concise accounts, are exceptions; KIRCHNER, LÖW, and SCHRÖTER [1908] also frequently provide detailed anatomical data. Another useful textbook is LINSBAUER [1922]. Introductions to the anatomy of plants are given by ESAU [1953] and HUBER [1961]. The plant cell wall is dealt with by FREY-WYSSLING [1959] and ROELOFSEN (in LINSBAUER [1959], who refer to its morphology, biochemistry, and biophysics. The microscopy of wood and bark may be found in FREUND [1951, Vol. 5]. SCHRÖDER [1952] describes the roots of about 100 bog and grassland plants. The morphology of fungi, the plectenchyma of which are often seen in soil thin sections, is presented by LOHWAG [1941] and BARNETT [1953].

The *taxonomic determination* of the plant species, from which an organ residue or tissue residue originates, is, in general, only of interest secondarily or indirectly—as, for instance, for comparisons with the parent material. If one investigates the vegetation of the site concerned and its undecomposed or slightly decomposed litter, it is usually very easy to determine the plant species related to the residues.

More difficult cases are the older humus formations, which have been produced from other than recent vegetation. This happens frequently in the deeper layers of peat. There the taxonomic determination of the plant residues is done by so-called *coarse residue analysis*. Here, as in pollen analysis, the scientist attempts to reconstruct the former vegetation. If the typical characteristics of the residues are not recognizable macroscopically or under the magnification of a hand lens, then coarse residue analysis has to depend, among other things, on the microscopic structure of the wood, the shape of the epidermal cells or of the cuticula ("cuticular analysis"), the stomata, hairs, or on other characteristic microscopic features (see the botanical literature quoted above). Decomposition phenomena are never considered more than superficially in such investigations, which are orientated toward vegetation history.

For identification after mechanical isolation and purification, the coarse residues are clarified chemically with sodium hypochlorite, potassium hydroxide, or oxalic acid (HOFMANN [1934], WILLERDING [1960]). Methods for the light microscopic determination of fossil charcoals are found in MÜLLER-STOLL [1936]. FIRBAS and BROIHAN [1937] have determined the spruce fraction in the formation of the different horizons of a raw humus cover by enumeration of the stomata from spruce needles. GROSSE-BRAUCKMANN [1964] has described and determined coarse residues from peats. Literature on coarse residue analysis with reference to vegetation history is given by FIRBAS [1949].

In humus research the successive stages of decomposition of the plant residues are of interest. The plant residues concerned can be isolated under the stereomicroscope from the successive horizons of the humus profile. For further investigation they are sectioned by microtome (BABEL [1964a]).

1. *The Parent Material*

In forests, *leaves* and *needles* are the most important parent material for humus formation. As a rule they amount to more than 80% of the litter. The remainder comprises pieces of *twigs*, *bud scales*, *fruits*, and *stems* from ground vegetation. (A detailed analysis of the litter of an oak forest is given by BURGES [1967]). Brushwood and timber lying on the soil in undisturbed natural forests equal the amount of litter. In forests under management, little brushwood and timber accumulate. Beside the litter, roots constitute an important part of the parent material. Root production in forests normally may be about a fifth of the total litter production (SCHLENKER [1962]) for southwestern Germany. But this amount varies greatly with environmental conditions. Sometimes roots predominate as parent material (Figure 12); WILDE [1958] then speaks of "rhizogenic humus."

In the upper H horizon of a beech raw humus in northwestern Germany, dead roots constituted 6.3% and living roots 9.6% of the total mass (VON BUCH, personal communication; see also VON BUCH [1962]). In this profile the maximum density of living root tips (which was in the F horizon) was 90 times as high as the maximum density in a mull profile (MEYER [1967]).

In grassland, DAHLMAN [1968] found the root production to be about as high as the shoot production. The same author determined the average life duration of roots in a grassland ecosystem to be about four years; similar results are found by ORLOW [1968] for the life duration of roots in a pine forest.

The organs mentioned consist essentially of three groups of tissues, namely, parenchyma, phlobaphene-containing tissues, and lignified tissues. The decomposition of the tissues shows

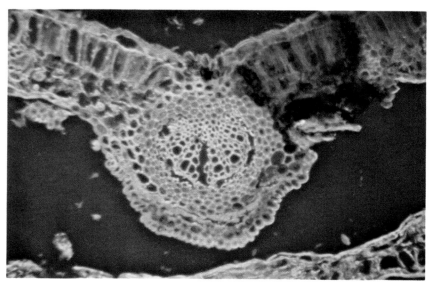

Figure 4. Beech leaf vein (cross section), isolated from L_n horizon, fluorescence illumination. Blue ring: sclerenchyma (primary fluorescence of lignin). In the xylem are distinguishable undecomposed, whitish blue fluorescing cell wall parts from decomposed, yellow appearing ones. Strong fluorescence of the lower cuticula. Top left, fluorescence of parenchyma cell walls readily visible; top right, it is masked by browning substances and fungal pigments. Raw humus (Northwest Germany). Sliced section (glycerine, 10 μm), unstained, fluorescence illumination, natural size 300 μm.

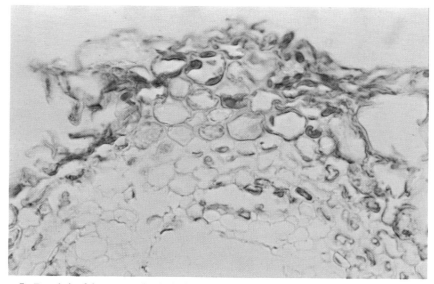

Figure 5. Beech leaf (cross section), isolated from F_m horizon of a raw humus. Upper half of the leaf vein. In the middle of the picture from the top downward, the following tissues: Epidermis (one cell layer, badly focused), collenchyma (about four layers, remaining cell walls yellow; brown bodies and yellow granular substances are remains of the secondary wall), sclerenchyma (about three cell layers, the colorless middle layers of the cell walls are more or less intact, the yellow substances on them as well as the brown rings are remains of the secondary wall), phloem (the small hole with yellowish brown substances on the edges, somewhat under the center of the picture), xylem (1 or 2 layers, remaining only the colorless middle layers), on the under edge of the picture the disappearance of the medulla-like central part of the xylem has left a great hole. (Northwest Germany) .Sliced section (10 μm), bright field, natural size 100 μm.

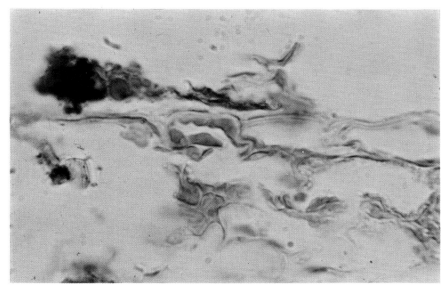

Figure 6. Dark bodies of probably microbial origin on the residues of the epidermis of a beech leaf. Epidermis outer wall colorless, epidermal cells from the inside encrusted with yellowish brown leaf-browning substances. Isolated from raw humus, F_m (Northwest Germany). Sliced section (10 μm), bright field, natural size 45 μm.

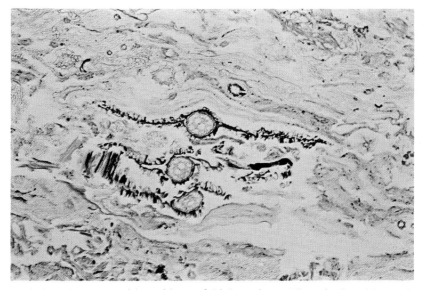

Figure 7. Sphagnum peat with residues of higher plants. Three-leaf residues of *Eriophorum vaginatum* with vascular bundles which are hardly altered, the parenchyma has changed to dark brown substances. Between the almost colorless sphagnum residues are yellow substances, part of them formed out of the solution state (compare Figure 29). Sliced section (Cremolan method, 10 μm), bright field, natural size 1.1 mm. (From a preparation by D. Puffe, Bremen.)

common features within each of these groups (BABEL [1965b]). (Brief details on the chemistry of the plant substances cited in the following are to be found in Section III B 2 bγ, cγ and dδ).

Parenchyma forms the green tissues of leaves, the cortex of young twigs and fine roots, and the cortex and medullary parts of the stem. They are built up from living cells; these have thin cellulose walls, a protoplast consisting mainly of protein, and a vacuole.

Phlobaphene-containing tissues give external protection to many organs. They include older or outer portions of the cortex, and most of the varied tissues of the bark of roots, twigs, and stems, especially cork tissues (periderms). The cork tissues are easily recognized by the regular parallel row arrangement of the rectangular cells. The tissues of many bud scales and fruit cases and seed cases also contain phlobaphenes. These tissues sometimes have thin cellulose walls, but usually they have thickened walls that, in the cork, are formed as a rule from middle lamella, suberin lamella, and carbohydrate lamella. Middle lamella and carbohydrate lamella can store lignin, but never suberin (MADER [1958b]). The phlobaphene-containing tissues contain dead cells while the plant is still living. The cell lumina contain yellow brown, or red tannin oxidation and polymerization products which have been formed following the death of the cell. These are termed *phlobaphenes*. In the so-called tannin cork the whole lumen of the cells is filled with phlobaphenes. In other outer tissues smaller amounts of phlobaphenes occur. The cell walls, especially the suberinized ones, are likewise occasionally colored by phlobaphenes (Figure 15).

Lignified tissues not only form the main mass of the woody parts of the wood growths but occur also as consolidating elements (sclerenchyma) in stalks, leaf epidermis, leaf hypodermis, leaf veins, and bark. The xylem of the vascular bundles is also lignified. The lignified cells are dead even in the living plant. Their lumina are nearly always empty. The cell walls are about 1 to 4 μ thick—much thicker therefore than those of the parenchyma. Under the light microscope at least two wall layers, the middle layer and the secondary wall, can be distinguished (Figure 4).

The middle layer may be differentiated electron microscopically into the middle lamella and the primary wall. The secondary wall, which forms by far the thickest wall layer, often shows a laminated structure. On it another thin wall, the tertiary wall, is deposited. The middle lamella consists of pectin. The primary wall consists of cellulose fibrils in a loose scattered texture, and the secondary wall consists of cellulose fibrils in a dense parallel texture. In all these layers lignin is deposited as an incrustation, and distinctly more so in the middle lamella and primary wall than in the secondary wall (RUCH and HENGARTNER [1960]). The lignin content is highest in the intercellular spaces. The tertiary wall apparently consists mainly of pectin, carbohydrate fibrils, and denatured residues of the cell contents (FREY-WYSSLING [1959]). There are chemical differences between the lignin of different plants and different kinds of tissues. They may be detected, for example, by the different methoxyl contents, and they often show different behaviors on degradation (SCHEFFER and ULRICH [1960], THOMPSON *et al.* [1964]).

In addition to these three groups, there are some other tissues of lesser importance quantitatively as parent materials for humus formation. The *epidermis* in leaves and stalks warrants consideration. It is similar to the parenchyma; however, the cells have a thickened outer wall which, starting from the cell lumen, consists in general of a rather thick cellulose layer, a pectin layer, and a cuticular layer. Across these three layers the cuticula can be separated as an outermost thin layer. The two last-named layers, which may also together be termed cuticula in the broader sense, are usually only $\frac{1}{2}$ to 1 μ thick combined. At the needles of coniferous trees, however, their thickness amounts to several microns. The cuticular layer consists of cutin, pectin, and cellulose; the cuticula proper consists of cutin only. Furthermore, waxes can occur in the cuticular layer, and also on the surface of the cuticula (FREY-WYSSLING [1959]).

Collenchyma are thick-walled tissues of living cells. They occur in leaf stems and leaf veins as well as in the stalks of herbaceous plants. The cell walls are distinguished by a high pectin content and by the poor order of the cellulose fibrils, as may be gathered from their shrinkage in dehydrating agents and from their weak birefringence. Under certain conditions it can be established that the secondary walls of the collenchyma consist of alternating lamellae of pectin and cellulose (FREY-WYSSLING [1959]).

The *phloem* of the vascular bundle and of bast contains, in addition to parenchymatous elements, the sieve tubes, which are thin-walled cells rich in hemicellulose (callose). Finally, *pseudotissues* (plectenchyma of fungi) should be included here. They are frequently found in soil thin sections—especially of dystrophic humus forms (see Part V)—in the form of dark-colored sclerotia and mycorrhiza mantles, and more rarely in the form of fruiting bodies. The framework substance of cell walls is chitin. The brown pigments are often collectively called melanins. They can constitute up to 54% of the mycelial weight (SCHEFFER and ULRICH [1960]) (Figures 20, 21).

Typical *chemical analyses* of humus parent materials might give, for straw, 5% crude protein, 40% cellulose, and 20 to 25% lignin. Forest litter might have about 5 to 15% crude protein, 15% cellulose, and 20% hemicellulose (SCHEFFER and ULRICH [1960]). The lignin content of the autumn foliage of different forest trees, according to STÖCKLI [1952], amounts to between 13% (*Castenea*, or chestnuts) and 27% (*Abies*, or firs), when determined with thioglycollic acid.

2. *The Decomposition of Tissues*

(a) General Phenomena

Changes in the decomposing tissues to the point of loss of coherence of the tissues will be discussed in this section. Even while the coherence of the tissue is maintained, manifold decomposition phenomena are shown in their interior as discolorations and deformations of the cell components. Later on in the train of breakdown processes, cracks, holes, and other flaws are formed, and finally the coherence of the tissues is completely lost. These symptoms arise in part through nonbiological processes, such as swelling, shrinkage, and the collapse of thin-walled tissues (for example, of the phloem). For the most part, however, they are formed by the activity of plant microorganisms and especially of animals (for further details, see Part III C 1).

The different tissue groups are destroyed at varying rates by these agents. The following order of increasing resistance, although somewhat schematic, is frequently valid: phloem, collenchyma, parenchyma, lignified tissue, epidermis, phlobaphene-containing tissue (Figure 12). This series is brought about essentially by different chemical and biotic activities in the tissues (different nutrient contents, special resistance to degradation of certain plant substances) and by mechanical differences (hardness and small water uptake by tissues of thick-walled cells). The difference in rapidity of destruction of parenchyma and lignified tissue is often striking, even macroscopically.

This is seen, for example, in the skeletonization of the leaves of deciduous trees. The lignified tissues of the leaf veins remain behind as the leaf skeleton after decomposition of the mesophyll (WITTICH [1939], FÜHRER [1961], ZACHARIAE [1965]). Similar phenomena are described by BARRATT [1965] in the decomposition of grasses and herbs. KULLMANN [1958], KONONOVA [1961], and JABLONSKI [1963] have made corresponding observations in decay experiments with fodder plant residues, and GRAFF [1955] in the case of destruction of straw by arthropods. The same conditions also apply to plant residues from peats; but in this case it should be noted that their decomposition has occurred not in the process of peat formation, but previously, under aerobic conditions (GROSSE-BRAUCKMANN [1963]).

The sequence outlined holds when conditions are favorable for breakdown, that means, in mull. Under other conditions deviations can occur. The observation by HANDLEY [1954] that, in raw humus, the mesophyll of *Calluna* leaves may remain intact longer than the veins is very important to the problem of raw humus formation.

The morphological decomposition phenomena which go on within the still coherent tissues seem, on the contrary, to proceed very similarly in different humus forms. Any variations are chiefly in rate of decomposition. (These ideas mainly have been obtained from investigations of the author on oligotrophic and dystrophic humus forms).

The breakdown products of tissues are solid, liquid (solutions), or gaseous. Only those decomposition products or residues that are solid can be examined microscopically. However, they may often be contaminated by organic substances, or rarely by mineral substances, from the solution phase, which have diffused into them or have been deposited on them as thin crusts. This possibility has to be considered with all humus forms, especially in the strongly decomposed horizons. As a rule, however, the contaminating substances cannot be detected morphologically with certainty. Only occasionally are rather thick organic crusts from the solution phase formed around plant residues, which are easily recognizable as such (Section III C 2 e). Frequently, of course, the dissolved substances are lost from the horizons investigated.

Experiments by NYKVIST [1963] show that the proportion of water-soluble substances in the litter may be assumed to be quite high. From needles, 1% of the dry weight may be leached out; from leaves, 4% (beech) to 17% (ash). Furthermore, there are the well-known brown seepage waters which often occur on the profile walls of podsols. ZIECHMANN and PRZEMECK [1964] have estimated in such a solution a concentration of 0.3 g organic matter per liter of solution. The uniform brownish-black coloration of a fossil wet moder is probably due to such solutions (Figure 14). KÜHNELT [1961] mentions a yellow coloring of wood as a result of infiltration of solutions from the dark droppings of diptera larvae.

The occurrence of *gaseous decomposition products*, arising during mineralization, can be inferred micromorphologically only indirectly, if at all, and in its morphological symptoms—namely, the disappearance of cell or tissue parts—cannot be distinguished from the dispersal of dissolved substances. However, sometimes solidified bubble-shaped, gas-filled pores that have been formed by putrefaction gases can be found in sections of sapropels.

MEYER [1960], from incubation experiments, has calculated that 60 to 80% of the organic matter is mineralized down to the lower limit of the fermentation layer of a mull. HOLLING and OVERBECK [1961] have computed a substance loss (in solution or gaseous form) of 33 to 80% for sphagnum peats, appearing macro- and microscopically only slightly or moderately decomposed.

(b) Parenchymatous tissues

(α) *Morphology.* Parenchymatous tissues often disappear rapidly and without residues (Figure 33). This was observed particularly in decay experiments, in which optimal conditions for decomposition are established—as, for example, by KONONOVA [1943] with the cortex and medulla of lucerne roots. In grass roots, which died off, the cortex sloughed off and disappeared, leaving only the central stele (GADGIL [1965]). GROSSE-BRAUCKMANN and PUFFE [1964] obtained corresponding results with the decomposition of parenchyma in peat. GROSSE-BRAUCKMANN [1963] in a similar study found differences in the rapidity of decomposition of parenchyma of various groups of mosses. Hepaticae decompose most quickly, followed by musci and finally sphagnum. The central parts of the moss stem decompose more quickly than the outer, cortex-like portions. This might be attributable to chemical differences in the composition of the cell wall. In mosses from peats small amounts of newly formed brown substances were found occasionally (GROSSE-BRAUCKMANN [1963]). KONONOVA [1961], in decay experiments with clover leaves, showed that brown water-soluble substances had been formed in the interior of parenchyma cells by autolysis of myxobacteria, which previously had shown an abundant development in the parenchyma, along with the consumption of the cell content substances. TRIBE [1963], in decay experiments with lettuce leaves, found dark spherical bodies which had apparently been produced from the contents of the mesophyll cells.

In the mesophyll of leaves and needles of forest trees, on the contrary, water-insoluble encrustations of the cell walls are formed (GROSSKOPF [1935], HANDLEY [1954], BABEL [1965a, b, 1972b]). Objects of intensive study were leaves of *Fagus* and *Calluna*, as well as needles of *Picea* and *Abies*. The crusts are brown, partly yellowish and partly reddish (Figures 10, 6). When present in large amounts, these substances can form spherical bodies (Figure 9). They appear immediately after the death of the cells, hence frequently even before leaf fall. During the first months on the soil, they develop further. They are here termed *leaf-browning substances*. The formation of browning substances may also be observed in other parenchyma as well as in plants other than those named. It seems to be the first step in the normal case of parenchyma decomposition.

Leaf-browning substances such as lignin can be stained with gallocyanine (GROSSKOPF [1935]). They are soluble in sodium hypochlorite (eau de javelle). After their removal, the remains of the cell walls are revealed, which, in early stages, scarcely show any decomposition symptoms and in which cellulose can still be detected by birefringence and reaction with chlorozinc iodide. This detection is possible to a limited degree or not at all unless the leaf-browning substances are removed.

Even before the loss of tissue coherence, a color change of the browning substances to dark brown may take place (Figure 8). This phenomenon was associated with a chemically detectable cellulose degradation in experiments by FALCK [1930]. It would be analogous, therefore, to the brown rot of wood. On the contrary, HANDLEY [1954] was able to show that cellulose is protected mechanically and possibly chemically from degradation in the soil by "masking" with browning substances. In raw humus, residues of cellulose walls encrusted with leaf-browning substances persist in the humus substance layer (HANDLEY [1954], BABEL [1965a]).

A part of the leaf-browning substances, however, disappears in the lower part of the L horizon (L_v, see Section II C). This is shown macroscopically in a striking clarification of the leaves. (Compare BABEL [1972b] for illustrations). The fully preserved leaves change to a clear yellow color, sometimes on the entire blade, and sometimes only in restricted areas. Microscopically somewhat the same picture is found as after treatment with sodium hypochlorite. All the colored substances in the mesophyll disappear and the slightly altered cell wall residues become clearly visible (Figure 8).

In other cases, however, the cell walls seem to disappear simultaneously with the decomposition of the leaf-browning substances, so that the entire leaf or the whole needle collapses, leaving only the epidermis, hypodermis, and vascular bundles. (Figure 33). Macroscopically such leaves appear to have become thinner and the needles appear shrunken (MEYER [1962], BABEL [1972b]. In the breakdown of the cell walls of the parenchyma, net-like and perforated decomposition phenomena have been observed in a few cases (ALTEMÜLLER and BANSE [1964] with straw, GROSSKOPF [1935] with pine needles). Whether the cell walls themselves are colored is difficult to establish clearly in the presence of the encrusting browning substances; after sodium hypochlorite treatment, they are colorless. The light refraction and birefringence of the morphologically fully intact cell wall regions apparently remain unchanged. In addition,

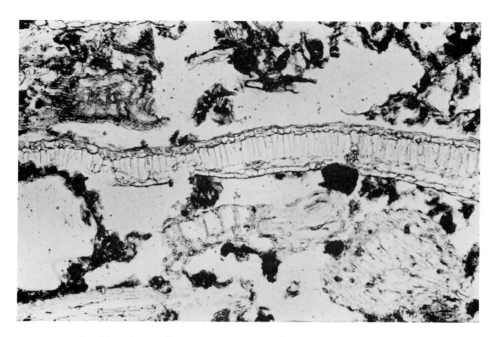

Figure 8. Leaf residues (occurring transversely), partly with unchanged leaf-browning substances (right part of the center leaf, piece of leaf below center), partly with darkened browning substances (top left), partly after decomposition of the browning substances and then still showing well the colorless unchanged cell walls (center). The very dark to opaque granular bodies on the leaves probably are collembola droppings (center below). Raw humus, F_m (Northwest Germany). Ground thin section (20 μm), bright field, natural size 380 μm.

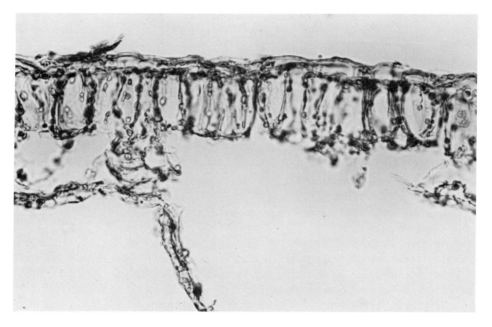

Figure 9. Beech leaf (cross section), L_v horizon. The nearly colorless upper cuticula is covered with dark substances; it separates from the epidermal outer wall (left). Epidermis and mesophyll are colored yellowish brown by leaf browning substances, which form in part very thin crusts in part conspicuous spherical bodies. Sponge parenchyma missing in some places. Lower epidermis broken up. Raw humus (Northwest Germany). Sliced section (20 μm), bright field, natural size 180 μm.

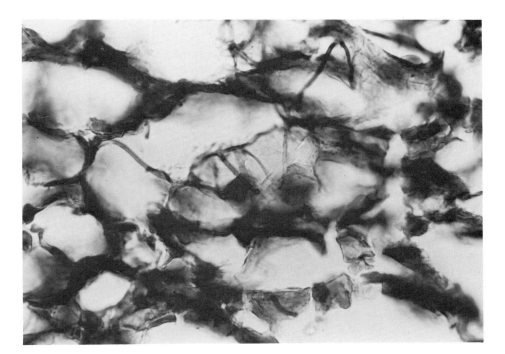

Figure 10. Leaf-browning substances in the assimilation parenchyma of an *Abies* needle. The cell walls are still well preserved (only partly torn), but on both sides irregularly encrusted by the browning substances. Shrinkage cracks in the browning substances. Some fungal hyphae. Moder, F_m (Black Forest, Germany). Sliced section (40μm), bright field, (strong green filter), natural size 180 μm.

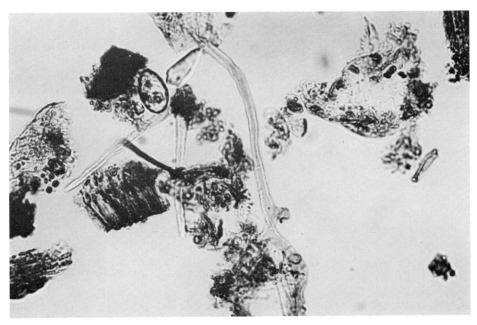

Figure 11. Debris preparation with fine substance particles of raw humus, F_m. Identifiable components, especially a thecamoeba shell (colorless, small, strongly contrasting oval, top left); double hair of a beech leaf (nearly colorless, immediately below); fragment of palisade parenchyma (strongly colored, below it); branched, septate fungal hypha (light brown, center); epidermis fragment (above right). Bright field, natural size 300 μm.

however, feebly colored parts were found, which appeared to have been formed from cell walls. Light refraction of these parts was greater than that of the unchanged cell walls (ALTE-MÜLLER and BANSE [1964]). In partially decomposed Ericaceae roots and Sphagnum, a weaker cellulose reaction with iodine-potassium iodide and sulfuric acid was found (FRERCKS and PUFFE [1964]).

(β) *Microbiology* In ground thin sections of decomposing parenchyma, *fungal hyphae* can quite often be found (Figure 10). But even where this is not the case, the participation of fungal degradative processes cannot be excluded. Colorless or very thin hyphae are often very difficult to recognize in ground thin sections among strongly colored irregularly shaped degradation products (Section III C 2 c). In addition, the hyphae can exercise activity over distances of several cell diameters by the excretion of ectoenzymes. It is true that some authors disagree with this (compare Section III B 2 d γ). Finally the hyphae may be already decomposed at the time of investigation.

As an example of morphological observations on bacteria that have participated in parenchyma decomposition, the works of KONONOVA [1961] have already been cited. In the decomposition of root cortex, mycorrhizal fungi which have become parasitic can, under certain conditions, take part (SCHAEDE [1962], MEYER [1964]. First, they degrade the pectin of the middle lamella, leading to a maceration of the tissue. KAUNAT and BERNHARD [1969] found pectinolytic species of Bacillus and Actinomycetes in the rhizosphere. Pectin decomposers are also regularly found on fallen leaves (*e.g., Pullularia pullulans* [WIERINGA, 1955]).

Fungi can also be observed as producers of brown pigments in the mesophyll of leaves. This applies especially to the congestive zones of mycelial growth. These can often be found between bleached and unbleached regions of the leaf blade and may be recognized even macroscopically as brownish-black demarcations. They are the result of extracellular fungal pigments (Figure 48).

The microbial degradation of the leaf-browning substances is the central point of interest in the microbiological phenomena in mesophyll decomposition. HANDLEY [1954], in numerous experiments with semisynthetic products and different fungi, established that the leaf-browning substances are, as a rule, difficult to degrade. But lignin-decomposing fungi seem to be capable of promoting this decomposition. This does not mean, however, that in bleached leaves the lignified tissues would likewise be already decomposed. It is true, however, that they are more decomposed than in brown leaves (BABEL [1972b]). FALCK [1930] observed a clearing of beech leaves by the white rot fungus *Agaricus nebularis*, and KÜHNELT [1963] by *Clitocybe infundi-buliformis*. In natural mull and moder formations, MEYER [1962] found macroscopically visible white aerial mycelium of *Mycena sanguinolenta* between bleached beech leaves and *Picea* needles. For some other species, see LEHMANN, MEYER, and SCHONLAU [1969]. Some of these fungi, as well as other lignin-decomposers (LINDEBERG [1955]), were found by MÜLLERSTÄEL [1962] only on mull and closely related moder formations. This seems to be in agreement with an hypothesis of FALCK [1930] that the lack of lignin decomposers is essential for raw humus formation. However, MIKOLA [1956] found bleached leaves as well as lignin decomposers in raw humus. Such contradictory observations are not surprising, since a whole series of factors are responsible for the much-discussed differences between mull formation and raw humus formation (see also the following section).

(γ) *Chemistry* According to morphological results on the decomposition of cell walls, cellulose does not seem to be concerned directly in the neoformation of brown substances. It is probably mainly mineralized, but partly converted into microbial substance.

Of chief interest here is the mode of formation and the chemical nature of the leaf-browning substances. According to HANDLEY [1954], it involves the formation of tannin-protein complexes. The tannins in living cells are localized in the vacuole, the proteins in the protoplast. On the death of the cell, the plasma membranes are broken down, and proteolytic enzymes and polyphenol oxidases are liberated from the protoplast. In addition, the redox potential of the cell increases. Polyphenolic tannins are thereby oxidized to quinones, which polymerize and, at the same time, take up amino acids released from the proteins. During these processes a color change—from no color to yellow and finally to brown—takes place. Chlorophyll may also be incorporated in the polymerisates, resulting in even a greater deepening of color (BÄBLER [1957]). Furthermore, an incorporation of monosaccharides into these mixed polymerisates is not improbable; for, according to NYKVIST [1963], monosaccharide-amino acid reactions in aqueous leaf extracts must be considered. Thus, purely autolytic postmortal processes lead to the formation of leaf-browning substances. They occur before decomposition of the cellulose walls begins. That means that extensively transformed as well as completely unchanged constituents exist in partially decayed plant residues. This is also indicated by absorption studies with fluorochromes (PAULI [1958, 1960]).

Differences in the quantity and stability of the tannin-protein complex in litter of different sites were found by HANDLEY [1954, 1961], COULSON et al. [1960a, b], and DAVIES et al. [1964a, b], and were thought to be responsible for the differences between mull and raw humus (see also DAVIES [1971]). Thus, the tannin content of beech leaves from a raw humus site was about 50% higher than of beech leaves from a mull site. In addition, typical raw humus plants, such as *Calluna*, are characterized by particularly high tannin contents (about twice as high as in beech). Finally, in weakly to strongly acid media, aqueous leaf extracts showed maximal precipitation action in gelatin solutions as well as maximal resistance of the precipitate to degradation by fungi.

In the degradation of leaf-browning substances by fungi, water-soluble tannins appear, and in addition, a fall in pH of about 2 units may be observed (KÜHNELT [1963], MIKOLA [1956]).

Apparently, leaf-browning substances are frequently falsely determined to be lignin by chemical analysis in the determination of lignin as an acid-insoluble residue. This explains the high lignin contents often given for litter (*e.g.*, birch litter 39% in comparison with pine wood 29%, (SCHEFFER and ULRICH [1960]). WITTICH [1943] found that a considerable proportion of the lignin fraction of needles in the litter in contrast to that in living needles was insoluble in acetyl bromide. This result might also be interpreted to be due to the formation of leaf-browning substances after the death of the needles and their erroneous inclusion in the lignin fraction. The same holds for the increase in nitrogen content of the lignin fraction, which is regularly found in the course of decomposition. GROSSKOPF [1935], even on the basis of morphological analyses, erroneously interpreted the leaf-browning substances as lignin degradation products, which were thought to have been produced in the decomposition of mesophyll cell walls. This interpretation was part of the basis for further discussions in the older humus literature (MAIWALD [1939]).

(c) Phlobaphene-Containing Tissues

(α) *Morphology* Many of the strongly colored "rotting products" often attract attention in sections of dystrophic or water-influenced humus forms (*e.g.*, of pseudogleys and gleys). They may be found on more precise morphological analysis to be residues of phlobaphene-containing tissues. The conspicuousness and wide distribution of such residues have even led to the separation and separate description of this group of tissues.

Their resistance to decomposition can be observed particularly well in fairly old cortex and cork tissue of roots. These tissues, as a rule, are preserved much longer than the central cylinder and inner parts of the cortex, which are formed from phlobaphene-free parenchyma (VAN DER DRIFT [1964]). The residues of cortex certainly show thereby a progressive loss of

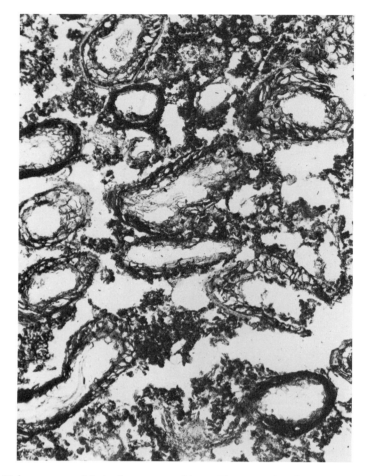

Figure 12. H layer, very rich in fine root residues. The central cylinders of the transversely or obliquely occurring root residues are partially or wholly decomposed, the cortex usually entirely preserved (partly collapsed) and colored reddish brown by phlobaphenes. Wet moder over peat, H horizon (Black Forest, Germany). Ground thin section (20 μm), bright field, natural size 1.5 mm.

the cell structures, but the color of the tissue remains unchanged for a long time (Figure 12). The residues of the cork tissue, which are cast off in the course of the thickening of the root are which originate from twigs, the wooden parts of which is decomposed, often accumulate in considerable quantities in the humus substance layer (Figure 43) (BABEL [1972b]). Particularly striking are the brownish-red cork residues of roots of the dwarf pine (*Pinus mugo*) in the tangel (Figure 13). On such residues a maceration often takes place which leads to the liberation of the individual cells, which then enter into the fine substance as intensely colored small bodies (Figure 15) (see also HOLLING [1958] and GROSSE-BRAUCKMANN [1963] (observations on residues of Ericaceae roots in peats). In peats and sometimes to some extent in other humus forms, even before the maceration, the phlobaphenes found in the cell lumen show an intense coloration, whereby the whole tissues may be finally transformed into coal-like masses (Figure 16) (HOLLING [1958], BABEL [1964b]). Fungal hyphae always seem to be present in this process. As yet, a complete disappearance of phlobaphenes, analogous to the disappearance of leaf-browning substances in the bleaching of leaves, does not seem to have been found anywhere.

JABLONSKI [1963], in decay experiments, noticed the appearance of fluorescent products in the vicinity of decomposing root cork. These products were found only at a certain intermediary stage of the decomposition.

Resistance to decomposition, similar to that of phlobaphene-containing tissues, is shown according to morphological observations, also by other outer tissues that are free of phlobaphenes but show suberinized cell walls. This pertains to some species of cork tissue as well as to exo- and endodermis of many roots. Where suberinization of the cell walls and phlobaphene content of the tissues coincide, as in the case of a normal tannin cork, the tissues that are most resistant to decomposition are formed. (For suberin, see also Section III B 2 e). Barks contain mostly phlobaphene, but they also contain some suberin tissue. ALLISON and KLEIN [1961] and ALLISON and MURPHY [1963] have shown that they decompose more slowly than wood.

(β) *Microbiology* It is well known that polyphenols, from which the phlobaphenes are formed, often exert an antibiotic effect (*e.g.*, WENZEL, KREUTZER, and ALCUBILLA [1970], who

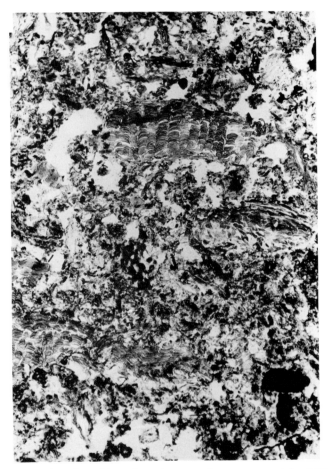

Figure 13. Brownish-red cork residues (mostly occurring transversely) of *Pinus mugo* in the H_r horizon of an oligotrophic tangel. Three larger cork residues are easily recognizable, some smaller ones are placed obliquely (*e.g.*, the black appearing in the center) Fairly dense fine substance: the black-appearing bodies, 5 to 10 μm in size, are single, reddish-brown cork cells (Alps). Ground thin section (15 μm), bright field, natural size 450 μm.

Figure 14. Uniform brownish black staining both of the plant residues as well as of the fine substance, probably by dissolved humus substances (thin parts appearing bright, thick parts opaque). Many leaf residues (transverse, center and top). Fossil (postglacial) wet moder of a heath (North Germany). Ground thin section (15 μm), bright field, natural size 450 μm.

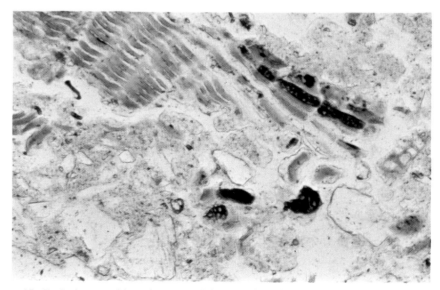

Figure 15. Cork tissue residues (transverse) in the mineral soil, cast off from root bark. The parts that are still coherent show the parallel row arrangement of cells characteristic of cork. Cell lumina reddish brown from phlobaphenes, cell walls colorless. Small groups and individual cells liberated by maceration. In places, cell contents transformed into dark fungal material. B/(B) horizon of a podzolbrown earth (Black Forest, Germany). Ground thin section (20 μm), bright field, natural size 350 μm.

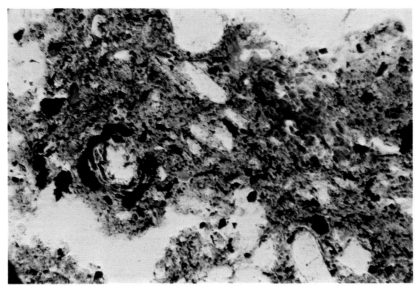

Figure 16. Two root cortices: right, oblique, without distinct outlines, brownish red; left, transverse, dark brown to opaque because of color change of the phlobaphenes (probably because of oxidative processes), central cylinder decomposed. Fine substance with high mineral content, some quartz. Wet moder with transition to peat formation. A_h (Black Forest, Germany). Ground thin section (25 μm), bright field, natural size, 0.5 mm.

investigated polyphenols from fir bast). The phlobaphenes themselves are difficult to degrade by microorganisms. Thus, phlobaphenes and the related monomers may protect from decomposition tissues which are impregnated by them. A maceration, however, may take place by pectinolytic microorganisms, as occur in the rhizosphere (*e.g.*, KAUNAT and BERNHARD [1969]).

Two cases of the blackening of phlobaphenes can be distinguished at moderate to fairly high magnification. The cell lumina can be densely filled with blackish-brown fungal mycelium (Figure 15). That means that the phlobaphenes are transformed into structural substances of fungi, mainly melanins and chitin. This is a parallel to the dense colonization of parenchyma cells by myxobacteria (KONONOVA [1961]), mentioned in the previous section. Under different conditions the phlobaphene bodies may darken with no other recognizable morphological changes occurring, but fungal hyphae are then often found on the surface of the tissue. In any case, probably only fungal groups such as basidiomycetes, which can also degrade lignin (HOLLING [1958], BABEL [1964a]), are capable of attacking phlobaphenes. According to BURGEFF [1961], catechin tannins of Ericaceae roots are degraded by the mycorrhiza-forming fungi.

(*γ*) *Chemistry* The formation of phlobaphenes shows a certain parallelism with the formation of the leaf-browning substances. In the degenerating metabolism, oxidases and peroxidases cause the oxidation and polymerization of monomeric tannins still in the living plant. Probably an incorporation of proteins is also involved therein (STAMM [1958]). Like the leaf-browning substances, phlobaphenes appear in the acid-insoluble residue designated as the lignin fraction (cf. Section III B 2 b *γ*) of substance groups analysis. The occasional blackening of phlobaphene-containing residues in the soil might indicate a higher oxidation.

From the morphological study of the decomposition of phlobaphene-containing tissues it becomes apparent that the phlobaphenes, and the whole tissues as well, are highly resistant to degradation. Frequently observed maceration indicates only a degradation of the pectin.

(d) Lignified Tissues

(α) *Morphology* As has already been mentioned, lignified tissues are decomposed more slowly than parenchyma under generally favorable decomposition conditions. In raw humus, on the contrary, they are often decomposed more quickly than parenchyma. For example, HANDLEY [1954] has found that in raw humus the parenchyma of the medullary rays is often better preserved than the lignified parts of the wood. BABEL [1972b] found rather a fast-disappearing of lignified tissues from twigs and tree roots in moder.

In ground thin sections of soil, *charcoals* are frequently found. Where they occur in fairly large pieces, they show very well-maintained structures. When very thin, they are brown and translucent, but otherwise opaque. Very small pieces (from 5 to 30 μ), which are often encountered as opaque constituents in superficial humus forms, can be recognized as charcoal with some certainty on the basis of their sharp-edged, straight-faced outlines and their very slow decolorization with sodium hypochlorite (Figure 17). By treatment with sodium hypochlorite,

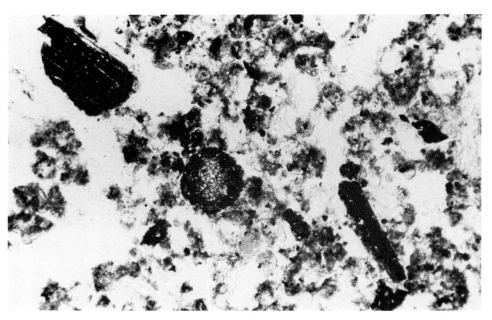

Figure 17. Two charcoal fragments, recognizable by their sharp-edged, in part parallel contours. A sclerotium (in the center). The very small, opaque particles probably are charcoals. Wet moder, A_h (central uplands of Germany). Ground thin section (15 μm), bright field, transmitted and incident light, natural size 450 μm.

which can also be carried out on ground thin sections, wood structures that previously were invisible because of the opacity of the particles become recognizable (BABEL [1964b, c]). Somewhat larger pieces of fossil charcoals may be determined taxonomically in the incident light microscope (MÜLLER-STOLL [1936], HOLDHEIDE [1941]). Details of structures, such as pits, remain well preserved. Only a slight shrinkage is liable to occur during carbonization. It is mainly in water-influenced humus forms that other plant residues also occur, even residues of leaves, which are dark brown to black and more or less opaque but, at the same time, show tissue structures that are scarcely changed (see Part V B).

The decomposition of wood by fungi has been very well investigated on account of its technological significance (*e.g.*, BAVENDAMM, MÜLLER-STOLL, and VON PECHMANN in Freund, volume V/2 [1951], RYPACEK [1966]). Electron microscope studies are reported by MEIER [1955], RYPACEK [1960, 1966], LIESE and SCHMID [1962], LIESE [1964], and SCHMID and LIESE [1964].

The two main types of fungal decomposition are called *white and brown rot*, corresponding to the discoloration of the wood. The tissue remains intact for a very long time, even after severe rotting. The individual cell wall layers behave quite differently, however. The secondary wall is in most cases attacked first. The most resistant to decomposition are the primary wall, the tertiary wall, and the substance filling the spaces between the cells consisting of almost pure lignin.

Fungal hyphae in both types of rot are generally found only in the lumen of the cells. Often they are almost colorless and, therefore, in the unstained preparation escape cursory observation. Under the electron microscope, microhyphae can be detected which branch out from the hyphae in the cell lumen and penetrate the walls. They are only 0.1 to 0.4 μm thick (LIESE [1964]). The hyphae lying on the walls are often surrounded by a narrow zone (up to about 3 μm wide) of lysis. Furthermore, degradation patterns are found in the secondary wall, the orientation and rhombic shape of which correspond roughly to the fine structure of this wall layer. These changes can also be recognized under the light microscope as indentations and flaws.

In brown rot, decomposition is often unevenly advanced in patches. Places with more severe decomposition can be easily recognized under the light microscope with crossed polarizers: the birefringence of the cell walls has largely disappeared there, before severe morphological changes are detectable by transmitted light (Figures 18, 19) (KUBIENA [1943], VON PECHMANN [1951]). In the electron microscope, however, in this stage it may be seen that the cell walls show sponge-like porosity in the portions no longer birefringent. In advanced stages they can become thinner by collapsing.

These microscopic investigations correspond to the chemical results, according to which brown rot is mainly characterized by cellulose degradation. A porous lignin skeleton remains after loss of birefringence due to cellulose. New formations in the place of the cellulose are not found, if one disregards the occasional occurrence of residues of excretions or of humification products of fungi.

In the case of white rot the electron microscope reveals a loose structure of cellulose fibrils in the secondary wall. Later the secondary wall disappears completely; SCHMID and LIESE [1964] were able to establish therein a layer-by-layer degradation. In contrast to brown rot, the middle lamella is often more readily attacked than the secondary wall.

These observations show—again in harmony with the chemical results—that in the first stages white rot breaks down mainly lignin and in later stages also cellulose. Brown colorations, which were found in intermediary stages (MEIER [1955]), may probably be interpreted as degradation products of lignin. In later stages, however, as indicated simply by the white color even macroscopically, there are no colored degradation or transformation products.

The decomposition of lignified tissues in the soil was studied by BABEL [1964a], with the light microscope, in the sclerenchyma and xylem of beech leaf veins. The material had been isolated from a natural raw humus formation. As a rule, the secondary wall is attacked first, as in the decomposition processes just described. It becomes yellow. Then it is detached from the middle layer, becoming further discolored. Finally, when seen in cross-section, it forms a

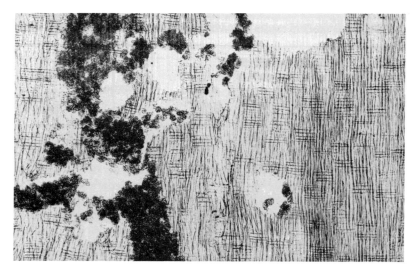

Figure 18. Brown rot *Pinus* wood, longitudinal section (cf. Figure 19). Tissues still intact (but stained yellow), apart from large feeding cavities, which are partly filled with droppings and decomposed dropping material. From moder (North Germany). Ground thin section (70 μm), bright field, natural size 1.5 mm.

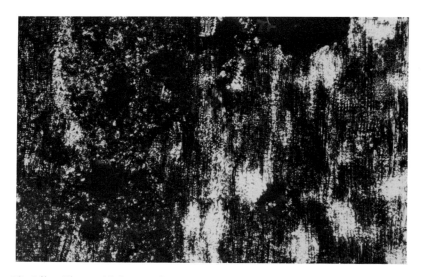

Figure 19. Like Figure 18 but under crossed polarizers. The morphologically well-preserved tissue is strongly birefringent only in patches; in other patches the cellulose is largely degraded.

brown lump, which lies in the lumen of the still completely intact middle layer. In a final stage which, however, is not often found because the veins are usually consumed by animals before then, this lump has disappeared completely (Figure 5). In rarer cases, conversely, a discoloration and subsequent disappearance of the middle layers occurs before changes in the secondary wall are observed. In these investigations fungal hyphae were found within the tissues only occasionally.

As far as is known, the decomposition of lignified tissues of leaves of other tree species or in other humus forms proceeds similarly (BABEL [1972b]). In mull, however, the formation of brown substances from the secondary wall is seldom found. In mull as well as in moder, the slightly lignified hypodermis of *Abies* needles formed only a few yellow decomposition products, which then disappeared without residues.

Woods that have decomposed under water show very similar morphology (MÜLLER-STOLL [1951]). The secondary wall in the cross section is transformed into rings and later to lumps. Ultimately, a homogeneous mass is formed from the whole cell material.

(β) *Histochemical and histophysical investigations* on the decomposition of lignified tissues in natural humus formations were made by BABEL [1964a]. The most important of the results are listed in Table 1, arranged according to the morphological decomposition stages of the secondary wall that have been outlined in the previous section.

For cellulose, birefringence was used as a qualitative criterion. For lignin detection, the blue primary fluorescence was used, as well as the absorption maximum at 280 nm and the phloroglucinol reaction. Sodium hypochlorite was used for the characterization of the newly formed substances.

Cellulose degradation is shown in Table 1 to run parallel to the morphological changes. The decrease in birefrigence indicates a destruction of the fine structure of the cellulose fibrils. From stage 3 onwards the morphologically unchanged middle layers can be stained with Ruthenium red. Since carboxyl groups are detected by Ruthenium red, this may be interpreted as the loosening of the fine structure or even as the incipient oxidation of the cellulose of the middle layer.

Lignin degradation begins at the side chains of the molecule, as shown by the early loss of the positive phloroglucinol reaction, which is specific for the side chains of the lignin units. Staining of the yellow and brown secondary wall residues with Schiff's reagent indicates the formation of aldehyde groups in the lignin degradation. Both observations are consistent with the results of macrochemical investigations (BRAUNS and BRAUNS [1960]). The absorption

Table 1. Morphological, histochemical, and histophysical investigations of beech leaf veins from a raw humus—degradation of the secondary walls in the sclerenchyma. (From BABEL [1964a]).

	1	2	3	4	(5)
Age	Fresh After $\frac{1}{2}$ year	After 1 year and later . . .			
Color	Colorless	Yellow	Brown-yellow	Yellow-brown	Sec. wall disappeared
Double refraction	+	+	(+)	(−?)	
Fluorescence	+	+	(+)	(−?)	
280 nm max.	+	(+)	(+)	—	
Phloroglucinol	+	?	—	—	
NaOCl	Not dissolved	Decolorized; partly dissolved		Decolorized; mainly dissolved	

maximum at 280 nm disappears after the loss of the positive phloroglucinol reaction. The specific lignin fluorescence disappears at last, and at about the same time as the birefringence. The fluorescent color seems to have been previously transformed to yellow. This effect, however, is due to the real color of the more strongly decomposed secondary wall (Figure 4). Accordingly, in the case investigated, simultaneous degradation of lignin and cellulose is involved (MEIER [1955]).

Decrease in the blue fluorescence of lignified tissues in the course of decomposition was observed also in straw by ALTEMÜLLER and BANSE [1964] and in lucerne roots by JABLONSKI [1963]. (For further details see Part VI B 3.)

Decolorization and solubility by sodium hypochlorite are properties which the secondary wall residues have in common with other brown humic substance materials, namely, the leaf-browning substances and the phlobaphenes already discussed. The transformation of lignin and the formation of humic substances can be followed still more closely by recording *ultraviolet absorption spectra of the microscopic preparation*. A publication on these investigations by the author is in preparation. The absorption spectrum of lignin is obtained in a pure form from fresh secondary walls. The above-described morphological transformation stages give progressively continuous transitions from the absorption spectrum of lignin to the nonspecific spectra of humic acids, which descend rather uniformly toward the longer wavelengths. Extractions of the secondary wall residues, which likewise were performed on the microscopic preparations, and subsequent repetition of the spectral determinations afforded no essential changes in the absorption. From this, together with the light microscopic investigations, it may be concluded that the conversion of the lignified cell walls to humic substances takes place without any deep-seated intermediate depolymerization of the lignin. It seems rather that oxidations, cleavage of monomers, and polymerization processes, in the sense of a formation of humic substances, take place on the lignin skeleton which is still little changed in its basic structure.

(γ) *Microbiology* The importance of fungi to the decomposition of lignified tissues emerges from the morphological observations on brown and white rotted woods. According to various culture experiments, however, there are also bacteria, especially those of the *Pseudomonas* group, that are able to degrade lignin (PREVOT *et al.* [1954], SORENSON [1962]). Up to now, however, it is not sure they can bring about a complete degradation of the lignin skeleton or peripheral changes only (HAIDER and SCHETTERS [1967]). Reliable indications on this subject have not yet been obtained from morphological investigations. The scarcity of fungal hyphae in the degradation of lignified tissue in the soil, however, could point in this direction (see also Section III B 2 b β). Dense colonization of slightly decomposed or undecomposed lignified tissues or of pure lignin preparations by myxobacteria was occasionally found by KONONOVA [1961] and JABLONSKI [1963].

The lytic zones around fungal hyphae indicate a certain remote effect of the fungal ectoenzymes. Another fact already mentioned is of interest here. The middle lamella or the secondary wall is decomposed before the tertiary wall, although the decomposition in question is caused by fungi that grow in the lumen of the wood cells. This leads to the conclusion that the enzymes diffuse rather extensively. The quoted observations on leaf veins and the degradation of leaf-browning substances in the mesophyll, in which hyphae are seldom found, suggest perhaps that the ectoenzymes diffuse through several cells. In this connection one should bear in mind the possibility that fungal hyphae, in later decomposition stages of the tissues, can be decomposed without leaving any trace, as may be often observed in the degradation of wood (BAVENDAMM [1951]). Lignin-degrading fungi are characterized by the formation of phenol oxidases (LINDEBERG [1955], HAIDER, LIM, and FLAIG [1964]). The most efficient lignin decomposers are Basidiomycetes, but in soil, lignin degrading is done also by Ascomycetes (HAIDER and DOMSCH [1969]). For lignin degradation see the section on leaf-browning substances (Section III B 2 b β), which show similarities to lignin not only in their analytical behavior but also in the microbiology of their decomposition.

(δ) *Chemistry* As suggested by the morphological findings, the resistance of lignin to degradation is not so high as is often assumed from chemical analysis (see Part III C 3). When specific reactions for quantitative determination of lignin are employed, however, they give results that correspond to the morphological findings. KICKUTH *et al.* [1969] found only traces of lignin in the A_h horizons of mull and mull-moder intergrades, but cellulose in amounts of 7 to 11 % of organic matter.

From the morphology of the decomposition of lignified tissues, conclusions may be drawn about the role of cellulose and lignin as parent materials for the production of humic substances. The topochemical neoformation of humic substances at the site of degradation of cellulose and lignin can be observed in many instances, as pointed out above. Participation of cellulose degradation products in the formation of brown substances is not unequivocally clear, however, from these investigations. The major part of the cellulose is evidently mineralized (compare Section III B 2 e). The lignin degradation, similarly, often leads to a direct mineralization by white rot (SCHEFFER and ULRICH [1960]), but often also to the formation of brown substances, which in soil may be transformed further to stable humic substances, especially by reaction with ammonia or amino compounds. This is also supported by chemical degradation experiments on humic substances. Vanillin and closely related compounds can be produced from lignin by chemical degradation. According to JAKAB *et al.* [1962], they can be "cleaved from those parts of the humic substance molecules that still bear characteristics of lignin, in which case one considers at least a partial formation of humic substances through peripheral changes of lignins" (see also GRABBE and HAIDER [1971].) This is the same idea of the humification products of lignin as was deduced above from histophysical investigations. Moreover, chemical investigations show that in the microbial degradation of natural lignin fission products the same intermediates are formed (*e.g.*, hydroxy-*p*-benzoquinone) as occur in the preparation of model humic acids from hydroquinone (FLAIG, SALFELD, and HAIDER [1963]). According to GADD [1957], vanillin is oxidized in the presence of fungal ectoenzymes (laccase) to catechol and *o*-benzoquinone, and these compounds can then polymerize to brown substances. When incorporation of nitrogen is not possible, however, the polymerizates can be degraded anew by the fungi as far as mineralization (HAIDER [1965]). This case might be realized in the quoted intermediary occurrence of brown substances in white rots and also in the case of the disappearance of brown substances in beech leaf veins.

The formation of *lignin-protein complexes* (WAKSMAN [1936]) in the conversion of lignin to humic substances is not very likely, according to morphological investigations. This formation could begin only after the mechanical disintegration of the lignified tissues, followed by intimate mixing with residues of proteinaceous tissues. The lignified tissues are largely free of protein and also do not contain any significant amounts of microorganisms which, on autolysis, might be possible sources of protein. The lignin-protein complex might be largely identical with the tannin-protein complex discussed in Section b.

(e) Other Tissues

The cell contents and the side and inner walls of the *epidermis* behave on decomposition like parenchyma. However, the thickened outer wall and the cuticula deposited on it deserve special consideration. Both often remain intact for a long time. As the thin section observations show, they are not readily consumed by the smaller soil animals and they are attacked by microbes only with difficulty (Figures 33, 22, 6) (PUFFE and GROSSE-BRAUCKMANN [1963] and (GROSSE-BRAUCKMANN and PUFFE [1964]). With more strongly decomposed leaf residues the cuticula (here understood in the broader sense—cuticula and cuticular layer) becomes wavy and cracked and, in places, separates from the epidermal outer wall. The histochemical and histophysical features investigated do not alter during these changes (observations on beech leaves by the author) (Figure 9).

The high resistance to decomposition of the epidermal outer wall with its deposits is due to the cutin. The resistance to decomposition of the cuticula on dead plant residues is consistent with its function in the living plant, to which it offers protection against pathogens (MADER [1958a]). In soils, however, there exist quite a lot of microorganisms which produce cutinase, as has been detected by HEINEN and DE VRIES [1966].

The exodermis of roots, which remains intact often longer than the primary cortex (see also KONONOVA [1961]), behaves similarly to the epidermis. Its resistance is caused by the suberin content of the walls.

Suberin and cutin, occurring on the cell wall as incrustations, are very similar to each other. They are essentially highly polymerized derivatives of fatty acids, but details of their composition are not yet precisely known. Sporopollenin, which forms the exin of pollen, is related to them. The degree of polymerization increases from suberin to cutin to sporopollenin. The resistance to decomposition appears to increase in the same order. Sporopollenin is the most resistant substance in the plant kingdom (see Section III C 2 c).

Collenchyma often is decomposed very quickly. Sometimes, however, masses of yellow-colored substances and, simultaneously, warty granular deformations of the inner wall parts may be observed (Figure 5), from a leaf vein. As a rule, the whole tissues disappear later

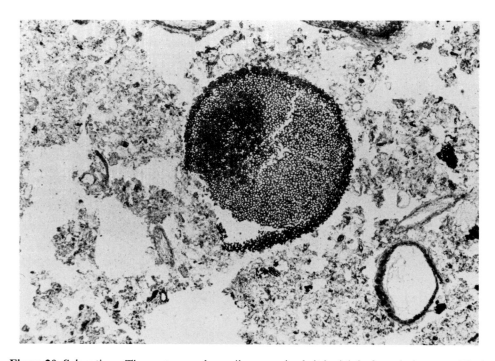

Figure 20. Sclerotium. The cortex can be easily recognized; it is rich in fungal pigments. The lower part is broken. In addition numerous small opaque particles, some of them are sclerotium fragments (*e.g.*, quite at the right). Below at the right a root residue, cork tissue still preserved. Raw humus, H (central uplands of Germany). Ground thin section (5 μm), bright field, strong blue-green filter, natural size 1.5 mm.

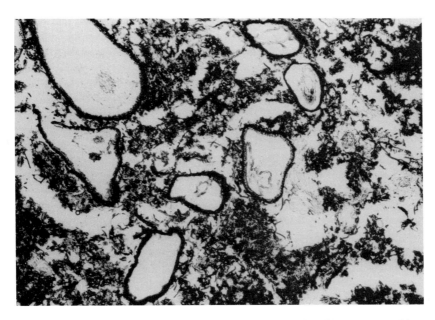

Figure 21. H horizon, very rich in fungal hyphae and hyphal residues. Mycorrhiza mantles as irregular dark rings; from these emerge hyphae, which together with hyphal residues constitute more than half of the fine substance. Only a few root residues are retained in the mycorrhiza mantles. Wet moder (Northwest Germany). Ground thin section (12 μm), bright field, natural size 1.5 mm.

without a trace (BABEL [1964a]). The relative ease of decomposition is induced by the looser structure and the high pectin content of the cell walls. If polyphenols play a role in the production of the yellow substances is not known. Otherwise, the discoloration should be taken as an indication of the production of colored substances from aliphatic compounds.

The *phloem* of the vascular bundles and the sieve strands of the bast are certainly the most easily decomposed tissues. In the upper litter layer, holes that are formed by the collapse of the *phloem* appear in the vascular bundles of leaves. The *phloem* residues form dense yellow masses with scarcely any structure. They disappear completely in the lower litter layer (BABEL [1964a], KONONOVA [1961]). This rapid decomposition is consistent with the thin-walled character of the sieve tubes and the phloem parenchyma cells and with the large proportion of pectin and hemicelluloses, especially callose, in the structure of the cells walls. Whether the disappearance of the phloem residues is due to removal in solution or to complete mineralization is not known.

The *fungal pseudotissues* of the sclerotias and mycorrhiza mantles are decomposed only with great difficulty (Figure 20). This is shown particularly clearly with mycorrhiza mantles, which often remain behind as nearly empty shells, while the roots surrounded by them are already decomposed (Figure 21). This resistance to decomposition might be largely due to fungal pigments. Their humic substance-like properties have been known for a long time. On the contrary, chitin, which is not impregnated with melanin, can be decomposed relatively easily (Section III C 3).

C. Organic Fine Substance

The small fragments which arise from the disintegration of plant residues, are called organic fine substance. In addition, under the morphologically collective group of the organic fine substance are included other small constituents such as fungal hyphae and spores, pollen grains and newly formed bodies from the solution phase. This discussion refers particularly to the organic fine substance of moder and raw humus. In these cases the organic fine substance is quantitatively prominent and it is horizon-forming in the humus substance layer (Figures 25, 23, 22). Very similar is the organic fine substance of the upper part of the A_h horizon of moder and mull-like moder (Figure 26). (For information on organic fine substance in mineral soil, see Section D 2 b.)

1. *Mechanical Disintegration of the Plant Residues*

In the disintegration of plant residues, soil zoological as well as microbiological and non-biological processes are active. The kind of disintegration that predominates differs markedly in the different horizons and different humus forms.

The feeble disintegration phenomena in the L layer are attributable mainly to non-biological processes. Because of the marked change between moistening and drying out, swelling and shrinking stresses occur. Also the brittleness of many dry plant residues facilitates mechanical disintegration. On the contrary, disintegration by the feeding activity of animals in the L layer is small. However, in the F layer this becomes dominant. The activity of fungi also contributes to the disintegration of the plant residues in the F layer. It can outweigh the disintegration by animals in raw humus. In the H layer a further reduction in size of the particles is not morphologically immediately apparent, although the mineralization and solution processes that take place there do in fact work in this direction.

(a)   Disintegration by the Activity of Animals

Most soil animals consume preferentially the softer and more hydrated plant parts. They are preferred because they are more palatable and because they contain microorganisms. The preliminary microbial decomposition of the plant residues and the microorganisms themselves provide an important source of food for soil animals (*e.g.*, ZACHARIAE [1965]). FÜHRER [1961] was able to show that dead Artemisia roots are preferred by an oribatid species to other plant residues because these roots were colonized by certain bacteria. For the nutrition of soil animals, see the summarizing accounts on soil zoology (*e.g.*, KÜHNELT [1961] and DUNGER [1964]).

By the selective uptake of food, there is formed, *e.g.*, the typical skeletonization of leaves, in which only the network of leaf veins remains intact (see also Section III B 2 a). That is chiefly the work of diptera larvae, diplopods, and larger oribatids, and less frequently of small arthropods. Earthworms also have similar feeding habits (EDWARDS and HEATH [1963]). Pieces of the leaf veins and leaf stalks remain behind.

The disintegration activity of soil animals becomes obvious also by analysis of their gut contents or excrements. The latter usually are called "Losungen" (droppings) in the German

Table 2.   Bite sizes of soil animals.   (After DUNGER [1963].)

Animal group	Bite sizes of the species examined
Diplopods	350–1000 μm
Isopods	250– 400 μm
Diptera larvae	400– 600 μm
Collembola	40– 170 μm
Lumbricidae	800–8500 μm

written micromorphological literature, fecal pellets in the English written literature. They are an important feature of the fabric of most humus soil horizons. Their size, shape, and arrangement, therefore, are dealt with in Part IV A 3.

The size of the tissue and cell fragments, which may be recognized in droppings, corresponds to the *bite size* of the soil animals. DUNGER [1963], by measurements of the fore gut content, has determined the bite sizes for the most important groups of litter-decomposing soil animals (see Table 2). The animals investigated had been recently caught in their natural habitat.

In the same work the hind gut contents were analyzed. In all the animal groups mentioned, several kinds of leaf residues were found, namely, parts of the vascular tissue and parts of the mesophyll and of epidermis, in addition to pieces of fungal hyphae and wood (the latter not in diptera larvae). Similar results to Dunger's were obtained by SCHUSTER [1956] (see also KÜHNELT [1961]) in very exact microscopic analyses of mite droppings. In these, animal residues (*e.g.*, chitin hairs), algal threads, and moss residues were also found. MACMILLAN and HEALEY [1971] have analyzed the gut contents of collembola.

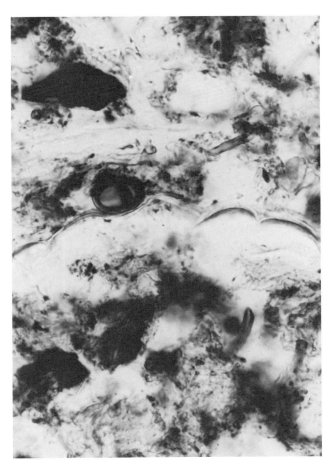

Figure 22. F$_r$ horizon of a raw humus. Outer wall of a beech leaf epidermis well-preserved (horizontal band in the middle). Fine substance strongly differentiated by color. A fungal spore (left center) and some pieces of colored hyphae. (Northwest Germany). Ground thin section (20 μm), bright field, natural size 110 μm.

Morphological changes that shed light on essential chemical changes, such as loss of structure, color changes, or a distinct decrease in certain components of the food, cannot be found in fresh droppings (Figures 34, 48). This appears to hold for earthworm droppings also. There are two ecological groups of animals: Primary decomposers, which consume plant residues; secondary decomposers, which consume droppings of other animals. Both groups bring about mainly a disintegration of the organic matter. ZACHARIAE [1965] assumes that many litter-feeding arthropods utilize only the slime of microorganisms and the degradation products produced from them. Droppings, which consist of more strongly decomposed material, are always derived from more strongly predecomposed food, as indicated in the above-mentioned investigations of DUNGER [1963].

Comparative measurements on extracts of soil animals food with extracts of their droppings showed only non-uniform feeble color changes (DUNGER [1963]). A small rise in nitrogen content in the droppings and a slight loss of substance, which were largely attributable to respiration of carbohydrates (BIWER [1961]), were not expressed morphologically.

Old droppings, on the contrary, in comparison with fresh ones show a deepening of color which corresponds macroscopically to the color intensification from the F to the H layer. It indicates chemical changes in the dropping material *after* deposition. In investigations of west European podsols a darkening of droppings has been found, especially under dry or very wet conditions (RIGHI [1969]).

(b)   Disintegration by Other Processes

After the decomposing action of fungi certain decomposition-resistant tissue parts often still remain. These disintegrate either as a result of mechanical stresses, such as swelling or shrinking, which always occur in the soil within small areas, or the parts remaining after fungal decomposition may be so finely divided that they are morphologically considered to be fine substance. An example of the latter case is the liberation of cells from the tissue association that is brought about by dissolution of the middle lamella (Figure 15). Detailed descriptions of disintegration by the decomposing activity of fungi are lacking in the literature.

The *mechanical disintegration procceses* due to alternate wetting and drying can be nicely observed especially on leaves. In the leaves of the L layer stresses occur between the lower epidermis and the vein network, which lead to a breaking up of the epidermis above the intercostal areas and finally to the liberation of small pieces of epidermis (Figure 9) (ZACHARIAE [1963]). Similar conditions cause the breaking up and breaking away of the lower parts of the leaf veins; these consist largely of collenchyma which, on wetting, swell more than the lignified parts of the leaf veins. Such processes, which are due generally to mechanical inhomogeneities and to swelling differences in the various tissues, are enhanced by partial microbial or animal decomposition.

The breaking up and piece-by-piece casting off of the root bark, which is a consequence of secondary thickening of the roots, leads similarly to the formation of small fragments of plant parts (Figure 15).

2.   *The Constituents of the Fine Substance*

For detailed morphological analyses of the fine substance it is most convenient to work not with sections but with debris preparations, which are prepared from suspensions (see Part VI A 3 c). When necessary, the inner structures of strongly colored pieces can be made visible by decolorization with sodium hypochlorite. Application of fluorescence microscopy sometimes helps in the microscopic analysis of organic fine substance (see Part VI B 3a).

(a)   Morphologically Identifiable Tissue Residues

Often among the conspicuous elements of the fine substance are reddish-brown homogeneous particles, spherical or longitudinal, about 20 to 40 μm in size. These are slightly deformed cells, which are released from phlobaphene-containing bark tissues and cork tissues by maceration. Still coherent groups of a few such cells also occur (Figures 15, 25, 13).

In debris preparations, moreover, very tiny pieces of the different parts of leaves can be recognized (see Table 3 and Figure 11). Easily recognizable in the example quoted were leaf

Table 3. Pieces of leaf in the fine substance of a beech raw humus.

Horizon*	Hairs	Pieces of			
		Cuticula	Epidermis	Mesophyll	Leaf veins
L_n	1–2	2–3	1	0–1	1
L_v	0 (−1)	2–3	2–3	1–2	1–2
F_r	0	x	3	5	1–2
F_m	0	x	1–2	5	4–5
H_r	0	x	0–1	3	3–4
H_f	0	x	x	3, x	2

Fine substance obtained with a 200 μm sieve; examination of debris preparations after sodium hypochlorite treatment. Estimation of the proportions in fine substance: 1 — rare; 5 — very frequent; x — hardly recognizable by purely morphological investigation. Investigations by the author. * = Symbols of horizons according to Babel [1971a] (see Section II C).

hairs, pieces of epidermis, the palisade parenchyma, and pieces of the leaf veins. Difficult to recognize were the colorless, irregular, thin, membranous pieces of the cuticula.

Table 3 shows the course of disintegration processes in the profile, which begins with the sloughing off of the cuticula already in the L_n horizon and ends with the disintegration of the leaf veins in the F_m horizon. The table shows also how, with increasing depth in the profile and therefore increasing transformation of the fine substance, these residues become increasingly difficult to recognize. For histochemical investigation of these particles, see Part III C 3. Pieces of soil animals, mainly chitin armour, may be found occasionally in debris preparations.

Corresponding investigations were performed in other cases, not on the total fine substance but on droppings or on the hind-gut contents of soil animals. The pertinent publications of SCHUSTER [1956] and DUNGER [1963] have already been discussed in Section III C 1 a. MÜLLER [1887] found in earthworm droppings pieces of moss leaves, beech leaves, and roots as well as hyphae.

(b) Morphologically Fully Destroyed Material

With increasing depth in the humus profile, an increasing proportion of the fine substance is changed so much that even on careful investigation only a strongly colored material with irregular structures may be recognized. In the F horizon this material, on microscopical examination, often shows bright colors, which in the H horizon recede in favor of darker and especially duller colors, and finally disappear altogether. Very dark to almost opaque parts, which often show granular internal structures, occur in different moder forms, particularly in the rendzina moder.

No clear features of the parent tissues can be recognized. The term "amorphous," however, which is often given to this material—particularly in the case of investigation at weak or medium magnifications—should certainly be avoided. Firstly, irregular structures can always

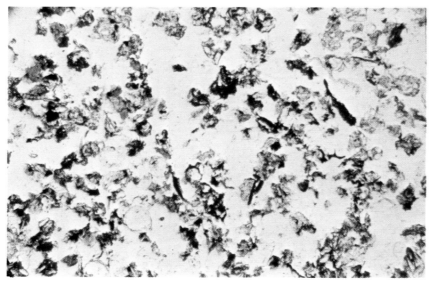

Figure 23. H/A$_{hh}$ of a moder (detail of Figure 39, see Figure 24 (fluorescence) and 27 (crossed polarizers)). The organic fine substance particles allow traces of cell wall structures to be recognized at best indistinctly; mainly dull and dark colors. A few brown hyphae. Bright field, natural size 600 μm.

Figure 24. Like Figure 23 but with fluorescence illumination. Numerous cell wall residues visible by yellowish or greenish primary fluorescence, some of them colorless in bright field. Fluorescence color frequently impaired by substances that are brown in bright field.

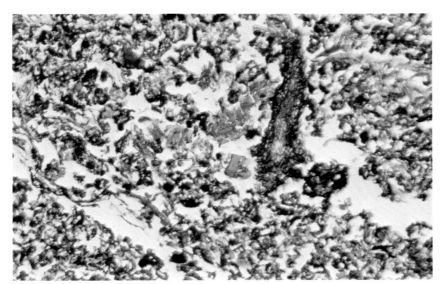

Figure 25. H$_f$ horizon of an alpine pitch moder (dense fine substance fabric, detail of Figure 41). Small, brightly colored, cell and tissue residues sporadically recognizable (center, immediately to the left of the root). Many hyphae and hyphal residues. A root, running perpendicular, occurring mainly in the mycorrhiza mantle. Bright field, natural size 450 μm.

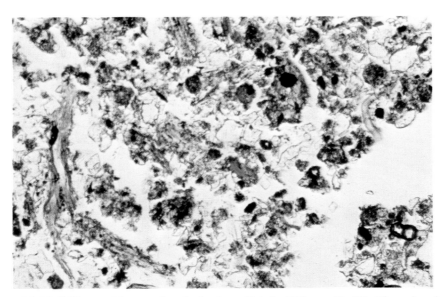

Figure 26. Mull-like rendzina moder, A$_h$ horizon (detail of Figure 52, 53). The colored particles are organic fine substance; some yellowish brown, small tissue residues; left, a fine root. The colorless grains are calcite. (Vienna woods, Austria). Ground thin section (20 μm), bright field, natural size 450 μm.

Figure 27. H/A_{hh} horizon of a moder under crossed polarizers (compare Figure 23 and 24). Quartz grains are visible because of birefringence; in addition organic parts (cell wall residues), which are sometimes not clearly distinguishable from small quartz grains. (Northwest Germany). Ground thin section (20 μm), natural size 600 μm.

be seen. Secondly, after treatment with sodium hypochlorite, it becomes evident that the material contains a large amount of cell wall fragments. These are coated with irregular, colored substances that are dissolved by sodium hypochlorite. In the fluorescence microscope the cell wall fragments may be recognized by their yellow and green or sometimes blue fluorescence, even without treatment with sodium hypochlorite (BABEL [1972a]). They are partly birefringent. The birefringence is, however, only very feebly apparent because of encrustation with colored substances. (Figures 24, 27, 23; see Section III C 3.)

(c)   Microorganisms and Their Residues; Pollen

Microorganisms are discussed only briefly here, as far as they can be recognized as morphological elements in microscopic humus preparations.

Colored hyphae and spores of fungi may often be found in thin sections as constituents of the fine substance (Figures 25, 22). Frequently their characteristic dark brown with slight violet color helps in their recognition. Colored hyphae are widespread, especially in organic layer formations. The colored substances, which are similar to melanin and humic substances (see Part III C 3), occur in the fungal cell walls, to lower amounts also in the cytoplasma. They can be formed also on the surface of the cell walls (sometimes in a mucilage layer, photo in DOMMERGUES and MANGENOT [1970]).

The study of fungal hyphae as constituents of soil organic matter is of interest for two reasons. On the one hand, fungi participate decisively in the degradation of plant residues; on the other, the dead residues of fungi are in turn a component of humus.

Hyphae are found in and on the plant residues, as well as in the pores (*e.g.*, Figures 10, 32). On mineral particles they occur to a much lesser extent (indirect studies in a podzol by

WILLIAMS and PARKINSON [1964]). In 1932 KUBIENA had already investigated the positions of fungi in the soil pores with the incident light microscope (KUBIENA [1932, 1938, 1943a]). Chewed pieces of hyphae often occur in droppings of mites and other soil animals (Figure 35) (DUNGER [1963]). Mycelial strands are sometimes found in moder and raw humus. They may be observed more easily under the stereomicroscope than in ground thin sections. Fungal hyphae, which are present for the most part in small fragments, can be found in abundance in the H layer of wet surface humus formations (ROMELL [1935] in greasy mor). In such cases it may often be morphologically established that the hyphae emerge from mycorrhiza mantles (Figures 28, 21). MEYER [1964] found the same in the upper F horizon of dry raw humus formations.

Colorless hyphae are easily overlooked in thin sections. Therefore, NICHOLAS and PARKINSON [1967] judged ground thin sections to be not quite suited for quantitative assessment of fungal mycelium. With phase contrast illumination, however, they may be found unequivocally in sufficiently thin sections (ALTEMÜLLER and BANSE [1964]; observations of the author). They are often filled with air and thereby more easily visible, but they may also be compressed. Colorless hyphae sometimes become visible in sections through their strong primary fluorescence (Figure 28; JABLONSKI [1963]). They are much less frequent than the brown hyphae, a fact that is quite in harmony with their experimentally established more rapid decomposition (see Section III C 3). Sometimes autolysis of hyphae may be observed, as well as lysis after colonization of bacteria (BURGES and RAW [1967]). On the average fungal hyphae in soil are thinner than in nutrition media (ZVAJGINCEV [1964]).

Bacteria can sometimes be made visible in sliced or ground thin sections of undisturbed humus samples by staining. Staining methods can be of advantage also for the investigation of

Figure 28. Colorless aerial mycelium with strong autofluorescence (in the bright field transmitted light it is hardly visible). Dark aerial mycelium without fluorescence, emerging from a mycorrhiza mantle that surrounds an obliquely sectioned root. In addition, several bright plant residues fluoresce that likewise were hardly visible in bright field (the central cylinder of the root at the right fluoresces bluish green: lignin). Raw humus, F horizon (Black Forest, Germany). Ground thin section (15 μm), unstained, fluorescence illumination, natural size 600 μm.

colorless hyphae. MINDERMAN [1956] was able to recognize bacterial colonies in gelatin sections of a moder. According to ALEXANDER and JACKSON [1955], hyaline fungal hyphae, algal filaments, and also bacteria can be observed in ground soil thin sections if the soil is first stained and afterward carefully infiltrated with synthetic resin. The technique has been developed further by JONES and GRIFFITHS [1964]. With this technique they could investigate the distribution pattern of colonies of bacteria in soil aggregates. On the other hand, ground humus thin sections in which bacteria are made visible give an important insight into the questions of localization of microorganisms in the soil and into the micromorphology of humus formation by microorganisms. These questions are discussed by BARTHOLOMEW [1963].

Observations on these questions may, of course, be made with the help of *other techniques*, some of which have been used in soil microbiology for a long time. Among these are fluorescent staining of bacteria with Acridine orange in debris preparations of soil (STRUGGER [1949]) and the plate growth method (CHOLODNY [1930], ROSSI and GESUE [1930]). In the latter, work can likewise be done with fluorescent dyestuffs (LEHNER and NOWAK [1959]). For further details on these techniques, see textbooks on soil microbiology (*e.g.*, MÜLLER [1965]). In such preparations filmy and clusterlike colonies of bacteria may be distinguished (ROSSI [1936]). The colonization of dead plant residues by bacterial colonies may also be observed (GLATHE [1955]). TROLLDENIER [1965a, b] has been able to observe the colonization of living roots by bacterial colonies with the aid of fluorescent staining. Direct observations of an A_h horizon of a moder found that only 0.02 % of the surface of the sand grains was covered by bacteria; by fluorescence microscopy the greatest part of the bacteria was found on organic particles (GRAY *et al.*, [1967]). The natural position of bacteria and fungi in soil was also shown by scanning microscopy (GRAY [1967]).

The occurrences of the shells of *diatoms* in ground thin sections from anmoor (Section V C) and of *thekamoebae* from moder and raw humus (Section V D 1) have already been mentioned (Section III A). *Green algae* and their residues have seldom been observed in ground thin sections (VAN DER DRIFT [1964]).

Residues of *pollen* were found in paraffin sections of peats in undisturbed samples (RADFORTH and EYDT [1958]). In other humus forms, pollen residues are also present, but in such small amounts that they can only be investigated microscopically after special enrichment methods (see textbooks and method books of pollen analysis, *e.g.*, FAEGRI and IVERSEN [1964]). Pollen analytical investigations on mineral soils were performed by WELTEN [1958]. Usually only the exine of pollens is preserved. It consists of sporopollenin (see Section III B 2 e). This substance, which is closely related to cutin and suberin, is degraded extremely slowly in acid soils (DIMBLEBY [1961]). VAN GIJZEL [1961] found a shift towards longer wavelengths in the fluorescence color of the pollen with increasing age. VAN GIJZEL [1961] interprets this shift by decomposition of some of the constituents of the sporopollenin. ERDTMAN [1971] observed hieroglyph-like corrosion structures in the exine, the outermost part of which seemed to be more resistant than the rest.

(d)   Opaque Particles

In thicker ground sections aggregates of strongly colored fine substance particles sometime appear opaque. Such material might have been envisaged by HARTMANN [1965] when he speaks of "coal-like fine humus." Also a dense cover of opaque, granular material on poorly colored cell wall residues, as in rendzina moder, gives, at lower magnification, the impression of completely opaque particles (see Figure 47).

In addition, however, there are also massive, clearly demarcated individual particles that are black in incident light, and completely opaque in transmitted light, even in very thin ground sections of about 10 μm. Such particles may be found in the fine substance of nearly all surface humus forms (BABEL [1964a]). They were also observed in water-logged mineral soils and in fossil peats (ALTEMÜLLER [1960], MOROZOWA [1963]; see also Part V C). They are often abundant enough to constitute an important fraction of the total material (up to some percent).

According to BABEL [1964b], some groups of opaque organic constituents, which are of very different origins, may be distinguished. Their identification is made possible partly by comparison with similar pieces having characteristic outlines or which are not completely opaque and thereby allow internal structures to be recognized. In other cases internal structures can be made visible by treatment with sodium hypochlorite (BABEL [1964c]).

Opaque formations from fungal *plectenchyma* frequently appear as fragments of sclerotial cortex as well as parts of mycorrhiza mantles and parts of perithecia (Figure 20). Other small fragments are derived from *charcoals* (Figure 17, cf. Section III B 2 d α). Blackened *cell parts* and *tissue parts* of higher plants occur preferentially in peats and wet moder formations. Sometimes, parts of them are somewhat transparent. Blackened wood pieces cannot always be distinguished with certainty from charcoals (see JONGERIUS and PONS [1962b]). Without doubt, in addition to carbonization, there are still other ways of rendering the residues of higher plants opaque. This is shown, *inter alia*, by fully opaque phlobaphenes in the cell contents of cork tissues, the cell walls of which are still colorless and birefringent. Also, opaque leaf residues with well-preserved mesophyll are found (Figure 14).

Chitin armour of animals can likewise be opaque, although this is very rare. More or less spherical, opaque formations of roughly 50 to 150 μm in diameter, which can occur in considerable amounts at the boundary between L and F layer, are enchytraeid and collembola droppings. They have apparently developed from very small-grained opaque impurities (grains of 1 to a few μm in diameter) which may be found in the L layer on the leaf surfaces (Figures 9, 6). These grains appear to be of microbial origin and are consumed together with the food by the animals named (ZACHARIAE [1963, 1964]). Easily confused with them are particles from technical fly ash, which can occur in densely populated and industrialized regions in the superficial horizons of soils. The differentiation of opaque organic constituents from opaque minerals is possible by means of sodium hypochlorite.

(e) Particles from the Solution Phase

In peats and related humus formations one finds morphologically homogeneous organic substances, which evidently represent pure formations from the solution phase. Morphologically similar constituents can occur in mineral humus horizons (see Section III D 2 b).

In holorganic terrestrial humus formations, in contrast, such constituents can be observed only seldom. The author has found red-brown, homogeneous organic substances forming 2 to 5 μm thick coatings on droppings (of diptera larvae probably) in spruce moder. In terrestrial humus formations, generally, dissolved organic substances are absorbed by the solid parts or they form more or less granular coatings on them, which are not distinguishable with certainty by micromorphological methods. Such coatings can contribute to the aggregation of individual particles. Spaces in rotting wood of stems and roots can be coated with films, which apparently are produced from soluble decomposition products of wood. This may be the case also in terrestrial humus forms. Such formations do not seem to have been accurately investigated.

Information on the formation of dissolved humus substances was presented in Section III B 2 a. It should be added that soluble substances are formed among others in the autolysis of bacteria (BARTHOLOMEW [1963]) and that KÜHNELT [1963] has detected water-soluble tannins among leaf residues of the upper F layer of a beech forest.

In peats there are precipitates of organic substances that have been formed from the solution phase and that have often been described and illustrated (KUBIENA [1943a], JONGERIUS and PONS [1962a], GROSSE-BRAUCKMANN and PUFFE [1964]; see Figure 29). These formations are orange-yellow to brown. Because their interiors are completely structureless, their formation from the solution phase can be deduced. Sometimes they form only very thin layers. But in more strongly decomposed peats or in humus-enriched horizons of peats they are usually very dark to opaque. In these cases they have greater thickness and completely envelope the plant residues. Often the interstices between the plant residues can then be entirely filled. In ground thin sections, however, as well as in cremolan sections, such crusts show shrinkage cracks. KOWALINSKI [1964b], who has investigated humus enrichment horizons in peats, has found that in terms of humus chemistry these crusts must be designated as hymatomelanic acids.

3. *The Chemistry of the Fine Substance*

The chemical composition of the fine substance may be deduced from morphological analyses.

A considerable proportion of the fine substance is not appreciably altered chemically as compared with the plant residues. This is true of fresh droppings, in which the parts of plant residues, apart from disintegration and sometimes deformation, show no morphological changes. It is also true for the small tissue fragments, the histological origin of which can still be identified microscopically. In the morphologically completely destroyed material there is still present a considerable amount of unchanged cell wall substances, as is shown by the significant amount of cell wall fragments visible after removal of colored substances. By test reactions, cellulose can be detected (HANDLEY [1954]). In the H layer of a beech raw humus, the content of cell wall residues amounted to a quarter of the fine substance (BABEL [1965a]). The histochemical and histophysical investigations of these cell wall fragments showed that apart from cellulose they contain also cutin, and in rarer cases, lignin (see Table 4). In the A_h horizon the amount of cellulose in the fine substance decreases appreciably, and lignin is found only in traces (BABEL [1972b]). (This is in agreement with chemical results of KICKUTH *et al.* [1969], quoted in Part III B 2d).

The brightly colored parts of the fine substance are, for the most part, either unchanged or only slightly changed derivatives of those colored substances that may be found in the plant residues before disintegration. They are browning substances from the materials of the parenchyma cell contents, phlobaphenes and degradation products of lignified cell walls (see Section III B 2).

The increase in dark and dull-colored constituents with increasing depth in the profile of

Table 4. Results of the histochemical and histophysical analyses of the cell wall residues from the fine substance ($< 200\ \mu$m) of a beech raw humus.

	Cutin-containing cell wall residues	Lignin-containing cell wall residues	Cellulose-containing cell wall residues
(Leaves from the litter)	(1)	(2–3)	(5)
Fermentation layer (F)	2	2	5
Humus substance layer (H)	3	1	4

Estimation scale of 1–5: 1 = little, 5 = very much.   (From BABEL [1965a].)

surface humus formations, however, indicates oxidative changes. These, apparently, involve mainly the colored components, since cell wall substances may be detected as before.

It has been frequently recognized from soil biological studies that the chemical composition of soil animal droppings is not greatly different from the composition of the food which the animal consumed. Rapid changes occur, however, after the droppings have been deposited (KÜHNELT [1961], DUNGER [1964]). The microbial count and the activity of microorganisms are considerably higher in fresh droppings than in the parent material (VAN DER DRIFT and WITKAMP [1960], PONOMAREWA [1962], BIWER [1961]). The most important reason for the later changes in the droppings is accordingly to be found in a stimulated activity of the microflora.

A rise in the content of true humus substances (content of acetyl bromide soluble matter) is found only in aged droppings (FRANZ and LEITENBERGER [1948], discussed in KÜHNELT [1957] and DUNGER [1958]).

Mineralization processes during passage through the gut (BIWER [1961], PARLE [1963]) and morphological changes of individual parts from digestive activity by the animals (SCHUSTER [1956]) are only seldom determinable. The essential activity of animals for humus formation is not the chemical alteration but the creation of new physicochemical and microbiological conditions and new surfaces for autoxidation and microbiological reactions (*e.g.*, SCHUSTER [1956]).

The considerable proportion of dark-colored spores, plectenchyma, and especially hyphae of fungi that may be found in many surface humus formations signifies chemically the presence of chitin and fungal pigments and their degradation products in the fine substances.

Colorless fungal hyphae are decomposed much more quickly than colored ones, as MÜLLER [1887] concluded from his morphological investigations and as PINCK and ALLISON [1944] have detected in experiments (see also HURST and WAGNER [1969] and MEYER [1970]).

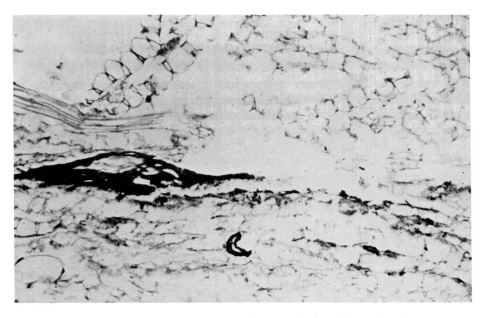

Figure 29. Sphagnum peat, slightly decomposed. Accumulation of strongly colored homogeneous substance from the solution phase (center left). Sliced section (Cremolan method, 10 μm), bright field, natural size 600 μm. (From a preparation by D. Puffe, Bremen.)

This difference, according to the last-named authors, is due to the content of humic substance-like pigments, which usually amounts to 20 to 30% in colored hyphae. The quantity of fungal pigments that is formed in a mycelium is dependent on various external conditions, *e.g.*, nutrition and microclimate (KÜSTER and HAIDER [1961], BUTIN [1961]). Chitinases are produced by many soil animals and soil fungi (DROZDOVA [1959], DEVIGNE and JEUNIAUX [1961]). They degrade chitin to acetylglucosamine and, from that, humic substances can be synthesized in melanoidin-like reactions. The relatively high amino sugar content, which was found by BREMNER [1955] in an acid soil, might be caused in particular by the high proportion of hyphae and hyphal residues that are usually found in such soils. Humus substance formation by microorganisms is treated extensively by SCHEFFER and ULRICH [1960]. They treat not only the formation of humus substance-like fungal pigments and fungal degradation products but also the formation of humus substances from aromatic compounds in the substrate by ecto-enzymes (see also HAIDER and MARTIN [1970], MARTIN and HAIDER [1971]). As is well known, the soil microorganisms are the most important producers of soil polysaccharides, which constitute often 10 to 30% of soil organic matter and which play an important role in soil aggregation (see MARTIN and HAIDER [1971]).

D. Organic Matter in the Mineral Soil

In the mineral humus horizon of mull as well as in deeper mineral soil horizons, the organic matter is more or less different from the organic matter in the holorganic humus horizons. These differences arise from fundamental differences in processes of formation. The morphology of the organic matter in mineral soil will accordingly be given special treatment here. This account can be composed briefly because reference can be made to the other sections, particularly to Parts IV and V. In addition, only few detailed investigations on this subject are available.

Sometimes organic constituents cannot be distinguished from mineral constituents purely by morphological means. There are, for example, opaque organic constituents and opaque minerals. There are further more or less homogeneous, reddish-brown formations, which may be phlobaphene-containing organic parts or iron oxide containing mineral parts. In such cases differentiation is possible by means of microchemical reactions which can be performed also on the ground thin sections such as the sodium hypochlorite and Prussian blue reaction (BABEL [1964c]).

1. *Blending of the Organic Matter into the Mineral Soil*

A considerable part of the organic matter is formed by roots and by microorganisms in the mineral soil itself. The amount of the root substance continually dying off is noted in Part III B1. The litter is mixed into the mineral soil by five processes:
1. Blending by animals.
2. Blending of solid particles by gravitation and rainwater.
3. Illuviation of dissolved substances by rainwater.
4. Processes of surface erosion on slopes.
5. Tillage.

These five processes are of very different importance in different humus forms. They are responsible for part of the most important characteristics of the humus forms.

Many animals in soil cause intermixing, but earthworms are considered to be the most important group. Depending on the species and the special conditions, their activities are extremely varied. Some earthworms (*e.g. Lumbricus terrestris*) draw plant litter into the soil consuming it later as food. On the other hand, droppings consisting mainly of mineral substances are often deposited in the F layer or even on the surface of the L layer (*e.g.*, ZAJONC [1967]). Some species make vertical burrows which are often lined with humus excrements. This results in blending of only small quantitative significance. Tipulidae larvae often deposit highly organic droppings in the upper soil horizon. (For details see the textbooks on soil biology by KÜHNELT [1961] and DUNGER [1964]; ZACHARIAE [1964].)

In forest soils earthworms usually consume material previously disintegrated by arthropods. They do not transport this material over large distances. Here the earthworms are of little significance in the intermixing of organic matter. Also by the simultaneous consumption of organic and mineral material they do not themselves bring about any mixing which would be of more importance to the macroscopic development of the soil profile (ZACHARIAE [1965]).

By the trampling and rummaging activities of game animals and the burrowing of mice and moles, in forests the litter can be intermingled in places to a considerable extent into the mineral soil (ZACHARIAE [1965]).

The mixing in of solid organic matter by gravitation and rainwater is a process by which great amounts are transported in all humus forms. In moder and raw humus organic fine substance from the H layer is washed into the small intergranular spaces of the upper 1 or 2 cm of the mineral soil (BAL [1970]). In mull, leaf and other small plant fragments fall into the larger interaggregate spaces which are continuously formed by the burrowing activity of earthworms in the upper 1 to 5 cm of mineral soils (ZACHARIAE [1966]). In Tirs and related soil formations (vertisols) shrinking fissures are of similar effect, whereby a self-mulching arises.

Movements of dissolved organic matter from the organic layers into the mineral soil take place in all humus forms. In mull they extend over distances of a few millimeters at most; in moder and raw humus over several decimeters (podzolization).

2. *Holorganic Constituents in the Mineral Soil*

(a) Plant Residues

Often root residues may be found. In the ground thin section it cannot be ascertained with certainty whether well-preserved roots were still living or dead. Frequently cast-off pieces of barks of old tree roots appear (Figure 15). Also, from the old grass roots the cortex sloughs off, leaving only the central stele (GADGIL [1965]).

The intermingled pieces of the litter can be still well preserved morphologically or they may show a progressive homogenization of the tissue structures.

Plant residues in pseudogleys and gleys are often infiltrated and encrusted to varying degrees by iron oxides (Figure 3). The infiltration extends in that case into the cell walls, as may be detected by the Prussian blue reaction (BABEL [1964c]). In semiterrestrial humus forms or those under water-logging conditions, often dark brown to opaque plant residues can be found (see Part V C).

The frequency of the plant residues changes greatly in the A_h horizon of a mull in the course of the year. It can be especially high in fresh earthworm droppings. Whether general differences in the frequency of the plant residues exist for different mull subforms cannot yet be definitely stated. In MÜCKENHAUSEN [1962], for the A_h horizon of pseudogleys, the humus form of which is, in general, a mull closely related to moder, a relatively large amount of plant residues is reported. In matted mull (BARRATT [1964]) under grassland large amounts of dead roots occur.

(b) Organic Fine Substance

In the upper centimeters of the mineral soil of moder and raw humus organic substances occur prevailingly as intrinsic fine substance particles. Morphologically it resembles very much the fine substance of the H horizon. But very small grains (about $1/2 \mu m$), which cover organic fine substance particles, are translucent here, whereas in the H horizon they are opaque (BABEL [1972b]). In the lower parts of the A_h horizon the whole material is lightened, probably by decomposition or displacement of organic substance (BAL [1970]).

Just as with plant residues, holorganic fine substance particles in the A_h horizon of mull are, as a rule, rare. They may be detected, rather than in thin sections, by fractionating by specific gravity. MCKEAGUE [1971], by those methods, found small cell and tissue residues in the silt fractions of Canadian chernozems (compare also BOTTNER [1970]).

Readily identifiable particles that are found relatively frequently are the reddish-brown cork and cortical cells of roots, described in Section III C 2 a. Dark hyphae of fungi occur especially at the upper boundary of the B_h horizon of posdols (ALTEMÜLLER [1962]). In acid- and water-influenced humus formations, there are remarkable amounts of colorless cell wall residues, which often become visible only in the fluorescence microscope by their blue or also yellowish primary fluorescence (Figure 30; BABEL [1972a] see also ALTEMÜLLER [1960]). In low amounts such particles also occur in other humus formations.

Opaque or dark-brown transparent plant fragments of the size of one or a few cells were observed likewise in water-influenced mineral soils (cf. Section III C 2 d; Figures 31, 49).

Very small dark particles, less than $1 \mu m$ in size, were observed in the agriculturally culti-vated humus horizon (A_p) of a mull-pararendzina (ALTEMÜLLER [1960]) and in the humus cutans of para-brown earths (ALTEMÜLLER [1956]). In the widespread forms of mull, particles of this kind, however, do not occur. They were found, however, in several soil types by PAR-FENOVA and YARILOVA [1967]; 2 to 3 μ brownish-black organic particles in southern chernozem soils seemed to be especially characteristic.

Around the mineral grains in the B_h horizons of podsols, coatings of organic matter with

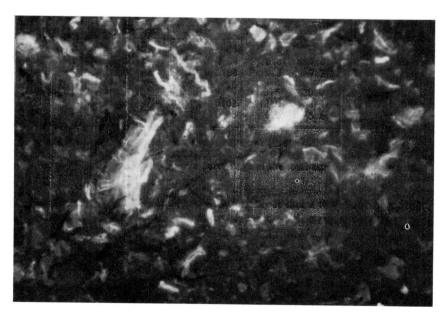

Figure 30. Numerous fluorescing, strongly crushed cell wall residues in the back water horizon of a pseudogley (dense silty loam; carbon content 1%). The cell wall residues are remains of fine roots. They are not visible in the bright field. Very dark: fungal hyphae (central uplands of Germany). Ground thin section (30 μm), primary fluorescence, natural size 400 μm.

clear yellow to brown colors are formed from the solution phase (covered fabric, see Part IV B3 and Figure 45).

3. *Organic Matter as Pigment of the Mineral Fine Substance*

In mull and related humus formations a substantial proportion of the organic matter is present as pigment of the mineral fine substance, which shows no holorganic particles, even with the strongest microscopic magnification.

Electron microscope pictures suggest that in this material a morphological separation of organic and mineral particles exists only at 0.01μm or below. (DUDAS and PAWLUK [1969, 70a] obtained particles with irregular and indistinct contours before and sharp contours after peroxide treatment.) For discussion of the arrangement of sesquihydroxides, humus, and clay minerals in the submicroscopic range, see HESS [1965]. From clay mineralogical investigations the localization of organic matter in the interlayers of montmorillonite can be concluded (GEBHARDT [1971]).

In mull practically the whole of the organic matter has been transported in the solution

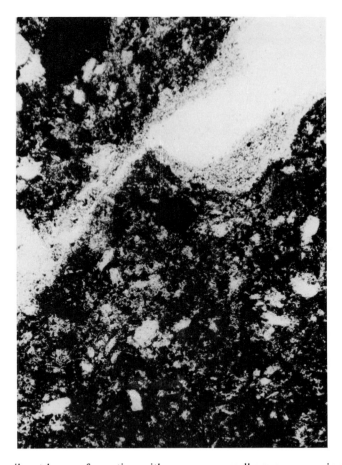

Figure 31. Fossil wet humus formation with numerous small opaque organic constituents in the strongly colored mineral fine substance (these cannot be distinguished in the photograph). The walls of a channel (top left) with humus-poorer, hence brighter cutan. Colluvial clay (central uplands of Germany). Ground thin section ($20\,\mu$m), bright field, natural size $600\,\mu$m.

phase, although only over the smallest distances, then to be precipitated at the surfaces of the clay minerals and the oxide hydrates. There, catalytic transformations of the organic compounds and the formation of clay-humus complex occur (see SCHEFFER and ULRICH [1960].)

The main object of humus microscopic investigation of mull, therefore, is the fabric analysis (see Part IV).

From the colors of the mineral fabric components, inhomogeneities in the distribution of organic matter can be recognized (Figures 31, 49; see also Part IV B3). Thus, cutans of channels can differ morphologically from the interior fabric by a darker color. According to ALTEMÜLLER [1956], the degree of their orientation birefringence decreases with increasing depth of color and hence increasing humus content.

IV. Fabrics in Humus Soil Horizons

The organic constituents dealt with in the preceding section, alone or together with mineral constituents, build up the fabric of the humus soil horizons.

The term fabric is used here, according to ALTEMÜLLER [1962], to be the arrangement of the constituents. According to KUBIENA [1935, 1938], elementary fabric (grouping of the single parts to each other) and aggregate fabric (formation and arrangement of the aggregates) may be distinguished. A comprehensive treatment of the terms "texture," "structure," and "fabric," as well as the bases of the fabric analysis, are given by BREWER [1964].

Here, only a brief treatment of those soil fabrics greatly affected by organic substances and organisms will be given. The most important phenomena of fabric formation and of typical fabric forms will be treated according to present knowledge in micromorphology. Too little is known about fabric systems to give a complete account of them (see Part IV C).

A detailed and systematic treatment of soil fabric is given by BREWER [1964]; but he gives little attention to organic matter and organisms. The same holds true for the treatment of soil fabric by ALTEMÜLLER [1966].

The fabric is determined by ratio, size, shape, and arrangement of the solid constituents and the pores. Fabric analysis of the horizons of the humus profile can provide information of biology and dynamics of humus formations.

The changing of fabrics from above downward in moder profiles has been investigated by macro- to micromorphological techniques and by estimating the changing amounts of some fabric features in the subsequent horizons (BABEL [1972b]). Micromorphometric measurements of those features have been carried out by several authors. KUBIENA, BECKMANN, and GEYGER [1962], by means of structure photograms, compared the fabrics of a brown loam, influenced by small animals, with the unaffected parts. JONGERIUS [1963], on the basis of such measurements, compared different humus forms, and BAL [1970] compared different subhorizons of a profile. GEYGER [1967] made comparative micromorphometric investigations on the biology and dynamics of alpine råmark.

A.   Processes of Fabric Formation and Stabilization

The fabric of the organic soil horizons is built up and stabilized by biological and non-biological forces. These forces often work against each other. They produce different types of fabric. Those formed biologically are often the more favorable from an ecological point of view (BENECKE and BABEL [1969]).

1. *Nonbiological Processes*

The nonbiological processes of fabric formation arise from swelling, shrinking, frost, and surface tension of water. These actions can be modified considerably by organic substances (Figures 46, 41, 42).

Experiments of JABLONSKI and RADOMSKA [1965] showed the formation of pores around plant residues that occur in mineral soils. This was explained by the differences of swelling and shrinking of organic and mineral substances. Soil aggregates are stabilized by organic substance (as pigment) against breakdown by nonbiological forces. Fissures nevertheless arising from shrinking will have rough walls (see soil fabric classification by BECKMANN and GEYGER [1967]).

By action of water meniscuses leaves are stratified very densely in the lower L layer (L_v according to BABEL [1971a]; Figure 33). Over the very close interstices, which are formed in this way, other binding agents become effective (very small amounts of fine substance, slimes of microorganisms, precipitations of leaf leachates (BABEL [1972b]) which bind the leaves even after drying. Action of surface tension plays a role also in the compaction of the dropping fabric of enchytraeidae. During drying, the loosely lying droppings are drawn together by the surface tension of the water (BABEL [1968, 69]; see Part IV B 3).

By displacement of humic clay material humus containing coatings are formed on pore walls in the lower soil horizons. Displacement of dissolved organic matter gives rise to formation of the covered fabric in podzols.

2. *Fabric Formation by Roots and Microorganisms*

Roots bind soil aggregates, especially where they are strongly branched and have developed root hairs or mycorrhiza. The root epidermis often forms a slimy "mucigel layer." This creates a very close union between roots and substrate (electron microscopic investigations by JENNY amd GROSSENBACHER [1963]). APINIS [1969], in a comprehensive review of literature, points out that there is also indirect influence of roots on soil aggregation, as roots promote the development of rhizosphere organisms and these promote a better aggregation.

Biopores develop in places where roots decay (SLAGER [1964], GADGIL [1965]). These

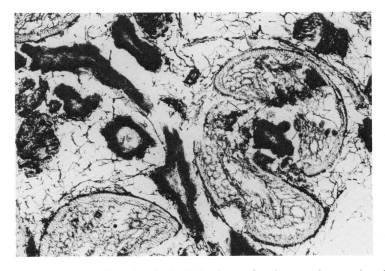

Figure 32. Dark aerial mycelium in the $L_v F_r$ horizon of a dry raw humus (produced under a *Pinus nigra* plantation on a former mull). Needle residues (transverse) and several roots (black-appearing structures, some transverse, some oblique); these parts are combined by the mycelium in a loose network (central uplands of Germany). Ground thin section (35 μm), bright field, natural size 1.5 mm.

are frequented by animals (*e.g.*, enchytraeidae), which loosen the surrounding mineral material.

On the other hand, growing roots cause a compaction of soil in their close environment (micromorphometric investigations of BLEVINS *et al.* [1970] on tree roots with a diameter of 4 to 7 mm).

Microbial filaments (bacterial colonies, actinomycete and fungal hyphae, algal filaments) contribute to living obstructions (SEKERA and BRUNNER [1943]). Numerous fungal hyphae may be observed microscopically, and often may be seen with the naked eye. Often they permeate both the plant residues of the lower L layer (L_v) and the upper F layer (F_r, according to BABEL [1971a]), as observed for instance by MEYER [1962] (Figure 32).

In A_h horizons fungal hyphae occur to combine sand grains (BARRATT [1962]) or primary aggregates (enchytraeidae droppings, BENECKE and BABEL [1969]). Rhizomorphae may act in the same way as hyphae. As in the case of roots, often a slime layer intensifies the binding action of the hyphae (ASPIRAS *et al.* [1971]); it consists of mycofibrillae (LIESE and SCHMID [1962]). With the electron microscope BEUTELSPACHER [1955] found microbial linear colloids combining mineral particles.

Pores may be formed by fermentation activities of microorganisms as was found by JEANSON [1964] in experiments in dense soil fabrics. (For further details on the influence of microorganisms on soil fabric, see textbooks of soil microbiology.)

3. *Fabric Formation by Animals; Droppings*

Animals living in soil exert a profound influence on the soil fabric by many of their activities. By their metabolic activities they change the organic constituents (see Part III C 1 a); by their blending activity they displace soil particles (see Part III D 1). By their movements in the soil they form channels. In biologically active soils the movements of the animals—including those which pass only a short period of life in the soil—can produce a strong loosening effect and the formation of roll aggregates (ZACHARIAE [1962]). The most important contribution, however, to fabric formation given by the soil animals are their excrements. These are the prevailing primary aggregates of many humus horizons.

The excrements of soil animals (at least if they are noticeable through their characteristic forms) are usually called droppings ("Losungen") in the German micromorphological literature; in the English literature, fecal pellets. They have been recognized since the studies of KUBIENA [1943a, b] as characteristic features for the biology of natural humus formations. Nearly all micromorphological publications on natural humus formations contain observations on droppings. Among them, however, detailed studies of these objects are rare.

Often droppings are found directly at the feeding places, especially in feeding cavities within plant residues (wood fragments, roots, needles, stem bases, and in dense packets of leaves). Those examples are usually produced by oribatid mites (Figure 33, further illustrations, *e.g.*, in PUFFE and GROSSE-BRAUCKMANN [1963]). In other cases droppings can accumulate to such an extent that they form whole horizons (*e.g.*, the H layer in pitch moder; see Part V D 3). Then the term "dropping fabric" is used (Part IV B; Figure 40).

The nature of the active animal groups may often be determined from the size, shape, internal structure, and arrangement of the droppings as well as from the food traces in the substrate. An unequivocal determination, however, is frequently possible only if the fauna of the humus formation under investigation was identified at least approximately. Taxonomically very different animal groups often produce very similar droppings, *e.g.*, enchytraeids and collemboles (ZACHARIAE [1964]) or diplopods, Tipulidae larvae, and the earthworm Dendrobaena (ZACHARIAE [1967]). The appearance of the droppings is also greatly dependent on the composition of the food. Finally, droppings are often partially or wholly destroyed very soon after

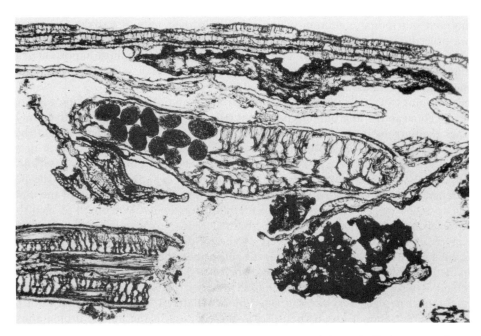

Figure 33. Mite droppings in a feeding cavity in the place of the assimilation parenchyma of an *Abies* needle (transverse; see also Fig. 34). To the right, the assimilation parenchyma is still well preserved, only browned. The droppings are of the same color (they seem to be darker because of the compaction of the material). The epidermis is intact throughout. Above that, a much shrunken *Abies* needle (likewise, transverse). Between these two the compressed residues of a needle, of which practically only the epidermis remains (thin, bright, leading out of the picture at the left; probably decomposition by fungi). At the top a densely compacted layer of three beech leaves, the uppermost collapsed in parts. Wet moder, F_r (Black Forest, Germany). Ground thin section (25 μm), bright field, natural size 1.5 mm.

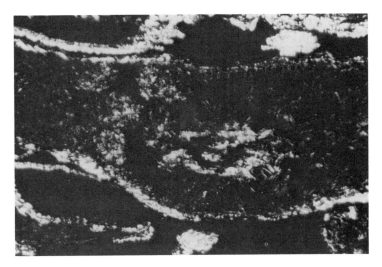

Figure 34. Detail of Figure 33, crossed polarizers. Also in the droppings the cellulose is still preserved (birefringence).

deposition. More precise investigations of dropping forms must therefore take into account the living and nutritional condition of the animals in their natural habitat; as an example of such a procedure, see the detailed work by ZACHARIAE [1965] already quoted.

In the following rough description of the dropping forms of the most important groups of soil animals (KÜHNELT [1958, 1961], DUNGER [1963], ZACHARIAE [1965]), the limitations mentioned should be taken into consideration.

The droppings of litter-consuming *diplopods* are among the most striking of droppings that one can come across in ground thin sections (Figures 37, 52). They are 0.5 to 4 mm in size. Sometimes they are built up by fairly large well-preserved plant fragments. Often they also contain mineral grains. They show a rather loose internal fabric. The contours of droppings are different in the individual subgroups of diplopods: The droppings of the large julids and glomerids are egg-shaped, those of the small julids more cylindrical, those of the polydesmids flattened. The droppings of the *isopods* are often very similar to those of the diplopods in respect to size and good preservation of the plant fragments. When well formed they are often flattened and sometimes show a longitudinal cleft.

Diptera larvae produce as many different types of droppings as there are different systematic groups on the one hand and different food on the other.

Detailed studies of gut contents of several groups of diptera larvae have been performed by HEALEY and RUSSEL-SMITH [1970]. They found different proportions of food residues (leaf fragments, fungal hyphae, very finely divided humus material, mineral material), not only in relation to different animal groups but also to the position in the humus profile and to the humus form.

Sometimes the droppings consist of very well-preserved tissue residues and therefore resemble diplopod droppings (Figure 38). When the larvae feed on more strongly predecomposed plant residues, the droppings contain much less recognizable tissue fragments. Mineral particles frequently occur in them. The droppings of Tipulidae larvae are egg-shaped. Bibionidae larvae produce cylindrical droppings, which sometimes fall apart easily. Investigations by SZABO *et al.* (1967) revealed that they contained leaf residues, some algal filaments and structureless organic substances and mineral particles. They were about 0.2 to 0.3 mm in diameter and up to 1 mm long. The droppings of lycoriidae larvae form, when wet, a pulp containing still easily recognizable cell residues. Under some circumstances it resembles fine substance formed out of enchytraeid droppings poor in mineral substance, which have subsequently fallen apart and welded together.

In ground thin sections the egg-shaped droppings of *oribatid* mites are recognizable with certainty (Figures 33 and 36). They are between 30 and 150 μm long and appear much denser than diplopod droppings—as the plant fragments contained, although still recognizable as such, are more disintegrated. They do not contain mineral particles. These droppings, as mentioned, often occur within feeding cavities according to the microhabitats of the animals (ANDERSON [1970]). There they form groups, a characteristic of which is that often a number of equally large droppings lie together, whereas neighboring parts consist of droppings of another size. These differences are probably caused by the progressive growth of the participating animals (ZACHARIAE [1965]).

Droppings of *collembola* are dark to black spheres, granular aggregates, or blotches of about 50 to 180 μm in size (Figure 35). They occur particularly in the lower L layer. Usually in ground thin sections they are opaque and no cell residues can be recognized. The food was slime films and microorganisms that had developed on the surface of plant residues. Partly, however, the droppings also show fungal hyphae or spores, or small birefringent cell fragments when the food was fungi or the mesophyll of leaves respectively (ZACHARIAE [1964]).

The droppings of *enchytraeids*, which live in the lower L layer on the same food as collembola, are very similar. A fairly reliable distinguishing feature from collembola droppings is that the enchytraeid droppings are often arranged in two parallel rows, thus forming channels among the leaves for the accomodation of the animals (ZACHARIAE [1964,1965]). In the F and H horizons, where the enchytraeids are active as secondary decomposers, they produce granular droppings (Figure 39). Here very small cell wall residues are easily recognizable, at least by fluorescence microscopy. In mineral soils the droppings are rich in mineral substance (mineral grains up to 60 μm in size) (Figure 44). The droppings are about 40 to 100 μm in size with most of them being 60 μm.

Granular to egg-shaped aggregates of about 60 μm in humus soil horizons probably will be in most cases enchytraeid droppings (observations of the author in central Europe). In the H horizon, as well as in the A_h horizon, great parts of the fabric often are built up of enchytraeid droppings (Part IV B) (BABEL [1968, 69]). In the mineral soil the enchytraeids often follow the channels of dead fine roots and leave their droppings there. They eat themselves into earthworm droppings, creating channels which they then fill with their own droppings (ZACHARIAE [1964, 1965]).

Among the *lumbricidae*, *Dendrobaena* species in the L layer form droppings which are similar to but larger than the dark enchytraeid droppings and, like them, are often arranged in channels. But if *Dendrobaena* in the F layer consume the material disintegrated by arthropods, their droppings are egg-shaped to cylindrical. They can be very similar to the droppings of Tipulidae larvae (ZACHARIAE [1967]). Leaf residues in the L layer can be glued together by the droppings (or "castings") of the great *Allolobophora* and *Lumbricus* species.

The earthworm droppings deposited in mineral soil horizons are usually strongly compressed. Skeleton grains and fine substance often are arranged in concentric lobes that are parallel to the former surface of the droppings (ZACHARIAE [1967]). Between the droppings, small wedge-shaped pores generally remain free. Where the droppings are deposited in the F or L layers or on the soil surface, they form the well-known knobbly racemose forms (see also Part V D 5). In the upper part of the A_h horizon earthworm droppings and their residues are often combined to form spongy fabric (Part IV B 3; Figure 2). Lumbricidae droppings can contain plant fragments, one to several millimeters in size; *e.g.*, well-preserved pieces of leaves. Still larger plant residues can be enclosed between the dropping material by the deposition of droppings in F and L layers. In earthworm droppings the proportion of plant fragments and of mineral material varies greatly.

Other important fabric features produced by earthworms are the channels, which are built in the mineral soil by the large species. There are detailed treatments of them in the textbooks of soil zoology. The walls of the channels often are covered by humus clay material, which originates from excrements (HENSEN [1882]).

The increase of pore volume of soils by earthworms is caused by deposition of droppings on the surface, by the permanent burrowing activity in the upper few centimeters of the mineral soil, and by the channels that go down to the C horizon. Deposition of droppings in the mineral soil, however, causes local soil compactions. JEANSON [1964, 1970, 1971] in laboratory experiments has investigated the fabric formation by earthworms, especially the earthworm channels in thin sections and by scanning microscopy.

The position and arrangement of droppings with respect to other droppings and to other fabric components can contribute to the indentification of animal groups. Moreover, the location of droppings near tissues can give a hint to the food consumed by the animal (Figure 33). The morphological comparison of the dropping material with the preserved tissues gives hints to chemical alterations (Part III C 1a).

Micromorphological investigations of droppings often have ecological objectives, concerning autecology of certain groups of soil animals as well as ecology of humus formation and

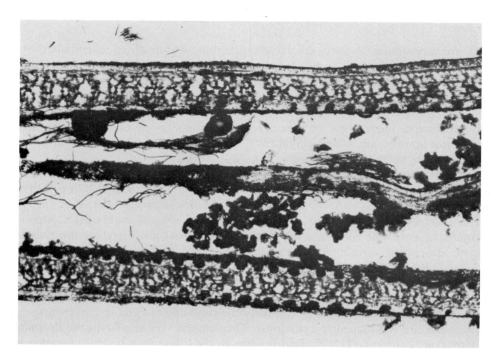

Figure 35. Collembola or enchytraeid droppings, consisting to a large extent of fungal material, which the animals have consumed from the surface of the (longitudinal) *Pinus mugo* needles. Left, dark aerial mycelium. Tangel, L_v horizon (Alps). Ground thin section (50 μm), bright field, natural size 1.5 mm.

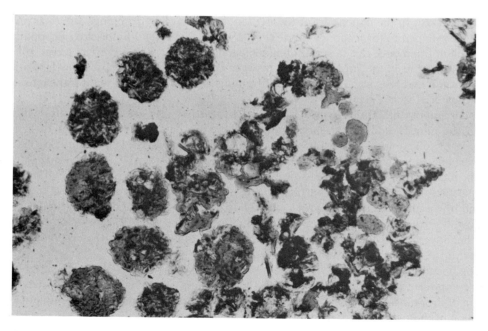

Figure 36. Mite droppings of different sizes and colors corresponding to different size of animals and different food. (Compare also Figure 33). *Picea* moder, F_r. (Northwest Germany). Ground thin section (12 μm), bright field, natural size 450 μm.

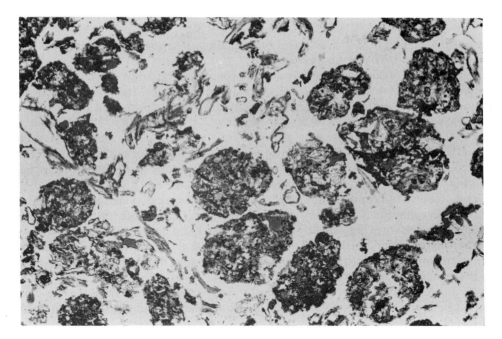

Figure 37. Droppings of isopods or diplopods (compare Figure 52). The stalked or ring-shaped parts between the droppings are for the most part fine roots of herbaceous plants. A moderate proportion of small mineral grains in the droppings is not visible in the bright field. H_r, moder to mull-like moder (central uplands of Germany). Ground thin section (15 μm), bright field, natural size 1.5 mm.

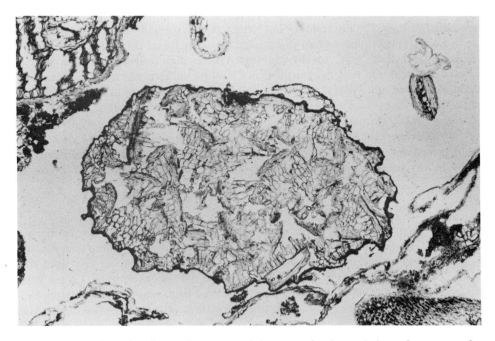

Figure 38. Dropping of a diptera larva, containing scarcely changed tissue fragments of spruce needles, encased by a peritrophic membrane. Spruce moder, F_r (central uplands of Germany). Ground thin section (12 μm), bright field, natural size 2 mm.

ecology of the whole site. Some examples of evaluation of dropping investigations for ecological aims are given by ZACHARIAE [1967].

Droppings of primary consumers usually are consumed (eaten) by secondary decomposers after a short time and thus are changed to other dropping forms. Mechanical influences cause formation of clusters out of the droppings or cause their breakdown (e.g., Figure 43). Breakdown is accelerated by processes of decomposition and dissolution (BAL [1970]). Also, fungal hyphae can destroy the droppings by growing into them. JONGERIUS and SCHELLING [1960] observed that the droppings fall apart more quickly in acid humus formations than in eutrophic ones. RIGHI [1969] observed that the droppings fall apart readily in wet humus formations on podzols but are preserved for a longer time in dry ones.

Great amounts of droppings do not necessarily indicate great animal activity. Even under very favorable conditions, when further transforming organisms are absent, droppings can accumulate appreciably in only slightly altered form, as ZACHARIAE [1965] has shown for droppings of enchytraeids and arthropods. This may also be the cause of the great amounts of droppings in alpine råmark (transitions from raw soil humus to moder and mull-like moder), which has been found by GEYGER [1967].

After the breakdown of droppings, a new formation of aggregates from the fine substance may often be found, but a compaction of the whole material may also occur (Figure 41; PUFFE [1965]).

B.   Fabric Types in Humus Soil Horizons

Here only the well-known typical fabric forms will be described. The present state of knowledge does not allow us to give a comprehensive picture.

1.   *Fabrics with Plant Residues Prevailing*

Laminated fabric of plant residues is characterized especially by a large proportion of more or less horizontally placed plant residues, particularly leaves (Figures 33 and 1). These are still preserved in rather large, macroscopically visible pieces. In some parts they can be compacted very closely, so that neither pores nor binding substances can be seen between them with a microscope. In other parts hyphae or organic fine substance are found between the plant residues. The fine substance regularly lies on the surfaces of the plant residues and joins them together in the form of bridges or larger masses. This fabric comes about by action of water films between the leaves in the L_v horizon, as well as by deposition of animal droppings which feed on microorganisms on the leaf surfaces. It offers suitable environmental conditions to the litter-disintegrating animals and is destroyed by them gradually (BABEL [1972b]).

2.   *Fabrics of Organic Fine Substance*

Dropping fabric consists of granular or egg-shaped droppings that are nearly equal in size and shape (Figure 40). Most frequently found sizes are about 60 μ. There are types with different degrees of compaction. In most cases the droppings are more or less welded and in places their contours are not well preserved. Figure 40 is a typical example. Figure 39 shows a highly porous example, where the granular aggregates of fine substance can scarcely be identified as droppings. In dense forms of this fabric, formation of fissures is observed. This fabric is frequent in the H horizon of moder. It is especially well developed in pitch moder. (It also occurs in mineral soils; see Part IV B 3.) Enchytraeids are especially important in building up this fabric. They act as secondary consumers of material, which has been formed in the F layer by litter-disintegrating animals.

In other cases organic fine substance can form a *dense fabric* with very small pores which originate by the contours of the single particles (single packing voids of BREWER [1964]). Besides, the fabric is divided by shrinking fissures (Figure 41). This fabric occurs above all in the H layer of wet moder forms; it develops by breakdown of the droppings.

3. *Fabrics in Humus Mineral Soil Horizons*

The differentiation of fabrics in mineral soils is greater than in the humus layers because of the addition of mineral components and because of the activity of larger soil animals, especially earthworms.

There are manifold forms of pores and aggregates. Channels (BREWER [1964]) are formed primarily by activity of organisms (roots or animals). If refilled, they are called pedotubules by BREWER. The material of the pedotubules often is darker (richer in organic matter). Sometimes it appears striated (striotubules: earthworm excrements), sometimes it is aggregated in the form of droppings or in similar forms (aggrotubules: excrements of earthworms and smaller animals). (See investigations of GHITULESCU and STOOPS [1970].) The walls of earthworm channels are covered by humus material deposited by the earthworms (see also Part IV A 3). Cutans of dark, peptized clay rich in humus are formed by deposition by seepage water (organo-argillan of BREWER). Thin humus clay cutans were found in chernozem-like soils and (more intensely developed) in para-brown earth (ALTEMÜLLER [1956], KUBIENA [1966], PARFENOVA and YARILOVA [1967]). Humus wall deposits are also described by KARPATSCHEWSKI [1960] and JONGERIUS and PONS [1962a] in muck mull.

The *agglomeratic fabric* (KUBIENA [1938], by BREWER [1964] called agglomeroplasmic fabric because of confusion with the definition in geology) is composed of loose-lying, more or less uncovered mineral grains between which loose-lying organic fine substance is deposited in small flaky to granular aggregates (Figure 43). Aggregates of a higher level are not formed out of these components. This fabric occurs when in the mineral humus horizon of raw humus and moder the mineral component is a sand. Besides, it occurs on the upper border of the B_h horizon of podzols. RIGHI [1969] supposes that it is formed there by transformation of the covered fabric by animals. Normally the agglomeratic fabric develops by illuviation of solid organic particles from the H horizon into the mineral soil. In this process dropping forms are more or less preserved, or they may be newly formed after depositing.

The *spongy fabric* (KUBIENA [1938]; for microphotos see also KUBIENA [1970]) is characterized by manifold racemous forms of aggregates that are connected to each other by irregular bridge formations. From this a manifold pore system results (Figure 2). Often bulbous and indented residues of coprogenous contours are found. By mixing aggregates of different color (by different humus and iron content) a mosaic-like pattern is often formed (ZACHARIAE [1965]). This is found mostly in dense forms (pore structure of BECKMANN and GEYGER [1967], see Part IV C). Sand grains, if they are present, do not occur free, but are incorporated into the aggregates. Plant residues are not frequent. In very loose forms (transitions to crumb structure (BECKMANN and GEYGER [1967], Part IV C) they lie in places free in the pores, or they are incorporated in the mineral aggregates. Particles of organic fine substance seldom occur. The main part of the organic substance occurs as pigment of the mineral fine substance (Part III D 3). The fabric is formed prevailingly by the burrowing activity and depositing of droppings of great earthworm species (in central Europe, above all, *Lumbricus terrestris* and *Allolobophora terrestris*). Sometimes the small droppings of enchytraeids are found in places especially rich in organic matter (ZACHARIAE [1965]).

A *dropping fabric* from enchytraeids in loess was described by BABEL [1968] (Figure 44). It is built up by enchytraeid droppings about 60 μ in size, which are the primary aggregates. These can lie very loosely, they can be welded weakly, or they can lie so compactly that the

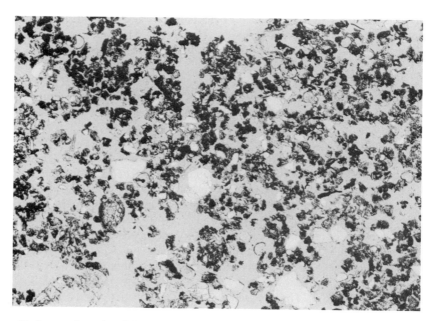

Figure 39. Loose dropping fabric (related to agglomeratic fabric; see also Figure 23). The small granular aggregates are enchytraeid droppings. Moderate proportion of quartz grains. *Picea* moder, H/A$_{hh}$. (Northwest Germany). Ground thin section (20 μm), half-crossed polarizers, natural size 1.5 mm.

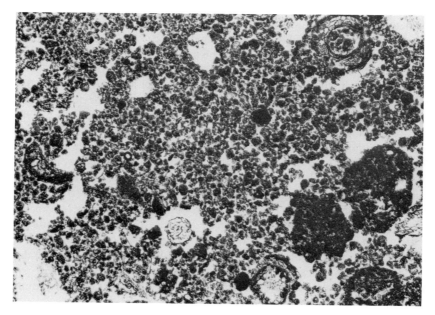

Figure 40. Dropping fabric. Nearly the whole material consists of moderate to well-preserved small droppings (of enchytraeids?) bottom right, two large droppings. Three transversely occurring roots. Pitch moder, H$_f$ (Alps). Ground thin sections (25 μm), bright field, natural size 1.5 mm.

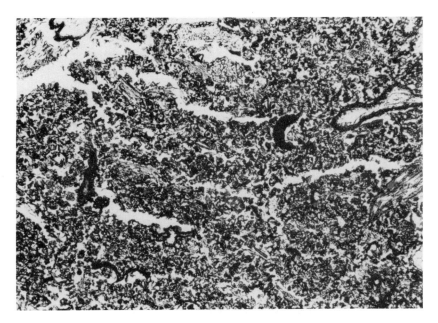

Figure 41. Dense fine substance fabric, with horizontal cracks (see also Figure 25). Of plant residues, only root residues of different sizes may be recognized. The fine substance has developed from droppings, which have fallen apart; the size of the particles corresponds to the bite size of the animals (compare Figure 40). Pitch moder, H_f (Alps). Ground thin section (25 μm), bright field, natural size 1.5 mm.

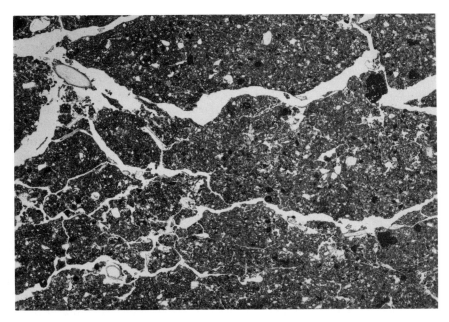

Figure 42. Mull, dense fabric with numerous cracks, which are often used by roots. *Terra fusca* A_h (4 cm) (Austrian limestone alps.) Ground thin section (25 μm), bright field, natural size 5.5 mm.

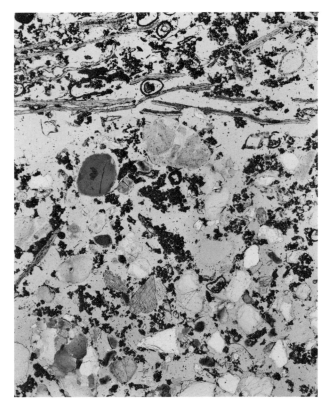

Figure 43. H and A_{hh} horizon of a raw humus. In the H (upper quarter of the picture) of plant residues foliated bark residues still remain; some root residues (occurring transversely, black rings: mycorrhiza mantles) may also be recognized. The A_h has an agglomeratic fabric: flocky to granular, dark small aggregates (loose clusters of droppings of enchytraeids) of organic fine substance between loosely arranged quartz grains. Beech raw humus. (Northwest Germany). Ground thin section (25 μm), half-crossed polarizers, natural size 5.5 mm.

pores are scarcely visible in the 30 μm thin section. The compact forms can be recognized more easily by stereomicroscopic investigation of the thin section; in the microscope only a slightly inhomogeneous distribution of the fine substance is seen in patches of 60 μm in size. Here the pores are about 10 μm in diameter and therefore important for retention of plant-available water. These fabrics up to now have been found in soils with loess-like grain size distribution and with transition forms from mull to moder. In those soils they often comprise more than half of the entire soil material. They are formed by activity of enchytraeids as probably the agglomeratic fabric also.

Spongy fabrics and dropping fabrics in the mineral soil are widespread. A well-developed spongy fabric usually can be found in chernozem and related soils (see *e.g.*, DUDAS and PAWLUK [1969, 1970b], GHITULESCU and STOOPS [1970]). The latter have performed micromorphometric measurements and found an amount of spongy fabric and pedotubules higher than 50 % down to 140 cm depth. The humus form of those soils is mull. Small areas of spongy fabric and dropping fabric, however, are found also in the B horizons of other soil types with humus-form moder, which have a denser fabric in the A horizon (BABEL [1972b]); also profiles and microphotos given by KOWALINSKI [1969/70], and PETTAPIECE and ZWARICH [1970] seem to show the same phenomena. At least in some cases they are the remains of a former animal activity, which has ceased after a change of environmental conditions, especially a change in vegetation by man.

Platy fabrics of the A_h horizon are formed in humus forms that are less favorable in biology (BECKMANN and GEYGER [1967]; Part IV C; Figure 42). An extreme formation found in arable soil was called puff-paste (flaky) structure (DeLEENHER [1967]). Other formations are the soil crusts in cultivated soils of the arid and semiarid regions, which show in thin sections a high density and sometimes horizontal fracturing (EVANS and BUOL [1968]) due to non-biological processes of fabric formation — swelling, shrinkage, and frost.

The fabrics quoted show transition formations according to mineral grain size distribution and activity of soil organisms.

In the *covered fabric* (compare KUBIENA [1953]) mineral grains are encased and cemented by organic matter (Figure 45). The covers often show a tangentially layered structure. The inner-most layer is brighter in color and consists of material rich in mineral matter; the outermost

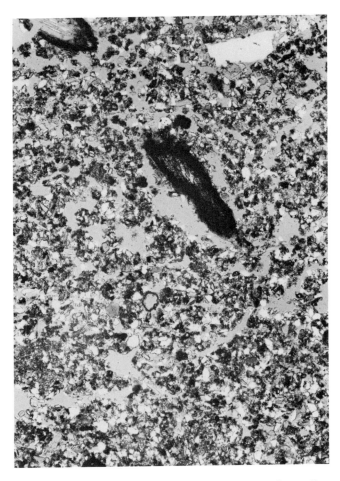

Figure 44. Dropping fabric of enchytraeids, in the A_{hh} horizon of a mull-moder (loess). Above middle to left, loose-lying droppings; above right, little compacted; beside, compacted dropping fabric; below, the droppings are hardly recognizable as primary aggregates in the compacted fabric, in this part some shrinking fissures. (Southwest Germany). Ground thin section (20 μm), half-crossed polarizers, natural size 1.5 mm.

layer is darker and nearly pure organic substance. Shrinkage fissures through the organic layer regularly are found in ground thin sections, but other irregular separations also occur (ALTEMÜLLER [1962a], ALTEMÜLLER and KLINGE [1964]; see also KARPATSCHEWSKI [1960]). This is the fabric of the B_h horizon of podzols. The covers are formed by precipitation of dissolved organic substances washed out from the humus layer (see Part III D 2b).

C. Fabric Classifications

KUBIENA [1938, 1953] describes the fabric formations of humus soil horizons which were known up to that time. He has named some of them (Part IV B). KUBIENA [1970], in part already [1953] names the fabrics from the soil types or the humus forms, for which they are characteristic, e.g., raw humus fabric (for the fabric often found in L_v and F_r, which here is named laminated fabric, Part IV B 1), moder fabric, and mull fabric (for the spongy fabric, Part IV B 3).

This nomenclature has provoked confusion in some cases, as the humus form comprises all horizons of the humus profile, and therefore, e.g., in a moder profile both, a raw humus fabric as a moder fabric may occur.

A systematic description of fabrics of mineral soil horizons is given by BREWER [1964]. But fabric formations caused essentially by organic substances or organisms are treated only incidentally. Brewer's terms "pedotubules" and "aggrotubules" have been quoted above (Part IV A, B). "Fecal pellets" are mentioned as a special group of "pedological features." Only "single fecal pellets" and "welded fecal pellets" are named.

BECKMANN and GEYGER [1967] give a simple classification of the void, aggregate, and structure forms of the mineral soils. There the fabric types of humus mineral soil horizons are taken into account. All quoted "structure forms" ("Spaltstruktur, Fugenstruktur, Bröckelstruktur, Porenstruktur, Schwammstruktur, Krümelstruktur") can occur in humus mineral soil

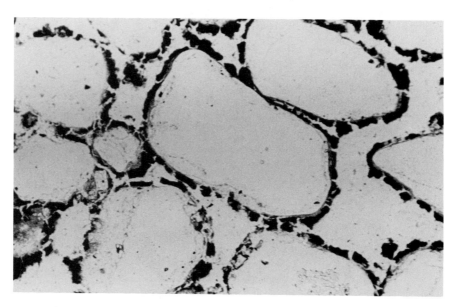

Figure 45. Covered fabric. Quartz grains enclosed by the covers and cemented by them. Inner, brighter cover layer mainly mineral, outer one, organic. Radial shrinkage cracks in the covers. Podsol (from ground moraine). B_h (Northwest Germany). Ground thin section (20 μm), bright field, natural size 0.5 mm.

horizons. The "Krümelstruktur" (= crumb structure) corresponds to a spongy fabric where the aggregates are not combined; it occurs in mull at the surface of the mineral soil. The "Porenstruktur" (= pore structure) corresponds to a compacted spongy fabric, where irregularly formed pores remain; it occurs in the lower parts of the A_h horizon of mull. "Spaltstruktur" (= fissure structure), "Fugenstruktur" (= joint structure), and "Bröckelstruktur" (= fragment structure) occur in humus formations where the nonbiological processes of fabric formation are stronger than the biological ones.

BARRATT [1968/69] proposes a comprehensive classification of microfabrics of humus soil horizons. The criteria of division are the following:

1. skeletal materials ("-skel") or plasmic materials ("-col");
2. organic particles ("humi-"), mineral grains ("lithi-"), organic substance as pigment, bound with ciay ("mulli-"), clay ("argilli-");
3. further subdivision according to kinds of tissues or minerals (*e.g.*, "calcitic," "silicic," "lignic," "mycetic");
4. shape of micro structure (*e.g.*, "platy," "pelleted," "spongy");
5. size of particles and aggregates (six classes from "very coarse" to "extremely fine");
6. fabric or arrangement of particles and aggregates.

From this it can be seen that this classification fundamentally is a short description, which has been standardized. Its advantage is that it makes possible a description of all existing microfabrics in a reproducible way. Its disadvantage is that typical fabric formations (typical by morphology as well as by their correlation with certain soil-forming processes) cannot be recognized by nomenclature. A very typical fabric such as the covered fabric is to be termed as humicol-silicic lithiskel complex. While the classification proposed by Barratt can be a useful aid for description of microfabrics, in the cases of well-known typical fabrics, true fabric names nevertheless may be used.

A proposition, taking into consideration the principles of Kubiena, Brewer, Beckmann and Geyger, and Barratt, has been presented by PARFENOVA and YARILOVA [1969].

JONGERIUS [1970] gives a nomenclature for "regrouping phenomena" in Dutch soils, where activities of soil fauna and vegetation are expressed by such terms as "fauna- pedoturbation" (the aggro- and striotubules of BREWER) and "fauna-" and "flora-pedocompaction" (compactions around earthworm channels or roots).

The classification of fabrics of humus soil horizons is still unsatisfactory. One of the reasons is the lack of knowledge about the manifold micromorphology and development of humus profiles.

V. The Micromorphology of the Most Important Natural Humus Formations

The most important terrestrial humus formations were classified by MÜLLER [1887] into three humus forms, which KUBIENA [1953] named mull, moder, and raw humus. These humus forms are defined morphologically. On the basis of micromorphological investigations, Kubiena was able to subdivide and to supplement them. Moreover, he has given short micromorphological descriptions of the submerged and semiterrestrial humus forms, some of which had been first described by VON POST [1862]. Accordingly, the only comprehensive classification of natural humus formations, which is based to a large extent on micromorphological properties, is due to KUBIENA [1953, 1970]).

The following account of the micromorphology of natural humus formations makes use essentially of the differentiation into humus forms given by Kubiena. It is not intended to modify or to supplement this classification. The purpose of this account is not to describe the system of classification but to describe the micromorphology of natural humus formations. At the present time, moreover, there is not enough micromorphological information material for firmly based changes or supplements, which are certainly desirable in some places.

The term "humus form" is used here as by Kubiena. It always refers to the humus formation as a whole; i.e., by it is meant not only chemical and physical features but, above all, a certain profile structure with all its horizons, its particular internal fabrics as well as the whole of the life in these (KUBIENA [1953, 1955a]). Most humus formations are typified by a special horizon. For raw humus this is the F horizon, or for mull, the A_h horizon. There can also be transition formations, e.g., in terra fusca under forest there is often an F horizon (as found in moder) occurring over the mull (A_h horizon) (KUBIENA [1953]). Still more marked deviations from the typical profile structure can often arise with rapid changes in the environmental conditions. This has induced HARTMANN [1952] to use the term "humus form" for the individual horizons.

A difficulty with the comprehensive appraisal of the literature also lies in the fact that even where the names of humus forms used by Kubiena are employed, there is no complete uniformity in their application. The same is true sometimes for the subhorizon designations, L, F, H. The various authors often do not clearly state how they use these terms (compare, however, BABEL [1971a]).

Other humus classifications that are not taken into consideration here are given by HARTMANN [1952, 1965], HOOVER and LUNT [1962], EHWALD [1956], BARRATT [1964], and WILDE [1971]. Except for Barratt, who gives a classification of grassland humus profiles, they concern forest humus forms. All are largely based on macromorphological features—Barratt and Hartmann on micromorphological ones also.

In recent years some detailed contributions to the micromorphology of peats and of raw humus and moder have been given (see references to Grosse-Brauckmann, Jongerius, Puffe, and Babel). Many other humus forms, on the contrary, have not yet been studied in detail—especially Dy, Gyttja, Sapropel, raw soil humus. Nevertheless, the attempt should be made here to discuss them, at least briefly. (A summary of the micromorphology of humus forms, based on Kubiena, is given by JONGERIUS [1961].)

The following descriptions are based largely on investigations of ground thin sections at low or medium microscopic magnifications. The text, within the individual humus forms, is arranged as follows: general features—micromorphological features (constituents, fabrics)—discussion of the micromorphology with regard to biology and chemistry—different formations (varieties) of the humus form in question.

It must again be emphasized that all humus forms should be defined strictly morphologically. Genetics and dynamics express themselves in morphology, but the connection is known, however, only in a fragmentary manner. Its investigation is plainly the object of morphology. Genetic and dynamic ideas should not therefore enter into the definitions of humus forms. Otherwise an objective, generally valid, basis for the classification of the humus forms will be missed.

A. Nonpeaty, Submerged Humus Formations

The nonpeaty submerged humus formations have as yet scarcely been investigated from the purely soil science side. In his soil systematics KUBIENA [1953] listed the humus forms Dy, Gyttja, and Sapropel. These terms are taken from limnology where they are designations for organogenic sediments (see also the textbooks of limnology).

The use of these terms for humus forms, at least in the case of Dy, is still problematical, since Dy is, as Kubiena himself states, "poor in life to harmful to life" and, therefore, can scarcely be conceived as a humus form. Gyttja, in contrast, possesses a rich fauna and microflora, Sapropel a very specific microflora of putrefaction bacteria. The delimitation of Dy, Gyttja, and Sapropel and their further subdivision is not quite uniform, either in the limnological or in the peat science literature (see a critical survey of the literature by GROSSE-BRAUCKMANN [1961]).

Dy is a macroscopic dark- to black-brown, acid humus formation occurring at the bottom of waters rich in humus, and, according to Kubiena, is characterized micromorphologically by a high proportion of humus gel precipitates. These do not show any regular contours and structures. Larger, pure accumulations of them occur macroscopically as so-called dopplerite. Recognizable plant residues, mainly of sphagnums, also of other mosses and water plants, exist usually only in small amounts; larger quantities characterize transition forms to Gyttja. Of microfossils, predominately sponge needles and diatoms can occur (GROSSE-BRAUCKMANN [1961]). Occasionally chitin residues of aquatic animals are found.

Gyttja is a muddy, organism-rich humus form at the bottom of eutrophic and oligotrophic waters. Micromorphologically it is characterized by numerous microorganisms and their residues, as well as by droppings and residues of droppings of aquatic animals. Recognizable plant residues (coarse and fine detritus of shore plants) can form a considerable proportion. In thin sections the appearance of diatom shells is characteristic. The organic fine substance often shows fissuring in ground thin sections (KUBIENA [1943a]).

The formation can vary in special cases depending on the variety. Von Post (see in GROSSE-BRAUCKMANN [1961]), who in 1862 introduced the term "Gyttja" into scientific usage, distinguished altogether 10 groups, each according to the inorganic impurity, the kind of waters etc. KUBIENA [1953] specified eight varieties. Many of them have an Äfja, an algal layer at the surface, which can contain, in addition to diatoms, especially blue-green algae, but also pellicles of bacteria (KRAUSE [1964]). Mineral components are usually present in considerable amounts; thus in the case of chalk gyttja (KUBIENA [1953]) often large quantities of chalk occur partly in the form of mollusk shells and partly in the form of tiny crystallites less than 1μ in size.

There are scarcely any investigations based on ground thin sections. Other microscopic investigations were made largely within the framework of hydrobiological and peat studies (for literature, see GROSSE-BRAUCKMANN [1961]).

Sapropel is a blackish to deep-black humus form, which is muddy, partly loose and partly dense. It is formed at the bottom of eutrophic waters under anaerobic conditions and is characterized by the occurrence of hydrogen sulfide and methane. It is produced from about the same starting material as Gyttja. According to KUBIENA [1953], reddish flocks of sulfur bacteria may often be found; droppings of small animals are absent. Occasionally elementary sulfur occurs in whitish-yellow films, but iron sulfide always appears as a bluish-black to black pigment. This often gives rise, on drainage and liming, to the formation of gypsum crystals, which may be recognized microscopically as prisms, pinacoids, lenses, or irregular grains. Sometimes bubble-like pores occur which are produced by the putrefaction gases. There are transition formations for Gyttja, as for Anmoor. Special thin section investigations are lacking.

B. Peats

Here, all peaty humus formations—low moor, transitional, and high moor peats (fen, carr, and moss)—will be dealt with together. In fact their micromorphology is generally very similar. Moreover, the terrestrial humus formations produced by cultivation of peats are relatively independent of the botanical composition of the parent material. Therefore, even in some of the works on the micromorphology of peats it is not clearly stated which kind of peat was investigated. Work can be done on ground thin sections. Often very severe shrinkages can, of course, interfere. It is better to use sliced thin sections, which offer, in addition, still other advantages. They can be readily prepared also from peat samples in natural arrangement (Cremolan method, see Part VI A 3).

1. *Primary Peats*

Primary peats here mean peats that have not been exposed, through a change in ecological conditions (particularly drainage), to secondary decomposition processes. The description here given is valid for the slightly and moderately decomposed forms mainly investigated hitherto.

As *constituents*, plant parts, many of which can still be identified botanically, can always be recognized with the naked eye (Figure 7; GROSSE-BRAUCKMANN [1963, 1964]). Correspondingly, most peats in thin sections show a large proportion of readily recognizable plant residues. These are sometimes almost colorless (sphagnum residues), usually intensely yellow, brownish-yellow, or brownish-red, and sometimes dark-brown to nearly opaque. Peculiar to more strongly decomposed peats is frequently a striated, juxtaposed arrangement, like nests of brightly and very darkly colored parts of tissue residues (Figure 50) (KUBIENA [1943]). Droppings of aquatic and also of land animals are very rare; in the superficial parts of natural sphagnum peat bogs they occur occasionally (GRACANIN [1962]), similarly also on the underwater surface of sedge bogs (KRAUSE [1964]). In deeper peat layers, droppings were still found by GROSSE-BRAUCKMANN and PUFFE [1964] in individual instances, but only if they had been protected from disintegration by their position in feeding cavities within woody parts, stalks, or leaves.

The same authors found fungal hyphae in places also, mainly in woods and in moss stalks, but also in the fine substance; fungal spores and sclerotia-like formations also occur—these more frequently in high moor than in low moor peats. Besides the droppings, greatly disintegrated material, in which however cell wall fragments are still clearly recognizable, may be found in variable but usually small amounts. In high moor peats it is derived predominantly from the woody and herbaceous plants of the high moor (Figure 7).

As a last group of constituents, homogeneous formations without definite structures should be mentioned (Figure 29). They are characteristic of peats. Often they are more or less fissured, yellow, brownish-yellow, not infrequently very dark to opaque. They often occur in layers, only a few micrometers thick, between the plant residues or as fillings of small spaces. Occasionally, however, they can become visible even macroscopically as dopplerite. They are accumulations of humus substances separated from the solution phase. These formations are, in part, not distinguishable with certainty from more or less topochemical transformation products of plant residues, which are recognizable only by the contours but no longer by the internal structures; such pictures can be presented, *e.g.*, by residues of cotton-grass-leaf sheaths (Eriophorum). Moreover, sometimes, as GROSSE-BRAUCKMANN and PUFFE [1964] have likewise shown, in dark parts seeming at first to be quite homogeneous, a still considerable proportion of well preserved but strongly disintegrated cell wall fragments may be recognized with the polarization microscope. In such forms it may be assumed that dissolved humus substances have precipitated in and between the still-intact structures.

The *fabric* that is built up from these constituents is determined by the high proportion of rather large plant residues. These form a laminated skeleton whose interstices are partly more, partly less filled in with finely divided residues and with homogeneous humus accumulations. In fresh, water-saturated peats the latter can thickly envelope the plant residues. In thin sections they are often shrunken characteristically so that cracks are added to the natural pore spaces. In slightly decomposed peats these humus portions without definite structures, form only, in places, thin horizontal layers (not more than 10 to 20 μ) between the plant residues.

As to the *decomposition of the plant residues* in primary peats, something more can be added from the works of GROSSE-BRAUCKMANN [1963, 1964] and GROSSE-BRAUCKMANN and PUFFE [1964].

The phenomena are, in principle, similar in high and low moors, although in low moors many plant species are decomposed until they are unidentifiable, depending on the more eutrophic conditions.

Sphagnums are preserved the best; they usually show distinct decomposition symptoms only in the stems (KOX [1954]). Herbaceous and woody plants, on the contrary, are considerably more strongly decomposed. Of these, the best preserved are subterranean shoot parts; the aerial parts are the worse preserved the drier the local site on which the plants in question are growing. Thus in peats of northwestern Germany, for example, leaves of the very common *Calluna vulgaris* are much more scarce than their basal stem parts and roots. The most easily decomposed tissues are generally parenchyma, followed by lignified tissues, although these are readily consumed by animals; barks and the outer parts of cortex and epidermis are the best preserved. This sequence corresponds exactly with the findings of plant residues in terrestrial humus forms (for details see Part III B). The degradation of cellulose, detected by double refraction, often parallels the decomposition of the structures. But there are also apparently not infrequent instances in which quite well-preserved structures no longer show double refraction. This, according to the author's observations, however, cannot be taken as in fundamental contrast to the conditions in the case of terrestrial humus forms (BABEL [1964]) (see also Part III).

On the whole it follows from these microscopic investigations, in agreement with already known notions, that the primary chemical properties of particular parent materials, on the one hand, and the anaerobic conditions, on the other, are the decisive factors leading to peat formation. Animals and fungi do not play any essential role. (Nothing can be said of the importance of bacteria, since they have not yet been made visible in peat thin sections.) A larger proportion of humus substances than with terrestrial humus formations is moved in solution and, on precipitation, forms fabric constituents of its own. Where, under dystrophic conditions the precipitated substances form continuous pure layers at the bottom of the waters, the formation of Dy is involved (see Section V A). No precise statement can be made as to the extent of the mineralization which, in addition to the removal of decomposition products in solution could have caused the complete disappearance of many tissue parts. The high values for the mineralization, which have been determined from comparisons with the dry matter production of living moors up to 80% (HOLLING and OVERBECK [1961]), could by no means have been expected, however, from the microscopic picture.

KOWALINSKI and KOLLENDER-SZYCH [1960] tried to characterize peaty muck soils by micromorphological analyses of constituents of humus and by chemical methods. They grouped the constituents into different groups according to BABEL [1965b].

2. *Secondary Humus Formations from Peats*

The decomposition and humification processes, which are promoted by drainage and cultivation of moors, may be followed very well micromorphologically. The account given here is based essentially on the works of PUFFE and GROSSE-BRAUCKMANN [1963], FRERCKS and PUFFE [1964] and JONGERIUS and PONS [1962a, b]; numerous illustrations may be found therein.

The increase in droppings is particularly striking. Feeding cavities with mite droppings may often be found down to a depth of some decimeters in sphagnum stems, in the stem bases of Eriophorum, in woody Ericaceae residues, and similar places. It may be very beautifully observed how first the thin-walled and later often nearly the whole material is converted into droppings, which remain well preserved in their egg-shaped contours for some time. Only the very resistant tissues usually remain behind undisintegrated. Similarly in Gyttja muck soils droppings have been found revealing recent biological processes (UGGLA *et al.*, [1969]). In addition, fungi occur both in the nondisintegrated plant residues as well as in the droppings and their breakdown products. Sclerotia also are frequently found. The decomposition of the droppings and the tissues decomposed by fungi lead to an enrichment of very fine-particled fine substance. This can become very dense, especially when formed by fungal degradative activity, and it tends to produce shrinkage cracks (PUFFE [1965]).

Intact tissues in secondary decomposed peats are often much darker than in primary peats; they can commonly appear opaque in 30 μm thick sections. Also, droppings can darken until opaque. Structureless substances, which likewise are very dark to opaque, are more frequent than in primary peats. In the subsoil they are often enriched (B horizon) and, as thick masses, surround the still quite well-preserved plant residues (JONGERIUS and PONS [1962a]).

According to JONGERIUS and PONS [1962a]), in Dutch moors microbiological transformations at first go on side by side or after physical "maturing processes," which are not described micromorphologically. The easily decomposable plant substances, such as pectin, cellulose, hemicellulose, and proteins are broken down; their humification products, which go into solution, are thought to cause the black color of the still-intact plant residues. Degradation by soil animals ensues usually only after this phase, but it is still accompanied by microbiological transformation processes. The droppings decompose, especially in the nutrient-poorer formations, after a short time.

On the whole, a humus form is thus produced which is related to moder. However, since the substrate—namely peat—is situated *below* the main sphere of action of the decomposition agents, it shows a fermentation layer that lies under the humus substance layer. From the top downward, however, organic material is likewise supplied by the living plant cover, so that one obtains also, as usual, above the humus substance layer, a fermentation layer. Thus there

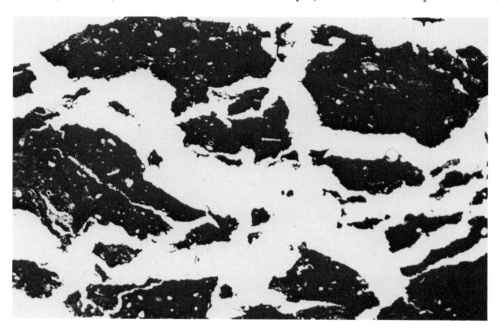

Figure 46. Pitch peat (strongly decomposed peat moder). Sharp-edged, dark aggregates with few plant residues. Cultivated peat over gley (South Germany). Ground thin section (20 μm), bright field, natural size 1.5 mm. (From a preparation by L. Menge, Stuttgart-Hohenheim.)

develops an L – F – H – F profile, as PUFFE and GROSSE-BRAUCKMANN [1963] and FRERCKS and PUFFE [1964] have shown in the case of drained high moors.

Under arable land the profiles of moder-like formations from peats are frequently characterized, in addition, according to JONGERIUS and PONS [1962a], by an organic B horizon which is formed by separations of dissolved humus substances migrated downward from above. In arable cultivation strong decomposition processes promote an unfavorable further development of peat moder, which is known agriculturally in German as "Vermullung," and which results in formation of pitch peat (KUBIENA [1953]) (Figure 46). It consists predominantly of fine substance, which forms a very dense fabric, which, on drying out, breaks down into hard, sharp-edged, scarcely wettable aggregates, which are deep brown.

According to JONGERIUS and PONS [1962], with larger proportions of mineral material and favorable physicochemical conditions but independently of the botanical composition of the peat, the development follows another direction. After similar initial stages, earthworms appear as secondary decomposers, and lead to the formation of a more or less stable spongy fabric, as is characteristic of typical mull. The lower parts of the profile can still have the characteristics of moder. This mull however always contains dark to opaque constituents, 25 to 50 μm in size, in the brown ground mass. As a rule, they exhibit plant structures, although some of them are without structure. The smaller the number of these inclusions, the more favorable seems to be the mull formation, as comparisons with agricultural experiences have shown. Peat-like moder formations also have small opaque constituents, as a rule (observations of the author; also HARTMANN [1952]; see also the following section on Anmoors).

Secondary humus formations from peats sometimes show transitions to Anmoor.

C. Anmoor

According to KUBIENA [1953], Anmoor is a gyttja-like humus form, produced under the temporary influence of ground water or back water. It is dark grey to blackish and has an inky smell. In the wet state it is muddy; in the dried condition it is transformed into an earthy fabric. The English term "Fenmull" appears to be synonymous with Anmoor (MADER [1953], EHWALD [1956]).

The micromorphological description is based on the works of KUBIENA [1953], JONGERIUS [1961], and KOWALINSKI [1964a, b]. Investigations of the aggregate fabric of Anmoor were made by MENGE [1965].

Plant residues do not usually play any quantitatively significant role as *fabric constituents* of the Anmoor; they are dull brown, dark brown, or opaque. Very dark or opaque plant fragments seem, in general, to be typical of humus formation under water or, at least, severe intermittent wetness (see also Figures 14, 16; also ALTEMÜLLER [1960]; this is true also for peats; Figure 50).

Fine substance is present among other things, in the form of strongly humified dropping material. The original dropping forms are usually no longer preserved. In addition, part of the organic substance always occurs in holorganic homogeneous bodies without definite structures. From these Anmoor resembles peats and nonpeaty submerged humus formations. This material can be developed in some Anmoor formations as illuviation cutans which cover pore walls and exhibit flow structures (JONGERIUS and JAGER [1964]). The cutans generally contain tiny opaque granules, scarcely more than 1 μm in diameter. Sometimes the cutans consist completely of opaque material.

Diatom shells are a characteristic feature present. Mineral matter is present in varying amounts. Often fine iron precipitates or iron concentrations occur as encrustations on roots. In some formations related to sapropel, gypsum crystals may be found as new formations.

Variations in hydrological conditions produce variations in *fabric*. Under the original semiterrestrial conditions a rather dense fabric usually forms, but if terrestrial conditions arise, a crumb fabric is formed. On the whole, dull and dark colors are dominant. According to JONGERIUS [1961], within the inner fabric gradually merging color transitions from very light brown to very dark and to opaque regions are typical (Figure 49). Mineral grains are loosely arranged. In organic B horizons, however, above all under plough land, the mineral grains are surrounded by black, homogeneous, often fissured, humus material. These results of *Jongerius*

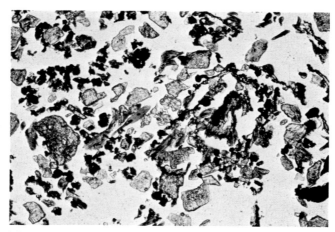

Figure 47. Rendzina moder, A$_h$ horizon. Reddish-brown plant residues, predominantly of mosses. Organic fine substance particles (dropping residues) dark to opaque. Calcite grains. Loose arrangement (central uplands of Germany). Ground thin section (35 μm), bright field, natural size 1.2 mm.

Figure 48. Tangel layer. The dark-colored fine substance which surrounds the needle residues, originates from earthworm droppings. It contains in part holorganic, in part mineral particles of sand and silt fraction (quartz, mica, calcite). The small, longish, brown particles are needle fragments. All the plant residues are of *Pinus mugo* needles. Above, right-hand side, a needle with well-preserved mite droppings; below that, a needle with densely lying residues of decomposed droppings; below, a needle, length-wise, with two fungal congestion zones. *Pinus mugo* tangel, F$_r$A$_{hh}$ horizon (Alps). Ground thin section (30 μm), bright field, natural size 4 mm.

have been obtained from Anmoors secondarily subjected to terrestrial conditions; Jongerius' definition of Anmoor is undoubtedly not completely identical with that of Kubiena.

In wet raw soils of the arctic regions KUBIENA [1970], found very typical "mud coatings,"

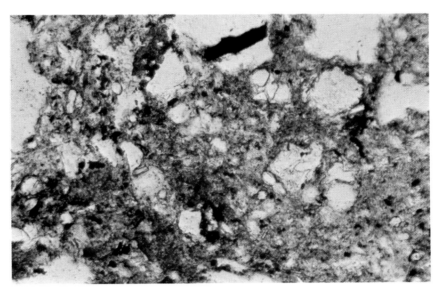

Figure 49. Anmoor. From the mineral fine substance, some organic fine substance particles stand out dull brown, dark brown, or opaque. Gradual color transitions in the mineral fine substance (larger quartz particles appear like holes in the fine substance). A_h (3 cm) (German Baltic coast). Ground thin section (15 μm), bright field, natural size 450 μm.

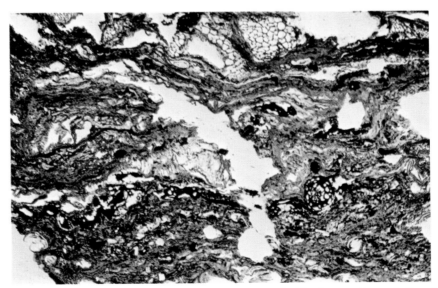

Figure 50. Strongly decomposed peat. Top, easily recognizable tissues; the majority of the plant residues compressed in bands; some nearly opaque bands and groups. Carr (wood peat, 25 cm) (Black Forest, Germany). Ground thin section (15 μm), bright field, natural size 1.1 mm.

which encase soil aggregates and isolated sand grains and which sometimes are very rich in humus, which has precipitated from the soil solution.

The dark-colored aggregates are usually very rich in iron, according to chemical studies by KOWALINSKI [1964a, b], so that this author speaks of aggregates of a silica-iron-humus type, which he contrasts with lighter-colored aggregates of a silica-aluminium-humus type.

The *variety* of Anmoor that develops depends on the amount and nature of mineral constitutents and upon hydrological conditions. Dystrophic varieties related to peats have a higher proportion of plant residues. Anmoor can pass over into Gyttja or into mull under certain hydrological conditions.

D.   Terrestrial Humus Formations

1.   *Raw Humus and Moder* (*Silicate Moder*)

The two most important terrestrial superficial humus forms, raw humus and moder, have many properties in common. In English and American classifications they are, therefore, grouped into one single humus formation mor (*e.g.*, HOOVER and LUNT [1952]).

Above all, both show macroscopically the characteristic horizon sequence (cf. Section II C): litter (L layer), decay layer or fermentation layer (F layer), humus substance layer (H layer). This humus profile represents the successive stages of decomposition of the litter (BABEL [1971]).

There is a principal theoretical difference between raw humus and moder. Fungi are dominant in the decay layer of raw humus, while soil animals are dominant in the decay layer of moder. Micromorphological investigations have often shown that this is only a quantitative difference. The definition of the two forms is not given by this genetical difference, but by differences in the macromorphology of their profiles. Transitions between the two forms, however, often occur. They have been investigated in the classical work of MÜLLER [1887] but further comparisons are needed.

(a)   Common Properties

The *litter layer* (L) consists of macroscopically slightly changed, only browned, loose-lying plant litter, mainly from leaves or needles. The parenchymatous tissues in them are browned, the others unchanged. Fine substance is present in only minimal proportions (in the L_n of a moder were found 2%, in the L_v 13% (BABEL [1971a]). It exists mainly in the form of a dark deposit on the surfaces of the leaves. This often consists chiefly of very dark to opaque droppings of collemboles and enchytraeids, which may sometimes accumulate in the lower part of the L layer (ZACHARIAE [1963, 1964], BABEL [1964b]).

In the *decay layer* (F) the residues of fine roots occur in addition to fragments of leaves and needles (Figure 1). They occur in very high amounts preferably under ecologically unfavorable conditions (for quantitative assessments see MEYER [1967]; see Figure 12). Leaf residues are more or less disintegrated. Changes, which point to chemical alterations, do not ensue at the same time as the mechanical disintegration, however, but usually occur later and gradually (for further details see Part III B and C). Most of the tissues in the upper parts of the decay layer are still readily recognized microscopically and as a rule easily interpreted. Their structures, however, are somewhat distorted. In addition to the parenchymatous tissues, the lignified tissues gradually become browned. Yellow, brownish-yellow, and brownish red colors predominate in the transmittent light. In the lower part of the F layer most of the parenchymatous and the lignified tissues are destroyed and phlobaphene-containing tissues (see Section III B 2 c α), are enriched.

Fungal hyphae are always to be found. They are usually stout-walled with a characteristic dark brown color. They grow most frequently on the surface of the leaf residues, but they are also found in the interior of the tissues. In addition, they occur in the fine substance, usually as small fragments. Frequently present are deep brown plectenchyma, which are opaque in places and ring-shaped in cross section. These are mycorrhiza mantels. The roots that were surrounded by them are often decomposed completely or to small residual amounts (Figure 21).

Droppings and dropping residues increase toward the lower parts of the decay layer. They are still often well preserved in their characteristic contours. This is particularly the case when they lie in feeding cavities that are frequently found in pieces of wood and in needles. Their color, as a rule, is that of the parent material. Free-lying droppings are often eaten again by other animals. If that is not the case, they sometimes fall apart only slowly (ZACHARIAE [1965]). This is true especially under dry conditions (RIGHI [1969]).

In the fine substance, in addition to fungal hyphae and dropping material, there are tissue and cell fragments left behind after the partial destruction of the tissues by fungi or animals. In debris preparations of the fine substance one occasionally finds chitinous parts of animals (MÜLLER [1887]).

The *humus substance layer* (H) is macroscopically black-brown to nearly black. Plant residues become rarer. These are mainly the residues of roots; sometimes residues of barks and leaf stalks occur (Figure 43; (BABEL [1972b]). In the phlobaphene-containing tissue residues, occasionally a blackening of the cell content substances may be observed (BABEL [1964b]). Sometimes also, the whole plant residues appear blackened as though carbonized; but the plant parts are structurally still well preserved (PARFENOVA and YARILOVA [1960]).

In the fine substance of the H layer, which consists mostly of the same consituents as those of the F layer, a darker coloration is usually shown than in that of the F layer. In thin sections dull and dark brown colors are predominant (compare RIGHI [1969]). Very small bright-colored tissue fragments occur occasionally. The fine substance is not homogeneous. Under stronger magnification, it still shows considerable amounts of cell wall fragments, which are enclosed in the browned, irregular granular flocky mass and which often become visible only by fluorescence illumination (Figures 24, 23, 22; for more details see Part III C 2). In un-favorable moder forms ("dysmoder") gelatinous accumulations of the browned granular masses were observed (DELECOUR and MANIL [1962]). In some, especially much wetter forms, fragments of fungal hyphae are numerous. They can constitute about half of the whole material (Figure 21) (ROMMELL [1955]). Frequently sclerotia may also be found (BABEL [1964b]); where they occur in large amounts, the microfabric is termed "sclerotial phase" by BARRATT [1968/69]. Finally, single quartz grains may be found.

The *mineral humus horizon* (A_h) is usually very dark black grey, or black brown. When the mineral matter is sand, it contains the same organic constitutents as the fine substance of the H layer (Figure 43). In the lower part of the A_h horizon the organic particles are brightened appreciably (BAL [1970]).

The *fabric* in the decay layer is composed of plant residues as a skeleton and the fine substance. The fine substance is deposited partly in separate, small, loose aggregates, and partly in larger, looser or dense masses between the skeleton. In ground thin sections one still observes many voids. Plant residues become rarer in the lower decay layer. They are often included in the fine substance of this layer.

In the humus substance layer exists a more or less dense fabric. It is usually fissured in ground thin sections. It is formed by the fine substance combining to give larger aggregates. But often a dropping fabric can be observed also.

In the mineral humus horizon (A_h) an agglomeratic fabric is usually formed. Occasionally brown fungal hyphae can be found as binding materials between the sand grains (BARRATT [1962], BURGES [1963]).

Traces of microbial and animal degradative activity occur particularly in the decay layer. The complete disappearance of whole tissue parts generally indicates microbial decomposition. With animal decomposition, droppings remain behind, which, in raw humus, frequently come from mites and enchytraeids. In addition to these in moder, many other animals are present.

The organic material that is present, especially in moder, enters the mineral horizon (A_h) mainly by particles being washed in with rain water. Part is also transported into the A_h horizon in the dissolved form (JONGERIUS [1961], HARTMANN [1952]). This, however, cannot be observed micromorphologically for certain. An indirect indication of the movement of dissolved organic substances in the poorer forms is the occurrence of uncovered quartz grains (KUBIENA [1953]). As to an indication of a chemical change of the organic substance by admixture into the A_h horizon, see Part III D 2b.

In BABEL [1972b], who has investigated a range from a raw humus-like moder to a mull-moder, the profile is subdivided into eight horizons. Their definition is macromorphological. Their dynamics has been concluded from macromorphological to micromorphological investigations. In the following the *processes* are quoted at those horizons, where their effects become visible.

L_n (definition: litter without macromorphological decomposition phenomena). Very low abiological disintegration and chemical alteration. Loose fabric, still moved by the wind.

L_v (definition: marked macromorphological alterations of the litter, but nearly without fine substance). High solution processes and microbiological actions. Formation of a laminated fabric.

F_r (definition: plant residues are predominant, low content of fine substance). High activity of primary decomposer animals. Breakdown of the laminated fabric. Penetration of roots.

F_m (definition: ratios of plant residues and fine substance are about the same). Further activity of primary decomposers, activity of secondary decomposers, resulting in high disintegration of plant residues. Beginning development of a fabric formed by fine substance only. Highly penetrated by roots.

H_r and H_f (definition: organic fine substance predominant, H_r (few) and H_f (no) above ground plant residues). Horizon with low decomposition processes. Further activity of secondary decomposers. Enrichment of plant substances which are resistant to decomposition and of newly formed stable substances. Dropping fabric or dense fine substance fabric.

A_{hh} (definition: upper part of the A_h horizon, very rich in humus). The horizon is formed by all sorts of blending processes. With exception of the high content of mineral material, only small changes in comparison to the H horizon. Low increase of chemical processes.

A_{hu} (definition: lower part of the A_h with lower humus content). Blending of organic substance only in solution form or by roots. Low turnover of the organic substance and very low animal activity. Droppings are sometimes enriched and form dropping fabric.

(b) Raw Humus

For the differentiation of raw humus from moder, the great thickness of the decay layer is decisive (KUBIENA [1953]). (In the literature written in German humus layers with a thick and dense H horizon often are also termed raw humus; here, however, the term raw humus is not applied for those humus formations.) Within the decay layer, the upper part predominates. It is composed predominantly of scarcely disintegrated but often soft or delicate and densely lying plant residues. They form a laminated, sometimes matted cover. Biologically, this profile formation involves low activity of the decomposing organisms, especially of the animals.

Special groups of fungi are capable of living in the frequently severely desiccated upper decay layer (MEYER [1964]).

Raw humus is produced from a vegetation, the litter of which is poor in nitrogen and rich in thick-walled tissues.

Among the plant residues of the *decay layer*, fine roots and their residues play a very big role, in addition to the material coming from the litter. The fine roots form a dense reticulum that permeates the remaining material. The root residues are preserved for a long time (MÜLLER [1887]). From microscopic investigations of the plant residues, the complete disappearance of whole tissues is often shown, which indicates fungal degradation. These phenomena, which are dealt with in more detail in Section III, occur also in other humus forms. In the upper F layer of raw humus, however, they indicate the only severe decomposition process. More precise investigations on the morphology of fungal degradation in raw humus are lacking. A part of the leaf parenchyma remains long preserved, however, in the browned state until disintegrated by animals.

In the fine substance of the decay layer brown fungal hyphae are common. MÜLLER [1887] describes three morphologically distinguishable groups. According to MEYER [1964], *Cenococcum graniforme* (Sow.) Ferd. et Winge, belonging to the fungi imperfecti, plays a significant role in raw humus formations, in northwest Germany. Strong development of an aerial mycelium penetrating the pores seems to occur particularly frequently in the upper F layer of fairly dry raw humus formations (Figure 32).

Droppings appear in the F layer in only small amounts, but they occur in all raw humus formations. The strongly disintegrated material, which remains after the droppings have fallen apart, is often not distinguishable with certainty from disintegrated residues from fungal decomposition.

In the *mineral humus horizon* (A_h) the quartz grains are without coatings. This is an indication of percolating dissolved humus substances. Precipitates of these cannot be recognized. (The humus illuvial horizon of podzols, B_h, showing such precipitates does not belong to the humus form raw humus. It often does not occur under raw humus. It is discussed in the treatment of the organic matter in the mineral soil horizons in Section III D 2 b.) Roots are normally very rare in the mineral humus horizon of raw humus.

The *fabric* of the upper decay layer (F_r) is determined by the laminated arrangement of the slightly disintegrated plant residues. These are bound by occasional bridge-forming fine substance aggregates, a great part of which is derived from enchytraeid and dendrobaena droppings (ZACHARIAE [1965]). The formation of solid felts, often striking in the fields, is caused by fungal hyphae, which in some forms can render the material matted and breakable, even in the F_m layer and in the H layer. Other differences in the fabric of the deeper horizons of moder and raw humus are not known for certain.

The *decomposition* of the plant residues is carried out mainly by fungi in the characteristic upper decay layer of raw humus. Animals also participate in the decomposition in the lower decay layer. The thickness of the decay layer shows that the processes take place slowly.

In the raw humus profile there are some *variations*. In the extreme case the F layer can become very firm and densely matted, while the H layer is nearly absent. This variety was earlier designated as "Auflagetorf" (peat humus) in the German literature (VATER [1904]) quoted after KUBIENA [1953]). This misleading term should not be used.

(c) Moder (Silicate Moder)

Litter, decay layer, and humus substance layer are about equally thick in the typical moder (KUBIENA [1953]). Often, however, litter and decay layers decline very markedly in comparison with the humus substance layer. The latter is often very rich in sand, so that in some cases in

its whole depth it is more properly designated as a mineral humus horizon (A_h). In other formations, H and A_h are clearly distinguished, however, even though the boundary is not so sharply
demarcated as with raw humus.

The *decay layer* consists of dropping materials and plant residues, whose particles are
usually smaller than in raw humus. Only a moderate proportion of the plant residues are fine
roots, in contrast to raw humus. Fungal decomposition of the tissues may also be observed
here (MEYER [1962]). Brown fungal hyphae are regularly found but rarely in large amounts.
Often various kinds of dropping forms suggest the occurrence of many species in the fauna.
Quantitatively, droppings and dropping residues regularly are very prominent (Figure 1).
MÜLLER[1887] reports that chitin residues of arthropods can often be found in the decay layer.

The *humus substance layer* shows no clear differentiating features in contrast to that of
raw humus.

The *mineral humus horizon* can be very thick in sandy soils. In such formations JONGERIUS
[1961] observed increasing dark coloration of the organic fine substance down to about 20 cm
deep, but between 20 and 40 cm a color change to lighter brown and sometimes yellow. Still
further down, the small organic aggregates fall apart completely and their residues settle on the
mineral soil skeleton. Jongerius interprets these phenomena as the results of gradual mineralization (compare also the detailed study of BAL [1970]). In contrast to the same horizon of raw
humus, the A_h horizon of moder contains varying amounts of roots.

The *fabric* of the F layer of moder is not matted and laminated as in raw humus. Fungal
hyphae are not numerous but droppings are very numerous, which indicates that animals are
very important in promoting disintegration of the plant litter. Collembola, enchytraeids, small
earthworms (SCHALLER [1950], DUNGER [1956], ZACHARIAE [1963, 1964], VAN DER DRIFT
[1964]) and insect larvae play a large role in these disintegration processes. According to
ZACHARIAE [1967], in spruce forests Tipulidae larvae often are the primary decomposers of
litter. In profiles richer in mineral nutrients, isopods and diplopods are also important.

Animals do participate in the transport of organic matter into the mineral soil horizons
(JONGERIUS [1961]) to a very small extent and distance only (a few millimeters). Above all

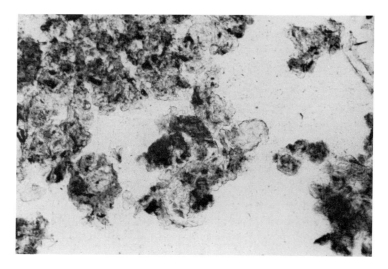

Figure 51. "Swollen moder" (see Barratt (1964)). Organic fine substance (partly with residues of
tissue structures) appearing swollen (perhaps an artifact that is characteristic of some wet moder
formations). Wet moder with transition to peat formation, H (Black Forest, Germany). Ground thin
section (20 μm), bright field, natural size 450 μm.

they effect a closer intermingling of the organic and mineral matter (*e.g.*, *Lumbricus rubellus*; see ZACHARIAE [1965]). Infiltrating of rain water was shown by VAN DER DRIFT [1964] to be largely responsible for the transport of organic matter from moder formations into lower sandy horizons. There, the organic particles in the A_h horizon often form horizontal bands, which may be interpreted only as an infiltration type of action.

Moder is much commoner than raw humus. It is the most widespread humus form of the forests of temperate climatic regions. There is a great deal of *variation* in silicate moder. MÜLLER [1887] and VAN DER DRIFT [1964] have done the only micromorphological comparisons of different varieties. Fluctuations in the thickness of the F and H layers and in the degree of disintegrations of the plant residues in the F layer can be seen macromorphologically. In the German literature the rather unfortunate terms "coarse" and "fine" moder have become established (KUBIENA [1953]), wherein coarse moder (according to DELECOUR and MANIL [1962]: dysmoder) resembles raw humus.

BARRATT [1964] mentions as a special "microfabric" a "swollen moder," which is formed at sites with temporary severe wetting. The structures of the fine substance appear swollen in thin sections. This impression is caused by a very light, lobular margin, sharply demarcated against the embedding medium. However, these structures may be artifacts which could just as well be typical of wet moder forms (Figure 51).

2. *Rendzina Moder*

Rendzina moder is the humus form of shallow soils on chalk or limestone (protorendzinas (KUBIENA [1943b, 1953]). It is a fine, macroscopically black moder, usually lacking a coherent decay layer.

A small proportion of disintegrated plant residues may be recognized microscopically in the H layer. These are often colored dark brown and their tissues are still rather well preserved. They often show traces of feeding. The main characteristic constituents are droppings showing well-maintained forms or only a partial breakdown. Their black color is the most striking characteristic of rendzina moder (Figure 47). In transmitted light they are opaque in the thicker places of thin sections. In the thinner places they are grey, brown, or colorless, and densely covered with opaque small-grained substances (calcium humates). Fresh droppings may be recognized partly by their shining brown colors; however, they appear to darken very quickly. In addition usually small calcite grains occur in varying amounts.

These constituents form a very loose *fabric*. Rarely, the droppings are slightly agglomerated. Aggregates of a higher order are never formed. Parts of the dark fine substance often encrust the clearer plant residues and the lower stem parts of living moss cushions. The calcite grains are not associated with other constituents. Because of this fabric, the whole humus formation is more or less dusty in the dry condition. KUBIENA [1938], described the fabric as rendzina fabric. The traces of feeding and the droppings originate from oribatids, probably partly from enchytraeids (ZACHARIAE [1964]).

3. *Alpine Pitch Moder*

This is the humus form of the alpine pitch rendzina (Alpine Pechrendsina [KUBIENA 1953, 1970]). It is formed at the alpine level under natural grassland on limestone, but is completely free of lime. While a decay layer is usually scarcely formed, the humus substance layer is more than 20 cm deep. The upper part (H_1) is strongly penetrated by roots, and it is rather loose, fine-crumbed, and blackish brown. The lower part (H_2) is dense, prismatically breakable, nearly black, and macroscopically homogeneous (see profile descriptions in SOLAR [1960]).

In the typical formations larger plant residues are found only in the H_1 layer and even there in small amounts. On the contrary, very small plant residues (100 μm and below) can be found in small or moderate quantities throughout (Figures 25, 40, and 41). They are usually reddish brown and often may be identified as residues of root cortex. The main quantities are a dirty brown, blackish brown in 30 μ sections, or reddish brown in thinner sections, very fine-particled material, which, in some pitch moder formations, is clearly formed as 30 to 50 μ large cylindrical droppings. This material consists of irregularly formed, strongly colored substances, partly with granular internal structures in which smaller amounts of difficult to recognize cell wall residues are included (Figure 25) (observations of the author). Under crossed polarizers the cell wall residues are visible by double refraction. Hyphae are rare. Calcite is lacking, quartz of aeolic origin is often present in small to moderate amounts.

The *fabric* appears moderately loose to moderately dense in thin sections. Between the small loosely cemented droppings a fine-pored space system remains free (Figure 40). A similar fabric is formed in other cases in which the primary aggregates show no clear dropping forms, but are irregularly delimited and loosely cross-linked with each other. In sections of the H_2 layer, shrinkage cracks always appear (Figure 41). The droppings most probably are produced by enchytraeidae (presumption of ZACHARIAE [1964] because of their great resemblance to enchytraeidae droppings in moder), in part perhaps also by collembola, which often occur abundantly.

According to KUBIENA [1953, 1970], pitch moder develops from rendzina moder under humid conditions by dissolution of the lime. There are dystrophic varieties, which contain more plant residues and more fungal hyphae and sclerotia.

4. *Mull-like Moder*

This humus form occurs especially in soils with advanced physical but slight chemical weathering and under grassland vegetation. It is very similar to mull macroscopically. It is characterized by the capacity for forming rather large aggreagtes. A true mull formation is not brought about, however, because of the lack of clay (KUBIENA [1943b, 1953]). Differentiation from mull is sometimes possible only with the help of the microscope. An F layer is seldom formed and an holorganic H layer is absent.

The mull-like moder of KUBIENA [1953], to which the humus formations described here should be assigned, probably frequently corresponds to the fine mull of HOOVER and LUNT [1952]. It is a true humus form and not a complex formation ("twin" formation) of moder and mull. In contrast, the mull-like moder of HARTMANN [1965] (see also EHWALD [1956]) is closer to the twin mull.

On microscopical investigation, plant residues can scarcely be found in the A_h horizon. The organic substance occurs almost entirely as a very fine-particled material which is similar to that in the H layer of moder. In addition, a high proportion of mineral grains of the sand fraction and predominantly of the silt fraction is always present (Figures 52, 53, and 26).

The *fabric* is characteristic. The organic fine substance particles are very intimately mixed with the mineral fine substance. Both components, which can be readily separated from each other microscopically, are contained in droppings, which are present in various formations and usually in sizes above 0.5 mm. They are evidently quite stable. Locally, nearly the whole fabric is composed of droppings.

The intimate mechanical mixing of organic and mineral substance is due to a rich soil fauna. These animals consume mineral material together with the organic food. These are julids, glomerids, insect larvae, small earthworms, and enchytraeids. Soil biological and electron microscopical studies by SZABO et al. [1962] and SZABO et al. [1964] show the morphology of the dropping and of the plant microorganisms, which mostly appear to be colony-forming.

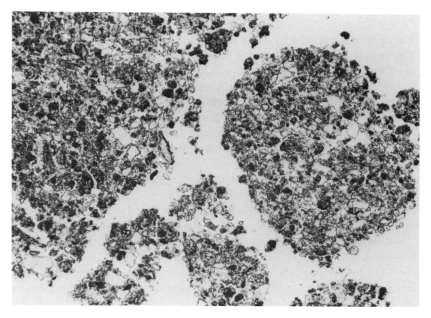

Figure 52. Mull-like rendzina moder, A_h horizon (see also Figures 26 and 53). Organic and mineral fine substance (calcite grains) are intimately mixed, but both are present in their own constituents that are readily distinguishable from each other; plant residues almost completely absent. Distinct formation of large aggregates. The great rounded aggregate to the right is a diplopod dropping. A_h (Vienna woods, Austria). Ground thin section (20 μm), bright field, natural size 1.5 mm.

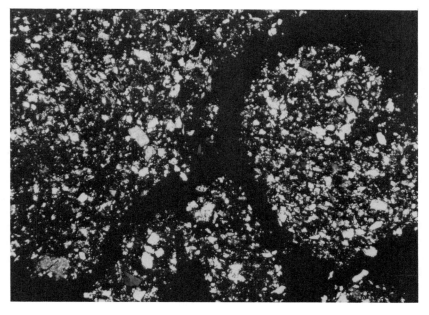

Figure 53. Like Figure 52, but under crossed polarizers. The calcite grains appear bright (fine sand and silt fraction, clay fraction missing).

5. *Mull*

Typical mull is characterized by a mineral humus horizon with a distinct clay content and a more or less pronounced crumb structure. It is colored brown or grey brown by organic matter. Apart from the color, the organic substance cannot be recognized macroscopically. It is more stable than in the humus forms discussed so far, through close attachment to the clay substance, namely forming of the clay humus complex (SCHEFFER and ULRICH [1960]). This humus form occurs in nutrient-rich soil formations with high biological activity. Mull formations under forest have sometimes a shallow decay layer under the litter; however, an holorganic humus substance layer is always absent.

The *decay layer* insofar as it is at present, resembles those of nutrient-rich moder formations. In addition to disintegrated plant residues, large amounts of droppings in manifold formations occur, especially large droppings of diplopods and the larger insect larvae. But signs of fungal degradative activities on the leaf residues are also found. Macroscopically, the most striking is a bleaching of the not yet disintegrated leaves, which is found frequently in forest soils. This, as the microscopic preparation shows, is a consequence of the disappearance of the leaf browning substances (for details, see Part III B 2 b).

In the *mineral humus horizon*, solitary intensely colored plant residues are sometimes found. Their amount varies with the season of the year; in late summer they have usually completely disappeared. They are often enveloped by mineral substances. Holorganic fine substance constituents are present only in exceptional cases. The main quantity of the organic substance exists as pigment of the mineral substance and is not recognizable as intrinsic particles, even at the highest microscopic magnifications (Figure 2).

The *fabric* of the mineral humus horizon is spongy fabric in the ecologically most favorable mull formations (Figure 2). If it is very loose at the surface of the mineral soil, the aggregates are hardly combined with each other and show their bulbous contours. Plant residues lie loosely between the aggregates. A spongy fabric occurs to usually no more than 5 cm depth. In the lower part of the A$_h$ horizon, mostly a coherent fabric occurs, which is divided more or less by fissures (Figure 42). In ecologically less favorable mull formations this fabric extends to the surface of the mineral soil. Traces of activity of earthworms or enchytraeids also can be found in thin sections of such mull formations (see Parts IV A 3 and IV B 3).

The primary disintegration of the plant residues is due to the work of small arthropods, whose droppings are quickly consumed by earthworms. The earthworms, in addition, also attack soft plant residues directly. As they consume mineral substances with organic substances at the same time, the intimate blending leads to the formation of the clay-humus complex in the gut of the earthworms. Termites (GENNART *et al.* [1961], HARRIS [1955]), enchytraeids, and probably also julids (JONGERIUS [1961], DUNGER [1963]), as well as Diptera larvae act partly in a similar manner (see Part IV A 3).

The very small proportion of plant residues and the usually complete absence of holorganic fine substance particles indicate the intensity of the transformation in mull, which takes place here presumably because of the close admixture with the clay-rich mineral substance and terminates in the stable binding of the organic matter with clay. This is in complete contrast to the mineral humus horizon of raw humus, moder, and mull-like moder, where, as already mentioned, the organic constituents are not essentially different from those in the humus substance layer.

On the basis of micromorphological investigations, particularly of arable soils from bog and Anmoor, JONGERIUS [1961] and JONGERIUS and SCHELLING [1960] distinguish different varieties of mull. They name "protomull" a humus formation in which the organic and mineral parts are still incompletely mixed and numerous large and small plant residues may still be recognized. This is at the same time the extreme example of a "heterogeneous mull." Classical mull, on the contrary, is an "homogeneous mull," in which the organic matter occurs only as pigment. "Postmull" is a form that exhibits breakdown of the crumb structure, decrease of the porosity, and brightening of the colors because of mineralization.

The subform occurring in the Tirs also shows low crumb stability and tendency to form cracks (KLINGE [1960]). The relationship of Tirs humus to Anmoor may also be recognized by the brownish-grey color shades in thin sections, and occasional occurrence of diatom shells and small, very dark to opaque pieces of tissue (KUBIENA [1957]). There are also cracks in the A_h horizon of most mull formations rich in clay. HOOVER and LUNT [1952] (further American literature quoted) list as macroscopically distinguishable mull subforms: firm mull, sand mull, coarse mull, medium mull. (A similar subclassification is also given by EHWALD [1956].) Fine mull, which has also been described by Hoover and Lunt, might in most cases turn out to be similar to mull-like moder on microscopic examination. On the other hand sand mull, in spite of very high sand content and low nutrient status, might often be a true mull formation which owes its origin largely to a favorable microclimate (VON ZEZSCHWITZ [1965]).

6. *Tangel*

Tangel (KUBIENA [1953, 1970]) is a complex humus form with features of raw humus and of mull or mull-like moder. The decay layer, however, shows a characteristic formation which makes it necessary to treat it as a distinct humus form. It is up to several decimeters thick and consists of slightly disintegrated plant residues in which mineral aggregates are intermingled (KUBIENA [1948], ZÖTTL [1965]). A holorganic humus substance layer is only present in oligotrophic or dystrophic forms (Figure 13). Tangel occurs mainly in high mountains and only under a vegetation that produces litter not easily decomposable (in the Alps, above all *Pinus mugo* and alpine dwarf bushes). The mineral soil must always show a certain clay or at least silt content.

The *decay layer* of tangel is also called tangel layer. It shows slightly to moderately disintegrated plant residues. The characteristic is the deposit of mineral aggregates, which contain numerous chewed plant residues (Figure 18). Formations on limestone, called tangel rendzinas contain, in the mineral aggregates, calcite fragments easily recognizable by polarization optics.

The decay layer does not pass abruptly but continuously into the *mineral humus horizon*. In particular, in its upper part, considerable quantities of fairly large plant residues are intimately mixed with the mineral material. Rather large amounts of only slightly changed fragments of root cortex are often noticeable, especially under *Pinus mugo* cover.

In mineral soil horizons more abundant in clay, the mineral aggregates in the decay layer and the whole mineral humus horizon have the same features as described for earthworm droppings and spongy fabric in the previous section (mull). In soils poorer in clay, in contrast, these parts show more of the features of mul-llike moder.

By the admixture of mineral material (especially if it contains much clay or lime) into the decay layer, acid humus substances, formed during the decomposition of the plant residues, cannot become active. Indeed, their effects (*e.g.*, dissolution of films of iron oxides from sand grains) can never be observed.

Dystrophic tangel shows a decay layer that resembles raw humus. But even here the mineral humus horizon is formed as mull-like moder or mull. In the case of shallow decay layers there are transitions to mull, which may occasionally be found at the mountain level.

VI. Methods

A short survey will be given here of the most important methods employed in work on humus microscopy. The applicability of the methods to materials and problems will be indicated. Details of working techniques, however, are not presented here; refer to the literature quoted.

All the methods are derived originally from the fields of mineralogy, biology, and medicine, in which microscopic work has been done much longer than in soil science. The methods have been partially modified for use in humus research. In many cases, however, methodical procedures can be exactly the same as in the original fields of study. It is then simply a question of the application of mineralogical and histological methods to problems of humus research. Relatively few special methods of this kind have been adopted. For the future there are further significant possibilities. In the following, only those methods that have in fact already been used in humus microscopy are quoted.

Older techniques of micromorphological soil investigations, especially incident light and isolation techniques, are treated by KUBIENA [1938]. The soil thin section technique is presented by BREWER [1964]. A short review of humus micromorphological methods is given by BABEL [1971b]. For macromorphological and micromorphological techniques in humus research, see also BABEL [1972b]. For biological microtechnique JOHANSEN [1940] and ADAM and CZIHAK [1964] may be mentioned. An extensive older account of staining and detection reactions in botany is given by MOLISCH [1921]. A more recent, brief presentation of staining methods is that by BAKER [1958]. Optical procedures are to be found in books on microscopy, of which FREUND (volume I/1 [1957]), RINNE and BEREK [1953], and OSTER and POLLISTER [1955/56] (some parts newly revised) may be mentioned here. Further literature references are to be found in the books quoted here.

A. Preparatory Work

1. *Sampling in the Field*

If fabric analyses are intended, samples must be taken in their natural arrangement. For this purpose different appliances such as tin boxes or cylindrical augers are used (KUBIENA [1953], ALEXANDER and JACKSON [1955], BREWER [1964]). Recently sampling after freezing by liquid nitrogen or carbon dioxide ice has been proposed especially for loose humus layers (BLEVINS *et al.* [1968]) or for soil zoological investigations (ANDERSON and HEALY [1970]). For investigations of the decomposition of plant residues and of humus formation, separate bag samples of the individual subhorizons should be taken in the finest possible subdivision.

2. *Methods for the Isolation and Enrichment of Certain Components*

The normal and most straightforward method of fractionation is the mechanical isolation of the desired humus components. For this, the work can be done with dissecting needles and forceps under the stereomicroscope. KUBIENA [1938] has also used the micromanipulator for such purposes.

Enrichments of particular constituents may be obtained additionally in various indirect ways. Thus, KROLL and MEYER [1959] worked with grain size fractionations. SEKERA [1956] has obtained organic residues from mineral soil by means of a density fractionation. In pollen analysis the accumulation of pollen by chemical means is commonly used (*e.g.*, FAEGRI and IVERSEN [1964]).

Since a great number of special methods can be applied to the various morphologically differentiatable particles directly in the microscopic preparation of the whole sample, the isolation of the constituents can often also be disregarded.

3. *Production of Microscopic Preparations*

(a) Ground Thin Sections

Just as for investigations of mineral soil, the ground thin sections are used as standard preparations for morphological humus investigations. They are usually prepared from samples in natural arrangement.

To be able to produce thin sections of humus samples, the samples must be impregnated with a material that hardens completely. The same impregnating agents are used as for sections of mineral soils. Today, polyester synthetic resins are generally used (ALTEMÜLLER [1956b, 1962], BREWER [1964]). As these are not miscible with water, drying of the samples before impregnation is necessary. Some authors use the milder freeze-drying in place of the usual air-drying (ALEXANDER and JACKSON [1955], GADGIL [1963], STEPHAN [1969]). For the further preparation procedures—sawing up of the impregnated material, grinding, cementing to the microscope slide—reference should be made to BREWER [1964]. Recently several modifications and improvements of the embedding and preparation techniques have been published in the journals of soil science (*e.g.*, GILE [1967], GUERTIN and BOURBEAU [1971]).

Thin sections may be made from samples of all humus forms. Real difficulties do not thereby arise. However, some slight disturbances and artifacts may be sometimes produced. The following data are applicable to impregnation with the German synthetic resin, Vestopal H (ALTEMÜLLER [1962]). Occasionally the embedding agent is colored yellow by humus samples. Also, fluorescing substances can penetrate, in rare cases, from the sample into the synthetic resin. The polymerization of the synthetic resin can sometimes be somewhat delayed (ALTEMÜLLER and BANSE [1964]). Particularly with inadequately dried samples, strain birefringence can occur in the synthetic resin.

Parts that have not been impregnated may be lost during grinding. This happens sometimes with the interior of needle residues, if their epidermis and hypodermis are well preserved and prevent impregnation of the interior.

The morphology of the organic constituents is altered appreciably only in rare instances. The shrinkage by the previous drying can be considerable with strongly decomposed material. Shrinkage could be avoided by impregnation of wet samples. Methods of wet impregnation fully satisfactory for any kind of soil sample do not exist. They would be of special importance for fabric analysis of strongly decomposed material. Some experiments for wet impregnation of peat samples have been performed with carbowax 6000. This is an artificial resin miscible with water. Because of its softness (similar to talc), special cutting, grinding, and polishing techniques must be used (MACKENZIE and DAWSON [1961], LANGTON and LEE [1965]). Carbowax 6000 cannot be used for soil samples rich in sand. Preparation of those samples without too much shrinkage may be possible by dehydration in an alcohol series previous to the transfer to a polyester resin (LUND and BEALS [1965]). Shrinkage is diminished significantly by freeze-drying (measurements by MENGE [1965]). On the other hand, under higher magnification the structures of tissue residues occasionally seem somewhat swollen as compared with corresponding material in sliced thin sections. Very bright lobular margins around the organic constituents which sometimes occur especially in sections of wet moder formations, are likewise probably artifacts (Figure 51; Section V D 1 c).

Thin sections serve largely for the analysis of the aggregate and the elementary fabric (cf. Section II B). Furthermore, they provide a survey of the kind and proportions of the constituents. More detailed investigations of the constituents, in contrast, are performed more advantageously on sliced thin sections or debris preparations (see below).

Usually sections of 20 μm in thickness are suitable. Where mainly fabric analysis at low magnifications is involved, even 30 or 40 μm sections suffice. Very strongly colored material, on the contrary, must be ground thinner so as to become transparent. With Vestopal H as the impregnating agent, sections of 6 to 8 μm thick may be easily produced. It is strongly recommended that in micromorphological publications the thickness of sections and the microscopic magnification used should be given, since many observed morphological properties, such as opacity or amorphousness, are dependent on these data.

(b)   Thin Sections by Slicing Techniques

Sections can also be sliced from naturally arranged humus samples. A prerequisite of most methods is that the samples should be mineral-free.

Various substances which are used in biology or medicine have been tested as impregnating agents. *Cremolan* has proved to be very good in the preparation of peat sections (PUFFE and GROSSE-BRAUCKMANN [1963]). It also has the advantage of being miscible with water, so the peat samples need not be dried. Sections 10 μm thick are cut with the microtome.

Paraffin has been likewise used in making peat sections (RADFORTH and EYDT [1958]). *Agar-agar* was used by HAARLØV and WEIS-FOGH [1953, 1955] for the impregnation of organic layer samples. Before impregnation, fixing with formaldehyde vapor was carried out because mainly microbiological investigations were intended. In this method cutting is done with razor blades. Sand-containing samples also can be cut, but the sand grains are pushed out by the razor blade. In this way, only sections of several 100 μm thickness can be obtained. MINDERMAN [1956] has developed this method further in that, after impregnation, for which he uses *gelatine*, the sand is dissolved out of the impregnated material with hydrofluoric acid. Sectioning is likewise performed with razor blades; even from originally sand-containing samples, section thicknesses may be obtained down to $7 \cdot 5$ μm. Another improvement of the gelatine embedding technique to get large, but rather thick sections (1 mm) for studies of animal positions in the soil, has been proposed by ANDERSON and HEALY [1970].

The subsequent techniques of mounting and covering of the sections are the same as used in biology and medicine.

Sliced sections have several advantages over ground sections. Instead of single sections, series of immediately successive sections may be prepared, and this does not require appreciably more effort. For impregnation with cremolan, agar-agar, or gelatine, drying is not necessary, so that corresponding artifacts are avoided. For staining and other manipulations, sliced thin sections are more easily accessible than sectioning by grinding.

Sliced thin sections are prepared, in addition, for the microscopic investigation of isolated plant residues (BABEL [1964a]). In this case, in contrast to the preparation of fresh tissues, fixing can be omitted. Paraffin may usually serve as the embedding medium. With such thin sections (sliced) there is another advantage over ground sections: the ease with which desired orientations may be obtained.

(c)   Debris Preparations

For the investigation of the fine substance debris preparations can be made (Figure 11; BABEL [1965a]). In this case one can use dry material, or suspensions. On the microscope slide the flat or elongated parts orient themselves in a manner advantageous for observation.

For direct investigation of the soil microflora, several other preparation methods, such as growth plates, are employed (see Section III C 2 c).

B.   Microscopic Investigations

The object of the microscopic investigation of natural humus formations is the constituent analysis and the fabric analysis (Part II). For constituent analysis (Part III) ground thin sections, sliced thin sections, and debris preparations are used. The interpretation of constituents according to their histological origin is given prevailingly by comparing morphological investigations. The chemical information obtained in this way is supplemented by microscopic detection reactions. For fabric analysis (Part IV) ground thin sections are used prevailingly. Besides comparing morphology, quantitative microscopic methods play an important role in fabric analysis.

1. Techniques of Qualitative Microscopic Investigation

With the stereomicroscope, work can be done directly on the sample. This also gives the connecting link between the macroscopic and the microscopic investigations. Constituents, which are indistinct in ground sections, are often easily recognized in the three-dimensional image of the stereomicroscope.

The microscopic investigation of ground sections, sliced sections, and debris preparations is usually made in *transmitted light*; *incident-light* investigations are only rarely employed, perhaps for opaque constituents. *Dark-field* illumination can be of value in the recognition of colorless structures, as well as for the clearer perception of color differences in organic parts, which show varying degrees of humification. *Phase-contrast* illumination also helps to make visible colorless parts, as for example, hyaline fungal hyphae. For the determination of the light refraction of humus constituents, the phase-contrast method has been used only in preliminary studies (ALTEMÜLLER and BANSE [1964]).

The *polarizing microscope* is also employed for qualitative investigations. Colorless cell walls and colorless mineral grains become visible by their birefringence (for detection of cellulose by birefringence see Part VI B 3). Fabric analysis investigation at half-crossed polarizers are often used, whereby pores (grey) can be differentiated from mineral grains (darker or brighter). *Fluorescence illumination* is also of great importance. Nearly all cell walls show primary fluorescence, so that tissue structures can be seen more distinctly in the fluorescence microscope. Small cell wall residues, which often occur in the organic fine substance and in humus mineral horizons can be found by fluorescence; the fluorescence microscope gives much more distinct pictures of such objects than the polarizing microscope (BABEL [1969]). (For lignin detection by fluorescence microscope see Part VI B 3.) The *interference microscope* has not been used directly in humus micromorphology.

Electron microscopic investigations have still scarcely been applied to humus micromorphological problems (BEUTELSPACHER [1955a, b]). Electron microscopic studies in neighboring fields, for example, in wood protection, however, also offer direct humus micromorphological information (LIESE and SCHMID [1962]).

For morphological comparison, besides a collection of permanent microscopic preparations, *microphotographs* are a very important aid (HEUNERT [1959], MICHEL [1962]). Moreover, microphotographs can serve for the prosecution and documentation of optical and chemical investigations that are performed on the microscopic preparations. Color photographs are by far superior to black and white photographs in most cases. The object under investigation nearly always is strongly differentiated in color. The color is an important feature of the organic constituents. Only in investigations of the void fabric are black and white photographs fully satisfactory.

2. Techniques of Quantitative Microscopic Investigation

A book edited by KUBIENA [1967] comprises several papers on techniques of micromorphometric soil analysis. The works of KUBIENA, BECKMANN, and GEYGER [1967], JONGERIUS [1963], JONGERIUS and JAGER [1964], GEYGER [1967], and GEYGER and BECKMANN [1967] contribute also to the application of such methods on problems of microscopic humus investigation.

Measurements of *length* and *thickness* as well as *countings* supplement the qualitative descriptions. The main subject of micromorphometric soil analysis are measurements of areas in the thin section. Under normal conditions they are measurements of *volumes* simultaneously. The most important method is point counting, which can be applied for many purposes, and requires but very simple equipment (NEUER [1966]). By point counting the volume ratio of pores and of any interesting groups of constituents or aggregates can be determined (*e.g.*, plant residues, organic fine substance, root residues, droppings). Structure photograms may be used for determination of volumes of pores and great mineral grains (KUBIENA, BECKMANN, and GEYGER [1962]). Also the *surfaces* of constituents and aggregates can be determined on structure photograms (GEYGER and BECKMANN [1967]). Measurements of *size of pores and particles* can be made by the particle size analyzer (Zeiss; *e.g.*, GEYGER and BECKMANN [1967]). Countings and measurements of sizes and volumes can now be made by the electronic quantimet-apparatus, when the groups of constitutents to be measured show enough difference in brightness (JONGERIUS [1969].

Estimation methods have proved to be of great value when the material under investigation consists of a great lot of samples. Such methods were developed in investigations applying humus micromorphology to ecological problems (BABEL [1967, 1972b]). Any micromorphological features which seem to be of interest can be estimated in scales from 0 to 5 according to their intensity or frequency. The estimation figures may be evaluated by simple statistical methods.

3. *Microscopic Detection Methods*

Optical and chemical methods so far applied to microscopic preparations of humus may serve to render cell and tissue structures visible. They may also give information about kind and degree of decomposition of the plant substances involved.

(a) Optical Methods

The positive *birefringence of cellulose* is one of the most important aids to microscopic humus investigations (principles in RUCH [1956], FREY-WYSSLING [1959]; first applications in humus investigation in GROSSKOPF [1935], KUBIENA [1943a]). Birefringent structures in plant residues indicate that cellulose is still present (*e.g.*, Figures 34, 27, 19). Naturally, it must be borne in mind that the brightening of cellulose residues under crossed polarizers can be almost completely masked by encrustations with strongly colored material. Especially in the mesophyll of leaves, very careful observations are necessary, since the cellulose walls are very thin. Besides the cellulosic cell walls, there are few birefringent organic constituents in humus. A smaller and negative birefringence may be often observed in the cuticular layer. It seems to be due to membrane waxes. The chitin walls of fungi usually do not show any distinct birefringence, but the chitin parts of soil animals are often strongly birefringent.

The synthetic resins of ground thin sections, when too rapidly polymerized, may show enough birefringence to impair observation of the birefringence of the organic parts.

The *autofluorescence* of many organic substances likewise has been used in humus microscopic investigations. A review on basics, techniques, and applications in humus micromorphology is given by BABEL [1972a]. For the application of fluorescence microscopy to biological objects, see PRICE and SCHWARTZ [1956]. In the field of coal petrography, fluorescence microscopic methods are also used (MALAN [1962], JACOB [1969]). MELHUISH [1968] determined living roots in polished blocks by their fluorescence. Single cell wall parts may often be detected in ground thin sections by fluorescence (Figures 24, 30). Sometimes they can be determined by their characteristic fluorescence colors. Thus, cuticulae usually fluoresce quite strongly yellowish or bluish green, cell walls of the mesophyll weakly yellowish, lignified cell walls strongly light blue to greenish blue (Figure 4). After removal of the yellowish decomposition products by means of sodium hypochlorite, even the cell walls in the mesophyll often appear bluish. Colorless hyphae sometimes show whitish or bluish autofluorescence (Figure 28).

Autofluorescence can result from traces of accompanying substances in the fluorescing particle. The fluorescence of lignified cell walls, however, is due to lignin; therefore the degradation of lignin can be followed by means of the fluorescence. In the course of decomposition of lignified tissues, the blue fluorescence decreases (JABLONSKI [1963], ALTEMÜLLER and BANSE [1964], BABEL [1964a]). Likewise the autofluorescence of the other organic residues decreases in the course of decomposition (observations by the author). Strongly fluorescing decomposition neoformations, on the contrary, have been found only in two instances, which may well be considered as not yet fully established (JABLONSKI [1963], ALTEMÜLLER and BANSE [1964]). Leaf-browning substances sometimes show weak fluorescence (BABEL [1972a]).

The synthetic resins of the ground thin sections usually fluoresce very little so that the observations of the fluorescence of organic particles are not disturbed.

Secondary fluorescence, hence fluorescence after staining with fluorochromes, is used in the direct investigation of the soil microflora (see Section III C 2 c). In the secondary fluorescence of undecomposed and decomposed tissues, differences were found by PAULI [1958, 1960] which, however, were not very distinct in microscopic preparations.

Microspectrographic determinations of the ultraviolet absorption of humus constituents were instituted by the author in relation to the questions of lignin degradation (see Section III B 2 d β, for methods, see RUCH [1960]).

(b)   Chemical Methods

As a rule these methods may be performed on sliced thin sections and on debris preparations. With ground thin sections, they are of only limited applicability, since the reagents penetrate only poorly into the synthetic resin used as impregnation agent. (It is practically impossible to dissolve the resin out of the preparation.) GAUTIER and HOLD [1960] give staining times of more than 24 hours for 1 μm thick sections of Vestopal W impregnations. With soil ground thin sections, however, some reactions may nevertheless be obtained in much shorter times (BABEL [1964c]). In 20 to 30 μm thick sections the organic parts were nearly completely decolorized in a 90-minute treatment with sodium hypochlorite. In a 5-minute treatment with yellow potassium ferrocyanide, the iron oxide-containing parts were colored blue and became thereby distinguishable from organic parts. The short times sufficed because here the reagents could attack particles that were in the surface of the thin sections (THOENES [1960]). In order to make microorganisms visible, especially bacteria, ALEXANDER and JACKSON [1955] as well as JONES and GRIFFITHS [1964] suggest staining the samples before impregnation with synthetic resin.

Various authors have carried out the usual *botanical detection reactions* for the most important cell wall substances in sliced thin sections, in particular, the methods for detection of lignin with phloroglucinol, of cellulose with chlorozinc iodide, and of cutin with Sudan III (HANDLEY [1954], HOLLING [1958], KONONOVA [1961], FRERCKS and PUFFE [1964], BABEL [1964a, 1964b]; for literature on the methods, see introduction to Section VI.). BABEL [1972b] has applied these reactions—in combination with optical methods—also to debris preparations of the fine substance. Indications of chemical changes in the decomposition of lignified cell walls may possibly be obtained also by staining with Ruthenium red and with Schiff's reagent (BABEL [1964a]).

Contrast stainings have likewise been performed on sliced soil sections (HANDLEY [1954], PUFFE and GROSSE-BRAUCKMANN [1963]). HAARLØV and WEIS-FOGH [1955] have stained living cells or cells that had just died with Violamine in agar-agar sections. For humus microscopic investigations one should always leave control preparations unstained, as the natural colors are a very important morphological feature.

In the investigation of colored organic neoformations, *sodium hypochlorite* has proved useful in various respects. This reagent is frequently used in botanical microtechniques for clearing, and was applied by HANDLEY [1954] for the first time to humus microscopic work. The colored organic substances are distinguished characteristically by the rapidity of decolorization and the speed of solution in sodium hypochlorite. Above all, however, the colored organic parts are distinguished from the uncolored cell walls, because the former are quantitatively soluble but the latter insoluble. Thus, fresh lignified secondary walls are not dissolved, but their brown-colored residues are dissolved rather rapidly (BABEL [1964a]). For the purely morphological investigation, sodium hypochlorite is important as a clearing agent. It often can reveal tissue structures that had been masked by colored substances (BABEL [1964b]). With sodium hydroxide, certain solution effects on colored organic residues are likewise obtained, but the results are by no means so clear as with sodium hypochlorite (HANDLEY [1954], VON BUCH [1962]).

In the search for characteristic stainings for humus substances, *Safranine T* was used by HOLLING [1958]. This dyestuff, however, also stains undecomposed lignified cell walls. In fact, the best microscopic characteristics of humification products are their own yellow, brown, or red colors.

REFERENCES

Adam, H., and G. Czihak [1964]. *Arbeitsmethoden der makroskopischen und mikropischen Anatomie* Stuttgart: Fischer.

Alexander, F. E. S., and R. M. Jackson [1955]. Preparation of sections for study of soil microorganisms. In D. K. Mc. E. Kevan ed. [1955]. *Soil Zoology*. Proc. Univ. Nottingham, 2nd. Easter School Agri. Sci. London: Butterworth.

Allison, F. E. [1968]. Soil aggregation—some facts and fallacies as seen by a microbiologist. *Soil Sci.* 106: 136–143.

———, and C. J. Klein [1961]. Comparative rates of decomposition in soil of wood and bark particles of several softwood species. *Proc. Soil Sci. Soc. Am.* 25: 193–196.

———, and R. M. Murphy [1963]. Comparative rates of decomposition in soil of wood and bark particles of several species of pine. *Soil Sci.* 27: 309–312.

Altemüller, H. J. [1956]a. Mikroskopische Untersuchung einiger Löss-Bodentypen mit Hilfe von Dünnschliffen. *Z. Pfl.-Ernähr. Düng. Bodenk.* 72: 152–167.

———. [1956b]. Neue Möglichkeiten zur Herstellung von Bodendünnschliffen. *Z. Pfl.-Ernähr. Düng. Bodenk.* 72: 56–62.

———. [1960]. Mikromorphologische Untersuchungen an einigen Gipskeupebröden im Raum Iphofen. *Bayer. Landw. Jb.* 37:70–85.

———. [1962a]. Beitrag zur mikromorphologischen Differenzierung von durchschlämmter Parabraunerde, Podsol-Braunerde und Humus-Podsol. *Z. Pfl.-Ernähr. Düng Bodenk.* 98: 247–258.

———. [1962b]. Gedanken zum Aufbau des Bodens und seiner begrifflichen Erfassung. *Z. Kulturtechnik* 3: 323–336.

———. [1962c]. Verbesserung der Einbettungs- und Schleiftechnik bei der Herstellung von Bodendünnschliffen mit Vestopal. *Z. Pfl.-Ernähr. Düng. Bodenk.* 99: 164–177.

———. [1966]. Die morphologische Untersuchung des Bodengefüges. In *Handbuch der Pflanzenernährung und Düngung*, Band II/1, Wien, New York: Springer.

———. und H. J. Banse [1964]. Die Bedeutung der Mikromorphologie hinsichtlich der organischen

Düngung. In A. Jongerius, ed. *Soil Micromorphology*. Proc. 2nd. Intern. Working Meeting on Soil Micromorphology, Arnhem. Amsterdam: Elsevier.

————. und H. Klinge [1964]. Mikromorphologische Untersuchungen über die Entwicklung von Podsolen im Amazonasbecken. In A. Jongerius, ed., [1964]. *Soil Micromorphology*. Proc. 2nd Inter. Working Meeting on Soil Micromorphology, Arnhem. Amsterdam: Elsevier.

————, and C. Tietjen [1967]. Some observations on animal activity in compost. In Graff and Satchell, *Progress in Soil Biology*. Proc. Coll. Dynamics Soil Communities, 1966. Vieweg: Braunschweig.

Anderson, J. M. [1970]. Observations on the vertical distribution of Oribatei [Acarina] in two woodland soils. 4. *Coll. Pedobiol.*, Dijon, 1970. Paris [1971]: I.N.R.A.

————, and J. N. Healey [1970]. Improvements in the gelatine-embedding technique for woodland soil and litter samples. *Pedobiol.* 10: 108–120.

Apinis, A. E. [1969]. Biocoenotic relationships of grassland soil fungi. *Mitt. flor.-soz. A. G., N.F., H.* 14: 68–85.

Aspiras, R. B., O. N. Allen, R. F. Harris, and G. Chesters [1971]. Aggregate stabilization by filamentous microorganisms. *Soil Sci.* 112: 282–284.

Babel, U. [1964a]. Dünnschnittuntersuchungen über den Abbau lignifizierter Gewebe im Boden. In A. Jongerius, ed. [1964]. *Soil Micromorphology*. Proc. 2nd Inter. Working Meeting on Soil Micromorphology, Arnhem. Amsterdam: Elseveir.

————. [1964b]. Opake organische Gemengteile in Auflagehumusformen. Scheffer-Festschrift 1964. Institut für Bodenkunde, Göttingen.

————. [1964c]. Chemische Reaktionen an Bodendünnschliffen. *Leitz-Mitteilungen* 3 (Nr. 1): 12–14.

————. [1965a]. Humuschemische Untersuchung eines Buchen-Rohhumus mittels mikrokopischer Methoden. *Mit. Ver. forst. Standortskunde* 15: 33–38.

————. [1965b]. Die Ansprache von Pflanzenresten im mikrokopischen Präparat von Humusbildungen. *Z. Pfl.-Ernähr. Düng. Bodenk.* 109: 17–26.

————. [1967]. Vergleich von Mikrogefügemerkmalen einiger Humusbildungen mit Hilfe einer Schätzmethode. *Geoderma* 1: 347–357.

————. [1968/69]. Enchytraeen-Losungsgefüge in Löss. *Geoderma* 2: 57–63.

————. [1971a]. Gliederung und Beschreibung des Humusprofils in mitteleuropäschen Wäldern. *Geoderma* 5: 297–324.

————. [1971b]. Methods of investigating the micromorphology of humus. In H. Ellenberg, ed., *Integrated Experimental Ecology. Ecological Studies, Analysis and Synthesis*, Vol. 2, Berlin, Heidelberg, New York: Springer.

————. [1972a]. Fluoreszenzmikroskopie in der Humusmikromorphologie. Third Intern. Working Meeting on Soil Micromorphology, Wroclaw, Poland, Sept. 1969. Warszawa: Polska Akademia Nauk.

————. [1972b]. *Moderprofile in Wäldern—Morphologie und Unsetzungsprozesse*. Hohenheimer Arbeiten 60, Stuttgart: Ulmer.

————. and R. Straub [1969]. Beziehungen makromorphologischer Merkmale von Humusprofileln zu der Behandlung und der Vegetation von Kahlschlagflächen. *Göttinger Bodenkundl. Ber.* 9: 125–169.

Bäbler, S. [1957]. Über die Verbräunung des Tabaks. Diss., Eidgenössische Technische Hochschule. Zürich: Juris-Verlag.

Baker, J. R. [1958]. *Principles of Biological Microtechnique*. London: Methuen, and New York: Wiley.

Bal, L. [1970]. Morphological investigation in two moder humus profiles and the role of the soil fauna in their genesis. *Geoderma* 4: 5–36.

————. [1973]. Micromorphological analysis of soils. Lower levels in the organization of organic soil materials. Thesis. University of Utrecht.

Barnett, H. L. [1958]. *Illustrated Genera of Imperfect Fungi*. Minneapolis: Burgess Publ.

Barratt, B. C. [1962]. Soil organic regime of costal sand dunes. *Nature* 196: 835–837.

————. [1964]. A classification of humus forms and micro-fabrics of temperate grasslands. *J. Soil Science* 15: 342–356.

————. [1965]. Decomposition of grass litters in three kinds of soil. *Plant Soil* 23: 265–269.

462   U. Babel

Barrett, B. C. [1967]. Differences in humus forms and their microfabrics induced by long-term top-dressings in hayfields. *Geoderma* 1: 209–227.

———. [1968a]. Micromorphological observations on the effect of land use differences on some New Zealand soils. *N.Z. J. Agric. Res.* 11: 101–130.

———. [1968b]. Macro- and micromorphological changes in topsoils under intensified grassland management. Ninth Intern. Congr. Soil Sci., Adelaide, Austr., Vol. 3, 211–219.

———. [1968/69]. A revised classification and nomenclature of microscopic soil materials with particular reference to organic components. *Geoderma* 2: 257–271.

Bartholomew, W. V. [1963]. The role of microorganisms in the formation and decomposition of organic compounds in soil. In R. A. Silow: The use of isotopes in soil organic matter studies. Rep. FAD/IAEA Techn. Meeting, Brunswick-Völkenrode, Sept. 1963.

Bavendamm. W. [1951]. "Mikroskopisches Erkennen und Bestimmen von holzbewohnenden und holzzersetzenden Pilzen." In H. Freund, ed. *Handbuch der Mikroskopie in der Technik 5/2.*

Beckmann, W., and E. Geyger [1967]. Entwurf einer Ordnung der natürlichen Hohlraum-, Aggregat- und Strukturformen im Boden. In W. Kubiena: *Die Mikromorphometrische Bodenanalyse.* Stuttgart: Enke.

Benecke, P., and U. Babel [1969]. Untersuchungen über die Auswirkung des Fichten-Reinanbaus auf Parabraunerden und Pseudogleyen des Neckarlandes. 3. Humusmorphologie und Bodengefüge. *Mitt. Ver. Forstl. Standortskunde* 19: 81–89.

Beutelspacher, H. [1955a]. Wechselwirkung zwischen anorganischen und organischen Kolloiden des Bodens. *Z. Pfl-Ernähr. Düng. Bodenk.* 69: 108–115.

Beutelspacher, H. [1955b]. Natürliche Fadenkolloide und Krümelbildung. *Landbauforschung Völkenrode* 5 (Nr. 4): 90–92.

Biwer, A. [1961]. Quantitative Untersuchungen über die Bedeutung der Asseln und der Bakterien für die Fallaubzersetzung unter Berücksichtigung der Wirkung künstlicher Düngemittelzusätze. I und II. *Z. Angewandte Entomologie* 48: 307–328, 377–394.

Blevins, R. L., G. M. Aubertin, and N. Holowaychuk [1968]. A technique obtaining undisturbed soil samples by freezing in situ. *Soil Sci. Soc. Am. Proc.* 32: 741–742.

———, N. Holowaychuk, and L. P. Wilding [1970]. Micromorphology of soil fabric at tree root-soil interface. *Proc. Soil Sci. Soc. Am.* 34: 460–465.

Blume, H. P. [1965]. Die Charakterisierung von Humuskörpern durch Streu und Humusstoffgruppen-analyse unter Berücksichtigung ihrer morphologischen Eigenschaften. *Z. Pfl.-Ernähr. Düng. Bodenk.* 111: 95–113.

Bottner, P. [1970]. La matière organique des principaux types de sols sous l'étage bioclimatique du chêne vert dans le midi de la France. *Sci Sol* No. 1: 3–18.

Brauns, F. E., and D. A. Brauns [1960]. *Chemistry of Lignin.* Suppl. vol. New York and London: Academic Press.

Bremner, J. M. [1955]. Recent work on soil organic matter at Rothamsted. *Z. Pfl.-Ernähr. Düng. Bodenk.* 69: 32–38.

Brewer, R. [1964]. *Fabric and Mineral Analysis of Soils.* New York and London: Wiley.

———, and J. R. Sleeman [1963]. Pedotubules: their definition, classification and interpretation. *J. Soil Sci.* 14: 156–166.

von Buch, M. [1962]. Vergleichende chemische und mikromorphologische Untersuchungen bei der Extrahierung von Huminstoffen aus Waldböden. *Z. Pfl.-Ernähr. Düng. Bodenk.* 97: 255–265.

Burgeff, H. [1961]. *Mikrobiolgie des Hochmoors.* Stuttgart: Fischer.

Burges, A. [1963]. The microbiology of a podzol profile. In Doeksen and van der Drift, eds., *Soil Organisms.* Proc. Coll. Soil Fauna, Soil Microflora and Their Relationships, Oosterbeek, Amsterdam: North Holand Publishing Company.

———. [1967]. The decomposition of organic matter in the soil. In Burges and Raw, eds, *Soil Biology.* London, New York: Academic Press.

———, and F. Raw, eds. [1967]. *Soil Biology.* London, New York: Academic Press.

Butin, H. [1961]. Versuche zum künstlichen Verblauen von Kiefernsplintholz mit dem Pilz Pullularia pullulans [de Bary] Berkh. *Angewandte Botanik* 35: 94–106.

Cholodny, N. [1930]. Über eine neue Methode zur Untersuchung der Bodenmikroflora. *Archiv Mikrobiologie* 1.

Coulson, C. B., R. J. Davies, and D. A. Lewis [1960a, b]. Polyphenols in plant, humus and soil. I. Polyphenols of leaves, litter, and superficial humus from mull and mor sites. II. Reduction and transport by polyphenols of iron in model soil columns. *J. Soil Sci.* 11: 20–29, 30–44.

Dahlman, R. C. [1968]. Root production and turnover of carbon in the root-soil matrix of a grassland ecosystem. In USSR Academy of Sciences, Intern. Symp. USSR 1968. Leningrad: Nauka.

———, and C. L. Kucera [1965]. Root productivity and turnover in native prairie. *Ecology* 46: 84–89.

Davies, R. J. [1971]. Relation of polyphenols to decomposition of organic matter and to pedogenetic processes. *Soil Sci.* 111: 80–85.

———, C. B. Coulson, and D. A. Lewis [1964a, b]. Polyphenols in plant, humus and soil. III. Stabilization of gelatin by polyphenol tanning. IV. Factors leading to increase in biosynthesis of polyphenols in leaves and their relationship to mull and mor formation. *J. Soil Sci.* 15: 229–309, 310–318.

Delecour, F., and G. Manil [1962]. Mikromorphologischer Beitrag zur Kenntnis der sauren Braunerden der belgischen Ardennen. *Z. Pfl.-Ernähr. Düng. Bodenk.* 98: 219–224.

Deuel, H. [1960]. Interactions between inorganic and organic soil constituents. Trans. 7th Intern. Congr. Soil Sci., Madison, Wisconsin, Vol. 1, p. 38–53.

Devigne, J., and C. Jeuniaux [1961]. Origine tissulaire des enzymes chitinolytiques intestinaux des lombrics. *Arch. Intern. Physiol.* 69: 223–234.

Dimbleby, G. W. [1961]. Soil pollen analysis. *J. Soil Sci.* 12: 1–11.

Doeksen, J., and J. van der Drift, eds. [1963]. *Soil Organisms*. Proc. Coll. Soil Fauna, Soil Microflora and Their Relationships, Oosterbeek. Amsterdam: North Holland Publishing Company.

Dommergues, Y., and F. Mangenot [1970]. *Ecologie microbienne du sol*. Paris: Masson.

van der Drift, J. [1964]. Soil fauna and soil profile in some inland-dune habitats. In A. Jongerius, *Soil Micromorphology*. Proc. 2nd Inter. Working Meeting on Soil Micromorphology Arnhem. Amsterdam: Elsevier.

———, and M. Witkamp [1960]. The significance of the breakdown of oak litter by Enoicyla pusilla Burm. *Arch. Néerl. Zool.* 13: 486–492.

Drozdova, T. V. [1959]. Chitin and its transformations in natural processes. Formation of melanoidins. *Usp. Sovrem. Biol.* 47: 277–296.

Dudas, M. J., and S. Pawluk [1969/70a]. Naturally occurring organo-clay complexes of orthic black chernozems. *Geoderma* 3: 5–17.

———. [1969/70b]. Chernozem Soils of the Alberta Parklands. *Geod.* 3: 19–36.

Dunger, W. [1956]. Untersuchungen über die Laubstreuzersetzung durch Collembolen. *Zoologisches Jahrbuch, Abt. Systematik, Ökologie und Geographie* 84: 75–98.

———. [1958]. Uber die Veränderung des Fallaubs im Darm von Bodentieren. *Z. Pfl.-Ernähr. Düng. Bodenk.* 82: 174–193.

———. [1963]. Leistungsspezifität bei Streuzersetzern. In Doeksen and van der Drift, eds., *Soil Organisms*. Proc. Coll. Soil Fauna, Soil Microflora and Their Relationships, Oosterbeek. Amsterdam: North Holland Publishing Company.

———. [1964]. *Tiere im Boden*. Wittenberg: Ziemsen.

Edwards, C. A., and G. W. Heath [1963]. The role of soil animals in breakdown of leaf material. In Doeksen and van der Drift, eds. *Soil Organisms*. Proc. Coll. Soil Fauna, Soil Microflora and Their Relationships, Oosterbeek. Amsterdam: North Holland Publishing Company.

Ehwald, E. [1956]. Über einige Probleme der forstlichen Humusforschung, insbesondere die Entstehung und die Einteilung der Waldhumusformen. Sitzungsberichte der deutschen Akademie der Landwirtschaftswissenschaften zu Berlin 5, (Nr. 12). Leipzig: Hirzel.

Erdtman, G. [1971]. Notes on the resistance and stratification of the exine. In J. Brooks *et al.*, *Sporopolenin* Proc. Symp. Geol. Dept., Imperial College, London. London, New York: Academic Press.

Esau, K. [1953]. *Plant Anatomy*. New York: Wiley, and London: Chapman and Hall.

Evans, D. D., and S. W. Buol [1968]. Micromorphological study of soil crusts. *Soil Sci. Soc. Am. Proc.* 32: 19–22.

Faegri, K., and J. Iversen [1964]. *Textbook of Pollen Analysis*. 2nd rev. ed. Oxford: Blackwell.

464   U. Babel

Falck, R. [1930]. Nachweise der Humusbildung und Humifizierung durch bestimmte Arten höherer Fadenpilze im Waldboden. *Forstarchiv* 6: 366–377.

Firbas, F. [1949]. Spät- und nacheiszeitliche Waldgeschichte Mitteleuropas nördlich der Alpen. 1. Band, Jena: Fischer.

————, and F. Broihan [1937]. Das Alter der Trockentorfschichten im Hils. *Planta* 26: 291–302.

Flaig, W., J. C. Salfeld, and K. Haider [1963]. Zwischenstufen bei der Bildung von natürlichen Huminsäuren und synthetischen Vergleichssubstanzen. *Landwirtschaftliche Forschung* 16: 85–96.

Franz, H., and L. Leitenberger [1948]. Biologisch-chemische Untersuchungen über Humusbildung durch Bodentiere. *Österr. Zool. Z.* 1: 498–518.

Frercks, W., and D. Puffe [1964]. Chemische und mikromorphologische Untersuchungen über den Zersetzungsgrad auf entwässertem und verschieden hoch aufgekalktem Hochmoor unter Grünland sowie in unkultiviertem, vorentwässertem Hochmoor. *Z. Kulturtechnik Flurbereinigung* 5: 149–171.

Freund, H. [1951]. *Handbuch der Mikroskopie in der Technik.* 8 Bände, Umschau-Verlag, Frankfurt/Main. Band I: Die optischen Grundlagen, Instrumente und Nebenapparate, Teil 1 [1957]: Allgemeines Instrumentarium der Durchlichtmikroskopie.

Frey-Wyssling, A. [1959]. *Die pflanzliche Zellwand.* Berlin, Göttingen, Heidelberg: Springer.

Führer, E. [1961]. Der Einfluss von Pflanzenwurzeln auf die Verteilung der Kleinarthropoden im Boden, untersucht an Pseudotritia ardua (Oribatei). *Pedobiologia* 1: 99–112.

Gadd, O. [1957]. Wood decay resulting from rot fungi. *Papper och trä* 8: 363–374.

Gadgil, P. D. [1963]. Soil sections of grassland. In Doeksen and van der Drift, eds., *Soil Organisms.* Proc. Coll. Soil Fauna, Soil Microflora and Their Relationships, Oosterbeek. Amsterdam: North Holland Publishing Company.

————. [1965]. Distribution of fungi on living roots of certain gramineae and the effect of root decomposition on soil structure. *Plant Soil* 22: 239–259.

Gäumann, E. [1951]. *Pflanzliche Infektionslehre.* 2. Aufl. Basel: Birkhäuser.

Gautier, A., and C. Hold [1960]. Technique de coloration histologique de tissus inclus dans des polyesters. *Experientia* 16: 124.

Gebhardt, H. [1971]. Zur Tonmineralzusammensetzung und Ton-Humus-Bindung in der Sequenz Schwarzerde-Parabraunerde (Griserde) der Hildesheimer Börde. *Göttinger Bodenkundl. Ber.* 19: 183–190.

Gennart, M., M. Kurbanovik, A. Nanson, and G. Manil [1961]. Contribution micromorphologique à l'étude de la biodynamique des sols forestiers équatoriaux. *Pedologie* 11: 110–122.

Geyger, E. [1967]. Die Strukturbildungen in den Humushorizonten von Böden der hohen Sierra Nevada, Spanien. *Geoderma* 1: 229–247.

————, and W. Beckmann [1967]. Untersuchungen zur Struktur eines Braunlehms unter montanem, immergrünem Regenwald in Peru. *Zeiss-Mitt.* 4: 185–207.

Ghitulescu, N., and G. Stoops [1970]. Étude micromorphologique de l'activité biologique dans quelques sols de la Dobroudja (Roumanie), *Pedologie* 20: 339–356.

van Gijzel, P. [1961]. Autofluorescence and age of some fossil pollen and spores. Proceedings K. Nederlandse akademie van wetenschappen Amsterdam Sect. B 64 (Nr. 1): 56–63.

————. [1971]. Review of the UV-fluorescence of fresh and fossil exines and exosporia. In J. Brooks et al. Proc. Symp. Geol. Dept., Imperial College, London. London, New York: Academic Press.

Gile, L. H. [1967]. A simplified method for preparation of soil thin sections. *Proc. Soil Sci. Soc. Am.* 31: 570–572.

Glathe, H. [1955]. Die direkte mikroskopische Untersuchung des Bodens. *Z. Pfl.-Ernähr. Düng. Bodenk.* 69: 172–176.

Goethe, J. W. [1807]. *Einleitung zur "Metamorphose der Pflanzen."* Jena.

Grabbe, K., and K. Haider [1971]. Die Huminstoffbildung und die Stickstoffverteilung bei der Strohrotte in Beziehung zur mikrobiellen Phenolbildung. *Z. Pfl.-ern. u. Bodenkunde* 129: 202–216.

Gračanin, Z. [1962]. Zur Genese, Morphologie und Mikromorphologie der Hangtorfbildung auf Kalksteinen in Kroatien. *Z. Pfl.-Ernähr. Düng. Bodenk.* 98: 264–272.

Graff, O. [1955]. Kleintiere bei der Strohrotte im Ackerboden. *Landbauforschung Völkenrode* 5: 81–83.

Graff, O., and J. E. Satchell, eds. [1967]. *Progress in Soil Biology.* Proc. Coll. Dynamics Soil Communities, 1966. Braunschweig: Vieweg.

Gray, T. R. G. [1967]. Stereoscan electron microscopy of soil microorganisms. *Science* 155: 1668–1670.

——, and T. F. Bell [1963]. The decomposition of chitin in an acid soil. In Doeksen and van der Drift, eds. *Soil Organisms.* Proc. Soil Fauna, Soil Microflora and Their Relationships, Oosterbeek. Amsterdam: North Holland Publishing Company.

——, P. Baxby, J. R. Hill, and M. Goodfellow [1967]. Direct observation of bacteria in soil. In Gray and Parkinson. *The ecology of soil bacteria.* Liverpool Univ. Press.

Grosse-Brauckmann, G. [1961]. Zur Terminologie organogener Sedimente. *Geologisches Jahrbuch* 79: 117–144.

——. [1963]. Zur Artenzusammensetzung von Torfen. (Einige Befunde und Überlegungen zur Frage der Zersetzlichkeit und Erhaltungsfähigkeit von Pflanzenresten.) *Berichte der Deutschen Botanischen Gesellschaft* 76: 22–37.

——. [1964.] Einige wenig beachtete Pflanzenreste in nordwestdeutschen Torfen und die Art ihres Vorkommes. *Geologisches Jahrbuch* 81: 621–644.

——, and D. Puffe [1964]. Untersuchungen an Torf-Dünnschnitten aus einem Moorprofil vom Teufelsmoor bei Bremen. In A. Jongerius, ed. *Soil Micromorphology.* Proc. 2nd Intern. Working Meeting on Soil Micromorphology, Arnhem, Amsterdam: Elsevier.

Grosskopf., W. [1935]. Stoffliche und morphologische Untersuchungen forstlich ungünstiger Humusformen. *Tharandter Forstliches Jahrbuch* 86: 48–111.

Guertin, R. K., and G. A. Bourbeau [1971]. Dry grinding of soil thin sections. *Can. J. Soil Sci.* 51: 243–248.

Haarløv, N. [1955]. Vertical distribution of mites and collembola in relation to soil structure. In D. K. McE. Kevan, ed. *Soil Zoology,* Proc. Univ. Nottingham, 2nd Easter School Agric. Sci. London: Butterworths.

——, and T. Weis-Fogh [1953]. A microscopical technique for studying the undisturbed texture of soils. Oikos 4, 44–57. See also in D. K. McE. Kevan ed. *Soil Zoology.* Proc. Univ. Nottingham, 2nd Easter School Agric. Sci. London: Butterworths.

Haider, K. [1965]. Untersuchungen über den mikrobiellen Abbau von Lignin. *Zbl. Bakteriol. I,* 198: 308–316.

——, and K. H. Domsch [1969]. Abbau und Umtsezungen von lignifiziertem Pflanzenmaterial durch mikroskopische Bodenpilze. *Archiv Mikrobiol.* 64: 338–348.

——, and J. P. Martin [1970]. Humic acid-type phenolic polymers from Aspergillus Sydowi culture medium. *Soil Biol. Biochem.* 2: 145–156.

——, and C. Schetters [1967]. Mesophile ligninabbauende Pilze in Ackerböden und ihr Einfluss auf die Bildung von Humusstoffen. In Graff and Satchell, eds. *Progress in Soil Biology.* Proc. Coll. Dynamics Soil Communities, 1966. Braunschweig: Vieweg.

——, S. Lim, and W. Flaig [1964]. Experimente und Theorien über den Ligninabbau bei der Weissfäule des Holzes und bei der Verrottung pflanzlicher Substanz im Boden. *Holzforschung* 18: 81–88.

Handley, W. R. C. [1954]. Mull and mor formation in relation to forest soils. Forestry Commission, Bulletin 23.

——, [1961]. Further evidence for the importance of residual leaf protein complexes in litter decomposition and the supply of nitrogen for plant growth. *Plant Soil* 15: 37–73.

Harris, W. V. [1955]. Termites and the soil. In D. K. Mc. E. Kevan ed. *Soil Zoology.* Proc. Univ. Nottingham, 2nd Easter School Agric. Sci. London: Butterworths.

Hartmann, F. [1952]. *Forstökologie.* Wien: Fromme.

——. [1965]. *Waldhumusdiagnose.* Wien: Springer.

Healey, J. N., and Russel-Smith, A. [1970]. Abundance and feeding preferences of fly larvae in two woodland soils. *Fourth Coll. Pedobiol.,* Dijon 1970. Paris [1971]: I.N.R.A.

Heinen, W., and H. de Vries [1966]. Stages during the breakdown of plant cutin by soil microorganisms. *Arch. Microbiol.* 54: 331–338.

Hennig, A. [1958]. Kritische Betrachtungen zur Volumen- und Oberflächenmessung in der Mikroskopie. *Zeiss-Werkzeitschrift* 30: 78–86.

Hensen, V. [1882]. Über die Fruchtbarkeit des Erdbodens in ihrer Abhängigkeit von den Leistungen der in der Erdrinde lebenden Würmer. *Landwirtsch. Jahrbuch* 11: 661–698.

Hess, C. [1965]. Über die Natur der Bindung des Humus in Schwarzerden und schwarzerdeartigen Böden, besonders in den Tirsen Marokkos. Diss., Göttingen.

Hesselmann, H. [1926]. Studier över barrskogens humustäcke. *Meddel. Stat. Skogsförsöksanst.* 22: 169–552.

Heunert, H. H. [1959]. *Praxis der Mikrophotographie.* 2. Aufl. Berlin, Göttingen, Heidelberg: Springer.

Hofmann, E. [1934]. *Paläohistologie der Pflanze.* Wien: Springer.

Holdheide, W. [1941]. Über zwei Funde prähistorischer Holzkohlen. *Ber. Deutschen Botanischen Gesellschaft* 59: 85–98.

Holling, R. [1958]. Untersuchungen zur Feinmorphologie und zum Chemismus einiger Torfe. Diss., Kiel.

———, and F. Overbeck [1961]. Über die Grösse der Stoffverluste bei der Genese von Sphagnum-torfen. *Flora* 150: 191–208.

Hoover, M. D., and H. A. Lunt [1952]. A key for the classification of forest humus types. *Proc. Soil Sci. Soc. Am.* 16: 368–370.

Huber, B. [1961]. *Grundzüge der Pflanzenanatomie.* Berlin, Göttingen, Heidelberg: Springer.

Hurst, H. M., and G. H. Wagner [1969]. Decomposition of ^{14}C-labelled cell wall and cytoplasmic fractions from hyaline and melanic fungi. *Soil Sci. Soc. Am. Proc.* 33: 707–711.

Jablonski, B. [1963]. The application of microscopical soil sections to investigations on the decomposition of plant residues in the soil. *Roczniki gleboznawcze* 13: 35–50.

———, and M. Radomska [1965]. The influence of different vegetable remainders upon the porosity of soil. Part I: Formation of soil pores under the influence of vegetable remainders. *Zeszyty Naukowe Wyzszej Szkoly Rolniczej we Wroclawiu, Rolnictwo* 19, Nr. 60.

Jacob, H. [1969]. Luminiszenz-Mikroskopie der organopetrographischen Bestandteile von Sedimentgesteinen. *Leitz-Mitt.* IV/8: 250–254.

Jakab, T., P. Dubach, N. C. Mehta, and H. Deuel [1962]. Abbau von Huminstoffen. I. Hydrolyse mit Wasser und Mineralsäuren. *Z. Pfl.-Ernähr. Düng. Bodenk.* 96: 213–217.

Jeanson, C. [1964]. Micromorphologie et pédozoologie expérimentale: Contribution à l'étude sur plaques minces de grandes dimensions d'un sol artificiel structuré par des Lombricides. In A. Jongerius, ed. *Soil Micromorphology.* Proc. 2nd Intern. Working Meeting on Soil Micromorphology, Arnhem. Amsterdam: Elsevier.

———, [1970]. Structure d'une galérie de lombric à la microsonde électronique. IV. *Coll. Pedobiol.*, Dijon. Paris [1971]: I.N.R.A.

———, [1971]. Nouvelles données microscopiques sur la contribution de la faune à l'élaboration de la structure du sol. *C. R. Acad. Sci. Paris*, 272: 422–424.

Jenny, H., and K. Grossenbacher [1963]. Root soil boundary zones as seen in the electron microscope. *Proc. Soil Sci. Soc. Am.* 27: 273–277.

Johansen, D. A. [1940]. *Plant Microtechnique.* New York: McGraw-Hill.

Jones, D., and E. Griffiths [1964]. The use of thin soil sections for the study of soil microorganisms. *Plant Soil* 20: 232–240.

Jongerius, A. [1961]. De Micromorphologie van de Organische Stof. In Bodemkunde, voordrachten, gehouden op de B-cursus "Bodemkunde," September 1959. s-Gravenhage.

———. [1963]. Optic-volumetric measurements on some humus forms. In Doecksen and van der Drift, eds. *Soil Organisms.* Proc. Coll. Soil Fauna, Soil Microflora and Their Relationships Oosterbeek. Amsterdam: North Holland Publishing Company.

———, editor [1964]. *Soil Micromorphology.* Proc. 2nd Intern. Working Meeting on Soil Micromorphology, Arnhem. Amsterdam: Elsevier.

———. [1969]. Application of quantimet-apparatus in soil micromorphometry. Third Intern. Working Meeting Soil Micromorphology Wroclaw, Poland, Sept. 1969. Warsaw [1972]: Polska Akademia Nauk.

———. [1970]. Some morphological aspects of regrouping phenomena in Dutch soils. *Geoderma.* 4: 311–331.

Jongerius, A., and A. Jager [1964]. The morphology of humic gley soils (orthic haplaquolls) under different land use. In A. Jongerius, ed. *Soil Micromorphology*. Proc. 2nd Intern. Working Meeting on Soil Micromorphology, Arnhem. Amsterdam: Elsevier.

————, and L. Pons [1962a]. Soil genesis in organic soils. *Boor en Spaade* 12: 156–168.

————. [1962b]. Einige mikromorphologische Bemerkungen über den Vererdungsvorgang im niederländischen Moor. *Z. Pfl.-Ernähr. Düng. Bodenk.* 97: 243–255.

————, and J. Schelling [1960]. Micromorphology of organic matter formed under the influence of soil organisms, especially soil fauna. Seventh Intern. Congress Soil Sci., Madison, Wisconsin, Vol. 2.

Karpatschewski, L. O. [1960]. Micromorphological investigation of leaching and podzolization of soils in a forest. *Pochvovedenie* 1960 5: 43–52.

Kaunat, H., and K. Bernard [1969]. Pektinhydrolyse durch rhizosphärenspezifische Bakterien und Actinomyceten. Zentralbl. Bakt. etc. II, 123: 464–466.

Kevan, D. K. Mc. E., ed. [1955]. *Soil Zoology*. Proc. Univ. Nottingham, 2nd Easter School Agric. Sci. London: Butterworths.

Kickuth, R., B. Meyer, and H. Schonlau [1969]. Die divergierende Humus-Metabolik benachbarter Sauer-Braunerden und Rendzinen unter Wald im Licht organischer Stoffgruppenuntersuchungen. *Göttinger Bodenkundl. Ber.* 8: 1–61.

von Kirchner, O., E. Loew, and C. Schröter [from 1908]. *Lebensgeschichte der Blutenpflanzen Mittleuropas.* Stuttgart: Ulmer.

Klinge, H. [1960]. Beiträge zur Kenntnis tropischer Böden. 1. Die schwarzerdeartigen Böden auf der Oberen Serie der Schichten von San Salvador, El Salvador, Zentralamerika. *Z. Pfl.-Ernähr. Düng. Bodenk.* 89: 102–114.

Kononova, M. M. [1943]. Use of microscopical method for study of the origin of humic substances. *Pochvovedenie* 1943, 6–7.

————. [1961]. *Soil Organic Matter.* New York, Oxford, London, Paris: Pergamon Press.

Kowalinski, S. [1964a]. Moorsh soils and their metamorphoses effected by plough tillage. *Travaux de la Société des Sciences et des Lettres de Wroclaw Seria B*, Nr. 124.

————. [1964b]. Micromorphological properties of "moorsh" soils. In A. Jongerius, ed. [1964]. *Soil Micromorphology.* Proc. 2nd Intern. Working Meeting on Soil Micromorphology, Arnhem, Amsterdam, Elsevier.

————. [1969/70]. Interdependence between micromorphological and chemical properties in some zonal soils of the Karkonosze Mountains. *Geoderma* 3: 89–115.

————. and A. Kollender-Szych [1969]. Micromorphological and physicochemical examination of the decomposition degree of organic matter in some peaty muck soils. Third Intern. Working Meeting Soil Micromorphology, Wroclaw, Poland, Sept. 1969. Warszawa [1972]: Polska Akademia Nauk.

Kox, E. [1954]. Der durch Pilze und aerobe Bakterien veranlasste Pectin- und Cellulose-Abbau im Hochmoor unter besonderer Berücksichtigung des Sphagnum-Abbaus. *Archiv für Mikrobiologie* 20: 111–140.

Krause, W. [1964]. Zur Technik der mikromorphologischen Untersuchung nicht getrockneter Unterwasserböden. In A. Jongerius, ed. *Soil Micromorphology*. Proc. 2nd Intern. Working Meeting on Soil Micromorphology, Arnhem, Amsterdam: Elsevier.

Kroll, W., and B. Meyer [1959]. Die Charakterisierung von Humusformen durch Siebanalyse und mikromorphologische Untersuchung am Beispiel einer Bodenserie auf mittlerem Buntsandstein bei Göttingen. *Landw. Forschung* 12: 12–17.

Kubiena, W. [1932]. Über Fruchtkörperbildung und engere Standortwahl von Pilzen in Bodenhohlräumen. *Archiv für Mikrobiologie* 3: 507–542.

————. [1935]. Über das Elementargefüge des Bodens. *Bodenkundliche Forschungen* 4: 380–412.

————. [1938]. *Micropedology.* Ames, Iowa: Collegiate Press.

————. [1943a]. Die mikroskopische Humusuntersuchung. *Z. Weltforstwirtschaft* 10: 387–410.

————. [1943b]. Beiträge zur Bodenentwicklungslehre. Entwicklung und Systematik der Rendsinen. *Z. Pfl.-Ernähr. Düng. Bodenk.* 29: 108–119.

————. [1944]. *Suelo y formacion del suelo desde el punto de vista biologico.* Publ. del Inst. Español de Edafologia, Ecologia y Fisiologia Vegetal.

Kubiena, W. [1948]. [1948]. *Entwicklungslehre des Bodens*. Wien: Springer.

———. [1953]. *Bestimmungsbuch und Systematik der Boden Europas*. Stuttgart: Enke.

———. [1955a]. Die Bedeutung des Begriffes Humusform für die Bodenkunde und Humusforschung. *Z. Pfl.-Ernähr. Düng. Bodenk*. 69: 3–7.

———. [1955b]. Animal activity in soils as a decisive factor in establishment of humus form. In D. K. Mc. E. Kevan, ed. [1955]. *Soil Zoology*. Proc. Univ. Nottingham, 2nd Easter School Agric. Sci. London: Butterworths.

———. [1956]. Zur Mikromorphologie, Systematik und Entwicklung der rezenten und fossilen Lössböden. *Eiszeitalter und Gegenwart* 7: 102–112.

———. [1957]. Neue Beiträge zur Kenntnis des planetarsichen und hypsometrischen Formenwandels der Böden Afrikas. *Stuttgarter Geographische Studien* 69: 51–64.

———. [1967]. *Die mikromorphometrische Bodenanayse*. Stuttgart: Enke.

———. [1972]. New advances and applications of soil micromorphology. Third. Int. Working Meeting Soil Micromorphology, Wroclaw, Poland, Sept. 1969. Warszawa: Polska Academia Nauk.

———. [1970]. *Micromorphological features of soil geography*. N. Brunswick, N. J.: Rutger's Univ. Press.

———, W. Beckmann, and E. Geyger [1962]. Zur Untersuchung der Feinstruktur von Bodenaggregaten mit Hilfe von Strukturphotogrammen. *Zeiss-Mitteilungen* 2: 256–273.

Kühnelt, W. [1957]. Die Tierwelt der Landböden in ökologischer Betrachtung. *Verhand. Deutschen Zool. Ges. Graz*, 39–103.

———. [1958]. Zoogene Krümelbildung in ungestörten Böden. *Deutsche Akad. der Landw. wissenschaften Berlin, Tagungsberichte* 13: 193–199.

———. [1961]. *Soil Biology*. London: Faber.

———. [1963]. Über den Einfluss des Myzels von Clitocybe infundibuliformis auf die Streufauna. In Doeksen and van der Drift, eds. *Soil Organisms*. Proc. Coll. Soil Fauna, Soil Microflora and Their Relationships, Oosterbeek. Amsterdam: North Holland Publishing Company.

Küster, E., and K. Haider [1961]. Untersuchungen über die Farbstoffbildung bei Streptomyces longisporus ruber. *Landbauforschung Völkenrode* 11 (Nr. 3): 61–62.

Kullmann, A. [1958]. Über die Verrottung von Futterpflanzenwurzeln an Hand von Modellversuchen. *Z. Pfl.-Ernähr. Düng. Bodenk*. 84: 127–132.

Kurotori, T., and H. Matsumoto [1958]. Microscopical study of forest soils. I. On the method of making thin sections of soils for microscopic observation and the application of this technique to the study of forest soils. *Forest Soils of Japan, Report 9*.

Langton, J. E., and G. B. Lee [1965]. Preparation of thin sections from moist organic soil materials. *Proc. Soil Sci. Soc. Am*. 29: 221–223.

De Leenheer, L. [1967]. Variation in time of the soil porosity and pore size distribution—(general picture of a six-years study on a loamy soil). *Pedologie* 17: 341–374.

Lehmann, H., B. Meyer, and H. J. Schonlau [1969]. Die divergierende Humus-Metabolik benachbarter Sauer-Braunerden und Rendzinen unter Wald im Licht standortsbiologischer Untersuchungen. *Göttinger Bodenkundl. Ber*. 9: 1–63.

Lehner, A., and W. Nowak [1959]. Neuere Ergebnisse des Direktnachweises von Bakterien im Boden mittels der kombinierten Aufwuchs-Fluorochromierungsmethode. *Zentbl. Bakt. Parasitkde*., 2nd Abt. 113: 32–34. 10/12.

Liese, W. [1964]. Uber den Abbau verholzter Zellwände durch Moderfäulepilze. *Holz als Roh- und Werkstoff* 22: 289–295.

———, and R. Schmid [1962]. Elektronenmikroskopische Untersuchungen über den Abbau des Holzes durch Pilze. *Angewandte Botanik* 36: 291–298.

Lieth, H., editor [1962]. *Die Stoffproduktion der Pflanzendecke*. Intern. Okol. Symp. Stuttgart-Hohenheim, Mai 1960. Stuttgart: Fischer.

Lindeberg, G. [1955]. Ligninabbau und Phenoloxydase-Bildung der Boden-Hymenomyceten. *Z. Pfl.-Ernähr. Düng. Bodenk*. 69: 142–150.

Linsbauer, K. [from 1922]. *Handbuch der Pflanzenanatomie*. 10 Vol. Berlin: Borntraeger.

Lohwag, H. [1941]. Anatomie der Asco- und Basidiomyceten. In Linsbauer [from 1922]. *Handbuch der Pflanzenanatomie*. Vol. 4.

Lund, Z. F., and H. O. Beals [1965]. A technique for making thin sections of soil with roots in place. *Proc. Soil Sci. Soc. Am.* 29: 633–635.

McKeague, J. A. [1971]. Organic matter in particle size and specific gravity fractions of some A$_h$-horizons. *Can. J. Soil. Sci.* 51: 499–505.

Mcmillan, J. H., and J. N. Healey [1971]. A quantitative technique of the analysis of the gut contents of Collembola. *Rev. Ecol. Biol. Sol* 8: 295–300.

Macfadyen, A. [1963]. The contribution of the microflora to total soil metabolism. In Doeksen and van der Drift, eds. *Soil Organisms.* Proc. Coll. Soil Fauna, Soil Microflora and Their Relationships, Oosterbeek. Amsterdam: North Holland Publishing Company.

Mackenzie, A. F., and J. E. Dawson [1961]. The preparation and study of thin sections of wet organic soil materials. *J. Soil. Sci.* 12: 142–144.

Mader, D. L. [1953]. Physical and chemical characteristics of the major types of forest humus found in the United States and Canada. *Proc. Soil Sci. Soc. Am.* 17: 155–158.

Mader, H. [1958a]. Cutin. In Ruhland (Herausgeber): *Handbuch der Pflanzenphysiologie.* Vol. 10. Berlin, Göttingen, Heidelberg: Springer.

———. [1958b]. Kork. In Ruhland (Herausgeber): *Handbuch der Pflanzenphysiologie.* Vol. 10 Berlin, Göttingen, Heidelberg: Springer.

Maiwald, K. [1939]. Beschaffenheit des organischen Bodenanteils. In E. Blanck, ed. *Handbuch der Bodenlehre* Ergänzungs-Vol. 1. Berlin: Springer.

Malan, O. [1962]. Die Entwicklung der optischen Methoden zur Beurteilung von Braunkohlen. *Freiburger Forschungsheft A* 253: 5–21.

Martin, J. P., and K. Haider [1971]. Microbial activity in relation to soil humus formation. *Soil Sci.* 111: 54–63.

Meier, H. [1955]. Über den Zellwandabbau durch Holzvermorschungspilze und die submikroskopische Struktur von Fichten-Tracheiden und Birken-Holzfasern. *Holz als Roh- und Werkstoff* 13: 323–338.

Melhuish, F. M. [1968]. A precise technique for measurement of roots and root distribution in soils. *Ann. Bot.* 32: 15–22.

Menge, L. [1965]. Untersuchungen über das Gefüge und die Wasserbindungsintensität einiger Anmoor- und Moorböden. Diss., Landw. Hochschule Stuttgart-Hohenheim.

Metcalfe, C. R. [1960]. *Anatomy of the Monocotyledones.* I. Gramineae. Oxford: Clarendon Press.

———, and L. Chalk [1950]. *Anatomy of Dicotyledones.* 2 vol. Oxford: Clarendon Press.

Meyer, B., and E. Kalk [1964]. Verwitterungs-Mikromorphologie der Mineral-Spezies in mitteleuropäischen Holozän-Boden aus pleistozänen und holøzänen Lockersedimenten. In A. Jongerius, ed. *Soil Micromorphology.* Proc. 2nd Intern. Working Meeting on Soil Micromorphology, Arnhem. Amsterdam: Elsevier.

Meyer, F. H. [1960]. Vergleich des mikrobiellen Abbaus von Fichten- und Buchen-Streu auf verschiedenen Bodentypen. *Archiv. Mikrobiologie* 35: 340–360.

———. [1962]. Vergleichende Mikrobiologie und Mikromorphologie der Humusbildung eines Buchen- und eines Fichten-Bestandes auf Basaltbraunerde. *Z. Pfl.-Ernähr. Düng. Bodenk.* 97: 234–243.

———. [1964]. The role of the fungus *Cenococcum graniforme* (Sow. Ferd. et Winge in the formation) of mor. In A. Jongerius, ed. *Soil Micromorphology.* Proc. 2nd Intern. Working Meeting on Soil Micromorphology, Arnhem, Amsterdam: Elsevier.

———. [1967]. Feinwurzelverteilung von Waldbäumen in Abhängigkeit vom Substrat. *Forstarchiv* 38: 286–290.

———. [1970]. Abbau von Pilzmyzel im Boden. *Z. Pfl. Ernahr. Dung. Bodenk.* 127: 193–199.

Michel, K. [1962]. *Mikrofotografie.* 2. Aufl., Wien: Springer.

Mikola, P. [1956]. Studies on the decomposition of forest litter by Basidiomycetes. *Comm. Inst. For. Fenn.* 48.2.

Minderman, G. [1956]. The preparation of microtome sections of unaltered soil for the study of soil organisms in situ. *Plant Soil* 8: 42–48.

Molisch, H. [1921]. *Mikrochemie der Pflanze.* 2. Aufl., Jena: Fischer.

Morozowa, T. D. [1963]. Micromorphological investigation of fossil soils. *Pochvovedenie* 9: 49–56.

Mückenhausen, E. [1962]. *Entstehung, Eigenschaften und Systematik der Boden der Bundesrepublik Deutschland.* 60 Tafeln, Frankfurt/Main: DLG-Verlag.

Müller, G. [1965]. *Bodenbiologie.* Jena: Fischer.

Müller, P. E. [1887]. *Studien über die natürlichen Humusformen.* Berlin: Springer.

Müllerstaël, H. [1962]. Humusbeschaffenheit und Pilzvegetation in einigen Buchen-und Fichten-Forsten des nordwestdeutschen Altdiluviums. Diss., Hamburg.

Müller-Stoll, W. [1936]. Untersuchungen urgeschichtlicher Holzreste nebst Anleitung zu ihrer Bestimmung. *Prähist. Z.* 27: 3–57.

———, [1951]. Mikroskopie des zersetzten und fossilierten Holzes. In Freund [from 1951]. *Handbuch der Mikroskopie in der Technik.* 8 Bände, Frankfurt/Main: Umschau-Verlag. Band 5.

Neuer, H. [1966]. Mengenanalyse mit dem Mikroskop. *Zeiss-Informationen* 60: 65–69.

Nicholas, D. P., and D. Parkinson [1967]. A comparison of methods for assessing the amount of fungal mycelium in soil samples. *Pedobiologia* 7: 23–41.

Nykvist, N. [1963]. Leaching and decomposition of water-soluble organic substances from different types of leaf and needle litter. *Studia Forestalia Suecica Nr. 3.*

Okafor, N. [1966]. The ecology of microorganisms on, and the decomposition of, insect wings in the soil. *Plant Soil* 25: 211–237.

Orlov, A. J. [1968]. Development and life duration of the pine feeding roots. Intern. Symp. USSR. USSR Acad. Sci., Leningrad: Nauka.

Oster, G., and A. W. Pollister [1955–1956]. *Physical Techniques in Biological Research.* New York: Academic Press.

Parfenova, Y. J., and Y. A. Yarilova [1960]. The problem of lessivage and podzolization. *Pochvovedenie* 9: 1–15.

———. [1967]. Humus microforms in the soils of the U.S.S.R. *Geoderma* 1: 197–207.

———. [1969]. Classification of soil fabric components. Third Intern. Working Meeting Soil Micromorphology, Wroclaw, Poland, Sept. 1969. Warszawa [1972]: Polska Akademia Nauk.

Parle, J. N. [1963]. Microorganisms in the interior of earthworms. *J. Gen. Microbiol.* 31: 1–11.

Pauli, F. W. [1958]. Fluorochrome Adsorption Studies on Decomposing Plant Residues. Diss., Potchefstroom.

———. [1960]. The fluorescence microscope in humus research. *Mikroskopie.* 15: 139–142.

———. [1961]. Flurochrome adsorption studies on decomposing plant residues. I. Decomposition studies. *South Afr. J. Agric. Sci.* 4: 123–134.

———, and J. H. Grobler [1961]. Characterisation of soil humus by acriflavine adsorption. *South Afr. J. Agr. Sci.* 4: 157–169.

von Pechmann, H. [1951] Die Mikroskopie in der Holzverarbeitungstechnik. In Freund (from 1951). *Handbuch der Mikroskopie in der Technik.* 8 Bände, Frankfurt/Main. Umschau-Verlag. Band 5.

Pettapiece, W. W., and M. A. Zwarich [1970]. Micropedological study of a chernozemic to grey wooded sequence of soils in Manitoba. *J. Soil Sci.* 21: 138–145.

Pinck, L. A., and F. E. Allison [1944]. The synthesis of lignin-like complexes by fungi. *Soil Sci.* 57: 155–161.

Ponomareva, S. J. [1962]. Die Bedeutung der biologischen Faktoren für die Steigerung der Fruchtbarkeit rasenpodsoliger Böden. *Z. Pfl.-Ernähr. Düng. Bodenk.* 97: 205–215.

von Post, H. [1862]. Studier öfver Nutidens koprogena Jordbildningar, Gyttja, Dy, Torf och Mylla. *Kgl. Svenska Vet.-Akad. Handl.,* N.F. 4, 1–59.

Prévot, A. R., M. Raynaud, G. Fischer, and B. Bizzini [1954]. Recherches sur la ligninolyse bacterienne dans le sol. Trans. 5ème Congrès International de la Science du Sol, Léopoldville 3:13.

Price, G. R., and S. Schwartz [1956]. Fluorescence microscopy. In Oster and Pollister, eds. *Physical Techniques in Biological Research.* Vol. 3. New York: Academic Press.

Puffe, D. [1965]. Gefügeuntersuchungen an Torfen. *Z. Kulturtechnik Flurbereinigung* 6: 301–312.

———, and G. Grosse-Brauckmann [1963]. Mikromorphologische Untersuchungen an Torfen. *Z. Kulturtechnik Flurbereinigung* 4: 159–188.

Radforth, N. W., and H. R. Eydt [1958]. Botanical derivatives contributing to the structure of major peat types. *Can. J. Bot.* 36: 153–163.

Righi, D. [1969]. Aspects morphologiques et physico-chimiques de la podzolisation en forêt de Rambouillet. Thèse de Doctorat 3ème cycle en Ecologie Végétale, 116p., Grignon.

Rinne, F., and M. Berek [1953]. *Anleitung zur optischen Untersuchung mit dem Polarisations mikroskop*· 2. Aufl. Stuttgart: Schweizerbart.

Rommell, L. G. [1935]. Ecological problems of the humus layer in the forest. Cornell-Univ. Agric. Exp. Sta. Memoir 170:2–28.

Rossi, G. [1936]. Direct microscopic and bacteriological examination of the soil. *Soil Sci.* 41: 53–66.

————, and G. Gesuè. Di un nuovo indirizzo nello studio biologico del suolo. *Ann. Technica Agr.* 3.

Roulet, N., P. Dubach, N. C. Mehta, M. Müller-Vonmoos, and H. Deuel [1963]. Verteilung der organischen Substanz und der Kohlenhydrate bei der Gewinnung "wurzelfreien" Bodenmaterials durch Schlämmsiebung. *Z. Pfl.-Ernähr. Düng. Bodenk.* 101: 210–214.

Ruch, F. [1956]. Birefringence and dichroism of cells and tissues. In Oster and Pollister, eds. *Physical Techniques in Biological Research.* Vol. 3. New York: Academic Press.

————.[1960]. Ein Mikrospektrograph für Absorptionsmessung im ultravioletten Licht. *Z. wiss Mikr.* 64:453–468.

————, and H. Hengartner [1960]. Quantitative Bestimmung der Ligninverteilung in der pflanzlichen Zellwand. Beih. z. d. Zeitschr. des Schweizerischen Forstvereins Nr. 30 (Festschrift Frey-Wyssling).

Ruttner, F. (1962). *Grundriss der Limnologie.* 3. Aufl. Berlin: de Gruyter.

Rypacek, V. [1960]. Die Veränderungen in der Struktur der verholzten Zellwand im Verlaufe der Humifikation (Czech, German summary). *Acta Agrobotanica* 9: 53–61.

————, [1966]. *Biologie der holzzerstörenden Pilze.* 2. Ausgabe. Jena: VEB G. Fischer.

Schaede, R. [1962]. *Die pflanzlichen Symbiose.* 3. Aufl. (neubearbeitet von F. H. Meyer), Stuttgart: Fischer.

Schaller, F. [1950]. Biologische Beobachtungen an humusbildenden Bodentieren, insbesondere an Collembolen. *Zool. Jb. (Syst.)* 78: 506–525.

Scheffer, F., and B. Ulrich [1960]. *Lehrbuch der Agrikulturchemie und Bodenkunde.* 3. Teü: *Humus und Humusdüngung.* Band I: Morphologie, Biologie, Chemie und Dynamik des Humus. 2. völlig neu bearbeitete Aufl. Stuttgart: Enke.

Schlenker, G. [1962]. Ertragspotentiale verschiedener Waldgesellschaften Südwestdeutschlands. In Lieth, ed. *Die Stoffproduktion der Pflanzendecke.* Intern. Okol, Symp. Stuttgart-Hohenheim, Mai 1960.

Schmid, R., and W. Liese [1964]. Über die mikromorphologischen Veränderungen der Zellwandstrukturen von Buchen- und Fichten-Holz beim Abbau durch Polyporus versicolor *From. Arch. Mikrobiol.* 47: 260–276.

Schroder, D. [1952]. *Unterscheidungsmerkmale der Wurzeln einiger Moor und Grünlandpflanzen nebst einem Schlüssel zu ihrer Bestimmung und einem Anhang für die Bestimmung einiger Rhizome.* Bremen: Schünemann.

Schuster, R. [1956]. Der Anteil der Oribatiden an den Zersetzungsvorgängen im Boden. *Z. Morph. Ökol. Tiere* 45: 1–33.

Sekera, F. [1956]. Beobachtungen über die Humusneubidung im Boden. *Phosphorsäure* 16 (Folge 3/4) 188–199.

————, and A. Brunner [1943]. Beiträge zur Methodik der Gareforschung. *Bodenkunde und Pflanzenernährung* 29 (74), 169–212.

Slager, S. [1964]. A study of the distribution of biopores in some sandy soils in the Netherlands. In A. Jongerius, ed. *Soil Micromorphology.* Proc. 2nd Intern. Working Meeting on Soil Micromorphology, Arnhem. Amsterdam: Elsevier.

Smolikova, L., and V. Lozek [1969]. Mikromorphologie und Molluskenfauna des mittelpleistozänen Aubodenkomplexes von Brozany (NW-Böhmen]. *Vestnik Ustredniho ustavu geologickeho* 44: 107–114.

Solar, F.[1960]. Zur Kenntnis der Böden auf dem Raxplateau. Diss., Hochschule für Bodenkultur, Wien.

Sørensen, H. [1962]. Decomposition of lignin by soil bacteria and complex formation between autoxidized lignin and organic nitrogen compounds. *J. Gen. Microbiol.* 27, 21–34.

Stamm, J. [1958]. Über das Vorkommen eosinophiler Zellschichten bei der Wurzel-Suberogenese einiger Pflanzen. *Photogr. und Wissenschaft* 2: 25–26.

Stephan, S. [1969]. Gefriertrocknung und andere bei der Herstellung von Bodendunnschliffen benutzbare Trocknungsverfahren. *Z. Pfl.-Ernähr. Dung. Bodenk.* 123: 131–140.

Stöckli, A. [1952]. Über den Abbau von Lignin, Zellulose und Hemizellulose durch Pilze–Versuche mit Streuematerialien, Holz und Ligninsulfonsäuren. Diss., E. T. H. Zürich.

Strugger, S. [1949]. *Fluoreszenzmikroskopie und Mikrobiologie.* Hannover: Schaper.

Szabo, J., M. Marton, and G. Partai [1964]. Micro-milieu studies in the A-horizon of a mull-like rendzina. In A. Jongerius, ed. *Soil Micromorphology.* Proc. 2nd Intern. Working Meeting on Soil Micromorphology, Arnhem. Amsterdam: Elsevier.

———, M. Marton, L. Varga, and S. Schönfeld [1962]. Complex soil biological investigations of rendzina soils. *Pochvovedenie* 10: 85–95.

———, T. Bartfay, and M. Marton [1967]. The role and importance of the larvae of St. Mark's fly in the formation of a rendzina soil. In Graff and Satchell, eds. *Progress in Soil Biology.* Proc. Coll. Dynamics Soil Communities, 1966. Braunschweig: Vieweg.

Thoenes, W. [1960]. Giemsa-Färbung an Geweben nach Einbettung in Polyester ("Vestopal") und Methacrylat. *Z. wiss. Mikr.* 64: 406–413.

Thompson, S. O., G. Chesters, and L. E. Engelbert [1964]. Comparative properties of plant lignins and humic materials of soils. 1. Yields and cation exchange properties of plant lignins isolated by different techniques. *Proc. Soil Sci. Soc. Am.* 28: 65–68.

Tribe, H. T. [1963]. The microbial component of humus. In Doeksen and van der Drift, eds. *Soil Organisms.* Proc. Coll. Soil Fauna, Soil Microflora and Their Relationships, Oosterbeek. Amsterdam: North Holland Publishing Company.

Trolldenier, G. [1965a]. Fluoreszenzmikroskopie in der Rhizosphärenforschung. *Zeiss-Informationen* 56: 68–69.

———. [1965b]. Fluoreszenzmikroskopische Untersuchung von Mikroorganismenreinkulturen in der Rhizosphäre. *Zentbl. Bakt. Parasitkde.*, 2. Abt. 119: 256–258.

Uggla, H., Z. Róg, and T. Woclawek [1969]. Micromorphology of the gyttja-muck soil of Jawty Male. Third Intern. Working Meeting Soil Micromorphology, Wroclaw, Poland, 1969. Warszawa [1972]: Polska Akademia Nauk.

Waksman, S. A. [1936]. *Humus.* Baltimore: Williams and Wilkins.

Welten, M. [1958]. Pollenanalytische Untersuchung alpiner Bodenprofile: historische Entwicklung des Bodens und säkulare Sukzession der örtlichen Pflanzengesellschaften. *Veröff. Geobot. Inst. Rübel*, H. 33, 253–274.

Wenzel, G., K. Kreutzer, and M. Alcubilla [1970]. Beitrag zur Klärung des Zusammenhanges zwischen Standort und Pilzhemmstoffgehalt des Fichtenbastes. *Fortswiss. Cbl.* 89: 372–381.

Wieringa, K. T. [1955]. Der Abbau der Pektine; der erste Angriff der organischen Pflanzensubstanz. *Z. Pfl.-Ernähr. Düng. Bodenk.* 69: 150–155.

Wilde, S. A. [1958]. *Forest Soils.* New York: Ronald Press.

———. [1971]. Forest humus: Its classification on a genetic basis. *Soil Sci.* 111: 1–12.

Willerding, U. [1960]. Beiträge zur jungeren Geschichte der Flora und Vegetation der Flussauen. *Flora* 149: 435–476.

Williams, S. T., and D. Parkinson [1964]. Studies of fungi in a podzol. I. Nature and fluctuation of the fungus flora of the mineral horizons. *J. Soil Sci.* 15: 331–341.

Wittich, W. [1939]. Untersuchungen über den Verlauf der Streuzersetzung auf einem Boden mit Mullzustand. 1. *Forstarchiv* 15: 96–111.

———. [1943]. Untersuchungen über den Verlauf der Streuzersetzung auf einem Boden mit Mullzustand. 2. *Forstarchiv* 19: 1–18.

Zachariae, G. [1962]. Zur Methodik der Geländeuntersuchungen in der Bodenzoologie. *Z. Pfl.-Ernähr. Düng. Bodenk.* 97: 224–233.

———. [1963]. Was leisten Collembolen für den Waldhumus? In Doeksen and van der Drift, eds.

Soil Organisms. Proc. Coll. Soil Fauna, Soil Microflora and Their Relationships, Oosterbeek. Amsterdam: North Holland Publishing Company.

——. [1964]. Welche Bedeutung haben Enchytraeen im Waldboden. In A. Jongerius, ed. *Soil Micromorphology.* Proc. 2nd Intern. Working Meeting on Soil Micromorphology, Arnhem. Amsterdam: Elsevier.

——. [1965]. Spuren tierischer Tätigkeit im Boden des Buchenwaldes. Forstwiss. Forschungen (Beihefte zum forstlichen Zentralblatt), H. 20.

——. [1966]. Die Streuzersetzung im Kohlgartengebiet. In Graff and Satchell, eds. [1967]. *Progress in Soil Biology.* Colloquium Dynamics of Soil Communities, Braunschweig Sept. 1966. Braunschweig: Vieweg.

——. [1967]. Der Einsatz mikromorphologischer Methoden bei bodenzoologischen Arbeiten. *Geoderma.* 1: 175–196.

Zajonc, J. [1967]. Über die Saisondynamik der Humusbildung durch Regenwürmer in einem Buchenwald der Karpathen. In Graff and Satchell, eds. *Progress in Soil Biology.* Proc. Coll. Dynamics Soil Communities, 1966. Braunschewig: Vieweg.

von Zezschwitz, E. [1965]. Biologische Aktivität und Basengehalt des Bodens. *Mitt. der Deutschen Bodenkundl. Gesellschaft* 4: 281–286.

Ziechmann, W., and E. Przemeck [1964]. Untersuchungen über die physiologische Wirkung von synthetischen und natürlichen Huminstoffen auf das Wurzelwachstum und auf die Phosphataseaktivität im Wurzelmeristem. Festschrift für F. Scheffer. Inst. für Bodenkunde, Göttingen.

Zöttl, H. [1965]. Zur Entwicklung der Rendzinen in der subalpinen Stufe. 1. Profilmorphologie. 2. Chemisch-biologische Dynamik. *Z. Pfl.-Ernähr. Düng. Bodenk.* 110: 110–114, 115–126.

Zvjagincev, D. G. [1964]. Study of shapes and sizes of soil microorganisms with the fluorescence microscope. *Pocvovedenie* 3: 101–105.

Chapter 8

Humus of Virgin and Cultivated Soils

M. M. Kononova

translated

by

T. Z. Nowakowski, Ph.D. and N. Walker, Ph.D.

Contents

I. Introduction

The manifold importance of organic matter in soil formation and soil fertility has been demonstrated by the experience of agriculture over many centuries and by numerous investigations, in which the role of organic matter in soil processes (weathering, soil profile formation, soil structure formation, etc.) as well as in supplying plants with nutrients and biologically active substances has been elucidated.

Therefore, the great interest shown in studies of soil organic matter is understandable. Although many questions concerning the nature and properties of soil organic matter components remain obscure, this is not due to lack of energy on the part of research workers but rather to the complexity of the problem, which demands a multilateral approach.

The complex nature of the organic part of soil is shown by the scheme in which we have tried to classify the diverse organic compounds that are components of soil organic matter (Table 1).

Table 1.   Classification of organic substances in soil.

I. Fresh and incompletely decomposed plant and animal residues.	II. Humus	
	(a) Strictly humus substances* Groups: humic acids fulvic acids humins hymatomelanic acid	(b) Products of the advanced decomposition of organic residues and products of microbial resynthesis (substances of a protein nature, carbohydrates and their derivatives, waxes, fats, tannins, lignins, etc.).

In this scheme the organic part of soil is denoted as "organic matter" and includes the following groups of substances:

First group—fresh and incompletely decomposed plant and animal residues; they are primary sources of humus (see second group of substances), but cannot yet be considered as inseparable components of the soil; they may be separated from the latter by mechanical

* *Translators' note:* At the suggestion of Professor Kononova, the Russian term "gumusovy veschestva" has been translated in this paper as "humus substances" because it was felt that the commonly used translation "humic substances" might be confused with "humic acids."

methods (sieving or sorting of residues, isolation by means of liquids of different densities, ultrasonic dispersion, etc.).

Second group — soil humus, including two subgroups: (a) strictly humus substances (humic acids, fulvic acids, humins, hymatomelanic acids) and (b) diverse products of the advanced decomposition of organic residues and also products of microbial resynthesis. Both these sub-groups of substances are an integral part of soil and cannot be separated from soil by mechanical methods.

An important aspect of the problem of soil organic matter, which for a long time has attracted the attention of research workers, is the elucidation of the geographical characteristics of humus formation that are determined by conditions of soil formation (vegetation cover, the hydrothermal regime, the nature of biological activity, the parent rock). These characteristics are reflected in the humus and nitrogen contents, the amount of humus and nitrogen accumulations, the nature of their distribution in the soil profile, and also in consistent differences in the nature and properties of humus substances.

The problem of the geographical laws governing humus formation was outlined by Dokuchaev in his work *Russian Chernozem* [1883]. This work included a map of the chernozem regions of European Russia on which were drawn isohumic belts, *i.e.*, belts with the same content of humus in the upper soil layer. The distribution of these belts showed that the maximum humus content occurred in the central regions of the chernozem zone; the humus content gradually decreased toward the north in the direction of podzolic soils and toward the south in the direction of dry steppes and semideserts.

According to Dokuchaev, accumulation of humus in chernozems is favored firstly by perennial grassy vegetation and secondly by the climatic conditions prevailing in the central belt of Russia, which determine the moderate rate of decomposition of plant residues.

The idea of the zonal nature of changes in the amount of humus in various soils was developed in the works of followers of Dokuchaev and Sibirtsev; they succeeded also in establishing certain qualitative differences revealed in the varying mobility of humus (Kozlovskii, Lesnevskii, Shcheglov, Naletov, and others; see Tyurin [1949, 1965]).

It should be noted that Hilgard [1893, 1906] regarded the humus content and its nitrogen content as important characteristics in soil classification. Hilgard found that soils of arid regions have low humus content but that their humus has a high nitrogen content. On the contrary, soils of humid regions are characterized by high humus contents, but the nitrogen content of the humus is low.

Subsequently, the accumulation of data characterizing the contents and reserves of humus and nitrogen and their distribution in the soil profile was accompanied by attempts to explain peculiarities in the composition and nature of humus substances and to show causal relationships between the "humus state" of the soil and the conditions of soil formation.

In the U.S.S.R. credit for many of these attempts is given to Tyurin, who developed several methods for studying soil humus. Information obtained by Tyurin, supplemented by current data of a number of Soviet soil scientists (summarized by Kononova [1966]) make it possible to observe consistent changes in the percentage of humus and nitrogen, their total reserves, C:N ratios, and the composition of humus and changes in the nature of humus substances in the main soil types of the U.S.S.R.

With various modifications, Tyurin's methods for determining the content and composition of humus are used in Bulgaria, Hungary, Germany, Poland, Rumania, Czechoslovakia, Yugoslavia, and other countries. Because of this, it is possible to make wider generalizations regarding the natural laws governing humus formation. These data have been used as much as possible in the present work.

We consider it appropriate to begin the account of these data with an examination of the

Table 2. Productivity of main types of vegetation of the Northern Hemisphere of the Old World. (Data of RODIN and BAZILEVICH [1964].)

Characteristic	Unit of measurement	Tundra	Arctic Brushwood	Spruce associations of northern, central, and southern taiga	Oak groves	Moderately dry steppes	Dry steppes	Semi-brushwood deserts	Sub-tropical deciduous forests	Dry savannahs	Savannahs	Humid tropical forests
Biomass	q/ha	50	280	1000–2600–3300	4000	250	100	43	4100	268	666	5000
Roots	q/ha	35	231	220–600–735	960	205	85	38	820	113	39	900
	% biomass	70	83	22–23–22	24	82	85	87	20	42	6	18
Leaf fall (aerial mass + roots)	q/ha	10	24	35–50–55	65	112	42	12	210	72	115	250
	% biomass	20	9	3.5–2–1.5	1.5	45	42	30	5	27	17	5
Litter or steppe tomentum	q/ha	35	835	300–450–358	150	62	15	—	100	—	13	20
Ratio of litter to leaf fall of green parts		14	92	17–15–10	4	1.5	1.0	—	0.7	—	0.2	0.1

importance of natural conditions, determining organic matter transformations in soils (characteristics of vegetation cover, soil microflora, hydrothermal regime, and chemical and physicochemical soil properties).

2. Natural Conditions of Humus Formation

(a) The Role of Vegetation Cover in Humus Formation

Decaying constituents of the vegetation cover are the main source of soil humus and, in addition, the main source of energy for numerous microorganisms and animals inhabiting the soil; the formation of humus substances is largely due to the activities of these.

The total reserves of aerial plant mass and root systems vary widely, depending on the type of vegetation, climatic conditions, and soil properties. Not only the total quantity of biomass, but the amount of plant residues entering the soil annually is of essential importance in humus formation. The amount of biomass (aerial part and roots) and the amount of annual leaf fall for various plant associations are given in Table 2.

Maximum accumulation of biomass is observed in forest associations; in humid tropical forests the accumulation is somewhat higher than in subtropical forests or in deciduous forests of the temperate zone.

The reserves of roots in deciduous forests of all zones are similar (820 to 960 q/ha). The total reserve of biomass is least in coniferous forests of the temperate zone. The proportion of subterranean organs, however, in the total biomass of forests of different types varies within fairly narrow limits (18 to 23%).

The root reserves in savannahs depend on the presence of woody vegetation in the plant cover. In savannahs with a predominantly grassy cover the reserves of roots are smaller than in dry savannahs, where woody vegetation is present, together with grassy vegetation.

The amount of biomass and its share of roots decreases from steppes of moderate moisture toward dry steppes and deserts, but the proportion of roots in the total biomass in this series remains practically the same (82 to 85%).

The amount of annual leaf fall (aerial mass) and roots in different associations varies widely. Grassy vegetation of steppe loses annually approximately half of the whole biomass. In forest associations, leaf fall does not exceed 5%, but in savannahs it constitutes 17 to 27%.

The ratios of the litter reserves to the leaf fall of the green part of the vegetation are of interest; these ratios characterize to some extent the rate of mineralization of the fallen plant material. The widest ratio of litter to leaf fall is observed in tundra, indicating a slow mineralization of the leaf fall and its accumulation in the form of litter. More intensive decomposition of the fallen plant material occurs in spruce associations of taiga. In broad-leaved forests (oak groves) the ratio of litter to leaf fall, which amounts to 4, indicates fairly intensive decomposition of the aerial leaf fall. Still more intensive decomposition of the leaf fall occurs on the surface of steppe soils. The ratio of litter to leaf fall, which in subtropical deciduous and humid tropical forests is 0.1 to 0.7, shows that vigorous mineralization of an enormous mass of leaf fall takes place on the surface of the soil in these plant associations.

No direct relationship between humus content in the soil and the total reserve of plant mass can be established (Table 3). As has been pointed out above, plant cover is only one of the factors of soil formation determining the extent of humus accumulation in soils. Nevertheless, a comparison of the data in Table 3 shows that the percentage of humus in the upper soil layer corresponds to the amount of leaf fall and to its nitrogen content.

(b) Microorganisms and Organic Matter Transformations in Soil

In mature soils—podzolic soils, chernozems, and serozems—the microbial protoplasm,

Table 3. Percentage humus content, amount of biomass, and constituents (aerial mass and roots) decaying annually in various soil vegetation zones of the U.S.S.R. (Data of Bazilevich [1962].)

Biocenoses, soils	Percent humus in 0–20 cm layer	Biomass, tons/ha		Decaying mass (aerial mass and roots)		Nitrogen added to the soil with the decaying mass, kg/ha
		Total	Roots	Percent of biomass	q/ha	
Coniferous forests on sod-podzolic soils	2.5–4.0	100–380	25–80	1–3	20–64	11–44
Deciduous forests on gray forest soils	4.0–6.0	100–500	25–125	1–4	24–64	25–72
Steppe-like meadows and meadowlike steppes on meadow-chernozem soils	9.0–10.0	11.5–32	7–20	50–55	60–80	90–230
Meadowlike steppes on typical leached chernozems	—	21–26	16–20	45–55	100–130	125–160
Mixed grass (*Festuca sulcata— Stipa*) steppes on ordinary and southern chernozems	8.0–6.0	18–25	15–20	40–45	80–110	90–120
Dry *Artemisia* (*Stipa–Festuca sulcata*) steppes on chestnut soils	4.0–2.0	10–23	9–18	35–45	40–80	45–70
Ephemeral (*Artemisia*) deserts on light serozems	1.0	11–14	9–12	60–85	60–80	80–110
Semibrushwood (*Salsola– Artemisia*) deserts on gray-brown soils	1.0	4–5	3.5–5.0	23–30	8–10	12–18

calculated on dry matter, constitutes no more than 1.2 to 1.3 tons/ha and therefore does not exceed 3% of the total reserves of humus. However, in some soils (*e.g.*, takyrs) microorganisms (mainly algae) are almost the only source of organic matter.

The enormous role played by microorganisms in the transformation of organic residues and in the formation of humus substances cannot be questioned.

To a large extent the further fate of newly formed humus substances depends on the activity of microorganisms. Newly formed humus substances may be either subjected to new biological processes and decomposed to end products of mineralization or preserved in the soil for longer or shorter periods (mainly linked to the mineral part of the soil). Thus, the humus reserve in the soil can be represented as follows:

Humus (reserve) = initial humus + newly formed humus − decomposed humus.

In the U.S.S.R. classified data on the number and group composition of the microbial population in various soil types are available (MISHUSTIN [1956]). Some of these data are included in Figure 1 The number of microorganisms is calculated per gram of soil and per gram of humus. The first value is an index of the total number of microorganisms in soil ("density of microbial population"); the second value, to a certain extent, shows the possible intensity of humus transformation by microorganisms. On the whole both indices characterize the "biogenic properties" of the soil.

Comparison of the data on the number of microorganisms and humus reserves allows the following conclusions to be drawn. The greatest reserves of humus correspond to a moderate number of microorganisms in 1 gram of soil humus. Evidently, a new formation of humus substances is effected under these conditions, and at the same time, the rate of their decomposition is not vigorous.

Neither the high "biogenic properties" of soil (in serozems) nor the weak "biogenic properties" of soil (in tundra-gley soils) favor the accumulation of humus.

It can be seen from the data in the following section that a similar relationship exists between humus reserves in soils and their hydrothermal regime, which influences the rate of microbiological processes and, in particular, the rate of organic matter transformations in soil.

(c) Hydrothermal Conditions and Organic Matter Transformations in Soil

Moisture and temperature are very important in humus formation, mainly because of their effects on the development of plant cover and on the activities of microorganisms, which are very important factors in soil formation.

Detailed investigations on the intensity of decomposition of plant residues and humus under various combinations of temperature and moisture were made by Kostychev [1886], Wollny [1886], and Waksman and Gerretsen [1931].

The results of these studies reveal the following relationships:

(1) The highest intensity of organic matter decomposition is observed under conditions

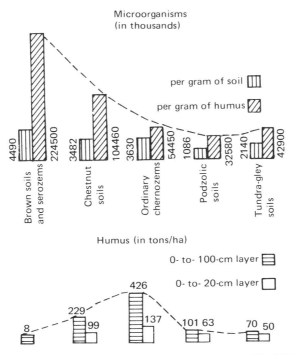

Figure 1. Microorganisms and humus in the soils of U.S.S.R.

of moderate temperature (about 30°C) and soil moisture of about 60 to 80% of its maximum water-holding capacity.

(2)   Simultaneous increase or decrease of temperature and moisture beyond the optimal values brings about a decline in the rate of organic matter decomposition.

(3)   With increasing value of one of the above two factors coincident with a decrease of the other, the intensity of organic matter decomposition, as with any biological phenomenon, is subject to that factor which is at a minimum. For instance, with low moisture content (<10 to 15%), the intensity of decomposition of plant residues remains low at all temperatures and therefore does not depend on the temperature factor. On the contrary, with adequate moisture content (50 to 60% or more) the intensity of decomposition depends essentially on temperature.

The influence of hydrothermal conditions on plant cover and microbiological processes, associated with the new formation and decomposition of humus, explains the relationship between these conditions and the humus reserves in soil.

TYURIN [1937], comparing the percentage content of humus in the main soils of the U.S.S.R. with mean values of annual temperature and precipitation, established the following important points: The largest amount of humus is observed at moderate values (in deep chernozem); it decreases both in soils with increased moisture regime and low temperatures (podzolic soils) and in soils with inadequate moisture and high temperatures (soils of dry and desert steppes) (see Figure 2).

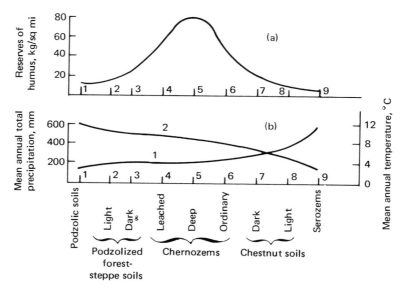

Figure 2. Humus reserves and climatic conditions of the main soil types and subtypes in the U.S.S.R. (after TYURIN [1937]).

A—humus reserves in soils to a depth of 100–120 cm in kilograms per square meter of surface area; B 1—mean annual temperature; B2—mean annual total precipitation.

On the basis of Tyurin's data, VOLOBUEV [1948] tried to derive a functional relationship between the humus content and the value of the hydrofactor (Hf) characterizing a change in moisture conditions for different values of precipitation (P) and mean annual temperatures (T). Volobuev calculated the hydrofactor from the empirical equation

$$T = 43.2 \log P - Hf$$

where T and P are variables, and Hf is the parameter which has a definite value for each hydroseries (Figure 3).

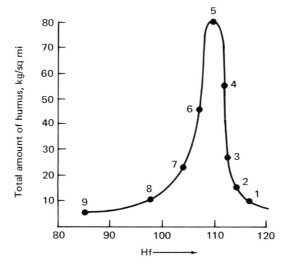

Figure 3. Change in humus reserves in the main soil types and subtypes in relation to the hydrofactor (VOLOBUEV [1948]). Numbers on the curve correspond to the order of soils in Figure 2.

Soils with high humus contents have moderate values of the hydrofactor, from 105 to 112; for the maximum humus content (in deep chernozems) the hydrofactor is 110. For hydrofactor values smaller than 110 or exceeding 112, the humus content decreases sharply. Volobuev [1958] stressed that, as a rule, the index hydrofactor is associated with the moisture content:

$$K_n = \frac{\text{precipitation}}{E_n}$$

Where K_n is an index of effective moistening and E_n is the natural evaporation rate.

A moderate value of the hydrofactor (105 to 110), characteristic for soils with high humus content, corresponds to $K = 0.8$ to 1.0 (*i.e.*, to the conditions of natural "equilibrated" moisture). The derivation of a mathematical relationship between the hydrothermal factors and the humus reserves in the soil is a complicated task, an example of which is the work of Volobuev. We should mention also the work of JENNY [1941], who tried to establish a relationship between humus and nitrogen contents in the soil and values of annual temperatures and moisture. In accordance with Van't Hoff's law, Jenny assumed that for each decrease of 10°C in annual temperature, the rate of humus decomposition and of nitrogen mineralization decreases by two to three times, and therefore, their contents in the soil increase in the same proportion. Jenny expressed this relationship by the following equation:

$$N = C_e{}^{-KT}$$

where N is the total content of nitrogen or humus in the upper soil layer, T is the temperature, and C and K are constants.

He pointed out that this equation is valid only for soils with the same conditions of moisture and with identical plant cover. This consideration is, moreover, justified because in microbiological processes (multiplication, respiration, enzymic activity) the value of Van't Hoff's temperature coefficient (equal to 2 to 3) either remains constant or changes only insignificantly within a comparatively narrow range (15 to 20°C). At high temperatures the coefficient decreases and at low temperature it increases (IMSHENETSKII [1944]).

Later on Jenny continued his study on American, Indian, and Egyptian soils and found that carbon and nitrogen contents depend upon climatic conditions and the way of soil utilization (generalized data are given in JENNY [1965]).

JAGNOW [1971], from data on 20 meadow, 13 forest, and 21 ploughed soils from East Africa found a mathematical relationship between carbon and nitrogen contents and site altitude, amount of precipitation and the length of soil cultivation.

The conclusion reached from the data of Tyurin and Volobuev, according to which soils with the highest humus contents are those that develop under a moderate hydrothermal regime, is no doubt logical. Such conditions are favorable to the development of perennial mixed stand grasses and other herbaceous vegetation, which is the best source of humus. At the same time these conditions favor the rhythmical character of the activities of microorganisms, when periods of rainfall during summer stimulate the formation of humus substances; the alternating periods are characterized by moisture deficit, because the depression of the activity of microorganisms protects newly formed humus substances from decomposition. The latter substances, by interacting with the mineral part of the soil, become less available to micro-organisms, even during subsequent periods of their activation.

A different phenomenon occurs in serozems, for which long periods of high temperature are characteristic; with favorable moisture conditions (particularly irrigation) an intensive new formation of humus substances is associated with an equally high rate of their decom-position. In northern soils, both new formation and decomposition of humus substances proceed slowly, because of the feeble activity of microorganisms, which is limited, even during the summer season, by unfavorable temperature conditions.

The connection therefore between humus reserves and the microflora (see Figure 1) as well as hydrothermal conditions (see Figures 2 and 3) becomes understandable.

While examining the influence of hydrothermal conditions on organic matter transforma-tions in soil, it should also be noted that the moisture regime has an essential influence on the humus composition and the nature of humic acids. The data given in the following sections show convincingly that under conditions of adequate and especially of excessive moisture, under leaching regime (and acid soil reaction), fulvic acids and humic acids similar to them in nature and properties predominate in the composition of humus. This characteristic is observed under the severe climatic conditions of the north (in podzolic soils) as well as under the tem-perature regime of the humid subtropics and tropics (krasnozems, ferrallitic soils).

In this case the absence of association with temperature conditions shows that the regu-larity found in the composition of humus (predominance of fulvic acids) is explained not so much by biological processes as by the mechanism of condensation and polymerization of humus substances, which are retarded under conditions of excessive moisture and acid soil reaction.

The conditions of alternating water regime (periods of moistening alternating with periods of drying) and neutral soil reaction favor mainly the formation of humic acids possessing fairly condensed aromatic rings. Soils of the chernozem type are an example of this.

In the further discussion of the data characterizing humus of various soil types, we shall see, from a number of examples, the multilateral influence of hydrothermal conditions on both biological and abiotic processes associated with the formation of different types of humus.

(d)   Chemical and Physicochemical Soil Properties and Organic Matter Transformations

The role of chemical and physicochemical soil properties in the transformation of organic matter is determined on the one hand by their influence on the development of vegetation cover and microbiological processes, and hence on the processes of new formation and decomposition of organic matter. On the other hand, the essential role of chemical and physicochemical soil properties in the fixation of humus substances and thus their retention in soil is not in doubt.

Lime plays an important role in the transformation of organic matter. Numerous investigations have demonstrated the promoting effect of calcium on the decomposition of fresh plant residues, which apparently is mainly associated with increased soil pH.

Turning now to the importance of the exchange of sodium in the transformation of organic matter it should be pointed out that a small amount of exchangeable sodium increases the rate of decomposition. This can be explained by the partial conversion of humus substances into a soluble form, more available to microorganisms, and also possibly, by desorption of microorganisms present in the soil in an absorbed and inactive state. However, with large amounts of exchangeable sodium, as a result of impaired aeration, microbiological activity is distinctly suppressed.

There are indications that the processes of organic matter transformation are impeded by the presence of iron and aluminum (see KONONOVA [1963, 1966]).

Soil reaction is of great importance not only because of its influence on microbiological processes, but also (in acid soils) because of its influence on the hydrolysis of humus substances. This phenomenon is a major reason for humus in acid soils under leaching regime being represented primarily by fulvic acids and forms of humic acids similar to them.

Chemical and physicochemical soil properties play an important role in the fixation of organic matter in soil. This phenomenon is complex in character and depends on the nature of the substances and the chemical and physicochemical soil properties. Clay minerals are important for the absorption of organic matter. Several authors have shown that this group of minerals interacts not only with humus substances, but also with many compounds of an individual nature—proteins, organophosphorus compounds, cellulose, hemicelluloses, and others; these compounds, when linked with clay minerals, become less available to microorganisms. Many publications are concerned with these problems; the review made by GREENLAND [1965] may serve as an example.

VINTILĂ, CHIRIAC, and BĂJESCU [1963], from a survey of many analytical data, established a direct relationship between the clay content and humus content in the arable layer of the main soil types of Rumania. However, they reported that the zonal character of humus accumulation retains its importance independently of the mechanical soil composition.

The direct relationship between the content of the mineral colloidal fraction (possessing high exchange capacity) in the soil and the amount of humus is explained by fixation of organic matter by clay minerals and hence its limited availability to microorganisms. Table 4 shows that the exchange capacity of the mineral part of the soil and, at the same time, the humus content increase from podzolic soils to fertile chernozems. In southern chernozems, chestnut soils, and serozems the exchange capacity and humus content decrease.

The existence of this kind of relationship makes it possible to define more accurately the factors that favor the accumulation of different amounts of humus in various soils. For instance, it becomes clear that the low humus content of serozems is explained not only by the high rate of organic matter decomposition but also by the fact that soils have a low exchange capacity and so the fixation of organic matter is negligible. At the same time, accumulation of large amounts of humus in chernozems depends to a great extent on the high exchange capacity of these soils.

The influence of soil chemical properties is shown not only in the total humus content but also in the humus composition. An important and interesting survey of data on the main

Table 4. Relationship between the exchange capacity and humus content of different soils of the U.S.S.R. (See data of Vinokurov, Madanov, and others in GORBUNOV [1948].)

Soil	Horizon	Humus, %	Exchange capacity in meq per 100 g of		Percentage of exchange capacity in	
			Mineral part	Whole soil	Mineral part	Organic soil
Strongly podzolic soil	A cultivated	2.82	7.22	10.08	71	29
	A₂	0.66	5.24	6.16	90	10
Moderately podzolic soil	A cultivated	3.24	7.07	12.56	56	44
	A₁	2.74	6.74	12.02	56	44
Slightly podzolic soil	A cultivated	5.29	14.65	26.54	55	45
	A₁	5.34	15.07	23.09	65	35
Slightly podzolic dark-gray soil	A cultivated	6.13	13.69	35.29	39	61
	A₃	4.97	15.46	34.09	45	55
Leached chernozem (high humus content)	A cultivated	8.02	20.16	48.35	42	58
	A₁	5.50	23.24	47.88	48	52
Fertile chernozem	A cultivated	11.10	23.82	62.24	38	62
	A₁	11.30	23.53	63.98	37	63
Ordinary chernozem	A cultivated	7.90	25.68	56.86	45	55
	A₁	5.96	25.00	52.91	47	53
Southern chernozem	A cultivated	4.95	24.66	42.50	58	42

zonal typical soils of Rumania and other countries, made by CERNESCU and CICOTTI [1964], showed that the humic/fulvic acid ratio in humus depends on the soil pH and increases with an increase of the degree of soil saturation by bases (see Figure 7).

The possibility of polymerization of organic compounds on the catalytically acting surface of inorganic compounds is an extremely interesting problem which recently has been raised more and more often in the literature.

The phenomenon of polymerization of polyphenols (which, as is well known, serve as structural units for humic acids) on iron hydroxide, silicic acid, or ion-exchange resins is described in a number of papers (see SCHEFFER and ULRICH [1960]) and may serve as an example of the possible nonbiotic synthesis of humus substances.

Further examination of the data will convince us of the importance of these phenomena in the formation of humus in some soils (*e.g.*, black tropical soils).

The information in this section shows that the state of the organic part of soil is determined by a combination and interaction of factors and conditions of soil formation: vegetation cover, activity of microorganisms, hydrothermal conditions, chemical properties of soil and of its parent rock.

On the basis of the above discussion we shall now proceed with an examination of the data characterizing the features of humus formation in soils.

3. General Features of Humus Formation in Various Soil Types

Fairly extensive data have been accumulated recently in the U.S.S.R. on the contents and reserves of humus and nitrogen, the group composition of humus, and the nature and properties of humus substances. They define the general features of humus formation in the main soil types. In addition, we have also used, in the present work, a number of review papers by foreign authors devoted to the above-mentioned topic.

(a)   Humus and Nitrogen Reserves in the Main Soil Types; C:N Ratio

In Figure 4 are given data by TYURIN [1949, 1951, 1965] summarizing the results of many Soviet soil scientists. The reserves were calculated for the 1-meter layer and for the 0 to 20 cm horizon. These results have been supplemented by data for a number of soils not included by Tyurin (tundra soils, light chestnut soils, brown semidesert soils and takyrs, mountain-meadow soils, and brown mountain-forest soils).

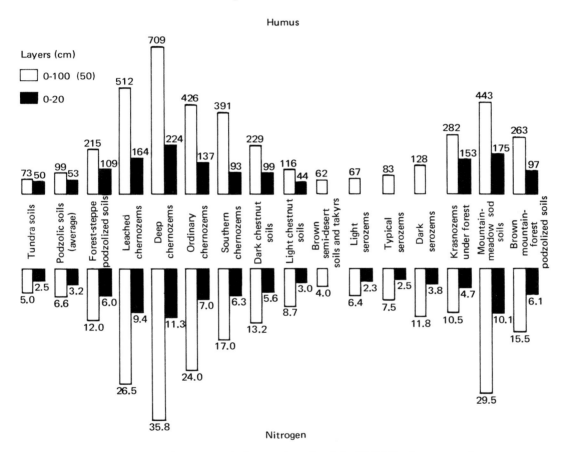

Figure 4.   Reserves of humus and nitrogen in soils of the U.S.S.R., in tons per hectare.

The richest in humus are deep chernozems, where the 1-meter layer soil contains more than 700 tons/ha humus and about 40 tons/ha of nitrogen. The reserves of humus and nitrogen in soils decrease in regions that are characterized by increasing amounts of precipitation and decreasing temperatures (leached chernozem, gray forest soils, and podzolic soils), as well as in regions where arid conditions occur (southern chernozems, chestnut soils, brown soils, and semidesert soils). Light serozems adjoin the latter group; the increase in the reserve of humus and nitrogen from light serozems to ordinary serozems and further to dark serozems is due to a change in the conditions of soil formation caused by vertical zonality.

The data for mountain-meadow soils and brown mountain-forest soils given in Figure 4 show that these soils are rich in humus and nitrogen.

We turn now to the nature of the distribution of humus in the profiles of various soils. In tundra soils humus is concentrated mainly in the surface layer; in podzolic soils and gray forest soils, for example, half of the whole reserves of humus and nitrogen is in the 0 to 20 cm layer. In chernozem soils, on the contrary, both humus and nitrogen are distributed over a greater depth of the soil profile, and therefore, the upper 0 to 20 cm layer contains only 25 to 30 % of the total reserves in the 1-meter layer. In chestnut soils and partly in serozems the humus and nitrogen is again observed to be more concentrated in the 0 to 20 cm layer. Soils of the humid subtropics (krasnozems) contain a large amount of humus, more than half of which is to be found in the upper (0 to 20 cm) layer.

The distribution of humus in the soil profile is explained in the majority of cases by the character and amount of plant residues, which are added to and generated in the soil. Under moss-lichens tundra, decaying components of the vegetation are concentrated in the topmost layer. In podzolic soils and gray forest soils, litter, grassy cover, and fine roots of woody vegetation, which are distributed in the surface horizons, serve as a main source of humus substances. Grassy cover of meadow-steppe and steppe vegetation of chernozems with roots penetrating into deep layers favors the formation of a deep profile with a high humus content. In soils of dry steppes and deserts the root systems of the vegetation are concentrated more and more in the upper soil layers, and the humus horizon in these soils is shallow.

We shall return to a discussion of the distribution of humus in the soil profile when dealing with the characteristics of humus of individual soils types (see Section 4).

The C:N ratio in soils, which is an index of the relative richness in nitrogen of the humus, is a characteristic value. Figure 5 is compiled from Tyurin's data referring to cultivated soils.

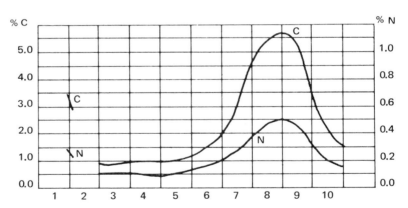

Figure 5.   Percentage of carbon and nitrogen in soils of the U.S.S.R. (for the upper 0–20 cm layer).
   1—krasnozems; 2—light serozems; 3—typical serozems; 4—brown semidesert soils; 5—light chestnut soils; 6—dark chestnut soils; 7—ordinary chernozems; 8—deep chernozems; 9—gray forest soils; 10—podzolic soils.

Deep and ordinary chernozems have the widest C:N ratio (over 10); to the north (in forest-steppe soils and podzolic soils) and to the south (in chestnut soils and particularly in brown semidesert soils) the ratio becomes narrower. The narrowest C:N ratio (8 or less) occurs in serozems, the humus of which is richer therefore in nitrogen than the humus of podzolic soils and particularly of chernozems. Humus of krasnozems is the poorest in nitrogen, as shown by the wide C:N ratio (18 to 19).

The reasons for such characteristics in the C:N ratios of various soils are not fully understood. The narrow C:N ratio in serozems can be due to the high population of microorganisms in these soils and therefore may be explained by the richness of the soil humus in microbial protein. In chernozems the C:N ratio, which is equal to 11 or 12, is probably caused by the high content of humic acids in the humus of these soils (the C:N ratio of humic acids is 14 or 15). The C:N ratio in cultivated podzolic soils (approximately 10) can be explained by a predominance in the humus composition of these soils of fulvic acids in which the C:N ratio is narrower than in humic acids. As has been pointed out above, all established C:N values refer to cultivated soils. It should be noted that in land of the same soil type or subtype but utilized differently, the C:N ratios may vary widely and depend, to some extent, on the care in removing semidecomposed plant residues during preparation of the soil sample for analysis.

Characteristics concerning the humus and nitrogen reserves and C:N ratios are applicable to the same soil types in other countries. TYURIN [1949] summarized the literature data on soils in the U.S.A. (Table 5).

Table 5. Humus, carbon, and nitrogen reserves, and C:N ratio in some soils of the U.S.A. (tons/ha in the 0 to 100 cm layer).

Soil group	Humus	C	N	C:N
Podzolic soils	109	64	5.6	11.3
Brown forest soils	221	128	10.5	12.2
Prairie soils	265 (220–300)	—	—	—
Chernozems of northern group	437 (400–500)	254	21.2	12.0
Chernozems of southern group	221	128	13.6	9.4
Chestnut soils	172	100	11.2	9.0
Dark serozems	125	72.5	9.3	7.8
Laterites (krasnozems) U.S.A.	68	40	3.8	10.4
Laterites (Hawaiian Islands, Mexico)	247	144	10.1	14.3

Tyurin reported that the chernozems of the northern group (Canada and North Dakota) are similar in humus and nitrogen reserves to ordinary and degraded Soviet chernozems, but they are somewhat poorer than deep and leached chernozems of the European part of the U.S.S.R.

The group of prairie soils approaches, in their humus and nitrogen reserves, the Soviet dark-gray soils of forest-steppes; similar to them is a group of brown (and gray-brown) forest soils. Podzolic soils of the U.S.A. scarcely differ from podzolic soils of the U.S.S.R. in their humus and nitrogen reserves.

Lateritic soils of the Hawaiian Islands, which have been cultivated only recently, are characterized by considerable humus and nitrogen reserves, similar to those in krasnozems of the U.S.S.R. and, like the latter, their humus has a low nitrogen content. The low hhmus content of lateritic soils of the U.S.A. is, according to Tyurin, due to erosion.

However, recent data obtained by FRIDLAND [1964] for lateritic (ferrallitic) soils of northern Vietnam show high humus and nitrogen contents, the reserves of which correspond to those in gray forest soils and comprise: Humus in the 0 to 100 cm layer, up to 200 tons/ha; and nitrogen, 7 to 12 tons/ha. In the 0 to 20 cm layer the reserves are: humus, 50 tons/ha; and nitrogen, 4 tons/ha.

Data with respect to humus and nitrogen reserves, confirming the pattern of humus formation characteristic for the U.S.S.R., can be found in the work of VINTILĂ, CHIRIAC, and

BAJESCU [1963] for the following Rumanian soils: podzolized soils and podzols, reddish-brown soils, and brown forest soils, calcareous, neutral, and leached chernozems, and light soils. From a survey of many analytical results, the authors have been able to compile a map showing humus and nitrogen reserves in the 0 to 20 and 0 to 100 cm layers for regions where these soils occur. Variations in the reserves for various types of Rumanian soils were generally found to be similar to those in corresponding soils of the U.S.S.R.

The zonal character shown in humus and nitrogen accumulations is also valid for soils of vertical zonality. A detailed survey of extensive analytical data on this subject for soils of vertical zonality of Azerbaidzhan was made by ALIEV [1956] and for Czechoslovakia by PELIŠEK [1962]

(b)   Humus Composition in Various Soil Types

The experience of soil scientists shows convincingly that the determination of the group composition of humus can be used successfully for studying soil genesis.

The most widely used method is that developed by TYURIN [1951, 1965]; it is also described

Table 6.   Composition of humus in soils of the U.S.S.R. (for upper humus layer).

Soil	Average per cent humus	Carbon of individual groups, as per cent of total soil carbon content				H.A. Carbon* / F.A. Carbon	Humic acids, free and combined with mobile R_2O_3†
		Substances extracted during decalcification	Humic acids	Fulvic acids	In soil residue		
	1	2	3	4	5	6	7
Tundra soils	About 1.0	20–30	10	30	30–40	0.3	75–100
Strongly podzolic soils	2.5–3.0	10–20	12–15	25–28	30–35	0.6	75–95
Sod podzolic soils	3.0–4.0	About 10	20	25	30–35	0.8	90–95
Gray forest soils	4.0–6.0	5–10	25–30	25–27	30–35	1.0	20–30
Deep chernozems	9.0–10.0	5–10	35	20	30–35	1.7	20–15
Ordinary chernozems	7.0–8.0	2–5	40	16–20	30–35	2.0–2.5	10–15
Southern and Cis-Caucasian chernozems	5.5–6.0	3–5	30–35	20	About 40	1.5–1.7	—
Dark chestnut soils	3.0–4.0	2–5	30–35	20	30–35	1.5–1.7	10–15
Light chestnut soils	1.5–2.0	8–10	25–29	20–25	30–38	1.2–1.5	Less than 10
Brown semidesert soils	1.0–1.2	3–5	15–18	20–23		0.5–0.7	About 10
Typical serozems	1.5–2.0	About 10	20–30	25–30	25–35	0.8–1.0	10 or less
Light serozems	0.8–1.0		17–23	25–35	25–35	0.7	10 or less
Takyrs	About 1.0	5–10	7–10	20–25		0.3–0.4	About 5
Krasnozems	4.0–6.0	10–20	15–20	22–28	35–38	0.6–0.8	90–100
Mountain meadow soils	6–15	10	15–30	28–35	20–40	0.4–0.8	—
Mountain forest brown soils	4.0–8.0	—	25–30	30–35	30–35	0.7–0.9	10–20

* This column is:
the ratio of humic acid (H.A.) carbon to the fulvic acid (F.A.) carbon = Humic acid carbon divided by fulvic acid carbon or $\dfrac{\text{Humic acid carbon}}{\text{Fulvic acid carbon}}$

† As percent of the total amount of humic acids (see column 3).

by KONONOVA [1963, 1966]. Extensive Soviet analytical data obtained by this method are given in Table 6.

The principle of Tyurin's method is as follows: A soil sample, after decalcification (by means of 0.1 N H_2SO_4), is treated repeatedly with 0.1 N NaOH for the extraction of humus substances. Humic acids are then precipitated from this alkaline solution by acidification; the acid solution contains fulvic acids. Carbon determinations in the extracts indicate the humus components: (1) substances retained in the 0.1 N H_2SO_4 extract, (2) humic acids, (3) fulvic acids (4) residual humus substances (so-called humins).

According to Tyurin, when decalcified soil is treated repeatedly with 0.1 N NaOH, humic acids linked to calcium and mobile forms of R_2O_3 are isolated. By treatment of a separate soil sample (without decalcification) with 0.1 N NaOH, humic acids combined with mobil (nonsilicate) forms of R_2O_3 are extracted. The difference between these two values gives the amount of humic acids linked to calcium. However, it should be pointed out that separations into fractions according to the form of link with the mineral part of the soil is, in Tyurin's method, hypothetical.

We note from Table 6 that the group of humic acids extracted from previously decalcified soil by repeated treatment with 0.1 N NaOH is characteristic of soil humus (column 2). In the majority of cases a change in the humic acid content corresponds to a change in the total amount of humus. The percentage content of this group in the composition of humus increases from podzolic soils to chernozems and again decreases in the direction of chestnut soils and further toward brown desert-steppe soils and also serozems.

The fulvic acid content (see column 4 of the table) changes less regularly, but on the whole it is in inverse relationship with the humic acids. The ratio between humic acid/carbon and fulvic acid/carbon (column 6), which in chernozems is about 2.0 and in chestnut soils 1.5 to 1.7, is characteristic. In podzolic soils, brown semidesert soils, serozems, and krasnozems this relationship is less than unity. The fulvic acid content exceeds that of the humic acids also in mountain-meadow soils.

Column 7 gives a rough idea of the amount of soil humic acids, free and hypothetically combined with nonsilicate forms of R_2O_3. As has already been mentioned, in Tyurin's analytical procedure for the determination of humus composition, this humic acid fraction is extracted from a separate soil sample (nondecalcified) by a single treatment with 0.1 N NaOH. In podzolic soils and in krasnozems this fraction constitutes 90 to 100% of the total amount of humic acids, whereas in soils of the chernozem series it comprises 10 to 15%. It is evident that in chernozem soils the humic acids are combined with calcium in complex forms that can be disrupted only by preliminary treatment with mineral acid.

Gray forest soils occupy an intermediate position; in these soils, humic acids, free and combined with mobile forms of R_2O_3, constitute 20 to 30% of the total humic acid content in the humus. From semidesert soils and also from serozems, about 10% of the total humic acid content is extracted by a single treatment with 0.1 N NaOH.

The amount of substances extracted during soil decalcification with dilute HCl or H_2SO_4 follows a fairly constant pattern (see column 2). This group includes various individual compounds (resin acids, carbohydrates, products of protein decomposition, etc.); in podzolic soils and in krasnozems this group includes a certain amount of fulvic acids and forms of humic acids similar to them. This explains why the amount of substances extracted during decalcification from these two soils reaches a maximum.

With regard to the carbon of the residue (column 5), which constitutes 30 to 40% of the total soil carbon content, no clear pattern of changes in its content in various soils can be observed. The absence of regular changes in this organic matter fraction can be explained by the fact that its fixation is mainly determined by the mechanical and mineralogical composition

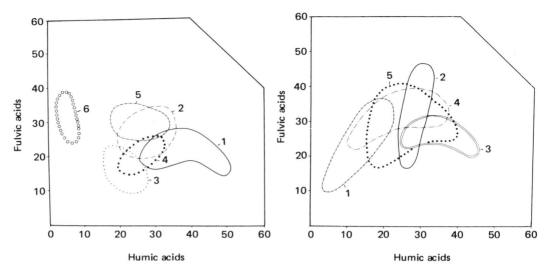

Figure 6. Limits of variation in the group composition of humus from soils of various genetic types.
(a) 1—chernozems; 2—chestnut soils; 3—serozems; 4—smolnitzes; 5—mountain meadow soils; 6—tundra soils.
(b) 1—podzolic soils; 2—brown forest soils; 3—gray forest soils; 4—krasnozems and shelto-zems of the subtropics; 5—cinnamonic forest soils (VOLOBUEV [1962]).

of the soil. In meadow-forest soils the high carbon content of the soil residue is partly due to the presence of semidecomposed plant material.

A consistent characteristic of the changes in the humus composition of soils of the U.S.S.R. is very well illustrated in Figures 6a and b, constructed by VOLOBUEV [1962], who used the method of triangular coordinates. The content of humic acids (as a percentage of the total humus carbon) is given on the horizontal axis and the percentage content of fulvic acids on the vertical axis. The difference between the sum of these values and 100% represents the value of the nonhydrolyzable residue (humin). Since, in natural soils the content of humic acids and fulvic acids does not exceed values of 50%, the apices of a triangle with acute angles are cut off, and then the triangular system practically assumes the form of a compact pentagon. The distribution of points on the graph revealed characteristic humus areas for the soil types considered.

Determinations of the humus composition in soil studies are commonly used in many countries, and currently there are numerous works confirming the characteristic pattern reported for soils of the U.S.S.R.: some of these will be quoted.

Figure 7 (taken by us from the above-mentioned paper by CERNESCU and CICOTTI [1964]) shows humic/fulvic acid ratios for the main soil types in Rumania. Comparison of the values in the figure with the data given in Table 6 shows that they have the same order of magnitude and confirms the stable character of the regular variations in humic acid/fulvic acid ratios which have been established for soils of the U.S.S.R.

A similar conclusion can be drawn from Table 7, which contains data on the composition of soil humus from various countries.

Investigations have shown that determinations of the group composition of humus are useful for revealing the genetic characteristics of soils. However, we consider it helpful to include along with the determination, a description of the nature and properties of humus substances. Information on this subject is given below.

(c)   Nature and Properties of Humus Substances in Various Soil Types

At present, many investigators are concerned primarily with the study of the nature and structure of soil humus substances.

The development of experimental techniques and new methods (spectroscopical, thermal, E.P.R., electrophoresis, X-ray, etc.) opens many new opportunities to study soil organic.

This is evidenced by the transactions of the Eighth (Rumania, 1964) and Ninth (Australia, 1968) International Soil Science Congresses and the "Humus and Plant" Symposiums

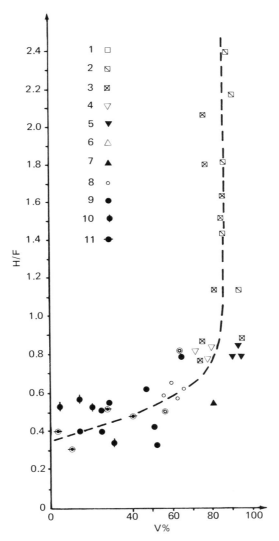

Figure 7.   Humic acid/fulvic acid ratios in various soil types in relation to the saturation with bases (V).
1—chernozem; 2—leached chernozem; 3—podzolized chernozem (and dark-gray soil); 4—podzolized gray forest soil; 5—gray-cinnamonic podzolized forest soil; 6—reddish-brown forest soil; 7—reddish-brown podzolized forest soil; 8—brown forest soil; 9—moderately and strongly podzolized soils; 10—podzolized forest soil (illuvial horizon B); 11—podzols and podzolic soil (CERNESCU and CICOTTI ,1964,).

Table 7.   Content and composition of humus in other European soils (for upper horizon).

| Soil | Percent humus | Percent C | Percent N | C:N | Carbon of individual groups (as percent of total soil carbon) | | | H.A. Carbon / F.A. Carbon | Author, country |
					Humic acids	Fulvic acids	In soil residue		
Podzolic loamy soil under forest, 0–20 cm	1.09	0.63	—	—	27	59	—	0.46	MUSIEROWICZ [1965] (Poland)
Chernozem overlying loess, arable land, 0–10 cm	3.74	2.17	—	—	32	13	—	2.46	MUSIEROWICZ [1965] (Poland)
Podzolized chernozem, 0–30 cm	2.14	1.24	—	—	26	21	29	1.23	KR'STANOV [1963] (Bulgaria)
Brown podzolic soil	2.85	1.49	0.13	14	22	26	32	0.8	DANILIUC [1957] (Rumania)
Ordinary chernozem	5.30	2.85	0.27	11	42	13	31	3.2	DANILIUC [1957] (Rumania)
Brown chernozem	3.61	1.90	0.21	10	34	19	25	1.8	DANILIUC [1957] (Rumania)
Chernozem overlying loess 5–25 cm	3.3	1.90	—	—	33	15	—	2.2	NĚMEČEK and POSPIŠIL [1966] (Czecho-slovakia)
Podzolic illuvial humic-ferrugi-nous soil									NĚMEČEK and POSPISIL [1966] (Czecho-slovakia)
10–20 cm	1.5	0.89	—	—	22	26	—	0.8	
40–45 cm	9.5	5.5	—	—	9	35	—	0.3	

(Czechoslovakia, 1967, 1971), as well as by voluminous literature published in periodicals.

However, these data are available for only a few soils and are not yet sufficient for establishing differences in the nature of humus substances resulting from varying soil-forming conditions. This is why the review is limited to those data that characterize the nature of humus substances available for the most soil types and subtypes, namely, elementary composition, measurements of optical properties, and coagulation (precipitation) threshold.

The presence of two basic structures in the molecules of humus substances can be asserted with confidence: (1) the aromatic carbon skeletons, and (2) the side chain carbon. The aromatic parts of the molecules possess hydrophobic properties and the side chains contain hydrophilic groups. The proportions of these structures determine a number of important properties of humus substances (hydrophilic nature, behavior towards electrolytes, the nature of the organometallic compounds, etc.).

The following discussion illustrates the regular differences in the nature and properties of humus substances (in particular, the ratio of the basic structures in the molecules) that are determined by the conditions of soil formation.

Elementary Composition of Humus Substances

In the elementary composition of humic acids, isolated from the main soil types of the U.S.S.R., the following characteristic differences can be detected. As can be seen from the data in Table 8, the percentage of carbon in humic acids consistently increases from podzolic soils

Table 8. Elementary composition of humic and fulvic acids from the main soils in the U.S.S.R. (as a percentage of absolutely dry ash-free substance).

Soil		C	H	O	N	C:H
Northern podzol under forest; humus-						
illuvial horizon, 16–24 cm,	I*	58.11	5.37	32.00	4.52	10.82
Arkhangel'sk region	II	52.37	3.53	42.89	1.21	14.84
Mountain-taiga ferruginized soil; 1–7 cm,	I	58.84	6.02	29.70	5.44	9.8
Western Trans-Baikal	II	50.64	3.95	43.48	1.93	12.9
Sod-podzolic soil; arable land, 0–20 cm,	I	57.63	5.23	32.33	4.81	11.02
Moscow region	II	46.23	5.05	44.60	4.12	9.15
Dark-gray forest soil under oak; 12–19 cm,	I	61.20	3.60	31.32	3.88	17.00
Shipov forest, Voronezh region	II	47.46	3.64	45.87	3.03	13.04
Ordinary chernozem; arable land, 0–20 cm,	I	62.13	2.91	31.38	3.58	21.35
Kamennaya steppe, Voronezh region	II	44.84	3.45	49.36	2.35	13.00
Chestnut soil; virgin land, 0–20 cm, Valuisk	I	61.74	3.72	30.62	3.92	16.60
experimental station, Volvograd region	II	43.19	3.61	51.43	1.77	11.96
Light serozem; arable land, 0–20 cm,	I	61.94	3.93	29.46	4.67	15.76
Pakhta-Aral, Kazakhs S.S.R.	II	45.80	4.30	46.00	3.90	10.65
Krasnozem under fern; 0–20 cm,	I	59.65	4.37	31.54	4.44	13.65
Anaseuli, Gruzinskaya S.S.R.	II	49.82	3.35	44.33	2.50	14.87

* I denotes humic acids; II denotes fulvic acids

toward dark-gray forest soil and chernozems; a decrease of this value is observed in humic acids from chestnut soils and serozems. An inverse relationship is observed in the percentage content of hydrogen. In accordance with these data, the C:H ratio in the elementary composition of humic acids changes very markedly.

Studies on humic acids of mined coals, peats, and soils (KASATOCHKIN [1951, 1953], KUKHARENKO [1955]) show that the percentage of carbon in these acids increases on "maturing" with a simultaneous decrease in the percentage of hydrogen and nitrogen. These features are considered indicative of the increasing degree of condensation of the aromatic rings in the humic acid molecules.

Correspondingly, the consistent increase of the C:H ratio in humic acids in the series from podzolic to gray-forest soils and further toward chernozems indicates an increasing proportion of aromatic carbon skeletons in the molecules of humic acids in this series of soils. A narrower C:H ratio in humic acids from chestnut soil and serozem shows the lower degree of condensation of the aromatic rings in these acids (as compared with those of chernozems).

This conclusion agrees with the data given below, characterizing the structure of humic acid molecules.

Optical Density of Humus Substances

At present, determinations of optical density, made in the visible and ultraviolet regions of the spectrum, are used for characterizing the nature of soil humus substances.

For these substances the continuous absorption spectrum, which is determined by the aromatic carbon ring, is characteristic; components of an aliphatic structure or of a saturated cyclic nature, forming the side radicals, do not absorb light in the regions of the spectrum examined (KASATOCHKIN [1964]).

Optical properties of solutions of humus substances at a given concentration are usually determined with a full range of filters and are represented in the form of curves (see Figure 8).

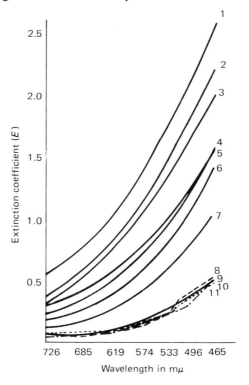

Figure 8. Optical density of humus substances.
Humic acids; 1—ordinary chernozems; 2—dark-gray forest soil 3—chestnut soil; 4—light sero-zem; 5—sod-podzolic soil; 6—krasnozem; 7—strongly-podzolic soil.
Fulvic acids: 8—krasnozem; 9—light serozem; 10—strongly podzolic soil; 11—ordinary cherno-zem.

However, it has been shown (SCHEFFER [1954], WELTE [1955]) that the ratios of the values of optical densities with filters in the vicinity of 400 and 600 nm (Q_4/Q_6 or E_4/E_6) do not depend on the concentration of substances in the solutions examined and are characteristic for humic acids of various types. For brown humic soils (isolated from podzolic soils) this ratio is equal to 5 or 6, for gray humic acids (from soils of chernozem type) E_4/E_6 equals from 2.5 to 3.0. Subsequently, a great diversity of values for optical density of humic acids from various soils has been explained.

The results of our comparative determinations of optical density of sodium humate solutions (pH 7.2 to 9.8) at a concentration of 0.136 g humic acid carbon per liter of solution are given in Figure 8. In a series of soils (strongly podzolic, sod-podzolic, dark-gray forest soil, ordinary chernozem), the optical density of the humic acids increases, thus indicating the increasing degree of condensation of the aromatic carbon skeletons. The humic acids of krasnozem are similar in this respect to those of podzolic soils.

Values of optical density of the humic acids of chestnut soils and serozems show a lower degree of condensation of the aromatic carbon skeletons compared with the humic acid from chernozems. On the whole the data of optical densities of humic acids from various soils agree with the values of the C:H ratios in the elementary composition (see Table 8). With regard to fulvic acids, the degree of condensation of their aromatic rings, as judged by the values of optical density, is extremely small.

The values of the E_4/E_6 ratios for humic acids from various soil types are as follows: podzolic soils, about 5.0; dark-gray forest soil, 3.5; ordinary chernozem, 3.0 to 3.5; chestnut soils, 3.8 to 4.0; serozems, 4.0 to 4.5; krasnozems, about 5.0. For fulvic acids this E_4/E_6 ratio varies from 6.0 to 8.5.

Similar values of optical density for humic acids of various soil types are given by other authors (CECCONI and TELLINI [1954], HOFFMAN [1964], HARGITAI [1955]).

Recently SALFELD [1971] suggested the determination and calculation of optical density by using three wavelengths and the $EQ_{45}:EQ_{56}:EQ_{67}$ ratios. The representation of these ratios in the three-dimensional system of axes is convenient for processing mass materials. The procedure is interesting and needs further testing.

Precipitation (Coagulation) of Humus Substances from Various Soils by Electrolytes (CaCl₂); The Nature of Their Complexes with Iron

As stated above, the ratio of aromatic to aliphatic structures in the molecules of humus substances determines a number of their properties, especially hydrophilic properties. The latter become apparent in connection with the precipitating action of electrolytes.

We have made comparative determinations of the threshhold of precipitation (coagulation) of humus substances from various soils by means of $CaCl_2$. The data in Table 9 (expressed in

Table 9. Threshold of precipitation (coagulation) of humus substances from various soils (in meq $CaCl_2$).

Preparations	Soil						
	Strongly podzolic	sod-podzolic	Krasno-zem	Dark-gray forest	Ordinary chernozem	Dark chestnut	Light serozem
Humic acids	40	20	18	16	5	8	14
Fulvic acids			Weak precipitation				

meq $CaCl_2$, referring to 1 liter of solution containing 0.136 g humic acid carbon, in an experiment lasting 2 hours) show regular differences between individual samples. In fact, humic acids from chernozems are precipitated even by the addition of a small amount of $CaCl_2$ (5 meq per liter of the above solution). Coagulation (precipitation) of humic acids from strongly podzolic soil occurs with the addition of 40 meq $CaCl_2$. All the remaining humic acids (from sod-podzolic soil, krasnozem, dark-gray forest soil, dark chestnut soil) are intermediate between the above-mentioned extreme values.

No precipitation of fulvic acids was observed, even after the addition of large amounts of $CaCl_2$.

Comparison of the data characterizing the threshold of coagulation (precipitation) of humic acids from various soils with the data on their elementary composition (Table 8) and optical densities (Figure 8) shows that with an increasing degree of condensation of the aromatic rings (possessing hydrophobic properties) and with a corresponding decrease of the aliphatic structures (containing hydrophilic groups) humic acids are precipitated by smaller amounts of electrolytes.

Different degrees of condensation of the aromatic rings of humic acids are reflected also in other physicochemical properties, especially in different viscosities. The viscosity is lower in the less-hydrated humic acid from chernozem than in the humic acid from podzolic soils (ORLOV and GORSHKOVA [1965]).

Another important property, the tendency to form intracomplex compounds (chelates) with iron, aluminum, and other polyvalent cations, can be explained by the presence of hydrophilic groups in the side-radicals of humus substances. The importance of these pheno-

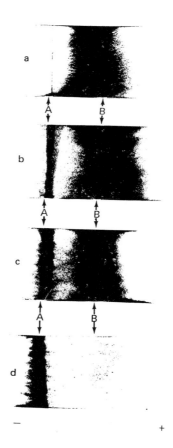

Figure 9. Electrophoretograms of humus substances treated with acidified potassium ferrocyanide solution.
(a) fulvic acids from krasnozem; (b) humic acid from humus-illuvial horizon of strongly podzolic soil; (c) humic acid, saturated with iron, from sod-podzolic soil; (d) humic acid, saturated with iron, from chernozem.
A—start; B—position of negatively charged iron-humus complex.

mena in soil-forming processes (in particular, in weathering and podzol formation) cannot be questioned.

Our data (KONONOVA and TITOVA [1961]) show the varying character of the iron complexes which depends on the nature of the humus substances. Even during the fractionation by electrophoresis (in borate buffer solution, pH 8) of fulvic acid preparations, purified by electrodialysis, negatively charged iron humus complexes appear as characteristic greenish-blue spots on electrophoretograms treated with a 4% solution of potassium ferrocyanide acidified with sulfuric acid. Similar complexes were also detected in humic acids isolated from the humus-illuvial horizon of strongly podzolic soil. (These humic acids resembled fulvic acids.)

Equally interesting are the results of electrophoretic studies of the complexes, obtained artificially when humus substances isolated from soils were saturated with iron. It was found that humic acids from sod-podzolic soil form both an immobile complex and a mobile (negatively charged) chelate-type complex. Humic acids from chernozem form only immobile complexes with iron (Figure 9).

The results given in Figure 9 illustrate the lack of complete identity in the nature and properties of humic acids from various soils and their tendency to approach fulvic acids in character. This aspect becomes clear when it is considered that humic and fulvic acids are heterogeneous and can be separated by various methods (*e.g.*, fractional precipitation at various pH values, ultracentrifugation, partition chromatography, electrophoresis) into several fractions.

A comparative study of the fractions isolated from humic and fulvic acids showed that in some characteristics (carbon and hydrogen contents, optical density) these fractions are similar. The ratio of these fractions determines to a large extent the differences in the nature and properties of various representatives of these groups. Undoubtedly, humic acids, similar in nature to fulvic acids, are also similar in the functions determining their participation in soil processes. Judging from all the characteristics (elementary composition, optical density, behavior toward electrolytes, fractional composition) humic acids in strongly podzolic soils resemble most closely in their functions fulvic acids.

(d) Soil Conditions Determining Regular Differences in Humus Content, Its Composition, and the Nature of Humus Substances from Various Soil Types

The causes of regular differences in the content, composition, and nature of humus substances from various soil types are still not clear because of the complicated dependence of the humus formation process on the complex conditions of the soil medium. Therefore, at the present time the investigator can only state to what natural conditions one or other type of humus corresponds.

A comparison of the above data characterizing the natural conditions (plant cover, microflora, hydrothermal conditions, chemical and physicochemical soil properties) under which humus formation takes place, with the data on humus reserves, humus composition, and the nature of humus substances leads to the following conclusions.

The highest humus content is observed in chernozems, due to the perennial meadow and steppe vegetation, moderate hydrothermal regime, neutral soil reaction, and adequate microbial activity. Under these conditions humus substances, mainly of the humic acid type, are formed which possess condensed aromatic rings and a small number of side radicals. We consider that alternate periods of wetting and drying (characteristic of chernozem soils) favors the condensation and polymerization of the humus substances (KONONOVA [1951]). A similar view was expressed by DUCHAUFOUR and DOMMERGUES [1963] on the basis of an examination of the humus of black tropical soils (see Section 4).

In soils with adequate or excessive moisture, leaching regime, and acid reactions, fulvic acids are predominant in the composition of humus. In the humic acids of such soils the aromatic skeleton becomes less condensed, with a simultaneous increase in the number of hydrophilic groups in the side chains.

If fulvic acids are regarded either as the initial forms or as the degradation products of humic acids, then it can be presumed that under the conditions of podzolic soils and krasnozems the formation of humus substances ceases at the fulvic acid stage. Further, it may be seen that lateritic (ferrallitic) soils of the tropics belong to this group.

The predominance of fulvic acids in the humus composition is also observed in brown semidesert soils and unirrigated serozems, where biological activity and the process of formation of humus substances are limited by inadequate moisture. In irrigated serozems vigorous biological activity prevents the "maturing" of humus substances by involving them in new biological cycles.

It is clear from the above discussion that questions concerning the mechanism of formation of humus substances in various soils require further experimental study.

4. Humus of Different Soil Types and Subtypes

The data given in the preceding section, characterizing humus of different soil types, are the mean values of many determinations and refer to only upper soil horizons. In the present section characteristics of humus from soil profiles of the main soil types occurring in different zones are presented. In addition to numerous data in this review, the works of various authors are also used (a bibliography is given in the book "Soil Organic Matter" by KONONOVA [1966]).

As criteria for characterizing the nature of humus, the percentage of carbon, the percentage of nitrogen, and the C:N ratio were used. There are also data on the contents of humic and fulvic acids and their ratios. Most analyses of humus composition were done by Tyurin's method. In individual cases the results on humus composition have been obtained by a rapid, simplified method of KONONOVA and BEL'CHIKOVA [1961], in which humus substances are extracted by a single treatment of nondecalcified soil with a mixture of 0.1 M $Na_4P_2O_7$ and 0.1 N NaOH. Determinations of humus composition by the two methods gave similar results. (see also MUSIEROWICH and SYTEK [1964], WIESEMÜLLER [1965]).

Data are also given on the amount of humic acids (the group of free humic acids and presumably of humic acids combined with nonsilicate forms of R_2O_3) extracted by a single treatment with 0.1 N NaOH of nondecalcified soil. For the characterization of the nature of humic acids we used results of the determination of optical density expressed as the coefficient E_4/E_6. In many instances the values of the coagulation threshold of humic acids are also given.

Humus of Tundra Soils

The tundra zone, because of low annual temperatures and high moisture (resulting from low evaporation), has unique conditions of soil formation. Excessive moisture favors reducing processes, in particular, soil gleying. Permafrost is of great importance and influences the processes of transformation of organic residues, the nature of humus substances, and their distribution in the soil profile.

The soil-forming process in the tundra zone develops mainly under moss-lichen and brushwood cover, with negligible participation and even absence of grassy vegetation. The microflora of tundra soils is scanty—low in number of species and in biochemical activity. Leaf fall is humified slowly, as shown by the wide ratio of litter to leaf fall (Table 2).

All these conditions—limited amount of fresh plant material entering the soil (Table 2), scanty microflora (Figure 1), and hydrothermal conditions unfavorable for microflora activity— contribute to the small humus reserves, usually concentrated in the 0 to 50 cm layer in tundra soils.

Studies on the nature of the humus of tundra soils show marked variations in both the total humus content and its composition, which can be explained by the diversity of natural conditions under which these soils are formed.

Table 10 contains data characterizing the humus of virgin tundra soils. Thus, for the tundra soils of Vorkuta and Salekhard, a low humus content and low humic acid content in the humus are characteristic. In the tundra of northern Yakutiya, even at a depth of 35 to 40 cm, a high content of organic carbon, equal to 4.7%, has been recorded. The fairly high content of humic acids in the humus of this soil is remarkable; in addition, at a depth of 35 to 40 cm, the humic acid, carbon/fulvic acid, carbon ratio increases and exceeds unity, as compared with that in the 6 to 10 cm horizon. Such a phenomenon can be explained by the migration of humus substances from the horizons of their formation toward the cold front of

Table 10. Content and composition of humus in tundra soils.

Soil	Depth of sampling, cm	Organic C, %	N %	C:N	Carbon of individual groups, as % of total soil organic carbon		H.A. Carbon / F.A. Carbon	Author
					Humic acid	Fulvic acid		
Tundra sur-	0–8	26.7	1.05	25.4	6.5	24.2	0.27	BARANOV-
face gley	10–16	2.2	0.154	14.3	8.7	27.8	0.31	SKAYA
loam	16–27	0.8	0.049	16.3	3.1	38.8	0.08	[1952]
Vorkuta	27–45	0.9	0.068	13.3	3.0	31.0	0.09	
Gley-podzolic	0–5	0.95	0.08	11.9	14.2	25.2	0.58	KOSHELEVA
tundra soil								and
Salekhard	10–15	0.76	—	—	5.9	21.7	0.27	TOLSTUKHINA [1957]
Tundra soil	6–10	12.17	0.95	12.8	25.0	36.2	0.69	KARAVAEVA
								and
Northern	35–40	4.74	0.35	13.5	25.3	23.7	1.07	TARGUL'YAN
Yakutiya								[1960]

permafrost, with their subsequent conservation under conditions of anaerobiosis, low temperature, and the heavy mechanical composition of the layer above the permafrost.

Humus of Podzolic and Sod-Podzolic Soils

The taiga-frost zone where these soils occur is characterized by a great diversity of climatic conditions, vegetation, and parent rocks, as shown by the soil cover. The latter develops under a combination of podzol-forming and sod processes; the relationship between these processes is very clearly reflected in the nature of the humus. There are many data characterizing the nature of humus in podzolic soils of the U.S.S.R. which occupy a vast area in the European and Asian parts of the Soviet Union. Some examples are given below.

(a)   Strongly podzolic soils of northern taiga subzone

Weak biological activity that is limited by the short period of relatively favorable combinations of temperature and moisture is typical of this subzone. During only 90 days in the year does the temperature exceed 10°C and are moisture conditions favorable for biological activity.

As Figure 1 shows, podzolic soils have a scanty microflora, and under these conditions, humification of large amounts of plant residues (Table 2) takes place slowly. This is evident from the thick litter; the ratio of litter to leaf fall in the northern and central taiga amounts to 15 to 17 (Table 2). Excessive moisture, acid soil reaction, and low temperature favor the formation of products of the incomplete decomposition of plant residues and humus substances of the fulvic acid type.

Soils of the Komi ASSR, for example, are acid (pH 3.5 to 4.2), and are developed on brown and red-brown loam; the sum of exchangeable bases in the mineral horizons amounts to 1-4 meq per 100 g soil.

Table 11.   Content and composition of humus in strongly podzolic soils of the northern taiga subzone (Komi A.S.S.R.)*.

| Soil, land use | Horizon | Depth of sampling, cm | Organic C, % | N, % | C:N | Carbon of individual groups as percent of total soil organic C | | H.A. Carbon / F.A. Carbon | Humic acids, free and combined with mobile forms of R_2O_3† |
						Humic acids	Fulvic acids		
Strongly podzolic soil	A_2	5–10	0.50	0.05	10.0	10.0	20.0	0.50	100
under scotch pine forest	A_2B	15–20	0.43	0.04	10.8	7.0	18.7	0.38	100
with moss cover	B	30–40	0.28	0.03	9.3	7.2	17.9	0.40	100
Sod-strongly podzolic	A_1A_2	7–12	1.35	0.10	13.5	20.0	29.6	0.68	100
soil. Dry valley	A_2B	20–30	0.24	0.02	12.0	8.4	16.7	0.50	100
meadow	B_1	45–55	0.18	0.03	6.0	5.6	22.2	0.25	—

* In Tables 11 through 14 are contained data of BEL'CHIKOVA [1961].
† In this table, as in Tables 12, 15, 16, 20, 21, 22, 27, 29, and 30, the amount of humic acids, free and combined with mobile forms of R_2O_3, are expressed as a percentage of the total amount of humic acids.

As may be seen from Table 11, the soil under forest has a low humus content, and fulvic acids predominate in the humus composition. Humic acids are represented by the fraction extracted by single treatment with 0.1 N NaOH solution, without preliminary soil decalcification (*i.e.*, the fraction of free humic acids and humic acids combined with mobile forms of R_2O_3).

Soil in meadows differs from forest soil, in having a fairly well defined but shallow humus horizon. A relatively high humic acid content is observed in the composition of the humus in this horizon and the humic acid carbon/fulvic acid carbon ratio is somewhat wider than in forest soil. However, the lower horizons of meadow soil differ neither in humus content nor in composition from those under forest. Consequently, in the northern taiga subzone the meadow grassy vegetation on strongly podzolized soil which has previously been under forest, favors the development of the sod process only in the upper soil horizon, and signs of this process are generally slight.

More intensive development of the sod process takes place in the deciduous-coniferous subzone, where it proceeds not only under meadow but also under forest vegetation.

(b) Sod-Podzolic Soils of the Deciduous-Coniferous Subzone

As an example of these soils, data of the Yaroslav region will be given. In this region hydrothermal conditions are more favorable to biological activity than in the northern taiga subzone. The period with temperatures higher than 10°C and a moisture supply favorable for biological activity lasts from 4 to $4\frac{1}{2}$ months of the year. This intensive microbiological activity is shown by the shallower (when compared with the northern taiga subzone) litter, whose ratio to the leaf fall is 10 (Table 2).

The sod-podzolic soils (under forest, meadow, or arable land) discussed have developed on a heavy blanket loam deposit; the soils are acid (pH∼3.5 to 4.9). The sum of exchangeable bases ($Ca^{++} + Mg^{++}$) in the mineral horizons constitutes approximately 3 to 8 meq per 100 g soil.

The data on the content and composition of humus in different soils are given in Table 12.

Table 12.   Content and composition of humus in sod-podzolic soils in deciduous-coniferous subzone (Yaroslav region).

Soil, land use	Horizon	Depth of sampling, cm	Organic C, %	N, %	C:N	Carbon of individual groups as percent of total soil organic C		$\dfrac{\text{H.A. Carbon}}{\text{F.A. Carbon}}$	Humic acids, free and combined with mobile forms of R_2O_3
						Humic acids	Fulvic acids		
Sod-Strongly pod-	A_1	4–7	4.03	0.26	15.5	22.3	28.3	0.79	94.4
zolic soil under	A_1/A_2	7–15	0.98	0.08	12.3	15.3	26.5	0.58	100.0
mixed forest	A_1/A_2	15–23	0.58	0.06	9.7	13.8	24.1	0.57	100.0
	A_2	23–38	0.30	0.03	10.0	13.3	26.7	0.49	—
Sod-moderately pod-	A_1	0–10	5.10	0.47	10.8	22.3	29.2	0.77	89.5
zolic soil; dry valley	A_2	10–23	0.83	0.05	16.6	16.9	28.9	0.58	100.0
meadow surrounded	A_2B	23–35	0.48	0.03	16.0	10.4	31.3	0.33	—
by forest									
Sod-moderately pod-	A_{arable}	0–10	1.16	0.14	8.3	22.4	28.4	0.78	92.3
zolic soil, arable land	A_{arable}	10–23	1.21	0.13	9.3	15.7	29.7	0.53	100.0
	B_1	23–29	0.52	—	—	15.4	28.8	0.54	—

The higher humus content and deeper humus-accumulating horizons indicate a more intensive development of the sod process under forest, meadow, and arable land in the deciduous-coniferous subzone as compared with corresponding areas in the northern taiga subzone; in the composition of humus, the content of humus acids increases and the humic acid carbon, fulvic acid carbon ratio becomes slightly wider, irrespective of the simultaneous increase in the amount of fulvic acids.

It should be stressed that the humus composition of soil under meadow and under arable land is generally similar to that under forest. Evidently this is because under the bioclimatic conditions of the deciduous-coniferous subzone the development of the sod process takes place even in soil under mixed forest.

Figure 10 illustrates the distribution of roots and humus in the profile of sod-podzolic

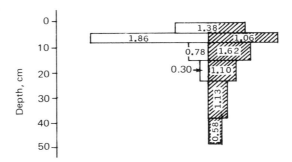

Figure 10.   Reserve of organic matter in sod-podzolic soil (Yaroslav region). Spruce forest: I—roots; II—humus. Total amount of roots in the soil profile per square meter of surface area—3.08kg, of which fine roots amount to 0.47 kg and litter to 1.38 kg. Total amount of humus in soil profile per square meter of surface area—5.49 kg.

soil under forest, the main mass of which is found in the upper horizon. It declines rapidly down the soil profile. A more uniform distribution of humus in the 0 to 20 cm layer of arable land (Table 12) is caused by mixing of the soil during ploughing.

The development of the sod process in soil formation in the deciduous-coniferous subzone is reflected not merely in the increased contents of humus and humic acids, but also in the change of the nature of the latter. The values of optical density (Figure 8) and coagulation threshold (Table 9), like the data given in Table 13 referring to the soils of the Komi ASSR and Yaroslav regions, indicate the increase in the degree of condensation of the aromatic rings in humic acids of sod-podzolic soils (compared with those of strongly podzolic soils) with a simultaneous decrease of hydrophilic side chains in the molecules.

In comparing the data characterizing the nature and properties of humic acids from various soil types, we made the assumption that the humic acids in strongly podzolic soils resemble the group of fulvic acids in a number of properties. There is every reason to assume that in podzolic soils, especially in the northern taiga subzone, humic acids (like fulvic acids) participate actively in the decomposition of the mineral part of the soil and in podzol formation.

Table 13.   Optical density and coagulation threshold (precipitation) of humic acids in podzolic soils.

Soil	Strongly podzolic soil (Komi ASSR)			Sod-podzolic soil (Yaroslav region)				
Characteristic	Coniferous forests, cm		Dry valley meadow, 7–12 cm	Mixed forests, cm		Dry valley meadow, cm		
	5–10	15–20		4–7	7–15	0–10	10–23	23–35
Optical density								
E_4/E_6	5.1	5.0	5.4	4.6	4.5	4.6	4.7	4.9
Coagulation threshold*	Complete coagulation not observed		40	13	—	15	—	16

* In this table, as in Tables 17, 23, and 24, the thresholds of coagulation are expressed in meq $CaCl_2/l$. of solution containing 0.136 g humic acid/carbon for a time experiment of 2 hours.

The examples quoted do not exhaust the diversity of podzolic soils and the nature of their humus. This diversity is explained by the peculiarities of climate, parent rock, and other factors of soil formation. The importance of the parent rock in the formation of humus of podzolic soils is illustrated below.

(c) Sod-Carbonate Soils of the Central Taiga Subzone

Two soils occurring under spruce–whortleberry–green moss association (Arkhangel'sk region), the age of which was 100 to 150 years, were studied. The first soil was strongly podzolic, overlying moraine loam; carbonates were present at a depth of 130 cm. The second soil was a sod-carbonate soil, strongly podzolized overlying carbonate moraine; effervescence with HCl was obtained at a depth of 40 cm.

The differences in the parent rock were reflected in the chemical composition of these soils; the strongly podzolic soil overlying moraine loam is very acid (pH 3.1 to 4.0) and has a negligible content of exchangeable bases (the sum $Ca^{++} + Mg^{++}$ amounts to 1.5 meq for 100 g soil). The sod-carbonate soil is less acid, the pH increasing from the upper horizon to the lower horizon from 4.5 to 6.7. The sum of the exchangeable bases Ca^{++} and Mg^{++} in the A_1 horizon amounts to 35.5 meq, and in the A_2 horizon 8·8 meq per 100 g soil. The content and composition of humus in these soils differ substantially (Table 14).

Table 14. The effect of soil-forming rock on content and composition of humus in podzolic soils.

Soil, land use	Horizon	Depth of sampling, cm	C, %	N, %	C:N	Carbon of individual groups, as percent of total soil carbon		H.A. Carbon / F.A. Carbon
						Humic acids	Fulvic acids	
Strong podzolic soil, overlying	A_1A_2	8–12	2.04	0.12	17.0	15.2	25.0	0.61
moraine loam, under forest	A_1A_2	12–21	0.71	0.05	14.2	11.3	22.5	0.50
	A_2	21–35	0.45	0.03	15.0	6.7	15.5	0.48
	B_1	35–45	0.28	0.03	9.3	3.6	17.9	0.20
	B_2	45–55	0.20	0.03	6.7	5.0	15.0	0.33
Sod-carbonate, strongly	A_1	7–15	7.23	0.46	15.7	32.2	27.2	1.18
podzolized soil, overlying	A_2	15–24	0.43	0.03	14.3	16.3	25.6	0.63
carbonate moraine, under forest	B_2	40–50	0.29	0.02	14.5	3.4	20.7	0.17

The data in Table 14 show that the upper, A_1, horizon of sod-carbonate soil contains a high percentage of humus, and humic acids predominate in its composition, a deviation therefore from typical podzolic soils under forest. However, in the lower horizons of both soils no substantial differences are observed either in the humus content or in its composition.

Incorporation of calcium from the parent rock into the biological cycle promotes the accumulation of humus and the preferential formation of humic acids in the upper biogenic soil layer.

Humus of Gray Forest Soils

Gray forest soils occurring in the deciduous forest zone under conditions of moderate moisture differ markedly from sod-podzolic soils by the deeper humus horizons and the absence of a continuous podzolic horizon The data of Figure 4 show the fairly high reserves of humus and nitrogen in gray forest soils.

The leaf fall of trees and of grassy vegetation supplies a large mass rich in nitrogen (Tables 2 and 3) and serves as the initial material for the formation of humus substances in these soils.

The development of the sod process in various groups of gray forest soils varies in intensity and this is reflected in the character of the humus formation; the sod process is most advanced in dark gray forest soils.

Humification of organic residues proceeds under conditions of fairly high biological activity. In this zone of the European part of the U.S.S.R. the period in which the temperature exceeds $10°C$ and the moisture supply is sufficient lasts for five to six months of the year. The narrow ratio of litter to leaf fall, equal to 4, shows the fairly active process of humification of the leaf fall on the surface of the soil (Table 2).

As an example, the data characterizing the humus of dark-gray soils of the Shipov forest (Voronezh region) will be examined. The vegetation consists of oak with an admixture of maple, with a cover of mixed grasses. The soil of the mineral horizons is neutral. The sum of exchangeable bases in these horizons is fairly high: about 50 meq per 100 g soil, of which Ca^{++} constitutes 85 to 90%, the remainder being exchangeable Mg^{++}.

Table 15.   Content and composition of humus in gray forest soils.*

Soil	Horizon	Depth of samp- ling, cm	Organic C, %	N, %	C:N	Carbon of individual groups as percent of total soil organic C		H.A. Carbon ———— F.A. Carbon	Humic acids, free and combined with mobile forms of R_2O_3
						Humic acids	Fulvic acids		
Dark-gray forest soil	A_0	3–10	6.11	0.60	10.2	28.2	25.4	1.11	33.1
under oak forest;	A_1	12–19	3.23	0.31	10.4	36.6	23.2	1.57	17.8
Shipov forest	A_2	23–40	2.29	0.23	10.0	47.1	19.7	2.40	6.5
	B_1	34–42	1.88	—	—	44.7	18.1	2.47	4.8
Gray podzolized soil	A_0	0–10	2.87	0.27	10.6	27.5	29.6	0.93	—
under oak forest;	A_1	14–24	1.49	0.14	10.6	39.6	29.5	1.34	—
Tul'sk region	A_2	26–32	0.68	0.07	9.7	33.8	32.3	1.05	—

* In Tables 15 through 17 data of BEL'CHIKOVA [1951] are used.

The data in Table 15 illustrates the nature of humus formation in dark-gray forest soil; namely, high contents of humus and nitrogen; predominance of humic acids over fulvic acids and in the A_0 and A_1 horizons, the humic acid carbon/fulvic acid carbon ratio is 1.11 to 1.57 and in the lower horizons, about 2.5. In contrast to podzolic soils, a large proportion of humic acids in dark-gray forest soil occurs in complex forms of combination with calcium; preliminary treatment of the soil with solutions of mineral acids (soil decalcification) is necessary to disrupt this link. Only in the upper 3 to 10, and 12 to 19 cm, horizons a single treatment of nondecalcified soil with 0.1 N NaOH extracts 33.1 and 17.8%, respectively, of the total amount of humic acids; these are apparently free forms of humic acids leached from the litter; in the remaining horizons this fraction constitutes only 5 to 6% of the total amount of humic acids.

The nature of humic acids in gray forest soils differs substantially from that in podzolic soils. The wide C:H ratio in humic acids from dark-gray forest soils (Table 8) and the fairly high optical density (Figure 8) show the higher degree of condensation of the aromatic rings in humic acids of this soil compared with that from podzolic soils. In addition, humic acids of this soil are precipitated more easily with $CaCl_2$ (Table 9), which indicates a decrease of the hydrophilic side radicals in the molecule.

It should be noted that in the lower layers of dark-gray forest soil of the Shipov forest, the content of humic acids in the composition of humus increases. This phenomenon might be explained by the history of the Shipov forest, the area of which has apparently been occupied, in the past, by chernozem. During afforestation the upper soil horizons were subjected to degradation by the action of organic matter leached from the litter.

It follows from the data given in Table 15 that the contents of humus and nitrogen in gray pozolized forest soil (Tul'sk region) are lower than in the dark-gray forest soil of the Shipov forest; a higher content of humic acid is also observed in the composition of humus of the latter soil. Thus, the diversity of conditions in the formation of gray forest soils is reflected in the nature of the humus.

Humus of Chernozem Soils

The formation of chernozems occurring in the steppe zone is associated with the development of steppe vegetation consisting of grasses and herbaceous plants that produce vigorous, deeply penetrating root systems and abundant leaf fall with a high nitrogen content (see Tables 2 and 3).

The presence in chernozems of an adequate and varied microflora ensures sufficient microbiological activity. For the chernozems of the Kamennaya steppe (Voronezh region),

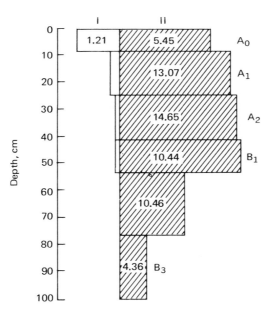

Figure 11. Reserve of organic matter in meadow-chernozem soil (Kamennaya steppe, Voronezh region), uncut steppe.
I—roots; II—humus. Total amount of roots in soil profile per square meter of surface area—2.0 kg; total amount of humus in soil profile per square meter of surface area in the 0–100 cm layer—58.43 kg.

for instance, the period in which the temperature is higher than 10°C lasts for $4\frac{1}{2}$ to 5 months. Fluctuations in moisture during the summer season causes the peculiar rhythm in biological activity when adequate microbial activity during the wet period alternates with depressed activity during soil drying. As has already been pointed out, such conditions promote the formation of humus substances of the humic acid type and protect them from rapid incorporation into new biological processes.

Fixation of humus substances in chernozems is also favored by a high exchange capacity, saturation with calcium, an abundance of mineral colloids, and a high content of minerals of the montmorillonite group.

In the Kamennaya steppe, the reserved uncut steppe (fallow since 1885), soils under agricultural crops, under 50-years-old oaks, and under maple plantations were investigated. All the soils studied were heavy loamy neutral soils overlying yellow-brown loam. The sum of exchangeable bases was 56 to 58 meq per 100 g soil, of which calcium constituted 56 meq and Mg^{++}, 2 meq.

Figure 11 (Kamennaya steppe, uncut steppe) shows the reserves of humus and roots and their distribution in the profile of the chernozem. It is noteworthy that, unlike podzolic soils (Figure 10), the large reserves of humus in chernozems are uniformly distributed down the soil profile.

The humus of chernozems is distinguished by its high content of humic acids, exceeding that of fulvic acids by two or more times (Table 16).

Table 16. Content and composition of humus in chernozems of Kammenaya steppe.

Soil, land use	Horizon	Depth of sampling, cm	Organic C, %	N, %	C:N	Carbon of individual groups, as percent of total soil organic C		H.A. Carbon / F.A. Carbon	Humic acids, free and combined with mobile forms of R_2O_3
						Humic acids	Fulvic acids		
Meadow-chernozem	A_0	0–7	7.35	0.54	13.6	36.3	22.4	1.62	25.9
soil under	A_1	10–20	5.16	0.44	11.7	38.8	22.5	1.72	12.9
herbaceous-mixed grass-vegetation	A_2	25–35	4.32	0.36	12.0	43.5	20.4	2.13	None
Ordinary chernozem,	A_1	0–20	4.79	0.37	12.9	40.5	19.6	2.07	4.1
arable land	A_2	20–35	4.82	0.44	10.9	42.7	20.1	2.12	5.8
Ordinary chernozem,	A_0	3–8	6.58	0.48	13.7	44.2	18.1	2.44	23.0
50-year-old oak +	A_1	10–16	5.79	0.50	11.6	50.8	16.9	3.00	27.9
maple plantation	A_2	17–22	5.44	0.45	12.1	50.0	18.4	2.72	22.4

Only a small fraction of the humic acids is extracted from nondecalcified soil by a single treatment with 0.1 N NaOH. A large part of the humic acids in chernozems is combined with calcium in complex forms which may be disrupted by a preliminary treatment of the soil with mineral acids.

The conditions of soil formation in chernozems (especially alternating periods of moistening and drying in summer and also a neutral soil reaction) favor the formation of even more complex forms of humic acids than in dark-gray forest soil. The high optical density is indicative of the degree of condensation of the aromatic rings in the humic acids of chernozems.

In full agreement with these data are the results of determinations of the coagulation (precipitation) threshold of humic acids: gel formation takes place rapidly with the addition of small amounts of $CaCl_2$, which shows a low content of hydrophilic side radicals in the molecules of humic acids from these soils (Tables 9 and 17).

Table 17. Optical density and coagulation (precipitation) threshold of humic acids from chernozems of the Kamennaya steppe.

Characteristic	Meadow-chernozem soil (uncut steppe), cm		Ordinary chernozem, arable land, cm		Ordinary chernozem under forest belt, cm		
	0–7	10–20	0–20	20–25	3–8	10–16	17–22
E_4/E_6	3.6	3.3	3.4	3.7	3.6	3.7	—
Coagulation threshold	10	5	4	—	5	—	5

In differently utilized land some variations in the content and composition of humus is observed, which is due to different plant cover, ploughing, and other factors. The differences, sometimes substantial, are observed in the humus of leached, deep and fertile, ordinary and southern chernozems (see Table 6).

However, the main features characterizing humus of the chernozem type of soil formation are preserved. These are great thickness of the humus horizon, high humus content, and predominance of the humic acids group in the composition of humus (the humic acid carbon/fulvic acid carbon ratio is 1.5 to 2.0 and sometimes higher). In all cases the humic acids are mainly represented by forms combined with calcium, which are extracted from soil usually after treatment with solutions of mineral acids.

The high degree of condensation of the aromatic rings and the low content of hydrophilic side radicals impart a quality of inertness in the humic acids; they are slightly mobile and coagulate rapidly in the presence of electrolytes. The latter property determines their importance in the formation of the agronomically valuable crumb structure.

Humus of Chestnut Soils

Chestnut soils are formed under conditions of limited precipitation, sharp daily and seasonal fluctuations in air temperature, and high evaporation coefficients.

A moisture deficit during the period of high temperatures is a factor that limits plant growth and the activity of microorganisms.

Less dense plant cover and, in consequence, a smaller amount of plant residues with a low nitrogen content (Table 3) entering the soil (Table 2) and less favorable conditions for microbial activity are the main causes for the smaller (in comparison with chernozems) humus content in chestnut soils (Figure 4).

Chestnut soils have a fairly high exchange capacity (approximately 30 to 35 meq/100 g soil), with a predominance of calcium and magnesium. This circumstance undoubtedly favors the humification of plant material and fixation of humus substances. At the same time the presence in these soils of exchangeable sodium brings about peptization of humus substances and their transformation into forms more available to microorganisms and capable of migration in the soil profile.

On the basis of the plant cover, humus content, and solonetz-like properties, chestnut soils are subdivided into dark-chestnut soils, chestnut soils, and light-chestnut soils. In the U.S.S.R.

dark-chestnut soils occur in the northern part of the zone; they were formed under *Festuca sulcata-Stipa* vegetation characteristic for dry steppes. The root systems penetrate to a depth of 30 to 40 cm and are the main source of humus in these soils; solonetz-like properties in dark-chestnut soils are only slightly developed.

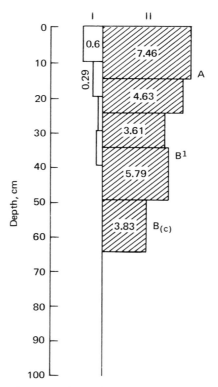

Figure 12.   Reserve of organic matter in dark-chestnut soil (Saratov region).
I—roots; II—humus. Total amount of roots in soil profile per square meter of surface area—1.18 kg; total amount of humus in soil profile per square meter of surface area—25.32 kg.

Figure 12 shows the reserves of humus and roots in dark-chestnut soil (Saratov region). The depth of the humus horizon and especially the uniformity of distribution of the humus in the soil profile of dark-chestnut soil resemble those in chernozems (see Figure 11).

Light-chestnut soils occur to the south of dark-chestnut soils; they are formed under *Artemisia-Festuca sulcata-Stipa* vegetation. Solonetz-like properties in these soils are more clearly defined.

In Table 18 data are given characterizing the content and composition of humus in dark- and light-chestnut soils of the European and Asiatic parts of the U.S.S.R.

Certain features, characteristic of the humus of chernozems, are present to some extent in chestnut soils. For instance, in these soils, as in chernozems, the percentage content of humic acids is higher than that of fulvic acids; accordingly, the humic acid carbon fulvic acid carbon ratio exceeds unity. Only a small fraction of the humic acids is extracted from nondecalcified chestnut soils by a single treatment with 0.1 N NaOH (see Table 6). This indicates the presence of humic acids in complex combination with calcium (and possibly with magnesium).

The data on the elementary composition and optical density indicate the lower degree of

Table 18.   Content and composition of humus in virgin chestnut soil.

Soil	Depth of samp- ling, cm	Organic C, %	N, %	C:N	Carbon of indi- vidual groups, as percent of total soil organic C		H.A. Carbon F.A. Carbon	Author
					Humic acids	Fulvic acids		
Dark-chestnut soil; Saratov region	0–15	1.82	0.17	10.7	33.5	19.8	1.69	Kononova [1966]
	15–25	1.66	0.15	11.1	32.5	24.7	1.31	
	30–45	1.06	0.10	10.6	25.5	20.8	1.23	
Dark-chestnut soil; Ala-Tau foothill zone	0–10	1.96	—	—	28.0	17.9	1.56	Emel'yanov [1953, 1956]
	11–20	1.15	—	—	28.3	22.3	1.27	
	30–40	0.92	—	—	26.4	23.1	1.14	
Light-chestnut soil; Volgograd region	0–20	1.41	0.14	10.0	26.9	20.7	1.30	Glotova [1956]
Light-chestnut soil; Karaganda region	0–8	1.35	—	—	18.6	13.5	1.38	Emel'yanov [1953, 1956]
	17–27	1.28	—	—	16.8	13.5	1.24	

condensation of the aromatic carbon skeleton in humic acids of chestnut soils as compared with those in chernozems (Table 8, Figure 8). The great stability of humic acids from chestnut soils to the precipitating action of electrolytes (see Table 9) reflects the increasing number of hydrophilic side radicals.

Differences in the humus composition and in the nature of the humic acids between chestnut soils and chernozems are more marked in light-chestnut soils than in dark, which is in part due to the direct and indirect influence of exchangeable sodium present in greater amount in light-chestnut soils.

The influence of sodium on humus is most pronounced in solonetzes, in which the exchangeable sodium content can reach 15% or more of the total exchangeable bases.

Humus of Brown Semidesert Soils and Takyrs

For soils of the semidesert zone, a low humus content is characteristic (Figure 4); this is due to the sparse grassy-*Artemisia* semiarid vegetation producing smaller leaf fall and root mass (see Table 2). Extremely low precipitation (the mean value for the semidesert zone of the U.S.S.R. is only 150 to 200 mm per year), occurring mainly during the period of low temperatures, does not favor humification of plant residues. In the humus of semidesert soils fulvic acids predominate: The humic acid carbon/fulvic acid carbon ratio is 0.8 to 0.5 and in gray-brown soils it decreases to 0.5 to 0.3 (Table 19). If fulvic acids are regarded as the initial forms of humic acids or as products of their destruction, then the predominance of fulvic acids in the composition of humus of brown semidesert soils may be explained by the decline in the rate of humus formation.

Takyrs are very peculiar soil formations. In the U.S.S.R. they occur in the southwestern part of the central Asian desert zone. They are characterized by low precipitation (75 to 130 mm per year), falling mainly during the period of low temperatures (from November to April). Only during the short spring are the hydrothermal conditions favorable to the growth of higher vegetation and to the activity of microorganisms.

Algae that grow abundantly on the surface of the soil during moist periods play an important role in takyr formation. In takyrs the organic mass is small; in the Kara-Kum area, on lichen-algae takyr, it amounts to 1 ton/ha, and of this, only about 30% is derived from the

Table 19. Content and composition of humus in brown, semidesert soils, and takyrs.

Soil	Depth of sampling, cm	Organic C, %	N, %	C:N	Carbon of individual groups, as percent of total soil organic C		H.A. Carbon / F.A. Carbon	Author
					Humic acids	Fulvic acids		
Brown semidesert-steppe soils; Gur'yev region	0–10	0.66	—	—	16.2	23.4	0.69	EMEL'YANOV [1953, 1956]
	13–23	0.46	—	—	10.3	23.1	0.45	
	23–33	0.45	—	—	15.9	24.7	0.64	
Gray-brown soil; Kara-Kum region	0–4	0.25	0.05	5.0	20.0	40.7	0.49	LOBOVA [1960]
	15–20	0.20	—	—	13.4	44.2	0.30	
Lichen algae-takyr; Kara-Kum region	2–3	0.43	0.064	6.7	10.0	21.7	0.46	PONOMAREVA [1956]
	3–9	0.30	0.048	6.3	6.5	24.4	0.27	
	9–25	0.27	0.046	5.9	4.6	18.4	—	

aerial parts and roots of higher vegetation (PONOMAREVA [1956]). The extremely limited resources of fresh organic material and unfavorable conditions for its humification are the main reasons for the lower amount of humus in takyrs (Figure 4).

The data characterizing the composition of humus in takyrs (Table 19) show a predominance of fulvic acids which, as in brown semidesert soils, may probably be regarded as the initial stages in humic acid formation.

Humus of Serozems

Serozems occur in the semidesert, the subtropical zone, and (in the U.S.S.R.) mainly in the regions of the central Asiatic republics (Uzbekskaya, Tadzhikskaya, Turkmenskaya, Kirgizskaya, and the southern part of the Kazakhskaya SSR). A small area is occupied by these soils in the Kura-Araksinsk lowland of the Caucasus.

Serozems develop under conditions of sufficiently high temperatures. The period having temperatures higher than 10°C lasts for seven months per year, but the total precipitation varies from 125 to 500 mm; the minimum temperature occurs in the lowest and southernmost parts of the zone; with increasing altitude the amount of precipitation increases. On the whole, continental characteristics prevail more in the zone of brown semidesert soils than in the zone of serozems.

The considerable amount of precipitation during March and April, together with the fairly high temperatures, promotes the vigorous development of an ephemeral vegetation and fairly intensive microbiological activity in these soils. The abundant microflora in serozems also contributes to this (Figure 1).

With the presence of fresh plant material during springtime, even on unirrigated serozems, both the new formation of humus substances and the decomposition of organic residues and humus take place.

Microbial processes are particularly active in irrigated serozems where, during a long season, hydrothermal conditions are especially favorable. However, at the same time, the high activity of microorganisms, which contribute both to the new formation of humus substances and to their decomposition, is one of the factors preventing the accumulation of humus in these soils.

The low humus content in serozems can also be explained to some extent by the distribution of plant residues. The grassy, ephemeral vegetation, characteristic of virgin serozems, is very diverse in its botanical composition and ecological adaptation to conditions of the soil medium. It has a limited density, but well-developed root systems, which are distributed in the upper soil layer, only a few centimeters deep. Therefore, in spite of the large reserves of roots, the absence of adequate contact with the soil mass does not contribute to the fixation of humus substances in the soil profile. This is illustrated by Figure 13, which shows roots and

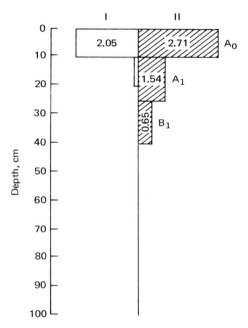

Figure 13. Reserve of organic matter in light serozem (State farm, Pakhta-Aral, virgin land). I—roots; II—humus. Total amount of roots in soil profile per square meter of surface area— 2.13 kg; total amount of humus in soil profile per square meter—4.9 kg.

humus reserves in the profile of virgin light serozem. Finally, the physicochemical properties of serozems (low content of the colloidal fraction and low exchange capacity, not exceeding 10 meq per 100 g soil) do not contribute to the fixation of humus.

Large differences are observed in humus content, humus composition, and the nature of the humic acids among dark, typical, and light serozems. The highest humus content, about 2%, is found in dark serozems. The lowest humus content and the lowest humic acid content occur in light serozems (Table 20).

The humus content in serozems increases when there is subsurface moisture; for example, in the upper horizon of the ordinary serozems of Azerbaidzhan the percentage of carbon reaches 1.45 and that of nitrogen, 0.19. The rapidly proceeding new formation and decomposition of organic matter in serozems favors the formation of humus substances of the fulvic acid type and relatively simple forms of humic acids. The low values of optical density indicate the low degree of condensation of the aromatic ring; according to this criterion, the humic acids of serozems are intermediate between those of chestnut soils and sod-podzolic soils (see Figure 8). They occupy a similar position in Table 9 regarding their behavior toward electrolytes ($CaCl_2$).

Humus of Krasnozems and Lateritic (Ferrallitic) Soils

Krasnozems occur in the subtropical humid-forest zone of all continents. Luxuriantly

Table 20.   Content and composition of humus in serozems.

Soil, land use	Depth of sampling, cm	Organic C, %	N, %	C:N	Carbon of individual groups, as percent of total soil organic C		H.A. Carbon F.A. Carbon	Humic acids, free and combined with the mobile forms of R_2O_3	Author
					Humic acids	Fulvic acids			
Light sero-zem, arable land; Kazakhskaya SSR	0–15	0.72	0.09	8.0	16.6	22.5	0.74	6.5	BEL'CHIKOVA [1948]
Typical sero-zem, arable land; Uzbekskaya SSR	0–20	0.78	0.08	9.8	20.5	23.0	0.89	5.0	BEL'CHIKOVA [1948]
Dark sero-zem, arable land; Tadzhikskaya SSR	0–20	1.48	—	—	28.8	30.0	0.96	—	ILOVAISKAYA [1959]
Ordinary serozem, arable land; Azerbaid-zhanskaya SSR	0–17 17–53	1.45 0.96	0.19 0.12	7.7 8.0	16.3 15.8	13.8 13.4	1.18 1.18	— —	EDIGAROVA [1961]

developing forest vegetation, which includes evergreen plants, produces massive amounts (in excess of 20 tons/ha per year, see Table 2) of plant residues, both leaf fall and roots.

In krasnozems organic matter transformations occur under conditions of prolonged periods of high temperatures. This explains the high activity of microorganisms and the soil fauna on the surface of krasnozems; in fact, as may be seen from Table 2, the litter/leaf fall ratio is only 0.7. In the whole soil, however, the rate of microbiological processes including humification is only moderate because of excessive moisture.

As an example, data pertaining to the organic matter of krasnozem of Chakva (in the vicinity of Batum) may be cited. The soil is characterized by pronounced decomposition of the mineral part; the exchange capacity does not exceed 20 meq/100 g soil, exchangeable hydrogen and aluminum predominating. The soil is acid.

The total reserve of humus in the krasnozem profile examined is fairly high; in the 1-meter layer it amounts to 250 tons/ha (Figure 14), slightly exceeding the corresponding value for chestnut soils (Figure 12). At the same time, as can be seen from Figure 5, the humus of krasnozems in the U.S.S.R. has a low nitrogen content and therefore a wide C:N ratio which in the mineral horizons is 17 to 18.

The unique conditions—formation of humus substances in the litter, excessive moisture, leaching regime, acid reaction—impart to the humus of krasnozems the features associated

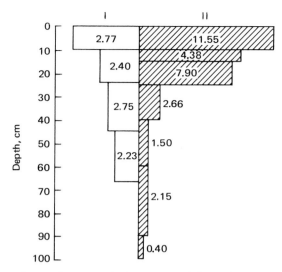

Figure 14. Reserve of organic matter in krasnozem (Chakva).
I—roots; II—humus. Total amount of roots in soil profile per square meter of surface area—10.15 kg, of which large roots amount to 2.94 kg; total amount of humus in the soil profile per square meter of surface area—30.54 kg.

with the humus of podzolic soils (Table 21). In the composition of the humus krasnozems, fulvic acids predominate over humic acids; the latter are present in forms that are extracted completely by a single treatment of nondecalcified soil with 0.1 N NaOH.

It may be seen from the values of optical density (Figure 8) and the coagulation threshold (Table 9) that the humic acids of krasnozems resemble those of podzolic soils. Evidently,

Table 21. Content and composition of humus in krasnozem (under forest vegetation).

Soil	Depth of sampling, cm	Organic C, %	N, %	C:N	Carbon of individual groups, as percent of total soil organic C		H.A. Carbon / F.A. Carbon	Humic acids, free and combined with mobile forms of R_2O_3	Author
					Humic acids	Fulvic acids			
Krasnozem overlying andesite-basalt; Chakva	0–15	7.01	0.39	17.9	23.5	28.5	0.83	97.0	BEL'CHIKOVA [1951]
	35–40	1.15	0.06	19.2	5.2	22.6	0.23	100.0	
Krasnozem overlying geniss; Chinese Peoples' Republic	0–10	1.54	0.15	10.3	12.3	30.5	0.40	79.0	NYU TSZI-VEN' [1961]
	50–60	0.55	0.06	9.2	None	27.3	—	—	

as in the latter, the aliphatic portion of the humic acids from krasnozems is more developed than the aromatic ring.

In Table 21 data are given on the content and composition of humus from krasnozems in China. This soil has a lower humus content than the krasnozem of Chakva, which may be explained by erosion. At the same time the krasnozem of China has a higher percentage of nitrogen, and therefore the C:N ratio is narrower than that in the krasnozem of Chakva (9 to 10). All the remaining parameters (humic acid carbon/fulvic acid carbon ratio, optical density of the humic acids, and behavior toward electrolytes) are similar for both soils.

The nature of the humification of plant residues in lateritic (ferrallitic) soils occurring in tropical forest zones appears to be similar to that in krasnozems. In ferrallitic soils there is a very large biomass and abundant leaf fall, the decomposition of which takes place very intensively on the soil surface; this is shown by the narrow litter/leaf fall ratio, which equals 0.1 (Table 2).

Data on the composition of humus in ferrallitic soils and in krasnozems are similar; the predominance of fulvic acids over humic acids is more pronounced in these soils than in krasnozems (Table 22).

Table 22. Content and composition of humus in ferrallitic soils of northern Vietnam. Analyses by Pankova and Bel'chikova (unpublished).

Soil character and vegetative cover	Depth of sampling, cm	Organic C, %	N, %	C:N	Carbon of individual groups, as percent of total soil carbon		H.A. Carbon / F.A. Carbon	Humic acids, free and combined with mobile forms of R_2O_3
					Humic acids	Fulvic acids		
Ferrallitic soils	0–20	2.25	0.23	9.8	6.2	34.7	0.18	100
overlying basalt	20–30	1.59	0.15	10.6	—	—	—	—
Forest	40–50	1.16	0.10	11.6	—	—	—	—
Ferralitic reddish-	0–5	3.42	0.24	14.3	8.5	27.2	0.31	100
yellow soils over-	12–17	0.86	0.07	12.3	4.7	41.8	0.11	100
lying weathering	25–30	0.72	0.08	9.0	2.8	43.0	0.06	100
products of	55–60	0.44	0.05	8.8	2.3	43.2	0.05	100
basaltic porphyrite Coffee plantation	95–100	0.29	0.05	5.8	None	48.3	—	—
Ferrallitic-margalitic soil	0–13	3.33	—	—	28.4	16.6	1.70	100
overlying ash and tuff	13–25	2.89	—	—	22.1	19.5	1.13	100
Herbaceous vegetation with occasional palms	25–40	1.20	—	—	4.5	24.6	0.18	100

The conditions of soil formation, as in krasnozems, favor the formation of humic acids in which the degree of condensation of the aromatic carbon skeleton is slight; this is shown by the values of optical density (Table 23), the E_4E_6 ratio of the humic acids being 4.8 to 5.0.

Ferrallitic-margalitic soils, developed on outcrops of volcanic tuffs (FRIDLAND [1964]),

Table 23. Optical density and coagulation threshold of humic acids from krasnozems and ferrallitic soils.

Characteristic	Krasnozems		Ferrallitic soils of northern Vietnam		Ferrallitic-margalitic soil of northern Vietnam. Herbaceous vegetation with occasional palms		
	Chakva, forest, 0–15 cm	China, forest, 0–10 cm	Forest, 0–20 cm	Coffee plantation, 0–5 cm	0–13 cm	13–25 cm	25–40 cm
E_4E_6	4.2	4.5	5.0	4.8	3.5	3.4	5.0
Coagulation threshold	13	20–25	—	—	—	—	—

are interesting. These soils have a higher humus content than ferrallitic soils; in the composition of humus, the humic acid content exceeds that of fulvic acid, and the humic acid carbon/fulvic acid carbon ratio is greater than 1 (Table 22). The humic acids have more condensed aromatic rings; the E_4E_6 ratio is 3.4 to 3.5 (see Table 23). However, all these peculiar features have been recorded only for the upper horizons.

Differences between the humus as of ferrallitic-margalitic and ferrallitic soils may be explained by the conditions of soil formation, amongst which a less acid soil reaction plays an important role. This reaction is due to the neutralizing action of alkaline products of weathering of volcanic ashes and tuffs. Probably no less important is the more moderate moisture regime due to evaporation from the soil surface under cover of the grassy vegetation.

Similar characteristics of humus have been observed in black tropical soils; some information on this is given below.

Humus of Black Soils of the Tropics and Humid Subtropics

This group of soils represents intrazonal formations in the zones of the tropics and humid subtropics of Asia, Africa, and America (GERASIMOV [1959], DUDAL [1963], FINK [1963]). Black soils are characterized by a high clay fraction content (with minerals of the montmorillonite type predominating), saturation with bases, and hence a neutral reaction.

Presumably the water regime plays an important part in the formation of humus in these soils; in spite of the large amount of precipitation (1500 to 2000 mm per year) infiltration in black soils is slight because of the low soil permeability. These soils often occur in regions with very long hot and dry periods.

These conditions (neutral soil reaction, alternation of soil wetting and drying, abundance of clay minerals) promote the formation of humus substances of the humic acid type with a moderate degree of condensation of the aromatic ring, which distinguishes them from the humic acids of zonal soils, indigenous to the tropics and humid subtropics.

Table 24. Content and composition of humus in black tropical soils (regurs) of India, arable land. Analyses by Pankova (unpublished data).

Depth of sampling, cm	Organic C, %	Carbon of individual groups, as percent of total soil carbon		H.A. Carbon F.A. Carbon	E_4/E_6 of humic acids	Coagulation threshold of humic acids
		Humic acids	Fulvic acids			
0–10	0.63	23.8	23.8	1.0	4.8	7
10–20	0.47	31.8	29.8	1.07	4.4	7
150–160	0.22	27.3	27.3	1.0	4.4	—

Examples of such a distinction are the ferrallitic-margalitic soils of northern Vietnam referred to in Table 22. The typical features of the humus of the black tropical soils of India (so-called regurs) described by Gerasimov [1959], are similar. The soil overlying the weathering products of traprocks is clayey, with a neutral reaction (pH about 7); carbonates are present in the surface layers.

From Table 24 it may be seen that, notwithstanding the low humus content, a comparatively high content of humic acids with fairly condensed aromatic ring systems and low stability toward electrolytes have been reported in the composition of the humus. All these features are peculiar to the humus of regurs and make these soils similar to the ferrallitic-margalitic soils of northern Vietnam and distinguish them from ferrallitic soils typical of the tropical forest zone.

The data of Duchaufour and Dommergues [1963] on the humus of black clayey magnesium tropical soils of southern Senegal are interesting and agree with our results. The soils studied were black soils (vertisols) overgrown by sparse savannah; annual precipitation varied from 400 to 700 mm; the soils were occasionally hydromorphous.

From a comparison of black tropical soil and ferrallitic soil Duchaufour and Dommergues concluded that the soils differed in the nature of the humus. Ferrallitic soil contained more humus (3.5%) than black soil, which had only 1.7%. However, the humic acid carbon/fulvic acid carbon ratio in the former soil was less than 1, while in black tropical soil this ratio was greater than 4. Furthermore, they found that in ferrallitic soil the humic acids had a less condensed aromatic ring ("brown" humic acids) than those in black tropical soil, where "gray" humic acids predominated.

The main reason for such a peculiarity is considered by the above authors to be the alternate wetting and drying of soil under savannah, which promotes condensation and polymerization of the humic acids. This hypothesis agrees with our opinion concerning the importance of alternate wetting and drying of chernozems during the summer period in the formation of humic acids with condensed aromatic rings (see Section 2). These ideas are supported by results obtained by Zonn [1964] in studies of the main types of soil formation in the tropical mountain regions of southeast Asia. He pointed out that in soils under a canopy of savannah forest the content of humic acids in the humus is higher than in soils under humid-tropical forests, and he regarded this phenomenon as due to the more moderate moisture regime and neutral soil reaction (under savannah forest).

There have been attempts to confirm experimentally the suggestion that alternate wetting

Table 25. Content and composition of humus in volcanic and nonvolcanic soils of Japan (Kawaguchi and Kyuma [1959]).

Soil	Organic C, %	Carbon of individual groups, as percent of total soil carbon		H.A. Carbon / F.A. Carbon
		Humic acids	Fulvic acids	
Volcanic soils:				
Fujigun	11.5	28.6	28.1	1.02
Sanpei	9.13	37.0	30.3	1.22
Camouno	12.2	49.5	21.2	2.33
Shimizu	9.04	39.5	27.0	1.46
Nonvolcanic soils:				
Nomoyana	3.16	24.2	26.1	0.93
Shizuhara	2.99	37.8	22.6	1.67
Nishikamo	2.13	33.4	24.1	1.39

and drying affects the polymerization of humic acids (BACHELIER [1963], BURGES [1960]). Burges established the possibility that under such a water regime "brown" humic acids could be converted into "gray" humic acids.

SINGH [1954, 1956] showed that the dark color of black tropical and subtropical soils with low humus contents is due to the formation of humus substances on the catalytically active surface of montmorillonite. Apparently humus substances penetrate into the crystal lattice and acquire high stability, as shown during digestion.

The above evidence concerning the peculiarities of humus in ferrallitic/margalitic soils and the black soils of the tropics and humid subtropics is very interesting as an example in which physicochemical soil properties, parent rock, and a number of other nonzonal factors of soil formation depart from the zonal regularities of humus formation that are determined by bioclimatic conditions.

GERASIMOV [1959] is of the opinion that black tropical soils resemble smolnitsas in a genetic respect. At present, however, available data indicate a great diversity in content and composition of the humus in smolnitsa soils, and so the problem of the genetic similarity of black tropical soils and smolnitsas is still unresolved.

Humus of Volcanic Soils

This group of soils, occurring in various natural zones, is formed with the active participation of volcanic products which often disturb the regularities established by bioclimatic conditions.

Detailed investigations of volcanic soils have been made in Japan; the results of these investigations include data on the humus of these soils (JAPANESE MINISTRY OF AGRICULTURE [1964]).

Basic materials of volcanic soils in Japan are allophane and aluminum hydroxide; the soils are acid. Japanese volcanic soils are related to black soils of the humid subtropics and resemble the margalitic soils of Indonesia (KANNO [1962]).

In their surface horizons these soils have a high percentage of humus (8 to 30%) and a wide C:N ratio (equal to 15 to 30 or more). In many instances the content of humic acids exceeded that of fulvic acids in the composition of humus, as determined by the method of TYURIN and PONOMAREVA (Table 25). However, this feature is not absolutely invariable.

Humic acids of volcanic soils were found to be more complex than those from soils of nonvolcanic origin. KOSAKA [1953] obtained some interesting results during the fractionation of humic acids into "gray" and "brown" (Table 26).

While humic acids of volcanic soils contained a considerable percentage of gray (*i.e.*, more complex) humic acids, only brown humic acid fractions were present in nonvolcanic soils.

Table 26. Results of fractionation of humic acids.

Soil	Humic acid fractions as a percent of total humic acids	
	Gray humic acids	Brown humic acids
Volcanic soils:		
Kataji	48.3	51.7
Kawasaki	20.0	80.0
Nonvolcanic soils:		
Nihondaira	None	100

Determination of the optical density of the humic acids fully confirmed this. The extinction coefficient for the humic acids from volcanic soils was considerably higher than for humic acids from nonvolcanic soils. The view that the more complex nature of the humic acids from volcanic soils is to some extent due to catalysis by minerals is not without foundation. Some interesting studies were made by Kyuma and Kawaguchi [1964], who found that allophane has catalytic properties promoting the oxidative polymerization of polyphenols, which are known to be structural units of humic acid molecules.

Conditions of formation of volcanic soils are diverse, and the features of humus formation differ. The volcanic soils of Kamchatka serve as an example (Sokolov and Belousova [1964]). These soils are formed under sparse birch forests with grassy cover; abundant leaf fall amounts to 10 tons/ha annually. Climatic conditions resemble the northern taiga regions of the European part of the U.S.S.R. Nevertheless, decomposition of the leaf fall proceeds intensively: approximately 50% of the total leaf fall is mineralized annually. The more intensive decomposition of the leaf fall is due to the influence of volcanic ashes, during the weathering of which, bases are liberated that neutralize the acid products of decomposition and the humification of plant residues.

Volcanic ashes largely determine the nutrient status, water, air, and heat regimes of soil and the character of soil formation. The soil profile is interstratified by buried layers, as illustrated by Figure 15, compiled by us from the data of Sokolov and Belousova [1964].

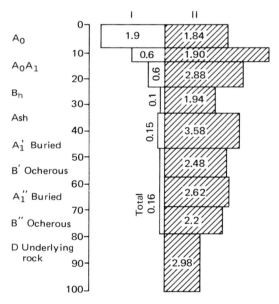

Figure 15. Reserve of organic matter in forest volcanic soil (Kamchatka).
I—roots; II—humus. Total amount of roots in soil profile per square meter of surface area—3.51 kg; total amount of humus in soil profile per square meter of surface area—22.42 kg.

However, in spite of the neutralizing influence of volcanic ashes, the humus of these soils preserves features peculiar to soils of the taiga zones, *i.e.*, its nature is determined by the bioclimatic conditions.

In the composition of humus of volcanic soils of Kamchatka fulvic acids predominate; humic acids are extracted completely by a single treatment of soil with 0.1 N NaOH (see Sokolov and Belousova [1964]).

Table 27. Content and composition of humus in volcanic soil of Kamchatka. Analyses by Bel'chikova (unpublished)

Depth of sampling, cm	Organic C, %	N, %	C:N	Carbon of individual groups, as percent of total soil carbon		H.A. Carbon / F.A. Carbon	E_4/E_6 ratio of humic acids
				Humic acids	Fulvic acids		
5–10	3.92	0.31	12.6	25.8	29.0	0.89	4.7
10–20	2.54	0.18	14.1	14.4	50.0	0.28	4.0
30–35	2.11	0.15	14.1	10.0	41.2	0.24	4.7
40–45	3.92	0.28	14.0	4.3	50.8	0.09	4.6
47–52	3.14	0.24	13.1	7.6	51.0	0.15	4.7
52–62	2.88	0.20	14.4	5.6	54.5	0.10	4.7
77–88	1.55	0.12	12.9	3.9	53.5	0.07	—

As an example, data characterizing the humus of one of the Kamchatka soils are given in Table 27.

Judging from the low values of the optical density, the humic acids have only a slightly condensed aromatic ring system. Data on the elementary composition of humic acids confirm this statement (Table 28).

The high content of hydrogen in the composition of humic acids should be noted, and in consequence, the C:H ratio is very narrow (10 to 10.1); it is similar to that in humic acids from podzolic soils (see Table 8).

Humus of Brown Forest Soils

The demarcation line of brown forest soils is not clearly established at present. These soils occur under diverse bioclimatic conditions and on various soil-forming rocks, under deciduous and coniferous forests. Therefore, they differ in nitrogen content, humus content, and composition.

Common features typical of brown forest soils are: high humus content, predominance of fulvic acids, and relatively low degree of condensation of aromatic rings in the humic acids.

The development of the podzolization process in brown forest soils proceeds differently from that in podzolic soils; some authors (PONOMAREVA [1964]) explain this by the rate of organic matter decomposition. (A summary of data characterizing the humus of brown forest soils of the U.S.S.R. is given in Figure 4 and Table 6.) At this point our data on two profiles of brown forest soils can be mentioned.

The first soil (Moldavia) is a podzolized slightly loamy brown forest soil. The profile was established under an ash-lime-hornbeam forest at an altitude of 380 m. Total precipitation amounts to about 600 mm per year. The soil is acid (pH 5.7 to 4.5), the sum of exchangeable

Table 28. Elementary composition of humic acids of volcanic soil (as a percentage of the dry, ash-free matter).

Humus substances	C	H	N	O	C:H
Humic acids from soil samples:					
5–10 cm	58.24	5.87	4.05	31.84	10.0
40–45 cm	55.60	5.49	5.28	33.63	10.1

Table 29. Content of composition of humus in brown forest soils of Moldavia. Analyses by Bel'chikova (unpublished).

Depth of sampling, cm	Organic C, %	N, %	C:N	Carbon of individual groups, as percent of total soil carbon		H.A. Carbon / F.A. Carbon	Humic acids, free and combined with mobile forms of R_2O_3	E_4/E_6 ratio of humic acids
				Humic acids	Fulvic acids			
0–4	4.14	0.39	10.6	13.5	21.3	0.64	77	4.6
4–13	2.02	0.22	9.2	18.3	25.7	0.71	50	4.9
13–20	0.87	0.10	8.7	19.5	35.6	0.55	47	4.2
20–30	0.68	0.09	7.6	10.3	33.8	0.30	85	4.7
30–36	0.58	0.08	7.3	8.6	31.0	0.28	—	—
40–50	0.40	0.06	6.7	10.0	35.0	0.28	—	—

bases ($Ca^{++} + Mg^{++}$) in the upper 1 to 7 cm layer is 39.3 meq, and in lower layers 18 to 20 meq/per 100 g soil. Data characterizing the humus of this soil are given in Table 29.

The second profile of a brown forest soil was examined at Stina de Vale (Rumania), in the western Carpathians, during an excursion organized by the International Congress of Soil Scientists in 1964. Stina de Vale is situated at 1600 to 1700 m above sea level; the annual precipitation is about 900 mm. The profile was established under spruce-beech forest with a grassy cover. The soil is acid, the pH in the upper horizon is 3.6 to 4.7. The content of exchangeable bases ($Ca^{++} + Mg^{++}$) amounted to about 1 to 2 meq per 100 g soil. Data characterizing the humus of this soil are given in Table 30.

Table 30. Content and composition of humus in brown forest soil of Rumania. Analyses by Bel'chikova (unpublished).

Depth of sampling, cm	Organic C, %	N, %	C:N	Carbon of individual groups, as percent of total soil carbon		H.A. Carbon / F.A. Carbon	Humic acids, free and combined, with mobile forms of R_2O_3	E_4/E_6 ratio of humic acids
				Humic acids	Fulvic acids			
5–10	6.22	0.41	15.2	14.3	34.6	0.41	100	4.6
10–20	3.14	0.23	13.7	8.6	43.9	0.20	100	5.0
25–35	1.51	0.14	10.8	6.6	51.7	0.13	100	5.0
45–55	1.12	0.11	10.2	4.5	59.8	0.07	100	5.8
65–75	0.55	0.06	9.2	5.5	54.5	0.10	100	5.0
90–100	0.52	0.08	6.5	3.8	53.9	0.07	100	—
170–180	0.40	0.06	6.7	2.5	55.0	0.05	100	5.0

It can be seen from a comparison of the data in Tables 29 and 30 that the features characteristic of the humus of podzolic type (predominance of fulvic acids over humic acids, mobility of the latter, low degree of condensation of the aromatic rings) in brown forest soil of Rumania are more distinct than in soil of Moldavia. These differences seem to be due to bioclimatic conditions of soil formation, and in particular, to the larger amount of precipitation and more severe climatic conditions associated with the brown forest soils in Rumania.

5. Conclusions

The information in this review shows that the state of the soil organic matter is determined by a variety of factors and conditions of soil formation (plant cover, activity of microorganisms, hydrothermal conditions, chemical and physicochemical soil properties, and parent rock).

From the material surveyed in Section 3 it follows that an outstanding role in humus formation is played by biochemical conditions which determine the zonal character of the composition and nature of the humus substances. However, in many cases the importance of chemical and physicochemical properties of soil and parent rock outweighs that of the bioclimatic factor. The sod-carbonate soils of the forest taiga zone, ferrallitic-margalitic soils, black tropical soils of the tropical forest zone, and also certain volcanic soils are all examples of this.

In these soils which develop on rocks and are rich in bases, transformations of organic matter proceed with the formation of more "mature" forms of humus substances of the humic acid type and often with more condensed aromatic rings than in corresponding zonal soils (podzolic, ferrallitic soils).

In these soils chemical and physicochemical properties influence not only the direction of the biological processes but, because of the catalytically active surfaces of some clay minerals, allophane, and others, probably also favor the abiotic phenomena of condensation and polymerization that accompany the formation of molecules of humus substances.

In view of the consistent character of the differences in the state of humus in various soils, attention should be paid to the fact that manifestation of the functions inherent in humus substances depends on the conditions of the soil medium (especially on moisture regime and soil reaction). For instance, the decomposing action of humic and fulvic acids on the mineral part of the soil is shown fully in soils with an acid reaction (podzolic soils, krasnozems, ferrallitic soils) and is checked in neutral soils (chernozems, chestnut soils, serozems, and others).

In this review are given data characterizing the humus of both virgin and reclaimed soils. There is no doubt that by cultivation and application of mineral and organic fertilizers, man brings about essential changes in the state of the organic part of the soil due to an intensification of the processes of the new formation and decomposition of organic matter. However, the information available indicates that, in spite of soil reclamation and soil cultivation, the humus still preserves features peculiar to the humus of the particular soil type.

Further studies of the patterns of humus formation should be continued by making investigations in greater detail on the basis of the results obtained from studies of the nature and properties of humus substances, by modeling the processes of their formation experimentally, and by widening the scope of investigations to include new soils. Only then can one be sure that the "blank areas" in humus studies will disappear from the world soil map.

REFERENCES

Aliev, S. A., 1956. Reserves of humus and nitrogen in soils of Azerbaidzhan. *Pochvovedenie*, 9:55.

Bachelier, G., 1963. Influence du climat sur les processus pedobiologiques de l'humification et de la deshumification. *Pedobiologia B.2*, H-2:153.

Baranovskaya, A. V., 1952. Characteristics of humus accumulation and humus composition in soils of the Komi ASSR. *Trudy Komi Fil. Acad. Nauk SSSR, Ser. Geogr.* No. 7.

Bazilevich, N. I., 1962. Exchange of mineral elements in various types of steppes and meadows on chernozem soils, chestnut soils and solonetses. In: "Problems of soil science" ("*Problemy pochvovedeniya*") Izd. Akad. Nauk SSSR.

Bel'chikova, N. P., 1948. Changes in the content and composition of organic matter of serozem with long-term application of farmyard manure and mineral fertilizers. *Pochvovedenie*, 1:39.

————. 1951. Some regularities in the content and composition of humus and in the properties of humic acids in the main soil groups of the U.S.S.R. *Trudy Pochv. Inst. Dokuchaeva*, 38:33.

Burges, A., 1960. Physico-chemical investigations of humic acids. *Trans. 7th Intern. Congr. Soil Sci.*, Madison, 2:128.

Cecconi, S., and M. Tellini, 1954. The relationship between humic acids and fulvic acids in various soil types. *Ann. Chim. Roma* 44: 943.

Cernescu, N., and M. Cicotti, 1964. The H/F ratio in relation to V and pH values of zonal soils, *Stiinta Sol.*, 2: 27.

Daniliuc, D., 1957. Humus composition in different types of soil. *Comun. Acad. Rep. Pop. Rom.* 7:979.

Dokuchaev, V. V., 1883. Russian Chernozem (*Ruskii chernozem*) SPb (1883).

————. 1949. Collected papers (*Sochineniya*). V. 1 Izd. Akad. Nauk SSSR.

Duchaufour, P., and Y. Dommergues, 1963. A study of the humic compounds of some tropical and sub-tropical soils. *Afr. Soils* 8, No. 1, 5.

Dudal, R., 1963. Dark clay soils of tropical and subtropical regions. *Soil Sci.* 95:264.

Edigarova, N. N., 1955. Soil organic matter in western part of Shirvanskaya steppe. *Trudy Inst. Pochvoved. Agrokhim. Akad. Nauk Azerbaidzhan SSR* 7.

Emel'yanov, I. I., 1953. Geographical characteristics of humus formation in soils of Kazakhstan. *Pochvovedenie*, 9:40.

————. 1956. Composition and characteristics of organic matter in soils of Kazakhstan. *Trudy Inst. Pochvoved. Akad. Nauk Kazakh. SSR* 6:180.

Finck, A. 1963. Tropische Böden. *Einführung in die bodenkundlichen Grundlagen tropischer und subtropischer Landwirtschaft*. Hamburg—Berlin: Verlag Paul Parey.

Fridland, V. M., 1964. Soils and weathering crusts of humid sub-tropics (*Pochvy i kory vyvietrivaniya vlazhnykh subtropikov*) Izd. Nauka.

Gerasimov, I. P., 1959. An outline of physical geography of foreign countries (*Ocherki po fizicheskii geografii zarubiezhnykh stran*). Geografiz.

Glotova, T. V., 1956. The organic matter of chestnut and estuarine soils of the arid southeast of the U.S.S.R. *Pochvovedenie* 6:45.

Gorbunov, N. I., 1948. The absorption capacity of soils and its nature (*Pogoltitel'naya sposobnost' pochv i ee priroda*) Sel'khozgiz.

Greenland. D., 1965. Interaction between clays and organic compounds in soils. *Soil Fertilizers*, 28:415, 521.

Hargitai, L., 1955. Comparative investigations by optical methods of organic matter in different soil types. *Agrártud. Egyet. Agron. Kaz Kiad.* 2, No. 10, 27.

Hilgard, E., 1893. Über den Einfluss des Klimas auf die Bildung und Zusammensetzung des Bodens, *Forschn Geb. Agrik Phys. B.16* Heidelberg.

————, 1906. *Soils, their formation, properties, composition and relations to climate and plant growth in the humid and arid regions*, New York-London: Macmillan.

Hoffmann, F., 1964. Farbmessungen an Humin—und Fulvosäuren. *Arch. Forstw. B 13*, H. 11.

Ilovaiskaya, N. N., 1959. Organic matter of the main soil types in Tadzhikistan. *Pochvovedenie* 8:15.

Imshenetskii, A. A., 1944. Microbiological processes at high temperature (*Mikrobiologicheskie protsessy pri vysokoi temperature*). Izd. Akad. Nauk SSSR.

Jagnow, G., 1971. The effect of altitude, precipitation and cultivation on the organic carbon and nitrogen content of East African soils. *Transactions of the Intern. Symposium "Humus et Planta V,"* p. 381, Prague.

Japanese Ministry of Agriculture and Forestry, 1964. *Volcanic ash soils in Japan.*

Jenny, H., 1941. *Factors of soil formation.* New York: McGraw-Hill.

———, 1965. Bodenstickstoff und seine Abhangigkeit von Zustandsfaktoren. *Z. Pflanzenernahrung, Düng. u. Bodenkunde* 109, 97.

Kanno, I., 1962. Genesis and classification of humic allophane soil in Japan. *Trans. Intern. Soil Sci. Conf. Comm. IV & V.* New Zealand.

Karavaeva, N. A., and V. O. Targul'yan, 1960. Characteristics of humus distribution in the tundra soils of Northern Yakutiya. *Pochvovedenie* 12:36.

Kasatochkin, V. I., 1951. The structure of carbonized substances. *Izv. Akad. Nauk SSSR, Otd. tekh. Nauk* No. 9 & *ibid.* No. 10 (1953).

———, M. M. Kononova, N. K. Larina, and O. I. Jgorova, 1964. Spectral and X-ray investigations of the chemical structure of humic substances in soils. *Trans. 8th Intern. Congr. Soil Sci.* 3:81 Bucarest.

Kavaguchi, K., and K. Kyuma, 1959. Quantitative investigation of the soil humus composition. *J. Sci. Soil Tokyo* 29:527.

Kononova, M. M., 1951. The problem of soil humus and contemporary tasks of its study (*Problema pochvennogo gumusa i sovremennye zadachi ego izucheniya*). Izd. Akad. Nauk SSSR.

———. 1963. Soil organic matter. Its nature, properties and methods of investigation. (*Organicheskoe veshchestvo pochvy. Ego priroda, svoistva i metody izucheniya*). Izd. Akad. Nauk SSSR.

———. 1966. *Soil organic matter.* 2nd English edition. Oxford: Pergamon.

———, and N. P. Bel'chikova, 1961. Rapid methods of determining the humus composition of mineral soils. *Pochvovedenie* 10:75.

———, and N. A. Titova, 1961. Use of paper electrophoresis in fractionating soil humus substances and studying their complex compounds with iron. *Pochvovedenie* 11: 81.

Kosaka, J., 1953. Studies on the relationship between soil types and forms of soil humus. *Nat. Inst. Agric. Sci. Tokyo Bull.* 2:49.

Kosheleva, I. T., and A. S. Tolstukhina, 1957. Improvement of the soils adjoining the northern reaches of the Ob River. *Pochvovedenie* 2:72.

Kostychev, P. A., 1949. Soils of chernozem region in Russia, P. 1. Formation of chernozem (*Pochvy chernozemnoi oblasti Rossii, ch. 1. Obrazovanie chernozema*) (1886). See also Sel'khozgiz (1949).

Kr'stanov, S., 1963. Organic matter in the main soil types. Jubilee collection of papers on the 50th anniversary of Bulgarian Soil Science.

Kukharenko, T. A., 1955. The present state of our knowledge on the structure and properties of humic acids of mineral coals. *Trudy Inst. goyruch. Iskopaem.* 5:11.

Kumada, K., 1955. Elementary composition of humic acids. *Soil and Plant Food* 1: 75.

Kyuma, K., and K. Kawaguchi, 1964. Oxidative changes of polyphenols as influenced by allophane. *Proc. Soil Sci. Soc. Amer.* 28:371.

Lobova, E. V., 1960. Soils of the arid zone of the U.S.S.R. (*Pochvy pustynnoi zony SSSR*) Izd. Akad. Nauk SSSR.

Mishustin, S. N., 1956. Micro-organisms and soil fertility (*Mikroorganismy i plodorodie pochv*). Izd. Akad. Nauk SSSR.

Musierowicz, A., Z. Czerminski, J. Skorupska, and J. Sytek, 1965. Fractionation of soil humus compounds using Tyurin's method and pyrophosphate method of Kononova and Bel'chikova. *Roczn. Nauk. rol.* **92**, Ser. A, **1**.

———, and J. Sytek, 1964. Fractionation of humus compounds from forest soils developed from löss. *Roczn. Nauk rol.* 89, Ser. A., 3:1.

Němeček, I., and F. Pospišil, 1966. Humus content and fractional composition of humus substances of the main soil types in Czechoslovakia. *Pochvovedenie* 8:45.

Nyu Tszi-ven', 1961. The nature of humus in soils of the tropics and humid sub-tropics. *Pochvovedenie* 5:34.

Orlov, D. S., and E. I. Gorshkova, 1965. Size and form of the particles of humic acids from chernozem and sod-podzolic soils. *Nauch. Dokl. vyssh. Shkloly, Biol. Nauki* No. 1.

Pelišek, J., 1962. Die vertikale Zonalität der Humusverhälnisse in den Waldgebieten der Tshechoslovakischen sozialistischen Republik, "*Studies about humus*" Symp. "*Humus and Plant.*" Naklad Ceškoslov. Akad. věd, Prahe.

Ponomareva, V. V., 1956. Humus of takyrs. In "Takyrs of Western Turkmeniya and ways of reclaiming them for agriculture" (*Takyry Zapadnoi Turkmenii i puti ikh sel'skokhozyaistvennogo osvoeniya*"). Izd. Akad. Nauk SSSR.

———, 1964. The theory of podzol-forming process (*Teoriya podzoloobrazovatel'nogo prozessa*) Izd. Nauka.

Rodin, L. E., and N. I. Bazilevich, 1964. Biological productivity of the main types of vegetation of Northern Hemisphere of the Old World. *Dokl. Acad. Nauk. SSSR* 157:215.

Salfeld, J., 1971. Optical measurements on humic systems. *Transactions of the Intern. Symposium* "*Humus et Planta V*," p. 257, Prague.

Scheffer, F., 1954. Neuere Erkenntnisse in der Humusforschung, *Trans. 5th Intern. Congr. Soil Sci.*, *Léopoldville* 1, 208.

———, and B. Ulrich, 1960. "Humus und Humusdüngung, Bd. 1, Morphologie, Chemie und Dynamik des Humus". Stuttgart (1960).

Singh, S. 1954. A study of the black cotton soils with special reference to their coloration. *J. Soil Sci.* 5:289.

———. 1956. The formation of dark-colored clay-organic complexes in black soils. *J. Soil Sci.* 7:43.

Sokolov, I. A., and N. I. Belousova, 1964. Organic matter in soils of Kamchatka and some problems of illuvial-humus soil formation. *Pochvovedenie* 10:25.

Tyurin, I. V., 1937. Soil organic matter and its role in soil formation and soil fertility (*Organicheskoe veshchestvo pochvy i ego rol' v pochvoobrazovanii i plodorodii*). Sel'khozgiz.

———. 1949. Geographical characteristics of humus formation (*Geograficheskie zakonomernosti gumusoobrazovaniya*). Proceedings of the Jubilee session commemorating hundred anniversary of Dokuchaev's birth. Izd. Akad. Nauk SSSR.

———. 1951. Analytical procedure for a comparative study of soil humus. *Trudy pochv. Inst. Dokuchaeva* 38:5.

———. 1965. Soil organic matter and its role in soil fertility ("*Organicheskoe veshchestvo pochvy i ego rol' v, plodorodii*"). Izd. "*Nauka.*"

Vintilă, J., A. Chiriac, and J. Bajescu, 1963. Date privind reservele de humus, azot si fosfor ale soluriror cu folosinta agricola din sudal R.P.R. *Anal. Inst. Cerc agric. Sec. Pedol.* 21:101.

Volobuev, V. R., 1948. Changes in the humus content in soils of the U.S.S.R. in relation to climatic conditions. *Dokl. Akad. Nauk. SSSR* 60:1.

———. 1958. Studies of the relationship between soil and hydrology. *Izv. Akad. Nauk SSR, Ser. geogr.* 6.

———. 1962. Application of graphical methods in studying the composition of humus of the main soil types in the U.S.S.R. *Pochvovedenie* 1:3.

Waksman, S. A., 1936. *Humus, Origin, Chemical Composition, and Importance in Nature.* London: Baillie're, Tindall and Cox.

———, and F. Gerretsen, 1931. Influence of temperature and moisture upon the nature and extent of decomposition of plant residues by micro-organisms. *Ecology*, 12:33.

Welte, E., 1955. Neuere Ergebnisse der Humusforschung. *Agnew. Chem.* 67:153.

Wiesemüller, W., 1965. Untersuchungen über die Fraktionierung der organischen Bodensubstanz, *Albrech-Thaer-Arch. B.9, H.5*:419.

Wollny, E., 1886. Untersuchungen über die Zersetzung der organischen Substanzen. *J. landw. Vers Sta.* 34:213.

Zonn, S. V., 1964. The main types of soil formation in mountain tropical regions of south-eastern Asia. In "Genesis and geography of soils in foreign countries according to investigations of Soviet geographers" ("*Genezis i geografiya pochv zarubezhnykh stran po issledovaniyam sovetskikh geografov*"). Izd. Nauka.

Index

Index